5-7-96

D0919535

PRINCIPLES OF FOREST PATHOLOGY

PRINCIPLES OF FOREST PATHOLOGY

F. H. TAINTER
Clemson University

F. A. BAKER
Utah State University

JOHN WILEY & SONS, INC.
New York • Chichester • Brisbane • Toronto • Singapore

This text is printed on acid-free paper.

Library of Congress Cataloging in Publication Data:
Tainter, Frank H.
    Principles of forest pathology / F.H. Tainter and F.A. Baker.
      p.   cm.
    ISBN 0-471-12952-6 (cloth : alk. paper)
    1. Trees—Diseases and pests.   2. Trees—Diseases and pests—
Identification.   3. Trees—Wounds and injuries.   4. Trees—Wounds
and injuries—Identification.   I. Baker, Fred A.   II. Title.
    SB761.T35   1996
    634.9′63—dc20   96-17500

Printed in United States of America

10 9 8 7 6 5 4 3 2 1

*This textbook is dedicated
to Dr. David W. French,
our teacher, friend, and mentor.*

Contents

16   Canker Diseases

Preface

This text differs from its American predecessors (Baxter 1952, Boyce 1961, Smith 1970, Verrall 1970, Hepting 1971, Tattar 1978, Manion 1981, 1991; these are listed in the bibliography at the end of Chapter 1) by including an introduction to the principles of plant pathology (Part One) followed by a systematic description of 85 representative tree diseases (Part Two). Both the principles and the diseases are discussed within the framework of contemporary forest management. Despite the absence of a text that follows such an organizational format, this approach to teaching forest pathology has long been used in most forest pathology courses throughout the United States and Canada. With this endeavor, we intend to fill that need.

Forest pathology—the art and the science—derives its foundation from two major fields, namely plant pathology, under which it is one of many crop-oriented divisions, and forestry, where it abides as one of many professional disciplines. Its main distinction within the plant-disease arena centers on the nature and economics of the crop. Most agronomic crops are annuals, from which yields are realized within a single growing season, whereas forests, as natural or planted communities of trees, are the longest termed of all perennial crops. Their culture is primarily extensive rather than intensive simply because the cost amortization of properly maintaining a growing forest, even at minimal rotations of 15–20 years, imposes strict limits on what management practice can be applied. Even so, significant areas of forest land are under intensive management and are besieged with pest problems similar to those of intensively managed agricultural crops. Within these economic and physical constraints, forest pathology strives to integrate the science and technology of both fields in developing practical and effective methods of disease prevention and management.

Accordingly, we dedicate this text to the requirements of undergraduate students and their teachers by offering a portrayal of tree diseases that is cognizant of current trends in both forestry education and forest practice.

In curricula where undergraduate courses in pest management are general in nature and specialization is concentrated at the graduate level, this book will serve in both capacities; however, it is not intended to be a research review. Adequate documentation is furnished throughout, but only in Part One, dealing with principles, are citations generally given in context. In describing representative tree diseases in Part Two, emphasis is given to the unique features of each particular type of disease and to the methods of diagnosis and management that have evolved in accordance with established phytopathological principles. The principles and diseases are interlinked as appropriate, and the pertinent literature supporting both is summarized in placing due accord upon the most appropriate management strategies for each disease. Consequently, the text also provides considerable utility to the practicing forester.

The mechanics of melding principles and disease biologies for such a diverse audience has necessarily imposed selectivity and restraint. This is particularly evident in Part Two where, instead of a lengthy compendium of all diseases, only 85 of the most representative are each succinctly described in a standard profile format. Expanded treatment may follow some profiles according to the significance of specific profile headings. Disease diagrams incorporate the latest information into an easy-to-follow and informative format. Statement of principles and scientific terms are highlighted for definition in context and are listed in the index.

F. H. TAINTER

F. A. BAKER, JR.

PRINCIPLES OF FOREST PATHOLOGY

PART ONE

PRINCIPLES OF FOREST PATHOLOGY

The first part of this textbook discusses the principles of plant pathology and their application in forest pathology. We begin with an introduction and history of forest pathology in North America. We then discuss the concept of disease and its causes. We generalize the disease cycle and then delve deeply into specific events common to disease cycles. With this foundation, we address epidemiology, the quantitative aspects of the disease cycle. Epidemiology and biology are integrated at the conclusion of Part One in a discussion of the principles of disease management.

1

INTRODUCTION

Synopsis: The objectives, scope, and development of forest pathology in North America are described within the context of forest protection and chronology of tree disease epidemics.

1.1 PLANTS AND DISEASE

Plants, as renewable sources of food and fiber, are essential to the survival and well being of humans. Lacking the mobility of humans and animals, the higher plants are products and victims of the environment to which they are physically bound. Whether cultivated or naturally seeded, crop plants of various sorts are governed in their adaptability and yield by the interaction of biotic and abiotic factors that constitute a given habitat. When one or more of these factors limit plant growth, disease may be manifested as growth retardation or as death of entire plants or their parts.

The impact of plant disease upon crop production in the United States is reliably estimated at 1% of the gross national product or about 10% of annual agricultural production (Young 1968). Reduction of such losses is the special challenge and practical objective of plant pathology. Reducing plant disease losses in the forest is the special concern of the forest pathologist.

1.1.1 Forest Pathology

The word *pathology* derives from the Greek *(pathos + logos)* meaning a "discourse on sickness or disease." By strict definition, forest pathology is the study of forest tree diseases. It is both a science and an art within the professions of forestry and plant pathology. As a science, it is one of many crop-oriented divisions of plant pathology that are collectively dedicated to understanding the nature of disease in plants. As an art, it is a discipline in forestry serving the public interest by applying scientific principles to the prevention and control of tree diseases.

Maintaining the health of a forest from seedling stage to timber class is a time-consuming process. Most agronomic crops are annuals from which yields are realized within one growing season, whereas forest rotations may range from a minimum of 10–50 years for pulpwood harvests to 30–100 years or more for timber harvests.

3

Through these years of growth and development, trees are subject to biotic and abiotic stresses that may invoke sudden to long-standing forms of debilitation. Tree pathology, however, is but one of many economic constraints that govern forestry practices. Consequently, pathology in forest practice is directed at disease prevention according to measures that are compatible with management objectives and are justified by economic constraints.

1.1.2 Forest Protection

The agents that cause losses in forests are diseases, insects, fire, weather, and animals (Fig. 1.1) as ranked by their impact on the growth of sawtimber in 1952. Reduction of these losses, which amounted to 92% of net growth and 90% of the harvest in that year alone, is the overall objective of forest protection. In reality, however, protection emphasizes the prevention and control aspects of diseases, insects, and fire.

Insects. Forest insects account for 20% of the total growth impact (Fig. 1.1) and are second only to the 45% proportion attributed to disease. Insect outbreaks, like fire, create the illusion of being more devastating than disease because the effects are usually more immediate, often spectacular, and certainly more visible. In contrast, disease is more insidious in that the effect on the tree is often hidden from view or not readily evident to the untrained observer. For example, heart decay contributes 73% of the total disease impact (Fig. 1.1), but its effect in destroying the internal heartwood column of trees goes unnoticed unless increment core samples are taken or external clues, if present, are recognized.

Insects that degrade trees and their products are the proper concern of forest entomology. They are not treated as causes of disease, either in the literature or in this text; however, the exclusion is based on tradition rather than on fact because impairment of

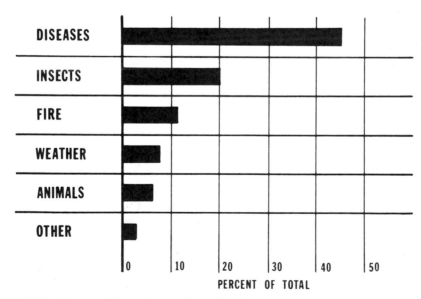

FIGURE 1.1   Agents of forest destruction and their associated proportion of growth loss in sawtimber. Redrawn from Hepting and Jemison (1958).

tree growth and function by insects is comparable to that caused by forest pathogens. Certain forest insects will be discussed, however, in at least two important relationships to disease, namely:

1. Insect and disease attacks are sometimes sequential and compounding, that is, one is primary or secondary to the other.
2. Insects may enhance the efficiency of disease-causing organisms; for example, insects that transport plant pathogens may be responsible for their dissemination, and insects, during feeding and breeding activities, may provide wounds that allow ingress of plant pathogens. Some insects, the more efficient vectors, perform both functions.

Fire. Prior to the 1940s, fire damage was the top offender among the agents of forest degradation; in fact, fire prevention and control were synonymous with forest protection. It ranked third in accounting for 12% of the total growth impact in 1952 (Fig. 1.1). Although fire is still an omnipresent threat, modern techniques of surveillance and suppression have reduced fire losses to a tertiary position in the total impact scale. Fire, however, often leaves injured survivors that are highly predisposed to disease. For example, a major portion of the heart decay in living sawtimber, especially in hardwoods, can be traced to earlier fire-wound invasion by wood-destroying organisms (Hepting 1941, Toole 1959). In addition, the sapwood of fire-killed timber is immediately subject to degradation by early successional organisms, and this weakness imposes initial time limits upon effective salvage operations.

Weather. Weather-induced losses (Fig. 1.1) consist of instantaneous injury occurring during storms, namely wind, hail, glaze, snow, and lightning, or the more subtle physiological stresses invoked by extremes or deficits in temperature and moisture, for example. The mechanical injury from storm damage, if not lethal in its direct effect on trees, is a major source of wound courts for subsequent disease establishment. Sudden or sustained deviations in weather (on a local scale) and climate (on a broader scale) can cause disease. In both cases, temperature, precipitation, and their interaction are the main offenders. Such stress conditions on tree growth, especially if prolonged and sublethal, will often predispose trees to the secondary complications of disease from biotic causes.

Vertebrate Pests. Damage (Fig. 1.1) stemming from various kinds of forest animals and domestic livestock constitutes 6% of the total impact on sawtimber. The major portion of this impact is caused by regeneration losses and growth-increment reduction. For example, browsing of tree reproduction by game animals and livestock can cause serious understocking and site deterioration of forests in various parts of the country. Smaller animals, such as birds, rabbits, porcupines, and beavers, can cause localized damage in some forest types and regions. Prevention of animal damage, whether from wild or domesticated sources, is largely a matter of regulation of population density as coordinated by forest, range, and wildlife managers. Animal injury in relation to forest diseases circumstantially implies roles as wounding agents and vectors. The few positive animal–tree disease correlations noted to date seem of little significance.

Disease. In summary, disease not only exerts a priority threat to forest productivity but also acts in concert with other agents of forest destruction, in particular, insects, fire, and weather. These interactions will be thoroughly explored as the principles and specifics of forest pathology are unfolded.

1.2 HISTORICAL PERSPECTIVES

Any attempt to document principles automatically implies that a body of knowledge exists from which established truths emerge. At this point, therefore, a capsule view of the development and progress of forest pathology seems appropriate.

1.2.1 European Origin

Forest pathology, like forestry, had its origin in Europe. That beginning is epitomized by the name of Robert Hartig (1839–1901), a German forester (Fig. 1.2). He is recognized as the father of forest pathology for pioneering the field and contributing its first text in 1874, which later was translated into English (Hartig 1894). Every

FIGURE 1.2   Robert Hartig, the father of forest pathology. Courtesy of the American Phytopathological Society.

serious student of forest pathology should read those portions of Hartig's book that bear on contemporary problems in order to appreciate fully the perception of his original accounts as tested by the passage of time. In his day, fully a century ago, plant pathology had just emerged from the falsifying influence of the autogenetic theory of plant disease (Whetzel 1918), and techniques for culturing and testing microorganisms in proving causation of disease had yet to be developed.

1.2.2   North American Development

The inception of forest pathology in the United States occurred at the turn of the century with the establishment in 1899 of the Mississippi Valley Laboratory at Washington University, St. Louis, Missouri, by the U.S. Department of Agriculture (Moore 1957). Dr. Hermann Von Schrenk was assigned there as a field agent. He had just completed his doctoral dissertation on heart decay. In 1902 G. G. Hedgcock and Perley Spaulding were added and were later joined by Ernst Bessey and Caroline Rumbold. While these scientists were concerned with tree diseases, Von Schrenk had an intense interest in the chemical preservation of timbers. A basic difference between Von Schrenk and the Division of Forestry in Washington, D. C., on research priorities resulted in the closing of the Mississippi Valley Laboratory in 1907. On that same day the task force was transferred to Washington, D.C., where it evolved into the Laboratory of Forest Pathology and then became known as the Office of Forest Pathology within the Bureau of Plant Industry. Von Schrenk left the Laboratory before the transfer and entered industrial employment. During those early years, additional scientists were hired and given specific responsibilities (Fig. 1.3). Hedgcock was made

FIGURE 1.3 Around 1910, the staff of the Office of Forest Pathology, from left to right, Carl Hartley, J. F. Collins, E. E. Carter, N. E. Watkins, T. C. Taylor, W. H. Long, C. J. Humphrey, G. G. Hedgcock, E. P. Meinecke, Haven Metcalf, J. R. Weir, Della Ingram, and Perley Spaulding. Reprinted from Buchanan (1976).

responsible for a survey of diseases on National Forests. Spaulding investigated nursery and plantation diseases and in 1909 was involved in the discovery of the introduction of white pine blister rust in the northeast. Carl Hartley was assigned to nursery diseases. C. J. Humphrey worked on the microbiological deterioration of wood. E. P. Meinecke was assigned the task of reducing the incidence of heart decay and improving salvage of infected stands. He had served a 3-year assistantship to Robert Hartig, founder of forest pathology. J. R. Weir and W. H. Long were soon hired to assist Meinecke in other regions. J. F. Collins was appointed to investigate shade tree diseases. Hedgcock and Rumbold continued their research in wood stains. Haven Metcalf was in charge of the operation. In 1931, the name was changed to the Division of Forest Pathology.

When the Division of Forest Pathology was transferred to the Forest Service in 1954, the name was changed to the Division of Forest Disease Research. In commemorating this transition, a bibliography of forest disease research was published (Moore 1957). The Division was credited with more than 2,200 publications in its then 53-year period of existence.

During the next two decades, the growth of the Division of Forest Disease Research was perhaps best indicated by a nearly three-fold increase in annual publication rate. This growth was facilitated by pathology's key role in the interdisciplinary staffing and research productivity of regional forest science laboratories.

A major impetus in the implementation of research into a national program of applied forest insect and disease control was provided by the Forest Pest Control Act of 1947. This legislation ultimately led in 1961 to the transfer of pest control programs from the Forest Experiment Stations to the newly created branch known as Forest Pest Control in State and Private Forestry at the regional level. This organization, which was renamed Forest Insect and Disease Management and then Forest Pest Management, and finally Forest Health, is responsible for the prevention, detection, evaluation, and suppression of forest insects, diseases, and weeds on federal lands and, indirectly through similar cooperative efforts, for the protection of state and private lands.

Through these cooperative efforts, many state forestry agencies now have forest protection units as a result of the technical and cost-sharing provisions of the 1947 legislation and serve as funding and administrative channels for federally funded–state cooperative projects. These field applications are often complemented by the research and educational programs of the associated state and private universities. Undergraduate courses in forest pathology or integrated forest pest management are now offered at 33 universities in 30 states. In addition, research and graduate degree programs are active at some 25 universities. During the past decade, however, severely restricted budgets and limited employment opportunities for graduates have caused a downsizing in state and federal forest pathology programs.

In Canada the first government forest pathologist was appointed in 1920 (Nordin 1961). In 1950 the Canada Department of Agriculture initiated a multimillion dollar laboratory building program for plant science and forestry research. By 1961 these new provincial laboratories were staffed by 51 research officers in forest pathology, approximately 100 laboratory and field technicians, and 71 forest biology rangers assigned to forest insect and disease survey projects. While some downsizing and redirection of emphasis has since occurred, forest pathology is still an important component of the research.

TABLE 1.1 Major Epiphytotics of North American Forest Trees

Year[a]	Disease	Cause	History
	Introduced Diseases		
1890	Beech bark disease	Insect-fungus	Ehrlich (1934)
1904	Chestnut blight	Fungus	Anderson and Rankin (1914)
1906	Blister rust (east)[b]	Fungus	Spaulding (1922)
1920	Blister rust (west)[b]	Fungus	Miller et al. (1959)
1927	Larch canker	Fungus	Tegethoff (1965)
1930	Dutch elm disease	Fungus	Clinton and McCormick (1936)
	Native Diseases		
1918	Elm yellows	Insect-mycoplasma	Swingle (1942)
1925	Ash dieback	Abiotic	Ross (1966)
1929	Pole blight[c]	Abiotic	Anonymous (1952)
1930	Birch dieback	Abiotic	Clark (1961)
1932	Littleleaf[d]	Abiotic-fungus	Campbell and Copeland (1954)
1940	Fusiform rust	Fungus	Dinus and Schmidt (1977)
1942	Oak wilt	Fungus	True et al. (1960)
1951	Sweetgum blight	Abiotic	Toole and Broadfoot (1959)
1951	Oak decline	Insect-abiotic	Staley (1965)
1954	Annosum root disease of southern pines	Fungus	Powers and Verrall (1962)
1955	Lethal yellowing of palms	Insect-mycoplasma	Sinclair et al. (1987)
1957	Maple blight	Abiotic	Anonymous (1964)

[a] Year introduced or first recognized.
[b] affects five-needled pines.
[c] affects western white pine.
[d] affects shortleaf and loblolly pines.

1.2.3 Major Disease Impacts

The development of organized efforts in forest pathology in the United States and Canada can be attributed to a steadily growing recognition that disease is a major deterrent to the national economies and to disease crises requiring emergency response. Both countries have experienced serious tree disease epidemics (syn. epiphytotic) of native and foreign origin (Table 1.1). The introduced diseases (or, more properly, the diseases caused by introduced organisms) have been especially damaging because native trees have had no opportunity to evolve resistance to new disease agents. Similarly, the introduction of exotic tree species has also resulted in unforeseen disease problems from formerly unimportant native pathogens.

Introduced Pathogens/Indigenous Tree Species

Chestnut Blight. Chestnut blight was the first pathogen to appear (Table 1.1) and proved to be one of the most devastating plant disease epidemics (Fig. 1.4). Originally detected in 1904 at the New York Zoological Park, the disease spread within a span of 50 years to virtually eliminate the American chestnut throughout its botanical

BEECH BARK SCALE

CHESTNUT BLIGHT

FIGURE 1.4 Maps of North America, showing dates and points of introduction of five major tree pathogens and the present distributions of the diseases.

10

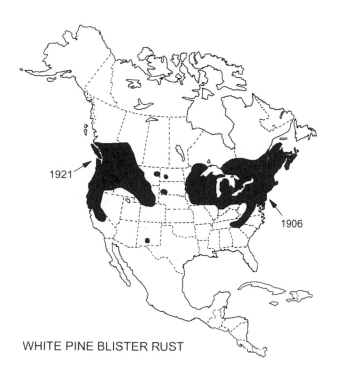

WHITE PINE BLISTER RUST

LARCH CANKER

FIGURE 1.4 *(Continued)*

11

DUTCH ELM DISEASE

FIGURE 1.4   *(Continued)*

range. The loss was inestimable because chestnut, with its many uses and values, was regarded as the elite species of the eastern deciduous forest. All efforts to halt the disease's 38-km/year progress failed. Even cutting a 1.6-km-wide barrier zone in advance of and traversing the disease perimeter in Pennsylvania could not check its progress (Beattie and Diller 1954).

In the midst of the battle against chestnut blight, a fledgling pathology work force accurately predicted the outcome (Hepting 1970). As a result of these estimates, lead-time allowances permitted chestnut-based industries to convert to substitute raw materials. For example, the impending demise of the chestnut pole industry forced the creation of an ultimately superior product, the creosoted pine pole. Also, a field and mill study showed that through 1971 only 226,629 m^3 (8 million cords) of sound, blight-killed chestnut would be available for tannin extraction to support the heavy leather industry. In the aftermath of chestnut blight, these forecasts were amazingly accurate.

Today, the fact that chestnut sprouts of some size can be found throughout the natural range of the tree has raised speculation about its recovery from the disease. Because the fungus does not attack the roots of the chestnut, many of the root systems of trees killed years ago are living and capable of sprouting. The fungus that causes the disease is still present, however, and attacks and kills the sprouts often before they reach seed-bearing age. There are now new breeding programs underway to replace the poorly designed early efforts. These efforts and the practical application of hypovirulence derived from the dsRNA mycovirus offer much hope for the eventual return of the American chestnut in eastern hardwood forests.

Blister Rust. In 1906, following closely on the heels of chestnut blight, the white pine blister rust was found at Geneva, New York (Table 1.1). The introduction of this fungus (Fig. 1.4) involves a rather ironic series of events in the trans-Atlantic perambulations of the eastern white pine. It and many other species of North American gymnosperms were imported and planted widely in a conifer-deficient Europe beginning in the early eighteenth century. Eastern white pine was one of the more successful exotics, but it fell prey to blister rust starting in the Baltic provinces of Russia in 1854. The spread of white pine blister rust intensified in Germany and northern Europe through the late nineteenth century (Spaulding 1922). The European epiphytotic apparently originated in Asia on the *pumila* variety of Swiss stone pine, which was resistant but not immune to the rust fungus. The contiguous natural range of both pine hosts provided a natural avenue through which blister rust could spread to the exotic white pine in Europe (Spaulding 1922).

In the United States at the turn of the century, much of the virgin white pine had been depleted in colonial settlement and industrial development, and reforestation with white pine seedlings had begun (Hirt 1956). American nursery workers then held the erroneous belief that white pine seedlings could not be successfully grown in this country or at least not grown as cheaply as transplants imported from Europe (Metcalf 1918). As a consequence, demand for planting stock far exceeded the national supply, and the importation of white pine began in 1899 (Metcalf 1918), despite the knowledge that blister rust was then widely distributed in Europe and known to be present in certain German nurseries. Thus, through a disastrous bit of foreign exchange, the alien progeny of our native white pines returned home two centuries later to inflict deadly harm on their natural cousins. Presently, after spreading relentlessly for some eight decades, the disease has encompassed the range of the eastern white pine from New Brunswick, Canada, south to North Carolina, and west to Minnesota and western Ontario (Miller et al. 1959).

Legislative Action. "Twice burned, doubly cautious" could well have been the motto of the Plant Quarantine Act of 1912. This legislation provided the legal and administrative authorization for the inspection of plants and plant products to exclude potential disease and insect pests from foreign imports and interstate commerce. Because many of the organisms harbored by plants are microscopic and even submicroscopic in size, particularly disease agents, they can easily escape detection. Realistically then, plant quarantine inspection is an appropriate deterrent but not totally exclusive. Unfortunately, this fact is borne out by the record of tree disease introductions subsequent to 1912.

Blister Rust Again. In 1921 white pine blister rust made a repeat performance (Fig. 1.4, Table 1.1), affecting planting stock via a French nursery importation to Vancouver, British Columbia, in 1910 (Miller et al. 1959). The disease is now well entrenched on sugar and western white pines, ranging from Washington to central California and inland to Montana and Wyoming.

At one time, both epiphytotics of blister rust commanded major attention in expenditures for disease control. In 1952, for example, of the total 3.85 million dollars (worth approximately $21.9 million in 1995) of state and federal funds appropriated for direct control of forest diseases, fully 93% was allocated to blister rust (Hepting and Jemison 1958).

Beech Bark Disease. At the same time that blister rust was making its appearance on the west coast, the beech bark disease had entered the scene in Nova Scotia and the northeastern United States (Fig. 1.4, Table 1.1). This disease of the American beech, which is caused by the primary feeding wounds of a scale insect and the subsequent wound invasion by a fungus, which eventually girdles the tree, is an interesting case history of changing values. The insect and probably the associated fungus were introduced into North America on ornamental beech trees (Shigo 1972) undoubtedly from Europe where the disease was known prior to 1849. Although the beech scale was introduced into Halifax, Nova Scotia, about 1890, the first recorded outbreak of the disease was not until 1920 (Ehrlich 1934).

As the disease spread during the 1920s and 1930s, there was moderate concern over the fate of the beech in view of the spectacular mortality and unknown potential for spreading the fungus. At that time, there was no significant market demand for beech; as a consequence, foresters welcomed the disease as a natural aid in converting beech stands to plantings of more valuable hardwoods. There is currently, however, greater acceptance of beech as a raw material (Hamilton 1955) on the strength of the economic reappraisals of the beech utilization studies of the 1950s. Should a rising market for beech continue to materialize as predicted, critical attention may once again focus on the beech bark disease, which has continued to spread steadily southward to establish a killing front in Vermont and eastern New York and an advance front (i.e., only the insects) in central New York and northern Pennsylvania (Shigo 1972). More recently, the disease has been found in Virginia, West Virginia, Tennessee, and North Carolina.

Larch Canker. In 1927 the larch canker disease was discovered in Essex County, Massachusetts (Fig. 1.4; Table 1.1), on trees planted in 1904 from nursery stock originating in Scotland (Tegethoff 1965). Because it was known that plantings of western larch in Europe had experienced heavy damage from the disease, there was immediate concern about the potential for spread of the causal fungus via the connecting link of eastern larch (tamarack) across Canada into the valuable reserves of western larch in the Rocky Mountains. An intensive survey began at once and resulted in the destruction in 1927 of more than 5,000 diseased trees within an 11-km radius of the discovery site. Resurveys were conducted in 1935, 1938, and 1952 with the resultant eradication of an additional 41 diseased trees, six of which were found in the last survey. The area was reexamined in 1965, and no evidence of the disease was found (Tegethoff 1965). The larch canker is a singular example of the prevention of an almost certain epiphytotic by timely action and direct control.

Dutch Elm Disease. The final, and hopefully last, episode in this North American chronology of introduced tree diseases occurred in 1930 with the importation of the Dutch elm disease fungus (Fig. 1.4, Table 1.1). The causal agent was concealed in a shipment of burl logs presumably originating from disease-killed elms in Holland and destined for veneer plants near Cleveland, Ohio, via entry through the port of New York (Clinton and McCormick 1936). It was identified by Curtis May who had just returned from studying the disease in Holland. Within 2 years of the discovery of the disease in both Cleveland and Cincinnati, it was reported rampant in New York and New Jersey and present in Connecticut and Maryland. Today, the disease range encompasses all but Florida and Louisiana in the east and Arizona, Nevada, and New

Mexico in the west (Barger and Hock 1971, Stipes and Campana 1981). In Canada the disease was first observed in Quebec (Pomerleau 1961) and is now known to occur additionally in New Brunswick, Nova Scotia, and Ontario (Canadian Forestry Service 1972) and is well established in Manitoba.

The earlier history of the Dutch elm disease in Europe dates to 1921 when it was first diagnosed in Holland (Clinton and McCormick 1936), hence the common epithet by which it became known. The causal fungus, however, probably originated in Asia because only the Chinese and Siberian species of elm are tolerant of the disease. All other elm species, especially those widely planted as shade and street trees, have yielded to the onslaught of the disease as it has swept over both continents.

The disease spreads rapidly; the Quebec epiphytotic, for example, has progressed at an average rate of 3,072 square kilometers annually since its inception there in 1944 (Pomerleau 1961). This overland movement is a function of strict dependence of the causal fungus for transmission by insect flight, specifically through the feeding and breeding activities of two elm bark beetles, a native species and principal vector, and the introduced European species. Although the European beetle apparently accompanied the introduction of the fungus in the 1930 veneer log importation (Clinton and McCormick 1936), there is evidence that it entered the United States from Europe prior to 1904.

As a shade tree problem, the Dutch elm disease has posed a real threat to municipal aesthetics with concomitant financial burdens in control and tree removal. Damage and shade tree valuation estimates are available from a number of sources, but one of the most comprehensive examples is a questionnaire survey of the history of the disease in Michigan's Lower Peninsula during the period 1950–1964 (Hart 1965). In 1964 alone, municipal and private expenditures for control amounted to $5.25 million and $3.5 million, respectively. Total monetary loss from approximately 2% tree mortality was estimated at $30 million annually in terms of control expenditures and aesthetic valuation.

Indigenous Pathogens/Indigenous Tree Species

Some epiphytotic diseases indigenous to North American forests are listed in Table 1.1 according to year of discovery. Their significance will be summarized under infectious and noninfectious causes rather than chronologically.

Infectious Diseases. These diseases, which include elm yellows, littleleaf, fusiform rust, oak wilt, annosum root disease, and lethal yellowing of palms, are like the introduced diseases in that they are caused by living or infectious agents. They are presumed, however, to be of indigenous origin as judged by their characteristics; antecedent history, if known; and absence elsewhere.

Elm Yellows (formerly elm phloem necrosis). This disease of *Ulmus* species, discovered in Ohio in 1919 (Swingle 1942), is one of only a few mycoplasmalike, organism-caused (Wilson et al. 1972) forest and shade tree diseases of any consequence in this country. The range of the disease in 1946, which included 15 central and southeastern states (Bretz and Swingle 1946), has remained seemingly stable except for outbreaks in Alabama in 1959 (Curl et al. 1959), Pennsylvania in 1971 (Merrill and Nichols 1972), and New York in 1972 (Sinclair 1972). The mycoplasmalike organism is vectored by a leaf hopper, which is known to occur in at least 26 states, among which Minnesota and Maine are the northernmost extremes (Sinclair 1972). Whether the disease becomes as extensive as the insect carrier remains to be seen. Any further spread will greatly intensify the pressure on America's elm resource, which is already in double jeopardy in some 18 states where both elm yellows and the Dutch elm disease are coincidentally present.

Littleleaf. In 1932 the long-term effect of soil deterioration in the southern Piedmont region became evident in a fungus-culminated root disease of shortleaf pine, known as littleleaf (Campbell and Copeland 1954). The disease intensified to significant proportions on over 2,400,000 hectares of pine lands from Virginia to Mississippi. Currently on the wane, the disease was minimized in forest practice by converting littleleaf areas to other pine species and by avoiding potential littleleaf sites in the regeneration of shortleaf pine through a scheme of soil-littleleaf hazard classification. More recently, very similar littleleaf factors have been reported in epiphytotics of exotic pines in New Zealand and of native hardwoods in Australia (Newhook and Podger 1972).

Fusiform Rust. The fusiform rust pathogen alternates between pines and oaks, and both hosts are native to North America. When botanical explorers passed through the southeastern United States in the nineteenth century, fusiform rust was a rather rare disease. It coevolved over millennia with its hosts and for a variety of reasons occupied a relatively minor position in the natural forest ecosystem (Dinus 1974). In surveys conducted in Louisiana, for example, the incidence of fusiform rust was 5% or less in 1925–1926. The first regional survey in 1938–1939 showed an increased incidence (20–50%) in Louisiana, Mississippi, and Alabama and a moderate incidence (5–30%) in Florida, Georgia, and South Carolina. Elsewhere, rust incidence was still less than 5%. By 1950, however, diseased stands were abundant from Maryland to Florida, and to east Texas and Oklahoma. Surveys in the 1970s indicated an even greater incidence and an average infection increase of about 3% per year. The reasons for such a dramatic increase of a previously rare disease are now regarded as a classic illustration of how man's altering of natural forested systems can dramatically affect natural checks and balances on potential native pathogens. The changes must be viewed in light of changes in southern pine forests. Southern pine forests are fire subclimax, and in the absence of fire, hardwoods will replace pines on many sites. Before European settlement, pines and hardwoods tended to occupy different sites, although mixtures did occur. Natural forces were such that wholesale mixtures over wide areas were infrequent. The natural forests were a patch mosaic of pines and hardwoods. Today, pines still occupy the same general range, but patterns of occurrence and abundance have changed. Resistant longleaf is less prevalent, having been replaced by loblolly and slash pines after clear-cutting and abandoning agricultural lands. Both disturbances, but especially clear-cutting, allowed hardwoods to increase in number and mingle more frequently with the pines. Artificial regeneration of loblolly and shortleaf pines began to increase around 1920 and continued at a rapid rate at the same time that the range of slash pine was increasing. Infected nursery stock was planted in many areas where fusiform rust was not previously abundant or even present. Plantations were more susceptible to rust due to reduced variation in tree size, favorable microclimatic conditions, and, most importantly, increased succulent tissue availibility for infection. The exclusion of fire favored the invasion of hardwoods, especially oaks, the alternate host. As a result, due to early disturbance and rearrangement, the two most susceptible pine species became more widespread and abundant while the number of alternate hosts increased. Because of these changes, a previously innocuous component of the natural ecosystem has within less than 100 years become the single most serious tree disease of pines in the southeastern United States.

Oak Wilt. For two decades following the discovery of oak wilt in Wisconsin in 1942 (Henry 1944), this killing fungus-caused disease of oaks commanded serious attention in the 21 states that now comprise its range (True et al. 1960). Fortunately, the early fears of a severe and widespread epiphytotic never materialized, except for the recent epidemic in central Texas. Nevertheless, oak wilt has a potential for long-term attrition of oaks that should not be ignored (Merrill 1967).

Annosum Root Disease. Originally studied by Robert Hartig in the 1870s, this disease is now found in most northern temperate regions and some tropical and subtropical areas (Bega 1963). Losses in the southeastern United States have been especially severe. Losses originate in two different ways. On deep, sandy, well-drained soils, losses due to mortality are common (Froelich et al. 1966). The pathogen gains entrance into the stand through fresh stumps created during thinning. This exposure leads to the development of infection centers just as the stand is entering its most productive growth phase. A more insidious form of loss occurs on sites of lower disease risk. Although infected trees may be uprooted by wind, the relative amount of mortality is much less, and the major impact results from slowed diameter growth (Bradford et al. 1978). These trees, however, become attractive to bark beetle attack and may also be prematurely killed (Alexander et al. 1981).

Lethal Yellowing of Palms. This disease was restricted to the Caribbean and west Africa until the 1950s (Sinclair et al. 1987). In 1955 introduced palms, especially the "Jamaica Tall," began dying in large numbers in Key West, Florida. Within 5 years, it had spread to mainland Florida, and by 1980 it had been discovered in south Texas. As the epidemic progressed, other introduced palms succumbed while palms native to Florida remained unaffected. Symptoms began as a yellowing of the older fronds with death of the entire palm occurring in 4–6 months. In 1972 lethal yellowing was associated with mycoplasmalike organisms with a planthopper being the principal vector. It is not known what triggered the epidemic in Florida.

Noninfectious Diseases. These diseases are often called diebacks or declines (Table 1.1) and are characterized by the absence of an infectious agent and by attributes that implicate factors of environmental adversity in a causal relationship. Drought and rootlet mortality have been common features found through investigations following the discovery of ash dieback in 1925 (Ross 1966), pole blight of western white pine in 1929 (Anonymous 1952), birch dieback in 1930 (Clark 1961), and sweetgum blight in 1951 (Toole and Broadfoot 1959). Although all these diseases have been epiphytotically destructive (Hepting and Jemison 1958), they differ significantly from diseases of biotic origin in that, as environmental conditions return to normal, tree recovery can be expected and further attrition ceases.

Oak decline, recognized as a developing problem in 1951 (Staley 1965), and maple blight, reported from the Lake States in 1957 (Anonymous 1964), are caused by a complex set of factors associated with primary and often-repeated defoliation by insects, injury from late frosts, or other stresses that make the affected trees prone to attack by secondary insect and disease pests.

1.2.4 The Literature

In the final analysis, the progress of American forest pathology is measured by its literature base. Books, as compendia of this knowledge, serve as convenient timeposts in documenting the overall contributions of the profession.

The first American text in forest and wood pathology, entitled *An Outline of Forest Pathology,* was published in 1931 by E. E. Hubert (1887–1954) of the University of Idaho. In 1938 J. S. Boyce (1889–1971) of Yale University authored *Forest Pathology,* which, through subsequent editions in 1948 and 1961 became widely adopted in forest practice and forestry education. D. V. Baxter (1900–1967) of the University of Michigan used an age-class format in *Pathology in Forest Practice,* which was published in 1943 and revised in 1952. Contemporary to this period and in the related field of shade tree pathology, P. P. Pirone wrote *Maintenance of Shade and Ornamental Trees* in 1941. It was revised in 1948 and retitled *Tree Maintenance* in 1959. In 1967 *Diseases in Forest Plantations* by Baxter was completed after his death as a final tribute to his service to the profession. Verrall (1970), in writing very specifically to the needs of foresters, published *Diseases of Forest Trees and Forest Products* under the auspices of Stephen F. Austin State University where he served on the faculty after his retirement from the U.S. Forest Service. In 1970 W. H. Smith of Yale University wrote *Tree Pathology: A Short Introduction* in which he outlined biotic causes of disease but emphasized the more topical abiotic sources of pathological stress in forest trees. A comprehensive diagnostic index of the diseases of 214 species of North American trees was completed by G. H. Hepting (1971) after his retirement from the U.S. Forest Service. *Diseases of Shade Trees* was written by T. A. Tattar in 1978 and is a well-illustrated book especially useful to urban tree specialists. The latest entry is *Tree Disease Concepts* which was written in 1981 and revised in 1991 by P. D. Manion of Syracuse University; it is an introductory text for undergraduate students in forest pathology and focuses on the major concepts that foresters need to deal with tree disease problems effectively. The present effort recognizes the need for a modern text devoted to

1. the principles of plant pathology as applied to forest diseases (Part One) and
2. the biology of specific tree diseases, selected to represent regional problems, symptom types, research and literature priorities, and strategies for control (Part Two).

LITERATURE CITED

Alexander, S. A., J. M. Skelly, and C. L. Morris. 1981. Effects of *Heterobasidion annosum* on radial growth in southern pine beetle-infected loblolly pine. *Phytopathology* 71:479–481.

Anderson, P. J., and W. H. Rankin. 1914. Endothia canker of chestnut. *N.Y. (Cornell) Agr. Expt. Sta. Bull.* 347:530–618.

Anonymous. 1952. *Pole Blight, What Is Known About It.* USDA For. Serv., No. Rocky Mt. For. and Range Exp. Sta. Misc. Publ. 4. 17 pp.

Anonymous. 1964. *The Causes of Maple Blight in the Lake States.* USDA For. Serv., Lake States For. Exp. Sta. Res. Paper LS-l0. 15 pp.

Barger, J. H., and W. K. Hock. 1971. Distribution of Dutch elm disease and the smaller European elm bark beetle in the United States as of 1970. *Pl. Dis. Reptr.* 55:271–272.

Baxter, D. V. 1952. *Pathology in Forest Practice,* 2nd ed. John Wiley and Sons, New York. 601 pp.

Baxter, D. V. 1967. *Diseases in Forest Plantations.* Cranbrook Institute of Science, Bull. 51. 251 pp.

Beattie, R. K., and J. D. Diller. 1954. Fifty years of chestnut blight in America. *J. For.* 52:323–329.

Bega, R. V. 1963. Symposium on root diseases of forest trees: *Fomes annosus. Phytopathology* 53:1120–1123.

Boyce, J. S. 1961. *Forest Pathology,* 3rd. ed. McGraw-Hill Book Co., New York. 572 pp.

Bradford, B., S. A. Alexander, and J. M. Skelly. 1978. Determination of growth loss of *Pinus taeda* L. caused by *Heterobasidion annosum* (Fr.) Bref. *Eur. J. For. Path.* 8:129–134.

Bretz, T. W., and R. U. Swingle. 1946. Known distribution of phloem necrosis of American elm. *Pl. Dis. Reptr.* 30:156–159.

Buchanan, T. S. 1976. *Forest Disease Research in the Western United States and British Columbia, Canada.* USDA Forest Service, Asheville. 234 pp.

Campbell, W. A., and O. L. Copeland. 1954. *Littleleaf Disease of Shortleaf and Loblolly Pines.* U.S. Dept. Agr. Circ. 940. 41 pp.

Canadian Forestry Service. 1972. *Annual Report of the Forest Insect and Disease Survey, 1971.* Ottawa. 106 pp.

Clark, J. 1961. Birch dieback, pp. 1551–1555. In *Recent Advances in Botany.* Univ. Toronto Press, Toronto.

Clinton, G. P., and F. A. McCormick. 1936. Dutch elm disease, *Graphium ulmi. Conn. Agr. Exp. Sta. Bull.* 389:701–752.

Curl, E. A., L. L. Hyche, and N. L. Marshall. 1959. An outbreak of phloem necrosis in Alabama. *Pl. Dis. Reptr.* 43:1245–1246.

Dinus, R. J. 1974. Knowledge about natural ecosystems as a guide to disease control in managed forests. *Amer. Phytopathol. Soc. Proc.* 1:184–190.

Dinus, R. J., and R. A. Schmidt (eds.). 1977. Management of fusiform rust in southern pines. Southern Forest Disease and Insect Research Council meeting Dec. 7–8, 1976 at Gainesville, Florida (Proceedings). 163 pp.

Ehrlich, J. 1934. The beech bark disease, a *Nectria* disease of *Fagus,* following *Cryptococcus fagi* (Baer.). *Can. J. Res.* 10:593–692.

Froelich, R. C., T. R. Dell, and C. H. Walkinshaw. 1966. Soil factors associated with *Fomes annosus* in the Gulf States. *For. Sci.* 13:356–361.

Hamilton, L. S. 1955. *Silvicultural Characteristics of American Beech.* USDA For. Serv., N.E. For. Exp. Sta. Beech Utiliz. Ser. 13. 39 pp.

Hart, J. H. 1965. Economic impact of Dutch elm disease in Michigan. *Pl. Dis. Reptr.* 49:830–832.

Hartig, R. 1894. *Textbook of the Diseases of Trees.* Geo. Newnes, Ltd., London. 331 pp. (Translated from the German by W. Somerville).

Henry, B. W. 1944. *Chalara quercina,* n.sp., the cause of oak wilt. *Phytopathology* 34:631–635.

Hepting, G. H. 1941. Prediction of cull following fire in Appalachian oaks. *J. Agr. Res.* 62:109–120.

Hepting, G. H. 1970. How forest disease and insect research is paying off. The case for forest pathology. *J. For.* 68:78–81.

Hepting, G. H. 1971. *Diseases of Forest and Shade Trees of the United States.* USDA For. Serv., Agr. Handbook 386. 658 pp.

Hepting, G. H., and G. M. Jemison. 1958. Forest protection. Timber Resources for America's Future. USDA For. Serv., *For. Res. Rept.* 14: 185–220.

Hirt, R. R. 1956. Fifty years of white pine blister rust in the northeast. *J. For.* 54:435–438.

Hubert, E. E. 1931. *An Outline of Forest Pathology.* John Wiley & Sons, New York. 543 pp.

Manion, P. D. 1981. *Tree Disease Concepts.* Prentice-Hall, Englewood Cliffs, N. J. 399 pp.

Manion, P. D. 1991. *Tree Disease Concepts,* 2nd ed. Prentice-Hall, Englewood Cliffs, N. J. 402 pp.

Merrill, W. 1967. The oak wilt epidemics in Pennsylvania and West Virginia: An analysis. *Phytopathology* 57:1206–1210.

Merrill, W., and L. P. Nichols. 1972. Distribution of elm phloem necrosis in Pennsylvania. *Pl. Dis. Reptr.* 56:525.

Metcalf, H. 1918. The problem of imported plant disease as illustrated by the white pine blister rust. *Brooklyn Bot. Gard. Mem.* 1:327–333.

Miller, D. R., J. W. Kimmey, and M. E. Fowler. 1959. *White Pine Blister Rust.* USDA For. Serv., For. Pest Leaf. 36. 8 pp.

Moore, A. E. 1957. *Bibliography of Forest Disease Research in the Department of Agriculture.* USDA, Misc. Publ. 725. 186 pp.

Newhook, F. J., and F. D. Podger. 1972. The role of *Phytophthora cinnamomi* in Australian and New Zealand forests. *Annu. Rev. Phytopath.* 10:299–326.

Nordin, V. J. 1961. Research trends in forest pathology in Canada, pp. 1560–1565. In *Recent Advances in Botany.* University of Toronto Press, Toronto.

Pirone, P. P. 1959. *Tree Maintenance,* 3rd ed. Oxford Press, New York. 483 pp.

Pomerleau, R. 1961. History of the Dutch elm disease in the province of Quebec. *For. Chron.* 37:356–367.

Powers, H. R., and A. F. Verrall. 1962. A closer look at *Fomes annosus. For. Farmer* 21(13):8–9, 16–17.

Ross, E. W. 1966. *Ash Dieback, Ecological and Developmental Studies.* Syracuse Univ. Coll. For. Bull. 88. 80 pp.

Shigo, A. L. 1972. The beech bark disease today in the northeastern United States. *J. For.* 70:286–289.

Sinclair, W. A. 1972. Phloem necrosis of American and slippery elms in New York. *Pl. Dis. Reptr.* 56:159–161.

Sinclair, W. A., H. H. Lyon, and W. T. Johnson. 1987. *Diseases of Trees and Shrubs.* Cornell Univ. Press, Ithaca and London. 574 pp.

Smith, W. H. 1970. *Tree Pathology: A Short Introduction.* Academic Press, New York. 309 pp.

Spaulding, P. 1922. *Investigations of the White Pine Blister Rust.* U.S. Dept. Agr. Bull. 957. 100 pp.

Staley, J. M. 1965. Decline and mortality of red and scarlet oaks. *For. Sci.* 11:2–17.

Stipes, R. J., and R. J. Campana. 1981. *Compendium of Elm Diseases.* American Phytopathological Society, St. Paul. 96 pp.

Swingle, R. U. 1942. *Phloem Necrosis, a Virus Disease of the American Elm.* U.S. Dept. Agr. Circ. 640. 8 pp.

Tattar, T. A. 1978. *Diseases of Shade Trees.* Academic Press, New York. 361 pp.

Tegethoff, A. C. 1965. Resurvey for European larch canker in Essex County, Massachusetts, 1965. *Pl. Dis. Reptr.* 49:834.

Toole, E. R. 1959. *Decay after Fire Injury to Southern Bottomland Hardwoods.* USDA For. Serv., Tech. Bull. 1189. 25 pp.

Toole, E. R., and W. M. Broadfoot. 1959. Sweetgum blight as related to alluvial soils of the Mississippi River flood plain. *For. Sci.* 5:2–9.

True, R. P., H. L. Barnett, C. K. Dorsey, and J. G. Leach. 1960. *Oak Wilt in West Virginia.* W.Va. Agr. Exp. Sta. Bull. 448T. 119 pp.

Verrall, A. F. 1970. *Diseases of Forest Trees and Forest Products.* Stephen F. Austin State Univ., School For. Bull. 21. 130 pp.

Whetzel, H. H. 1918. *An Outline of the History of Plant Pathology.* W. B. Saunders Co., London. 130 pp.

Wilson, C. L., C. E. Seliskar, and C. R. Krause. 1972. Mycoplasma-like bodies associated with elm phloem necrosis. *Phytopathology* 62:140–143.

Young, R. A. (ed.). 1968. *Plant Disease Development and Control.* National Academy of Science, Washington, D.C. 205 pp.

2

DISEASE IN CONCEPT

Synopsis: Plant disease is defined in terms of symbiotic, ecologic, and environmental concepts. The scope of pathogen-suscept interactions in tree pathology is delineated. Fundamental terminology is introduced, and disease classification systems are summarized.

2.1 DISEASE DEFINED

Disease is an abnormality factor in populations of all living things. We have all experienced disease in one way or another, whether it be through the personal discomfort of the common cold or through the disappointment of gardens that do not yield. Disease is not always so obvious, however, and more often than not, its effect on living systems is very subtle. Thus, to cope with all the broad ramifications of distinguishing abnormality from normality in trees, disease must be clearly defined at the onset.

There are nearly as many definitions of plant disease as there are books on the subject. Most of the definitions are valid in their own way and in their own time. One of the earliest, that of Whetzel (1925), perhaps has weathered the passage of time better than others. He defined disease as "injurious physiological activity, caused by the continued irritation by a primary causal factor, exhibited through abnormal cellular activity, and expressed in characteristic pathological conditions called symptoms."

If we look at the key words and phrases in this definition, much in the manner of Horsfall and Dimond (1959), many of the common misconceptions about disease can be clarified. Some of these explanations follow.

1. Disease is injurious, but injury, in the sense of being instantaneous rather than a continuing irritation, is not disease. For example, a wound of any sort in a tree does not constitute disease; wounds are often associated with disease, however, because they provide a major means of access by which disease-causing agents gain entry to trees.

2. Disease is not the primary causal factor alone. Disease is the result of a dynamic interaction between the causal agent and the affected tree. Thus, it is not correct to say that the fungus *Cryphonectria parasitica* is chestnut blight but rather that *C. parasitica* is the cause of chestnut blight.

22

3. Disease is not solely the symptoms, or in medical parlance, the syndrome. Symptoms are the visible expressions or response of the affected tree to the actions of the causal agent.

4. Disease is not catching or contagious, in the sense that only the infectious cause of disease can be transmitted. Although not stated in the definition, Horsfall and Dimond (1959) give the correct version of this aspect of disease when they emphasize the common misconception. The purity of scientific expression, however, often yields to communicative conveniences. For example, the phrase "introduced disease" is commonly employed (Chapter 1) in the literature, even though precision would dictate that it is the infectious agent that is introduced.

Definition and critique are used here to introduce certain basic features of disease as a prelude to additional concepts that follow. No single definition is sacred; if a definition is good, it will promote understanding without memorization. Accordingly, we each have our own defined versions of disease, which the reader may adopt or reject on the strength of additional facts as they are presented. Plant disease is the detrimental interaction of a living plant with its environment. Disease is a sustained impairment in function, structure, or form of an organism as provoked by biological, chemical, or physical factors of the environment.

2.2 THE SYMBIOSIS CONCEPT OF DISEASE

The term *symbiosis* derives from the Greek *sym.* (with) and *bioun* (to live). Literally, symbiosis means the living together or intimate occupation of a single habitat by dissimilar organisms. Infectious disease (disease caused by living organisms) meets these specifications in the form of antagonistic symbiosis in which one symbiont, the parasite, benefits unilaterally at the expense of another, the host. Because the benefit to one and the loss to the other is primarily in nutrient or energy transfer, antagonistic symbiosis is perhaps better known as parasitism. Some endophytic fungi in leaves that survive surface sterilization may play a role in mediating insect attack. This example may be considered as a form of mutual symbiosis.

Disease, however, is not always the end result of parasitism. Figure 2.1 illustrates some of the varying degrees of parasitism. Certain symbioses involving a parasitic union have evolved a balance or mutualistic symbiosis whereby both symbionts directly benefit from the association. The host yields essential growth materials to the parasite, and the parasite reciprocates with off-setting physiological returns to the host. Disease is not expressed. Important and intriguing associations of this type are

1. the legume–bacterium (*Rhizobium* spp.) interaction, which balances host nourishment of the bacterium in root nodules in return for nitrogen via atmospheric fixation by the bacterium (Nutman 1965);
2. mycorrhizae (Chapter 13), which are formed by certain fungi in the parasitism of rootlets of higher plants, benefit the host in a number of ways, particularly through enhanced nutrient absorption (Marks and Kozlowski 1973); and
3. certain algae and higher fungi, which unite so perfectly that the resultant lichen thallus gives the appearance of a single taxon; through the mutual exchange of

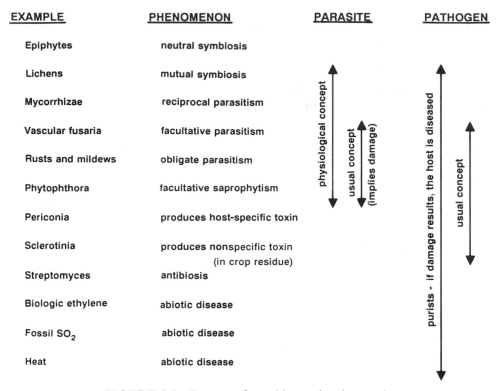

FIGURE 2.1   Degrees of parasitism and pathogenesis.

materials necessary to the nutrition of each, some combinations have even attained synergism whereby both symbionts grow better in the consortium than either does alone (Culberson 1970).

Finally, neutral symbiosis implies that the symbionts have no perceptible effect on each other. Neutral symbionts with trees are important here only from the standpoint that, unless carefully scrutinized, they sometimes produce an appearance that can be misinterpreted as disease. Certain epiphytes such as Spanish moss on the branches of live oak (Fowells 1965) and surface growths of fungi such as *Septobasidium* spp. on tree bark (Couch 1938) are strictly superficial in their growth habit. The latter example is of additional interest in that it also involves a symbiosis with insects.

Organisms growing in close proximity but not in symbiotic contact may exert ecological influences upon the course of disease by way of antibiosis, metabiosis, and synergism, as described in some detail by Stakman and Harrar (1957). Antibiosis between organisms involves the metabolic products (antibiotics) of one having inhibitory or toxic effects upon another. Antibiotic influences upon disease-causing organisms are implied primarily in relation to root diseases (Chapter 12) because *in vitro* antibiosis is so easily demonstrated with almost any sample of soil-microbial populations. There is evidence (Leben 1965), however, that antibiosis could likewise prevail between the established epiphytic microflora of aerial plant parts and disease-causing agents, if and when they are deposited. Gibbs (1980) showed that if *Ceratocystis piceae* were introduced to a fresh wound on healthy oak prior to its inoculation with *C. fagacearum,* no infection would result. *C. piceae* is very common in oak-wilt trees (Shigo 1958). Could the action of *C. piceae,* then, be considered as an example of antibiosis? In opposite function to the biological system that might serve to block disease development, as just described, metabiosis introduces the possibility that one kind of organism may create an environment that is favorable to another. Drawing just one example, such successions leading to the discoloration and decay of wood in living trees have been demonstrated and reviewed by Shigo (1967). Ecological synergism, as involved in disease causation, means that two, or possibly more, different disease agents produce greater effects conjointly than can one separately. Examples of pathological synergism discussed later include black root decay of tree seedlings (fungus + fungus) (Hodges 1963) and mimosa wilt (nematode + fungus) (Gill 1958).

2.3   THE DISEASE SQUARE

To this point, new terminology has been minimized as much as possible. Because the disease square concept is presented as a unifying view of disease, it is the most appropriate vehicle for introducing and emphasizing the basic vocabulary of plant pathology.

Disease Components

Our concept of disease would not be complete without stressing the constant regulatory role of the environment upon the disease agent (pathogen), the plant (suscept), and their interaction (disease), should it occur. Disease then is really a quadripartite entity; the so-called disease square is illustrated in Figure 2.2. Note, in particular, that abiotic, or physical, factors of the environment are not only conditioners of disease but also primary causes of disease when conditions such as nutrition, temperature, moisture, and air pollutants are deficient or excessive. These diseases caused by abiotic factors are the noninfectious diseases, sensu Smith (1970). The biotic environment includes competitors for a substrate as a multitude of antagonistic interactions.

The terminology of the disease square's components requires definition and clarification. Pathogen, as originally employed (Whetzel 1926), meant any organism capable of inducing disease. Contemporary use of the term (Agrios 1969, Strobel and Mathre 1970), however, has broadened its meaning to an entity or anything that can produce disease. Thus, as used hereafter, pathogen carries the same connotation as causal agent (i.e., any biological, chemical, or physical factor of the environment that is capable of causing disease). The term *suscept* was coined by Whetzel (1926) in an effort to avoid the pitfalls of equating pathogen with parasite–host in relation to disease. The parasite–host combination, by definition, denotes a nutritional relationship that may or may not produce disease as governed by the nature of the symbiosis. Therefore, if we define suscept as any plant that can be or is attacked by a given pathogen, a terminology that accommodates all types of disease is available and war-

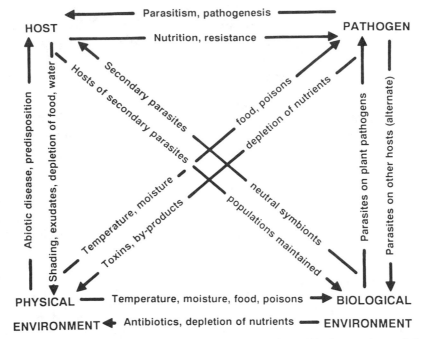

FIGURE 2.2 The disease square, showing the principal possible interactions of the four groups of variables that contribute to the occurrence and intensity of plant disease. This is a qualitative diagram; it does not include any quantitative information such as amount of inoculum, temperature, and moisture.

ranted. The use of the terms *parasite* and *host* is entirely appropriate when parasitism is involved because the majority of plant and tree diseases are caused by parasitic organisms. Be aware, however, that not all parasites are pathogens and vice versa; and that not all hosts are suscepts and vice versa; therefore, choose the proper terms accordingly.

Another version of disease (Walker 1957) would have *incitant* serve as a singular term for pathogen and environment in the sense that an organism incites or causes a disease under the influence of other factors. In our opinion, *incite* and *incitant* tend to project a more restricted view of disease, when, in fact, the basic components of the disease square are often a complex of multiple interactions. The environment in particular is a composite of factors that may differentially influence the suscept, the pathogen, and the suscept–pathogen interaction.

Therefore, disease is a coalition of suscept, pathogen, and physical and biological environments relative to space and time. In a single suscept, the ultimate severity of disease is primarily a function of its inherent susceptibility in relation to the aggressiveness of the pathogen and the modifying effects of the prevailing environment. The real impact of a forest disease, however, is measured in terms of its epidemiology (its generative capacity within a population of suscepts).

A closer look at each component of disease will permit a brief preview of pertinent functions, additional terminology, and scope of the subject matter.

2.4 THE DISEASED TREE—THE SUSCEPT

2.4.1 Suscept Lists

Some 845 species of native trees exist in the United States including Alaska but not Hawaii; 165 are listed as commercially important in forestry (Little 1949). Many of these are also indigenous Canadian trees. Grouped taxonomically and geographically, the American species number 58 conifers or softwoods (23 in the east and 35 in the west) and 107 broadleaf trees or hardwoods, of which 86, the greatest majority, are found in the east. Fowells (1965) documented the silvicultural characteristics of 122 species, furnishing habitat and life history data of considerable value in judging normality of growth relative to the effects of disease. A diagnostic index to the diseases of 214 species of trees and ornamental shrubs by Hepting (1971) provides another primary and comprehensive source of information. This descriptive sourcebook is complemented by a host index (Anonymous 1960), which lists known pathogens and other organisms circumstantially associated with disease on each and all plants in taxonomic order. In combination, these indices of tree and disease information essentially delineate the spectrum of suscepts most likely to be encountered in forest practice.

2.4.2 Vital Functions

Because symptoms are the ultimate indicators of abnormality in the suscept, comparison with a like but disease-free tree should reveal the internal attributes of the disorder. Consequently, a diagnostician relies heavily upon a basic knowledge of normal tree anatomy and physiology in order to interpret pathological alteration of tissue and function. For now, this complex of cells which compose the roots, stem, and crown of a living tree can be reduced to two basic factors to affect two critical determinants of infection by a given pathogen—its access and nutrition. Once the pathogen has gained entrance and begun to derive sustenance, host physiology is altered in various ways in response to specific pathogens.

At least six vital processes can be affected according to McNew's (1959) scheme of classifying associated symptom and pathogen types. The effects of disease on these critical physiological functions are illustrated in Figure 2.3 and listed as follows:

1. food storage in the form of seed, root, stem, and bud reserves;
2. juvenile growth as either seedling or shoot development;
3. root extension in the procurement of water and included minerals;
4. water transport;
5. food manufacture or photosynthesis;
6. translocation of photosynthate to sites of cell utilization; and
7. structural integrity.

The last function is not a physiological process, but it is exceedingly important in trees because many are weakened to the point of stem and root breakage from internal decay of physiologically inactive heartwood. If any of the first six functions is completely disrupted, the tree dies. Ordinarily, the function is impaired but not destroyed, so that only the efficiency and yield of the tree are reduced. On a tree population or

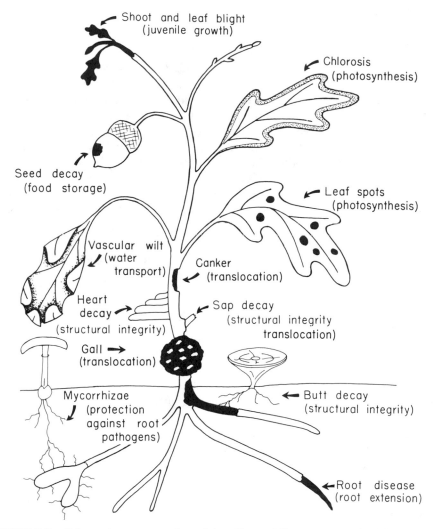

FIGURE 2.3 Schematic representation of the effects of diseases on tree health, showing the vital functions of a tree and their impairment by various types of pathogenic influences.

forest basis, however, growth retardation from disease can be highly significant in limiting the productivity potential of a given forest site.

2.4.3 Symptomatology

With the exception of immunity, all other suscept responses are ultimately expressed as microscopic and macroscopic symptoms. Naturally, susceptible responses exhibit a more pronounced symptomatology than do tolerant and resistant reactions. The suscept may and generally does display more than one symptom, so that a progression of symptoms (the symptom complex) usually characterizes a given disease. Identification or diagnosis of disease is based, in part, on visible alterations in plant morphology, the macroscopic symptoms. Some 40 or more specific symptoms (Table 2.1)

...ase Symptoms

...)eath of Cells or Tissues

...af cells
...hardwoods
...bar—on conifers
...perature extremes
...wood of roots and stems of living trees

...cells in wood of roots and stems of living trees

...ll or part of tree-due to vascular wilt, sudden
..., or drought

...—Overgrowth of Tissues

...ventitious buds—dwarf mistletoe
...ust/broom rusts

...causing puckering
...ue to insects, nematodes, or associated with

...ment or Growth of Plant or Some Organs

General, marginal, or intervenal chlorosis—due to absence of chlorophyll
    caused by pathogens, toxins, mineral deficiency, air pollution, drought, excess
        water, or chemical burns
Dwarfing—due to root disease, mineral deficiency, or mycoplasmas

are classified under three major types of tissue alterations, namely necrosis (death), hypertrophy (overdevelopment), and atrophy (underdevelopment). Symptoms are also indicative of the extent of suscept involvement in that some are consistently localized while others are systemic or expressed by the entire tree. For example, stem cankers (Chapter 16) and leaf spots (Chapter 14, and Boyce 1961, pp. 129–134) are localized necroses of respective tree parts while systemic wilts (Chapter 17) are symptomatic of a malfunction of the water-conducting system of a tree. Multiple local symptoms can coalesce, however, to give the illusion of systemic action by the pathogen. For example, a canker or coalescent cankers may completely girdle a stem causing death of stem, branches, and foliage beyond that point. In this and other like instances, the canker is a primary symptom associated with the direct influence of the pathogen while the distal necroses that follow are secondary symptoms, those arising indirectly and apart from the locus of the pathogen. Yellows, or chlorosis, is a very common symptom of plant disease used to describe any physiological disturbance or disease that results in the destruction or reduced synthesis of chlorophyll. Factors that can cause chlorosis include saline or alkaline soils, deficiencies or excesses of essential mineral elements, toxic concentrations of pesticides, reduction in number of mycorrhizal feeder roots, parasitic flowering plants, excessive or insufficient soil water, high or low soil and air temperatures, feeding by mites or insects, viruses and mycoplasmas, and parasitic fungi, bacteria, and nematodes. The very excellent book by Sinclair et al. (1987) contains many detailed color photographs of signs and symptoms of most North American tree diseases of importance.

2.4.4 Susceptibility

Suscepts that are amenable to the full potential of a pathogen are considered susceptible, or fully liable, to infection. Susceptibility is an absolute condition, just as immunity, the antonym of susceptibility, implies total exclusion from a given pathogen. Suscept responses intermediate to these extremes are qualified by the terms *tolerance* and *resistance*. Tolerance is the ability of a suscept to sustain a pathogen without suffering serious injury or reduction in yield, whereas resistance is the inherent capacity of a suscept to prevent or restrict the entry or subsequent activities of a pathogen under environmental conditions that favor disease development.

2.5 THE PATHOGEN

2.5.1 Types and Diversity

Tree pathogens show the same diversity that characterizes plant pathogens in general. Although *pathogen* is an inclusive term, a natural division requires that we distinguish between animate and inanimate agents of disease. Thus, we qualify pathogens accordingly or use the synonyms biotic and abiotic.

2.5.2 Biotic Pathogens

The living agents of disease encompass the fungi, bacteria, mycoplasmalike organisms, viruses, parasitic flowering plants, nematodes, and even some insects. They are all capable of infection, meaning that they can become established as parasites within a given suscept. The infectious nature of these kinds of pathogens is especially critical because they can spread from diseased to healthy trees by means of dispersal in growth or reproduction. These vegetative and reproductive structures of pathogens (the signs), in association with symptoms, characterize a given disease and facilitate its identification or diagnosis. However, noninfectious diseases, which are caused by abiotic agents, differ notably in lacking signs and, for that reason, are sometimes more difficult to diagnose.

Chemicals in disease causation may derive from either biotic or abiotic origin. Biochemicals, such as toxins, enzymes, and growth regulators, are, respectively, the mechanisms by which pathogens, especially fungi, kill cells of, obtain nourishment from, and modify tissues of the suscept. Some species of trees, such as black walnut and cherrybark oak, must even be regarded as pathogens on the basis that they release phytotoxins (allelopathy) into their local environment (DeBell 1970, 1971).

2.5.3   Substrate Classification of Biotic Pathogens

What follows now is only a brief characterization of each kind of biotic pathogen; they are covered in more detail in Chapter 4.

The fungi (Eumycetes) and bacteria (Schizomycetes) are lower, achlorophyllous plants, which reproduce by spores and simple cell division, respectively.

Mycoplasmas, discovered only recently in 1967 (Doi et al. 1967), are the smallest of living organisms, which apparently bridge the taxonomic gap between bacteria and viruses. The few mycoplasmas identified to date were previously regarded as viruses associated with phloem disorders in yellows-type diseases (Davis and Whitcomb 1971).

Viruses are included in the biotic group despite the fact that they do not possess all attributes of life; nevertheless, in common with the other kinds of biotic agents of disease, they reproduce their own kind and hence are transmissible from tree to tree. Viruses are submicroscopic particles composed of protein and nucleic acid. They reproduce only in living cells of animal vectors and suscepts and at the expense of the latter in eliciting disease by diversion of protein synthesis.

The parasitic flowering plants include the mistletoes (Loranthaceae), dodders (Convolvulaceae), broomrapes (Orobanchaceae), and figworts (Scrophulariaceae). In common, these plants establish rootlike haustorial connections with the vascular elements of suscept stems or roots in the withdrawal of water and nutrients.

Nematodes (Nematoda), or eelworms, are the only members of the animal kingdom that have received major attention as recognized plant pathogens. The plant parasitic types are small, wormlike creatures that feed primarily on plant roots by puncturing them with a hollow, retractable stylet.

2.5.4   Abiotic Pathogens

Noninfectious disease agents or abiotic pathogens are physical and chemical factors of the environment that are unfavorable for normal growth and development of a given suscept at extremes of deficiency or excess. In addition, trees in a stressed or weakened condition from abiotic causes are often more susceptible to biotic pathogens. Abiotic factors function concurrently as agents of predisposition.

The physical extremes of particular note are temperature and moisture. We will find, for example, that a prolonged moisture deficit is largely responsible for dieback and decline epiphytotics of a number of forest tree species (Chapter 3). Chemical effects from abiotic sources range from extremes in mineral nutrition of trees to direct toxicity in the form of air pollution impacts.

2.6   NUMERICAL SPECTRUM OF BIOTIC PATHOGENS

In documenting the fact that parasitism of plants is a common phenomenon, Agrios (1969) estimates that, in North America alone, some 8,000 species of fungi cause approximately 80,000 diseases; in addition, at least 180 species of bacteria, more than 500 different viruses, and over 500 species of nematodes also attack crops. No enumeration has been made for the tree component of this crop total; however, reference to the current and most comprehensive lists should provide a reasonable numerical estimate of pathogen affinities for native trees.

The fungi are by far the largest group as evidenced by the approximately 1,300 species indexed by Hepting (1971) in this disease compendium of 214 species of woody plants. In comparison, and from the same source, only 21 species of bacteria are listed.

To date, only one mycoplasma seems to be a definite cause of disease; elm yellows, formerly attributed to a virus, recently has been reassessed and reclassified (Wilson et al. 1972). Even at that, there are only 19 different viruses recorded from just nine species of native trees, all hardwoods (Thornberry 1966), and none is considered potent as a forest pathogen. They do, however, represent a considerable potential threat for vegetatively propagated trees. On a worldwide scale, conifers essentially appear immune to viruses. Only one disease, a virosis of spruce, in Czechoslovakia, has been substantiated to date (Cech et al. 1961); however, virus infection of red spruce has recently been reported (Jacobi and Castello 1991).

Nematode associations with forest trees, as compiled by Ruehle (1967), reveal that 94 plant-parasitic species have been found in the root zones of 73 species of timber trees in the United States. Only 30 of these tree species, however, are known to suffer appreciable root damage as caused primarily by species of *Meloidogyne, Pratylenchus,* and *Xiphinema,* the root knot, lesion, and dagger nematodes, respectively.

Of the more than 2,500 species of parasitic flowering plants estimated by Agrios (1969) and reviewed by Kuijt (1969), only the dwarf mistletoes (*Arceuthobium* spp.) are of major concern in forestry. Hawksworth and Wiens (1972) recognized 17 taxa of dwarf mistletoes in the United States, which collectively are principal pathogens of 30 species of conifers, primarily of the western forest types. Other kinds of tree pathogens in the category of parasitic flowering plants, such as dodder in forest nurseries (Latham et al. 1938), true mistletoes on ornamental trees, and, more recently, *Senna seymeria* on planted southern pines (Grelen and Mann 1973), are noteworthy because they may occasionally cause severe damage on a local scale.

Thus, the array of potential tree pathogens, both biotic and abiotic, is rather impressive in scope and may seem quite overwhelming at the moment. Fortunately, the establishment of principles (Part One) permits a systematic cataloging of facts that should dispel such concerns as we prepare the foundation for studying specific diseases in Part Two.

2.7   PARASITISM AND PATHOGENICITY

Because *parasitism* and *pathogenicity* are not necessarily synonymous, a scheme of classifying pathogens based upon substrate and nutritional relationships is appropriate by way of explanation. Like all living things, biotic agents of disease are differentiated nutritionally as either autotrophs or heterotrophs. Autotrophic, or independent, organisms synthesize food from inorganic sources, while heterotrophic, or dependent, organisms require organic materials for nutrient elaboration.

Because the majority of recognized plant pathogens, exclusive of viruses, nematodes, and abiotic factors, are members of the plant kingdom, the distinction between autophytes and heterophytes, respectively, is that of chlorophyllous and achlorophyllous, or higher flowering plants and lower plants, or phanerogams and cryptogams from older botanical classification. Except for the major group of parasitic flowering plants (Chapter 4), of which the dwarf mistletoes (Chapter 19) are the most important, additional autotrophic pathogens are relatively few in number, and, unlike the mistletoes and their kind, they typify pathogenicity without parasitism. This is evidenced by the allelopathy example (DeBell 1970, 1971) given previously and by the strangulation and shading effects of lianas or climbing vines such as poison oak, honeysuckle, and kudzu.

Heterophytic organisms, which are incapable of synthesizing carbohydrates out of carbon dioxide and water, are linked nutritionally to suscepts either as saprophytes on dead tissue, as parasites in living tissue, or most commonly as both in the course of their life cycles. When a heterophyte can function as both, its parasitic and saprophytic phases are qualified further by the modifying term *facultative,* meaning occasional. Hence, a facultative saprophyte is an organism that usually exists as a parasite but, on occasion, can live or survive as a saprophyte, whereas a facultative parasite usually functions as a saprophyte but, on occasion, can derive its nutrition from living tissue as a parasite. Some heterophytes are restricted to one life system or the other and therefore are identified by the adjective *obligate*. Hence an obligate saprophyte or saprobe exists solely on dead organic matter, while an obligate parasite is strictly dependent upon living tissue of the suscept for sustenance and survival. Although this terminology is most frequently applied to the fungi because of their preponderance in disease causation, other agents of disease are typically, and without exception, obligate parasites, namely, parasitic flowering plants, mycoplasmas, and viruses.

Obviously, pathogen does not always equate as parasite and vice versa. Notable examples of parasites that are not pathogens, as mentioned previously, are the legume nodule bacteria and the mycorrhizal fungi. Pathogens that are not parasites include epiphytes, as previously cited; sooty mold fungi, of questionable pathogenicity; heartwood decay fungi in living trees; and, quite definitely, abiotic causal agents.

Further elaboration of two of the examples of pathogens that are not parasites may clarify these rather perplexing disease relationships. Sooty mold is a sign that is not accompanied by visible symptoms. It is composed of any one or more of a number of species of obligately saprobic fungi that grow superficially and dependently on the sugary secretions or "honey dew" of aphids, which feed on the foliage of many tree species. The growth has the appearance of soot. When it occurs on Christmas tree crops in particular, its unsightliness causes considerable alarm. In fact, uninformed growers have destroyed whole blocks of affected trees thinking they were severely diseased and beyond recovery. Actually, such trees are merchantable because the fungus growth or sooty mold will dissipate with rain wash, or it can be removed from cut trees by spraying them with a directed stream of water. The question remains, is sooty mold a disease? Although most references to the subject, as cited in Hepting (1971), allude negatively to the question, there is the distinct possibility that the mass and opacity of the fungus may, by shielding, reduce the photosynthesis of affected leaves. In addition, the fungus growth may plug stomates and thus affect gas and water exchange. Therefore, it seems quite plausible that sooty mold is a disease, even though tree symptomatology may be evidenced only in slightly reduced growth.

The second example, heartwood decay fungi in living trees, accounts for 73% of the total disease loss in American forests (Hepting and Jemison 1958). Wood decay fungi, which number in the hundreds of species, run the gamut between parasitic and saprophytic types in living trees, dead trees and their debris, and wood in use. Heart decay, because it involves the central cylinder of physiologically dead wood in living trees, is a saprophytic activity if we consider only the site of action; it could be regarded as parasitism if we take the viewpoint of whole tree involvement. However, the latter interpretation fails to recognize that parasitism is really the yield of nutrients from living host cells and tissues to the demands of a parasite. Consequently, heart-decaying fungi are not parasites, but are they pathogens? They are not, at least according to most definitions of disease, including that of Whetzel (1925) used at the beginning of this chapter. With heart decay, there is no injurious physiological activity or continued irritation; there is only the activity of the fungus in its breakdown of inert, nonreactive heartwood. Yet, even with this in mind, it seems inconceivable that we would not consider heart decay as a disease when it ranks as an important agent of forest degradation. In our opinion, disease alters more than function in the suscept; it also alters structure and form, as defined earlier. By this concept, then, heart decay qualifies as a disease.

In concluding this section, only those attributes of pathogens pertinent to the conception of disease have been employed. Other characteristics, such as variability and infection biology, are more appropriate to later elucidation of specific pathogen types.

2.8 THE ENVIRONMENT

Environmental components of the disease square, in the broadest sense, include all biotic, physical, and chemical factors that influence the pathogen–suscept interaction, which by themselves are part of and influenced by their habitat, be it primarily edaphic, atmospheric, or both. Having already addressed the biotic and chemical factors of the environment, we will examine briefly the physical aspects of the atmospheric environment as viewed on three scales—climate, weather, and microclimate.

Weather and climate affect the biotic components of disease through the individual or interacting effects of precipitation, temperature, humidity, fog and dew, wind, and radiation, differing primarily in the time span of influence (Pirone 1959). In this regard, Hepting (1963) points out that annual crops reflect weather changes, whereas trees and other perennial flora will reflect, in addition, climate changes. The latter, as reviewed by Hepting (1963), is especially accountable for a number of regionally important diebacks and declines of forest trees in the United States during the 1930s through the 1950s.

Neither weather nor climate are static, but weather, being more changeable with the seasons, is a more localized determinant of whether a pathogen–suscept combination will develop into disease. Some of the more devastating crop diseases are conspicuously associated with periods of aberrant weather. Normally, disease is such a synchronization of so many biological factors subject to weather regulation that the probability of infection is relatively low. For example, weather influences a pathogen at its source, in terms of its reproduction and liberation, and in reaching the suscept, in terms of transport, deposition, and infection. If any one of several weather factors is unfavorable to the pathogen at any one of these stages in its transmission from source to suscept, disease does not occur. Waggoner (1962) expressed this observation in a different yet interesting way—"A severe outbreak is a rare removal by the weather of obstacles that ordinarily restrain the pathogen."

TABLE 2.2 Plant Disease Classification Systems with Examples and References to Each

Classification Type and Examples	Context	
	Major Emphasis	Chapter Topic

1. Crop Divisions
  Seedlings, saplings
  Forest trees
	Baxter (1952), Boyce (1961), Smith (1970)	Baxter (1952)

2. Specific Suscepts Oaks, pines, etc.	Hepting (1971)	Pirone (1959)

3. Suscept Culture and Use
  Nurseries, plantations
  Shade trees
  Wood products
	Baxter (1952, 1967), Cartwright and Findlay (1950)	Baxter (1952), Boyce (1961)

4. Suscept Age Class Seedlings, saplings	Baxter (1952)	Boyce (1961)
5. Suscept Part Affected Roots, stem, foliage	Boyce (1961)	
6. Symptom Types Decays, cankers, wilts	Boyce (1961)	
7. Suscept Function Affected Photosynthesis, etc.	Horsfall and Dimond (1959)	Agrios (1969)

8. Regional Suscepts
  Native
  Exotic
	Matuszewski (1973), Spaulding (1958)	Boyce (1961)

9. Pathogen Types Fungi, bacteria, etc.	Agrios (1969), Smith (1970), Strobel and Mathre (1970), Walker (1957)	

Another effect of weather is on the suscept itself. Favorable weather may produce increased growth rates for a time, but, when unfavorable weather is once again prevalent, dieback of roots and reduced photosynthesis may open a "window of susceptibility" to a given pest. Changing weather may also affect the subsequent development of the pathogen in the new host, thus altering the disease. Major environmental stresses, of which changing weather may be one component, may induce "cohort senescence," a concept developed during the study of dieback and decline of ohia forests in Hawaii (Mueller-Dumbois et al. 1983). A group of trees on a poor-quality site and exposed to frequent or prolonged stress may become senescent at an earlier age than the same tree on a good-quality site. This is discussed in more detail in Chapter 18.

The study of microclimate in relation to disease focuses attention upon the environment immediately surrounding the plant and draws upon the biological and physical sciences in the biometeorological measurement and interpretation of physical parameters that govern plant infection processes (Brooks 1963, Lowry 1963, Platt 1963). By means of sophisticated instrumentation and methodology, the importance of applying micrometeorological precision to the evaluation of plant disease epidemiology has been convincingly demonstrated (Van Arsdel 1965). Correlation of these kinds of data with synoptic weather patterns has permitted sufficient accuracy to recommend the use of synoptic weather charts in appraising at least one tree disease hazard, the important fusiform rust of the Gulf Coast area (Davis and Snow 1968).

The edaphic environment of pathogen and suscept is sufficiently complex and so oriented to root diseases that the topic is deferred to Chapters 10, 11, and 13, where fundamentals of root pathology will be developed.

2.9   CLASSIFICATION OF PLANT AND TREE DISEASES

As a final consideration in this chapter, we call attention to the various ways of classifying plant diseases. The orderly indexing of causes, suscepts and parts affected, symptoms, and the like is essential to the communication and retrieval of information pertinent to the diagnosis and control of disease. Most plant pathology texts use more than one type of classification, as shown by the cross references in Table 2.2.

Part Two of this text, which deals with the biology of specific forest diseases, is organized according to tree parts affected and their symptoms.

LITERATURE CITED

Agrios, G. N. 1969. *Plant Pathology*. Academic Press, New York. 629 pp.

Anonymous. 1960. *Index of Plant Diseases in the United States*. U.S. Dept. Agr. Handbook 165. 531 pp.

Baxter, D. V. 1952. *Pathology in Forest Practice*, 2nd ed. John Wiley and Sons, New York. 601 pp.

Baxter, D. V. 1967. *Disease in Forest Plantations: Thief of Time*. Cranbrook Inst. Sci. Bull. 51. 251 pp.

Boyce, J. S. 1961. *Forest Pathology*, 3rd ed. McGraw-Hill, New York. 572 pp.

Brooks, F. A. 1963. Biometeorological data interpretation to describe the physical microclimate. *Phytopathology* 53:1203–1209.

Cartwright, K. St. G., and W. P. K. Findlay. 1950. *Decay of Timber and Its Prevention*. Chemical Publishing Co., New York. 294 pp.

Cech, M., O. Kralik, and C. Blattny. 1961. Rod-shaped particles associated with virosis of spruce. *Phytopathology* 51:183–185.

Couch, J. N. 1938. *The Genus Septobasidium*. University of North Carolina Press, Chapel Hill, N.C. 480 pp.

Culberson, W. L. 1970. Chemosystematics and ecology of lichen-forming fungi. *Annu. Rev. Ecol. and System.* 1:153–170.

Davis, R. E., and R. F. Whitcomb. 1971. Mycoplasmas, rickettsiae, and chlamydiae: Possible relation to yellows diseases and other disorders of plants and insects. *Annu. Rev. Phytopath.* 9:119–154.

Davis R. T., and G. A. Snow. 1968. Weather systems related to fusiform rust infection. *Pl. Dis. Reptr.* 52:419–422.

DeBell, D. S. 1970. Phytotoxins: New problems in forestry? *J. For.* 68:335–337.

DeBell, D. S. 1971. Phytotoxic effects of cherrybark oak. *For. Sci.* 17:180–185.

Doi, Y., M. Teranaka, K. Yora, and H. Asuyama. 1967. Mycoplasma- or PLT group-like microorganisms found in the phloem elements of plants infected with mulberry dwarf, potato witches' broom, aster yellows, or paulownia witches' broom. *Ann. Phytopathol. Soc. Japan* 33:259–266.

Fowells, H. A. 1965. *Silvics of Forest Trees in the United States*. USDA For. Serv., Agr. Handbook 271. 762 pp.

Gibbs, J. N. 1980. The role of *Ceratocystis piceae* in preventing infection by *Ceratocystis fagacearum* in Minnesota. *Trans. Brit. Mycol. Soc.* 74:171–174.

Gill, D. L. 1958. Effect of root-knot nematodes on Fusarium wilt of mimosa. *Pl. Dis. Reptr.* 42:587–590.

Grelen, H. E., and W. F. Mann, Jr. 1973. Distribution of senna seymeria *(Seymeria cassioides)*, a root parasite on southern pines. *Econ. Bot.* 27:339–342.

Hawksworth, F. G., and D. Wiens. 1972. *Biology and Classification of Dwarf Mistletoes (Arceuthobium)*. USDA For. Serv., Agr. Handbook 402. 234 pp.

Hepting, G. H. 1963. Climate and forest diseases. *Annu. Rev. Phytopath.* 1:31–50.

Hepting, G. H. 1971. *Diseases of Forest and Shade Trees of the United States*. USDA, Agr. Handbook 386. 658 pp.

Hepting, G. H., and G. M. Jemison. 1958. Forest protection. pp. 185–200. In *Timber Resources for America's Future*. USDA For. Serv., For. Res. Rept. 14.

Hodges, C. S., Jr. 1963. Black root rot of pine. *Phytopathology* 53:1132–1134.

Horsfall, J. G., and A. E. Dimond. 1959. The diseased plant. pp. 1–17. In Horsfall, J. G. and A. E. Dimond (eds.). *Plant Pathology*. Vol. 1. Academic Press, New York. 674 pp.

Jacobi, V., and J. D. Castello. 1991. Isolation of tomato mosaic virus from red spruce in the Adirondack Mts. (abst.) *Phytopathology* 81:1173.

Kuijt, J. 1969. *The Biology of Parasitic Flowering Plants*. University of California Press, Berkeley. 246 pp.

Latham, D. H., K. F. Baker, C. Hartly, and W. C. Davis. 1938. Dodder in forest nurseries. *Pl. Dis. Reptr.* 22:23–24.

Leben, C. 1965. Epiphytic microorganisms in relation to plant disease. *Annu. Rev. Phytopath.* 3:209–230.

Little, E. L., Jr. 1949. Important forest trees of the United States, pp. 763–814. In *Yearbook of Agriculture*. USDA, Washington, D.C.

Lowry, W. P. 1963. Biometeorological data collection. *Phytopathology* 53:1200–1202.

Marks, G. C., and T. T. Kozlowski. 1973. *Ectomycorrhizae: Their Ecology and Physiology*. Academic Press, New York. 444 pp.

Matuszewski, M. 1973. *Forest and Shade Tree Pests of Kentucky*. Ky. Dept. Nat. Res. and Environ. Prot. 85 pp.

McNew, G. L. 1959. The nature, origin, and evolution of parasitism. pp. 19–69. In Horsfall, J. G., and A. E. Dimond (eds.). *Plant Pathology,* Vol. 2. Academic Press, New York. 715 pp.

Mueller-Dombois, D., J. E. Canfield, R. A. Holt, and G. P. Buelow. 1983. Tree-group death in North American and Hawaiian forests: a pathological problem or a new problem for vegetation ecology? *Phytocoenologia* 11(1):117–137.

Nutman, P. S. 1965. The relation between nodule bacteria and the legume host in the rhizosphere and in the process of infection. pp. 231–246. In Baker, K. F., and W. C. Snyder (eds.). *Ecology of Soil-Borne Pathogens*. University of California Press, Berkeley. 571 pp.

Pirone, P. P. 1959. *Tree Maintenance,* 3rd ed. Oxford Press, New York. 483 pp.

Platt, R. B. 1963. Axioms for biometeorological research. *Phytopathology* 53:1198–1199.

Ruehle, J. L. 1967. *Distribution of Plant-Parasitic Nematodes Associated with Forest Trees of the World.* USDA For. Serv., SE For. Exp. Sta. Rept. 156 pp.

Shigo, A. L. 1958. Fungi isolated from oak-wilt trees and their effects on *Ceratocystis fagacearum. Mycologia* 50:757–769.

Shigo, A. L. 1967. Successions of organisms in discoloration and decay of wood. *Int. Rev. Forest Res.* 2:237–299.

Sinclair, W. A., H. H. Lyon, and W. T. Johnson. 1987. *Diseases of Trees and Shrubs.* Cornell University Press, Ithaca and London. 574 pp.

Smith, W. H. 1970. *Tree Pathology, A Short Introduction.* Academic Press, New York. 309 pp.

Spaulding, P. 1958. *Diseases of Foreign Forest Trees Growing in the United States.* USDA For. Serv., Agr. Handbook 139. 118 pp.

Stakman, E. C., and J. G. Harrar. 1957. *Principles of Plant Pathology.* Ronald Press Co., New York. 581 pp.

Strobel, G. A., and D. E. Mathre. 1970. *Outlines of Plant Pathology.* Van Nostrand-Reinhold Co., New York. 465 pp.

Thornberry, H. H. 1966. *Index of Plant Virus Diseases.* U.S. Dept. Agr. Handbook 307. 446 pp.

Van Arsdel, E. P. 1965. Micrometeorology and plant disease epidemiology. *Phytopathology* 55:945–950.

Waggoner, P. E. 1962. Weather, space, time, and chance of infection. *Phytopathology* 52:1100–1108.

Walker, J. C. 1957. *Plant Pathology.* McGraw-Hill, New York. 707 pp.

Whetzel, H. H. 1925. *Laboratory Outlines in Plant Pathology.* Saunders, Philadelphia. 225 pp.

Whetzel, H. H. 1926. The terminology of plant pathology. *Proc. Intern. Congr. Plant Sci. 1st. Congr., Ithaca.* 2:1204–1215.

Wilson, C. L., C. E. Seliskar, and C. R. Krause. 1972. Mycoplasma-like bodies associated with elm phloem necrosis. *Phytopathology* 62:140–143.

3

ABIOTIC CAUSES OF DISEASE

Synopsis: This chapter emphasizes the regulatory role of the physical and chemical environment upon tree growth and the stress conditions that arise from extremes.

3.1 CLIMATIC CAUSES

The species-rich and geographically diverse flora of the North American forest is a result of a series of changing climatic influences in recent geologic time. Most of the present genera existed at least 100 million years ago, just before the end of the Cretaceous period. Forests similar to those now found in the southern Appalachians were widely distributed across a shrunken continent that contained no mountains. These forests were nurtured by a warm, moist climate with no seasonal variation.

The end of the Cretaceous saw the uplift of the Rocky Mountains and a shift to a harsher climate. This changing climate caused mesophytic deciduous trees to retreat, and the more ancient and more xerophytic conifers took their place. By the end of the Oligocene, most of the mountains had been eroded, and a very mild climate once again prevailed. The Miocene and Pliocene epochs, which followed, were periods of deteriorating climate accompanied by a renewal of mountain building.

The present Pleistocene epoch, the age of ice and humans, began after the violent uplift of the Pacific coastal range and marked the beginning of the coldest climate since the Permian period 200 million years earlier. Periodic snow accumulated and continental glaciers gradually developed. A great continental ice sheet advanced four times and melted away three times. The last glaciation, the Wisconsin, still persists but left most of continental North America about 12,000 years ago.

At their maximum extension, the glaciers extended southward to the Ohio and Missouri river courses. Except for portions of Alaska and the so-called driftless area in southwestern Wisconsin, the northern half of the continent was covered with ice. As the great ice sheets advanced, many forest species were faced with extinction. In the European Alps, which are oriented east-west, most preglacial forest species disappeared. In North America, however, there were several escape routes by which forest species could migrate ahead of the glaciers and escape extinction. So much ocean water was frozen in ice that the oceans were lowered and the continental shelves were colonized by forest species, serving as refuges. The low ocean levels also exposed a dry-land bridge between Alaska and Siberia, permitting species to migrate beween continents. North–south mountain ranges allowed southward migration.

North American forest species survived by migrating far southward. Jack pine and spruce grew as far south as Louisiana. The distinctive local climates created in the various escape areas likely caused much speciation. As the ice sheet receded, the new species migrated northward and occupied new sites and in new combinations. Although these forests are now contiguous, individual species still occupy distinct regions with dissimilar environments. Geography has had, and still has, a great influence on the present distribution of forest species. The present flora of the Rocky Mountains would be far more complex if not for the north–south mountain ranges that intercept the moisture-laden winds from the Pacific Ocean. In the more mesic Appalachians, the north-south mountains do not impede the prevailing weather patterns, which come from the Gulf of Mexico.

Temperature and soil moisture are two major factors controlling the composition and vigor of forests. Low temperature limits tree growth at high elevations and at high latitudes. Moisture must be sufficient to accumulate in the soil and in adequate quantities for tree growth. If either or both of these factors are limiting, the variety of species tends to be limited, and growth may be slow, as in the northern Great Plains. In the southern Appalachians and Puget Sound, where both of these factors are favorable, growth is luxurious.

3.1.1   Temperature

Almost all North American forests experience winter dormancy. At higher elevations and latitudes, there may be an average of only 30 frost-free days each year. On the Canadian–United States border, the average growing season is about 100–120 days. On the Pacific Coast and southern Atlantic Coast, the growing season exceeds 200 days.

Inland, the climate is typified by very cold winters and very hot summers. On the Pacific Coast, the moderating influence of the Pacific Ocean tends to prevent temperature extremes, an influence that would be felt farther inland if it were not for coastal mountain ranges. This maritime influence is far less significant on the Atlantic Coast because prevailing winds are westerly.

3.1.2   Moisture

Availability and patterns of moisture distribution depend on the oceans. Air masses moving over oceans pick up significant amounts of moisture, which may be deposited as rain and snow as they move over land masses. Precipitation is induced when saturated air is cooled by lifting. Air masses are lifted when they cross over mountains (orographic lifting) or when they are heated by land surfaces (convective lifting).

Precipitation patterns in North America result from atmospheric circulation around two great anticyclonic air masses that lie between latitudes 20° and 40° N. They form high-pressure areas with continual clockwise movement of air around themselves. The Azores-Bermuda High over the North Atlantic brings moist air from the Gulf of Mexico into the United States and Canada east of the Rocky Mountains. The air mass then moves east across the Atlantic Ocean to western Europe and south-

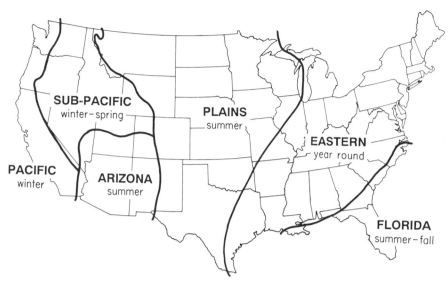

FIGURE 3.1   The major seasonal distribution zones of precipitation in the main portion of the United States. Redrawn from Barrett (1962), *Regional Silviculture of the United States.* Copyright © 1962 John Wiley & Sons, Inc. Reprinted by permission of John Wiley & Sons, Inc.

ward to Africa and then west across the Atlantic. As the air mass crosses the Atlantic, it picks up moisture and begins another cycle.

Over the Pacific Ocean, a similar situation exists, but because of the mountain barriers and prevalence of a condition in which warm air overlies cooler air, little release of rain or snow is possible, resulting in dry summers. During the rest of the year, rainfall is abundant, providing sufficient moisture for growth of the most productive forests in North America.

Six main types of seasonal distribution of precipitation have been recognized within the United States (Fig. 3.1). The *Eastern type* prevails over the East. Annual precipitation exceeds 89 cm and is evenly distributed during the year. The *Florida type* has additional amounts of heavy rain from tropical storms in late summer and early fall. The *Plains type* has a pronounced summer maximum of rainfall. Forests exist only where annual precipitation exceeds 51–64 cm. The *Pacific type* receives virtually all its precipitation from cyclonic storms in winter. The storms may be very light or extremely heavy, depending on latitude and altitude. The *Sub-Pacific type* receives precipitation from winter snows, but only enough to support forests on the mountains. The winter precipitation is prolonged into spring. The *Arizona type* occurs in Arizona and New Mexico and has an extended dry spring followed by intense shower and thunderstorm activity in summer as warm, moist air moves in from the Gulf of Mexico. Nevertheless, annual precipitation is sufficient to support forests only at high elevations.

3.1.3   Storm System Elements

Wind, snow, hail, glaze, and lightning can cause rather dramatic and often unforeseen losses to forests.

Wind has a variety of effects on tree health. Some wind is needed; trees that sway in the wind develop stronger stems. Less obvious and less understood is the shaking effect of light breezes, which cause a reduction in growth. Stronger winds cause obvious damage including uprooting, leaning, main stem breakage, bending, and broken tops or branches (Fig. 3.2). Hurricanes seldom occur but cause widespread destruction when they do. Hurricane Camille caused catastrophic destruction in Louisiana and Mississippi in 1969 (Fig. 3.3). On Steptember 21, 1989, Hurricane Hugo made landfall near Charleston, South Carolina (Fig. 3.4). It was the tenth strongest hurricane of this century and the strongest to make landfall since Hurricane Camille. With wind speeds in excess of 216 km/hr, it destroyed more than 1.6 million ha of trees, and an additional 36% of the state's forest area had more than 50% of the tree stems per hectare either broken or blown down. The volume of pine sawtimber damaged was equal to that normally harvested in 3 years and was worth more than $581 million. About 23% of the state's total hardwoods were damaged, representing 15 years of harvesting and a value of more than $231 million. Most of this timber was not salvageable for wood or fiber because of splitting, shakes, checks, and other internal strength defects and rapid deterioration caused by insects and fungi. The downed timber also posed a significant threat of fire.

Uprooted trees are colonized very quickly by stain and decay fungi, and secondary insects such as *Ips* bark beetles, borers, powder post beetles, and ambrosia beetles. Rapid drying causes checking, and solid wood products milled from these logs will have lower values. The longer salvage is delayed, the greater the amount of defect.

Leaning trees, also known as root sprung, may live for several years but will have symptoms of declined vigor and may eventually die. Leaning trees almost always have sprung roots. Trees with broken stems usually decline more quickly. Pines may be invaded by bark beetles, which introduce blue stain fungi. Hardwoods may be invaded by root pathogens and are also subject to insect attack. Oaks in the southeastern

FIGURE 3.2 Sporadic stem breakage of oak caused by strong winds in March 1990.

FIGURE 3.3   Extensive main stem bending and breakage caused by Hurricane Camille in 1969. Reprinted from Murphy (1978).

United States are particularly vulnerable to attack by the two-lined chestnut borer and colonization by *Hypoxylon atropunctatum,* which produces a rapidly advancing soft rot, or decay, of sapwood. Hardwoods frequently survive longer than conifers but are more severely attacked by decay fungi. Trees with broken branches may also be attacked by insects and stain fungi. However, wounds created by broken branches smaller than 8 cm in diameter, with no heartwood exposed, will often close with no serious long-term losses. High-value trees with major limb damage should be pruned to promote rapid wound closure.

Trees that are severely bent may resist attack by insects and fungi but should be removed in subsequent thinnings especially where wood quality is important. Leaning trees produce a large proportion of reaction wood (compression wood in conifers, tension wood in hardwoods), which may have undesirable properties for some applications. In addition, many large trees may have excessive ring shake and other internal defects, which may lead to considerable unforeseen economic losses. Planting tree species resistant to storm-related damage may help to reduce future losses in high-risk areas (Table 3.1).

Although air pollution has been suggested as a cause of crown dieback, growth decline and mortality of red spruce, and balsam fir on subalpine sites in the northern

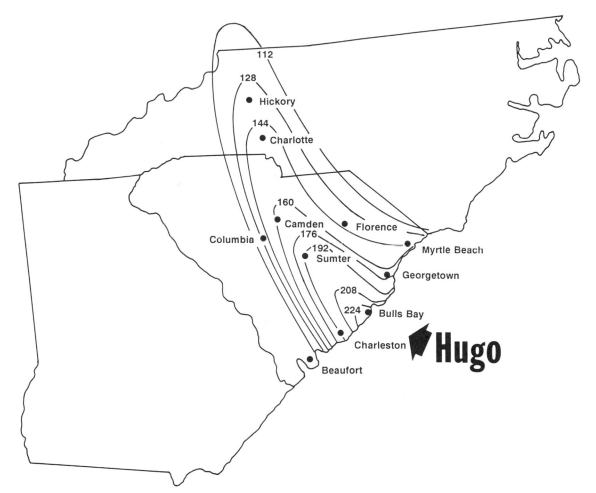

FIGURE 3.4 Gust wind speeds (km/hr) during Hurricane Hugo.

Appalachian Mountains, evidence suggests that wind is the most common form of disturbance on these sites. Excessive winds cause trees rooted in shallow, rocky soils to sway in the wind, producing breakage of small roots and collisions between rime ice-coated branches. Root and butt decay fungi may colonize the wounds and cause further injury. Trees that suffer a substantial reduction in crown size in an environment that is marginal may not be able to recover from the injury.

Although forests are generally well adapted to winter snows, occasional unseasonal, heavy, wet snows will cause injury. Heavy snow accumulations often bend young trees of many coniferous species, especially in the Rocky Mountains. Although snow bending can significantly reduce height growth the year after injury, even severely bent stems straighten after a few years. A photographic study of severely bent ponderosa and sugar pine seedlings showed no permanent stem crook 10 years later (Oliver 1970).

Movement of deep snowpack in steep mountainous country can, however, cause a variety of stem deformities in young conifer stands. Six classes of deformities have been identified (Leaphart et al. 1972) of which butt sweep was the most frequent and

TABLE 3.1 Relative Resistance of Tree Species to Hurricane-Related Damage

Breakage	Uprooting	Salt	Deterioration by Insects and Diseases
MOST RESISTANT			
Live oak	Live oak	Live oak	Live oak
Palm	Palm	Palm	Palm
Bald cypress	Bald cypress	Slash pine	Sweetgum
Pond cypress	Pond cypress	Longleaf pine	Water oak
Sweetgum	Tupelo gum	Pond cypress	Sycamore
Tupelo gum	Redcedar	Loblolly pine	Bald cypress
Mimosa	Sweetgum	Redcedar	Pond cypress
Dogwood	Sycamore	Tupelo gum	Southern red oak
Magnolia	Longleaf pine	Bald cypress	Magnolia
Sweet bay	Mimosa	Sweetgum	Tupelo gum
Southern red oak	Southern red oak	Water oak	Sweet bay
Water oak	Magnolia	Sycamore	Hickory
Sycamore	Slash pine	Sweet bay	Pecan
Longleaf pine	Loblolly pine	Southern red oak	Redcedar
Slash pine	Sweet bay	Hickory	Red maple
Loblolly pine	Water oak	Mimosa	Mimosa
Redcedar	Red maple	Pecan	Dogwood
Hickory	Dogwood	Magnolia	Longleaf pine
Red maple	Hickory	Red maple	Slash pine
Pecan	Pecan	Dogwood	Loblolly pine
LEAST RESISTANT			

Source: Barry et al. (1982).

dog leg and stem failure were the most injurious (Fig. 3.5). Most deformities originated within 1.5 m of ground line but caused the entire tree height to be deformed. Snow-bent western larch eventually recovers (Schmidt and Schmidt 1979).

In the southeastern United States, glaze injury is rather common. Glaze forms when rain comes in contact with branches whose temperature is below freezing. Ice

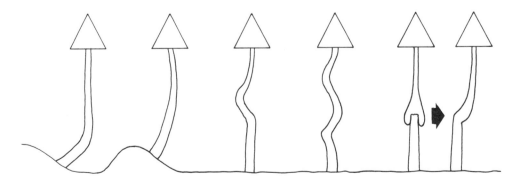

Butt sweep   Stem sweep   Dog leg   S-curve   Stem failure

FIGURE 3.5 Stem deformities in young trees caused by snowpack and its movement. Redrawn from Leaphart et al. (1972).

FIGURE 3.6   (a) Loblolly pine seedlings that have been bent over from glaze buildup. They will recover. (b) Young sawtimber loblolly trees bent and broken from glaze. They will not recover. Courtesy of C. Lee, Texas A & M University, College Station.

buildup causes a weight increase, which in turn causes breakage. If light glazing occurs, only twigs and small branches may be broken off. Heavy glaze buildups, especially if accompanied by strong winds, may cause complete breakage of crowns or uprooting. Throughout the southern Appalachian Mountains, glaze forms regularly and at higher elevations glaze damage is severe. It is not unusual to see a well-defined elevation above which glazing is common and below which glazing is seldom a problem. In the higher incidence areas, repeated limb and top breakage may deform trees, reduce growth, and provide entry to pests. Glaze damage may be a major reason for the poor form and condition of many hardwood stands in the southern Appalachians.

Long-needled conifers are especially susceptible to glaze. Large trees may be stripped of their branches, uprooted, or snapped off along the bole. Sapling and pole-sized trees may be bent to the ground and never regain an upright position (Fig. 3.6a, b). In the Ozark region of Missouri and Arkansas, the natural range of shortleaf pine is several hundred kilometers north of the natural range of loblolly pine. Severe glaze-related injury to the longer needled loblolly is suspected to be one of the reasons for this difference in their natural ranges. When loblolly pine is planted near the northerly range of shortleaf pine, it grows very well but is periodically devastated by glaze injury. Jones and Wells (1969) found great differences in glaze damage in 11-year-old loblolly pine provenances, with damage being much less among trees from the colder, more inland locations (Fig. 3.7). Although trees from the coastal areas sustained greater damage, their superior growth rate more than offset the greater glaze damage.

Lightning causes relatively less direct injury to forests because the damage is localized and response of an individual tree to lightning is highly variable. A lightning strike may have no visible effect, yet the struck tree may die unexpectedly weeks or

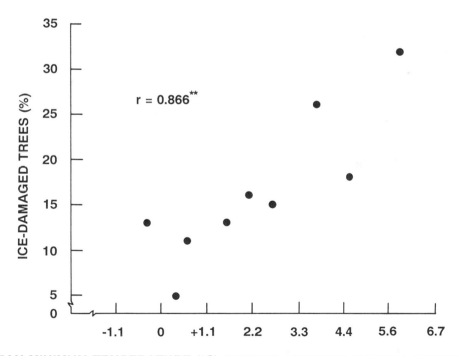

MEAN MINIMUM TEMPERATURE (°C) DURING JANUARY AT SEED SOURCE

FIGURE 3.7   The amount of ice damage sustained by loblolly pine from various seed sources is significantly correlated (at the 1% level) with the mean minimum temperature during January at the seed sources. Redrawn from Jones and Wells (1969).

months later. Often, a strip of bark from an upper branch to the root crown is blown off as the electricity is discharged into the soil (Fig. 3.8). Trees often recover from this type of injury. More extreme cases result when the tree is shattered (Fig. 3.9). Resistance in the wood causes instantaneous heating, and the wood literally explodes as the sap is converted to steam.

An important, often unrecognized, side effect of lightning strikes is that they often reduce the tree's vigor. This in turn makes the tree attractive to various other disease and insect pests. Bark beetles often infest lightning-struck coniferous trees. Many isolated southern pine beetle "spots" are associated with lightning strikes.

FIGURE 3.8   A pecan tree that has been struck by lightning. Note the strip of bark and wood that has been blown off.

FIGURE 3.9   An oak tree that has been shattered by lightning.

3.1.4   Climatic Change and Associated Stresses

Climate is the average of weather conditions at a given location. In the United States the 30-year averages of temperature and precipitation are frequently referenced as a climatic norm. Weather, conversely, includes a variety of predictable measures of the current state of the atmosphere including, but not restricted to, temperature and precipitation.

The energy in solar radiation is the driving force for climate. Natural factors that alter the amount of radiation that reaches the earth's surface may alter climate. Some of these factors include volcanic eruptions and interactions between the atmosphere and the oceans. More recently, human intervention is suspected of having an effect as well.

There is evidence that the total range of variation of average global temperature has been approximately 10°C (18°F) over the past 100,000 years. This is remarkably stable, but a short-term shift of a few degrees toward either extreme could have a significant effect on forest health. Hepting (1963) reported an increase in average global temperature during the first 40 years of this century in the United States and noted that in the early 1940s a cooling phase began. The cooling phase ended in the mid 1960s and warming resumed in about 1970 (Coakley 1988). A general warming trend of about 0.5°C (0.9°F) has occurred in the Northern Hemisphere over the past

100 years. Regional climate, however, may show a significant departure from the global trend. Figure 3.10 shows the variability in regional climate of Spokane, Washington. Note that the average minimum spring temperatures, based on a 9-year moving average, have dropped below the 87-year average beginning in the mid 1940s and have remained there to the present. During the same period, precipitation was more frequent. In the United States, precipitation decreased from the 1880s to the 1930s. Since then it has gradually increased, with a marked increase from autumn through spring for the last 30 years (Coakley 1988).

In the short term, from year to year, extreme variations in precipitation can stress trees and cause some mortality. Mainly, though, the stress is reflected in reduced growth. If moisture extremes persist, tree populations may not replace themselves and eventually die. Temperature extremes, however, have a potentially more serious effect, especially if they are colder. Lower than normal temperatures effectively shorten the growing season. Late frosts, which may affect reproduction, damage or kill flowers. Early frosts may have less of a deleterious effect because succulent tissues may have already hardened somewhat from late summer moisture stress. Extreme variations in either precipitation or temperature, which do not immediately kill trees, reduce their vigor, and they may become susceptible to diseases and insect pests.

Currently, there is much concern about the effects that humans are having on climate. Not only does climate directly affect crop yields, but it also has a largely unknown effect on pathogens, insects, and weeds that reduce crop yields. Of particular interest is the relationship between the increasing carbon dioxide concentration in the atmosphere, which traps solar energy and could lead to global warming. Carbon dioxide levels have increased from 275 ± 1 ppmv in the 1850s to 315 ppmv in 1958 to 343 ± 1 ppmv in 1984 (Coakley 1988). In addition, other gases are radiatively active. These other gases, which include ozone, nitrous oxide, chlorofluorocarbons, and methane, collectively have about the same heat-trapping ability as carbon dioxide. The chlorofluorocarbons are particularly troubling because they can also interact with the ozone in the stratosphere and may be responsible for a hole in the ozone layer over the Antarctic. The ozone layer plays an essential role in protecting life on earth from ultraviolet radiation. Methane is also increasing, at about 1.1–1.3%/yr, and is generated by marshes, swamps, and ruminant animals. As a result of anthropogenic increase in these "greenhouse" gases, global temperature could increase 1.5–5.5°C (2–10°F). Some scientists feel that the slight global warming trend observed over the past 100 years is an example of carbon dioxide interaction. Others consider the evidence inconclusive.

Some models estimate a decrease in summer moisture over middle and high latitudes if carbon dioxide levels continue to increase. This increase would tend to have a xeric influence on forests of North America but may allow longer growing seasons. Trees adapt to these changes slowly because of their long generation time. Disease and insects, however, could adapt very quickly by producing several generations each year. These pests might pose a potentially more serious threat to forest health than higher temperatures alone.

Several forest tree disorders have been induced by medium-term (10–200 years) changes in climate. One of these is pole blight of western white pine. This disease is potentially serious because pole blight occurred on nearly one seventh of the area presently occupied by western white pine. How much more will be affected in the future is unknown. Pole blight was named because it affects white pines that are in

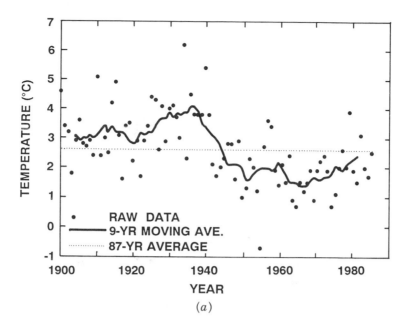

(a)

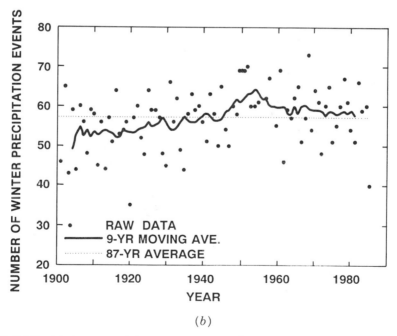

(b)

FIGURE 3.10 (a) Average minimum temperature for spring. (b) Precipitation frequency for winter at Spokane, Washington. Redrawn from Coakley (1988). Reproduced, with permission, from the *Annu. Rev. of Phytopathology*. Vol. 26, 1988, by Annual Reviews Inc.

the pole class, or those trees about 40 years old. At present, one fourth of the total white pine area under former blister rust control is in the pole class. In affected stands, trees representing 50% of the white pine basal area are diseased or dead because of pole blight.

The affected trees have slightly yellowed and shorter needles. Height growth is reduced, and resin flows from lesions on the main stem. Affected trees may die a few years after initial symptoms are apparent. Extensive rootlet mortality precedes the appearance of symptoms in the crown, and it has been suggested that the disease is systemic in nature, accompanied by a reduction in the vigor of associated mycorrhizae. Initial studies on the problem suggested that a fungus, *Europhium trinacriforme*, was a possible cause of pole blight, but subsequent studies show this fungus to be secondary, invading only after the tree has started to die.

At present, the best evidence indicates an edaphic–physiologic cause of pole blight. The disease has never been found on soils with more than 13 cm of available moisture storage capacity in the upper 0.9 m of soil. Pole blight is common on soils with storage capacities from 1.8 to 13 cm. As in birch dieback, pole blight developed during a period of years (1917–1940) when lower regional precipitation and higher temperatures prevailed.

Management of pole blight necessarily will depend on what the future research programs can develop. At present, though, it can be assumed that sites with >13 cm of available moisture storage capacity in the upper 0.9 m of soil will support white pine for 120 years and longer. Sites with less favorable moisture storage capacities may not be suitable for growing white pine.

Another example of climate-mediated forest decline is that of birch dieback which was a major problem in the northeastern United States in the decade 1938–1948. Yellow and paper birches died in large numbers, first in New Brunswick and then in the New England states. Birches, with their shallow root systems, are extremely sensitive to increases in soil temperature. The combination of circumstances that cause birch dieback continue to be responsible for the loss of many birch trees, especially birch planted as ornamentals. It is common for birch, wherever grown, to die when the other components of the stand are removed. Thus birch dieback is found occasionally wherever birch occur.

Studies of the birch dieback problem showed that the regional climate in the decade 1938–1948 was unfavorable not only for birch but also for other tree species in the area. Precipitation was less than normal, and average temperature during the growing season was higher than normal. Both yellow birch and white birch are shallow-rooted species and are sensitive to small increases in soil temperature that occur for any length of time during the growing season. Therefore, it is important to avoid sudden and extreme exposure of birch stands by removing adjacent trees that shade the birch roots. An average increase in soil temperature of just 2.2°C (4°F) can cause rootlet mortality, which is followed by a general decline in the crown and dieback of birch. In sugar maple, a less sensitive species, the soil temperature had to be raised 4.4°C (8°F) above normal before root mortality occurred. Other organisms may contribute to birch dieback. A virus may be involved. The bronze birch borer may increase the rate of birch mortality but is not considered a primary cause.

Management of birch dieback is possible to a certain extent. Increased use of birch would not leave these trees to die after logging. If more than 40% of the total volume of a fully stocked birch stand is to be cut, then all the birch should be cut. Improvement cuttings in young stands should be done to minimize exposure of root systems of residual trees to heating by sunlight. Partial cuttings at 6-year intervals have been suggested.

3.2 CHEMICAL CAUSES

The behavior of an introduced toxic chemical in the forest environment begins with its entrance and continues with its movement through the forest ecosystem until its form is modified to such a degree that it is no longer toxic. Depending on the specific chemical and its mode of action, there may be little or no effect, a delayed effect, or an immediate toxic effect on one or more components of the forest.

Harmful chemicals have either of two kinds of toxicity: acute or chronic. *Acute toxicity* is the relatively rapid response to a relatively large dosage of a particular chemical exposed over a short time period. *Chronic toxicity* is a much slower or delayed response to relatively small doses over a long period of time. There are, of course, gradations at all points between these two extremes.

3.2.1 Toxic Salts

Salt problems on trees may be divided into six arbitrary groups based on

1. soil salinity,
2. irrigation sprinkling with saline water,
3. highway deicing operations using salts,
4. ocean salt spray or cyclic salt,
5. cooling tower drift, and
6. spray canals (Curtis et al. 1977).

Salt injury results both from direct spray and from absorption through roots. Salts, mainly sodium chloride, come from two sources: the ocean and deicing compounds used on highway systems during winter months in northern North America. Tropical storms and hurricanes along the Atlantic and Gulf coasts of the United States can carry salt water as much as 80 km inland. Flooding of inland areas with salt water is more injurious than direct spray on foliage (Little et al. 1958). This salt water is trapped and cannot drain back into the ocean. Salt solution is taken up by roots and produces injury called salt burn, which may take weeks to develop and may ultimately kill the tree. Some 8,000 ha of forested land were damaged by Hurricane Hazel in North Carolina alone. Species vary in their sensitivity to salts. Loblolly pine is more sensitive than slash pine and other conifer species. Blue spruce, Austrian pine, and live oaks are quite tolerant. Salt concentrations in areas where trees died were 5,675–6,818 kg/ha in the upper 15.2–30.5 cm layer. Deposits in some localities following a storm were as high as 10,227 kg/ha. This salt is leached from the root zone rather quickly, returning to reasonably low levels within a few months after a storm.

In northern latitudes, sodium chloride and, occasionally, calcium chloride are used to keep highways in safe driving condition. One interstate highway in Connecticut received approximately 19,700 kg of sodium chloride per two-lane kilometer and approximately 840 kg of calcium chloride per two-lane kilometer in one winter (Smith 1970). Trees along highways, especially conifers, are often covered with salt spray blown from the highway surface by high-speed cars and trucks. During the following spring, affected trees are brown because of the damaged foliage. The lower portions of the trees, about 61 cm, usually escape injury because of the snow, which protects them from the salt, and the tops of the taller trees, about 4.6–6.1 m, are often not damaged. Susceptible trees such as white pine are more severely damaged, even when planted behind other species that are injured to a lesser degree. Spruce seem to be tolerant, while cedars are susceptible. The pines, other than white pine, are intermediate in susceptibility. The trees may appear to be dead in May and early June, but as the current year's needles develop, the trees seemingly recover until in September they again appear healthy. On close examination it is evident that all the older needles have been dropped and only the current year's needles survive. Inevitably these trees are not as vigorous as trees free of salt. Severely burned trees grow slowly and are often killed. To stop the use of salt is not a reasonable solution, but it is possible to minimize the damage by using resistant tree species, by planting further from the highway, and by not placing trees at road intersections where larger amounts of salt are used. In many cases the amount of salt used can be reduced. Within metropolitan areas the salts applied to the streets are taken up by the root systems of all species of trees and the damage does not appear until July or later of the following growing season (French 1959).

The margins of affected leaves turn yellow and then brown with these color changes progressing from the mid-vein toward the margin, where concentrations may become toxic. This marginal necrosis occurs because the tree translocates the salt to the leaf. Entire leaves may be killed. In severe cases, witches' brooms are formed, and twigs are stunted. The damage is typically on the side of the tree toward the street, and trees at an intersection, at the foot of a hill, or near major drainage from the street are more seriously affected than trees in less exposed locations. The symptoms are similar on all affected species, although some kinds of trees are probably more tolerant. This type of injury occurs as a result of the uptake of the salt. Affected leaves contain higher concentrations than healthy leaves. Damage is due to toxicity of the chloride ion along with other factors causing water stress (Walton 1969). Salts that contain chromium ions, which reduce vehicle rust, seem to be as damaging as plain sodium chloride or mixtures of sodium and calcium chlorides.

Injury due to roadside salt is not completely understood. Use of the most common deicing salts produces potentially excessive concentrations of sodium, calcium, and chloride, any of which can be phytotoxic. Although the chloride ion is considered the most toxic component of salt, the sodium ion readily enters the plant and may be toxic at levels less than the toxic concentrations of chloride. The sodium threshold of toxicity in eastern white pine needle and twig tissues is approximately 0.05–1.0% of the dry weight. Sugar maples have exhibited severe salt injury symptoms with foliar concentrations as low as 0.05–0.37% (Smith 1970). Sodium is readily absorbed by plant leaves and likely moves directly through the cuticle and is rapidly accumulated in leaf and twig cells and intercellular spaces until toxic levels are reached.

Shortle and Rich (1970) have determined the relative salt tolerance of 22 species of trees often used in roadside plantings (Table 3.2). Trees in the tolerant group had a significantly greater chloride content than unaffected woodlot trees, yet they exhib-

TABLE 3.2 Relative Tolerance of Roadside Tree Species to Salt Applied as a Deicing Agent

MOST TOLERANT

Red oak
White oak
Eastern redcedar
Black locust
Quaking aspen
Black birch
Paper birch
Gray birch
Yellow birch
Black cherry
White ash
Largetooth aspen
Basswood
Shagbark hickory
American elm
Red maple
Eastern white pine
Ironwood
Eastern hemlock
Sugar maple
Speckled alder
Red pine

LEAST TOLERANT

Source: Shortle and Rich (1970).

ited a very low incidence of salt injury. This study suggests that these trees can tolerate above-normal amounts of chloride in their leaves. Red oak, white oak, redcedar, and black cherry do not accumulate chloride as do the other salt-tolerant species in this group, suggesting that more than one mechanism may be involved in salt tolerance.

Salt injury to roadside trees via aerially borne salt splash, aerosol, and dust is particularly injurious to coniferous species. These have foliage that has a high surface-to-volume ratio and is particularly efficient in the interception of airborne particulates. Townsend and Kwolek (1987) have determined the relative susceptibility of many common roadside species of pine that should be taken into account when planning future plantings in high-risk situations (Table 3.3).

3.2.2 Herbicides and Silvicides

Herbicides are chemicals that are designed to kill or modify the growth of unwanted vegetation. In forestry they are usually applied to vegetation to reduce their competition with crop trees. They are often employed in site preparation for reforestation. Silvicides are herbicides that are used to kill woody or treelike vegetation. Silvicides are more likely to be used to kill or remove competing woody vegetation in an already established stand.

TABLE 3.3 Relative Susceptibility of 13 Pine Species to Sodium Chloride Spray

MOST TOLERANT

Japanese black pine
Ponderosa pine
Austrian pine
Southwestern white pine
Bristlecone pine
Japanese white pine
Red pine
Scotch pine
Eastern white pine
Jack pine
Swiss stone pine
Balkan pine
Japanese red pine

LEAST TOLERANT

Source: Townsend and Kwolek (1987).

Aerially applied chemicals may be distributed to the air, vegetation, forest floor, and water (Fig. 3.11). The movement, persistance, and fate of herbicides in the forest ecosystem may be quite variable. As much as 75% of aerially applied herbicides can move to locations other than the target area, often leading to unintended injury. In the great plains, where it is difficult to grow trees because of low annual rainfall, Siberian elms and boxelder were used extensively in the past. These trees are often planted as shelter belts, in close proximity to agricultural fields where herbicides are used. These trees have been damaged by 2,4-D and other agricultural herbicides (Otta 1974) (Fig. 3.12).

Some spray material is dispersed by the wind as drift. The amount of drift depends on release height, droplet size, and wind velocity. Some chemical may volatilize as it falls through the air or as it evaporates from plant or soil surfaces. Some may fall on surface water. Most of the herbicide not lost through drift or volatilization is intercepted by vegetation or the forest floor.

The amount of herbicide intercepted by vegetation depends on the nature of the herbicide, the rate of application, and the density of the vegetation. There is limited absorption of many herbicides and very little translocation if they are absorbed. Much of the herbicide is washed off by rain. Growth dilution and metabolism to nonbiologically active substances also reduce herbicide effectiveness.

The forest floor is a major receptor of sprayed herbicides and is a large, potential reservoir of stream pollutants. Any undegraded herbicide can dissolve in surface run-off or leach through the soil profile.

There is, thus, great potential for unplanned exposure of adjacent vegetation to herbicides. To use herbicides safely, we must take advantage of the toxic behavior characteristics of the chemical while maximizing exposure of target organisms and minimizing exposure of nontarget organisms.

Other methods of application such as direct injection of unwanted trees avoid most of the problems associated with drift but are much more costly to apply. Pelletized

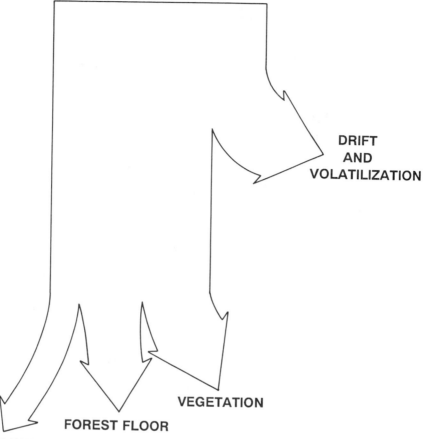

**DISTRIBUTION OF
AERIALLY APPLIED CHEMICALS**

**DRIFT
AND
VOLATILIZATION**

VEGETATION

FOREST FLOOR

SURFACE WATER

FIGURE 3.11   The initial distribution of aerially applied chemicals. Redrawn from Norris (1971).

formulations in slow- or controlled-release granules or grid balls permit aerial applications with minimal drift problems.

3.2.3 Air Pollutants

Air pollution has become a popular issue in recent years but is certainly not new. Compounds emitted from various manufacturing processes have been a part of our environment for many years, but as the world's population multiplies at an ever increasing rate, the problems of excessive air pollutants become more acute. Many toxic compounds are dispersed into the atmosphere or dumped for meaningful or other purposes into water systems and on the landscape. Only a few of the more important materials toxic to plants are discussed here.

FIGURE 3.12 Oak leaves and twigs with growth imbalance resulting from drift of 2,4-D herbicide.

The Nature of Plant-Pathogenic Air Pollution

Air pollutants affecting plants arise from a variety of natural and human-related phenomena. They can be divided into primary and secondary types. *Primary pollutants* are those that originate at the source in a form toxic to plants; sulfur dioxide (SO_2) and hydrogen fluoride (HF) are examples of primary pollutants. *Secondary pollutants* develop as a result of reactions among pollutants. The photochemical pollutants—

peroxyacetyl nitrate (PAN) and ozone (O_3)—are examples of this type. Ozone, SO_2, HF, PAN, oxides of nitrogen (NO_x), particulates, aldehydes, chlorine (Cl_2), hydrogen chloride (HCl), ethylene, hydrogen sulfide, and silicon tetrafluoride (SiF_4) are the most common air pollutants affecting plants in the United States today. Of these, the first four are by far the most important and cause the most damage. In addition, there are many other pollutants of rather minor importance.

Sources of Plant-Pathogenic Air Pollutants

Sulfur dioxide is emitted principally from the combustion of coal; the production, refining, and use of petroleum and natural gas; the manufacturing and industrial use of sulfuric acid and sulfur; and the smelting and refining of ores, especially copper, lead, zinc, and nickel.

The combustion of coal represents the major source of SO_2. The amount of SO_2 emitted depends upon, among other things, the sulfur content of the coal. In the United States, the sulfur content of coal ranges from less than 1% to as much as 6%, with 2% as an average. Coal-burning power plants are the most important single source of SO_2. Many instances of vegetation damage have been associated with coal-burning power plants.

Fluorine-containing compounds, such as HF and silicon tetrafluoride (SiF_4), originate principally from aluminum reduction processes, manufacture of phosphate fertilizer, steel manufacturing plants, brick plants, pottery and ferroenamel works, and petroleum refineries. Fluorides originate from the molten cryolite bath in the manufacture of aluminum and from impurities in the raw materials used by other industries. Fluorides are toxic at much lower concentrations than are most other air pollutants. Therefore, while the annual tonnages produced do not compare with those of pollutants such as SO_2, fluorides still represent a major problem.

Ozone (O_3), usually a secondary-type pollutant, has been long recognized as a phytotoxicant, but only recently has its effect on trees been appreciated. Sources of O_3 are the upper atmosphere, electrical storms, and photochemical reactions.

Tropospheric O_3 may be brought to the earth's surface during atmospheric disturbances such as hurricanes and violent storms. Ozone may also form during thunderstorms, when electrical discharge splits molecular oxygen (O_2) to form atomic oxygen (O), which subsequently reacts with molecular oxygen to form O_3. The concentration of O_3 at ground level increases during many thunderstorms, often enough to injure vegetation. Weather fleck of tobacco, a disease that was associated with atmospheric disturbances for many years, is caused by O_3. In this case, ozone is a primary pollutant.

Ozone, however, is usually considered a secondary pollutant because the most important source of O_3 today is photochemical reactions in polluted atmospheres. Oxides of nitrogen (NO_x), which are emitted into the atmosphere by automobiles and a variety of industries and utilities, may react in the presence of sunlight and oxygen to form O_3 as follows:

$$NO_2 + \text{light} \rightleftarrows NO + O$$

$$O + O_2 \rightleftarrows O_3$$

$$NO + O_2 \rightleftarrows NO_3$$

$$NO_3 + O_2 \rightleftarrows NO_2 + O_3$$

Likewise, the irradiation of mixtures of NO_x and various hydrocarbons often results in O_3 formation. Hydrocarbons emitted by motor vehicles and NO_x emitted by motor vehicles, industrial processes, and the generation of electrical power can react to form the secondary pollutant O_3, which is a major phytotoxic ingredient of urban smog.

Peroxyacetylnitrate has been recognized as a common phytotoxic constituent of the smog over cities in recent years. Peroxyacetylnitrate is one member of an homologous series of compounds that originate principally from the reaction of olefin-type hydrocarbons and NO_x in the presence of light.

$$\text{Olefin} + NO_x \xrightarrow{\text{light}} > CH_3\text{--}COONO_2$$

Peroxyacetylnitrate, its relatives, and a number of other compounds, such as aldehydes, can form. Their formation depends on the type of olefin involved, on whether NO or NO_2 is present, and on the duration of the time of irradiation. The automobile and other forms of transportation represent the major sources of hydrocarbons and NO_x that react to form PAN.

Terpenes released by coniferous vegetation can also react photochemically with NO_x to form ozone and PAN. Consequently, natural sources of hydrocarbons in the vicinity of human-related sources of NO_x, such as power plants, could result in the formation of photochemical air pollutants and thus complicate control efforts.

Recently, scientists have recognized that NO_x compounds have the potential to be primary phototoxic pollutants. Large amounts of these compounds are emitted annually. Oxides of nitrogen originate from a variety of sources that include gasoline combustion in motor vehicles; refining petroleum; combustion of natural gas, fuel oil, and coal; and incineration of organic wastes.

Hydrogen chloride was an important phytotoxic air pollutant in Europe approximately 100 years ago. However, HCl emissions were reduced when technological changes replaced the Le Blanc soda process. In recent years there has been a reappearance of diseases attributed to HCl. Chlorine, which is also highly toxic to plants, has been implicated as the cause of injury to vegetation in several instances. These two chemicals are increasing in importance as phytotoxicants. Current major sources of HCl and Cl_2 are refineries, glass-making, incineration and scrap burning, and accidental spillage. Polyvinyl chloride (PVC) is used in large quantities in manufacturing packaging materials and wire insulation. Combustion of PVC releases HCl and a myriad of other compounds. The increased use of PVC suggests that HCl could again become important as an air pollutant.

Particulates have been recognized as phytotoxic air pollutants for many years; however, there has been relatively little research on the problem. The major sources of atmospheric particulates include combustion of coal, gasoline, and fuel oil; cement production; lime kiln operation; incineration; and agricultural burning and other agricultural-related activities.

Transport of Plant-Pathogenic Air Pollutants

The transport and dispersion of air pollutants is determined by the concentration of the pollutant at the source and meteorological factors such as wind speed and atmospheric stability. Dispersion of pollutants is maximum when wind velocity is high and the atmosphere is unstable. In an unstable atmosphere, temperature decreases with altitude, and vertical and horizontal dispersion are maximal due to thermal and dynamic turbulence. These conditions exist most frequently during the day, especially cloudless days. In contrast, in a stable atmosphere, temperature increases with altitude, and there is little turbulence. Stable conditions occur most frequently during nighttime hours. Pollutant dispersion is minimal under stable conditions.

Other factors such as source height are also important in pollutant dispersion. When the source is at ground level, maximal ground-level pollutant concentrations occur near the source, and the ground-level concentration diminishes rapidly with distance from the source. An elevated source, such as a smokestack, results in a low ground-level concentration near the source. However, ground-level concentration increases rapidly with an increase in distance from the source to a point at which a maximum occurs, and then ground-level concentration decreases with a further increase in distance from the source. In an area where a pollution source exists at ground level, we would find maximal injury to vegetation near the source and a decrease in the amount of injury as distance from the source increased. Because of the high reactivity of halogens, this relationship is especially pronounced (Fig. 3.13). With elevated sources, there is often little or no injury near the source; maximal injury may occur near the point at which maximal ground-level concentrations occur. Also, there is usually no gradient of injury to vegetation in the vicinity of elevated sources but rather "pockets" of injury in varying directions and at different distances from the source.

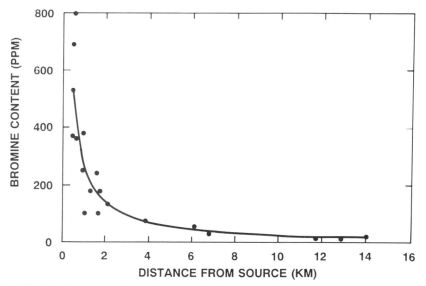

FIGURE 3.13   Concentrations of bromine in first-year loblolly pine needles growing around a bromine extraction and production plant. Note the exponential decrease of bromine in plant tissues farther from the source. Redrawn from Tainter and Bailey (1980). Adapted with permission from *Environmental Science and Technology,* Vol. 14. Copyright 1980 American Chemical Society.

Effects of Plant-Pathogenic Air Pollution

In general, the symptoms of air pollution injury to plants can be classified as either necrotic, chlorotic, or atrophic. Specific effects include changes in cell wall permeability, plasmolysis, changes in tissue pH, interference with cell wall synthesis, acceleration of respiration, inhibition of enzymes, and reduction in the rate of photosynthesis. Effects are often classified as either acute or chronic, with acute implying some degree of necrosis. In contrast, chronic refers to nonnecrotic, usually more subtle effects.

Sulfur dioxide causes an interveinal chlorosis and necrosis of the leaves of broad-leaved plants and usually tip necrosis or banding of the leaves of evergreens. On broad-leaved plants such as pinto bean and alfalfa, the interveinal necrosis is usually ivory colored; on broad-leaved plants such as red maple and hawthorn, the interveinal necrosis may be reddish brown. The tip necrosis of evergreens is usually characterized by a dead needle tip with a chlorotic band between the necrotic area and the healthy tissues. On many evergreens, older needles are shed prematurely, which ultimately reduces growth rate.

Hydrogen fluoride, in contrast to SO_2, causes a marginal chlorosis and necrosis of the leaves of broad-leaved plants (Fig. 3.14) and tip chlorosis and necrosis of evergreens. Fluorides are highly mobile within the plant. Instead of causing injury at the point of every entry, for example in the interveinal areas, they are usually translocated to the leaf margin where they accumulate and eventually cause injury. Because of their mobility and ability to accumulate in certain plant tissues and because of their extreme toxicity (atmospheric concentrations of one part per billion (ppb) may injure some plants), fluorides present a special type of problem. The ability of fluorides to accumulate in vegetation and their toxicity to animals have caused additional problems. Cattle

FIGURE 3.14   Injury to leaf margins of dogwood due to hydrogen fluoride emitted from a nearby fiberglass insulation plant.

grazing on vegetation near fluoride sources often develop a disease known as fluorosis, caused by the excessive fluoride taken in during foraging. Fluorides interfere with blood and bone metabolism, which results in a blackening of the teeth and overall debilitation of the animal. Serious outbreaks of fluorosis have occurred in cattle herds in the western United States.

Ozone damage to plants is often characterized by a metallic flecking or pigmented stippling of the upper leaf surface of broadleafed plants. The upper leaf surface effect is due primarily to the fact that the palisade cells are much more sensitive to O_3 than the spongy mesophyll cells. Ozone causes tip necrosis, tip chlorosis, banding, and chlorotic mottling of the leaves of conifers. With species such as ponderosa pine, the older needles are most sensitive to mottling. Older leaves of broadleafed plants are also more sensitive.

Ozone that develops photochemically in the atmosphere over cities is not necessarily confined to the city. In recent years, the chlorotic decline of ponderosa pine in the San Bernardino Mountains east of Los Angeles has been attributed to O_3 that originates in the smog cloud over Los Angeles. Researchers have demonstrated this association by measuring ozone concentrations in affected areas. Then, they exposed plants in fumigation chambers using similar levels of ozone and were able to duplicate the chlorotic mottle observed on the older needles in the field. Hence, photochemical air pollutants, such as O_3, that develop in the atmosphere over urban centers may be carried to adjacent forest areas and cause damage as far as 80, 96, and 112 km away.

Peroxyacetylnitrate, in contrast to O_3, causes symptoms on the lower surfaces of the leaves of vegetables. The symptom that occurs most often is glazing or silvering of the underleaf surface, which may develop into a necrotic-type symptom. This glazing or silvering is due primarily to a destruction of spongy mesophyll cells and the subsequent movement of air into the area of the leaf between the lower epidermis and the palisade cells. Peroxyacetylnitrate causes interveinal necrosis, premature defoliation, and other symptoms on woody plants such as white ash. Also, young, developing leaves are most sensitive. Although PAN is an important pollutant, we know relatively little about it because it is explosive and few researchers will work with it. Also, measuring PAN concentrations is quite difficult.

As far as relative toxicity is concerned, SO_2 is the least toxic and HF is the most toxic of the four compounds—O_3, SO_2, HF, and PAN. PAN is slightly more toxic than O_3. Many factors are important in determining the severity of plant response, including the degree of sensitivity of a given plant; the nature of environmental conditions before, during, and after fumigation; the concentration of the pollutant during fumigation; and the length of the fumigation.

Management of Plant-Pathogenic Air Pollution

Air pollution diseases of plants can be controlled by

1. eliminating the source,
2. using resistant varieties,
3. using chemical protectants, and
4. forecasting air pollution episodes.

Eliminating the Source. If we establish air-quality standards, we can eliminate the source of air pollution. Once an acceptable standard is established, emissions from the source are controlled to meet the standards. Compliance with established standards is enforced through legal channels.

Using Resistant Varieties. Because tremendous variation exists in species populations with respect to pollutant sensitivity, we can use standard breeding techniques to develop resistant selections. Because pollutants do not vary, pathogen variability is eliminated; hence, resistant selections are more likely to survive than are selections resistant to biological pathogens.

Using Chemical Protectants. Chemicals applied as sprays or soil drenches can provide protection against pollutants such as ozone. For example, a soil drench of benomyl protected pinto beans from injury at O_3 exposures of 25 ppm for 4 hours. There is much research interest in this area at the present time.

Forecasting Air Pollution Episodes. Forecasting the occurrence of air pollution episodes that might result in vegetation injury is, as the phrase indicates, nothing more than predicting the occurrence of ground-level concentrations that coincide with favorable environmental conditions and the presence of susceptible plant populations. Modified forecasting systems have been used to reduce pollutant emissions during periods of maximum plant sensitivity. Coal-burning power stations have curtailed emissions by reducing operating load or using low-sulfur coal. Efforts are presently underway to model air pollution-induced diseases with the ultimate goal of developing simulators to be used as a basis for prediction. This approach to air pollution control appears to be highly feasible, especially with greenhouse crops.

3.3 EDAPHIC CAUSES

3.3.1 Disturbed Mineral Nutrition

Minerals can be defined as any naturally occurring substance that is neither animal or vegetable. Minerals are often more narrowly defined as those naturally occurring elements that plants must absorb to live and grow. The study of mineral nutrition is an important subscience of plant physiology. The disruption in mineral physiology of host trees is an important cause as well as a result of disease.

To determine which elements are essential for plant growth, healthy plants are dried to remove the water and then the principal elemental components are scientifically determined. For most plants, these include oxygen and carbon (45% of each on a weight basis) and much lower concentration of hydrogen, nitrogen, potassium, phosphorus, sulfur, calcium, magnesium, chlorine, iron, manganese, copper, boron, zinc, and molybdenum (Table 3.4). In all, 16 elements are considered essential for most angiosperms and gymnosperms. To be considered essential, the plant must require that element to complete its life cycle, or the element must be a part of an essential molecule or constituent (such as the magnesium in chlorophyll). The listing of adequate concentration of elements in tissues (Table 3.4) is useful to forest manag-

TABLE 3.4 Adequate Internal Concentration of Essential Elements Typical of Most Higher Plants

Element	Concentration in Dry Tissue	
	ppm	%
MICRONUTRIENTS		
Molybdenum	0.1	0.00001
Copper	6	0.0006
Zinc	20	0.0020
Manganese	50	0.0050
Boron	20	0.002
Iron	100	0.010
Chlorine	100	0.010
MACRONUTRIENTS		
Sulfur	1,000	0.1
Phosphorus	2,000	0.2
Magnesium	2,000	0.2
Calcium	5,000	0.5
Potassium	10,000	1.0
Nitrogen	15,000	1.5
Oxygen	450,000	45
Carbon	450,000	45
Hydrogen	60,000	6

Source: Salisbury and Ross (1958).

ers and arborists because an analysis of the concentration of elements in certain tissues (such as leaves or fruits) often indicates whether the tree is suffering from a deficiency. Figure 3.15 shows the generalized range of element concentration and plant growth response. Above the critical concentration, the relative amount of a given element has little effect on plant health. If the concentration of the element is below the critical concentration, growth may be reduced, and as a result, deficiency symptoms may be present. Excessive concentrations of an element may lead to toxic effects and reduced growth or death.

Another useful method for determining the role of specific macronutrients and micronutrients is to grow plants in an otherwise complete, balanced nutrient solution and to delete specific nutrients, one at a time. Nutrient deficiencies often produce characteristic visual symptoms. If deficiency symptoms for two or more elements are similar, a chemical analysis of the tissues will usually identify the deficient element. An excellent application of this technique for black walnut, eastern cottonwood, black locust, sweetgum, and Scot's pine was done by Hacskaylo et al. (1969). Essential elements either are part of an important compound or have an enzyme-activating role. Some essential elements may play both roles.

The first seven elements listed in Table 3.4 are often called trace elements or micronutrients. They are needed in tissues at concentrations of less than 100 ppm of dry weight of tissue. The last nine are macronutrients, which are usually needed in tissues at concentrations of greater than 1,000 ppm of dry matter.

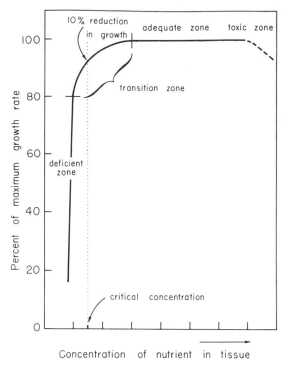

FIGURE 3.15   A generalized plot of plant growth as a function of tissue nutrient concentration. Redrawn from Epstein (1972).

Nutrient Deficiency Symptoms of Some Essential Elements

A deficiency of an essential element causes a characteristic symptom, which may also be associated with stunting chlorosis or necrosis of various plant parts (Fig. 3.16). A knowledge of characteristic nutrient deficiency symptoms helps foresters, arborists, seed orchard managers, and tree nursery managers to diagnose problems and take corrective action. However, nutrient deficiencies can be very difficult to diagnose accurately because the symptoms are often similar to those produced by many pathogenic organisms. A characteristic of some nutrient deficiencies, however, is that the young, growing parts of plants, and especially flowers and seeds, can have a pronounced sink effect and can withdraw mobile elements from older parts. Hence, older tissues may show a nutrient deficiency first.

Some Functions of Essential Elements

Molybdenum.   The only known function of molybdenum is in the nitrate reductase system, an enzyme involved in nitrogen fixation by legumes. Only trace amounts are required by plants. Molybdenum deficiencies are rare, especially in basic soils.

Copper.   Copper is associated with chloroplasts and proteins and is present in several enzyme systems involved in oxidation and reduction. Only small amounts are required by plants. Excesses frequently become toxic.

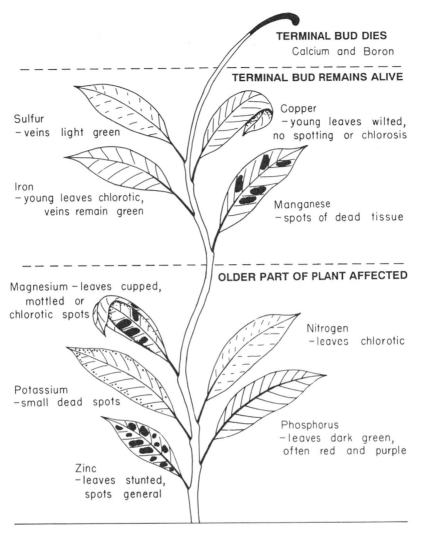

FIGURE 3.16 A guide to plant nutrient deficiency symptoms. Drawn from data presented in Salisbury and Ross (1985).

Zinc. Many enzymes require zinc, and it is necessary to produce the growth regulator indoleacetic acid. Zinc deficiency causes a rosette type of disorder, which results from a stunting of leaves and stem internodes. Zinc deficiency has been recognized as the cause of rosette disease of pecans in the coastal plain of southeastern United States. In the early stages the newer leaves at the top of the tree become yellow, mottled, or wrinkled. Later, the leaves become narrow and twisted, and the shoots die back from the tips.

Boron. Physiological functions of boron are not clear, but it appears to be involved in nucleic acid synthesis, sugar translocation, respiration, reproduction, and water relations. Terminal twigs and root tips fail to elongate normally when boron is deficient, resulting in stunting, poor survival, and stand irregularities. Boron deficiency is common in many countries, especially in exotic plantations of eucalypts and pines (Stone 1991). Application of slowly soluble boron fertilizer in intensively managed plantations on susceptible soils is becoming routine.

Manganese. Manganese plays a role in chloroplast formation and oxygen evolution in photosynthesis; it also activates many enzymes involving oxidation-reduction reactions. Manganese deficiency has caused a serious growth-retarding physiological disorder of the plumy coconut palm in Florida. Leaves of affected plants become increasingly chlorotic with the veins remaining darker green. New leaves are stunted, and much of the leaf area may be killed. In final stages, growth stops and the plant dies. Effective control is achieved by application of manganese to the soil or as a foliar spray.

Chlorine. The chloride ion is widely available and is absorbed in far greater quantities than needed. It plays a role in photosynthesis and in cell division.

Iron. Iron forms parts of certain enzymes, is involved in respiration, and is closely related to chlorophyll formation. Iron deficiency results in a rapid breakdown in the formation of chlorophyll. Some plants are particularly sensitive to iron deficiencies. Iron is relatively immobile.

Calcium. Calcium is needed during cell division and occurs in the middle lamella as calcium pectate. It is essential for proper structure and function of cell membranes. Because calcium is almost immobile, deficiency symptoms are most evident in young tissues.

Magnesium. This element activates many enzymes and is essential in photosynthesis, respiration, and formation of nucleic acids. It is readily translocated, and older tissues are the first to exhibit deficiency symptoms.

Sulfur. Sulfur is an important component of many proteins and is found in the amino acids cystein, cystine, and methionine. Excess sulfur can be absorbed from the atmosphere and has toxic effects. Sulfur deficiency is usually not a problem, but there are sulfur-deficient regions. Sulfur is not readily redistributed in the plant.

Potassium. Potassium is present in plants in large amounts, is an activator of many enzymes in photosynthesis and respiration, and is an important contributor to maintain internal osmotic pressure. Potassium deficiency symptoms develop quickly, especially in older tissues because the element is readily retranslocated.

Phosphorus. Phosphorus is a component of the high-energy phosphate bonds of adenosine triphosphate (ATP), which is responsible for energy transfer in biochemical systems. It forms a portion of ribonucleic acids (RNA) and deoxynucleic acids (DNA). These molecules mediate protein synthesis and transfer of genetic information. Phosphorus is also involved in many other metabolic processes. It is second only to nitrogen as the element most limiting in soils. Phosphorus deficiencies appear first in older tissues. Phosphate is low in solubility and is fixed in unavailable forms by iron and aluminum in soil.

Nitrogen. Nitrogen deficiencies are common (Fig. 3.17). Nitrogen is an important component of chlorophyll, enzymes and structural proteins, nucleic acids, and other organic compounds. Nitrogen-containing compounds may make up 40–50% of the dry weight of protoplasm. For this reason, it is required in relatively large amounts. Nitrogen is mobile and moves from older tissues to younger growing areas. Hence, deficiency symptoms appear first in the older tissues.

FIGURE 3.17   Chlorosis and partial failure of loblolly pine planted in an old field in South Carolina (a and b). Greatest mortality is in areas formerly in alfalfa (arrows). Soil pH in these areas ranged from 6.7 to 7.0, which is much too high for the pines to use the available nitrogen.

Many pine species exhibit a foliar discoloration during the dormant season. In Scot's pine this discoloration is due primarily to the loss of chlorophyll, accompanied by an increase in carotenoid concentration (Jones 1971). The loss of chlorophyll was not shown to be related to a nitrogen deficiency. A sharp increase in chlorophyll concentration accompanied the return to summer green.

Chlorosis of shade-tolerant species can also result from exposure to full sunlight under severe growing conditions. It appears first in older foliage but eventually even the newer growth becomes chlorotic. This phenomenon, known as solarization, is caused by exposure to extreme light intensities that inhibit photosynthesis. It is recognized by the chlorotic appearance of foliage, which results from the destruction of chlorophyll (Ronco 1970). It may be confused with nitrogen deficiency.

Role of Soil pH

Soil pH has an important influence on availability of some mineral elements. Iron, zinc, manganese, and copper are less soluble in alkaline than in acidic soils because at the higher pH they precipitate as hydroxides. Phosphate is less readily absorbed at pH of greater than 6.5 or lower than 5.5. At a high pH phosphate is usually present as insoluble calcium phosphate and at low pH, as aluminum phosphate.

Iron chlorosis affects many tree species on alkaline soils. Chlorosis is a common disease in some prairie regions of the central United States, Colorado, and other locations because of the high alkalinity of the soil. At pH levels of 7.8 or greater, iron is too tightly bound to soil particles for optimum plant uptake. Early symptoms include a general yellowing of younger leaves, which is most intense between the veins and at the margins. Eventually, shoot growth is reduced, and within a few years the tree may die. Iron chlorosis is easily treated by foliar or soil application of iron sulfate, iron ammonium citrate, and synthetic iron chelates. Iron chelates are more effective in soil because their buffering ability helps them maintain iron in a soluble form. Iron compounds may also be treated by injecting them directly into the tree trunk. Iron chlorosis has been temporarily corrected by adding sulphuric acid to the soil.

Role of Pathogen

Nutrient deficiencies can produce a variety of symptoms that greatly resemble those caused by some pathogens. Pathogens often cause nutrient-deficiency-like symptoms when they injure roots. Roots are the plant's absorbing structures, which absorb water and mineral elements from the soil. Pathogenic injury to the root system may cause foliar symptoms, which may be very difficult to diagnose as to cause. The littleleaf disease of shortleaf pine is a good example. Root-tip infections by species of *Phytophthora* and *Pythium* cause foliar symptoms typical of a nutrient deficiency. In the early stages a slight yellowing and shortening of the needles is evident. Later, the new needles are discolored and much shorter than normal, and only the current needles are retained. Early research to determine the cause(s) identified a nitrogen deficiency in affected trees (Roth et al. 1948), and symptoms could even be alleviated temporarily by adding nitrogen. Later, when Koch's postulates were successfully used to show the involvement of root disease organisms, the disease scenario was clarified.

Mycorrhizae form symbiotic partnerships with most tree species and facilitate the extraction of water and mineral nutrients from soil. Conifer seedlings, which were planted in the great plains states during the 1930s failed to grow well, or even survive, until infested soil was inoculated into the mycorrhizal-deficient soils. Many similar instances have been reported around the world. Mycorrhizal fungi are essential in increasing absorption of nutrients that normally diffuse very slowly toward roots yet are needed in high demand. These include NH^{4+}, K^+, NO^{3-}, and especially phosphate.

Role of Nutrition in Disease

The nutritional condition of a tree can dramatically affect its response to pathogen attack. Excess of certain nutrients can affect the physiological condition of a plant, which in turn alters response of the plant to disease. For example, seedlings grown with excess nitrogen are spindly, and much of their tissue is succulent and slow to harden off. Damping-off disease in forest tree nurseries is an example. Nursery managers often withhold nitrogen during the early growth, which causes seedlings to harden off quickly and makes them more resistant to attack by soilborne pathogens. Another example in the nursery is the steps taken to encourage proper mycorrhizal development by limiting soil phosphorus. Infection of pine roots by mycorrhizal fungi is difficult to achieve unless the seedlings are grown under a phosphorus stress. If phosphorus is plentiful, the seedlings will not form mycorrhizal associations. Seedlings that are not mycorrhizal when outplanted are less likely to survive the stress of the first growing season in the field. The fungus partner is not present to quickly colonize soil around the planting site and the seedling will die from lack of water.

These examples illustrate some of the effects of mineral nutrition on disease susceptibility. Although the specific examples discussed previously illustrate situations when excess nutrients favor disease, a vigorously growing tree is generally more likely to be a healthy tree. The effects of disease on mineral nutrition in the host have not been well studied, but they certainly have an effect. There is circumstantial evidence that nutrient deficiency in aspen increases susceptibility to Hypoxylon canker. Colorado blue spruce is often planted off-site, especially in urban settings, and is unable to develop an adequate root system so that it often suffers from nutrient deficiency. As a result, it is susceptible to attack by *Cytospora,* which first colonizes lower branches and later higher branches as the tree generally declines in vigor.

3.3.2   Temperature Effects

Growth of trees is extremely sensitive to temperature. Some species can grow well over a wide range of temperatures; others require a much narrower range. When temperature maximums or minimums are approached, we can expect some type of injury to result. Exposure to elevated temperatures often accompanies drought conditions, which alone are an important source of stress. A discussion of noninfectious diseases caused by temperature extremes follows.

High Temperature

Heat defoliation and leaf scorch may occur as a result of excessive temperatures coupled with severe dry winds. These conditions cause a rapid loss of water, especially in maple leaves. Leaf margins turn yellow or brown, and leaves fall prematurely. Leaf scorch can be prevented by planting trees, especially maples, in locations protected from long exposure to sun and wind. Smaller trees in high-risk locations can be mulched and then watered during periods of hot weather to compensate for the excessive water loss.

Temperature stresses on urban trees can be especially severe. Extreme soil temperatures, which occur under the more open stands or adjacent to pavement, can themselves be lethal to vegetation. Soil temperatures under trees planted in holes in asphalt in a parking lot were up to 10°C (18°F) higher than in uncovered soil at a depth of 15 cm below the surface (Halverson and Heisler 1981). Higher soil temperatures may increase evapotranspiration, which can deplete soil moisture and lead to water stress. Pavement covers a significant portion of urban areas.

Sunscald describes localized death of the cambium resulting from overheating of cambial tissues after sudden exposure to the sun such as when surrounding trees are removed. Solarization is a photochemical reaction that reduces photosynthetic activity of plants exposed to extremely high light intensities (Ronco 1975). The light inhibits photosynthesis followed by oxygen-dependent bleaching of chloroplasts. Plant mortality may occur. Certain carotenoid pigments normally protect against solarization by absorbing excess light energy. Some species, in particular Engelmann spruce, do not have adequate protection and may have a low survival rate unless they are shaded. This type of damage is most common at high elevations in mountainous areas.

Winter sunscald is caused by a combination of above-freezing temperatures during the day and freezing temperatures at night. Winter sunscald usually occurs on the southwest side of the tree during the late winter or early spring. The cambial tissues warmed by the sun during the day become active and then are killed when they freeze at night. An elongated canker results. Thin-barked species such as young apple and maple trees are most susceptible to winter sunscald (Fig. 3.18). The injury can be avoided in newly planted shade trees by wrapping them with crepe paper, sisal kraft paper, or other materials that will prevent excessive warming of the cambial tissues.

Heat injury results from high soil temperatures. Seedling mortality is common on cutover areas in the central Cascade range of Oregon (Hallin 1968). Maximum soil temperatures of 68–73°C were observed in August on the south face of 100% slopes (Fig. 3.19). A relationship of approximately 0.9°C change in maximum temperature per 10% change in slope was found on south-facing slopes.

Heat injury is also common in nurseries when the soil surface temperature becomes high enough to kill cells in the stem, which in turn results in a swelling of the stem above the injured tissues (Barnard 1990). This swelling is near the ground line (Fig. 3.20) and can be confused with symptoms of damping-off. Soil temperatures in nurseries have been as high as 71–79°C, and temperatures of 54°C are not unusual. Heat injury can be prevented by shading the seed beds. Dark soils pose more of a threat to high-temperature buildup than light soils as dark soils absorb more solar radiation.

Low Temperature

Frost injury occurs in the fall (early frost), in the winter, and in the spring (late frost). Frost damage occurs when temperatures suddenly drop below freezing, before plants are hardened (Fig. 3.21).

Plants have differing tolerances to low temperatures. Some plants can grow and even flower under snow. Low temperatures dramatically slow the metabolism of many life processes, but freezing has the most dramatic effect. Ice crystals may form in succulent tissues thus rupturing cell membranes and leading to cell death.

Many tree species avoid freezing injury by deep supercooling of their susceptible tissues, to temperatures as low as −40°C. Xylem tissue in hardwoods is generally too compact to permit formation of ice (Fig. 3.22). In succulent tissues of hardy species,

FIGURE 3.18   Young maple with elongated canker resulting from winter sunscald. Note bark cracks at base and fruiting bodies of *Schizophyllum commune* just beyond visible upper edge of the canker.

ice crystals form only in the intercellular spaces, outside the cell membranes. With extremely hardy tree species such as quaking aspen and the birches, large ice crystals that remove all but the bound water form. Cells of these species then must be able to withstand extreme dehydration. If properly acclimatized, these species can withstand temperatures as low as $-196°C$ (Salisbury and Ross 1985).

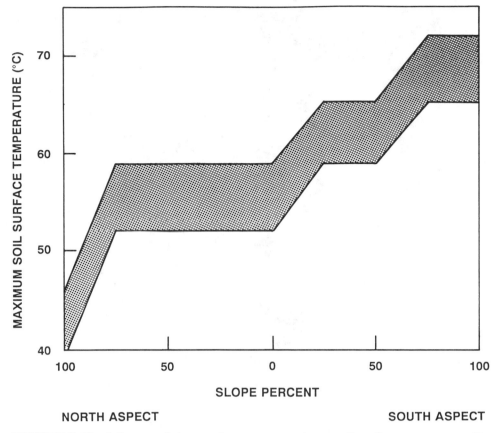

FIGURE 3.19   Influence of slope and aspect on maximum soil surface temperatures. Redrawn from Hallin (1968).

Intracellular (within cells) freezing is usually lethal, whereas plants often recover completely from intercellular (between cells) freezing if it is not extended too long. In succulent tissue, the intercellular water is low in solutes, and this water freezes at 0°C. The water in the cell vacuole contains solutes, and the freezing point is below 0°C. If the air temperature gradually drops below 0°C, ice crystals form initially in the intercellular spaces. Water diffuses out of cells and freezes between the cells, thus increasing the concentration of solutes in the cells and further depressing the freezing point. If this gradual drop in temperature does not continue for too long a period of time and the temperature slowly rises above freezing, severe damage to the cells is unlikely to occur. During a more rapid drop in temperature below freezing, the water may not move out of the cells quickly enough, and ice crystals can form in the vacuoles of the cells. Then the ice crystals rupture the cell membranes. Upon thawing, the semipermeable membranes are destroyed, and the affected plant tissues die. Damage also can result from gradual temperature drops if the cell dehydrates following extensive movement of water out of the cell into intercellular spaces or if tissues thaw quickly and the rapid movement of water into the cells ruptures their membranes.

Many species of trees introduced to more northerly areas (e. g., fruit trees, Siberian

FIGURE 3.20 Necrosis and possible stem enlargement in seedlings resulting from heat injury.

and Chinese elms, and lombardy and bolleana poplars) are commonly injured by frost (Fig. 3.23). In 1951 a drop in temperature from 2.2°C on October 29 to −17°C on November 4 killed approximately 80% of the weeping willows in southern Minnesota. Many of these trees began to produce leaves or, in the case of Siberian elm, heavy crops of seed in the spring of 1952 and then died. The cambium on the main stem was already brown on some of these trees that produced leaves in the spring of 1952. Some of the willow trees with injured branches reacted by producing roots that grew toward the ground.

Temperature extremes during the winter of 1950–1951 also damaged plants over wide areas in southeastern United States (Hepting et al. 1951). The autumn of 1950 was unusually warm. At 7:30 A.M. on November 24, the temperature was 8.3°C. Twenty-three hours later, it was −17°C. Temperatures remained at or below freezing until November 29, and the following winter was unusually cold. As a result of the warm fall, plants were not hardened sufficiently to withstand the 25°C drop on November 24–25 and the subsequent subfreezing temperatures. Many tree species developed frost cracks, and there was an unusually high incidence of bacterial wetwood, or slime flux, the following spring, especially in white oaks. Many large trees died in 1951.

Excessive frost damage can be avoided by planting native tree species from a local seed source and by not planting susceptible trees in areas where frosts commonly occur. Frost or freezing injury occurs when nursery stock lifted for shipment is not

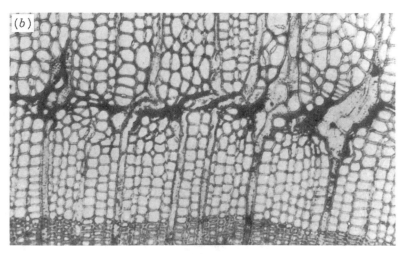

FIGURE 3.21 (a) Young tree of grand fir repeatedly injured by late frosts, showing an abnormal compact and bushy growth form. (b) Cross section of stem, showing formation of frost ring. Reprinted from Rhoads (1923).

FIGURE 3.22 Quaking aspen in northern Wyoming that was cold injured in December 1988 when a cold air mass known as the Alaskan Clipper unexpectedly moved southward out of Canada.

protected from cold (Hodges 1961). Roots are more susceptible to freezing injury than tops of trees.

In frost-injured Douglas-fir seedlings, some injury remains hidden and cannot be identified visually for months afterward (Edgren 1970). Subsequent growth was most severely affected when seedlings were lifted soon after being frosted. Multiple tops was the most common symptom.

FIGURE 3.23   Immature gingko leaves injured by late frost. The larger leaves were unaffected and have resumed growth.

Plants with relatively low levels of hardiness may escape minor frost injury by supercooling. Unlike other plants, which can supercool water many degrees, these slightly hardy plants can cool water only a few °C below the freezing point. Ice nucleation results when something provides nuclei for ice crystal formation. Some species of bacteria (especially *Pseudomonas syringae*) have the ability to initiate ice formation at supercooled temperatures. The gene in the bacterium responsible for ice nucleation has been identified and can be removed. There is presently considerable research interest in developing ice nucleation-free bacterial strains for commercial application to leaves of fruit and vegetable crops at risk from frost injury. These bacteria colonize tissue surfaces and physically outgrow the native strains. Because the introduced bacteria cannot induce ice crystal formation, the plants are protected from minor frost injury. These bacteria could be especially useful in areas where temperatures rarely drop below freezing.

Chilling injury occurs in some plants even at temperatures either slightly above freezing or sometimes at temperatures far above freezing. It is most prevalent in tropical or subtropical plants grown in southern regions of North America. Sensitivity may be related to the temperature at which the lipids in cell membranes solidify. This temperature is determined by the ratio of saturated to unsaturated fatty acids. Below the critical chilling temperature, these lipids are more or less solid, and the membranes tend to leak, upsetting solute balances and disrupting the plant's physiology.

Frost crack is a common phenomenon in both hardwoods and softwoods growing in northern latitudes. A sudden and pronounced drop in winter temperatures can cause the outer layers of wood to contract more rapidly than the inner layers, which are more insulated. The contraction of the wood apparently is due in part to rapid loss of water from the wood cell walls. More rapid desiccation of the outer cells in comparison with inner cells may be involved as well. An elongated narrow split results; it may be up to a meter or more long and extend deep into the tree. Frost cracks usually occur on the main stem near the base of the tree. If the cracks are kept open, successive layers of callus tissues result in the formation of so-called frost ribs. The initial cause of frost crack is usually some internal defect, which subjects the tree to frost crack under favorable conditions. Improper pruning is often associated with frost crack.

Frost shake is a crack or separation of wood following the growth ring and is formed as a result of the sudden warming of the outer layers of wood while the inner tissues are still cold. More rapid expansion of the outer layers occurs, and the frost shake results. The shake often develops in a growth ring, which contains a barrier zone layer of cells formed in response to injury. Some species, eastern hemlock especially, are very susceptible to frost shake.

Winter browning is a tree injury, common on some species of ornamental conifers in northern latitudes, and is assumed to be caused by desiccation of the foliage during warm periods in the winter or dormant season. It is now known that the foliage of northern white cedar (arborvitae) can be injured by a rapid drop in temperature (9°C/min), a temperature change that can occur at sunset or when the sun is blocked by obstructions such as other trees or buildings (White and Weiser 1964). Reducing desiccation of these arborvitae does not reduce winter browning of the foliage, and there is no relationship between moisture content of leaves and the degree of injury. Whether this is the only explanation of winter browning is not known. A disease of western conifers called red belt may be due to loss of moisture during the dormant season.

3.3.3 Moisture

Water plays a vital role in the metabolism of plants. Water and substances dissolved in water directly influence many plant functions. These functions include mineral transport and optimal turgidity. Transpiration of water from the plant is an important means of dissipating heat to the environment. There is much evidence that some water deficit is necessary for optimal plant growth. In order for plants to be able to absorb moisture from the soil, the water potential of the cytoplasm of its cells must be more negative than of the soil–water solution. Substances such as the polyols (arabitol, glycerol, and mannitol) and the amino acids (aspartic acid, glutamic acid, and proline) increase in cells that are subject to drought or other stresses such as freezing or high salt concentration. These compounds can exist at relatively high concentrations in the cytoplasm without damaging the enzymes there.

Water Stress

Insufficient water directly affects cell growth; soon other processes such as photosynthesis and synthesis of proteins and cell wall materials are also affected. Some plants, known as xerophytes, are adapted to survive under extreme drought conditions. But they may cope with drought in different ways. Some plants, such as palms and mesquite, grow down to the water table and avoid the drought. Ephemeral plants escape the drought by becoming dormant during the dry period. Succulents, such as cacti, resist drought by storing water when it is available and are able to minimize unnecessary water loss. The true drought-resistant species are able to regulate the osmotic potential of protoplasm without limiting its functionability. Some extreme xerophytes can lose large quantities of water, yet they are not killed. They are often exceptionally efficient in using dew, rain, or even moisture in the atmosphere when the relative humidity is high. Among the many adaptations of desert plants, the allelopathic chemicals are often produced to eliminate the competition of adjacent plants. These plants are discussed in Chapter 4.

While the evolutionary reactions of trees to desert conditions is of interest to forest ecologists, most forest managers are more concerned about the effects of water stress on trees that are growing under normally mesophytic conditions.

Water stress appears to have its first noticeable effects on cellular growth (Fig. 3.24). If the tissue water potential, which is a measure at atmospheric pressure of the tendency to absorb additional water molecules, is decreased by only −0.1 MPa, it may affect cell growth (Salisbury and Ross 1989). This relationship may be one reason why plants grow mainly at night when water stress is lowest. Reduced cell growth leads to a slowing of shoot and root growth, followed by a reduction in cell-wall synthesis. Eventually chlorophyll formation is inhibited. Many enzyme systems also show decreased activity and nitrogen fixation, and reduction drops.

As stress becomes more severe, abscisic acid levels increase in leaves and roots, and the production of ethylene increases. Both of these growth regulators are affected by other kinds of stress so it is difficult to know their exact role in water stress.

Water stress can have several effects on tree health and growth. Stressed trees may die prematurely. They may suffer reduced growth, or they may become disposed to attack by other organisms. Predisposition is discussed in detail in a number of other chapters. It is not always easy to distinguish damage due to drought from other causes. Deciduous trees appear to die from the top down, have small off-colored foliage and narrow growth rings (Fig. 3.25), and may be invaded by many secondary fungi and insects, which seem to be causing the trees to decline. Pine trees appear to die from the bottom up when subjected to artificial drought. Moisture stress may have caused drought cracks in western white pine (Hungerford 1973). Trees vary in their resistance to drought. Green ash, hackberry, locust, and ponderosa pine are

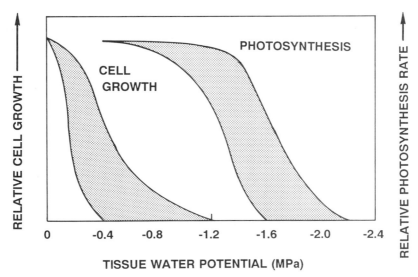

FIGURE 3.24 Cell growth and photosynthesis as a function of decreasing tissue water potentials. Redrawn from Salisbury and Ross (1985).

FIGURE 3.25   Drought-induced mortality of oaks in northern West Virginia.

some of the best trees for areas that periodically experience low rainfall. Tree species that are particularly susceptible to drought should not be planted on southwest-facing slopes or in droughty soils.

Winter drying, or winter browning, is assumed to be the result of excessive losses of moisture during warm days in the winter when the soil is frozen. It is primarily a disease of evergreens and occurs in winter or early spring. Red belt is a specific name given to winter drying of conifers in the Rocky Mountains, and it occurs at an elevation where the roots are in frozen ground and warm winds cause excessive transpiration. From a distance, red belt appears as a band of reddish brown foliage following that particular contour where the injury occurred. Portions of plants that are beneath the snow are usually not affected.

Winter drying is common where species are at the northern edge of their range. In Virginia, seed source has been demonstrated to affect the degree of susceptibility of loblolly pine to winter drying. Trees grown from seeds collected from southern sources were highly susceptible to winter drying, whereas seeds from the same species that were collected from northern areas survived very well.

The results of winter drying usually do not become evident until after the trees start to grow in the spring. Winter drying can be avoided to some extent by protecting evergreens with a mulch to prevent deep freezing. More susceptible species such as northern white cedar and its horticultural varieties should not be planted on exposed southwest-facing slopes. Some varieties of a species are more resistant to winter browning—the Techny variety of arborvitae is much more resistant than other varieties. Wrapping susceptible evergreens in burlap is helpful. As mentioned previously, rapid drops in temperature may sometimes be more important than desiccation.

Leaf scorch is a disease that resembles leaf spot, blotch, or blight diseases caused by infectious agents such as fungi or bacteria. The exact causes of leaf scorch are not

FIGURE 3.26   Construction of a parking lot and resulting soil compaction have injured this oak tree.

well understood. It results, however, when water is evaporated from the leaves and is not replaced by water in the soil. This loss of water may be due to a prolonged drought or some form of diseased or injured roots coupled with a period of continuous warm dry wind. Measures that alleviate loss of moisture from the soil, such as mulching, may prevent leaf scorch. This practice is often employed in nurseries, especially with evergreens. Portable shields consisting of wooden slats or burlap, placed above the plants to allow air circulation, can help to prevent damage from direct sunlight and drying winds.

Symptoms of leaf scorch consist of brownish or reddish dead areas on leaf margins or interveinal areas. Scorched evergreens have needles with dead tips. Slightly different symptoms are associated with needle droop, which results in the abnormal drooping and eventual death of the current needles. Needle droop of red pine following a

FIGURE 3.27   Declining live oak with severe restrictions to its root system.

particularly dry summer in 1974 was found to be associated with root deformities induced during planting, such as J-rooting and root clumping (Bergdahl and French 1976).

Excessive water, or flooding, may cause waterlogging of soils. Flooding is followed by a rapid reduction in transpiration and the water-absorbing capacity of roots. Within several days, wilting develops followed by chlorosis of the lower leaves. Water fills the air spaces. As oxygen is depleted around the roots, ethylene synthesis is also reduced, but the ethylene that is synthesized is trapped near the root. Ethylene precursors accumulate and are transported to the shoots. Here ethylene is readily produced, which in turn causes leaf epinasty, retardation of stem elongation but increased stem thickness, leaf senescense, abscission, adventitious root formation, and increased susceptibility to pathogens. Along with decreases in oxygen, carbon dioxide levels increase, and permeability of the cell membranes decreases.

Slash pine absorbs nutrients and water best in spring and early summer because soil moisture, soil aeration, and temperature are apparently optimal at that time (Shoulders 1976). Fertilization may partially substitute for drainage on some sites because poor root aeration may reduce the plant's ability to absorb water and nutrients.

Tree roots tend to seek a level in the soil that best supplies their needs for both oxygen and water. Spring floods, if they last long enough, smother the root system of trees and cause dieback in the top of the trees. Secondary fungi and insects often invade these trees.

Filling around trees, especially with heavy soils such as clay, can produce similar symptoms (Figs. 3.26, 3.27). Symptoms normally will not appear until later in the same year or the year following root damage. Conifers such as black spruce can be killed by changes in drainage, which results in raising the water table in a stand.

As a result of filling and flooding, certain anaerobic microorganisms may increase rapidly, and they can cause the accumulation of toxic materials such as nitrates. These toxic compounds can cause root mortality in well-aerated soils. In general, flood-tolerant species tolerate these toxic materials, or the damage they cause, better than less-tolerant trees. Tolerant species that experience root deterioration during flooding may quickly develop adventitious roots. Vigorous dominant hardwoods of many species benefit from flooding during the dormant season, but relatively few species can tolerate much flooding during the growing season (Broadfood and Williston 1973). Some species can tolerate intermittent flooding during the growing season (Table 3.5). Flood tolerant tree species that will survive intermittent flooding in the south-

TABLE 3.5 Relative Tolerance of Various Trees to Intermittent Flooding During the Growing Season

MOST TOLERANT

Silver maple
Green ash
Deciduous holly
American sycamore
Eastern cottonwood
Overcup oak
Pin oak
Black willow
Bald cypress
Boxelder
River birch
Pecan
Shagbark hickory
Hackberry
Eastern redbud
Common persimmon
Honeylocust
Osage-orange
Red mulberry
Bur oak
American elm
Bitternut hickory
Black walnut
Eastern redcedar
Black cherry
Blackjack oak
Chinkapin oak
Northern red oak
Post oak
Buckthorn
Black locust

LEAST TOLERANT

Source: Loucks (1987).

eastern United States include green ash, sycamore, cottonwood, willow, sweetgum, American elm, pecan, mulberry, silver maple, red maple, bald cypress, river birch, and persimmon (Barry et al. 1982). Species that are relatively tolerant may experience root deterioration during flooding but adventitious roots develop quickly. Intolerant species lack this ability.

LITERATURE CITED

Barnard, E. L. 1990. *Groundline Heat Lesions on Tree Seedlings.* Fla. Dept. Agric. and Consumer Services, Pl. Path. Circ. No. 338, 2 pp.

Barrett, J. W. 1962. *Regional Silviculture of the United States.* The Ronald Press, New York. 610 pp.

Barry, P. J., R. L. Anderson, and K. M. Swain. 1982. *How to Evaluate and Manage Storm-Damaged Forest Areas.* USDA For. Serv., For. Rept. SA-FR 20. 15 pp.

Bergdahl, D. R., and D. W. French. 1976. Needle droop: An abiotic disease of plantation red pine. *Pl. Dis. Reptr.* 60:472–476.

Broadfood, W. M., and H. L. Williston. 1973. Flooding effects on southern trees. *J. For.* 71:584–587.

Coakley, S. M. 1988. Variation in climate and prediction of disease in plants. *Annu. Rev. Phytopathol.* 26:163–181.

Curtis, C. R., T. L. Lauver, and B. A. Francis. 1977. Foliar sodium and chloride in trees: Seasonal variations. *Environ. Pollut.* 14:69–80.

Edgren, J. W. 1970. *Growth of Frost-Damaged Douglas-Fir Seedlings.* USDA For. Serv., Res. Note PNW-121. 8 pp.

Epstein, E. 1972. *Mineral Nutrition of Plants: Principles and Perspectives.* John Wiley and Sons, New York. 412 pp.

French, D. W. 1959. Boulevard trees are damaged by salt applied to streets. *Minnesota Farm and Home Sci.* 16(2):2.

Hacskaylo, J., R. F. Finn, and J. P. Vimmerstedt. 1969. *Deficiency Symptoms of Some Forest Trees.* Ohio Agri. Res. and Dev. Center, Res. Bull. 1015. 68 pp.

Hallin, W. E. 1968. *Soil Surface Temperatures on Cutovers in Southwest Oregon.* USDA For. Serv., Res. Note PNW-78. 17 pp.

Halverson, H. G., and G. M. Heisler. 1981. *Soil Temperatures under Urban Trees and Asphalt.* USDA For. Serv., Res. Pap. NE-481. 6 pp.

Hepting, G. H. 1963. Climate and forest diseases. *Annu. Rev. Phytopathol.* 1:31–50.

Hepting, G. H., J. H. Miller, and W. A. Campbell. 1951. Winter of 1950–51 damaging to southeastern woody vegetation. *Pl. Dis. Reptr.* 35:502–503.

Hodges, C. S. 1961. Freezing lowers survival of three species of southern pines. *Tree Planters Notes* 47:23–24.

Hungerford, R. D. 1973. Drought crack on western white pine in northern Idaho. *For. Sci.* 19:77–80.

Jones, J. K. 1971. *Seasonal Recovery of Chlorotic Needles in Scotch Pine.* USDA For. Serv., Res. Pap. NE-184. 9 pp.

Jones, E. P., Jr., and O. O. Wells. 1969. *Ice Damage in a Georgia Planting of Loblolly Pine from Different Seed Sources.* USDA For. Serv., Res. Note SE-126. 4 pp.

Leaphart, C. D., R. D. Hungerford, and H. E. Johnson. 1972. *Stem Deformities in Young Trees Caused by Snowpack and its Movement.* USDA For. Serv., Res. Note INT-158. 10 pp.

Little, S., J. J. Mohr, and L. L. Spicer. 1958. Salt-water storm damage to loblolly pine forests. *J. For.* 56:27–28.

Murphy, P. A. 1978. *Mississippi Forests—Trends and Outlook.* USDA For. Serv., Resour. Bull. SO-67. 32 pp.

Norris, L. A. 1971. The behavior of chemicals in the forest. pp. 90–106. In *Pesticides, Pest Control and Safety on Forest Range Lands.* Proceedings from a Short Course for Pesticide Applicators. Oregon State University, Corvallis.

Oliver, W. W. 1970. *Snow Bending of Sugar Pine and Ponderosa Pine Seedlings—Injury Not Permanent.* USDA For. Serv., Res. Note PSW-225. 4 pp.

Otta, J. D. 1974. Effects of 2,4–D herbicide on Siberian elm. *For. Sci.* 20:287–290.

Rhoads, S. A. 1923. The formation and pathological anatomy of frost rings in conifers injured by late frosts. *USDA, Bull.* 1131:1–15.

Ronco, F. 1970. Chlorosis of planted Engelmann spruce seedlings unrelated to nitrogen content. *Can. J. Bot.* 48:851–853.

Ronco, F. 1975. Diagnosis: "Sunburned" trees. *J. For.* 73:31–35.

Roth, E. R., E. R. Toole, and G. H. Hepting. 1948. Nutritional aspects of the littleleaf disease of pine. *J. For.* 46:578–587.

Salisbury, F. B., and C. W. Ross. 1985. *Plant Physiology,* 3rd ed. Wadsworth, Belmont, California. 540 pp.

Schmidt, W. C., and J. A. Schmidt. 1979. *Recovery of Snow-Bent Young Western Larch.* USDA For. Serv., Gen. Tech. Rept. INT-54. 13 pp.

Shortle, W. C., and A. E. Rich. 1970. Relative sodium chloride tolerance of common roadside trees in southeastern New Hampshire. *Pl. Dis. Reptr.* 54:360–362.

Shoulders, E. 1976. *Poor Aeration Curtails Slash Pine Root Growth and Nutrient Uptake.* USDA For. Serv., Res. Note SO-218. 6 pp.

Smith, W. H. 1970. Salt contamination of white pine planted adjacent to an interstate highway. *Pl. Dis. Reptr.* 54:1021–1025.

Stone, E. L. 1991. Boron deficiency and excess in forest trees: A review. *For. Ecol. and Mgmt.* 37:49–75.

Tainter, F. H., and D. C. Bailey. 1980. Bromine accumulation in pine trees growing around bromine production plants. *Environ. Sci. Tech.* 14:730–732.

Townsend, A. M., and W. F. Kwolek. 1987. Relative susceptibility of thirteen pine species to sodium chloride spray. *J. Arbor.* 13:225–228.

Walton, G. S. 1969. Phytotoxicity of NaCl and $CaCl_2$ to Norway maples. Phytopathology 59:1412–1415.

White, W. C., and C. J. Weiser. 1964. The relation of tissue desiccation, extreme cold, and rapid temperature fluctuations to winter injury of American Arborvitae. *Amer. Soc. Hort. Sci.* 85:554–563.

4

BIOTIC CAUSES OF DISEASE

Synopsis: This chapter serves to introduce the major biotic (living) disease-causing agents of forest and shade trees in North America. The major types of infectious pathogens are characterized; and, for each, representative life forms and life cycles are illustrated. How they cause disease is also briefly discussed.

4.1 FUNGI

4.1.1 The Nature of Fungi

Fungi are small and structurally very simple plants that lack chlorophyll (Bessey 1965, Burnett 1970). The vegetative portion (thallus) is a mass of threadlike structures collectively called hyphae or mycelium (Fig. 4.1). Hyphae range in diameter from less than 1 μm (one millionth of a meter) to more than 20 μm. As long as food is available

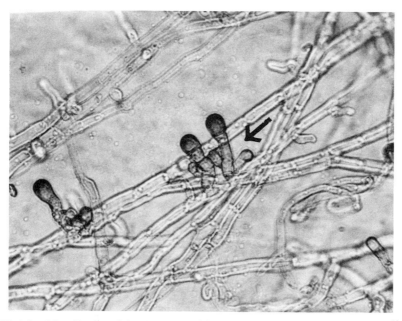

FIGURE 4.1 Mycelium and immature conidia (arrow) of *Epicoccum* spp. (Moniliales), a cause of degradation and staining of conifer lumber.

and the environment is not limiting, hyphal tips are potentially immortal and have great growth capacity. One hyphal tip cell in 24 hours may produce 1 km of mycelium, and in 48 hours, hundreds of kilometers of mycelium. The ability to grow rapidly and completely colonize a substrate contributes to the success of fungi in the biological world.

The hyphae, which develop initially from a single strand called a hypha, may grow together to form large, fleshy or hard fruiting structures such as mushrooms or conks (Fig. 4.2). Fungi do not develop true roots, stems, or leaves as do higher plants, and they reproduce by means of microscopic spores. Because fungi do not have chlorophyll, they cannot fix atmospheric carbon; hence, they must depend upon preformed organic materials as their source of energy. Fungi produce many specialized enzymes, however, that can break down virtually all organic substances, no matter how complex, from the most decay-resistant wood to the latest synthetic plastics to paint (Fig. 4.3). Complex carbohydrates are broken down into simple sugars that will then diffuse through the hyphal wall and be used by the fungus protoplasm to produce new mycelium, fruiting structures, and spores.

Many fungi degrade organic debris and wastes and, so, are beneficial. Slash from tree-harvesting operations, for example, would soon accumulate to unmanageable proportions if it were not rapidly decayed by fungi. Atmospheric carbon would soon be tied up in undecayed organic matter, and in about 20 years a global carbon dioxide deficiency would result. Other beneficial fungi, such as species of *Penicillium* (Hyphomycetes), are used in the production of antibiotics, including penicillin, and in cheese manufacture, such as Roquefort and Camembert, to impart special flavors or to produce commercial acids and enzymes. Root systems of almost all terrestrial higher plants depend on mycorrhizal fungi to provide water and certain nutrients such as phosphorus.

FIGURE 4.2 Mushrooms of *Coprinus micaceus* (Coprinaceae), one of the inky cap fungi, growing from a dead elm root.

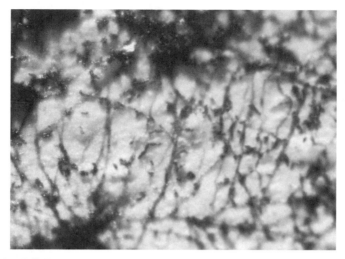

FIGURE 4.3   A light microscope view of hyphae and spores of *Cladosporium* spp. (Moniliales) discoloring a painted wood surface.

Although most of the 100,000 known species of fungi use only dead organic matter—these are known as *obligate saprophytes*—many fungi can invade and obtain their nourishment from living plants. These are known as facultative parasites. Fungi that are normally parasitic on living plants but that can also grow in dead plant material are called *facultative saprophytes*. Some fungi have evolved to the extent that they can live only on or in a living host, such as the rust fungi. These are known as *obligate parasites*. If the host plant dies, then these fungi die as well. With obligate parasites, specialized fungal-absorbing structures called haustoria enter host cells and extract nutrients (Fig. 4.4).

4.1.2 Growth Requirements

Water is needed for growth of all fungi, but different species of fungi can tolerate wide ranges of water availability. Some of the water molds spend their entire lifetime in water. Other fungi, such as *Aspergillus restrictus,* which is a low-water requirer (xerophytic), can grow in stored seeds with a moisture content as low as 12–15%. Wood decay fungi, such as *Poria incrassata* require at least some free moisture in the wood cells. If wood to be placed in service is dried below the fiber saturation point, which is approximately 28%, and maintained at that moisture content, it will never become decayed. At the other extreme, species of *Trichoderma* can grow in wood that is nearly saturated with water.

Because fungi do not have chlorophyll, light is not a prerequisite for growth, although at least some light is necessary to initiate fruiting and spore production in many fungi. Fungi, especially those that produce large sporocarps, respond to gravity and orient their spore-bearing hymenium to ensure maximum spore dissemination. Oxygen is also necessary for growth, although fungi can survive in the presence of extremely small amounts. Some fungi grow best in almost pure carbon dioxide while others can tolerate only small amounts. Although fungi can grow at a range of pH from 2 to 12, they generally prefer a slightly acidic environment (pH 5–7). Fungi may secrete enzymes that can change the pH of an unsuitable substrate to their pre-

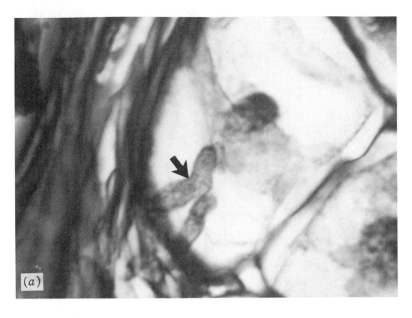

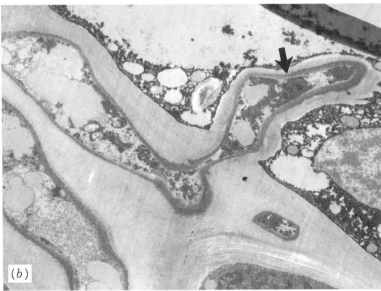

FIGURE 4.4   Haustoria (arrows) of the blister rust fungus, *Cronartium comandrae* (Melampsoraceae), in cortical cells of pine; (a) light microscope and (b) transmission electron microscope views.

ferred optimum. Fungi can survive extremely low temperatures but generally grow only when the temperature is above 0°C. Snow molds grow best at temperatures only slightly above freezing and may not grow above 15°C. Optimum temperatures for growth (20–30°C) for most fungi are similar to those of higher plants and animals. Temperatures approaching 70°C are fatal in minutes and those near 90–100°C, kill fungi immediately.

4.1.3   A Typical Life Cycle

Most fungi, but not all, reproduce both sexually and asexually. The ability to do either, depending on the need, imparts in them a powerful survival ability in a dynamic environment. A generalized life cycle of most disease-causing fungi is shown in Figure 4.5. As long as a food source is plentiful, most fungi will grow vegetatively with little tendency to reproduce.

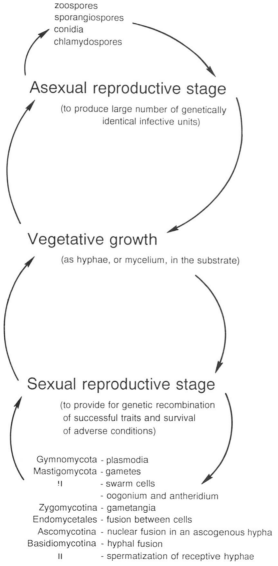

zoospores
sporangiospores
conidia
chlamydospores

Asexual reproductive stage

(to produce large number of genetically
identical infective units)

Vegetative growth

(as hyphae, or mycelium, in the substrate)

Sexual reproductive stage

(to provide for genetic recombination
of successful traits and survival
of adverse conditions)

Gymnomycota - plasmodia
Mastigomycota - gametes
!I   - swarm cells
               - oogonium and antheridium
Zygomycotina - gametangia
Endomycetales - fusion between cells
Ascomycotina - nuclear fusion in an ascogenous hypha
Basidiomycotina - hyphal fusion
II   - spermatization of receptive hyphae

FIGURE 4.5   A schematic life cycle of fungi showing important features of asexual and sexual reproduction.

As hyphal growth becomes extensive, the asexual reproductive stage is initiated, and very soon large numbers of asexual spores are usually produced. These propagules are genetically similar to those of the parent strain, which successfully colonized the substrate. Hence, they have an advantage if the same substrate is encountered again. Asexual spores are usually produced in great quantities, the purpose being to colonize the available substrate rapidly and to ensure that at least a few spores find a suitable substrate. If they are effectively disseminated, many new colonies far from the initial colony can be initiated. Zoospores can swim for only short distances under their own power, but encysted spores, chlamydospores, or conidia may be carried much farther by moving water. Airborne spores such as sporangiospores and conidia can be carried for many kilometers. Sometimes chlamydospores and even hyphae can be transported long distances in infested plant debris or in contaminated soil, seeds, or host plants. Most fungi must depend on wind, flowing water, vectors, or people to assist in disseminating asexual spores.

Asexual spores are small, usually single-celled, rather delicate structures and may live for only a few minutes. Other spores can remain viable for days, months, or even years. Successful colonization depends upon their encountering a susceptible substrate and penetrating it before their energy reserves are expended. Most spores depend upon chance to encounter a suitable substrate. Some spores, however, can enhance their chances of colonization. Zoospores, for example, can detect very small gradients in leached sugars and actively swim toward root tips. Germ tubes of germinating conidia may preferentially grow toward open stomata.

With time, the substrate or the environment changes. As the growing season progresses, host tissues often become less susceptible to infection and colonization. Most fungi adapt to these changes by switching their metabolism toward production of the sexual reproductive stage. "Male" and "female" sexual structures form, and nuclei are exchanged in some fashion followed by nuclear fusion. This change is followed by reduction division (meiosis) and the formation of one or more sexual spores, depending on the species. Meiosis provides for recombination of genetic factors and may produce daughter nuclei with superior genetic qualities.

Following formation of the sexual stage, the vegetative thallus often dies, although this may not occur with obligate parasites. Some sexual spores are thick-walled and are very resistant to dessication and extreme temperatures. They can survive years of harsh conditions. Most, however, will germinate the following spring or upon return of favorable growing conditions and colonize a new substrate. Some will not germinate unless a suitable host is in close proximity.

Although specific details vary with different fungi, most fungi can reproduce either sexually or asexually and will do whichever is to their advantage. An intimate knowledge of the pathogen, its life cycle, and the environmental factors that affect it are crucial to its successful control. As you will learn in later chapters, successful control of many pathogens is not possible because much of this information is not known.

4.1.4 Reproduction

Some genera such as *Sclerotium* survive quite well without producing spores of any kind. These fungi have special structures called sclerotia, which allow the fungus to survive (Fig. 4.6) harsh conditions. Most fungi, however, are able to reproduce by means of spores. Spores are usually produced in large numbers and may be dissemin-

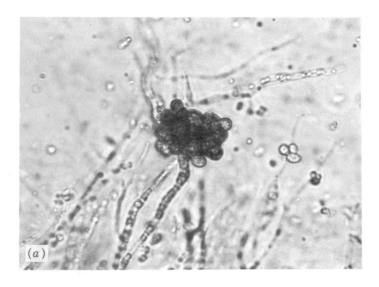

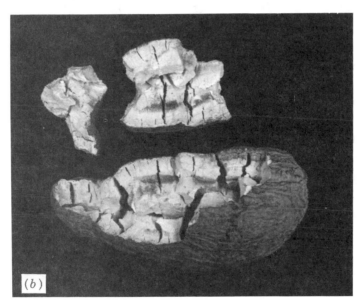

FIGURE 4.6   (a) Microsclerotium of the root rotting and vascular wilt fungus, *Verticillium albo-atrum* (Moniliales). (b) Macrosclerotia of *Poria cocos* (Polyporaceae). These latter, also known as Tuckahoes, are frequently unearthed by fire plows.

ated by wind, splashing raindrops, or certain insect or animal vectors. Once dispersed into the upper atmosphere, spores may be carried for thousands of kilometers and can be brought down to host plants by rain. Spores have been trapped at altitudes of over 12,000 m. Dark-colored spores with thick walls are more likely to resist the extreme temperatures and harmful ultraviolet light encountered at high altitudes.

Spores may be unicellular or multicelled; most are hyaline (colorless), while others are colored or black. Spores are produced in or on special reproductive structures called sporophores, and tremendous numbers may be produced. A single fruiting body, or conk as they are sometimes referred to, of *Ganoderma applanatum (Fomes applanatus)* produced an estimated 30,000,000 spores/day for 6 months. This estimate amounts to 350,000 spores/sec and appears to the spectator almost as smoke emanating from the pore surface. Spores range in length from 1 to more than 200 μm, but most are 5–25 μm long, slightly smaller than an average-sized parenchyma cell of higher plants. Some spores are produced in a sticky gelatinous mixture that oozes out of the sporocarp. These spores are adapted not only to dissemination by splashing rain but also by insect vectors. The Dutch elm disease fungus and blue stain fungi depend on bark beetles for their dissemination. Animals and birds also can carry spores. Over 757,000 spores of *Cryphonectria parasitica,* the chestnut blight fungus, were counted on one downy woodpecker (Heald and Studhalter 1914).

In addition to the natural factors involved in disseminating spores, people have acted as vectors and have disseminated such notorious pests as the Dutch elm disease fungus, chestnut blight fungus, and white pine blister rust. More recently, in 1967, a more virulent form of the Dutch elm disease fungus was reintroduced into Great Britain with devastating consequences. In spite of quarantine restrictions, elm logs from Canada were simultaneously shipped into three ports of entry.

Although small and simple in form, fungi are unusually well adapted to survive adverse conditions, and because of their rapid growth and ability to produce large numbers of spores and a large variety of enzymes, they can rapidly take advantage of favorable conditions. Largely for these reasons they are able to cause so great concern to managers of forest and shade trees and cause more losses than do all the other biotic causes combined.

4.1.5   Summary Classification of the Fungi

This section summarizes the classification of fungi and lists some key characteristics and observations (adapted from Alexopoulos and Mims 1979).

Kingdom: Myceteae (Fungi)

Members of the fungal kingdom are nonvascular plants, lacking chlorophyll, and are usually associated with reduced vegetative forms. They are often only microscopic in size and usually reproduce by means of spores.

As suggested by this summary classification, a prevailing opinion of biologists is that fungi should be placed in a separate kingdom, the Myceteae. Fungi are quite distinctive from both animals and plants, especially in regard to the manner in which fungi obtain nutrients. Fungal cell walls have cellulose as well as chitin. Chitin is similar to the cellulose of green plants and fungi but has an *N*-acetyl amino group attached to the parent glucose molecule (Fig. 4.7). Interestingly, chitin also forms the exoskeleton of arthropods, suggesting that plants, animals, and fungi may all have evolved from a common ancestor.

Division I: Gymnomycota. The first of three major divisions of fungi contains the Gymnomycota, a group of phagotrophic organisms whose vegetative structures lack cell walls (Martin and Alexopoulos 1969). Because of this structure, they are believed to be primitive forms of fungi. Although the cellular slime molds, myxamoebae, and

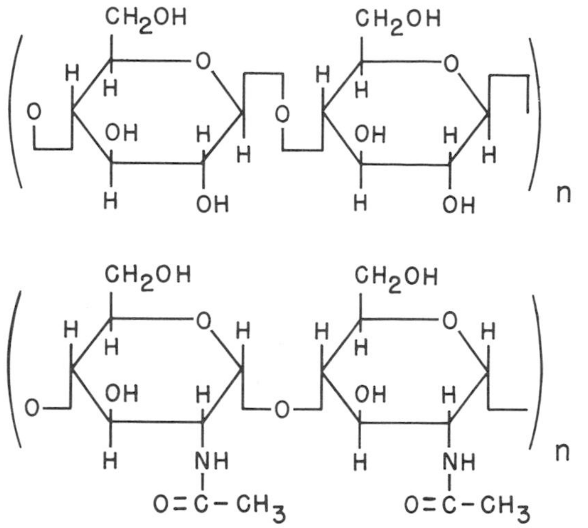

$$
\left(\begin{array}{c} \end{array} \right)_n \quad \text{CELLULOSE}
$$

$$
\left(\begin{array}{c} \end{array} \right)_n \quad \text{CHITIN}
$$

FIGURE 4.7   Cellulose and chitin, two basic cell-wall components of fungi.

slime molds are included in this division, the latter are most frequently encountered by foresters.

The vegetative portion of the slime molds is a plasmodium of naked protoplasm that grows over or through the substrate on which it occurs, deriving its nourishment by ingesting bacteria, fungi, and other microorganisms (Fig. 4.8). In some genera, spores are borne in a sporangium, which is surrounded by a delicate crusty wall. A netlike mass of sterile hyphae, called a capillitium, may form along with the spores. If present, the capillitium expands at maturity, probably assisting in the controlled release of spores. Sporangia are often borne upon stalks. Other sporangia are not stalked

FIGURE 4.8   Fruiting bodies of the slime mold *Hemitrichia* spp. (Gymnomycota).

FIGURE 4.9   Aethalia of the slime mold *Fuligo* spp. (Gymnomycota) on a decayed tree stump.

and, when in clusters, may resemble insect eggs. Some species that are not stalked have fruiting bodies known as aethalia or plasmodiocarps (Fig. 4.9) which simply disintegrate upon maturity, releasing spores into the air. Slime molds damage higher plants only when they occur in such masses that they partially smother the plants by preventing sunlight from reaching the leaves. Although they cause no tree diseases, they are extremely common in the forest on logs and other decaying vegetation, and the brightly colored (often yellow or white) plasmodia may be very visible during moist conditions.

The remaining divisions of fungi contain large numbers of fungi of great importance not only to forestry but to agriculture and humankind as well.

Division II: Mastigomycota (The Now Obsolete Class Phycomycetes).   The second division, the Mastigomycota, is analogous to the old class, Phycomycetes, or algal fungi. Most members are aquatic, and asexual reproduction is typically by zoospores. These zoospores have one or two flagella and depend upon free water for their dissemination. There are two types of flagella. Whiplash flagella are whiplike appendages that assist in movement in water by transmitting a succession of undulating waves. Tinsel flagella have tiny filaments that extend laterally from the main axis. They assist in movement by extending into the water and then retracting. The filaments cause friction, which assists in pulling the zoospore along. Aquatic fungi range in size and complexity from single-celled individuals in the class Chytridiomycetes, which infect algae and grow saprophytically on pollen grains and hair, to highly specialized parasitic forms in the most advanced class, the Oomycetes. Here, species such as *Phytophthora* develop sexual oospores that form following the fusion of an antheridium with a large fertile cell called an oogonium. Following fertilization, one or more oospores form within the oogonium. The classes are identified based on the number

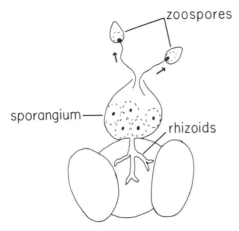

FIGURE 4.10   A chytrid (Chytridiomycetes) infection of a pine pollen grain.

and kind of flagella on the zoospores. As a group, these fungi cause many damping-off diseases and can be especially destructive in tree nurseries.

Class 1: Chytridiomycetes. Chytridiomycetes are popularly known as chytrids. Zoospores have a single posterior whiplash flagellum (Fig. 4.10). These fungi have a greatly reduced growth form, usually consisting of rhizoidallike hyphae and simple sporocarps. They are parasitic on algae and higher plants but also are saprophytic on cast pollen, other aquatic fungi, microscopic animals, and plant debris.

Class 2: Hyphochytridiomycetes. Zoospores have a single anterior tinsel flagellum (Fig. 4.11). This is a small class whose members have much the same habits as do the chytrids.

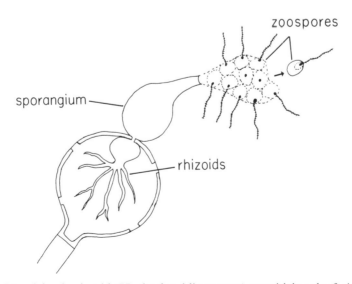

FIGURE 4.11   A hyphochytrid (Hyphochytridiomycetes) parasitizing the fruiting structure of an alga.

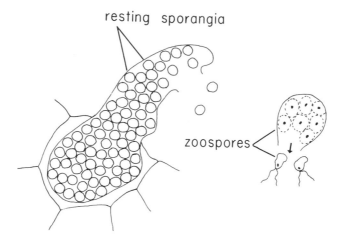

FIGURE 4.12   *Plasmodiophora* spp. (Plasmodiophoromycetes) infecting a plant root. Note sporangia that have formed inside an infected root cell and an enlarged view of one sporangium in the process of releasing zoospores.

Class 3: Plasmodiophoromycetes. Zoospores have double unequal-sized flagella, both of the whiplash type (Fig. 4.12). These fungi are obligate parasites of algae, aquatic fungi, and higher plants and often deform (hypertrophy) infected tissues. There are no hyphae. A naked plasmodium is produced within host cells. The plasmodium cleaves into resting sporangia, which are released as host tissues decay. The sporangia release zoospores, which may become intermittently amoeboid until they come in contact with the host plant and initiate new infections.

Class 4: Oomycetes. Zoospores have double equal-sized flagella (Fig. 4.13). One flagellum is of the tinsel type and is directed forward. The other is of the whiplash type and is directed backward. Some reduced species are parasitic on algae and other aquatic fungi. Some more advanced and complex species are saprophytes or parasites

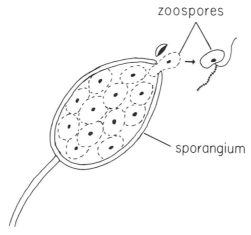

FIGURE 4.13   A zoosporangium of *Phytophthora* spp. (Oomycetes) releasing zoospores.

on plant and animal debris in water, on animals and fish, and on roots of higher plants. They are important as a cause of damping-off in forest tree nurseries.

Division III: Amastigomycota. In this division no zoospores are produced.

Subdivision 1: Zygomycotina (The Now Obsolete Class Zygomycetes). The third major division, Amastigomycota, is characterized by a lack of zoospores. Within this division, the subdivision Zygomycotina comprises the old, outdated class Zygomycetes. This group is large and diverse. Its members are mostly saprophytic, or parasitic on or in arthropods. Some are parasitic on other fungi, higher plants, and some animals. Many species are commercially beneficial because of valuable metabolic byproducts. Members depend upon wind for dispersal of their dry, asexually formed sporangiospores (Fig. 4.14). These are borne within a sporangium on a specialized aerial structure known as a sporangiophore. Each sporangiophore may be supported by a prominent set of rhizoids, which penetrate the substrate.

In sexual reproduction two gametangia fuse to eventually form the zygospore (Fig. 4.15). If gametangia from one thallus type fuse, the fungus is called *homothallic*. If two thalli of opposite mating types are required, the fungus is *heterothallic*. The two mating types are often designated as plus and minus. Most of the Zygomycotina are heterothallic. After nuclear fusion, a zygote that develops into a thick-walled zygospore is formed. Zygospores often permit the fungus to survive winter or other adverse conditions. Inhibitors prevent the zygospore from germinating immediately; these inhibitors often break down after exposure to extreme conditions.

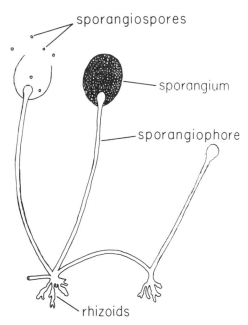

FIGURE 4.14 Asexual fruiting structures of *Rhizopus nigricans* (Zygomycotina), the common bread mold.

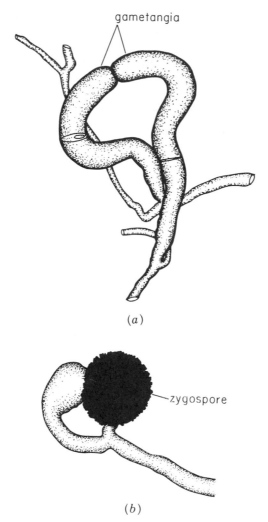

gametangia

zygospore

(a)

(b)

FIGURE 4.15   Sexual reproductive structures in (a) *Phycomyces blakesleeanus,* a heterothallic zygomycete, and (b) a developing zygospore of *Zygorhynchus heterogamus,* a homothallic zygomycete.

Subdivision 2: Ascomycotina (The Now Obsolete Class Ascomycetes).   The second major subdivision, Ascomycotina, comprises the old class Ascomycetes. This is a very large group that includes a diverse range of largely terrestrial fungi. Most are saprophytes on decaying organic matter, but some are highly specialized and grow only on parts of certain plant hosts. Many species are destructive parasites of forest trees and cause such notable diseases as the Dutch elm disease, chestnut blight, and oak wilt.

Ascomycotina are also known as the sac-fungi. Sexual reproduction in this group is characterized by production of sexual spores (ascospores), in a saclike cell (ascus) by a process known as free-cell formation (Fig. 4.16). In this process the ascus at first contains a diploid (2N) nucleus, which undergoes meiosis to form haploid (1N) ascospores. Eight ascospores is the usual number resulting from this process, al-

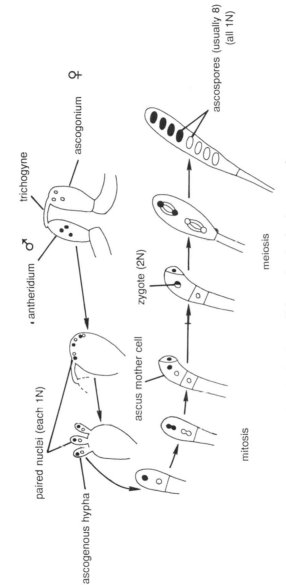

FIGURE 4.16   Free-cell formation in the Ascomycotina.

though as few as four and larger numbers in multiples of eight are characteristic of certain species, depending on the number of mitotic divisions that take place. Asci may be produced upon or within complex multicellular fruiting bodies. Illustrations of some of these are shown in the following discussion of subclasses and families.

Asexual reproduction in the Ascomycotina is typically by production of spores called conidia, although not all members reproduce asexually. In its simplest form, asexual reproduction is represented by budding of cells, as in the yeasts. Production of conidia is very common. A conidium is by definition any asexually produced spore. Conidia are often produced in large numbers continuously as long as substrate is available and environmental conditions are favorable. The asexual spore is often referred to as the imperfect stage of the fungus. In late summer or early autumn, the fungus may shift to production of sexual fruiting bodies, also known as the perfect stage, which mature and discharge ascospores in the following spring.

Class: Ascomycetes. There is one class, Ascomycetes, and five subclasses of importance to forest disease specialists.

Subclass 1: Hemiascomycetidae. Most hemiascomycetes normally produce neither hyphae nor an ascocarp. They include the yeasts, which are common inhabitants of leaf and bark surfaces. Some yeasts are useful for the production of bread and alcohol. Other hemiascomycetes produce primitive hyphae and are obligate parasites of vascular plants.

ORDER: ENDOMYCETALES (THE YEASTS). The yeasts produce no mycelium and reproduce by budding (Fig. 4.17). None of the yeasts are important plant pathogens, but they are well known for their ability to ferment sugars. Their nearly ubiquitous presence on plant surfaces suggests that they are an important component of the biological environment.

ORDER: TAPHRINALES. All members of this order are obligate parasites of vascular plants where they cause various leaf and fruit malformations in the form of blisters.

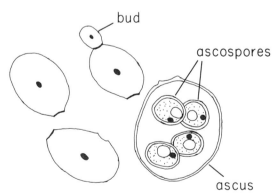

FIGURE 4.17 Yeast (Endomycetales) showing vegetative cells and an ascus and ascospores. Note bud scars.

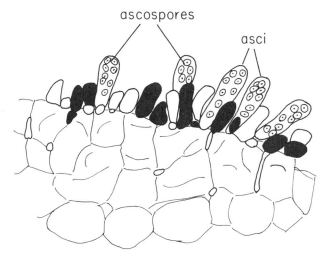

FIGURE 4.18   Asci of *Taphrina* spp. (Taphrinales) on a leaf surface.

They have limited mycelium, which is restricted to the epidermis (Fig. 4.18). Unprotected asci are organized into a palisadelike layer and are borne on plant surfaces. These fungi overwinter as conidia, which form from ascospores.

Subclass 2: Plectomycetidae. In the Plectomycetidae, the ascocarp is either a closed sphere (cleistothecium) or a globose sphere with a long neck (perithecium). In the two families included here, the asci dissolve as the ascospores mature, and either the ascocarp wall disintegrates to release the ascospores (in the Eurotiaceae) or the ascospores are oozed out in a sticky, gelatinous mass for insect dissemination (in the Ophiostomataceae).

Family: Eurotiaceae. Members of the Eurotiaceae are very widely distributed. The ascocarp is a cleistothecium, which is a spherical structure with no openings through which ascospores may be released (Fig. 4.19). Release of ascospores is eventually

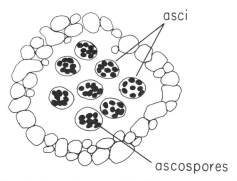

FIGURE 4.19   A cleistothecium, the sexual fruiting structure of *Eurotium* spp. (Eurotiaceae).

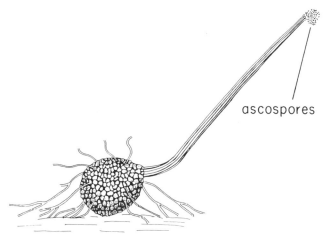

ascospores

FIGURE 4.20   A perithecium, the sexual fruiting structure of *Ceratocystis* spp. (Ophiostoma-taceae).

successful after the cleistothecial wall weathers away. Two important form genera of conidial stages are *Aspergillus* and *Penicillium,* which, among many other attributes, cause deterioration of improperly stored agricultural and forest tree seeds. They also cause black, blue, and green surface molds and stains on lumber and veneer and are an important cause of value reduction in these products.

Family: Ophiostomataceae. These fungi have globose perithecia with long necks (Fig. 4.20). The family contains many species of *Ceratocystis* and *Ophiostoma,* which cause vascular wilt diseases and blue stain of wood.

Subclass 3: Hymenoascomycetidae. In the third subclass, unitunicate (single-walled) asci are formed on a basal hymenium in various types of ascocarps.

In the family Xylariaceae, the ascocarp is also a perithecium, but it is embedded in a stroma, with only the perithecial ostioles protruding. Some fungi that cause cankers and soft rot of wood are found in the Xylariaceae. The remaining families in this subclass also contain canker-causing fungi, and some species cause anthracnose and leaf spots.

ORDER: ERYSIPHALES. Erysiphales contains the powdery mildews, which attack a wide range of living hosts. The mycelium is entirely superficial but is attached to the epidermis by haustoria through which nourishment is obtained from the host. The ascocarp is a cleistothecium covered with distinctively ornamented appendages (Fig. 4.21). Although these are obligate parasites, their cleistothecia survive the winter on dead leaves. The number of asci produced and the form of ornamentation of the appendages are used to identify the powdery mildews.

FIGURE 4.21 A cleistothecium of *Microsphaera* spp. (Erysiphales). Note the asci that have been released in the ruptured area, the ascospores that they contain, and the sculptured appendages.

ORDER: XYLARIALES. Ascocarps of the Xylariales are characterized by being dark colored and leathery or hard. Most are saprophytic on dead woody materials, but several families cause important economic losses.

Family: Chaetomiaceae. The ascocarp is a perithecium covered with hairlike appendages (Fig. 4.22). These fungi are cellulolytic; that is, they are preferentially saprophytic on cellulosic products such as paper and fabrics and cause important losses.

Family: Xylariaceae. In the family Xylariaceae the ascocarp is also a perithecium, but it is embedded in a stroma, with only the perithecial ostioles protruding (Fig. 4.23). The stroma consists entirely of fungal tissue. Its morphology is rather characteristic for each of these common genera. In *Hypoxylon* the stroma tends to be appressed to the substrate and appears as a crustlike growth just under the bark. In *Daldinia* the stroma is a large, black, hemispherical structure with internal dark-colored distinctive zone lines. *Xylaria* has stromata shaped somewhat like human fingers, and those of *Nummularia* are club-shaped. Members of this family produce a type of decay of woody tissues known as soft rot, and some are weakly parasitic. *Hypoxylon mammatum* is an economically important species that causes a destructive canker of aspen.

Family: Diatrypaceae. In the family Diatrypaceae ascocarps are embedded in a stroma that is composed of both host and fungal tissues (Fig. 4.24). Members are chiefly saprophytic on dead bark and wood, although some cause cankers on living trees. The most important genera are *Diatrype* and *Eutypella.*

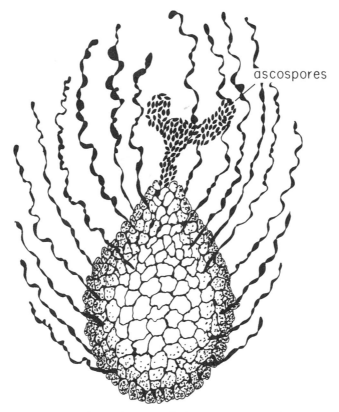

ascospores

FIGURE 4.22   A perithecium of *Chaetomium* spp. (Chaetomiaceae).

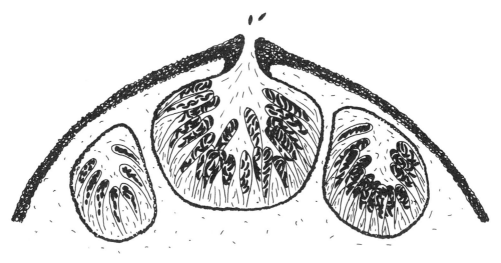

FIGURE 4.23   Cross section of portion of a stroma of *Xylaria* spp. (Xylariaceae) showing embedded perithecia. Note asci and ascospores. Note also that the black outer rind is rather thin and covers a strikingly white inner stromatic tissue.

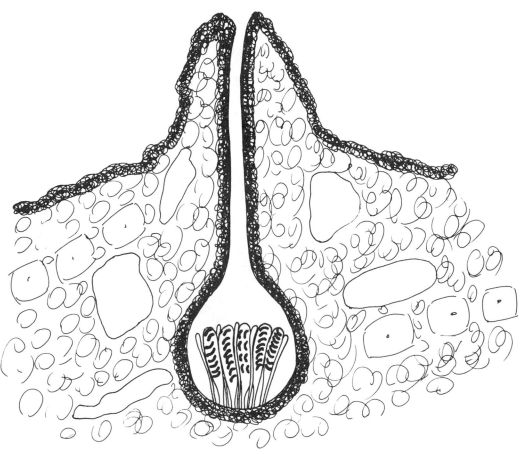

FIGURE 4.24 Stroma and embedded perithecium of *Eutypella parasitica* (Diatrypaceae). Note the allantoid (weiner-shaped) ascospores.

Family: Sordariaceae. The Sordariaceae is not an economically important family, but some species are of much value as experimental fungi. They are easy to grow on artificial media, and because they produce distinctive perithecia, asci, and ascospores (Fig. 4.25), they are also useful for educational instruction. They are all saprophytic on dung and decaying plant materials.

ORDER: DIAPORTHALES

Family: Gnomoniaceae. In the family Gnomoniaceae perithecia are buried in host tissues, and each has a prominent beak (Fig. 4.26). The asci have a small refractive ring at the apex. There are many species that cause leaf spots of many woody and fruit tree species.

Family: Diaporthaceae. In the family Diaporthaceae the perithecia are immersed in a stroma (Fig. 4.27). The perithecia have long, often sinuous, beaks. Asci have an apical refractive ring. The most notorious species in this family is *Cryphonectria parasitica,* the cause of chestnut blight.

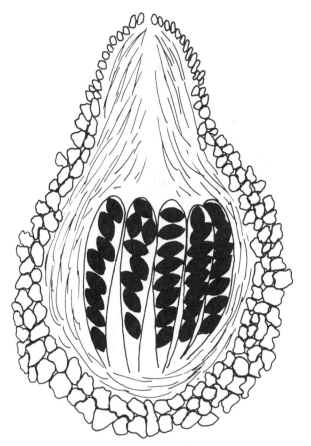

FIGURE 4.25   Perithecium of *Sordaria* spp. (Sordariaceae).

ORDER: HYPOCREALES

Family: Nectriaceae. Species in the Nectriaceae produce superficial perithecia and are important saprophytes and parasites of trees (Fig. 4.28). Perithecia are typically fleshy in texture when moist and are brightly colored, usually a shade of yellow, orange, pink, or red.

Discomycetes. Discomycetes is an artificial grouping of ascomycetous fungi that have the common feature of producing asci in an apothecium, which is a saucer or cuplike structure (Fig. 4.29). For this reason, they are often known as the cup fungi.

The order Phacidiales contains two families: the Rhytismataceae and the Hypodermataceae. The former has irregular apothecia and causes a group of distinctive diseases known as the tar spots. The latter contains many needle-cast fungi and has a slitlike apothecium (a hysterothecium), which may be quite short or nearly as long as the conifer needle.

The orders Pezizales and Tuberales include the cup fungi, morels, and truffles. These fungi are not destructive but are common in the forest. Some are mycorrhizal.

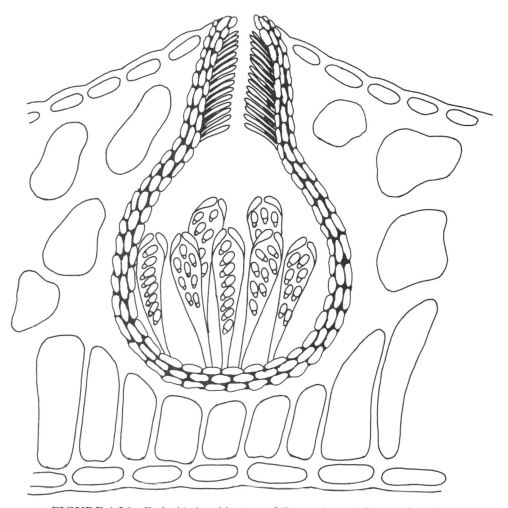

FIGURE 4.26   Embedded perithecium of *Gnomonia* spp. (Gnomoniaceae).

Others are prized for their edibility.

ORDER: PHACIDIALES. The Phacidiales produce apothecia with outer layers that make them appear black. Many are leaf or needle pathogens of trees.

Family: Rhytismataceae. In the Rhytismataceae the stroma is either superficial or slightly buried in the substrate (Fig. 4.30). The most important genus, *Rhytisma,* has stromata that are somewhat bulging and circular and that appear black, producing a tar spot-appearing structure on the leaves of broadleaf species. Apothecia are borne within and, when mature, split into a series of irregular radiate fissures from which the needle-shaped ascospores are forcibly ejected in great numbers by a puffing action.

Family: Hypodermataceae. The Hypodermataceae have stromata with a single apothecium. The apothecium is elongated and is referred to more specifically as a hystero-

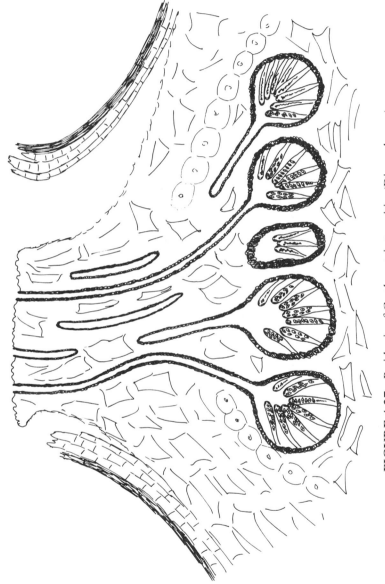

FIGURE 4.27 Perithecia of *Cryphonectria parasitica* (Diaporthaceae).

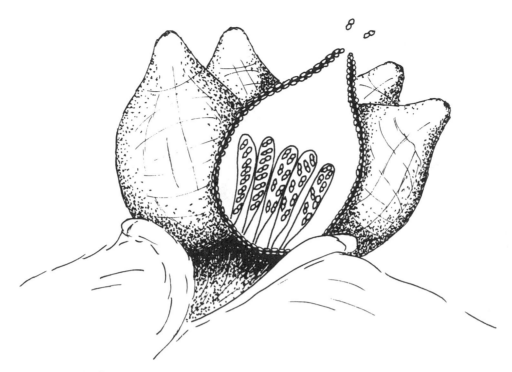

FIGURE 4.28   Superficial perithecia of *Nectria* spp. (Nectriaceae).

thecium (Fig. 4.31). The apothecium is covered by the dark edges of the stroma to form a slitlike opening. Most members are weak parasites of conifers, causing needle-cast disease; some have caused serious losses.

ORDER: PEZIZALES. Pezizales is a large order of members that have asci that open by means of a lidlike structure known as an operculum. This feature allows the asco-spores to be forcibly discharged when subjected to a mechanical disturbance and is visible as a small cloud that is produced with a hissing sound. All are saprophytes on plant debris in soil or on dung.

Family: Pezizaceae. Pizazaceae are known as cup fungi because of the characteristic shape of the ascocarps (Fig. 4.32). They may be small or quite large. Most are rather drab in color, but some are brightly colored and may be shades of red, orange, or yellow. Many members grow on decaying wood.

Family: Morchellaceae. The morels are characterized by a large, distinctive fruiting structure with a well-defined stalk (Fig. 4.33). The ascocarps are irregularly shaped apothecia that cover a swollen, spongelike pileus, or cap. All are saprophytic on or-ganic matter. Most of this family is very tasty and very much sought after by mush-room hunters.

ORDER: TUBERALES. Tuberales is a small order containing mostly subterranean mem-bers. They produce ascocarps, commonly known as truffles. In Europe they are con-

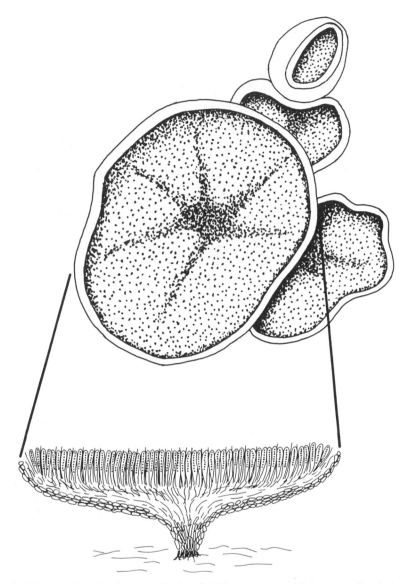

FIGURE 4.29   Typical apothecia of a Discomycete, with cross section below.

sidered a delicacy, even more widely appreciated than the morels. They may be an important food source for rodents living in forested environments. The ascocarps range in size from the size of a match head to that of a walnut. They usually remain closed at maturity with the ascospores being released by decay of the ascocarp or its being broken by animals (Fig. 4.34). All Tuberales are probably mycorrhizal fungi, living in a symbiotic association with roots of trees. For this reason they are especially important components of forested ecosystems.

FIGURE 4.30   Apothecium of *Rhytisma* spp. (Rhytismataceae).

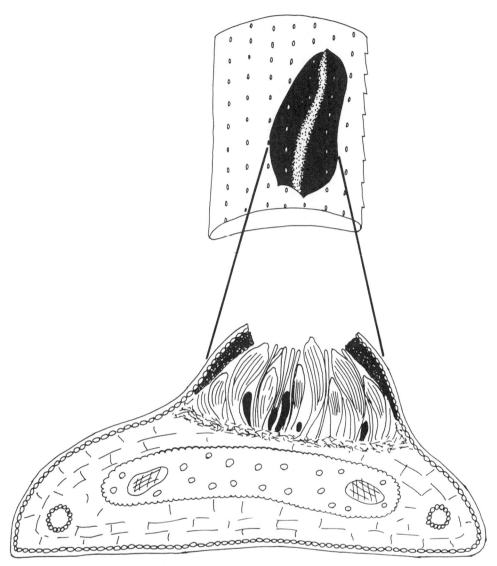

FIGURE 4.31   Hysterothecia of *Davisomycella fragilis* (Hypodermataceae) on jack pine. The drawing on top is a macroview of a hysterothecium on a needle. The bottom is of a cross section through a hysterothecium on an infected needle. Redrawn from Darker (1967).

Subclass 4: Laboulbeniomycetidae.   A fourth subclass of the Ascomycetes, the Laboulbeniomycetidae, contains fungi that lack true mycelium. They are primarily parasites of arthropods and, in the southern hemisphere, of marine red algae. These fungi are highly host-specific and, therefore, grow only on certain unique substrates or organs such as the legs of certain species of cockroaches; some are even restricted to one sex of an insect species. Whether they cause diseases of their hosts is not known.

FIGURE 4.32 Apothecia in the Pezizaceae, showing the range in size and shape.

Subclass 5: Loculascomycetidae. The major features of the Loculascomycetidae are the bitunicate asci, which are produced in stromatic locules. *Bitunicate* means that the asci are double-walled. During release of ascospores, the outer ascal wall splits open on the end, and the inner wall extends upward immediately before the ascospores are ejected. This feature may assist in more effective release and dissemination of the spores. Stromatic locules are hollow areas within the stroma; their inner walls are lined with asci. Members cause leaf spots, needle casts, and cankers of trees.

ORDER: DOTHIDIALES

Family: Dothidiaceae. In this large family the stromatic locule is known as a pseudothecium, and these are immersed or erumpent from either a stroma or host tissues (Fig. 4.35). Many species are parasitic on agricultural crops. Several species of *Guignardia* cause leaf spots and fruit rots of woody plants.

Family: Capnodiaceae. The Capnodiaceae are known as the sooty molds. They are epiphytic fungi, which live on the sugary excretions of insects, particularly aphids. While not a parasite of plants, the dark-colored mycelium covers the aerial parts of plants and may reduce their ability to photosynthesize. The mycelium may become so dense as to form a black, carbonaceous, or spongy covering (Fig. 4.36).

FIGURE 4.33 Fruiting body of *Morchella* spp. (Morchellaceae), and cross section of hymenium.

ORDER: PLEOSPORALES

Family: Venturiaceae. The Venturiaceae are largely plant parasites that produce their ascostromata either subepidermally or subcuticularly. The most important genus, *Venturia,* produces shoot blight, leaf blight, and fruit scabs of many fruit tree species, and maple, aspen, and willow.

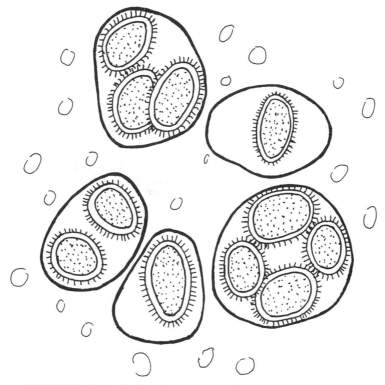

FIGURE 4.34   Asci and ascospores of *Tuber* spp. (Tuberales).

ORDER: HYSTERIALES

Family: Hysteriaceae. The Hysteriaceae, which has an ascocarp resembling the hysterothecium found in the Hypodermataceae, contains species primarily saprophytic on wood and bark. They are of little economic importance.

Subdivision 3: Basidiomycotina (The Now Obsolete Class Basidiomycetes). The major feature of the Basidiomycotina is that their sexual spores (basidiospores) are produced on the surface of various types of specialized club-shaped structures known as basidia. Many Basidiomycotina lack the ability to produce asexual spores, although some can produce buds (such as in some of the yeasts), chlamydospores, conidia, and oidia. Oidia usually originate from haploid mycelium and may serve as spermatia. Spermatia are specialized spores involved in diploidization.

Class 1: Basidiomycetes. There is only this one class, Basidiomycetes.

Subclass 1: Holobasidiomycetidae. In the subclass Holobasidiomycetidae the basidia are nonseptate (single-celled) and often produced in a basidiocarp (Fig. 4.37). There are two major subgroups, the Gasteromycetes, which have no distinct hymenium, and the Holobasidiomycetes, which have basidia in a well-defined hymenium. The hymenium is usually produced on a well-developed basidiocarp.

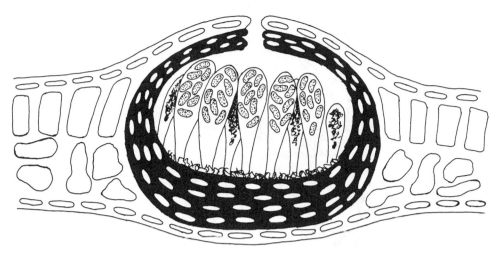

FIGURE 4.35   Pseudothecium of *Guignardia* spp. (Dothideaceae).

ORDER: APHYLLOPHORALES.   The Aphyllophorales are known as the pore, coral, and tooth fungi. The order includes both saprophytic species, which are common in the forest on decaying organic matter and wood, and a large number of parasites that cause heart and root rots of living coniferous and deciduous trees.

Family: Thelephoraceae.   In the family Thelephoraceae the basidiocarp is usually thin, leathery or hard, with a smooth hymenium (Fig. 4.38).

Family: Clavariaceae.   In the family Clavariaceae the basidiocarp is usually erect, fleshy, and club- or coral-shaped with a smooth hymenium (Fig. 4.39).

FIGURE 4.36   A well-developed example of a sooty mold (Capnodiaceae) growing from a twig. The black, feathery structures are spongelike masses of mycelium and perithecia.

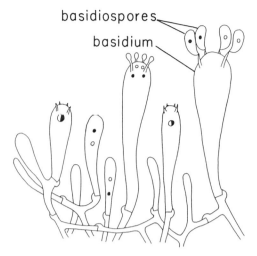

FIGURE 4.37 Schematic drawing showing the formation of nonseptate basidia in the Holo-basidiomycetidae.

Family: Cantharellaceae. In the Cantharellaceae the basidiocarp is fleshy and fun-nel-shaped with a smooth hymenium (Fig. 4.40). These are highly prized as edibles.

Family: Hydnaceae. In the family Hydnaceae the basidiocarp is fleshy or woody with the hymenium covering spines or teeth that hang down from the basidiocarp (Fig. 4.41).

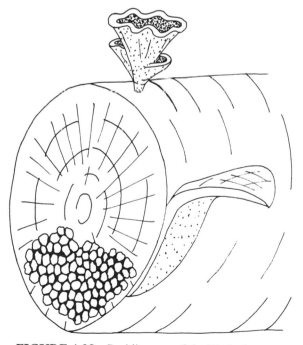

FIGURE 4.38 Basidiocarps of the Thelephoraceae.

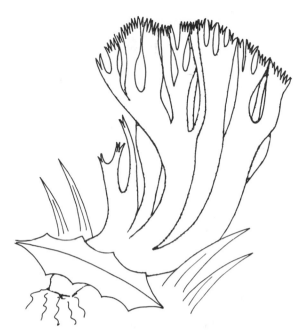

FIGURE 4.39   A basidiocarp of the Clavariaceae.

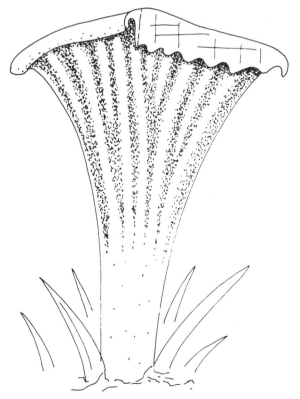

FIGURE 4.40   A basidiocarp of the Cantharellaceae.

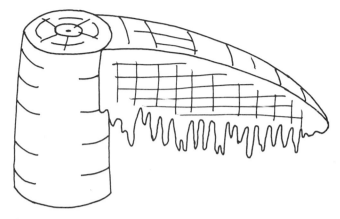

FIGURE 4.41   A cross section of a basidiocarp of the Hydnaceae.

Family: Meruliaceae. In the family Meruliaceae the basidiocarp is woody, fleshy, or leathery with the hymenium lining shallow pits or tubes of unequal depth (Fig. 4.42).

Family: Polyporaceae. In the family Polyporaceae the basidiocarp is woody, fleshy, or leathery, with the hymenium lining deep tubes of equal depth (Fig. 4.43).

ORDER: AGARICALES. Closely related to the Aphyllophorales, the Agaricales also contain many species that can produce heart decay in living trees. Other species are mycorrhizal. Included in this order are the mushrooms and boletes. The basidiocarps are usually fleshy, with a distinct stem and cap (pileus) and with the hymenium lining gills or pores.

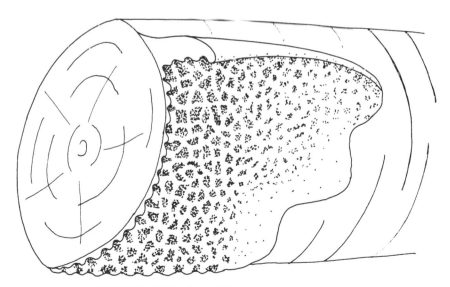

FIGURE 4.42   A basidiocarp of the Meruliaceae.

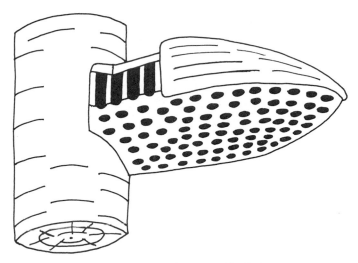

FIGURE 4.43   A basidiocarp of the Polyporaceae.

Family: Boletaceae. The boletes have vertically arranged tubes on the underside of the pileus with the hymenium lining the inside of each tube (Fig. 4.44). Many are mycorrhizal, and some species are found only in association with certain tree species.

Families: Russulaceae, Hygrophoraceae, Amanitaceae, Tricholomataceae, Volvariaceae, Strophariaceae, Agaricaceae, Cortinariaceae, and Coprinaceae. In older classification

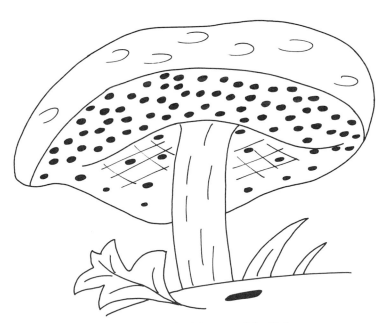

FIGURE 4.44   A basidiocarp of the Boletaceae.

FIGURE 4.45   A basidiocarp of the Agaricaceae.

systems, all these families were lumped together in the Agaricaceae. As taxonomists better understand their relationships, it has been possible to separate them into more or less natural families. All, however, have in common a hymenial layer that covers vertically arranged gills on the underside of the pileus (Fig. 4.45).

ORDER: EXOBASIDIALES. The order Exobasidiales contains a small group of obligate parasites that infect *Azalea* and *Rhododendron*. They produce galllike overgrowths on twigs and leaves (Fig. 4.46). The most important genus is *Exobasidium*.

ORDER: DACRYMYCETALES. The order Dacrymycetales contains some of the jelly fungi, which assist in decaying slash. Basidiocarps are waxy or jellylike and are usually bright yellow or orange in color (Fig. 4.47).

Gasteromycetes. This somewhat artificial catch-all grouping includes the common puff-balls, earthstars, bird's nest fungi, and stinkhorns. All are saprophytes on decaying organic matter.

FIGURE 4.46   Gall-like growth caused by infection by *Exobasidium* spp. (Exobasidiales).

ORDER: SCLERODERMATALES. Members of the Sclerodermatales order resemble puffballs but have a hard outer rind (Fig. 4.48). *Pisolithus tinctorius* is an important ectomycorrhizal fungus, especially in poor soils, and has been used commercially in tree nurseries to increase tree growth and protect against pathogens.

ORDER: LYCOPERDALES. The Lycoperdales are the common puffballs and earthstars. Fruiting bodies are characterized by having an outer wall that surrounds the *gleba,* which is the fertile portion where basidiospores are produced (Fig. 4.49). Spores are puffed out following a blow that fractures the outer wall.

ORDER: PHALLALES. The Phallales are known as the stinkhorns, a name that is derived from the rather fetid odor released when the gleba is exposed. Upon exposure, the gleba autodigests and the basidiospores are trapped in a foul-smelling gelatinous matrix (Fig. 4.50). Flies are attracted by the odor, feed upon the matrix, and, thus, inadvertently assist in dissemination of the spores.

FIGURE 4.47   Basidiocarps of *Dacrymyces* spp. (Dacrymycetales).

ORDER: NIDULARIALES. The Nidulariales order contains the bird's nest fungi and gets it name from the basidiocarp that resembles a miniature bird's nest containing several tiny egglike structures known as peridioles (Fig. 4.51). The peridioles contain the basidiospores and are forcibly ejected when a drop of water falls into the hollow outer nestlike structure.

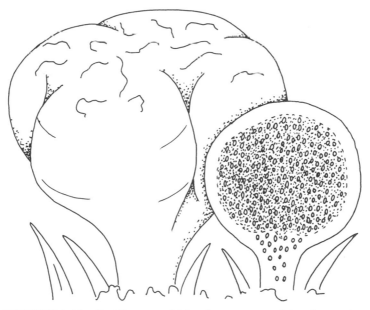

FIGURE 4.48   Basidiocarps of *Scleroderma geaster* (Sclerodermatales).

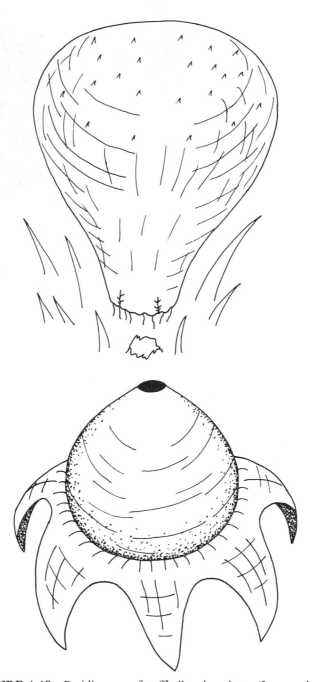

FIGURE 4.49   Basidiocarps of puffball and earthstar (Lycoperdales).

Subclass 2: Phragmobasidiomycetidae. The subclass Phragmobasidiomycetidae has basidia, which are septate and four-celled within its two orders. This group includes a few jelly fungi.

ORDER: TREMELLALES. Fungi in the Tremellales order produce a specialized basidium known as a metabasidium, which is divided vertically into four cells (Fig. 4.52).

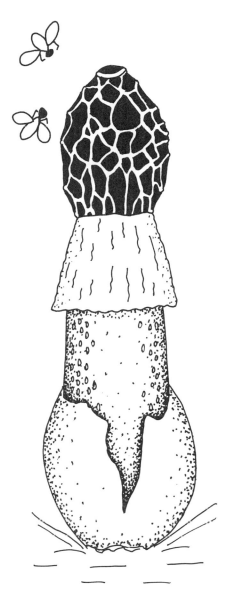

FIGURE 4.50   Basidiocarp of a common stinkhorn (Phallales).

Each of these cells produces a sterigma at the tip of which is produced a basidiospore. Members are known as jelly fungi because of the gelatinous consistency of the fruiting bodies and are primarily saprophytes.

ORDER: AURICULARIALES. The order Auriculariales is very similar to the Tremellales, but the metabasidium is divided transversally into four cells (Fig. 4.53). Each of these cells produces a separate sterigma and basidiospore.

Subclass 3: Teliomycetidae. The subclass Teliomycetidae contains a large number of obligately parasitic fungi collectively known as the rusts or the smuts (Fig. 4.54). There is no basidiocarp produced by either group. A thick-walled resting structure

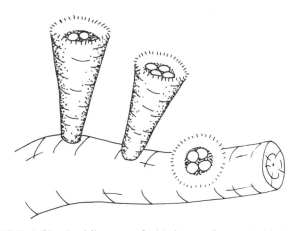

FIGURE 4.51   Basidiocarps of a birds' nest fungus (Nidulariales).

known as a *teliospore* is characteristic. The teliospore germinates to produce a promycelium on which four basidiospores are formed. The many rust fungi are of economic importance on both coniferous and deciduous forest trees. The smuts are of consequence only on cereal crops.

Most forest tree rusts typically have five spore stages. The pycnial and aecial stages are usually found on the primary host (the one that is economically more important), which is usually coniferous. Aeciospores infect leaves of the alternate host and give rise to uredinia and eventually telia. Teliospores germinate and produce basidiospores, which infect new or succulent needles on the conifer host to complete the life cycle.

ORDER: UREDINALES. The Uredinales are the rusts and include a large number of economically important plant parasites. White pine blister rust is perhaps the most damaging and well-known example of a tree rust.

Mycologists classify organisms by virtue of their sexual or perfect stage, which in rusts occurs in the germinating teliospore. Consequently, mycologists designate the

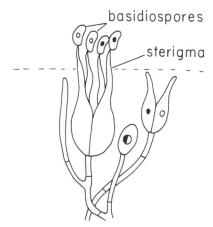

FIGURE 4.52   Typical basidia of Tremellales.

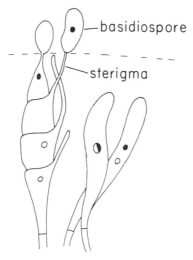

FIGURE 4.53 Typical basidia of the Auriculariales.

telial host as the primary host; the pycnia and aecia form on the alternate, or second-ary, host. By definition then, the alternate host of white pine blister rust is white pine! Many students in forest pathology have learned this on examinations! The common names for tree rusts often include the name of the tree. Forest pathologists refer to the other host as the alternate host. Because most rusts on conifers produce pycnia and aecia on the conifer, the common usage of alternate host is mycologically incor-rect. While we strive to be accurate in our use of mycological terms, the information conveyed by our misuse of alternate and primary host outweighs the benefits of being correct. Therefore, we will use primary host when referring to the economically im-portant host, and alternate or secondary host when referring to the other host.

A brief review of how the rusts are believed to have evolved may help put the development of the various spore stages into a clearer perspective. According to fossil

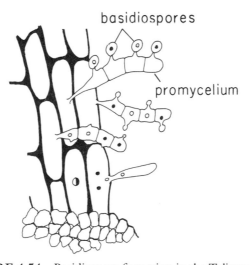

FIGURE 4.54 Basidiospore formation in the Teliomycetidae.

records, the oldest rust spore stage discovered appeared on ferns in the Devonian. The nuclear condition of the spores was dikaryotic, or N + N. This means that the fungus has two haploid nuclei that have not yet fused (*di* = two; *karyo* = nuclei). As this taxon developed and spread, the greater distance between infections produced a strong selection pressure for development of monokaryotic (N) conidia, which would allow the fungus to more rapidly colonize plants in the intervening space. Pycnia may have developed as both monokaryotic and dikaryotic types and initially projected from stomata. Most became extinct, but some persisted in tropical fern swamps and produced small conidia. The simple conidiophores became grouped together and were associated with a nectar that attracted insects. These conidia, or pycniospores, then lost the ability to germinate, but they could pass through pores into receptive hyphae. The aecium was the last to evolve and resulted from the initiation of heteroecism or the ability to infect quite different hosts, which is a response to a severe population decrease twice each year. In this case, rust or host population decreased due to dramatic annual fluctuations in temperature and precipitation. With the dry season, fern fronds died back each year. The only alternative to extinction for the rusts was to infect adjacent progymnosperms, which could survive through the seasonally dry weather. Modern relics of some of these old associations can still be seen in the northern temperate forests where the rusts are believed to have originated. Tree rusts are mainly a phenomenon of the northern hemisphere. They coevolved with their coniferous hosts, which for a variety of reasons did not migrate into the southern hemisphere.

At lower elevations, the several generations of uredinia each year helped increase the rust population on widely scattered hosts; and pycnia aided in dikaryotization of the scattered monokaryons. As one approaches arctic conditions or the alpine tree line, rusts occur only when the primary and secondary hosts are in increasingly close proximity because few uredinia can form during the short growing season before they give way to telia.

Family: Pucciniaceae. The teliospores of the Pucciniaceae are generally stalked and embedded in a gelatinous matrix. Upon germination of the teliospore, these fungi form a septate basidium known as a promycelium. The genus *Gymnosporangium* causes an important group of diseases known as the cedar or cedar apple rusts.

Family: Melampsoraceae. The Melampsoraceae have teliospores that are united into columns and form a septate promycelium upon germination. Most of the tree rusts are found in this family.

Family: Coleosporiaceae. In the family Coleosporiaceae the teliospores are united to form crustlike layers. As germination proceeds, the teliospore becomes four-celled and produces basidiospores without forming an external promycelium. Many of the conifer needle rusts are included here.

ORDER: USTILAGINALES. The Ustilaginales are commonly known as the smuts. All are extremely important plant pathogens; but since none cause tree diseases, they are not discussed further.

Subdivision 4: Deuteromycotina. The last subdivision of the Mycetae is the Deuteromycotina. This grouping is highly artificial with the common feature that sexual reproduction is unknown. For this reason, they are also known as the Fungi Imperfecti, or imperfect fungi. Three simple form subclasses are based on how the asexual spores are borne.

1. *Blastomycetidae*. These fungi are yeastlike.
2. *Coelomycetidae*. Conidia are produced in fruiting structures called pycnidia or acervuli.
3. *Hyphomycetidae*. Conidia are produced on special hyphae but not in pycnidia or acervuli.

These fungi can grow indefinitely by remaining in the mycelial state and producing conidia. The asexual, or conidial, state is commonly encountered, but the sexual state may be encountered rarely, if at all. These fungi are practically identified, then, based upon the reproductive state in which they happen to be found. If the asexual stage is more prevalent or we cannot induce formation of the sexual stage in culture, then, the first description is likely to be of the asexual stage. But, the traditional taxonomic one is based on the sexual stage; it was reviewed previously in this chapter. The Deuteromycotina was constructed to accommodate fungi with an uncommon or absent sexual stage and is based solely on the morphology of the conidial state. For convenience and ease of identification, species with similar conidia are placed in the same form genus. We use the term *form* genus because a true genus is based on the morphology of the sexual stage. When a sexual or perfect stage is found, the fungus is classified on the basis of its perfect state and named accordingly. The perfect name becomes the only validly recognized scientific name; the imperfect name is then used to refer to the asexual state. For example, *Spiniger meineckellus* is the anatomical name of the asexual stage of *Heterobasidion annosum (Fomes annosus)*. Several of the fungi in this subdivision are responsible for tree diseases including species causing Septoria canker of aspen, pitch canker of pines, and Cytospora canker of many tree species.

4.2 BACTERIA

In spite of their microscopic size, bacteria have a remarkable potential for rapid multiplication and biochemical activity. Following the discovery of life forms such as bacteria, it is not strange that men doubted that these minute organisms could destroy a human being or even a plant. Robert Koch, a German country doctor, presented the results of his brilliant studies on the dreaded anthrax disease of cattle to a skeptical audience of scientists in 1876 and began to break mental barriers to accepting the germ theory of disease. It remained for an American scientist named Burrill to show for the first time that bacteria also caused disease in plants. This was in 1884. From 1890 to 1920 Erwin F. Smith significantly advanced our understanding of bacterial diseases of plants.

4.2.1 Nature of Plant Pathogenic Bacteria

Bacteria are in the class Schizomycetes of the Thallophyta. They are single-celled, microscopic organisms without chlorophyll. They have neither mitochondria nor the typical membrane-bound nucleus (they are prokaryotic) of higher plants and animals. The single cells may be held together in chains, branched or unbranched filaments, or loose masses, but each cell is a unit and is capable of independent existence. Bacteria

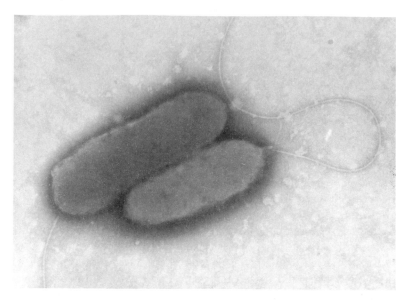

FIGURE 4.55   Bacteria (*Xanthomonas* spp.), each with a single polar flagellum.

range in size from 0.4–0.7 μm in diameter to 1.0–3.0 μm in length. They are just large enough to be seen with the light microscope.

A bacterial cell divides by simple fussion. This process of one individual dividing to form two new individuals may take place once every 20–30 minutes. At the end of 12 hours, the population arising from a single cell may be about 20 million bacteria if all cells remain alive. The ability to produce tremendous numbers of individual cells in a relatively short time partially explains why these organisms can be so destructive. They also produce a wide variety of enzymes that are capable of digesting many kinds of food materials.

Bacterial cells are either spherical (as in *Streptococcus*), spiral, or rod-shaped. Many bacteria have long, threadlike flagella that are attached to the wall. The manner of arrangement of the flagella serves as a basis for classifying bacteria in one or more different genera. For example, the genus *Erwinia* is characterized by perithichous flagella (flagella all over the cell). The genus *Xanthomonas* is characterized by a single flagellum (polar) attached at one end (Fig. 4.55).

Bacterial genera and species are not easily differentiated solely on the basis of colony characteristics or cell morphology, so certain biochemical or physical tests are used for their differentiation, including serology, homology, and percentage of guanine-cytosine. However, one of the most significant physiological tests in the identification of plant pathogenic bacteria is their effect on specific host plants.

Approximately 170 species of plant pathogenic bacteria have been described. Fortunately, only a few of these attack trees, including orchard trees, and only one or two cause disease in forest trees.

4.2.2   Dissemination

Bacteria that cause plant disease do not produce spores, so they are not adapted for wind dissemination. Although they are capable of movement in water, they depend, rather, on wind-splashed rain and on people for their spread and dispersal. Insects are also responsible for spread of some bacterial diseases.

Protection from desiccation is important in the survival of bacteria. Because they are susceptible to desiccation, most plant pathogenic bacteria must survive unfavorable periods in association with host plants or tissues. In apple or pear trees infected with fire blight, the bacteria surround themselves with a gummy substance that prevents desiccation (Schroth et al. 1979). In general, bacteria that cause foliar diseases do not survive for long periods of time in the absence of suitable host plants. Although these bacteria can survive as saprophytes, other microorganisms are antagonistic to them, and plant pathogenic bacteria do not survive in the competition for survival that is so characteristic of microbial life in the soil.

4.2.3   Mechanism of Pathogenesis

Bacteria are incapable of direct penetration through the cuticle, epidermis, or bark of the host; they must enter through wounds or stomata. Thus we consider the infection process passive. Once inside the host, bacteria produce a multitude of enzymes and toxic materials that upset the metabolism of the host, causing injury or death. Because the surfaces of most plant parts are covered with large populations of free-living bacteria, it is amazing that so few have evolved to become parasites of plants. Bacteria can be easily isolated from the sap of apparently healthy trees, yet only a few species can cause a disease called wetwood (Hartley et al. 1961) (Figs. 4.56 and 4.57).

FIGURE 4.56   Elm, showing discoloration of bark caused by exudations from bacterial wetwood in the pruning wound.

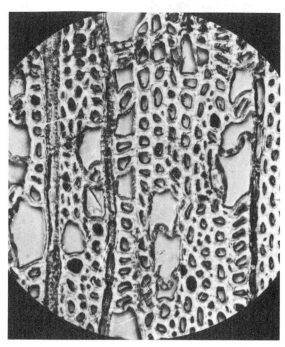

FIGURE 4.57   Collapse of Eucalyptus wood due to effects of bacterial wetwood. Note the angularity of the collapsed vessels.

If the plant pathogenic bacterium is incompatable with the host, a hypersensitive reaction quickly develops, usually within a few hours, and further reproduction of the bacterium ceases (Horsfall and Cowling 1979). Infection of a susceptible host apparently involves a "recognition" phenomenon. Some evidence suggests that interwoven fibers of branching sugar molecules enable the bacterial cell to tightly adhere to susceptible host cell walls. After entering a susceptible plant, a series of events ensues: respiration usually increases or is altered, phenolic substances accumulate, permeability of cell membranes increases, and peroxidases become more active. In crown gall and other bacterial gall diseases the induction of plant tumors represents a compatible reaction (Fig. 4.58).

Few bacteria can use starch, and few can degrade cellulose at a very rapid rate. No bacterial plant pathogens are known to degrade lignin. Bacteria are dependent on macerating enzymes that degrade the pectic portion of the middle lamella mainly by hydrolysis. These pectic enzymes tear holes or weaken areas in the cell wall, which in turn weakens areas in the plasmalema. Cell death results from rapid electrolyte loss. An exception is *Erwinia amylovora*, the fire blight pathogen, which produces none of the major cell wall component-degrading enzymes. How this bacterium spreads within tissues is not known.

Some bacteria cause vascular wilting when the stem xylem tissue is plugged with bacterial products and host substances formed in response to the bacteria. Both substances are composed of highly viscous polysaccharides. Their deposition on pit membranes could be a major factor restricting water movements.

Some phytopathogenic bacteria disrupt metabolism by producing toxins. Symptoms appear as a chlorosis resulting from inhibition of some important metabolic

FIGURE 4.58 A bacterial gall on redbud.

process. Although most toxins are not host specific, *amylovorin,* the fire blight toxin, apparently is.

Some bacteria can enhance the formation of auxins, cytokinins, and ethylene, growth regulators whose roles are not yet clearly understood. Alteration of the balance of these growth regulators, however, is suspected to be involved in hypertrophic growth, a characteristic of some bacterial diseases such as crown gall. The crown-gall bacterium, *Agrobacterium tumefaciens,* contains a specific large plasmid, an extrachromosomal, circular piece of DNA that is present in all virulent strains (Lippincott and Lippincott 1975). Isolated plasmids transferred to avirulent strains of *A. tumefaciens* make these strains become virulent. A contiguous segment of DNA, approximately 5–10% of the plasmid, is transferred to the host cell and is responsible for inducing the tumor.

4.2.4 Ice Nucleation Bacteria

Other bacteria, although not plant pathogenic, can have a direct influence on plant health by increasing damage due to frost injury. Some plants cannot withstand ice crystal formation in their cytoplasm and hence suffer greatly from frost damage. Water in plant cells below freezing is supercooled and needs a catalyst to start crystallization. Some bacteria can serve as this catalyst.

Leaves are colonized by many nonplant pathogens, or saprophytic bacteria. These include certain ice-nucleation bacteria in the *Pseudomonas syringae* group. These bacteria can catalyze ice formation at temperatures as high as $-1°C$. As their population increases, the chance and degree of injury likewise increases. There are large seasonal increases in spring in populations of ice-nucleation bacteria. They are normally found in anticlinal wall junctions between epidermal cells. There is little or no invasion of leaves, nor is there secondary invasion of frost ruptured tissues by these or other bacteria.

Not much is known about the ice-nucleation phenonemon. Ice nucleation bacteria that are killed shortly before frost retain the ice-nucleation characteristic but lose it when they are killed far in advance of frost. There is a microsite on the outer nuclear membrane of the bacterium that governs ice nucleation. But, ice nucleation apparently starts on the leaf surface.

Water stress influences ice nucleation. Plants growing under water stress are resistant to frost injury apparently because the ice-nucleation bacteria are not as numerous.

Other Pseudomonads antagonistic to the ice-nucleation bacteria have been isolated and used to control frost injury biologically. Applied just after bud break, they increase rapidly on expanding leaf tissue. This has also been done experimentally with potato, tomato, strawberry, and almond. Scientists have also tried to identify the most effective antagonistic mechanisms and have tried to induce mutants without ice-nucleation ability.

4.2.5 Beneficial Effects

Another interesting bacteria–plant interaction has recently been discovered. Some bacteria extract iron from the soil and bind it tightly. This iron is then unavailable to other soil microorganisms, some of which could be harmful. In sugar beets, the only such association investigated so far, the reduced competition for other essential nutrients results in significantly higher yields.

4.3 MYCOPLASMAS

Mycoplasmas are bacterialike organisms that cause many diseases in humans and animals and in 1967 were discovered to cause diseases in plants as well (Agrios 1975). Since then, many of the yellows diseases (Fig. 4.59) formerly thought to be virus-caused have been associated with mycoplasmas (Wilson and Seliskar 1976). As research progressed, it became obvious that these plant disease agents did not belong to the true mycoplasma family, which contained the human and animal pathogens. Hence, the more current term *mycoplasmalike organism* (MLO) has come into use. MLOs are known to be associated with over 50 yellows disorders in plants.

The MLOs differ from bacteria in being smaller and in having an exterior membrane instead of a cell wall. MLOs are the smallest known prokaryotes and can be seen only with the aid of the electron microscope. A closely related group, the rickettsialike organisms (RLO) have cell walls and are probably small bacteria. Because MLOs have no rigid cell wall, they may assume various shapes. Most appear oval or spherical, but many appear filamentous as well (Fig. 4.60). Mycoplasmalike organisms appear to divide by fission. Most MLOs apparently require insect vectors in order to infect plants. These vectors, sucking insects, can also act as hosts, or at least as reservoirs, for MLOs. After an MLO is ingested, approximately 5 days must elapse before transmission can occur. These MLOs can also be spread through natural root grafts, by budding, and by cuttings.

The MLO-caused disorders are characterized by an overall yellowing and dwarfing of foliage (Fig. 4.61) or by the development of witches' brooms (Fig. 4.62). How

FIGURE 4.59 Yellows disease of oak caused by mycoplasmas.

MLOs cause disease is not well understood. Some MLOs produce one or more toxins that injure leaf cells. Some of the disease symptoms have been produced by injecting healthy plants with the purified toxin(s). As invading MLOs become established inside a plant, a large quantity of cholesterol is needed for manufacture of their own membranes. Because cholesterol uptake is a purely physical absorption process and does not depend on metabolic energy, cholesterol usage probably causes little harm

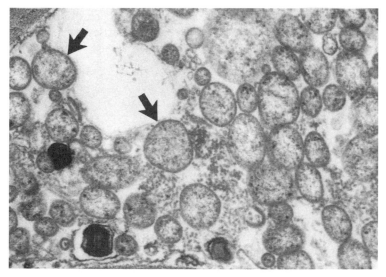

FIGURE 4.60 Mycoplasmas in a portion of a phloem cell. Courtesy of J. L. Dale, University of Arkansas, Fayetteville.

FIGURE 4.61   A branch of *Aralia spinosa* infected with mycoplasmas (right) compared with a healthy branch (left). Courtesy of J. L. Dale, University of Arkansas, Fayetteville.

to the plant. Severe photosynthetic and hormonal imbalances soon result, however. They are manifested by witches' broom production, shoot proliferation, big buds, leafy flowers, and yellows symptoms.

Tree diseases associated with MLOs include black locust witches' broom, pecan and walnut bunch disease, witches' broom of dogwood and sassafras, and both ash and elm yellows.

In the past, control of some MLOs has been partially successful by controlling the insect vectors. Recently, symptoms of some yellows diseases have been suppressed following injection of trees with tetracycline (oxytetracyline hydrochloride) antibiotics.

4.4   VIRUSES

Viruses cause some of the most destructive diseases, not only of plants but also of man and animals. The list of virus diseases is an imposing one. In people they cause infantile paralysis, encephalitis, smallpox, yellow fever, the common cold, and even warts. In animals they cause foot and mouth disease, infectious anemia, swine fever, rabies, distemper, and equine encephalitis.

Viruses usually stunt the growth of plants and reduce their yields, but a few also cause mortality. At times these diseases enhance the value of ornamental plants by producing a variegation in flowers or leaves. Among tulips, broken color of the blossoms is regularly due to viruses and was formerly sought after as a horticultural oddity. Even bacteria suffer from viruses known as bacteriophages.

FIGURE 4.62 Witches' broom on willow caused by mycoplasmas.

4.4.1 Nature of Viruses

Viruses are extremely small, obligate, intracellular parasites that cannot multiply on synthetic culture media (Agrios 1975). The contagious nature of viruses was first recognized at the end of the nineteenth century, when they were considered to be either filterable bacteria (Iwanowsky) or a "contagious living fluid" (Beijerinck). Most scientists thought that viruses were living organisms too small to be detected with the microscope but were in most other respects analogous to bacteria.

In 1935 Stanley prepared a partially crystalline proteinaceous material from the juice of tobacco plants infected with the tobacco mosaic virus. We now know that this was the virus itself in pure form. We have since learned to crystallize many other viruses and that the crystals are composed of protein and also some nucleic acid.

Although virus particles are too small to be seen by light microscopy, they can easily be made visible by transmission electron microscopy (Fig. 4.63). Viruses assume various morphological forms, including rigid rods (tobacco mosaic virus) and bacteriophages that have hexagonal heads and tiny tails. The nucleic acid portion of the virus particle is usually protected and surrounded by an outer sheath composed of

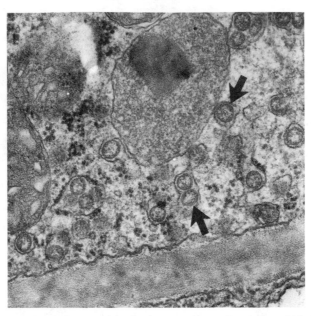

FIGURE 4.63   Spherical virus particles in the cytoplasm of an infected black locust leaf cell. Courtesy of J. P. Fulton, University of Arkansas, Fayetteville.

small, uniform protein molecules. The nucleic acid component of most plant-infecting viruses contains only RNA, which is the infectious entity.

A related group, known as viroids, consists of a small loop of infectious RNA without any protective protein. Citrus exocortis and avocado sun blotch are the only viroids of trees and shrubs known in North America.

Viruses are named according to the nature of symptoms they produce. Elm leaf mosiac, for example, vividly describes a virus infection in this host.

4.4.2 Transmission

Many viruses are highly infectious and may be transmitted by simply rubbing healthy plants with infected plant parts. Handling of plants during transplanting and pruning is a common means of spreading viruses.

The majority of viruses, however, are spread in nature by insects. In some cases, insects feed on virus-infected plants, and their mouth parts become contaminated with virus particles in the plant sap. When they feed on healthy plants, they infect them in a more or less mechanical manner. In many other instances, the insect is the only natural means by which the virus can pass from one plant to another. The insect feeds on a diseased plant, the virus passes into its digestive tract, and only after an incubation period of several days or weeks does the insect begin to transmit the virus to healthy plants. Thereafter, throughout its life, it continues to transmit the virus to every plant on which it feeds.

Some viruses can be transmitted by vegetative propagation such as grafting and budding. Natural root grafting sometimes results in the spread of viruses from one tree to another. A few viruses are borne in seed or tubers.

4.4.3   Symptoms

Many virus diseases of forest trees probably exist, but, unfortunately, little is known about them (Kim and Fulton 1973). Virus diseases of agricultural crops have justifiably received the most research attention in the past. Because of the lack of definite information on viruses in forest trees, considerable survey work and research needs to be done. Some forest and shade tree viruses are well known, but very little information exists on the majority of suspected or reported forest tree viruses. We have recently learned that some alleged "virus-infested" trees, especially those with "yellows" disease, are infected with mycoplasmas or MLOs.

A mosaic appearance is probably the most common type of symptom associated with virus infections in forest and shade trees (Fig. 4.64). Mosaic symptoms may include yellow mottling, vein clearing, or gradate into chlorotic rings. Other common symptoms include reduction in leaf size, leaf curling, stunting of growth, dieback, and generalized chlorosis.

In the United States, probably the most common virus affecting shade trees is elm mosaic virus, which apparently is a strain of tomato ringspot virus. This virus is distributed throughout the midwest and some eastern states but apparently occurs infrequently. Symptoms include dieback of upper branches, sparse foliage, small leaves that have chlorotic ring patterns, and small overgrowths interveinally on undersides of leaves. Mild to moderate brooming may occur. Affected trees lose vigor and become unsightly because of thin foliage and dead branches. The causal virus is a small sphere that can be transmitted with tree sap or by grafting. In addition to elm mosaic, elm ringspot, elm scorch virus, and elm zonate canker virus have been reported in the United States.

FIGURE 4.64   Mosaic virus symptoms on boxelder.

Because of their higher crop value, *Prunus* spp. have received much attention and study by plant virologists. But symptoms suggestive of virus infections have also been observed on a number of other forest trees. A bright yellow ringspot commonly occurs on several species of oak. Redcedar often shows evidence of peculiar types of irregular yellowing. A virus that occurs in black locust (Fig. 4.62) is similar to the virus causing tomato spotted wilt, but the spherical virus particles are larger in black locust than in tomato. Tobacco streak virus also occurs in black locust. Virus symptoms have also been noted in black gum, sassafras, hard maple, and redbud.

Poplar mosaic is found in certain cultivated varieties of *Populus deltoides* in Canada, Europe, and the United States. In infected trees, light green or yellow blotches and veinal necrosis occur in leaves. Leaves may curl, and reddish brown tumors are often found on petioles. Some stunting of current season's shoots may occur. This virus is transmitted in propagative cuttings and grafts. The causal virus is filamentous and flexuous (650 nm average length). Eradication of infected stock is the best method of control.

Both white and yellow birch have been graft-inoculated with the apple mosaic virus, which causes birch line pattern. In Europe, ash infectious variegation is believed to be virus-caused, as is ash necrotic curl in Italy. Maple variegation of boxelder and sycamore maple is striking in appearance and is common in parts of South America. Black locust mosaic is common in Europe. Ash ringspot, caused by tobacco ringspot virus, reduced tree height growth, leaf size, and root development (Hibben and Hagar 1975). Ashes in the United States are also affected by ash chlorotic–necrotic leaf spot.

4.4.4   How Viruses Induce Disease

Disease generally denotes some type of host malfunctioning. There are instances, however, of viral infections not causing any measurable harmful effect in the host even though the virus increased in the host plant (Horsfall and Cowling 1979). Wilting and death of leaves can result from vascular plugging in the stem or the effects of many confluent local lesions. Except for the local lesion reaction in hypersensitive hosts, viral infection seldom kills infected host cells.

Neither the site(s) in the cell where virus and viroid disturbance first occurs nor the nature of the disturbance is known. There is some evidence that virus- or viroid-engendered RNA molecules are the agents that interact with host DNA. They may serve as primers for aberrant DNA or RNA synthesis. Thus the metabolic machinery of the host cell can be diverted from its normal functions to the synthesis of new virus particles. The very close similarity between viruses and the genetic material of the nucleus of host cells explains why studies of virus infections may ultimately lead to a more complete understanding of life itself. The capacity of self-replication is the unique feature of living systems. Viruses have this capacity, but lack the complexity of the various membranes, mitochondria, and other cell organelles of their hosts.

4.5   PARASITIC FLOWERING PLANTS

Parasitic flowering plants are higher plants that parasitize other higher plants during most of their life cycle. There are approximately 3,000 parasitic angiosperms in 15 plant families. With one exception, *Podocarpus ustus,* they are all dicotyledons. Although many of them contain functional chlorophyll, they depend upon their hosts for all or most of their fixed carbon needs, as well as other nutrients and water.

The parasitic nature of a green plant on another green plant intrigued the ancients (Fig. 4.65). Several Greek legends emphasized the supposed mystical nature of the mistletoes. Pliny the Elder remarked about them in his *Natural History.* The ancient Germanic tribes and Druids held the mistletoe in reverence because it was able to stay green during the winter when other trees lost their leaves. In their language, mistletoe meant all-healing.

Parasitism of higher flowering plants by other higher flowering plants has an evolutionary advantage that is not completely understood. One mistletoe scientist (Kuijt 1969) has proposed that, because of the tremendous diversity of existing forms of parasitic flowering plants, parasitism has arisen on at least eight known different occasions in unrelated groups of dicotyledons. Such parasitism offers some advantages. Most parasitic plants are believed to have originated in the tropics, where the soils are

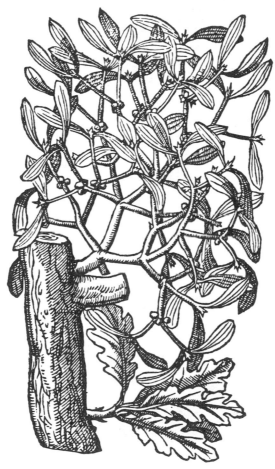

FIGURE 4.65   The leafy mistletoe, *Viscum album* on oak, from an old herbal published in 1581. From Kuijt (1969).

typically mineral-poor. By parasitizing already successful plants, such parasites were able to exploit an adequate supply of mineral nutrients and water.

Most parasitic flowering plants have a basically similar infection process. Upon contact with the host, the parasite produces a contact organ called a holdfast (Fig. 4.66), which is analogous to the appressorium of fungi. In some, a cutinlike substance

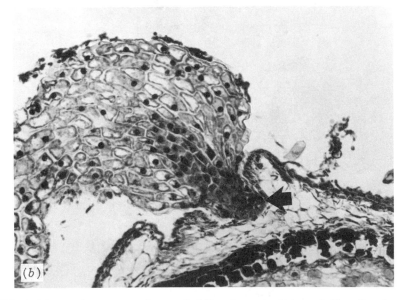

FIGURE 4.66   (a) An appressorium or holdfast (arrow) of *Arceuthobium* forming on black spruce. (b) A penetration peg (arrow) is evident in the section.

is produced on the host–parasite interface that helps hold the two together. A wedge-shaped structure called a penetration peg then grows from the holdfast and penetrates the host by means of mechanical and enzymatic action. Parasitic xylem unites with host xylem soon after penetration. The mistletoes also grow in phloem tissue, but form sinkers, which are initiated at the cambium, forming the endophytic system which grows outward with the cambium, leaving the sinker embedded in the xylem. This invariable xylem union suggests an early need for xylem-transported materials. In *Arceuthobium* the main body of the parasite develops in the phloem tissue, suggesting a further evolutionary development to perhaps eventual exclusive phloem parasitism.

Once xylem contact has been made, a burst of metabolic activity occurs. Respiration and transpiration increase in host tissues and, after a few days, photosynthesis decreases. The infection site then begins to act as a sink for organic compounds such as sugars, phosphorous, potassium, and sulfur.

In infected host tissues the size and number of cells increases, suggesting that the host is stimulated to produce an increase in size and number of cells suggesting that the metabolism of growth regulators is altered. Cytokinins, for example, are known to cause the conversion of cortical host tissue to xylem elements in *Orobanche,* which speeds up the host-to-parasite xylem "bridge." With *Arceuthobium* the inhibition of lateral buds may be reversed, and the familiar witches' broom begins to form (Fig. 4.67). Chlorophyll senescence is delayed. In old infections of *Arceuthobium,* witches' brooms still retain green needles after the rest of the foliage is lost. The infection area, or broom, acts as a nutrient sink. The dwarf mistletoes can photosynthesize at the rate of only 10% of the rate of the host, and they act as powerful sinks. On the other hand, the leafy mistletoe *(Phoradendron)* fixes its own carbon and extracts primarily

FIGURE 4.67   Witches' brooms on black spruce caused by *Arceuthobium pusillum*.

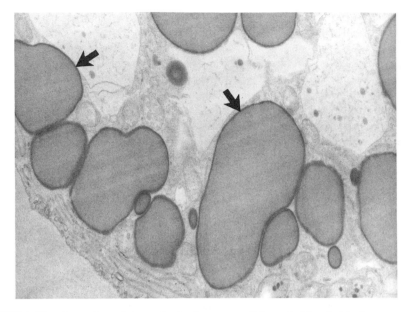

FIGURE 4.68   A portion of the endophytic system of *Arceuthobium* in black spruce showing an abundance of lipid (arrows). From Tainter (1971).

water and other nutrients. The parasites maintain water and mineral movement into the area by maintaining a high osmotic pressure. In *Arceuthobium,* large quantities of lipids accumulate (Fig. 4.68).

Recent evidence indicates that nitrogen starvation occurs in the host. The early association of parasite and host xylary cells allow the parasite direct access to the host-produced organic nitrogen compounds.

The most important families, genera, hosts, and geographic regions of occurrence are summarized here.

Summary of the Major Parasitic Flowering Plants Important to Forestry

  I. Root parasites

  A. Broomrapes (Orobanchaceae)—parasitic primarily on agricultural crops in India, Pakistan, Ceylon, south central Asia, Philippines, Mediterranean region, Europe, New Zealand, and South Africa.

  B. Parasitic figworts (Scrophulariaceae)—parasitic primarily on agricultural crops. Some serve as alternate hosts for tree rusts, in South Africa, Rhodesia, West Indies, Madagascar, East Africa, and the southeastern United States.

  C. Sandalwoods (Santalaceae)—worldwide in distribution, *Geocaulon, Buckleya, Comandra.*

  II. Stem or leaf parasites

  A. Cuscutaceae—dodder—on agricultural crops worldwide.

  B. Lauraceae—evergreen shrubs and ornamentals in India, Hawaii, East Indies, and Puerto Rico.

  C. Viscaceae

  Arceuthobium—conifers in the United States, Canada, Mexico, China, India,

Pakistan, Kenya, the Mediterranean region, and the Middle East.

Dendrophthora—rubber, mango, avocado, and cacao in South America.

Korthalsella—acacia and eucalyptus in Hawaii and Australia.

Notothixos—eucalyptus in Australia.

Phoradendron—coffee, avocado, teak, and various forest and shade trees in Bolivia, Central America, Mexico, United States, and the West Indies.

Viscum—rubber, conifers, fruit trees, deciduous trees in Europe, Asia, and Africa. One species introduced to California in the United States.

  D. Loranthaceae

Amyema—eucalyptus in Australia.

Dendrophthoe—southeast Asia.

Elytranthe—rubber and cashew in Malaya.

Loranthus—Europe and Asia.

Phthirusa—rubber in Brazil.

Psittacanthus and *Struthanthus*—citrus and acacia ranging from Mexico to Chile.

Tapinanthus—in Africa.

Tristarix—cactus in Chile.

  E. Myzodendraceae—on *Nothofagus* spp. in Patagonia.

  F. Eremolepidaceae—a rare South American group.

The parasitic figworts, dodder, dwarf mistletoes, and leafy mistletoes are among the most important parasitic flowering plants causing economically important damage to forest trees in North America.

Parasitic Figworts

The parasitic figworts (Scrophulariaceae) comprise a large and widespread group of root parasites in the southeastern United States. There are 11 genera and approximately 30 species. The largest genera are *Agalinis, Aureolaria,* and *Seymeria.* Some species are quite widely distributed, but, except for *Seymeria,* very little is known of the potential for causing damage.

Damage to pines by senna seymeria *(Seymeria cassioides)* is, however, well documented. This is one of the few root parasites in this group that exhibits any host specificity. It will grow well only when grown with pines. At least three species of pines—slash, loblolly, and longleaf-are susceptible, especially on poorly drained moist sandy sites of the lower Gulf and Atlantic coastal plains. The parasite occurs in nine states from Virginia through Louisiana. Once established, the parasite increases its numbers and, because of its pine host requirement, only the pines are affected. If the pines can survive a fire, the parasite can be managed with a prescribed burn after seeds germinate in the spring and before flowers develop.

Dodder

The Cuscutaceae (or dodders) are more of a problem in tropical and semitropical regions, but in the United States they have caused minor losses in nurseries on black locust, green ash, and poplar. Over 50 species are found in North America, but most are on field crops. Dodder occurs in patches that increase in size as the yellowish

FIGURE 4.69   Plants of dodder (*Cuscuta* sp.).

vinelike stems grow around the aerial portions of its host (Fig. 4.69). Haustoria are produced in host tissue, and through these structures nutrients are extracted during the growing season.

Dodder is best managed by preventing its introduction into nurseries. This can be accomplished by using cleaned seed. Using methyl bromide to fumigate soil before planting will kill seeds of dodder. Preemergence and early postemergence herbicides such as 2,4-D can be used. Drastic treatment of an established infestation to eliminate hosts and dodder has involved using chemicals such as pentachlorophenol.

Dwarf Mistletoes

Of the parasitic seed plants, the dwarf mistletoes (*Arceuthobium*) of the Viscaceae, are of the greatest concern in the North American conifer forests and cause the single most important disease problem in the western conifer forests of the United States (Hawksworth and Weins 1972). As the older timber has been harvested, dwarf mistletoe has become even more important than heart decay. Because of their importance, the dwarf mistletoes are discussed in detail in Chapter 19.

Leafy Mistletoes

The leafy, or so called true or green, mistletoes have been known since ancient times. They were considered then, perhaps because of their parasitic habit and evergreen nature, to have unusual medical and magical value; and they were frequently used in the religious rites of the ancient Germanics and Druids. At Midsummer Eve, a mistletoe plant growing on the sacred oak was cut from the tree by a Druid with a golden sickle and caught in a white cloth or by immaculate virgins. Only then was the sacrificial ceremony allowed to begin, and animals and human beings were slain and burned. The beginnings of the currently fading custom of kissing under the mistletoe at Christmas time are shrouded in mystery.

Leafy mistletoes are comprised of several genera and are common throughout the world. In the United States the leafy mistletoes do not extend north of a line drawn across the country from approximately New Jersey to Oregon. Their northward extension is probably limited by cold temperature. *Phoradendron* species do not occur in Canada. About a dozen species of *Phoradendron* (Viscaceae) are found in much of the forested areas of the United States but are most abundant in the arid and semiarid Southwest (Fig. 4.70). These mistletoes are more abundant in warmer climates and in some areas are extremely common on a wide range of hosts, primarily on hardwoods, but several species also occur on juniper, cypress, and incense cedar. *Phoradendron flavescens* causes a disease of pecans in Florida, of citrus in Texas, and of walnuts and persimmons in California. *Phoradendron juniperinum libocedri* and *P. bolleanum pauciflorum* have caused significant losses on incense cedar and white fir.

In Europe *Viscum album* is the common species, damaging many different hosts including apple, almond, cherry, pine, fir, and poplar in parks, orchards, forests, and plantations. *Viscum album* now occurs in California on 24 species including willow, alder, poplar *(Populus tremuloides* and *P. fremontii)*, elm, mountain ash, crabapple, and pear. In South America, Central America, Africa, and southern Asia, various species of the leafy mistletoes are destructive on citrus, tea, mango, rubber, cashew, cacao, cactus, avocado, teak, and coffee. In Australia, *Eucalyptus* species are the most important hosts.

Most leafy mistletoes have well-developed leaves. In one species the stems are as large as 38 cm in diameter, although, for most species, the stem diameters are 2.5 cm

FIGURE 4.70   A severe infestation of *Phoradendron* spp. (Viscaceae) on American elm.

or less. Pollination is accomplished by birds and insects, and birds disseminate the seed. Leafy mistletoes cause overgrowth in the host; the infected portion of the tree may swell, and a broom may result. Portions of the tree beyond the point of mistletoe infection may become deformed and die, but in many instances the host will live for years with only minor reduction in growth rate.

In comparison to the damage caused by the dwarf mistletoes, damage caused by *Phoradendron* spp. is slight, although they can cause economically important damage in certain areas. In the Australian Eucalypts, when the mistletoe infests 38% of the crown, increment is reduced by 38%. Species of *Phoradendron* are capable of photosynthesizing most of their own carbohydrates and depend on their hosts only for water and whatever mineral and organic nutrients are carried with water in the transpiration stream.

In many areas the leafy mistletoes are considered highly desirable and are used for decorations at Christmas time. In the southern states, principally in Oklahoma and Texas, sprigs of *Phoradendron* are collected and sold for the Christmas season. Pruning has been practical as a control measure, primarily in the tropics, but because birds can continually bring in more seed from other areas, pruning must be repeated. Fire has been used to remove mistletoe as an alternative for pruning. Formulations of 2,4-D have killed 70–100% of the infections on species of *Eucalyptus*. Such treatment causes some defoliation, and about 5% of the host plants are killed; however, it is a promising method in that large areas can be treated at low costs. Herbicide sprays have also been successful in controlling *Phoradendron* spp. in California walnut groves if applied during the dormant season.

4.6 NEMATODES

Plant parasitic nematodes (phytonematodes) are very small, elongate, tubular animals often called roundworms. Most nematodes are less than 2.5 mm in length. They are nonsegmented or superficially segmented. They are fairly primitive organisms yet have well-developed digestive, nervous, and reproductive systems. Nematodes apparently formed long ago and evolved into many different forms. Many species are parasitic on plants; others are parasitic on man and on domesticated animals. There are about 500,000 species, with 1,100–1,200 parasitic on plants. Zoologists regard nematodes as a distinct group and classify them in either the phylum Aschelminthes or, more commonly in recent years, the phylum Nematoda. Although nematodes were known to attack plants since the eighteenth century, it was not until the 1930s when workers in Hawaii proved that nematodes could be controlled with halogenated hydrocarbons such as chloropicrin, that attention began to be paid to phytonematodes and the diseases they cause.

Many nonplant–pathogenic nematodes feed on bacteria, other nematodes, or mites, which they macerate within a muscular structure containing teeth. All plant parasites, or phytonematodes, have an interesting structure called a spear, or stylet (Fig. 4.71). It is a hollow, tubelike structure located at the head end of the nematode and is used for feeding. Its sharp tip can be forced into plant tissues, then powerful muscles at its base inject enzyme-containing digestive fluids through the stylet and into the host. Then another muscular pump sucks host cell contents back through the stylet into the nematode's body. Most phytonematodes kill host cells by extracting

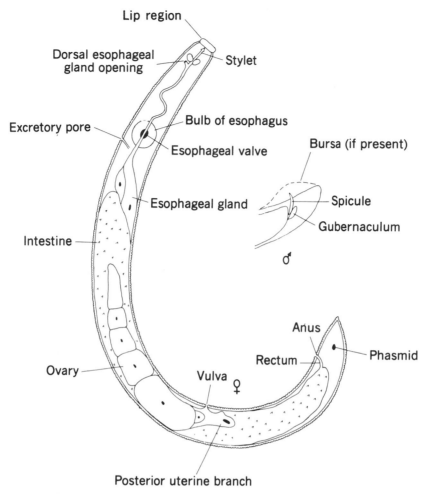

FIGURE 4.71 Drawing of a fictitious phytonematode, showing most morphological features needed for identification.

their contents. A single nematode can feed on dozens, hundreds, or even thousands of cells. Because a single plant may be attacked by hundreds or even thousands of nematodes, the efficiency of the host's root system can be considerably reduced, causing stunting, chlorosis, or even death. Aboveground symptoms may be rather nonspecific—irregular growth or wilting. For this reason, nematode diseases can be difficult to detect. The below-ground parts may contain galls or knots, stubby roots, reduced root growth, excessive branching, and/or lesions. Because nematodes reproduce rather more slowly than bacteria or fungi, and their spread is restricted unless aided by man, nematode-induced diseases tend to remain localized.

Phytonematodes damage plants in various ways. A small amount of feeding and removal of cell contents by a few nematodes may have no apparent permanent effects. Feeding can cause host cells and usually adjacent cells to die, however, resulting in lesion formation. Feeding by some species interferes with the normal growth regulator balance in plants, resulting in either a cessation of cell division or abnormal cell division and enlargement. The wounded area may also favor entrance and development of fungi.

Phytonematodes normally move only short distances in soil, usually less than 1/3 meter. They can, however, be carried long distances in soil adhering to shoes, machinery, or plant roots.

Most phytonematodes have a soil stage and live and move in water surrounding soil particles. Water is essential for their activity, but they are adversely affected by too much water and/or poor aeration. They can survive under dry conditions as eggs or as juveniles. Most phytonematodes take about 30 days for one generation to occur. The rapid increase in population of phytonematodes is at least partly due to the fact that the length of the life cycle decreases as the soil temperature increases, with many species completing their cycles most rapidly at temperatures of 25–30°C.

Nematodes have a rather simple life cycle. There may be as many as 200–500 eggs per female. Nematodes may molt several times to produce the adult. The first-stage larva develops in the egg and typically molts once before it hatches. The second-stage larva, which emerges from the egg, is in most forms capable of moving to and feeding on plant roots or other plant parts. Nematodes undergo three more molts before they become adults. They may overwinter as eggs, second-stage juveniles, or as adults.

The presence of nematodes can often be determined by visually examining the plants, or after incubating them to allow for symptom expression. The most common symptom is a gall. Nematodes can also be extracted from soil by various methods: the active method involves seiving out the live nematodes, and the passive method uses a high-density sugar solution to float out the nematodes. For identification, a nematode is mounted in water on a microscope slide and viewed, using a microscope with an oil-immersion objective. The taxonomy of nematodes is based on characteristics of the mature adult female.

Comparatively little is known of the effects of nematodes on forest trees (Ruehle 1972). The root-knot (*Meloidogyne* species) and root-lesion (*Pratylenchus* species) nematodes enter, feed, and reproduce within tissues of over 2,000 plant species including elms, mulberry, pecan, cherry, catalpa, and many other hardwood species. Other nematodes that remain in the soil and feed on the root surface include dagger nematode (*Xiphinema* species), lance nematode (*Hoplolaimus* species), sheath nematode (*Hemicycliophora* species), ring nematodes (*Criconema* and *Criconemoides* species), pin nematode (*Paratylenchus* species), and sting nematode (*Belonolaimus* species). All these nematodes have been observed on many hardwoods and conifers. The following list shows the parasitic habits and hosts of some forest tree nematodes.

Parasitic Habits of Common Forest Tree Nematodes

I. Ectoparasites—remain outside of roots and penetrate with only a small portion of their bodies, where they feed on the epidermal and cortical cells; occasionally may penetrate partially or wholly into the cortical tissue.
   A. *Belonolaimus* spp. (sting nematode)—associated with trees, primarily a problem in the southern United States.
   B. *Criconema* and *Criconemoides* spp. (ring nematodes)—associated with many woody plants, pines, and others.
   C. *Hemicycliophora* spp. (sheath nematodes)—affect many woody plants, especially lemon.

D. *Hoplolaimus* spp. (lance nematodes)—observed on and in roots of several forest trees including slash and longleaf pines.

E. *Paratylenchus* spp. (pin nematodes)—commonly found around roots of forest and shade trees.

F. *Xiphinema* spp. (dagger nematodes)—associated with many hardwoods and conifers.

II. Endoparasites—enter, feed, and reproduce within tree roots.

A. *Meloidodora* spp. (pine cystoid nematode)—often associated with the roots of pines, particularly in the southern United States.

B. *Meloidogyne* spp. (root-knot nematode)—infects over 2,000 plant species including catalpa, cherry, mulberry, pecan and willow.

C. *Pratylenchus* spp. (root-lesion nematode)—has a very wide host range including many hardwood and orchard species.

Most of these phytonematodes have caused problems in the southeastern United States, particularly in nurseries. Their role in established stands is largely unknown. The stable, relatively undisturbed environment found around the roots of perennial plants such as trees is very conducive to the buildup of large and potentially damaging populations of phytonematodes. Trees that appear to be declining or are unthrifty exhibit two of the major symptoms of heavy nematode infestation.

In the tree nursery the same soil sterilization or fumigation practices commonly employed to reduce the incidence of damping-off fungi are also very effective against nematodes.

The pine wood nematode *(Bursaphelenchus xylophilus)* is one of many nematodes that can survive in insects (Mamiya and Kujohara 1972). This nematode is suspected of damaging trees when introduced by feeding Cerambycids. Because of the important problems it causes, the pine wood nematode will be discussed in Chapter 18.

4.7   INSECTS

Insects cause a wide variety of injuries to plants, mainly as a result of their feeding activities. The effects of insects on forest and shade trees is usually studied in a separate course entitled Forest Entomology and cannot be summarized in any adequate manner in this chapter.

There is an increasing awareness, however, that insects can also cause disease, in addition to feeding injuries. These insect-caused diseases will be briefly discussed here.

Insects' role in disease is encompassed into two generally discrete realms:

1. the regurgitation, excretion, secretion, or injection of nonliving material onto, or into, plant tissues, which results in the development of disease symptoms and

2. insect interactions with microbes that result in disease.

4.7.1   Insect Materials That Result in Disease Symptoms

The first group contains the insect-derived phytoallactins (*allact* = Greek for change) and includes all insect-derived chemicals that cause changes in plants. There are two major types: phytotoxins and phytohormones.

Phytotoxins

Just as certain pathogens may produce a toxin that causes symptoms in a plant, so, too, insects or mites may also be toxin-producing agents. Most of the insects are in the orders Hemiptera and Homoptera (the sucking insects), with some mites (Acarina) also included. Many of these secrete a viscous material just before and during feeding, which gels to form a sheath around the sucking mouthparts. A more watery saliva that spreads from the sheath area is also injected. This saliva partially digests cells and contents and is sucked back into the insect's food canal. Exactly how this process causes injury is not known, but three theories exist:

1. Free amino acids in the watery saliva are toxic to host plant cytoplasm.
2. Pectinase enzymes dissolve the middle lamellae of cells, which allows more extensive diffusion of the salivary secretions.
3. An imbalance of the growth regulator indole-acetic acid (IAA) interferes with basic subcellular processes.

Perhaps all three occur. The resulting injuries (phytotoxemias) can be either local-lesion or systemic. Of the former, Aphididae (aphids) cause chlorotic spots, and Coccidae (scales) cause dieback of ash. Of the latter, Miridae (capsids) cause gnarled stem canker and systemic necrosis of deciduous trees.

Phytohormones

The phytohormones apparently cause an imbalance of growth regulators, especially auxins, in meristematic tissues and are manifest as production of adventitious buds, rosetting, shortening of internodes, galls, tumors, and witches' brooming.

Some gall-forming cecidogenic insects contain the commonly occurring natural phytohormone IAA in salivary glands or ovipositional glands. The tumerous growth caused by the balsam woolly aphid *(Adelges picae)* can be partially duplicated by applying IAA in lanolin to scarified bark surface.

4.7.2 Insect/Disease Interaction

Some insects commonly bear several either ecto- or endomicrobial organisms that contribute to disease development in the plant. Indeed, often the insect cannot survive on the host plant without the presence of the symbiotic microbes.

These ectosymbiotes are borne by insects in ectodermal structures called mycangia, mycetangia, or esophageal bulbs. They are common in larvae, nymphs, and adults. These structures enable immature insects to spread the ectosymbiotes along walls of their tunnels. The adults spread the ectosymbiotes from tree to tree and inoculate them through feeding scars, tunnels, and brood galleries.

There are extremely complex interrelationships with fungi. The nutrition, growth, metamorphosis, reproduction, and life span of the ambrosia beetle *(Xyleborus ferruginum)* are regulated by symbiotic-derived chemicals. Other systems are even more complex, including those of bark beetles, nematodes, and mites.

The endosymbiotes spend at least a part of their life cycle within the insect. Many viruses and mycoplasms are examples of phytopathogenic agents that can be both ecto- and endosymbiotes of insects, especially of leafhoppers.

4.8 ALLELOPATHY

The detrimental effect of released organic chemicals by one plant on the germination, growth, or metabolism of a different plant is known as allelopathy (Fisher 1980, Rice 1979).

Recognized as early as 1832 by De Candolle, the importance of allelopathy has been largely ignored until recently. For many years the phenomenon was recognized as a "soil-sickening" problem in agricultural fields. There are many examples in forestry, as well, and the list is rapidly growing.

Growth of blue grama in Arizona rangelands was reduced in the presence of juniper trees. In the South, cherrybark oak produces a phytotoxic substance that retards the development of reproduction under oak seed trees. Western bracken *(Pteridium aquilinum)* quickly invades disturbed sites in the Pacific Northwest. Water leachates from senescent bracken fronds reduce or delay germination of western thimbleberry *(Rubus parviflorus)* and salmonberry *(Rubus spectabilis)* and cause reduced height growth in Douglas-fir seedlings. Extracts from green foliage of Arizona fescue *(Festuca arizonica)* and mountain muhly *(Muhlenbergia montana)* reduce germination of ponderosa pine seeds, and retard elongation of radicles. Foliage extracts of bearberry *(Arctostaphylos uva-ursi)* and sheep laurel *(Kalmia angustifolises)* are only slightly inhibitory to growth of jack pine but are progressively more inhibitory to red pine, white pine, white spruce, and balsam fir. Because this is also the approximate successional sequence for these species, allelopathy may be one reason why jack and red pines may be the most successful pioneers on these sites.

Some of the phytotoxins causing allelopathy include

1. Phenolic acids—*p*-hydroxy benzoic, gentisic, benzoic, salicylic, ferulic, and cinnamic
2. Aldehydes—salicylaldehyde, benzaldehyde, and vanillin
3. Coumarins—coumarin, esculetin, and scopoletin
4. Glucosides—juglone, amygdalin, and phlorigin
5. Terpenes—camphor, cineate, and α-pinene

Hackberry produces ferulic, caffeic, and *p*-coumaric acid, which inhibit the growth of associated herbaceous species. In the chaparral-grassland type of southern California, *Salvia* shrubs produce volatile terpenes that prohibit normal grassland growth within 6–9 m. *Eucalyptus globulus* also produces volatile terpenes that prevent understory vegetation from growing beneath canopies. Various phenolics have also been isolated from the fog drip and litter leachates of *Eucalyptus*.

Juglone is produced by walnuts and is an effective allelopathic phytotoxin against many plants. It apparently functions as an electron acceptor for NADH dehydrogenase, which reduces oxygen uptake, or it regulates respiration by inhibiting coupled intermediates of oxidative phosphorylation.

Decomposing roots of *Prunus* are noted for their allelopathic effects. Living roots contain amygdalin and emulsin. These are compartmentalized away from each other in the living plant. Upon death and the loss of cell integrity, the amygdalin and emulsin react and release toxic HCN gas.

Foresters and agriculturists have learned to control many allelopathic interactions without even knowing of their existence. The planting of seedlings rather than seed, intensive site preparation, fertilization, and weed control all reduce the allelopathic effect that new trees must endure. If forests are to be established without these intensive silvicultural treatments, a great deal more must be learned about allelopathic effects.

4.9   MULTIPLE PATHOGEN COMPLEXES—SYNERGISTIC, NEUTRAL, AND ANTAGONISTIC

4.9.1   Synergistic Effects

Synergism occurs when two or more plant pathogens act together in a complementary manner, such that the mixed infection produces more extensive damage than either pathogen could produce by itself.

Synergistic effects have been shown for combinations of some bacteria, fungi, viruses, and nematodes. Some saprophytes, or nonpathogens, may not become pathogenic but must accompany another pathogenic agent for infection to be successful. There are many bacteria, for example, including anaerobes, that accompany decay fungi in the invasion of trees. Some bacteria can fix nitrogen and thus help supply nitrogen to the succession of fungi, which colonize wounded tissues in trees. Each successional stage modifies the substrate slightly and sets the nutritional stage for the next colonizer and eventually the decay fungi.

Many well-documented cases of synergism have implicated some pathogenic agents causing some type of wound on the host that allows the second pathogen to invade. Seed and storage fungi such as *Aspergillus* and *Rhizopus* can more easily infect if they enter through insect-feeding wounds. Certain root-rot fungi and nematodes cause very little injury to plants if only one or the other is present in the soil. Severe injury will develop, however, only if both are present; the nematode apparently wounds the root, and the fungus then enters through this wound.

There are other cases of nematode–fungi synergism, though, where simple wounding cannot explain the increased susceptibility of plants to attack by another pathogen. For example, attack by the root-knot nematode does not increase susceptibility to root-rot fungi until 3–4 weeks later. This tendency has also been demonstrated for nematodes and *Verticillium* and *Fusarium* wilt diseases. The nematode or the gall tissue may cause production of a translocatable signal or other growth-regulating substance.

Although the best examples of pathogen synergism often involve nematodes as one of the pathogens, there are fungus-only synergisms. Often these involve an alteration in the concentration of sugars in some parts of the plant. There are low-sugar diseases (such as early blight of tomato caused by *Alternaria solani*) and high-sugar diseases (such as leaf rust caused by *Puccinia recondita*). An interesting situation can develop in tomatoes infected with the potato leaf roll virus. Leaf sugar increases as a result of the virus infection. Because *Alternaria solani* does not require much sugar, it can infect the virus-infected tomato leaves and not produce any visible disease.

There are other examples in which the primary invader may improve the nutritional status of invading fungi. Nematodes produce enzymes that can stimulate the germination of dormant spores. Root-knot nematode galls are extremely rich in amino acids along with some growth-regulating substances. There is also an increase of phosphorus in the galls, which probably is from the increased DNA and RNA synthesis.

The initial pathogen may also cause a change in resistance to a secondary pathogen. Nematode attacks will often render normally resistant cultivars of tobacco and tomato susceptible to attack by *Phytophthora parasitica* and *Pythium ultimum*, respectively.

Pathogens may also cause normally nonpathogens to become pathogens. *Aspergillus, Penicillium,* and *Trichoderma* spp. are common soil inhabitants and normally are not pathogenic on tobacco and other crops. Infection of tobacco roots by root-knot nematode is often followed by extensive invasion by these fungi.

Nonpathogenic organisms also can influence the infection process. Many bacteria and fungi grow saprophytically over leaf and root surfaces. These bacteria and fungi decompose dead organic debris and in the process secrete various chemical compounds that may stimulate germination or enhance infection.

Infection by some viruses has changed the amino acid content of the host, which was reflected in different root exudates that stimulated infection by *Fusarium*.

Some pathogens may infect the host simultaneously and the resulting disease may be different from either of the two if acting alone. A joint infection by potato virus X and potato virus Y produces symptoms of rugose mosaic, but infection by either virus alone produces diseases that are much less severe. In some virus diseases, there appears to be an interaction between the pathogens rather than an effect of the one pathogen on the host that may alter its reaction to a second pathogen.

Just as there are synergistic effects between biotic causes of diseases, there are also synergistic effects between biotic and abiotic causes of disease. Exposure to air pollutants can increase the amounts of disease of grey mold, bean rust, powdery mildew, and tobacco mosaic virus, for example. In addition, direct synergism between various air pollutants can occur with the only biotic involvement being from the host. Acid rain may remove cutin, an important barrier to infection of leaves. Pathogens may benefit not only from the increased flow of exudates but also by an easier entry into the epidermis.

Usually the abiotic cause has an influence on the biotic cause, but the reverse may also occur. There are certain bacteria (*Pseudomonas* spp.) that have strains that are active in ice nucleation. If these bacteria are present on plants that are cooled to slightly below freezing temperatures, they can aggravate frost damage. If they are absent, there is no severe frost damage.

4.9.2 Neutral Effects

Multiple pathogen complexes that have neutral effects are less well known. Many major tree pathogens apparently have neutral effects on each other, but whether there really is a neutral relationship or that we simply have not yet identified synergistic interaction is not known. It is possible to aseptically remove either tree sap or increment cores from tree stems, plate them onto suitable nutrient media, and find a rich variety of bacterial and fungal organisms. There are presently not available sufficiently sensitive methods to detect or measure any deleterious harm to their tree host. Yet it is very likely that water and nutrients are taken from the tree. Perhaps small quantities of toxic substances are released as a result of normal metabolism. Because there is as yet no perceived hint of any possible economic injury to the hosts, it is likely to be a long time before these nutritional, metabolic, and pathogenic interrelationships are fully understood.

4.9.3 Antagonistic Effects

Infection and invasion of a plant by a biotic agent, which may or may not be a pathogen of that plant, may elicit a biochemical reaction by the host that will protect that plant against subsequent invasion by pathogens. Lesions produced on such protected plants, for example, may not only be reduced in number but in size as well compared to those produced on unprotected plants.

Specific antibodies protect animals from foreign proteins that enter their systems. Thus mammals can defend themselves against infection by many microorganisms. Although there is no conclusive evidence that plants produce antibodies against invading pathogens, many plants produce fungitoxic substances that inhibit their pathogens. These substances are mostly phenolics and are named phytoalexins. Not only can phytoalexin production be stimulated by pathogens but by nonpathogens and chemical and mechanical injury as well. Some phytoalexins include ipomeamarone, which is produced by sweet potatoes inoculated with *Ceratocystis fimbriata;* isocoumarin, which is produced in carrot roots inoculated with the nonpathogen *C. fimbriata;* orchinol, which is produced by *Orchis militaris* following infection by *Rhizoctonia repens;* and pisatin, which is produced by peas in response to injury or inoculation with many fungi.

4.10 THE DISEASE CYCLE

Most pathogens follow a predictable series of steps or stages during the disease cycle. Although certain details of some steps may differ somewhat between specific pathogens, each must be successfully achieved before the next can be attempted. As we will see in Chapter 9, detailed knowledge of each of these steps can play a decisive role in developing a control strategy for a specific pathogen.

4.10.1 Source of Inoculum

The disease cycle begins with a source of inoculum (Fig. 4.72). Inoculum may come from an active infection in a previously diseased host or alternate host growing in relatively close proximity to the suscept. Or, it may come from dormant fruiting bodies or spores that have survived in plant debris. In other instances, the pathogen may be growing saprophytically on organic matter and produces inocula that infect living hosts that are in close proximity or predisposed due to some unfavorable environmental factor(s).

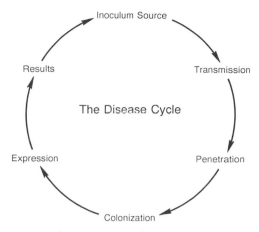

FIGURE 4.72   The disease cycle.

4.10.2   Transmission

Inoculum can do no harm, however, unless it comes into very close proximity of a suitable host. As you will learn in Chapter 5, most inoculum is in the form of spores that are adapted for chance dissemination by wind and to a lesser extent by wind-driven rain or by moving water such as natural overland flow or as irrigation water. For some pathogens, inoculum comprised of spores or microsclerotia is deposited in the soil by decaying plant debris and depends upon a potential host eventually growing into close proximity. Zoospores may swim for short distances and get close enough to susceptible tissues to cause infection. A few fungi such as *Armillaria mellea* produce specialized rootlike structures called rhizomorphs, which can grow several meters to find a suitable host. Many fungi have specialized spores or fruiting bodies that attract and assist insects in transmission. Others, like *Ophiostoma ulmi,* go one step further and depend upon the bark beetle vector to identify the correct host tree also and even to create a wound in order to successfully inoculate spores into receptive tissues. More importantly, however, humans can and do serve as unwilling vectors of pathogens and have been responsible for the introduction of several very destructive disease pathogens into new regions or countries. Apart from the direct transfer of pathogens, human activities have frequently changed natural environmental and biological checks and balances of previously insignificant pathogens resulting in some new and very destructive tree diseases. As a matter of fact, we could very persuasively argue that nearly all important forest and shade tree pathogens have become so as a direct result of human intervention or attempts to alter natural ecosystems.

Some plant pathogens, such as bacteria, are difficult to transmit because they do not produce spores or other resistant structures. They, like nematodes, must largely depend on movement of infected plant debris or soil for transmission. At the other extreme are the dwarf mistletoes. They produce seeds that are too large and heavy to be disseminated by wind or insects, although birds may carry the seeds into previously uninfested stands. The dwarf mistletoes, however, very effectively transport themselves by an explosive fruit, which shoots the seed several meters away. This simple but effective strategy has allowed the dwarf mistletoes to follow their conifer hosts very closely through most of their ranges as they expanded northward following recession of the last Pleistocene glaciation.

4.10.3   Penetration

Once the pathogen has contacted the host, it must breach the host's natural defenses and gain entrance. Some pathogens produce toxic substances that kill host tissues in advance of penetration. Others produce powerful enzymes that can degrade cuticular waxes and the epidermal cell wall. These pathogens require that wounds be present before they can successfully gain entrance. Still other pathogens enter the host more subtly by growing through natural openings such as stomates. Fungal pathogens use all the preceding strategies. Because bacteria are immobile, they must be carried into wounds by vectors or may be pulled into stomates or hydathodes by transpirational suction. Dwarf mistletoes penetrate largely through mechanical action, which tears apart outer host tissues and allows specialized invading cells or tissues to enter and become established.

The strategy of penetration is similar to that of transmission and is largely a matter of allocation of energy resources. Passive movement to and into a host does not require much of an energy expenditure, but it likewise does not carry with it much assurance that either will be successful. Pathogens with elaborate mechanisms for transmission and penetration usually produce relatively fewer propagules, and these pathogens are more successful.

4.10.4   Colonization and Disease Expression

Once inside the host, the pathogen exhibits one of two strategies. It may remain as a relatively small, or local, colony, or it may systemically colonize most or all of the host. Because of this characteristic, disease expression is often closely linked with the kind and degree of colonization.

Local infections may have a relatively minor affect on disease expression. For example, a few leaf spots scattered over many leaves may have only a minor deleterious effect on the host. Conversely, many thousands of spots that coalesce and reduce photosynthesis may cause dramatic symptom expression and result in the loss of the entire current growth of foliage, or possibly even death of the host. At the other extreme, a relatively small lesion, if it kills the cambium and girdles the tree, can have a much quicker and an equally fatal effect. A root pathogen that is causing a small number of lesions may not produce visible crown symptoms, but if a severe drought coincides with infection, symptoms could change from barely visible to severe. Obligate parasites can have either local or systemic infections. Rusts tend to be of a local nature, and symptoms may not be very dramatic or the host may be severely injured until sporulation. Then, the infested tissues may visibly decline in health rather dramatically. Viruses are obligate parasites and can have local or systemic colonies. Symptom development can be rather dramatic, but often there is no external evidence of infection, and except for a slight reduction in crop yield, little damage may be evident.

A particular symptom may not be characteristic of a given disease. Vascular wilt symptoms are characteristic of several vascular wilt-causing pathogens. Yet, very similar symptoms could result from a severe root-disease infection, for example, or even exposure to an air pollutant. The expression of symptoms is usually visible evidence that infection has occurred and that colonization is proceeding. Symptom expression is a direct result of pathogen-caused injury to the host.

4.10.5   Results

The entire plant may be killed, or only certain parts might be damaged. This could include virtually any part of the plant. Whether there is an economic loss depends upon the part of the plant affected, the time of infection, and the value of the crop. Heart decay may have little deleterious effect if nuts or leaves are the primary crop. Some of the vascular fusaria may have no measurable effect on tree health whatsoever, but the slight stain they produce in the wood may significantly reduce its commercial value. In a wilderness park, pathogens may be beneficial by increasing the rate of nutrient cycling. Irrespective of the economic goals to be achieved, the disease cycle is complete when the pathogen has completed its life cycle and successfully produced adequate numbers of propagules to begin another disease cycle.

4.11   GENERALIZED CYCLE OF A LEAF DISEASE

We focus our discussion on the disease cycle and illustrate some of the events that influence the success of a typical leaf disease caused by a fungus. During its life span a leaf pathogen faces a series of crucial events, any one of which can cause a decline in the pathogen or its outright demise. A leaf pathogen is typical of most pathogens in that it must be able to cope not only with the changing weather but with a changing host as well.

We begin our generalized leaf disease cycle in the spring when new growth begins. Assuming it survived the winter, the pathogen must be able to grow and produce propagules quickly enough to infect new and susceptible host tissues. Precise timing is critical. Fungi, which produce spores before new host growth is present, must make some provision for these spores to remain viable while lodged on or under bud scales. Ideally, the spores will be disseminated just as new growth is forming. If they are too late, either the host tissues are hardened off and resistant to infection, or the weather no longer favors germination of the spores and infection. With many host-pathogen combinations, only young, succulent, partially developed leaf or stem tissues are susceptible to infection. This susceptible period persists for at most several weeks, during which favorable weather conditions must occur. If environmental conditions are too favorable and a large percentage of spores cause infection, all new host growth may be attacked so severely that the leaves are killed and shed, and there is no subsequent infection in that season. This scenario often happens with sycamore anthracnose. With this disease, though, surviving twig lesions produce additional spores, which cause subsequent infections on regrowth.

This fairly typical series of events also occurs with the powdery mildews. Their ascospores mature in early spring. At about bud burst, or slightly before, the ascospores are forcibly discharged into the air and are carried about by the wind. Some become lodged under the bud scales of enlarging buds, and a few also become stuck to the emerging leaves just after bud burst. If there were no subsequent spore production after this initial set of infections, most pathogens would probably never amount to much. Due to random chance, predation, harsh effects of winter, and so on, relatively few spores survive to establish the fungus in the new growing season. This is largely true not only for leaf pathogens but also for any pathogen that must start afresh each year in a new growing season. For viruses, mycoplasmas, parasitic seed plants, and other obligate parasites, the pathogen can survive in a perennial host or in its vector in a considerably more protected fashion for much longer than one season but infection of new host individuals can be nearly as risky. A vector is any organism capable of carrying a disease-causing organism.

As the initial infections mature, the asexual spore stage predominates. Because the number of successful infections resulting from the overwintering sexual ascospores was likely limited, this secondary mechanism evolved to allow the pathogen to increase its numbers substantially and to enable it to infect healthy host tissue interspersed within and among the widely spaced initial infections. Many spores contain powerful germination self-inhibitors that must be leached away before germination can proceed, ensuring a wide dispersal of spores. The asexual spores, or conidia, are perfect genetic copies of the genetic material contained in those successful first lesions. The conidia are carried away and disseminated by wind, splashing rain, rodents, birds, and man's activities. During the spring, there is abundant succulent growth and plenty of free water (usually) to enhance spore germination and infection; hence, there are many new infections. Each successive wave of infection responds rapidly to rainy periods by producing another crop of spores.

As summer approaches, there are usually fewer rains, and conditions generally become less favorable for germination and infection. Some fungal spores, though, can germinate and cause infection in just a few hours using a relatively small amount of moisture such as that derived from dew.

By late summer to early fall, the pathogen no longer produces asexual spores. The mycelium in infection sites begins gearing up to produce the perfect, or sexual, stage. The sexual process ensures that favorable mutations will be more likely recombined to produce offspring better suited to survive changing environmental conditions, which comprise not only variable weather but genetic changes in the host as well.

Most leaf diseases overwinter as either mature or partially mature sexual reproductive units in or on the fallen leaves. Many sexual spores are extremely thick-walled and can withstand harsh winter conditions.

This leaf disease cycle is obviously not very detailed and did not address many obligate parasites (especially the rusts), mycoplasmas, viruses, bacteria, and parasitic flowering plants. All have certain aspects in common, such as overwintering, dissemination, infection of new hosts in the growing season, and except for the viruses, coping with sexual reproduction. Specific details of these, and many other pathogens, will be discussed in Part Two.

LITERATURE CITED

Agrios, G. N. 1975. Virus and mycoplasma diseases of shade and ornamental trees. *J. Arbor.* 1:41–47.

Alexopoulos, C. J., and C. W. Mims. 1979. *Introductory Mycology,* 3rd ed. John Wiley & Sons, New York. 632 pp.

Bessey, E. A. 1965. *Morphology and Taxonomy of Fungi.* Hafner Publishing Co., New York. 791 pp.

Burnett, J. H. 1970. *Fundamentals of Mycology*. Edward Arnold Ltd., London. 546 pp.

Darker, G. D. 1967. A new *Davisomycella* species on *Pinus banksiana. Can. J. Bot.* 45:1445–1449.

Fisher, R. F. 1980. Allelopathy: A potential cause of regeneration failure. *J. For.* 78:346–350.

Hartley, C., R. W. Davidson, and B. S. Crandall. 1961. Wetwood, bacteria, and increased pH in trees. *U.S. For. Prod. Lab. Rept.* 2215:1–34.

Hawksworth, F. G., and D. Weins. 1972. *Biology and Classification of Dwarf Mistletoes (Arceuthobium)*. USDA For. Serv. Agri. Handbook No. 401. 234 pp.

Heald, F. D., and R. A. Studhalter. 1914. Birds as carriers of the chestnut-blight fungus. *J. Agr. Res.* 2:405–422.

Hibben, C. R., and S. S. Hager. 1975. Pathogenicity of an ash isolate of tobacco ringspot virus. *Pl. Dis. Reptr.* 59:57–60.

Horsfall, J. G., and E. B. Cowling (eds.). 1979. *Plant Disease—An Advanced Treatise: How Pathogens Induce Disease,* Vol. IV. Academic Press, New York. 466 pp.

Kim, K. S., and J. P. Fulton. 1973. Association of viruslike particles with a ringspot disease of oak. *Pl. Dis. Reptr.* 57:1029–1031.

Kuijt, J. 1969. *The Biology of Parasitic Flowering Plants*. University of California Press, Berkeley. 246 pp.

Lippincott, J. A., and B. B. Lippincott. 1975. The genus *Agrobacterium* and plant tumorigenesis. *Ann. Rev. Microbiol.* 29:377–405.

Mamiya, Y., and T. Kujohara. 1972. Description of *Bursaphelenchus lignicolus* n sp. (*Nematoda Aphelenchoidae*) from pine wood and histopathology of nematode-infected trees. *Nematologica* 18:120–124.

Martin, G. W., and C. J. Alexopoulos. 1969. *The Myxomycetes*. University of Iowa Press, Iowa City. 561 pp.

Rice, E. L. 1979. Allelopathy—An update. *Botan. Rev.* 45:15–109.

Ruehle, J. L. 1972. Nematodes and forest trees, pp. 312–334. In J. M. Webster (ed.), *Economic Nematology,* Academic Press, New York. 563 pp.

Schroth, M. N., W. J. Moller, S. V. Thompson, and D. C. Hildebrand. 1974. Epidemiology and control of fire blight. *Ann. Rev. Phytopathology* 12:389–412.

Tainter, F. H. 1971. The ultrastructure of *Arceuthobium pusillum. Can. J. Bot.* 49:1615–1622.

Wilson, C. L., and C. E. Seliskar. 1976. Mycoplasma-associated diseases of trees. *J. Arbor.* 2:6–12.

5

INOCULUM*

Synopsis: The source, production, release, dissemination, deposition, dormancy and survival of forest disease inoculum are discussed. In addition, the concept of inoculum potential is discussed along with the quantity and quality of inoculum produced by specific forest pathogens.

5.1 INTRODUCTION

Forest diseases are caused by a variety of organisms. The dominant disease-causing agents are fungi and parasitic higher plants (e.g., dwarf mistletoes) (Manion 1991). Viruses, bacteria, mycoplasmas, and nematodes can also cause forest diseases, but they are generally considered to be of minor importance. The major concern with these organisms lies with their ability to spread and infect new hosts. If they didn't spread and damage large numbers of trees, there would be much less concern.

Disease-causing organisms can be dispersed in the soil or in the air with airborne dispersal being the most rapid mechanism of spread. Disease spread, however, cannot be understood simply by understanding dispersion processes in the soil or air. It is necessary to study production, release, dispersion, deposition, and viability of disease propagules and the effects of meteorological factors on these processes to obtain a fuller understanding (Edmonds 1979). Many viable spores may be released into the soil or atmosphere, but unless they arrive able to infect the host, disease will not spread. In other words, it is the quantity and quality of inoculum arriving at an infection court that is all important to disease spread and its economic impact.

5.2 INOCULUM

5.2.1 Inoculum Defined

Inoculum is defined as the organism itself or specialized cells of an organism that are capable of infecting a host (Stakman and Harrar 1957).

---

*Written by R. L. Edmonds, College of Forest Resources, University of Washington.

5.2.2   Sources of Inoculum

Forest pathogens produce several types of inoculum. No special types are produced by viruses, bacteria, and mycoplasmas. The individual virus particles and bacteria constitute the inoculum. In the case of nematodes, either the organism itself or its eggs can serve as inoculum. The seeds of parasitic plants are their inoculum.

On the other hand, fungi produce many different kinds of inoculum in many different ways. Examples of inoculum produced by fungi are actively growing mycelium, dormant mycelium in seeds or other plant parts, rhizomorphs or mycelial strands, sclerotia and other resting stages such as chlamydospores, and spores. Vectors, such as insects and birds, can spread inoculum but are generally not considered as inoculum sources in themselves.

Spores, seeds, bacterial cells, mycoplasmas, viral particles, and nematodes are considered to be primary inoculum, while the others are considered to be secondary sources. Spores are the most important source of inoculum because they are produced in large numbers and can spread more rapidly than other inoculum sources. Because fungi are the dominant forest pathogens, most of the subsequent discussion will concern them with less emphasis on the other organisms.

5.2.3   Production of Inoculum

Primary Inoculum. Only the production of fungal spores will be examined in detail. There are a number of mechanisms used by fungi to produce enough inoculum to be successful. Generally, they not only have to produce enough inoculum, but they also have to do it in small spaces. Thus they generally produce inoculum that is microscopic and easily and quickly carried by air and water currents.

Sizes and shapes of fungus spores, however, are variable. Very small spores tend to be spherical, whereas larger spores may be elongated. Most spores are $5-50$ μm in diameter but the range is from less than 5 μm in diameter to as much as 115×350 μm for some ascospores. Although most spores are dispersed as single units, aggregates do occur (Gregory 1973). In *Eutypella parasitica* (Diatrypales) the entire contents of a single ascus are discharged and remain together in flight (Johnson and Kuntz 1979).

The simplest method for producing large amounts of inoculum in a small space is found in bacteria. It is division. One cell divides to become another and the process goes on until large numbers of cells are formed. Some fungi fragment their hyphae into oidia and chlamydospores-forming asexual spores. They are, however, not produced in specialized structures or on spore-producing branches. A good example of oidia production is by the Basidiomycete *Phanerochaete gigantea (Peniophora gigantea)* (Aphyllophorales).

A third very simple method of inoculum production involves yeastlike budding, which is common among the yeasts and some smut fungi. These methods of producing inoculum, however, are generally not considered as efficient as the methods of spore production used by most of the fungi whereby specialized structures are produced to ensure the successful dissemination of propagules in either air, water, or soil. They range from simple hyphal branches or sporophores to the elaborate fruiting structures found in some Basidiomycetes.

Spores may be produced endogenously or exogenously. Endogenously produced spores are produced in specialized cells such as sporangia or asci. Exogenously produced spores, such as conidia, are produced externally. In the Deuteromycetes (Fungi Imperfecti) and some other groups, conidia are produced on conidiophores, which may be single or grouped to form coremia, sporodochia, acervuli, or pycnidia. They can be produced in chains and on single or branched conidiophores. Basidiospores and other specialized rust spores are also produced exogenously. All these structures enable the fungi to use vertical space efficiently.

Asexual reproduction in the fungi can be regarded as a mechanism for rapid multiplication of inoculum by using close aggregation of sporophores and efficient use of vertical space. This is not strictly true for sexual reproduction. In the Phycomycetes, sexual fusion results in the formation of oospores and zygospores, but these are generally produced in far fewer numbers than asexual zoospores or sporangiospores.

The sexual fruiting bodies of Ascomycetes and Basidiomycetes vary in size and shape and are more complicated than those of the Oomycetes and Zygomycetes. They may be considered as efficient spore production machines. Ascomycete fruiting bodies vary from simple nonaggregated asci in the yeasts to apothecia or cups, where asci are lined along the surface, to perithecia and cleistothecia developed in plant leaves. Such fruiting bodies produce spores abundantly, particularly after a dormant season. Many Ascomycetes also produce conidia, so they are particularly adept at producing inoculum under a variety of conditions.

The Basidiomycetes are the most diverse group of fungi with respect to spore production. Fruiting bodies vary from the elaborate in many of the wood-decaying fungi to very simple in the rusts, which generally do not have distinct sporophores. Aecia, however, may be produced in what could be considered as cuplike structures in some cases. In the gill fungi (or mushrooms) the arrangement of gills enables very efficient spore production. There may be as many as 1,000 gills on the lower surface of some mushrooms (Stakman and Harrar 1957). Without gills, mushrooms would have to be up to 20 times larger to produce the same number of spores (Stakman and Harrar 1957). The pore fungi may be even more efficient at producing spores in a small space. *Phellinus igniarius (Fomes igniarius)* (Aphyllophorales), which causes a heart decay on conifers, has about 2,000 tubes (4 mm deep and 0.15 mm wide) per square centimeter (Stakman and Harrar 1957). The specific increase in comparison to a flat surface in this case is 942 compared to 20 for gills.

Many environmental factors influence the production of fruiting bodies and their maturation. The most commonly invoked factors are temperature, light, and relative humidity (Ingold 1971). Similar factors have been invoked for seed production in dwarf mistletoe with a further factor being plant size or age (Strand and Roth 1976). Many of these factors are also involved with spore release and will be discussed later. However, it is sometimes extremely difficult to discriminate between the effects of these factors on spore maturation and spore release.

Secondary Inoculum.   Secondary inoculum such as rhizomorphs, sclerotia, and mycelial strands are important on a local basis. They are essential for maintaining the viability of fungi under stress circumstances and are extremely important in the case of decay fungi, which exist in the soil or in woody residues when not active. This type of inoculum is unable to traverse the long distances that spores are capable of covering, and typical growth rates may only be on the order of 1 m/yr. Nevertheless, they are important sources of inoculum. A good example of this is *Phellinus weirii*

(Aphyllophorales) which exists in Douglas-fir residues in Oregon, Washington, and British Columbia and spreads via root contacts to healthy trees when their roots contact old residues (Hansen 1979, Thies 1984, Nelson and Starrock 1993). In fact, this fungus rarely produces viable sporophores and relies primarily on secondary inoculum for spread.

5.2.4   Release of Inoculum

Spore release in the fungi can be either passive or active. This topic is covered well by Meredith (1973) and Ingold (1971). In general, passive release involves spores being blown away, removed by the action of water, or carried by agents such as insects.

Passive release generally involves members of the Zygomycetes and Deuteromycetes in the fungi and also bacteria, viruses, and mycoplasmas. Most conidial fungi are passively released by wind action. Some, however, release spores by hygroscopic twisting motions of conidiophores [e.g., *Botrytis cinerea* (Hyphomycetes), which causes grey mold of seedlings] (Jarvis 1962). Many rust spores, although they are Basidiomycetes, are released passively. Some of the Zygomycetes have developed a squirt gun mechanism that shoots sporangia from the sporangiophores (e.g., *Pilobolus* sp.). Spores may be shot vertically as far as 180 cm (Meredith 1973).

With wind-released spores, Stepanov (1935) found that the minimum wind speed required to release spores varied from fungus to fungus. The specific action of wind release and the forces required to liberate conidia are discussed further by Aylor (1978), Kramer (1979), and Gregory and Lacey (1963). Other means of dry liberation of spores include leaf flapping and abrasion due to leaves rubbing together (Aylor 1978).

In contrast to wind-released spores, spores dispersed by water are generally considered to have a restricted distribution from a source, but they can be splashed and resplashed in contrast to dry air spores which are usually deposited permanently (Fitt et al. 1989). Gregory et al. (1959) found that raindrops landing on a surface could carry spores to a height of 20 cm. However, to become truly airborne, the splash droplets must rapidly evaporate and leave the spores suspended.

Active release of spores occurs mainly in the Ascomycetes and Basidiomycetes. However, as mentioned previously, some members of the Zygomycetes and Deuteromycetes also release spores actively. The general mechanism of ascospore discharge involves the controlled bursting at the apex of a turgid ascus sending spores distances ranging from a few millimeters to 40 cm (Meredith 1973). Mostly, however, the distance is around 1 cm (Ingold 1979). Hayes (1980) showed that ascospore discharge in *Crumenulopsis sororia* (Helotiales), a discomycete causing stem cankers of pine, ceased at relative humidities below 65%, and at temperatures above 16°C. Relative humidity, net incoming radiation, and rainfall up to 24 hours prior to discharge also affected spore release. Spore discharge in many Ascomycetes occurs only after rain, wetting, irrigation, or dew fall. The active mechanism of discharge of basidiospores has been described in detail by Ingold (1971, 1979), but it is still not completely understood. When a basidiospore is mature, a drop of water exudes from the spore hilum forming a droplet. The spore is then jerked off violently.

Spore release in the Basidiomycetes and many Deuteromycetes commonly follows circadian patterns determined by one of the major environmental factors that also show rhythms of this kind (e.g., light, temperature, relative humidity, and wind velocity) (Kramer 1979). Some spores have maximum release at night or early in the morning (e.g., many Basidiomycetes). Other spores have release maxima in early afternoon (e.g., thick-walled spores of the Deuteromycetes, which exhibit passive release when wind speeds are high).

Heterobasidion annosum (Fomes annosus) (Aphyllophorales) (Wood and Schmidt 1966, Schmidt and Wood 1972), *Cronartium quercuum* f. sp. *fusiforme (Cronartium fusiforme)* (Snow and Froelich 1968), and *Cronartium ribicola* (Uredinales) (Van Arsdel et al. 1956, Van Arsdel 1967) are good examples of forest pathogens that release their spores with maxima at night. The time of release appears to be related to periods of high humidity and cooler temperatures. McCracken (1978) indicated that temperature is the most important periodic stimulus for basidiospore release of decay fungi.

Environmental conditions, again mostly moisture and temperature, also influence the length of time a fruiting body can sustain spore production. In the polypore fungi many species have tough perennial fruiting bodies and can produce spores throughout the year as long as temperature and moisture conditions are favorable. For example, basidiospores of *H. annosum* are produced throughout the year in Europe (Kallio 1970), Canada (Morrison and Johnson 1970), and the northern United States (Miller 1960) except during periods of drought or extreme cold. Maximal production occurs in late summer or early fall. In contrast, in the southeastern United States few spores are produced in summer, and maxima can occur in fall, winter, or spring (Hodges 1969).

Spore release is thus a very complicated process and varies markedly from one group of fungi to another. In contrast to fungi, plant pathogenic bacteria are released into the atmosphere passively, generally by the action of the wind blowing across surfaces (Lighthart 1979) or water-splash.

The seeds of dwarf mistletoes, like the spores of many fungi, are actively released into the atmosphere (Hawksworth and Wiens 1972, Hawksworth and Johnson 1989). Generally, release takes place in the fall months on warm days. Seeds are shot into the air considerable distances with velocities as high as 27 m/sec.

5.2.5 Dissemination

Dissemination of propagules can be by the action of wind, water, and animals (including insects, birds, and humans). Wind, insects, and humans are the most effective disseminators. Wind is the principal distributor of fungus spores, bacteria, and seeds of some parasitic plants. Insects can carry fungus spores (Harrington and Cobb 1988, Schowalter and Filip 1993) and viruses (Manion 1991). For example, spores of *Ophiostoma ulmi (Ceratocystis ulmi)* (Ophiostomatales), which cause Dutch elm disease, are carried by beetles (Manion 1991). Nematodes can also carry viruses (Ruehle 1973). Humans have spread many plant pathogens, including the important forest diseases— Dutch elm disease, chestnut blight, and white pine blister rust (Liebhold et al. 1995). Zadoks (1967) has thoroughly discussed the role of humans in the international transport of plant pathogens.

Most forest pathogens are carried by wind, and their spores are generally well adapted for this. They generally have small spores (between 5 and 50 μm diameter) but are subjected to gravity fall at a terminal velocity governed by Stoke's Law. Stoke's Law in simplified form states that the rate of fall per second of a spherical spore is proportional to the square of its radius. Table 5.1 shows dimensions and terminal

TABLE 5.1 Terminal Velocities and Dimensions of Representative Fungus Spores

Genus or Species	Spore Type	Dimensions (μm)	Terminal Velocity (cm/sec)
Alternaria sp.	Conidia	10×20	0.3
Cladosporium sp.	Conidia	4–5×10–14	0.07
Cronartium ribicola	Urediniospore	19×22	0.8
Heterobasidion annosum	Basidiospore	3.5–4.5×4.5	0.08
Puccinia graminis	Urediniospore	18×30	1.1

velocities of some common airborne spores. Spores of *Heterobasidion annosum,* for example, are well adapted for aerial dispersal because they are small, although larger spores (rust urediniospores) are also capable of traveling long distances. Spore size thus does not entirely determine travel distance. When turbulence is high, say during the day, differences in terminal velocities between different sized spores are minimized. Terminal velocities of elongated spores are less than those of spherical spores of the same volume.

Spore dispersion has been compared to the dispersion of a plume of smoke from a chimney or a small ground-based fire. The plume travels in the direction of the mean wind and is dispersed by both thermal and mechanical turbulence. This analogy is probably reasonable for spore clouds generated in agricultural fields during daylight. Gregory (1945) developed an equation for this dispersion pattern and included a term to describe spore deposition from the bottom of the cloud as it moves downwind. Although this model may work for short distances over agricultural fields, it generally does not work over longer distances or in the case of many forest pathogens.

Horizontal and vertical dispersion would be expected to greatly reduce spore concentrations at long distances from spore sources. This does not always occur because discrete clouds of spores have been identified over the North Sea between Britain and Denmark (Hirst and Hurst 1967). In forests, spores may be released at night where they move close to the ground in cold air drainage currents. However, at greater heights they may move in the opposite direction as a result of updrafts over bodies of water or swamps (Van Arsdel 1967). During daylight, clouds of spores may be ejected vertically from forest openings as a result of differential heating of tree canopies and openings. Just the presence of trees in the wind field strongly influences dispersion patterns in forests in comparison to open fields. However, close to the point of spore release (say less than 60 m) classical models of dispersion (e.g., Gaussian plume models) may work reasonably well beneath a forest canopy (Edmonds and Driver 1974). Beyond this distance dispersion patterns are complicated by shifting wind fields (Fritschen and Edmonds 1976), and traditional models are difficult to apply. Changes in vegetation density also markedly affect spore dispersion patterns.

Burleigh et al. (1978) also used a Gaussian plume model to predict downwind dispersion of aeciospores and urediniospores of *Cronartium ribicola* from a line source. The model closely approximated observed numbers of aeciospores at all downwind sampler locations (127–470 m) but was less accurate for urediniospores.

Because of their small spore size, forest pathogens are well adapted for dispersion, but many may be deposited on surfaces in the forest and not travel far. Relationships between spore size and deposition are discussed later. Some spores, however, manage to travel extremely long distances. Van Arsdel (1967) indicated that basidiospores of *C. ribicola* may be deposited 20 km from their source even in relatively calm conditions as a result of below-canopy dispersal from *Ribes* bushes at night, updrafts over bodies of water, and return flows at higher levels. Aeciospores of *C. ribicola* may travel even further. Christensen (1942) indicated that such spores could be carried 480–640 km by wind. Pennington (1925) found *Ribes* bushes infected with *C. ribicola* 176 km beyond the limits of white pine and infection centers 240–320 km apart.

Viable spores of *H. annosum* have been found 360 km from land in England (Rishbeth 1959). Kallio (1970) also trapped spores of *H. annosum* in Finland that had traveled between 50 and 500 km across the Baltic Sea.

Some forest diseases are transmitted by insects that carry spores or mycelia. Classic examples are Dutch elm disease (Manion 1991) and Leptographium disease (Harrington and Cobb 1988) by beetles. Mycelia is also carried on insects in specialized sacks, called mycangia. Insect dissemination is also controlled by environmental conditions, but it is generally more directed because many seek out a host. Wellington (1945) has shown that atmospheric pressure, wind temperature, and humidity strongly influence the dissemination of insects.

Soil pathogens are largely confined to the soil and move in soil water, but some possess spores that are dispersed by wind (Wallace 1978). Strong wind gusts or tornadoes may carry aloft soil containing such pathogens. *Pythium* sp. and nematodes may be dispersed by birds, and rodents may be important in dispersion of some subterranean fungi. *Phytophthora cinnamomi* (Peronosporales) may be dispersed by termites and wild birds (Keast and Walsh 1979). However, the wide dispersion of this fungus in Australia is thought to be attributed to human activity (Newhook and Podger 1972, Old 1979).

Birds as well as wind may also carry seeds of dwarf mistletoe plants (Hawksworth and Wiens 1972). Travel distances are generally greater by birds than by wind because the large size and heavy weight of seeds generally restricts their airborne spread to within 15 m of the source (Hawksworth and Wiens 1972).

5.2.6 Deposition

The principle methods of spore deposition in nature are sedimentation, impaction, turbulent deposition, rain washing and electrostatic deposition (Gregory 1973). Not all these mechanisms are of similar importance, and their relative importance tends to vary with spore size, wind speed, and distance from a spore source.

Close to a source, impaction may dominate except under calm conditions where sedimentation plays a large role, especially for large spores with high terminal velocities. Spore size is also important with impaction. Many stem and foliar pathogens have large spores, which may increase their likelihood of impaction on stems, leaves, and needles. Small spores tend to flow around such objects without impacting. Spores in the atmosphere may also be charged (Leach 1976) and could be attracted to plant surfaces bearing opposite charges.

As mentioned previously, when fungus spores are carried by insects, their flight and deposition are more directed and less dependent on the vagaries of wind transport. For example, Pomerleau and Mehran (1966) labeled spores of *O. ulmi* with P^{32} and found that the labeled spores accumulated at nodes of leaves, green shoots, and 1-year-old twigs where bark beetles had deposited them.

5.2.7   Dormancy and Survival

For a pathogen to infect a host after dispersal, its inoculum must be viable. In many cases fungus spores may have only a few hours to infect their host before they die. In other cases, inoculum may become dormant and survive for months or even years.

Most plant viruses occur in plants and insect vectors and thus are not exposed directly to the environment. However, they are generally sensitive to environmental conditions and have survival times varying from minutes to months (Stakman and Harrar 1957). Most plant pathogenic bacteria and fungi are also sensitive to external environmental conditions and are exposed to them directly. Even so, some bacteria may persist as long as 20 months on seeds (Stakman and Harrar 1957). Most plant pathogenic bacteria don't produce spores or other survival structures.

Fungi, unlike bacteria, possess specialized cells that enable them to survive for many years (e.g., sclerotia). Generally, spores quickly lose their viability, although some are capable also of surviving for years (Sussman 1968). Table 5.2 lists the longevity of fungus spores and other structures for selected forest pathogens and the conditions under which they survive. Most basidiospores have a short survival time, but some may survive for 170 days if kept dry.

Rust spores vary in their viability: aeciospores and urediniospores are the most durable spores of *Cronartium ribicola*. Aeciospores may survive several weeks under dry conditions and occasionally as long as 6 or 7 months (Van Arsdel et al. 1956, Siggers 1947). However, exposure to 8 hours of direct sunlight will reduce aeciospore viability to 1 in 2,000 (Van Arsdel et al. 1956). Urediniospores in storage may last for more than 270 days. However, in more normal environmental conditions only 50% of urediniospores may germinate with close to 0% after 2 weeks (Van Arsdel et al. 1956). Rust basidiospores have a very short survival time, as short as 10 hours in some circumstances (Spaulding 1922). Yarwood and Sylvester (1959) suggest that *C. ribicola* basidiospores have a half-life of 5 hours in the atmosphere. Ascospores and conidia of *Cryphonectria parasitica (Endothia parasitica)* (Diaporthales) also have variable survival times, depending on the conditions to which they are subjected (Table 5.2).

Survival times for spores of soil fungi are not vastly different from airborne spores. Hwang and Ko (1978) suggested the following survival times for zoospores (3 weeks), mycelium (2 months), and chlamydospores (1 year) for *Phytophthora cinnamomi* in soils. A similar value for chlamydospores (10 months) was found by Weste and Vithanage (1979). They also found that soil moisture stimulated chlamydospore formation while decreasing low organic matter content, and small numbers of microorganisms increased chances of survival.

In mycelial form, fungi can survive for long periods, especially in woody materials (Table 5.2). *Heterobasidion annosum* is a good example. Low and Gladman (1960) recorded *H. annosum* on a larch stump cut 63 years previously. Other estimates of survival times in residues include 6–62 years (Greig and Pratt 1976), 40 years (Miller 1960) and 10–15 years (Rennerfelt 1957).

High temperatures in woody debris may limit survival times (e.g., *H. annosum* in the southeastern United States) (Driver and Ginns 1969). Additional examples of long survivability in stumps are provided by *Phellinus weirii* in Douglas-fir, which may survive for 50 years (Hansen 1979) and *Armillaria* sp. (Agaricales) in citrus roots, which may survive for 6–14 years (Garrett 1956), in conifer stumps for at least 50

TABLE 5.2 Longevity of Fungus Spores, Mycelia, and Other Structures of Selected Forest Pathogens

Organism	Stage	Longevity	Treatment	Reference
Cryphonectria parasitica	Ascospores	1 yr	Dried in bark	Anderson and Rankin (1914)
	Ascospores	5 mo	Removed from bark	Anderson and Rankin (1914)
	Conidia	1 yr	Dry spore horns	Anderson and Rankin (1914)
	Conidia	1 mo	Separate dry	Anderson and Rankin (1914)
Cronartium ribicola	Basidiospores	10 min	Room temp., 90% RH	Spaulding (1922)
	Basidiospores	5–6 days	Air dried	Spaulding and Rathbun-Garrett (1926)
	Aeciospores	8 wk	—	Spaulding (1922)
	Urediniospores	7 mo	10°C	Siggers (1947)
	Urediniospores	223 days	4–10°C	Siggers (1947)
	Teliospores	2 mo	—	Siggers (1947)
Melampsora medusae	Urediniospores	5 yr	Freeze dried	Shain (1979)
Stereum hirsutum	Basidiospores	56–64 days	Dry, room temp.	Harrison (1942)
Stereum sanguinolentum	Basidiospores	131–137 days	Dry, room temp.	Harrison (1942)
	Basidiospores	162–170 days	Dry, room temp.	Harrison (1942)
Phaeolus schweinitzii				
Phellinus pini	Basidiospores	>65 days	Dry, room temp.	Harrison (1942)
Phellinus igniarius	Basidiospores	91–99 days	Dry, room temp.	Harrison (1942)
	Basidiospores	>10–<80 days	Open field laboratory	Good and Spanes (1958)
Fomitopsis pinicola	Basidiospores	>173 days	Dry, room temp.	Harrison (1942)
Heterobasidion annosum	Basidiospores	>18 mo	On bark in soil	Evans (1965)
	Basidiospores	<18 mo	Without bark in soil	Evans (1965)
	Mycelium	63 yr	Larch stump	Low and Gladman (1960)
Fusarium sp.	Conidia	>2 mo–8 yr	7°C on agar	Hesseltine (1947)
Phytophthora cinnamomi	Zoospores	3 wk	In soil	Hwang and Ko (1978)
	Mycelium	2 mo	In soil	Hwang and Ko (1978)
	Chlamydospores	1 yr	In soil	Hwang and Ko (1978)
	Chlamydospores	10 mo	In soil	Weste and Vithanage (1979)

TABLE 5.2 *(Continued)*

Organism	Stage	Longevity	Treatment	Reference
Armillaria sp.	Mycelium	>6–<14 yr	Citrus roots	Garrett (1956)
	Mycelium	460 yr	Pine roots and soil	Shaw and Roth (1976)
Phellinus weirii	Mycelium	50 yr	Douglas-fir stumps	Hansen (1979)

Source: Adapted from Sussman (1968).

years, and in Eucalypt stumps for 70 years (Redfern and Filip 1991). Shaw and Roth (1976) also reported the survival of a clone of *A. mellea* in root systems in a ponderosa pine forest for 460 years. Its average spread rate was 1 m/yr. Rhizomorph networks formed by some *Armillaria* sp. may be relatively long-lived, becoming supported by a succession of food bases as they become available to different parts of the network. However, rhizomorphs of *A. mellea* are short-lived and are produced in successive waves (Redfern and Filip 1989).

Viability of spores and other inoculum ultimately determines the effective spread of the pathogen, but the overall survivability of the pathogen determines its long-term impact. Sussman (1968) suggested that survivability of fungi results from resistance to deleterious agents, competitive saprophytic ability, disseminability, responsiveness (timing), and mutational capacity. Not all these factors are of equal importance for all organisms.

The most important deleterious agents are temperature, radiation, chemicals, and to some extent, moisture. The hazard of dessication is greatest during the day and in air layers near the ground surface. At higher altitudes and at night, conditions are more favorable.

The wavelengths of radiation that are most lethal are in the ultraviolet (UV) region. Ascent to the upper air greatly increases the UV dosage except in clouds. Fungus spores may be transparent and colorless (hyaline) or colored with yellow, red, brown and purple, which are the most common colors. Spore walls may be very thick, and the surface may be hydrophobic, hydrophilic, sticky, smooth, or shiny.

Pigmented spores have more protection against UV dosage than hyaline spores. This fact may explain the low viability of many basidiospores after a short time in the air. There is also some evidence of the lethal action of radiation in the visible region (Gregory 1973). Photoreactivation may reduce apparent radiation damage as spores return to lower altitudes. Interactions of temperature, moisture, and radiation are not well understood in terms of spore viability, but Carlisle (1970) has shown that low temperature and dessication may protect spores against radiation damage.

The impact of chemicals on spore viability is also not well understood. Air pollutants are generally thought to reduce viability, but Lighthart (1973), for example, working with airborne bacteria, found that high humidities protected them from loss of viability due to carbon monoxide exposure.

Soil organisms are exposed to many chemicals in the soil, including antibiotics. Many human-made chemicals have now been added to soils, but concentrations are low in most forest situations with the exception of forest nurseries. For an organism to survive, particularly in the soil, it must possess high competitive saprophytic ability

(Garrett 1956). This ability is also necessary for organisms to survive on leaf surfaces. Presence of conidia of *Cladosporium* sp. (Hyphomycetes) on poplar leaves was found to reduce the viability of urediniospores of *Melampsora larici-populina* (Uredinales) (Sharma and Heather 1978).

Timing or responsiveness, however, does need further discussion. Survivability is enhanced when an organism is able to respond to environmental stimuli when it is advantageous. Spore formation and dormancy, circadian rhythms, and mutational capacity are survival mechanisms.

5.3 INOCULUM POTENTIAL

5.3.1 Inoculum Potential Defined

Garrett (1956) defined inoculum potential as the "energy of growth of a fungus (or other microorganism) available for colonization of a substrate at the surface of the substrate to be colonized." From this definition it is apparent that there is more to disease development than just the number of propagules or spores threatening the host. Environmental factors and host properties must certainly be involved. Baker (1978) further elaborated the concept with the following equation:

$$\text{disease expression} = \text{inoculum potential} \times \text{disease potential}$$

where inoculum potential is inoculum density (virulent propagules) as modified by environmental factors, and disease potential is the host susceptibility during its life cycle as influenced by its proneness to disease. The latter involves the genetic resistance factors of the host. The genetic potential of the pathogen should also be considered as part of inoculum potential.

In the previous section the processes of production, dispersion, and deposition of inoculum were discussed, but to accurately assess disease development, estimates of quantities of inoculum are needed. Such estimates are needed especially where modeling approaches are used.

5.3.2 Quantities of Inoculum

Pathogen populations are measured directly, by determining the actual or relative inoculum level, and indirectly, by measuring host colonization or damage (Baker 1978). Determining the quantities of inoculum present is extremely important if simulation or other modeling approaches describing disease spread are used. Examples of such models are provided by Bloomberg (1979a,b) for *Fusarium oxysporum* (Hyphomycetes) in forest nursery beds, by Strand and Roth (1976) for dwarf mistletoe, by Edmonds and Sollins (1974) for foliage diseases, and by Bloomberg (1988), Shaw et al. (1991), and Shaw and Eav (1993) for root diseases.

Assessment of populations of soil organisms is usually expressed directly by sampling soils and expressing inoculum as number or mass of propagules per gram or cubic meter of soil. Airborne inoculum also can be determined directly by spore traps where inoculum density can be expressed as a concentration (e.g., number of propagules per cubic meter). Inoculum quantities also have been expressed as the number of spores released from a fruiting body or deposited on a substrate per unit surface area and time (e.g., number of spores per square centimeter per hour). Johnson and Kuntz (1979) have developed a relative spore discharge index (0 = no spores; 1 = few scattered spores; 2 = moderately heavy deposit on spore traps).

Inoculum density for foliar pathogens has been expressed indirectly as the number of lesions per unit area or diameter of leaf discs (Hamelin et al. 1994). Other indirect measures include sporocarp area or numbers per log length in the case of wood-decaying fungi.

In many cases, airborne inoculum concentrations are not well related to disease development because environmental conditions may strongly influence inoculum potential. In the case of soil pathogens, however, disease development is generally strongly related to inoculum quantities because the effects of environmental variables are buffered (Baker 1978). Mathews et al. (1978) also showed an additive relationship between basidiospore inoculum density and infection on susceptible, intermediate, and resistant families of slash pine to fusiform rust.

Table 5.3 shows inoculum quantities for selected forest pathogens. It is apparent from Table 5.3 that there are many ways of expressing inoculum quantities. Maximal quantities are recorded in the table, but you should recognize that inoculum quantities vary with time of day, month, or year largely in relation to changes in environmental conditions, and at times they may be zero. For example, Kallio (1970) found maximum deposition of *H. annosum* spores in July and August in Finland, but deposition fell to zero in January, February, and March when conditions were too cold for spore production. In a study by Pronos and Patton (1978), maximal production of *Armillaria* rhizomorphs in the soil around herbicide-killed oaks occurred after 10 years with a steady increase before this time and a decline after.

Thus to obtain exact quantities of inoculum at any given time, a complete understanding of the quantities produced in relation to environmental conditions is required. Mathematical modeling of inoculum quantities is not an easy task.

An interesting point in relation to inoculum quantities of soil pathogens was raised by Baker and McClintock (1965). They examined the relationship between inoculum density (in propagules per gram of soil) and the distance (in millimeters) between propagules of root-infecting fungi. An increase in the inoculum density decreases the distance between propagules most rapidly up to 2,000–3,000 propagules/g. Beyond this density very large inoculum densities are needed to decrease distances between propagules significantly, perhaps indicating an ecological limit to inoculum densities in the soil for this type of organism. Most published values of root-infecting fungi in forest nurseries are below 3,000 propagules/g.

5.3.3 Inoculum Quality

Not all propagules have an equal potential to induce disease. Genetic and physiological capabilities differ, and once inoculum is produced, it is subject to environmental stresses. The effective inoculum density encountered by a host is dependent on the survival characteristics of the pathogen. The longevity varies markedly as seen in Table 5.2, depending on the type of propagule.

Let us now consider these capabilities in more detail, in particular the ideas of Garrett (1956) concerning the energy of growth of a pathogen available for coloniza-

Table 5.3 Inoculum Quantities of Some Representative Forest Pathogens

Species	Inoculum Quantity	Comments	Reference
Armillaria mellea	1.3 g rhizomorphs/100 cm² soil	In area around pine trees	Pronos and Patton (1978)
	28 mg rhizomorphs/g stump	Rhizomorphs in stumps	Rishbeth (1972)
	0.54 g hyphae/g stump	Hyphal density in stump	Rishbeth (1972)
Cronartium comandrae	45 infected comandra leaves/plant	Uredinia and telia	Dolezal and Tainter (1979)
Cronartium ribicola	50 uredinia/9 mm diam. leaf disk		
	700 sporidia/telial column		Van Arsdel et al. (1956)
	10 aeciospores released/sec/m		Burleigh et al. (1978)
	5 urediniospores released/sec/m		
Chrysomyxa abietis	90 sporidia/m³		Collins (1976)
Heterobasidion annosum	125 cm² sporocarp/1-m log		Schütt and Schuck (1979)
	7 sporocarps/log m		
	96,000 spores released/cm²/hr	Maximum rate of spore release from sporocarp in field	Kallio et al. (1974)
	10,000 spores released/cm²/hr	Maximum rate of spore release in laboratory	Wood and Schmidt (1966)
	1,932 spores deposited/100 cm²/hr	Maximum deposition rate	Kallio (1970)
	27 spores deposited/100 cm²/hr	Maximum deposition rate	Morrison and Johnson (1970)
	1 propagule/9 mg soil	Spruce forest	Siepmann (1974)
Fusarium oxysporum	170 propagules/g soil	Nursery bed	Edmonds and Heather (1974)
	0.25–0.5 inoculum particles/cm³ soil	Nursery bed	Bloomberg (1979a)
Lophodermium pinastri	200,000 ascospores/m³	Maximum airborne concentration	Lanier and Sylvestre (1971)
Phytophthora cinnamomi	200 propagules/g soil	Maximum number	Weste and Vithanage (1979)
	320 propagules/g soil	Maximum number	Kliejunas and Nagata (1979)

tion of a host. Each spore has a degree of energy associated with it, but it may need assistance from other organisms to actually infect a host. For example, the spores of many decay fungi may not germinate unless surfaces are preconditioned by bacteria, fungi, or yeasts (Paine 1968, Brown and Merrill 1973). Inoculum potential can thus be affected by organism succession on substrates in forests.

In addition, forest pathogens may possess additional energy for infection from secondary inoculum, particularly rhizomorphs and mycelium in woody residues in the soil. Perhaps this factor makes forest pathogens different from fungal pathogens of annual crops.

Another aspect that should be considered with forest pathogens is that of diversity, because forest ecosystems are generally more diverse than agricultural ecosystems. Schmidt (1978) has covered this topic well, but it is useful to discuss some of this material. He has identified host, pathogen, and environmental factors that influence disease in natural forest ecosystems, and many of these reduce inoculum potential.

The host factors considered are

1. genetic resistance, including immunity and specific and general disease resistance,
2. limited infection sites (many tree pathogens require wounds for infection),
3. numbers and mixtures of tree species,
4. physiological age,
5. community structure, and
6. species succession.

Pathogen characteristics that influence the resistance of forest ecosystems to disease are

1. the quantity of inoculum and the latent and infectious periods,
2. the energy and genetic potential of inoculum, and
3. dependency on insects for spermatization, release, dissemination, and inoculation.

Latent periods may vary from 2 to 6 weeks for *Mycosphaerella dearnessii (Scirrhia acicola)* (Dothideales) but may be perennial for stem-decay fungi. The length of the primary disease cycle is also quite variable; from 1 year or less in the case of foliar pathogens and stem canker fungi to many years in the case of rusts and stem-decay fungi.

Another factor that strongly influences inoculum potential is pathogenic variability. It usually is very large and has been examined in *Cronartium quercuum* f. sp. *fusiforme* (Powers 1980, Mathews et al. 1978, 1979, Snow et al. 1975), *Cronartium ribicola* (McDonald and Hoff 1971, McDonald 1978), *Cronartium comptoniae* (Uredinales) (Tauer 1978), *Heterobasidion annosum* (Cowling and Kelman 1964), *Hypoxylon mammatum (Hypoxylon pruinatum)* (Xylariales) (Bagga and Smalley 1974), and *Armillaria mellea* (Raabe 1967). There are pathogenic and nonpathogenic species of *Armillaria* (Shaw and Kile 1991), and at least two strains of *H. annosum* occur in northern hemisphere conifer forests: the S or spruce strain, which occurs on spruces, firs, and hemlocks; and the P strain, which typically occurs on pines but is also found on S-type hosts (Chase 1989).

Environmental factors affecting inoculum potential are climatic factors (temperature and moisture), edaphic factors (nutrients, moisture, and pH), biotic factors (mycorrhizae and competitive saprophytes), and stress factors such as air pollution (Otrosina and Cobb 1989). The role of climatic factors in influencing inoculum potential is direct and fairly obvious. Inoculum potential is highest when temperature and moisture regimes are optimal and lowest when extremes of temperature and moisture occur. Edaphic factors may act more subtly. For example, soil nutrient levels may strongly influence the inoculum potential of foliar pathogens. *Dothistroma septospora (Dothistroma pini)* (Coelomycetes) has recently been defoliating radiata pine plantations in Australia. High levels of infection occur in areas where soil nitrogen and sulfur are high. Apparently this results in high levels of the amino acid asparagine in the needles, which favors the pathogen (Eldridge et al. 1981). Ozone injury to ponderosa and Jeffrey pines in California greatly increased stump colonization by *H. annosum* (Otrosina and Cobb 1989).

5.4   CONCLUSIONS

The concepts of inoculum and inoculum potential are complicated. With many diseases, there is a very strong relationship between inoculum quantity and disease amounts and progression (e.g., damping-off and root disease in forest nurseries). With other classes of diseases (e.g., stem rusts and decay fungi and perhaps foliage diseases), this relationship is not so strong. This relationship illustrates the need to consider not only inoculum quantities but also inoculum potential, because, in the latter cases, host and environmental resistances override the influence of numbers alone. This relationship is well illustrated with rust fungi, many of which have complicated life cycles involving two hosts and five spore stages, each with its own specific inoculum potential.

Efforts to model forest diseases (Bloomberg 1979a, b, 1986; Edmonds and Sollins 1974; Strand and Roth 1976; Shaw et al. 1991; Shaw and Eav 1993) have demonstrated the need for determining inoculum quantities and potentials for specific forest diseases. Simulation modeling of such diseases, however, is an extremely complicated and difficult task.

Consideration of inoculum potential for particular diseases may help forest managers to implement practices that will avoid epidemic situations. Schmidt (1978) has pointed out the importance of maintaining functional diversity in forests for this purpose. Forest practices tending to favor one species (either exotic or native) and generally selected seedlings may allow diseases that normally are in equilibrium with their host to assume epidemic proportions because of loss of tree diversity and an increase in inoculum potential. Schmidt (1978) suggests that the epidemics of fusiform rust and beech bark disease in the United States and *Dothistroma septospora* in New Zealand are examples of this concept. Thus, as our forest management practices become more sophisticated, it is evident that our knowledge of inoculum potential of forest diseases will assist in the safe application of those practices.

Forestry is becoming increasingly international. Logs move around the world following markets (e.g., from Chile, New Zealand, and Russia to North America). Understanding inoculum sources and potentials is extremely important in preventing disease outbreaks of introduced pathogens.

LITERATURE CITED

Anderson, P. J., and W. H. Rankin. 1914. Endothia canker of chestnut. *Cornell Univ. Agr. Exp. Sta. Bull.* 47:661–668.

Aylor, D. E. 1978. Dispersal in time and space: Aerial pathogens, pp. 159–180. In J. G. Horsfall and E. B. Cowling (eds), *Plant Disease. Vol. II. How Disease Develops.* Academic Press, New York. 436 pp.

Bagga, D. K., and E. B. Smalley. 1974. The development of Hypoxylon canker of *Populus tremuloides:* Role of ascospores, conidia, and toxins. *Phytopathology* 64:654–658.

Baker, R. 1978. Inoculum potential, pp. 137–157. In J. G. Horsfall and E. B. Cowling (eds.), *Plant Disease. Vol. II. How Disease Develops.* Academic Press, New York. 436 pp.

Baker, R., and D. L. McClintock. 1965. Populations of pathogens in soil. *Phytopathology* 55:495.

Bloomberg, W. J. 1979a. A model of damping-off and root rot of Douglas-fir seedlings caused by *Fusarium oxysporum. Phytopathology* 69:74–81.

Bloomberg, W. J. 1979b. Model simulations of infections of Douglas-fir seedlings by *Fusarium oxysporum. Phytopathology* 69:1072–1077.

Bloomberg, W. J. 1988. Modeling of control strategies for laminated root rot in managed Douglas-fir stands: Model development. *Phytopathology.* 78:403–409.

Brown, T. S. Jr., and W. Merrill. 1973. Germination of basidiospores of *Fomes applanatus. Phytopathology* 63:547–550.

Burleigh, J. R., P. D. Sebesta, and W. A. Wood. 1978. Predicting dispersal of aeciospores and urediospores of *Cronartium ribicola* Fisher from a live source. *Proc. Am. Phytopath. Soc.* 4:112 (abst.).

Carlisle, M. J. 1970. The photoresponses of fungi, pp. 309–344. In P. Halldal (ed.), *The Photobiology of Microorganisms.* John Wiley and Sons, New York. 479 pp.

Chase, T. 1989. Genetics and population structure of *Heterobasidion annosum* with special reference to western North America, pp. 19–25. In W. J. Otrosina and R. F. Scharpf (tech. coords.) Proc. Symposium on Research and Management of *Annosus* Root Disease (*Heterobasidion annosum*) in Western North America. USDA For. Serv., Gen. Tech. Rept. PSW-116. 177 pp.

Christensen, J. J. 1942. Long distance dispersal of plant pathogens, pp. 78–87. In F. R. Moulton (ed.) *Aerobiology.* Publ. No. 17. AAAS, Washington, D.C. 289 pp.

Collins, M. A. 1976. Periodicity of spore liberation in *Chrysomyxa abietis. Trans. Br. Mycol. Soc.* 67:336–339.

Cowling, E. B., and A. Kelman. 1964. Influence of temperature on growth of *Fomes annosus* isolates. *Phytopathology* 54:373–378.

Dolezal, W. E., and F. H. Tainter. 1979. Phenology of comandra blister rust in Arkansas. *Phytopathology* 69:41–44.

Driver, C. H., and J. H. Ginns, Jr. 1969. Ecology of slash pine stumps: Fungal colonization and infection by *Fomes annosus. For. Sci.* 15:1–10.

Edmonds, R. L. 1979. *Aerobiology: The Ecological Systems Approach.* Dowden, Hutchinson and Ross, Stroudsburg, PA. 386 pp.

Edmonds, R. L., and W. A. Heather. 1973. Root diseases in pine nurseries in the Australian Capital Territory. *Pl. Dis. Reptr.* 57:1058–1062.

Edmonds, R. L., and C. H. Driver. 1974. Dispersion and deposition of spores of *Fomes annosus* and fluorescent particles. *Phytopathology* 64:1313–1321.

Edmonds, R. L., and P. Sollins. 1974. The impact of forest diseases on energy and nutrient cycling and succession in coniferous forest ecosystems. *Proc. Am. Phytopath. Soc.* 1:184–190.

Edmonds, R. L., K. B. Leslie, and C. H. Driver. 1984. Spore deposition of *Heterobasidion annosum* in thinned coastal western hemlock stands in Oregon and Washington. *Pl. Dis.* 68:713–715.

Eldridge, R. H., J. Turner, and M. J. Lambert. 1981. *Dothistroma* needle blight in a New South Wales *Pinus radiata* plantation. *Aust. For.* 44:42–45.

Evans, E. 1965. Survival of *Fomes annosus* in infected roots in soil. *Nature* 207:318–319.

Fitt, B. D. L., H. A. McCartney, and P. J. Walklate. 1989. The role of rain in dispersal of pathogen inoculum. *Annu. Rev. Phytopathol.* 27:241–270.

Fritschen, L. J., and R. L. Edmonds. 1976. Dispersion of fluorescent particles into and within a Douglas-fir forest, pp. 280–301. In R. J. Engelmann and G. Sehmel (eds.), *Atmosphere Surface Exchange of Particulate and Gaseous Pollutants (1974)*. ERDA Tech. Info. Center, Oak Ridge, TN.

Garrett, S. D. 1956. *Biology of Root-Infecting Fungi*. Cambridge University. Press, London. 292 pp.

Good, H. M., and W. Spanes. 1958. Some factors affecting the germination of spores of *Fomes ignarius* var *populinus* (Neuman) Campbell, and the significance of these factors in infection. *Can. J. Bot.* 36:421–437.

Gregory, P. H. 1945. The dispersion of air-borne spores. *Trans. Brit. Mycol. Soc.* 28:26–72.

Gregory, P. H. 1973. *The Microbiology of the Atmosphere*, 2nd ed. Holstead Press, John Wiley & Sons, New York. 377 pp.

Gregory, P. H., E. J. Guthrie, and M. E. Bunce. 1959. Experiments on splash dispersal of fungus spores. *J. Gen. Microbiol.* 20:328–354.

Gregory, P. H., and M. E. Lacey. 1963. Liberation of spores from mouldy hay. *Trans. Brit. Mycol. Soc.* 46:73–80.

Greig, B. J. W., and J. E. Pratt. 1976. Some observations on the longevity of *Fomes annosus* in conifer stumps. *Eur. J. For. Path.* 6:250–253.

Hamelin, R. C., R. S. Ferris, L. Shain, and B. A. Thielges. 1994. Prediction of poplar leaf rust epidemics from a leaf-disc assay. *Can. J. Res.* 24:2085–2088.

Hansen, E. M. 1979. Survival of *Phellinus weirii* in Douglas-fir stumps after logging. *Can. J. For. Res.* 9:484–488.

Harrington, T. C., and F. C. Cobb, Jr. (eds.). 1988. *Leptographium root disease on conifers*. American Phytopathological Society Press, St. Paul, MN. 149 pp.

Harrison, C. H. 1942. Longevity of the spores of some wood-destroying hymenomycetes. *Phytopathology* 32:1096–1097.

Hawksworth, F., and D. Wiens. 1972. *Biology and Classification of Dwarf Mistletoe (Arceuthobium)*. U.S. Dept. Agric. Handbook No. 4. Washington, D.C. 234 pp.

Hawksworth, F. G., and D. W. Johnson. 1989. *Biology and management of dwarf mistletoe in lodgepole pine in the Rocky Mountains*. USDA For. Serv., Gen. Tech. Rept. RM-169. 38 pp.

Hayes, A. J. 1980. Spore liberation in *Crumenulopsis sororia*. *Trans. Brit. Mycol. Soc.* 74:27–40.

Hesseltine, C. W. 1947. Viability of some mold cultures. Mycologia 39:126–128.

Hirst, J. M., and G. W. Hurst. 1967. Long-distance spore transport, pp. 307–344. In P. H. Gregory and J. L. Monteith (eds.), *Airborne Microbes*. Cambridge University Press, New York, New York. 385 pp.

Hodges, C. S. 1969. Modes of infection and spread of *Fomes annosus*. *Annu. Rev. Phytopath.* 7:247–266.

Hwang, S. C., and K. H. Ko. 1978. Biology of chlamydospores, sporangia and zoospores of *Phytophthora cinnamomi* in soil. *Phytopathology* 68:726–731.

Ingold, C. T. 1971. *Fungal Spores—Their Liberation and Dispersal*. Oxford University Press, London. 302 pp.

Ingold, C. T. 1979. Dispersal of micro-organisms, pp. 10–21. In P. R. Scott and A. Bainbridge (eds.), *Plant Disease Epidemiology*. Blackwell Scientific Publications, Oxford. 329 pp.

Jarvis, W. R. 1962. The dispersal of spores of *Botrytis cinerea* Fr. in a raspberry plantation. *Trans. Brit. Mycol. Soc.* 45:549–559.

Johnson, D. W., and J. E. Kuntz. 1979. Eutypella canker of maple: Ascospore discharge and dissemination. *Phytopathology* 69:130–135.

Kallio, T. 1970. *Aerial Distribution of the Root-Rot Fungus Fomes annosus (Fr.) Cooke in Finland*. Acta For. Fenn. 107. 55 pp.

Kallio, T., J. Selander, and A. Uusi-Rauva. 1974. Labelling of *Fomes annosus* basidiospores with radioactive isotopes H^3, P^{32}, and I^{125}. *Silvae Fennica* 8:1–9.

Keast, D., and L. G. Walsh. 1979. Passage and survival of chlamydospores of *Phytophthora cinnamomi* the causal agent of forest dieback disease through the gastro intestinal tracts of termites and wild birds. *Appl. Envir. Microbiol.* 37:661–654.

Kliejunas, J. T., and J. T. Nagata. 1979. *Phytophthora cinnamomi* in Hawaiian forest soils: Seasonal variations in population levels. *Phytopathology* 69:1268–1272.

Kramer, C. L. 1979. Fungus, moss, and fern spores, pp. 24–41. In R. L. Edmonds (ed.), *Aerobiology: The Ecological Systems Approach*. Dowden, Hutchinson and Ross, Stroudsburg, PA. 386 pp.

Lanier, L., and G. Sylvestre. 1971. Epidemiologie du *Lophodermium pinastri* (Schrad.) Chev. *Eur. J. For. Path.* 1:50–63.

Liebhold, A. M., W. L. MacDonald, D. Bergdahl, and V. C. Mastro. 1995. Invasion by exotic forest pests—A threat to forest ecosystems. *For. Sci.* 41(2) Suppl. 1–49.

Lighthart, B. 1973. Survival of airborne bacteria in a high urban concentration of carbon monoxide. *Appl. Microbiol.* 25:86–91.

Lighthart, B. 1979. Airborne microbial models, pp. 361–364. In R. L. Edmonds (ed.), *Aerobiology: The Ecological Systems Approach*. Dowden, Hutchinson, and Ross, Stroudsburg, PA. 386 pp.

Leach, C. M. 1976. An electrostatic theory to explain violent spore liberation by *Drechslera turcica* and other fungi. *Mycologia* 68:63–86.

Low, J. D., and R. J. Gladman. 1960. *Fomes annosus* in Great Britain. An assessment of the situation in 1959. *Forest Rec. Lond.* 41. 22 pp.

Manion, P. D. 1991. *Tree Disease Concepts*. 2nd ed. Prentice-Hall, New Jersey. 402 pp.

Mathews, F. R., T. Miller, and C. D. Dwinell. 1978. Inoculum density: Its effect on infection by *Cronartium fusiforme* on seedlings of slash and loblolly pine. *Pl. Dis. Reptr.* 62:105–108.

Mathews, F. R., H. R. Powers, Jr., and L. D. Dwinell. 1979. Composite vs. single gall inocula for testing resistance of loblolly pine to fusiform rust. *Pl. Dis. Reptr.* 63:454–456.

McCracken, F. I. 1978. Spore release of some decay fungi of southern hardwoods. *Can. J. Bot.* 56:426–431.

McDonald, G. I. 1978. Segregation of red and yellow needle lesion types among mono aeciospore lines of *Cronartium ribicola*. *Can. J. Genet. Cytol.* 20:313–324.

McDonald, G. I., and R. J. Hoff. 1971. Resistance to *Cronartium ribicola* in *Pinus monticola*: Genetic control of needle-spots only resistance factor. *Can. J. For. Res.* 1:197–202.

Meredith, D. S. 1973. Significance of spore release and dispersal mechanisms in plant disease epidemiology. *Annu. Rev. Phytopath.* 11:313–342.

Miller, O. K., Jr. 1960. The distribution of *Fomes annosus* (Fries) Karst. In *New Hampshire Red Pine Plantations and Some Observations on Its Biology*. Fox Forest Bull. 12. 25 pp.

Morrison, D. J., and A. L. S. Johnson. 1970. Seasonal variation of stump infection by *Fomes annosus* in coastal British Columbia. *For. Chron.* 46:200–202.

Morrison, D. J., and A. L. S. Johnson. 1978. Stump colonization and spread of *Fomes annosus* 5 years after thinning. *Can. J. For. Res.* 8:177–180.

Nelson, E. E., and R. N. Sturrock. 1993. Susceptibility of western conifers to laminated root rot (*Phellinus weirii*) in Oregon and British Columbia field tests. *West. J. Appl. For.* 8:67–70.

Newhook, F. J., and F. D. Podger. 1972. The role of *Phytophthora cinnamomi* in Australian and New Zealand forests. *Annu. Rev. Phytopath.* 10:299–320.

Old, K. M. 1979. *Phytophthora and Forest Management in Australia*. CSIRO, Melbourne, Australia. 114 pp.

Otrosina, W. J., and F. W. Cobb. 1989. Biology, ecology and epidemiology of *Heterobasidion annosum*, pp. 26–33. In W. J. Otrosina and R. F. Scharpf (tech. coord.) *Proc. Symposium on Research and Management of Annosus Root Disease (Heterobasidion annosum) in Western North America*. USDA For. Serv., Gen. Tech. Rept. PSW-116. 177 pp.

Paine, R. L. 1968. Germination of *Polyporus betulinus* basidiospores on non-host species. *Phytopathology* 58:1062–1063 (abst.).

Pennington, L. H. 1925. Relation of weather conditions to the spread of white pine blister rust in the Pacific Northwest. *J. Agr. Res.* 593–607.

Pomerleau, R., and A. A. Mehran. 1966. Distribution of spores of *Ceratocystis ulmi* labelled with phosphorus 32 in green shoots and leaves of *Ulmus americana*. *Naturaliste Can.* 93:577–582.

Powers, H. A., Jr. 1980. Pathogenic variation among single aeciospore isolates of *Cronartium quercuum* f. sp. *fusiforme*. *For. Sci.* 26:280–282.

Pronos, J., and A. F. Patton. 1978. Penetration and colonization of oak roots by *Armillaria mellea*. *Eur. J. For. Path.* 8:259–267.

Raabe, A. D. 1967. Variation in pathogenicity and virulence in *Armillaria mellea*. *Phytopathology* 57:73–75.

Redfern, D. B., and G. M. Filip. 1991. Inoculum and infection, pp. 48–61. In C. G. Shaw, III and G. A. Kile (eds.). *Armillaria Root Disease*. USDA For. Serv., Agric. Handbook 691, Washington, D.C. 233 pp.

Rennerfelt, E. 1957. Untersuchungen uber die Wurzelfaule auf Fichte und Kiefer in Schweden. *Phytopath. A.* 28:259–274.

Rishbeth, J. 1959. Dispersal of *Fomes annosus* Fr. and *Peniophora gigantea* (Fr. Massee). *Trans. Brit. Mycol. Soc.* 42:243–260.

Rishbeth, J. 1972. The production of rhizomorphs by *Armillaria mellea* from stumps. *Eur. J. For. Path.* 2:193–205.

Ruehle, J. L. 1963. Nematodes and forest trees—Types of damage to tree roots. *Annu. Rev. Phytopath.* 11:99–118.

Schmidt, R. A. 1978. Diseases in forest ecosystems: The importance of functional diversity, pp. 287–315. In J. G. Horsfall and E. B. Cowling (eds.), *Plant Disease. Vol. II. How Disease Develops*. Academic Press, New York. 436 pp.

Schmidt, R. A., and F. A. Wood. 1972. Interpretation of microclimate data in relation to basidiospore release by *Fomes annosus*. *Phytopathology* 62:319–321.

Schütt, P., and H. I. Schuck. 1979. *Fomes annosus* sporocarps. Their abundance on decayed logs left in the forest. *Eur. J. For. Path.* 9:57–61.

Schowalter, T. D., and G. M. Filip (eds.). 1993. *Beetle–Pathogen Interactions in Conifer Forests*. Academic Press, New York. 252 pp.

Shain, L. 1979. Long term storage of *Melampsora medusae* urediospores after freeze drying. *Pl. Dis. Reptr.* 63:368–369.

Sharma, J. K., and W. A. Heather. 1978. Parasitism of urediospores of *Melampsora larici populina* by *Cladosporium* sp. *Eur. J. For. Path.* 8:48–54.

Shaw, C. G., III. 1981. Basidiospores of *Armillaria mellea* survive an Alaskan winter on tree bark. *Pl. Dis.* 65:972–974.

Shaw, C. G., III, and G. A. Kile (eds.). 1991. *Armillaria Root Disease*. USDA For. Serv., Agric. Handbook 691. Washington, D.C. 233 pp.

Shaw, G. C., III, and L. F. Roth. 1976. Persistence and distribution of a clone of *Armillaria mellea* in a ponderosa pine forest. *Phytopathology* 66:1210–1212.

Shaw, C. G., III, A. R. Stage, and P. McNamee. 1991. Modeling the dynamics, behavior, and impacts of *Armillaria* root disease, pp. 150–156. In C. G. Shaw, III, and G. A. Kile (eds.). *Armillaria Root Disease*. USDA For. Serv., Agric. Handbook 691. Washington, D.C. 233 pp.

Shaw, C. G., III, and B. B. Eav. 1993. Modeling interactions, pp. 199–208. In T. D. Schowalter, and G. M. Filip (eds.). *Beetle–Pathogen Interactions in Conifer Forests*. Academic Press, New York. 252 pp.

Siepmann, A. 1974. Occurrence of *Fomes annosus* in the soil of spruce stands (*Picea abies* Karst). *Eur. J. For. Path.* 4:74–78.

Siggers, P. V. 1947. Temperature requirements for germination of spores of *Cronartium ribicola*. *Phytopathology* 37:855–864.

Snow, G. A., and R. C. Froelich. 1968. Daily and seasonal dispersal of basidiospores of *Cronartium fusiforme*. *Phytopathology* 58:1532–1536.

Snow, G. A., R. J. Dinus, and A. G. Kais. 1975. Variation in pathogenicity of diverse sources of *Cronartium fusiforme* on selected slash pine families. *Phytopathology* 65:170–175.

Spaulding, P. 1922. *Investigations of the White Pine Blister Rust*. U.S. Dept. Agr. Tech. Bull. 957. 100 pp.

Spaulding, P., and A. Rathbun-Gravatt. 1926. The influence of physical factors on the viability of sporidia of *Cronartium ribicola* Fisher. *J. Agr. Res.* 33:297–433.

Stakman, E. C., and J. G. Harrar. 1957. *Principles of Plant Pathology*. The Ronald Press Co., New York. 581 pp.

Stepanov, K. M. 1935. Dissemination of infectious diseases of plants by air currents (translated title). *Bull. Plant. Prot. (USSR) Ser. II Phytopathology* 8:1–68.

Strand, M. A., and L. F. Roth. 1976. Simulation model for spread and intensification of western dwarf mistletoe in thinned stands of ponderosa pine saplings. *Phytopathology* 66:888–895.

Sussman, A. S. 1968. Longevity and survivability of fungi, pp. 447–486. In G. C. Ainsworth and A. S. Sussman (eds.), *The Fungi. Vol. III. The Fungal Population*. Academic Press, New York. 738 pp.

Tauer, C. G. 1978. Sweet fern rust resistance in jack pine seedlings: Geographic variation. *Can. J. For. Res.* 8:416–423.

Thies, W. G. 1984. Laminated root rot—The quest for control. *J. For.* 82:345–356.

Van Arsdel, E. P. 1967. The nocturnal diffusion and transport of spores. *Phytopathology* 57:1221–1229.

Van Arsdel, E. P., A. J. Riker, and R. F. Patton. 1956. The effects of temperature and moisture on the spread of white pine blister rust. *Phytopathology* 46:307–318.

Wallace, H. A. 1978. Dispersal in time and space: Soil pathogens, pp. 181–201. In J. G. Horsfall and E. B. Cowling (eds.), *Plant Disease. Vol. II. How Disease Develops*. Academic Press, New York. 436 pp.

Wellington, W. G. 1945. Conditions governing the distribution of insects in the free atmosphere. I. Atmosphere pressure, temperature and humidity. *Can. Entomol.* 77:7–15.

Weste, G., and K. Vithanage. 1979. Survival of chlamydospores of *Phytophthora cinnamomi* in several nonsterile host-free forest soils and gravels at different water potentials. *Aust. J. Bot.* 27:1–10.

Wood, F. A., and R. A. Schmidt. 1966. A spore trap for studying spore release by basidiocarps. *Phytopathology* 56:50–52.

Yarwood, C. E., and E. S. Sylvester. 1959. The half-life concept of longevity of plant pathogens. *Pl. Dis. Reptr.* 42:125–128.

Zadoks, J. C. 1967. International dispersal of fungi. *Neth. J. Plant Path. 73 Supp.* 1:61–80.

6

INOCULATION AND PENETRATION*

Synopsis: A sequence of critical events is involved in the initiation of disease in trees.

6.1 INTRODUCTION

The beginning of the disease sequence is the arrival and first contact of the pathogenic agent with the suscept, followed by entry into the suscept, and then development of a series of interactions between the pathogen and the suscept, which eventually results in disease. Here we are concerned with the first stages in this sequence—inoculation and penetration.

6.2 INOCULATION

Inoculation is the arrival of inoculum on the plant surface or the first contact of inoculum with the suscept at or near an infection court. The inoculation may be brought about in several ways.

1. There may be passive transfer of the inoculum through some agent of inoculation. Wind is a common agent of dissemination of fungal ascospores, such as ascospores from the apothecia of the fungus causing Scleroderris canker and shoot dieback. Splashing rain is responsible for the local spread of many pathogens, such as the brown spot needle blight fungus in plantations of susceptible pines. Insect vectors are a major means of overland spread by many pathogens, a notorious example of this being the tree-to-tree spread of *Ophiostoma ulmi* (Ophiostomatales) by the elm bark beetles. Various other agents of transmission could be listed.

2. Active movement of the pathogen may result in contact with the suscept. For example, zoospores produced by *Phytophthora cinnamomi* (Peronosporales) swim through water films in the soil to reach plant roots, and nematodes also move in water

---

* Written by R. F. Patton, Professor Emeritus, Plant Pathology Department, University of Wisconsin.

films in soil or on plant surfaces. Attraction of some spores to root surfaces by exudates from the roots may further assure pathogen contact with the suscept. Fungi may reach a new suscept by active growth through a substrate. The rhizomorphs produced by the root-disease fungus *Armillaria mellea* (Agaricales) are the major agent of inoculation by this fungus and they grow through the soil from a food base to a plant root. Other fungi may grow along the surface of roots or in the outer bark scales from a diseased tree to a healthy one at root contacts or root grafts. This behavior is typical of the spread of the root decay fungus *Inonotus tomentosus* (Aphyllophorales) in roots of pine and spruce. Root-disease fungi also may spread from a diseased root through the connected vascular tissue in a root graft to infect a root of a previously healthy tree. Trees may be inoculated by the movement of vascular wilt pathogens, such as those causing Dutch elm disease or oak wilt, from diseased to healthy trees through root grafts. Such instances may well be a combination of passive movement of spores in the transpiration stream along with a certain minimal growth of the fungus across walls between connecting vascular elements.

3. Growth of the suscept itself, usually roots in the soil, may result in accidental contact with an inoculum source already present in the soil. Roots of pine seedlings in a nursery bed may come into contact with the microsclerotia of *Cylindrocladium* (Hyphomycetes), for example, and these are stimulated to germinate and infect the roots.

Inoculation is subject to a variety of influences that may determine whether it occurs at all. Such factors include the presence or production of inoculum and its amount, the occurrence of environmental conditions that are favorable to release of inoculum or activity of its vector, the occurrence of meteorological conditions that affect propagule dissemination, the distance from a source of inoculum to a suscept, and the size and density of the suscept.

After inoculation occurs there is no guarantee that infection will follow. Subsequent interactions between the pathogen and the suscept are governed by a number of factors. It is essential, of course, that the inoculum arrive at or near a specific infection court, and so the location on the surface and the nature of the plant surface are important. Some other influences are the environmental conditions, the presence of stimulatory or inhibitory substances, and the presence of antagonists. Predisposition is primarily the physiological state of the suscept as affected by its age and the conditions under which it grew, also governs whether infection will occur. The physiological and ecological factors that affect tree vigor and, thus, the predisposition of a tree to infection are different influences from the effects of environmental factors on the pathogen.

6.3   GERMINATION

After the arrival of inoculum in a dormant or quiescent state on a plant surface, the next step in the initiation of disease is the start of growth, or germination. Germination includes the process and changes occurring during resumption of growth of a resting structure, such as a seed or fungal spore. The usual indication of germination is the emergence of the root and plumule from a seed or a germ tube from a fungal spore. The hatching of eggs might be considered an equivalent process for certain structures, such as the eggs of nematodes.

When a living seed or spore fails to germinate after it has been subjected to conditions considered adequate for germination, it is in a state of dormancy. Dormancy is normally the result of an interaction of imposed environmental conditions and hereditary properties of the organism. Dormancy in seeds may result from physiological, that is, endogenous or internal factors, and physical factors such as a hard, impermeable seed coat or the presence of a germination inhibitor in the seed coat. In nature, dormancy is usually broken if enough time elapses; otherwise, dormancy may be broken by chemical or mechanical treatments of the seed coat or by application of certain moisture, temperature, and light treatments alone or in combination. Fungal spores may have an environmental type of dormancy such as the presence of self-inhibitors formed at the time of sporulation, which might be removed by washing, or constitutional or physiological dormancy, which is broken by heat shock, chemical treatment, or application of germination stimulants.

The germination of seeds of higher plants is a familiar process and has implications in pathogenesis when we are concerned with diseases caused by parasitic higher plants such as the mistletoes. Nondormant seeds may germinate immediately or remain in a quiescent state if environmental conditions are not favorable for germination. This quiescent period is a state of persistent viability and may last from a few weeks to many years. The germination process usually incorporates three stages:

1. imbibition of water,
2. activation of metabolic processes, and
3. growth of the embryo and emergence of the radicle.

Seeds of most species of dwarf mistletoes germinate in late winter or spring, after undergoing a period of dormancy. Those of some other species germinate immediately after seed dispersal. During germination a radicle is produced, which, after limited growth, produces a holdfast on the host branch. The holdfast develops a penetrating wedgelike mass into the host, beginning the process of infection.

The germination of the fungal spore depends upon the breaking of dormancy, if any, and the presence of the necessary environmental and physiological conditions that evoke the events of germination. Often spores do not germinate, or they germinate only in very low percentages, in dense suspensions, or in masses upon a surface. Such behavior may be due to the presence of inhibitory substances formed at the time of spore formation; before germination can occur such blocks must be removed. Dilution of these self-inhibitors to a noneffective concentration usually occurs through the normal processes of spore dispersal.

Spore germination generally encompasses three phases of activity that are interrelated:

1. nuclear division and cell growth;
2. absorption of water and (usually) swelling of the spore; and
3. emergence of a germ tube.

Thus there is a change from a state of low metabolic activity to a highly active state and greater expenditure of energy with resulting morphological changes and formation of a vegetative thallus. One or more germ tubes protrude from the spore wall either at any place or from one or more particular sites called germpores. The basidio-spore of *Cronartium ribicola* (Uredinales), for example, usually forms one germ tube, but three or four are common, and as many as seven have been observed. In assessments of germination, a spore is usually considered to have germinated if the germ tube is one-half the larger diameter of the spore or the time required to produce a germ tube can measured. Occasionally, instead of a germ tube, additional spores are formed by budding from the original spore, and this also is considered germination.

Most germination occurs by the formation of a germ tube, often called direct germination. The germ tube may continue to grow and form a vegetative thallus, which in turn may produce additional spores or other reproductive structures. A few spores [e.g., basidiospores of the rust fungi such as *C. ribicola* and *Cronartium quercuum* f. sp. *fusiforme* (Uredinales)] may form an outgrowth that functions as a sterigma, on the tip of which is borne another spore of the same kind called a secondary basidio-spore. This type of germination has been called indirect germination or germination by repetition, and usually occurs under what might be considered adverse conditions. This process can be repeated several times if conditions so dictate, with each successive generation being formed at the expense of the preceding one. Such an alternative type of germination might provide a means of vegetative perpetuation, with the spore able to act either as an organ of infection or an organ of sporulation (Bega 1960). In some Phycomycetes, sporangia are formed, and these then either produce a germ tube, as in normal spore germination, or form zoospores. Although the formation of zoospores is really sporogenesis, the process is usually considered a type of germination. Whether germination of a sporangium occurs by formation of a germ tube or zoospores seems to be determined largely by the environmental conditions at the time. The zoospores formed may then encyst and germinate by a germ tube. Another type of spore, the oospore of the Phycomycetes, which results from a sexual process, may germinate, like the sporangium, either by a germ tube or by production of zoospores.

Elongation and further growth and development of the germ tube continue under favorable environmental conditions and provided that enough essential nutrients are available either from reserves stored in the spore or from the immediate external environment. Different species vary in their specific environmental requirements, but appropriate moisture, temperature, and other physical conditions must be present to allow the metabolic functions to proceed. Although many spores germinate in water, external nutrients or stimulating agents may be important for some. Exudates from seeds or plant roots may stimulate germination of some spores, or inhibit germination of other spores. Host exudates also may attract motile spores or organisms. Stimulation of hatching of cysts or eggs of nematodes (the "hatching factor") is also an influence of host exudates. The germination process is time-dependent, but the time required will vary with external conditions. Also, spores are inherently variable so that germination in a group of spores will not occur simultaneously.

Before the suscept can be invaded by certain fungi, the development of specialized hyphal structures from the germinating spore may be required. The formation of these infection structures, including vesicles, appressoria, and infection cushions, often occurs as a normal part of the germination sequence. Evidence indicates that some fungi form infection structures in response to substances exuded from the suscept hypocotyl or root, the physical surfaces of the suscept, volatile substances emanating from the suscept through stomata, or even some chemical agents in vitro.

6.4 PENETRATION

After inoculum arrives on the plant surface, it must penetrate or enter and establish the intimate contact or association with the suscept necessary to produce disease. But before penetration can be effected, the outer plant surfaces or protective barriers of the plant must be breached or bypassed. Thus penetration refers to the initial invasion of the suscept by a pathogen. Penetration may occur without infection; the resistance reactions of the suscept may result in death of the pathogen so that it does not proceed beyond the mere penetration stage. If the pathogen does become established within the host following penetration, then the suscept is infected, and disease is induced. The *infection court* is that part of a susceptible plant through which the pathogen penetrates the outer protective barrier and in which the pathogen establishes a disease relationship. The infection court may be a more specific site than the mere point at which the inoculum arrives at the time of inoculation. Sometimes growth of the pathogen from the site of arrival is necessary to place it at the actual infection court. For example, the germ tube of some fungi must grow on a leaf surface into a stoma, through which penetration occurs. Or a moisture film on the plant surface might provide a migratory pathway for movement of bacteria or other pathogens to a specific infection court such as a lenticel or to a wound that has ruptured the outer protective barriers such as the epidermis or bark and exposed the inner susceptible tissues that can serve as an infection court. But other pathogens may penetrate right at the initial site of arrival of the inoculum. Before we consider the details of penetration and processes at the infection court, however, it is appropriate to consider the prepenetration interactions between suscept and pathogen.

6.5 PREPENETRATION INTERACTIONS BETWEEN PATHOGEN AND SUSCEPT

6.5.1 Morphological Protective Barriers of Plants

All the outer plant surfaces form protective barriers of varying effectiveness against the entrance of pathogenic agents. Included among these barriers are elements such as the cuticle, surface waxes, trichomes, epidermis, root cortex, periderm of stems and roots, and thick seed coats. Some pathogens have developed ways of penetrating these barriers directly, whereas others depend on the various natural openings that exist in these covering layers or on breaks in the barriers caused by wounds. The natural openings include stomata, lenticels, hydathodes, leaf traces, nectaries, points where secondary roots emerge through the cortex, and openings in the ectodesmata, or walls of surface cells, through which strands of cytoplasm extend.

Some of the surface elements may act as barriers to penetration by keeping the pathogen away from the actual infection court or by altering the wetting characteristics of the surface, thus interfering with germination or growth on the surface. Trichomes extend from the epidermis and may prevent some propagules from reaching the plant surface. Some are connected to glands that exude toxic or offensive substances that may help defend the plant against insects. The *cuticle* is a protective noncellular membrane consisting of pectin and cutin layers and surface waxes over the outer surface of the epidermis. Although it probably does not present a strong mechanical barrier to penetration by fungal hyphae, it is important because of its influence on responses of bacteria, fungi, and nematodes, which may be governed by *hydrotropism* (response to water) and *thigmotropism* (response to surface contact phenomena). Surface waxes are exuded onto the cuticle surface in a great variety of forms and composition and are particularly important in affecting the wettability of the surface; they may affect spore germination and may even form plugs in stomata, thus decreasing the chances for penetration of a fungal germ tube through a stoma, as sometimes occurs in white pine blister rust, for example.

More important as mechanical barriers to some pathogens are the *epidermis,* the outer layer of cells on the primary surfaces of plants, and the *periderm,* a layer of corky cells that replaces the epidermis of many plant surfaces during secondary growth. These physical barriers may be important methods of defense against some plant pathogens, although they usually do not constitute the only factor in conferring resistance to attack by plant pathogens. Penetration through these physical barriers either directly or through natural openings or wounds will be discussed later.

6.5.2   Adherence of a Pathogen to the Plant Surface

Of major importance in the infection process is the ability of a pathogen to adhere to the outer covering layer of plant parts or to the surface of an individual cell. In some cases this initial interaction between surface components of a suscept and the pathogen may largely determine the final outcome of the relationship.

When spores are deposited on a plant surface, they must remain in position to be effective. Mucilaginous substances on the spore wall or exuded from the germ tube may enable the spore and germ tube to stick to the surface. During the stages of adhesion of spores and external growth of germ tubes or mycelium, a fungus pathogen is most vulnerable to changes in the environment, especially the microenvironment of the plant surfaces.

One important group of factors that either allow or prevent entry by a pathogen is the recognition and compatibility phenomena associated with plants and pathogens. More will be said of this in Chapter 7. When microorganisms come in contact with a plant, the host cells "recognize" some common component of the surface of the potential invaders and then respond in a predictable manner. An interaction of complementary macromolecules at cell surfaces, generally involving carbohydrates or glycoproteins or both, is involved in recognition. Recognition includes the information potential contained in surfaces that come in contact and host response to a complementary interaction. Many organisms may adhere to plant surfaces, however, without involving these recognition phenomena.

Most of our knowledge of recognition interactions pertains to bacterial pathogens. Although the epiphytic colonization of leaf surfaces by certain bacteria is specific, the nature of the adherence is not yet known. But the selective colonization of leaf surfaces by certain plant pathogenic bacteria depends on the type of recognition mechanisms already described. Limited amounts of leachates or exudates on leaf surfaces may contain lectins or lectinlike substances, which are proteins that have the capacity to bind to specific sugar residues on the surface of bacterial cells. The interaction with lectins or the formation of the so-called glycocalyx (a network of interwoven fibers of branching sugar molecules between the surfaces of bacterial and host cells) may be involved in the specific adherence of bacteria. This phenomenon also could be a determinant of specificity between certain plants and strains of *Pseudomonas syringae.* These hosts support ice-nucleation activity, which is responsible for frost injury and then often secondary invasion by other microorganisms. Plants which recognize the bacteria and allow them to adhere become susceptible to frost injury, which in turn permits invasion by other organisms.

Recognition and compatibility phenomena also play a role in the adsorption of soil microorganisms to root surfaces, whereby irregular aggregates or colonies may form. The nodulation of legumes by *Rhizobium* species is very specific, and considerable selectivity for homologous strains is shown before the complex series of reactions proceeds that eventually end in the distribution of bacteria throughout the tissues of the nodule that develops.

6.5.3   Germination and Growth

Often, before the process of penetration begins, the pathogen grows and moves on or around the plant. Fungal propagules may germinate near or on the plant surface, and germ tube growth might be necessary for the fungus to reach the actual infection court or to form certain structures necessary for direct penetration of the plant's protective barriers. Also, there may be multiplication of bacteria, the production and release of motile zoospores from an encysted sporangium, or the molting and movement of nematode larvae. Various physical factors of the environment may affect either the pathogen or the suscept and thus affect these prepenetration phases of activity.

6.5.4   Germ Tube Growth

The direction of growth of germ tubes and hyphae often is governed by certain external factors so that location of and entrance into suitable infection sites is most efficient. Germ tubes of some fungi grow toward stomata, for example, *Dothistroma septospora* (Coelomycetes) on pine needles (Peterson and Walla 1978). Although evidence regarding the nature of the controlling factors is generally lacking, the random growth of germ tubes of many fungi on pine needles in enclosures with a saturated atmosphere indicates that a gradient of some volatile stimulus such as moisture or CO_2 emanating from the stomata may be the directing influence. When *Dothistroma* spores grew on plastic needle replicates placed among branches of Austrian pines, germination was similar to that on real needles, but germ tube growth was randomly oriented. When germ tubes by chance entered stomatal antechambers, growth continued as on the surface, and no appressoria were differentiated, as occurs normally. Thus the response to a stomatal stimulus seen on living needles was not duplicated on the plastic surfaces. Another type of directed growth is orientation of the germ tube so that it has a maximum opportunity to encounter a stoma. Lewis and Day (1972) ascribed the growth of germ tubes of urediniospores of the wheat rust fungus at right angles to the epidermal ridges on wheat leaves to a thigmotropic response to the topography of the leaf surface.

6.5.5   Formation of Infection Structures

The invasion of a suscept often requires the development of specialized hyphal structures from the germinating spore. Such structures, regardless of morphology, are here included in the term *appressoria,* which incorporates all structures adhering to host surfaces to achieve penetration. There are two basic groups of appressoria: simple, developed from a single modified cell of a hypha or germ tube, and compound, developed from many cells. Simple appressoria vary from slightly swollen, hyaline germ tube apices or swollen hyaline tips separated by a septum to dark, usually thick-walled, sometimes swollen structures with one or more septa. Compound appressoria vary from a few cells to multicelled structures called infection cushions. A mucilaginous sheath or some type of binding substance adheres the appressorium to the plant surface. From the appressorium an infection peg or hypha is produced. It penetrates the surface barrier of cuticle and epidermis or enters a stomatal opening. Appressoria also may act in an auxiliary capacity that increases the chance of successful infection. If adverse conditions halt the germination process, the appressorium can remain dormant until conditions again favor germination and infection. The ability to form appressoria is determined genetically for those species that form such structures, but environmental influences are variable. With some species appressorial formation is only slightly influenced by the environment, but with others a fairly well-defined environment is necessary. Infection cushions are formed by many root-infecting fungi, and these are induced by substances exuded from noncutinized surfaces of the hypocotyl and root. On aerial surfaces, germ tubes may form appressoria often in response to the physical nature of the surface but perhaps also in response to certain chemical stimuli.

The formation of infection structures is a particularly important aspect of the life history of organisms that are *biotrophic,* that is, which must derive nutrients for growth and development from the living tissues of a compatible host. This includes the many rusts and downy mildews. The fungus grows on the plant surface to a site suitable for haustorium formation and forms an appressorium there. For example, the urediniospores of rusts produce an appressorium over a stoma, and from the appressorium an infection thread grows through the stoma into the substomatal chamber. Here a substomatal vesicle is formed from which grows the infection hypha. Other fungi, for example the powdery mildews, may form an appressorium on the epidermal surface, and an infection peg then penetrates directly through the cuticle and epidermal cell wall into the epidermal cell.

6.5.6 Chemical Exudates

Plants exude from their surfaces various chemicals that may be either stimulatory or inhibitory to germination or prepenetration growth of pathogens. Root exudates contain a wide diversity of compounds, and materials may exude into water from various aerial parts of plants. The cuticle sometimes contains substances that can be fungistatic. Wounds may result in increased exudates, which may encourage superficial growth of various microorganisms. Although water films on leaf surfaces are usually necessary for germination and early stages of infection, rain has a leaching action, and this may have negative effect, when various materials such as carbohydrates, amino acids, and organic acids are removed from the leaf surface. Germination and subsequent growth of a propagule may be affected by the final concentration of exuded or leached materials in a film or water droplet on a plant surface. Diffusion from nonpathogenic microorganisms present on plant surfaces may also or further modify the prepenetration environment.

Host exudates may contain nutrients such as sugars or amino acids that often favor germination, particularly with some organisms that produce spores unable to germinate in pure water. Root exudates often are stimulatory, whereas the normal phenomenon of soil fungistasis keeps sclerotia or spores from germinating. With an increase in exudates from wounds, growth is encouraged, as for example in the enhanced development of *Verticillium* (Hyphomycetes) and *Fusarium* (Hyphomycetes) seen around diseased or damaged roots.

Inhibitory materials also may exude from plant surfaces, and these materials may repel plant pathogens by direct toxic or lytic action. Root exudates may contain toxic phenolics or such substances as the hydrogen cyanide in exudate from sorghum roots. The toxicity of exudates may inhibit spore germination, germ tube growth, or infection-structure formation. Another type of direct influence is the infusion of toxic materials such as phenols, lignins, or proanthocyanidins in seed coats so that penetration by a pathogen is prevented. Exudates also may act indirectly by favoring growth of nonpathogenic microorganisms that are antagonistic to plant pathogens through antibiotic or competitive influences.

6.5.7 Biological Activities

Significant interactions occur between a pathogen on a plant surface and nonpathogenic epiphytic microorganisms. One mechanism was mentioned previously, that of antagonism to pathogens by antibiotic action or competitive interference. A classic example of biological control is the inhibition of the root-decay pathogen, *Heterobasidion annosum* (Aphyllophorales), by the fungus *Phanerochaete gigantea* (Aphyllophorales) on the surface of freshly cut pine stumps. Also, Bier and Rowat (1962, 1963) suggested that commonly associated microflora on stem bark of aspen are antagonistic to some canker-causing facultative saprophytes, probably by antibiotic action. In other situations there may be no antagonistic influences, but microorganisms in the immediate vicinity of the root (rhizosphere) or on stem or leaf surfaces may modify the environment within which pathogens germinate, explore the plant surface, and begin penetration. Also, normally saprophytic bacteria may form synergistic partnerships with plant pathogenic bacteria, fungi, or nematodes. A wide variety of bacteria accompany decay fungi in the invasion of trees. Many of these bacteria occur with the pathogen at the site of initial penetration by the primary pathogen or appear at a later stage in a succession of microorganisms at the infection court.

A number of field observations that trees with ectomycorrhizae were damaged less by root pathogens than trees with few or no mycorrhizae have led to suggestions that ectomycorrhizae protect roots against invasion by root pathogens. Several mechanisms for this protection have been proposed (Zak 1964). One proposed mechanism is that the mantle of ectomycorrhizae serves as a mechanical barrier to penetration. Marx (1970) demonstrated that ectomycorrhizae of shortleaf pine provided passive physical barriers to infection of the feeder roots by *Phytophthora cinnamomi*. Also, this protective barrier was broken by the feeding of certain plant parasitic nematodes, which created infection courts for *P. cinnamomi*. Another proposed mechanism is that ectomycorrhizae may alter normal plant root exudates and thereby change the interaction between host and potential root pathogen. Germination of *P. cinnamomi* zoospores is faster on nonmycorrhizal roots than on mycorrhizal roots, and the suggestion was made that germination-stimulating root exudates were used by the intervening fungal symbiont (Marx and Davey 1969b). The production of antibiotics by ectomycorrhizal fungi also may be a part of the protective effect of ectomycorrhizae. Antifungal diatretynes produced by the ectomycorrhizal symbiont *Leucopaxillus cerealis* var. *piceina* (Agaricales) seemed to be functional in resistance to infection by *P. cinnamomi* (Marx and Davey 1969a). Besides direct action on a pathogen, antibiotics, if important under natural conditions, might also act indirectly by influencing the microbial flora of the rhizosphere in which the pathogen must compete successfully in order to invade its host. A final suggested mechanism is the ectomycorrhizal induction of chemical inhibitors of host origin. Krupa and Fries (1971) showed that mycorrhizal roots produced more volatile terpenes than nonmycorrhizal roots and that these volatile terpenes inhibited *P. cinnamomi* in culture. Also Stack and Sinclair (1975) found that the ectomycorrhizal symbiont *Laccaria laccata* (Agaricales) protected Douglas-fir roots from damping-off caused by *Fusarium oxysporum* (Hyphomycetes), possibly by inducing host–root secretion of volatiles or extractives inhibitory to the pathogen. Although it is clear that mycorrhizal fungi may be biological deterrents to root pathogens by interference in the prepenetration phases of activity, more research is needed to define the specific mechanisms before their use as biological control agents will be possible.

6.6 PENETRATION THROUGH PROTECTIVE BARRIERS

Entry of a pathogen by way of the infection court is made either by direct penetration of the intact protective barriers or through an opening in these barriers, either a natural opening such as a stoma or a break or wound that has exposed the underlying susceptible tissue. Usually the character of the pathogen determines the type of penetration that is followed. At the infection court a certain minimal invasive force is required. It comes from nutrients usually supplied by the inoculum until a self-supporting infection is established. Probably more energy is required for direct penetration of the cuticle and cell wall, and lesser amounts for growth through natural or artificial openings. Even at the infection court, but also of course after infection is established, the pathogen and host are interacting in an ionic medium, where pH and ionic characteristics are important in determining the course of the resulting pathogenesis. At the point of penetration, nutrients, enzymes, toxins, and growth regulators all may have some role in the ensuing host–pathogen relationships.

6.6.1 Direct Penetration

Direct penetration through the cuticle and host cell wall is largely the result of a combination of mechanical and chemical forces by some pathogens, mostly fungi. Ectoparasitic nematodes do, however, probe the outer tissues, mostly roots, of plants by inserting their stylet, a hollow spearlike structure connected to a muscular esophageal pump. Endoparasitic nematodes may penetrate the outer tissues of their hosts probably by mechanically forcing their way into tissues softened by enzymatic degradation. But in most instances of direct penetration of the outer protective barriers of plants, we are concerned with the activities of fungi or parasitic seed plants.

Fungi may penetrate plant tissues directly, probably by a combination of mechanical pressure and enzymatic degradation of the cuticle and cell wall. There seems to be little evidence of penetration by mechanical force alone, although there have been a few reports of penetration of films by fungi where chemical degradation played no part. Forceful penetration accompanied by some physical distortion, however, perhaps of tissues softened by previous or concurrent chemical action, is an important means of entry for many fungi. The general process is similar with most fungi, although there may be variation in the details. The germ tube becomes attached to the cuticle either by intermolecular forces or a mucilaginous material. Some fungi form an appressorium, which also adheres to the surface. From this or from the germ tube, a small protuberance or new growing point appears in the adherent area that grows into a stylelike structure, often called an infection peg or a thread or a penetration peg. This penetration peg, constricted to a narrow tube, penetrates the cuticle and outer wall and then returns to the normal size of a hypha in the cell lumen. Enzymatic action occurs during this growth process whereby extracellular enzymes may be excreted into the immediate surroundings or into the cell wall matrix; wall-bound enzymes act locally at the growing tip of the germ tube or penetration peg and catalyze limited hydrolysis of the host cuticle and cell wall. After penetration has occurred, the pathogen begins to establish itself by colonizing susceptible cells and tissues, and the plant is then considered infected. Enzymatic degradation of cells and tissues is important in the processes of penetration as well as in the later stages of pathogenesis (disease development). Chemical action affects the waxes, cutin and suberin, cellulose, and proteins in breaching of the outer protective barriers. In fungi three main types of softening and solubilizing enzymes catalyze the breakdown of pectin, cellulose, and lignin.

Another mode of direct penetration is by hyphae acting in concert. Some fungi may form an infection cushion of massed hyphal cells, which then form appressoria and penetration pegs resulting in multiple penetrations. Or a hyphal aggregate, which is organized into a rhizomorph such as that produced by the common root-decaying fungus, *Armillaria mellea,* may enter. Entrance by a rhizomorph is partly by mechanical pressure and partly by chemical action. A rhizomorph growing on the surface of a root becomes attached at certain points by a mucilaginous substance, and also single side-hyphae from the rhizomorph tip ramify among the outer cork cells and help anchor the tip to the root surface. At this point of contact, one or more rhizomorph branches develop and enter the bark as a unit. When mechanical force is exerted, cells beneath the rhizomorph branch tip are compressed, and some of the collapsed cells are pushed aside. At the same time, chemical action degrades cells immediately surrounding the rhizomorph tip, and these are killed and dissolved. Thus death of cells precedes the advance of the rhizomorph into the root tissue (Thomas 1934).

Penetration by parasitic flowering plants, such as the mistletoes, is also directly through the bark tissues and by a combination of mechanical and enzymatic action. A typical example is invasion of a stem by dwarf mistletoe. The seed becomes attached to a twig by a viscous coating and upon germination forms a short radicle. The radicle forms a holdfast, or appressoriumlike attachment disc. From the underside of the holdfast, a wedge-shaped organ, the primary haustorium, penetrates the bark. The haustorium consists of xylem elements and a core of thick-walled parenchyma, the whole surrounded by parenchyma tissue. Usually host cells are killed and crushed in advance of the parasite. Branches develop in the living inner bark (phloem) tissue, and these cortical haustoria extend most rapidly in a longitudinal direction but ramify irregularly through the inner bark tissues of the host. Sinkers develop from the cortical haustoria and grow down to the cambium. With continued diameter growth of the host, and concurrent growth of the parasite, the sinkers gradually become embedded in the xylem, particularly in the rays, so that the parasite develops a close association with both the phloem and xylem tissues of the host.

An interesting and special type of penetration occurs in the nodulation of legumes by species of *Rhizobium*. When a bacterial cell of a compatible strain adheres to a root hair, softening and degradation of the cell wall results in the formation of a pore and then development of a so-called infection thread through the root hair into the root cortex. The infection thread then serves as a pathway to the host cortex; bacteria growing along the infection thread spread throughout the cortical host tissues in the localized infection area, which becomes a root nodule (Callaham and Torrey 1981, Newcomb 1976).

6.6.2   Penetration Through Openings

Superficial covering layers passively protect the plant body against invasion by many pathogens. Still, there are surface openings through which entry into susceptible tissues can be made, including several types of natural openings as well as wounds that have broken the protective barrier.

6.6.3   Natural Openings

Natural openings available for penetration by some pathogens include stomata, lenticels, leaf traces, hydathodes, nectaries, and cracks where secondary roots have emerged through the cortex. All these openings may be used by different pathogens, but usually an individual microorganism has become adapted to a particular mode of entry through a specific type of infection court.

Of all the natural openings, stomata are probably most commonly used as entry points for invading pathogens, especially fungi and bacteria, and also for air pollutants. Fungi enter through stomata in two major ways.

1. Growth of a germ tube or vegetative hypha into the stomatal antechamber, if present, and between the guard cells into the substomatal chamber. No accessory structures are formed prior to penetration, and there may or may not be a restriction in size of the germ tube or hypha as it passes between the guard cells. The pathogen probably can enter even through closed stomata.
2. Formation of an appressorium over or in a stoma followed by subsequent penetration by a hypha or infection peg between the guard cells.

Once in the substomatal chamber after penetration by either of these two methods, the "penetration" hypha may form a substomatal vesicle from which one or more infection hyphae are produced, or it may proceed to infect the tissue directly without the intervening formation of a vesicle. For some small-spored fungi, stomata also might serve as a means of passive entrance of spores in rain or dew. Various characteristics of stomata including number, spatial arrangement, structure, wax plugging, and time and degree of opening or closing may have some influence in the disease relationship between a suscept and a pathogen, but it is unlikely that stomatal characteristics alone are the absolute determinants of susceptibility or resistance to fungi.

Stomata also may be entry points for bacteria. They may be drawn in passively in water films, and it is assumed also that active movement of bacteria that possess flagella may also enable such bacteria to enter stomata or other openings. The structure of some stomata, however, may deter entry by bacteria and thus inhibit disease development, possibly through excluding water.

Lenticels are another major pathway for entrance by bacteria. These structures for gas exchange may be closed or partially closed by cork layers or layers of rather densely packed nonsuberized cells. Particular environmental conditions, often favoring stress of the plants, may result in rupture of the closing layers and provide a port of entry for waterborne bacterial cells.

6.6.4   Wounds

Wounds of all types and sizes may serve as infection courts, from microscopic insect punctures to tiny cracks or abrasions to large logging wounds or cut stump surfaces. Some pathogens may use a wound as an alternate site for entry, taking advantage of the opportunity to establish an infection in exposed susceptible tissue. Wounds are a common site for invasion by soilborne pathogens. Several factors are influential in the role of wounds as infection courts.

1. The availability of inoculum when the wound is made or during the period when the wound is susceptible to invasion.
2. The ability of a pathogen to germinate and grow on or in damaged tissue.
3. The wound-closure processes and host-resistance mechanisms set into action by the wound stimulus.
4. The ability of a pathogen to compete with other microorganisms whose germination or growth also might have been favored by the wound.

Wounds serve as the usual point of entry for many fungi and bacteria and probably as the only pathway into the host for viruses, mycoplasmas, rickettsialike organisms, and flagellate protozoans. Pathogens, particularly fungi and bacteria, may invade previously made wounds, or they may be injected into the host by the agent making the wound, usually during feeding by the vector.

Nematodes cause wounds in plants by their activities in feeding and in penetrating plant tissues. Such activities in themselves can result in plant diseases, but also with many nematode–host associations the wounds themselves may serve as points of entry for other pathogens. Both fungi and bacteria, many of which also can cause plant diseases, may become part of nematode–fungus or nematode–bacteria disease complexes through their entry into susceptible plant tissues via nematode wounds. Perhaps even more important are interrelationships between nematodes and viruses, whereby viruses are injected directly into host plants by nematode feeding. Reference was made previously to the protection afforded to roots by the mantle of ectomycorrhizae, but nematodes can breach this protective barrier and create a pathway for invading fungi into the cortex of the root (Baham et al. 1974).

Like nematodes, insects may be the direct causes of certain diseases in plants, but they may be of most importance in plant disease relationships through their role as components of symbiotic complexes. Microbial symbiotes may be borne outside or inside insects, and this relationship aids in dispersal, multiplication, and sometimes direct inoculation of the pathogen. Insects are important in relation to wounds as an entry point for pathogens in that they may visit wounds or create wounds. An example of the significance of visitation to wounds is the role of Nitidulid beetles as vectors of the oak wilt fungus, *Ceratocystis fagacearum* (Ophiostomatales). The beetles, which are sap and fungus feeders, are attracted by the fruity odor to fruiting pads of the fungus beneath the bark of trees killed by oak wilt. Spores of the fungus adhere to the insect's body and if after leaving this fruiting structure the insect is attracted to fresh wounds, such as pruning wounds or storm-broken branches on nearby oaks, the spores may be incidentally inoculated into the wounds, which immediately serve as an infection court for the fungus. The beetles that serve as the vectors for *Ophiostoma ulmi*, play a somewhat different role, however. Spores formed in the breeding galleries are carried on the insect's body when it emerges and flies to nearby healthy trees to feed. In the course of the insect's feeding on twigs, spores may be deposited in the exposed outer xylem tissues in the feeding wound made by the beetle. An even more direct role of insects in introducing pathogens into wounds is the direct injection into plant cells of viruses, mycoplasmas, and rickettsialike organisms. All such pathogens must be introduced into cells by wounding, and insects such as aphids, leafhoppers, and mites are important vectors that inject pathogens directly into host plants as they feed.

6.7 CONCLUSIONS

Forest diseases caused by biotic agents begin with the arrival and first contact of the pathogen with the suscept. Inoculation occurs by passive transfer of inoculum or through active movement of the pathogen. Subsequent interactions between pathogen and suscept are governed by a variety of factors, and the fact that inoculation took place is no guarantee that infection will follow. If inoculum arrives in a dormant or quiescent state, germination occurs as a result of conditions that induce the breaking of dormancy and the resumption of vegetative growth. Germination of fungal spores usually entails the formation of a germ tube, and associated infection structures may or may not develop. Before a pathogen penetrates the outer protective barriers of the plants, numerous prepenetration interactions between the pathogen and suscept influence further development of the pathogen. Morphological protective barriers vary in their effectiveness against the entry of pathogens. Adherence of a pathogen to a plant surface must precede penetration, and it is helped by the exudation of adhesive substances or the action of molecular forces involved with recognition phenomena. Growth of the pathogen on the surface often is necessary for it to contact the actual infection court, the site of penetration. Chemical exudates from plant surfaces may either stimulate or inhibit germination or prepenetration growth of pathogens. Also, significant interactions may occur between a pathogen and nonpathogenic epiphytes in the prepenetration stages of activity. Finally, entry of a pathogen occurs at the infection court either by direct penetration through the intact protective barriers or by growth or passive transfer through an opening in the barriers. Such openings may be natural features such as stomata or lenticels or artificial breaks such as wounds. Inoculum may enter wounds made prior to the time of inoculation, be deposited by vectors as an incidental accompaniment of the wounding process, or be injected directly into cells or tissues as a part of the feeding process by vectors such as nematodes or insects. If, after penetration, the pathogen grows and establishes itself in susceptible cells or tissue, the plant is deemed infected.

LITERATURE CITED

Baham, R. O., D. H. Marx, and J. L. Ruehle. 1974. Infection of ectomycorrhizal and nonmycorrhizal roots of shortleaf pine by nematodes and *Phytophthora cinnamomi*. *Phytopathology* 64:1260–1264.

Bega, R. V. 1960. The effect of environment on germination of sporidia in *Cronartium ribicola*. *Phytopathology* 50:61–69.

Bier, J. E., and M. H. Rowat. 1962. The relation of bark moisture to the development of canker diseases caused by native facultative parasites. VII. Some effects of the saprophytes on the bark of poplar and willow on the incidence of Hypoxylon canker. *Can. J. Bot.* 40:61–69.

Bier, J. E., and M. H. Rowat. 1963. Further effects of bark saprophytes on Hypoxylon canker. *For. Sci.* 9:263–269.

Callaham, D. A., and J. G. Torrey. 1981. The structural basis for infection of root hairs of *Trifolium repens* by *Rhizobium*. *Can. J. Bot.* 59:1647–1664.

Krupa, S., and N. Fries. 1971. Studies on ectomycorrhizae of pine. I. Production of volatile organic compounds. *Can. J. Bot.* 49:1425–1431.

Lewis, B. G., and J. R. Day. 1972. Behaviour of urediospore germ tubes of *Puccinia graminis tritici* in relation to the fine structure of wheat leaf surfaces. *Trans. Br. Mycol. Soc.* 58:139–145.

Marx, D. H. 1970. The influence of ectotrophic mycorrhizal fungi on the resistance of pine roots to pathogenic infections. V. Resistance of mycorrhizae to infection by vegetative mycelium of *Phytophthora cinnamomi*. *Phytopathology* 60:1472–1473.

Marx, D. H., and C. B. Davey. 1969a. The influence of ectotrophic mycorrhizal fungi on the resistance of pine roots to pathogenic infections. III. Resistance of aseptically formed mycorrhizae to infection by *Phytophthora cinnamomi*. *Phytopathology* 59:549–558.

Marx, D. H., and C. B. Davey. 1969b. The influence of ectotrophic mycorrhizal fungi on the resistance of pine roots to pathogenic infections by *Phytophthora cinnamomi*. *Phytopathology* 59:559–565.

Newcomb, W. 1976. A correlated light and electron microscopic study of symbiotic growth and differentation in *Pisum sativum* root nodules. *Can. J. Bot.* 54:2163–2186.

Peterson, G. W., and J. A. Walla. 1978. Development of *Dothistroma pini* upon and within needles of Austrian and ponderosa pines in eastern Nebraska. *Phytopathology* 68:1422–1430.

Stack, R. W., and W. A. Sinclair. 1975. Protection of Douglas-fir seedlings against *Fusarium* root rot by a mycorrhizal fungus in the absence of mycorrhiza formation. *Phytopathology* 65:468–472.

Thomas, H. E. 1934. Studies on *Armillaria mellea* (Vahl) Quel., infection, parasitism, and host resistance. *J. Agric. Res.* 48:187–218.

Zak, B. 1964. Role of mycorrhizae in root disease. *Annu. Rev. Phytopath.* 2:377–392.

7

COLONIZATION AND PATHOGENESIS

Synopsis: Colonization and pathogenesis are the processes by which pathogens obtain food and energy from their hosts.

7.1 INTRODUCTION

The processes of colonization and pathogenesis involve a constant physical and biochemical interaction between pathogen and host. Pathogens are, through mutations and sexual recombinations, constantly developing and testing new means of overcoming host resistance. When we consider the abundance of pathogens in nature and their arsenals of biochemical tools, we might logically conclude that there should be no disease-free plants. The host, however, is not waiting to become infected. Through mutation and recombination it is refining an already inhospitable host environment into an even more inhospitable one to discourage invading pathogens. If we examine the variety of histological and biochemical defenses that form in response to colonization, we could just as easily conclude that it is amazing that plants ever become infected. In this chapter, we will examine how pathogens colonize trees and cause disease and how trees resist pathogens.

7.2 LOCAL VERSUS SYSTEMIC INVASION

Pathogens differ greatly, but they usually differ predictably for a given species, in the general manner in which they colonize a host. Black spot of elm, for example, is always limited to discrete spots on leaves. Each individual spot does not exceed a certain size. If there are many infections, or if they coalesce, the entire leaf may be killed. As a rule, individual spots attain a certain size, and then sporulation occurs. Colonization of a host requires much energy. There may be no benefit to be derived to the pathogen by continued colonization if the spot is already large enough to support fruiting structures. Because of the transient availability of susceptible host tissues, the speed at which some inoculum is produced is often more important than is the relative amount.

200

In the case of facultative saprophytes or facultative parasites, nutrition is derived from dead host tissues killed by the pathogen. With obligate parasites, colonization cannot be so harsh so as to severely injure or kill host tissues because the pathogen will then die. In these cases growth regulator imbalances are created. They cause the infection and colonization site to act as a sink for nutrients needed by the pathogen. More will be written about this aspect later.

A different kind of local infection would be a canker caused by *Nectria* sp. where colonization is restricted to a very narrow zone around the canker margin. The host cannot quite outgrow the pathogen, but neither can the pathogen grow at will in host tissues. Other canker-causing pathogens such as the chestnut blight pathogen can grow so rapidly and aggressively in host tissues that the host is unable to initiate a callus response and succumbs quickly.

Although most forest tree infections tend to be of the local lesion type, the vascular wilts differ somewhat because they may begin as systemic infections as spores are carried through the tree's vascular system, but each germinating spore in essence forms a local lesion. The net effect of numerous internal local lesions is a rapid and generally irreversible decline and death of the host.

Relatively few pathogens can cause truly systemic infections. Yet, the mistletoes, rust fungi, bacteria, nematodes, mycoplasmas, and viruses all contain taxa, which can cause systemic infections. With some pathogens, though, there are limitations. Dwarf mistletoes and the rusts only systemically colonize host tissues formed after infection and only at the infection site. Some mycoplasmas and viruses form truly systemic infections and will colonize virtually every cell in the host. This form of colonization seldom kills the host as the pathogens are obligate parasites, and living tissues are essential for survival. Some pathogenic organisms have evolved along with complex vectoring systems that ensure their survival when in transit between hosts.

7.3   DESTRUCTIVE VERSUS BALANCED PARASITISM

Parasites are organisms that must obtain their nutrition from living hosts. Parasites can also be necrotrophs if they obtain their nourishment from host cells killed in advance of their colonization. They survive by continually invading new and living tissues. If the tissue or host they occupy dies, then they cease growth and either die or form a resting stage.

Obligate parasites, conversely, only attack and colonize living hosts. They do not kill host cells, but rather they depend upon host cells for their continued nutrition. Obligate parasites produce a rather limited amount of middle lamella and cell wall-degrading enzymes. Colonizing hyphae are intercellular and do not usually penetrate host cells. Penetration is accomplished by specialized cells known as haustoria. Precisely applied enzymatic action allows the haustorium to penetrate the host's cell wall. The haustorium does not, however, penetrate host plasmalemma, but this invaginates into host cytoplasm and in almost all cases separates the haustorium from host cytoplasm. The presumption is that the enlarged or delicately branched haustorium provides a large surface area of close contact between the host and fungus, but little is known of how nutrients are passed from host to pathogen or how this delicate balance is maintained.

7.4 BIOCHEMISTRY OF INFECTION

Necrotrophs produce disease in either of two ways. Usually they kill cells quickly by secretion of extracellular enzymes or toxins. As infection progresses, a soft, watery rot of colonized tissues develops. Hyphae then penetrate the area between host cells and secrete a variety of cell wall degradative enzymes. Host cells die because they lose turgor pressure and become "leaky." The pathogen uses both the degraded cell-wall materials and the dissolved materials that leak from the dead cells. Soft-rot disease symptoms usually progress rapidly once started. Damping-off disease of conifer seedlings is an example of a necrotroph, which produces soft-rot symptoms. Other necrotrophs produce infections that tend to remain dry. These infections are slow in spreading. In many cases the pathogen produces low molecular weight toxins that move about the plant and cause symptoms to form distant from the infected tissues.

There are some disadvantages but also some very definite advantages to being an obligate parasite. Because success of the pathogen depends upon the well-being of the host, there is initially not much harm generated to the host by infection. Toxins are not produced nor are large quantities of extracellular lytic enzymes, and their production may be switched off and on as needed. Host-produced nutrients are rapidly removed by the pathogen, which tends to create a nutrient sink into which host-produced nutrients are diverted. Eventually the lack of nutrients leads to debilitation and even death of the host, but usually not until after sporulation or fruiting in the pathogen has occurred.

One of the most noticeable physiological effects of colonization by obligate parasites is a two- to fourfold increase in the respiration rate. Most of this increase is due to the stimulation of the host's metabolism by the parasite. There is also a shift away from the major metabolic pathways of glycolysis and the tricarboxylic acid cycle to the pentose phosphate pathway. This shift provides pentose sugars for the eventual synthesis of nucleic acids and proteins and provides NADPH, which is used in the synthesis of sugar alcohols, and lipids, which are needed in fungal metabolism. The overall tendency is for the host to biosynthesize materials that the pathogen requires for metabolism and synthesis.

Starch is deposited in host leaf chloroplasts around the infection sites, and these deposits disappear as the fungus begins to sporulate. The rate of photosynthesis in infected tissues usually declines, and leaf tissues become chlorotic and senesce prematurely. With some obligate parasites such as the powdery mildews, the tissues immediately surrounding an infection site remain green and appear healthy while the rest of the leaf becomes chlorotic. Cytokinins and auxins usually sharply increase in infected tissues. Whether these growth regulators are produced by the pathogen or by the host is not clear, but they are believed to play an important role in pathogenesis by obligate parasites.

Symptoms of infection by obligate parasites vary greatly. Small discrete uredinial lesions on broad-leaved hosts infected by some of the tree rusts produce little or no measurable effect on the host, whereas hyphal growth of the pycnial and aecial stages in conifers causes an excessive overgrowth of phloem and xylem tissues. Oak leaf blister causes curling and swelling of portions of leaves. Yellow-fir rust causes formation of dense witches' brooms. Elytroderma needle cast grows systemically in host branches and needles and also causes overgrowth of the host, which appears as very loose, rather unorganized witches' broom-like growths of branches. Interestingly, one of the most important tree diseases in the United States, fusiform rust of southern pines, causes much of its impact by killing infected seedlings or young trees, which may cause unacceptable reduced stocking levels. Only infected trees that survive this early mortality will be deformed and subsequently be of much lower economic value.

All living cells, whether they are of pathogen or of host plants, respire continuously and during that process absorb oxygen (O_2) and release an equal volume of carbon dioxide (CO_2). In this overall process, which is known as respiration, organic compounds are oxidized to CO_2 and O_2 is reduced to form water (H_2O). Energy is an important additional product. The simplified equation for respiration of the common sugar glucose can be represented as

$$C_6H_{12}O_6 + O_2 \rightarrow 6CO_2 + 6H_2O + energy$$

While much of the energy released is in the form of heat, a much more useful form is the energy trapped in adenosine triphosphate (ATP). Adenosine triphosphate is used in many essential processes of life, including growth. The preceding equation is somewhat misleading, though, because it implies that free CO_2 and H_2O are released. In fact, only a portion of the substrate is fully oxidized, and much of the CO_2 and H_2O are diverted into synthetic processes to satisfy growth demands, producing carbon skeletons for a myriad of organic substances. The most common respiratory substrates include sugars, starch, fats, organic acids, and, under rare circumstances, even proteins.

Even though the process of respiration is similar for pathogens and hosts, there is a major difference in the sources of organic materials suitable for metabolism. Plants with chlorophyll, the host or suscept in most cases of concern to us, convert energy from sunlight into useful chemical bonds. Carbon, derived from CO_2 in the atmosphere, is a basic building block for most organic compounds, and O_2 is a byproduct of photosynthesis. The high-energy bonds from sunlight are used to build and maintain complex organic compounds culminating in the structure of the plant itself. In essence, these compounds, which form the substrate of the host, serve as a food base for the pathogen, either as a living food base in the case of parasites or as a dead food base for saprophytes.

The process of photosynthesis in the host is of direct consequence for us only to the extent that it might be influenced by disease and the effect that disease has on the production of economically valuable plant products, but the process of respiration is important for both the suscept and the pathogen. Respiration in pathogens is important because it provides the energy and carbon building blocks necessary for their growth and pathogenesis. These building blocks are required for synthesis of enzymes that break down relatively simple sugars in the host, and the complex enzyme systems needed to degrade cell-wall structures and toxic substances produced by the host. Respiration in the host is important because it reflects a response to infection by diverting energy and intermediate substances to the production of materials toxic to the invading pathogen and to materials needed to repair damage caused by the infection process. Our introduction to respiration will begin from the standpoint of the pathogen and concentrate on fungi because they are the most plentiful pathogens of higher plants.

7.4.1   Primary Metabolism

A generalized metabolic pathway illustrating some of the major biochemical events common to both pathogens and higher plants is shown in Figure 7.1. Photosynthesis in green plants synthesizes glucose, which is the energy-containing carbon skeleton starting point for the major metabolic pathways in both hosts and pathogens. In green plants, glucose is quickly converted and stored as starch until needed and is

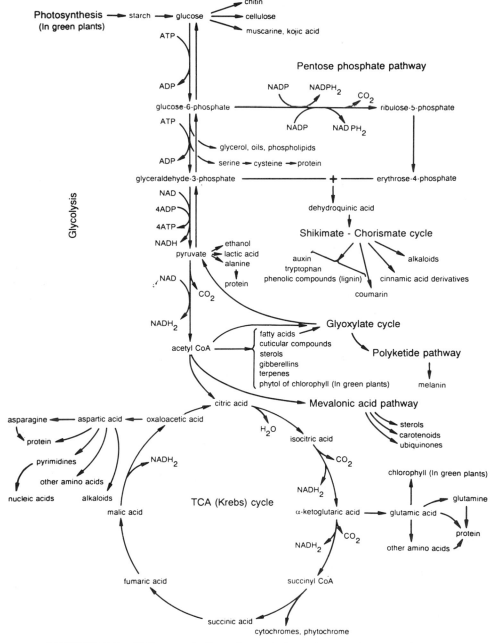

FIGURE 7.1   Major metabolic pathways in host plants and most pathogens.

eventually transformed into cell walls and a variety of other carbon-containing cell components. For pathogens, glucose is the energy-rich substrate that is used for synthesis of the same variety of cell components. There are two major forms of metabolism, primary and secondary, and they are active in both green plants and pathogenic organisms.

Primary metabolism is responsible for the degradation of organic molecules such as glucose for the production of energy and the synthesis of lipids, carbohydrates, nucleic acids, and proteins. Primary metabolism involves two basic aspects that are opposite in their actions but complimentary in their effects—the degradation of pre-formed carbon structures and the synthesis of new materials. The pathways of secondary metabolism usually are only synthetic and produce metabolites that have no obvious cellular function but do tend to be specific for a particular group of organisms. The metabolites are produced by both green plants and fungi.

In green plants the major first product of photosynthesis is glucose, most of which is immediately converted into and stored as starch and later used as needed. Complex polysaccharide structures such as cell walls that are being attacked by fungi are first hydrolyzed extracellularly by the fungi to subunits of a size that can be transported into the cell. This basic subunit is also glucose. The starting point in basic respiration for green plants and pathogens is known as *glycolysis* (Fig. 7.1). Here the aerobic conversion of glucose to pyruvate furnishes energy and carbon precursors for subsequent metabolic reactions. Glycolysis occurs in the cytoplasm of the respective organisms. Remember that the colonizing pathogen may be trying to degrade as quickly as possible as much of the host as it can while at the same time the host is trying to resist that effort.

For the present we will focus our attention on the two major glycolytic pathways in all plant and fungal cells—the Embden—Meyerhof (EM) pathway and the pentose–phosphate (PP) pathway. Later, when we examine specific interactions between host and parasite during pathogenesis, we will look at secondary pathways. In the EM pathway, glucose is phosphorylated to form glucose-6-phosphate. It is then cleaved to form glyceraldehyde-3-phosphate, which is then converted through a series of steps to pyruvate. In the PP pathway, glucose-6-phosphate is alternatively reduced to yield ribulose-5-phosphate and CO_2. Ribulose-5-phosphate then feeds a series of reactions from which intermediates are diverted for synthesis of other compounds. The PP pathway functions as a major source of nicotinamide adenine dinucleotide phosphate (NADPH), which is used for biosynthesis of new compounds. Glycolysis, conversely, produces nicotinamide adenine dinucleotide (NADH), which is used for degradation reactions.

In bacteria there is another glycolytic pathway that is less efficient, the Enter–Doudoroff pathway. In that pathway, glucose-6-phosphate is converted to 6-phosphogluconate and finally to glyceraldehyde-3-phosphate and pyruvate with a major difference from the EM pathway in that no energy is released as ATP.

Regardless of its mode of synthesis, pyruvate is transported from the cytoplasm to the mitochondria and is converted there to acetyl CoA, which then enters the tricarboxylic acid (TCA) or Krebs cycle and is ultimately degraded into CO_2-releasing electrons and hydrogen. In this process most of the energy of respiration is produced. In addition to providing energy, the TCA cycle also provides carbon skeletons for biosynthesis into other compounds. For example, many TCA intermediates are precursors for amino acid synthesis (Fig. 7.1).

The liberated electrons and hydrogens are carried to the electron transport chain,

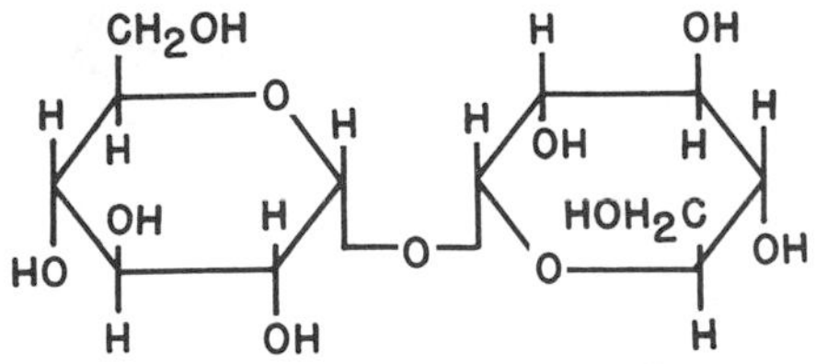

FIGURE 7.2   Trehalose, a major storage sugar of fungi.

which is the last step in respiration. Here oxidative phosphorylation occurs, and ATP is formed. The components of this chain release energy in a stepwise controlled manner through ubiquinone and cytochromes. These various coenzymes accept the hydrogen ion from each preceding intermediate carrier and then pass it along to the subsequent carrier. The last carrier passes the hydrogen to molecular oxygen and forms water. The hydrogen ions generate an electrochemical gradient across the mitochondrial membrane to provide the driving force for production of ATP. A theoretical potential of 38 moles of ATP is formed per mole of glucose. The energy produced as ATP is used to do all kinds of biosynthetic cellular work such as the synthesis of new cell walls, amino acids, and enzymes.

Formation of starch is a major way in which green plants store surplus energy. When this energy and the carbon building blocks contained within starch are needed for growth, it is converted to glucose, which then enters the glycolytic pathway. In fungi, in contrast, formation of starch is rather uncommon, and formation of polyols (lipids) and trehalose (a sugar) (Fig. 7.2) are more efficient ways for fungi to store energy for future needs. When sporulation occurs, for example, a large amount of energy, along with organic intermediates, is needed in a relatively short time. Lipids are synthesized in the cytoplasm in a series of reactions using the fatty acid molecule, acetyl CoA, as a starting point (Fig. 7.1). Recall that acetyl CoA is also a precursor for the TCA cycle. Lipids are stored in vacuoles. They are oxidized in the mitochondria to carbon dioxide and water and release considerable energy as ATP in the process.

The process by which cells use energy to synthesize glucose and other carbohydrates such as cell walls from smaller molecules is called *gluconeogenesis*. The major pathway of gluconeogenesis is basically a reversal of the EM pathway. This process requires an expenditure of energy, however. Many different compounds can be used by fungi as carbon sources if they are first converted into pyruvate or phosphoenol-pyruvate for entry into gluconeogenesis, which is accomplished in the glyoxylate cycle.

7.4.2   Secondary Metabolism

Pathways of secondary metabolism often are initiated after active growth slows. During this stage, active cell division and growth have ceased, but storage products such as lipids and polysaccharides are accumulated. It is during this stage that synthesis of secondary metabolites begins. Many secondary metabolites have no known physiologic function to the organism that produces them. These metabolites are produced either through the mevalonic acid pathway, the polyketide pathway, or the shikimate–chorismate pathway (Fig. 7.1). Secondary metabolism may be a way for the plant or fungal cell to remove excess intermediates before they become toxic, and these products may coincidentally convey significant survival value against attacks by predators or parasites.

An important secondary metabolite of the polyketide pathway is melanin (Fig. 7.1). Melanins are dark-brown to black pigments that enhance the survival and competitive abilities of fungal species in certain environments (Bell and Wheeler 1986). They occur in cell walls of hyphae, conidia, and sclerotia and absorb all visible wavelengths of light. Some species that produce large amounts of melanin include *Agaricus, Aureobasidium, Daldinia concentrica, Humicola, Thielaviopsis,* and *Verticillium dahliae*. Most soilborne fungal biomass is melanized, and this appears to have an antibiotic effect against antagonistic organisms. Melanins also protect against hydrolytic enzymes such as chitinase, cellulase, and glucanase produced by antagonists. In sporocarps, melanin helps impart structural rigidity. As fungal melanins are degraded in the soil, they slowly release some elements such as phosphorus into the environment.

Melanins are also found in abundance in the medium around fungal cells. Decay fungi secrete laccase, peroxidase, and tyrosinase that polymerize phenols liberated in decaying plant tissues. Extracellular melanins are formed by the oxidation of these secreted phenols and account for much of the soil humus produced by fungi. Foreign plant phenols such as catechol, dopamine, tannic acid, gallic acid, quinol, catechin, chlorogenic acid, and caffeic acid are thus oxidized by fungi into the black- or dark brown-pigmented organic residues known as humus. Some fungi can use melanin as their sole carbon source.

Some other important intermediate metabolites of fungi include sterols, gibberellins, patulin, cinnamic acid derivatives, penicillin, the mushroom toxins phalloidin and amanitin, and the alkaloids LSD and psilocybin. Some of these compounds will be discussed later in this chapter. In higher plants the emphasis in metabolism is on synthesis from the beginning. In the pathogen the initial emphasis is on degradation and then on synthesis. An examination of the pathway in Figure 7.1 reveals that they differ also in the emphasis placed on certain secondary pathways. Because of their relatively small size and need to store a large volume of high-energy substances, pathogens divert much energy into lipid synthesis. For example, higher plants certainly produce lipids, but they also divert much energy into the synthesis of cell walls or starch. Similarly, some pathogens produce little or no phenolic compounds, yet a tree under attack by a vascular wilt pathogen such as *Ceratocystis fagacearum* may produce relatively large amounts of phenolics as the shikimate–chorismate cycle is activated (Fig. 7.1). Depositions of these phenolics may be so great as to be visible to the unaided eye. They are especially visible in the xylem of trees affected by vascular wilts (Tainter and Fraedrich 1986). Simple phenols are colorless solids when pure, but they usually oxidize and become dark on exposure to air. The brown color of a partially eaten apple is due to oxidation of phenolic materials.

7.5   EFFECTS ON THE HOST

When pathogens penetrate a plant, either through stomata or through wounded tissues, the plant cells recognize the pathogen as an outsider in two ways:

1. the specific recognition of a pathogenic compound by particular receptive cells and
2. a nonspecific recognition of the physical and chemical injury and a response to that recognition.

The injury or death of parasitized cells often induces the oxidation of polyphenols or the formation of lignin in the adjacent noninfected cells.

Pathogens tend to produce either stimulatory effects on plant metabolism or repressive effects on certain aspects of plant metabolism. A plant infection may cause a continuous injury to plant tissues, and this injury will induce a continuous response. Pathogens that inhibit various metabolic processes may do so through the production of phytotoxins and toxic enzymes. Of the two kinds of phytotoxins secreted in plant tissues, these may be either host-specific or host-nonspecific. Some toxic enzymes produced during pathogenesis include pectinase and cellulase, which attack cell-wall materials. These enzymes severely disrupt the host's chemical and physical cell-wall structure, including membranes, and enable the pathogen to penetrate plant cells and colonize tissues.

We shall now examine some of the specific effects that pathogens have on their hosts. According to Scheffer (1983), pathogenic determinants produced by microorganisms that influence plants include

1. low molecular-weight toxins that interfere with metabolism or change the structure of protoplasm,
2. enzymes that break down cell walls or other cellular structures,
3. hormonal or antihormonal compounds that upset normal growth and development functions in the host plant,
4. genetic information passed into the host, as in *Agrobacterium tumefaciens,*
5. nontoxic determinants involved in host selectivity (such as cell-wall carbohydrates), and
6. products that interfere with normal water, nutrient, or metabolite movement.

Although the following discussion examines some of these specific influences on plants during pathogenesis, in very few cases do any of these influences act alone. Several of them usually act synergistically, and it can be very difficult to determine if any one of them has the greater influence over a particular set of symptoms. The effect that these determinants have on trees can also be classified according to the life-support processes, which are affected. These include

1. dysfunction in photosynthesis and respiration,
2. dysfunction in the flow of food,
3. dysfunction in the water system,
4. disturbed mineral nutrition,
5. alteration of growth,
6. dysfunction in symbiosis,
7. disrupted reproduction, and
8. loss of structural integrity.

In the next section we will discuss the pathogenic determinants identified by Scheffer (1983) and show how they affect the life-support processes of trees.

7.5.1   Toxins

Two types of toxins concern forest managers. The first type are produced by pathogens, affect tree health, and are directly or indirectly responsible for affecting all the life-support processes. The second type includes toxic metabolic byproducts in mushrooms or other fungal fruiting bodies. Because many people are interested in eating wild mushrooms, a brief discussion of some of these important toxins is included.

Some bacterial and fungal metabolites seriously injure or kill plant cells. These metabolites are known as toxins. They may be general and affect many species of plants, or they may be specific to a single plant species. Toxins are highly reactive, may be unstable, and are usually produced only in small amounts. For these reasons, relatively few toxins have been identified and characterized. Fungal toxins are substances that may cause a pathological disorder in other organisms by invasion and parasitism of the host, ingestion of poisonous mushrooms by humans or animals, and ingestion of agricultural products on which the fungus has grown causing food poisoning (the last are often referred to as mycotoxins) (Griffin 1981). If the toxins are active against other microorganisms, they are called *antibiotics*. A summary list of fungal toxins appears in Table 7.1.

Fungal toxins have been proposed to be involved in the disease syndrome of Dothistroma needle blight (Shain and Franich 1981), annosum root disease (Bassett et al. 1967), chestnut blight, and Hypoxylon canker of aspen (Stermer et al. 1984). Toxins have also been implicated in two wilt diseases of trees—Dutch elm disease and oak wilt.

Although there is some controversy regarding the definition of what is a toxin, it is generally regarded as a microbial product, other than an enzyme, that causes obvious injury to plant tissues and that is known with reasonable confidence to be involved in disease development (Scheffer 1983). Rudolph (1976) narrowed the definition somewhat by including toxicity at low concentrations in his criteria. Wheeler and Luke (1963) suggested use of phytotoxin for compounds that are nonspecific, have low activity, and incite few or none of the symptoms of the pathogen. In contrast, the term *pathotoxin* roughly corresponds to the requirements of a host-specific toxin. Host-specific toxins have selective toxicity to the host of the pathogen that produces them but are not toxic to resistant plants at normal physiological concentrations (Scheffer 1976). They cause all the visible and known physiological effects induced by the pathogen (Scheffer 1983). The term *vivotoxin* refers to a substance produced in vivo, which functions in disease development but is not the inciting agent (Dimond and Waggoner 1953). More will be said of these toxins in a later section.

There is considerably more evidence for the involvement of pathogen-produced plant growth regulators in the disease process. Auxins, cytokinins, ethylene, and gibberellin have all been produced by fungi. However, whether pathogens produce them during pathogenesis or the degree to which they contribute to the disease symptomatology is still not well known. Plant-growth regulators are very unstable, and it is difficult to extract them from diseased tissues and even more difficult to prove their origin. These regulators will be discussed in a later section.

Fungal toxins in the broad sense are compounds that cause a pathological condition in a plant or animal host (Griffin 1981). Possible modes of action of toxins include nuclear or ribosomal disruption, energy transfer dysfunction, cell-membrane disturbance, enzyme inhibition, and interference in water transfer. Growth inhibitory ef-

TABLE 7.1 Some Toxins Synthesized by Fungi

Precursor or Fungus Pathway	Toxin	Source	Function
Tryptophan	Psilocybin	*Psilocybe*	Poisonous mushrooms
Tryptophan	Bufotenine	*Amanita*	Poisonous mushrooms
Tryptophan	Lysergic acid	*Claviceps*	Mycotoxin
Tryptophan	Sporodesmin	*Aspergillus*	Mycotoxin
Tryptophan	Auxin	Many fungi	Plant growth regulator
Polyketide	Aflatoxins	*Aspergillus*	Mycotoxin
Polyketide	Zearalenone	*Fusarium*	Mycotoxin
Polyketide	Cytochalasins	*Phoma*	Mycotoxin, antibiotic
Polyketide	Griseofulvin	*Penicillium*	Mycotoxin, antibiotic
Polyketide	Ergochromes	*Penicillium*	Mycotoxin, antibiotic
Polyketide	Cercosporin	*Cercospora*	Phytopathogenic
Polyketide	Martisin	*Fusarium*	Phytopathogenic, antibiotic
Mevalonic acid	Trichothecenes	*Fusarium, Trichoderma, Cephalosporium*	Mycotoxin, phytopathogenic, antibiotic
Mevalonic acid	Fusidanes	*Aspergillus, Cephalosporium*	Mycotoxin, antibiotic, phytopathogenic
Mevalonic acid	Sesquiterpenes	*Fusicoccum, Cochliobolus, Cephalosporium*	Phytopathogenic Phytopathogenic Phytopathogenic
Mevalonic acid	Gibberellin	*Gibberella*	Plant growth regulator
Mevalonic acid	Abscisic acid	*Cercospora*	Plant growth regulator
Shikimate	Xanthocillin	*Penicillium*	Mycotoxin
Shikimate	Coumarins	*Phytophthora, Endothia, Nectria*	Phytopathogenic Phytopathogenic Phytopathogenic
Directly from glucose	Muscarine	*Amanita muscaria*	Poisonous mushrooms
Directly from glucose	Kojic acid	*Aspergillus*	Toxic
Directly from glucose	Cerato-ulmin	*Ophiostoma ulmi*	Phytopathogenic
Adenine	Cytokinin	*Exobasidium, Nectria, Taphrina*	Plant growth regulator Plant growth regulator Plant growth regulator
Methionine	Ethylene	Many fungi	Plant growth regulator
Phenylalanine	Gliotoxin	*Myrothecium*	Mycotoxin
Leucine	Pulcheriminic acid	*Candida*	Mycotoxin
Aspartic acid	Fusaric acid	*Fusarium*	Phytopathogenic
Cysteine	Penicillin	*Penicillium*	Antibiotic

Source: From Griffin (1981).

fects of some pathotoxins may be caused by interference with nucleic acid production. Some mycotoxins have been proven to affect directly DNA expression in animal cells (e.g., aflatoxin), but similar effects in plant cells have not yet been proven. Effects that appear to result from nucleic acid disruption may actually result from membrane disruption instead (Daly 1981, Rudolph 1976).

Plant mitochondrial or chloroplastic energy systems may be disrupted by toxins. In the mid 1950s, scientists suggested that high respiration in some diseased plants was caused by diffusible toxins. Since that time, explanations other than mitochondrial damage have been developed for increased respiration in many plant diseases. *Helminthosporium maydis* race T on susceptible corn may be an exception. T-toxin from *H. maydis* appears to affect mitochondria, perhaps through uncoupling of respiration (Daly 1981).

Tentoxin, produced by *Alternaria alternata,* has been proposed to affect photosynthesis by interfering with chlorophyll synthesis or plastid development or by uncoupling cyclic photophosphorylation (Rudolph 1976), possibly by binding noncompetitively with CF_1-ATPase in the thylakoid membranes (Gilchrist 1983). Whichever the case, these are two possible modes of action that would interfere with the energy functions of plants.

Plasma membrane disturbance may be one of the more predominant modes of action for toxins. Studies of host-specific toxins have shown that several of these change the properties of cellular membranes in susceptible plants, probably by binding to specific proteins in the membrane (Scheffer 1976).

Fusicoccin (produced by *Fusicoccum amygdali*) is an example of a nonspecific toxin that affects cell membranes. One proposed mode of action is stimulation of a plasmalemma ATPase, which leads to H^+ extrusion. The resulting effect on pH balance could theoretically cause all the metabolic disturbances induced by fusicoccin (Strobel 1982). Several other nonspecific toxins have been demonstrated to cause changes in permeability of membranes (Rudolph 1976, Owens 1969).

Most of the documented examples of specific enzyme inhibition are for toxins produced by nonfungal microorganisms. Rhizobitoxine (produced by *Rhizobium japonicum*) is an irreversible inhibitor of β-cystathionase. Tabtoxin (produced by *Pseudomonas tabaci*) has been proposed to inhibit an enzyme, possibly glutamine synthetase, by acylation of the enzyme's active sites. *Pseudomonas phaseolicola* toxin inhibits ornithine metabolism (Rudolph 1976).

7.5.2 Poisonous Mushrooms

A small number of the mushroom toxins and alkaloids are poisonous to humans. A brief description summarized from Griffin (1981) is included here because many readers may have an interest in them. More than 60 accidental and needless deaths occur each year because of ingestion of poisonous mushrooms.

The most deadly toxin is the amatoxin, consisting of α- and β-amanitin, which is a cycloprotein produced by *Amanita phalloides* (Fig. 7.3). This toxin is useful as a research tool in molecular biology. It inhibits RNA polymerase, which is responsible for transcribing DNA into messenger RNA in mammals and yeasts but not in bacteria. It binds to the enzyme, and the result is a dramatic reduction in protein synthesis. In humans a 3–4 day induction period ends with kidney and liver failure. A contributing factor is the fact that, because of the kidney damage, amanitin is reabsorbed in the kidneys. Hence, the blood is not cleaned quickly, and the toxin is recycled. Attempts to develop antibodies have not been effective because the toxin is extremely potent and it persists for a long time in the blood.

Another related toxin, phalloidin, is also a cycloprotein and affects the membranes of the liver. This toxin causes swelling of the liver and efflux of potassium ions. An

FIGURE 7.3   Fruiting bodies of poisonous mushrooms: (a) *Amanita phalloides;* (b) *Amanita muscaria;* (c) *Gyromitra esculenta;* and (d) *Coprinus atramentarius.* Courtesy of C. M. Christensen, University of Minnesota, St Paul.

interesting note is that, if the liver has been prepoisoned with carbon tetrachloride, this chemical ties up the binding sites, and phalloidin is not effective.

Another characteristic of *Amanita*-type toxins is that the victim often experiences delayed symptoms. In about 12 hours the first signs of ingestion appear, including nausea, vomiting, and diarrhea. The patient then seems to improve for 3–5 days. This recovery is followed by fatal hepatitis with kidney failure and coma.

FIGURE 7.3   *(Continued)*

The second major group of mushroom secondary metabolism toxic metabolites includes the psilocybe indoles, psilocybin, and bufotenine. Psilocybin produces effects similar to those produced by mescaline, LSD, and amphetamines. It attacks the central nervous system. Bufotenine is produced by *Amanita* and skin glands of toads. In humans it is effective only when injected into the brain. Both toxins have great therapeutic value in treating mental illness and have the potential for treating drug abuse. At low dosages they are hallucinogens and cause intellectual and bodily relaxation. Individuals being treated with hallucinogens feel a detachment from the environment. At higher dosages, they feel distortions of space and time perception and see illusions. Possession of either toxin in the United States is illegal without a permit from the Drug Enforcement Agency.

The third major toxin is muscarine. It was the first mushroom toxin to be chemically isolated. Kobert accomplished this task in 1869. His assistants peeled caps off 1 ton of fresh mushrooms in one night and several became ill before morning from the toxin taken up through their skin. Muscarine is produced in the fly agaric *Amanita muscaria* (Fig. 7.3b) and, to some extent, in *Inocybe* spp. Muscarine interferes with the neurotransmitter acetylcholine between nerve cells. Symptoms develop within 30–120 minutes after eating. These include blurred vision, excessive perspiration, slowing of heart beat, lowered blood pressure, and asthmatic breathing. Because of its structure, it is not removed from the receptor, and its action is not readily terminated. It affects smooth muscles and glands, but not ganglia of the autonomic nervous system nor skeletal muscles.

The fourth toxin is gyromitrin (or methylhydrazine), which is produced by *Gyro-*

FIGURE 7.3   *(Continued)*

mitra, the false morel (Fig. 7.3c). This mushroom, which has no insect pests, was once highly popular. By 1930 Germany and Poland produced 350,000 kg/yr. It was not lethal if it was parboiled and the water was discarded. Its sale is now illegal, however, because some deaths resulted from cooking water used in gravies and sauces. Humans have considerable variation in their tolerance. Methylhydrazine is extremely volatile and was hazardous to workers in mushroom canneries in Poland. It causes changes in kidney and liver functions. Methylhydrazine is the same material used as a rocket propellant.

The fifth toxin, coprine, is produced by the inky cap mushroom, *Coprinus atramentarius* (Fig. 7.3d). Little is known of its mode of action. Symptoms appear if victims drink alcohol while eating the mushroom. The symptoms cause a red-faced appearance and difficulty in breathing; it lasts up to 48 hours. The behavior of coprine is similar to that of *Antabuse,* which is a drug used in cases of alcohol abuse.

FIGURE 7.3   *(Continued)*

7.5.3 Toxins Produced by Forest Tree Pathogens

As early as 1964, Hubbes (1964) suggested that a toxin produced by the fungus *Hypoxylon mammatum,* causal agent of Hypoxylon canker of aspen, was capable of inhibiting wound callus formation in aspen. A host-specific toxic substance called mammatoxin was extracted from cell-free extracts of culture media. The toxic components of culture extracts are very soluble in polar substances but relatively insoluble in nonpolar solvents. Relative forms of the toxin vary in proportion according to time in culture. Younger cultures yield larger, more nonpolar components. Older cultures yield smaller, more lipophyllic molecules. Toxicity is not destroyed by autoclaving. Lesion size in a leaf bioassay was positively correlated with the percentage of infection in natural stands (Bruck and Manion 1980). Sensitivity to toxin injury was positively correlated with environmental predisposition induced by withholding moisture from potted rooted cuttings of aspen.

The role of this toxin complex is by no means clear. As Schipper (1978) found a virulent conidial isolate that produced little toxin activity, Griffin and Manion (1985) found an isolate that had lost virulence but retained the ability to produce toxin. Leaf bioassay may not be a true reflection of stem responses (Griffin and Manion 1985), however, and may be responding to other elements in the toxic culture filtrates.

These authors also suggested that the toxin bioassay response could be related to the ability of aspen clones to produce a resistance response rather than a measure of toxin in the culture filtrates (Fig. 7.4). Although much is known of some aspects of the toxin complex (Stermer et al., 1984), no toxic metabolite has been purified and identified.

Ceratocystis fagacearum has been shown to produce two phytotoxic substances in culture. White (1955) separated two toxic components and tentatively identified one as a polysaccharide. The nature of the other is unknown but is of smaller molecular weight than the polysaccharide toxin. It causes discoloration and necrosis of oak leaves. McWain and Gregory (1972) have shown the polysaccharide component to be a neutral mannan with a considerably branched structure composed of $\alpha(1,6)$ linked mannose and a molecular weight of 1.07×10^6. Although it is probably involved in blocking the vascular system, White (1955) did not consider it to be the principal toxic product of *C. fagacearum*.

Relatively more is known about the toxins produced by *Ophiostoma ulmi*. Dimond (1947) and Dimond et al. (1949) isolated a thermostable polysaccharide in a screening of toxic substances in culture filtrates. This and other large polysaccharide toxins

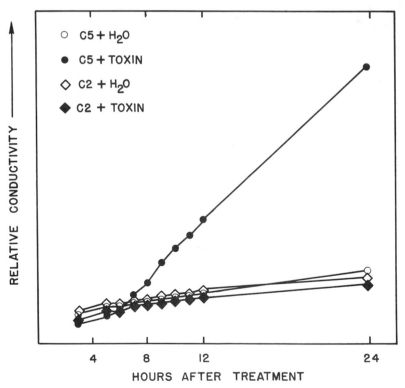

FIGURE 7.4 Effect of mammatoxin on electrolyte leakage from leaf disks of aspen, showing clone C5 to be susceptible to the toxin and C2 to be resistant. Redrawn from Stermer et al. (1984). Courtesy of B. A. Stermer, R. P. Scheffer, and J. H. Hart.

have not been well characterized, but a peptidorhamnomannan was shown to have a protein backbone comprising 6.9% of its mass and was determined to be toxic when it reduced conductance in elm stems (Van Alfen and Turner 1975). It was not species-specific, however, and caused wilting in a variety of plants (Strobel et al. 1978). Furthermore, other high molecular-weight dextrans (500,000) also reduced conductance in elm cuttings. A phytotoxic glycopeptide, probably a peptidorhamnomannan, was detected in xylem sap of infected elm tissue (Scheffer and Elgersma 1981).

A more credible wilting toxin, different from the peptidorhamnomannan (Stevenson and Takai 1981), known as cerato-ulmin, has since been discovered (Takai 1974). It elicits both external and internal symptoms of wilting, which closely resemble field symptoms of Dutch elm disease. It has a molecular weight of about 13,000 and is a single polypeptide chain possessing only 4.4% carbohydrate (Stevenson et al. 1979). The toxin has crystallike properties that cause it to form short polymers giving it a milky appearance (Takai 1974). Cerato-ulmin is a hydrophobic, surface-active molecule. It has low solubility and accumulates at an air/water interface in low concentration, so it could easily accumulate and be transported through vessel openings. In this way, it could affect vascular flow in low concentrations or be transported to another point of activity (Russo et al. 1981). Significantly, cerato-ulmin loses its toxic effect when its native tertiary structure is modified (Stevenson and Takai 1981). The use of immunocytochemistry and scanning electron microscopy has revealed that the toxin is more abundant with aggressive isolates of *O. ulmi* (Svircev et al. 1988).

Heterobasidion annosum produces at least two toxic substances in culture—oxalic acid and fomannosin. Fomannosin is a toxic sesquiterpene that causes drooping, water soaking, and browning of needles on treated loblolly pine seedlings. Fomannosin is not produced in quantity until fungal growth has reached maximum dry weight, which suggests it may be a product of autolysis. Fomannosin production does not seem to be correlated with pathogenicity of fungus isolates, and it has not been isolated from diseased tissue. However, the fact that symptoms are similar to those of pathogen infection suggests that fomannosin may have a role in disease development (Bassett et al. 1967). The chemical structure of fomannosin has been partially determined and is reviewed by Casinovi (1970).

The causal agent of chestnut blight produces at least two toxins in vitro. One of these is diaporthin, the empirical formula $C_{13}H_{14}O_5$ (Fig. 7.5). The other toxin is skyrin, a bianthraquinone with the empirical formula $C_{30}H_{18}O_{10}$. Both sub-

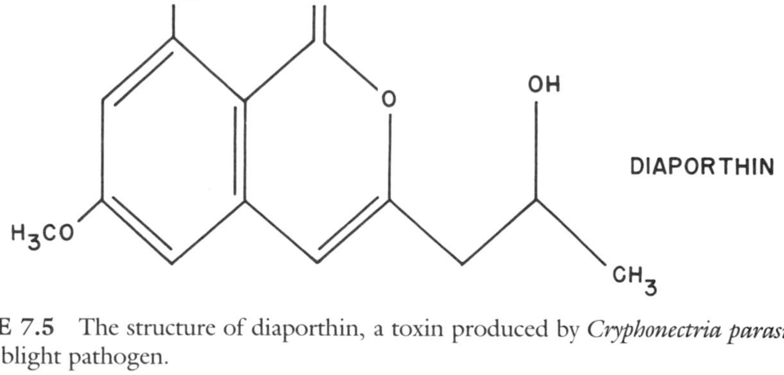

FIGURE 7.5   The structure of diaporthin, a toxin produced by *Cryphonectria parasitica*, the chestnut blight pathogen.

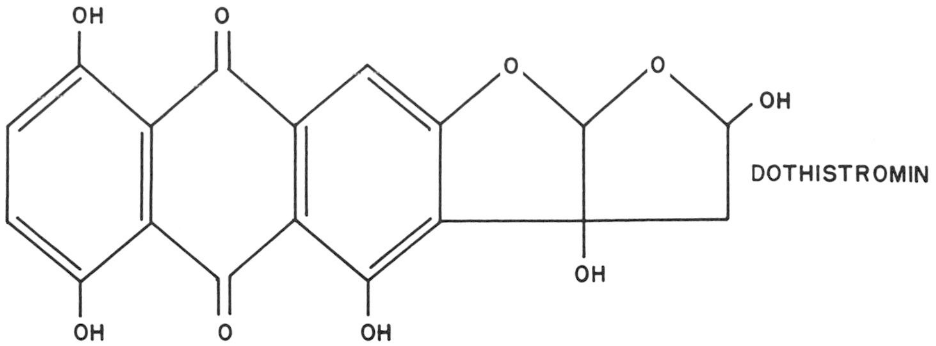

FIGURE 7.6   The structure of dothistromin, a toxin produced by *Dothistroma septospora*.

stances are nonspecific and cause necrosis of the leaf and collapse of stem tissue of tomato.

Dothistroma needle blight of pine is characterized by red bands and necrotic lesions. The toxin dothistromin has been isolated from natural lesions, and the epidemiology of the fungus shows that tissue is killed before invasion by fungal hyphae. The red bands can be induced by dothistromin in concentrations lower than those occurring in diseased needles. Lesion development from dothistromin treatment is favored by high light intensity. This development corresponds to natural lesion production being favored by high light. There also appears to be a correlation between host susceptibility to blight and sensitivity to the toxin (Shain and Franich 1981). The toxin dothistromin has been identified as a difuranoanthroquinone (Bassett et al. 1970) (Fig. 7.6).

7.5.4 Cell-Wall Degradative Enzymes

Host cell-wall structural materials are large or insoluble and must first be digested into smaller units before they can be absorbed through the cell wall and plasmalemma and into the cytoplasm of fungi. Most of this digestion must occur exocellularly. Fungi must, therefore, produce a wide variety of exoenzymes. The complement of enzymes produced by a particular fungus depends upon its genetic potential and the immediate environmental conditions that may induce or repress synthesis of particular enzymes.

The speed at which degradative enzymes can be synthesized may give certain saprophytic fungi a competitive advantage when attacking a certain dead organic substrate. Plant pathogens, on the other hand, must not only break down host tissues but may also need to overcome natural or induced resistance in the host.

One of the first cell-wall components with which an invading pathogen must contend is the cuticular wax coating on the epidermis. Cuticular waxes are deposited on the outer epidermal surface either as solid layers or as granular projections and are extremely resistant. There are no known pathogens that can degrade cuticular wax. It is believed that pathogens penetrate cuticular waxes by mechanical means. Many pathogens avoid this barrier by penetrating through open stomata.

Cutin is a major component of the cuticle, representing a transition from the cuticular waxes in its outer part and merging into a mixture of pectin and cellulose where it connects with the outer cell walls. Cutin is also relatively stable, but some fungi apparently do produce cutinases, which cause a softening of the cutin in the immediate vicinity of the invading hypha.

Once penetration of the cuticle has been achieved, the next preformed cell-wall barrier is that of pectin (Fig. 7.7). Pectin is an intercellular cement that holds plant cells together and acts as an amorphous gel that fills much of the intercellular space. Pectin is a complex polysaccharide composed of α-1,4-galacturonic acid with various sugar side chains. A variety of pectinases produced by many plant pathogens either cleave the chain-forming 1–4 glycosidic linkage, thus releasing chain portions containing residues of galacturonic acid, or remove methyl groups, which does not affect chain length but increases solubility and increases their accessibility to chain-splitting enzymes. Activity may be as endo- or exoenzymes. Endoenzymes cause cleavage of the pectin chain at random; exoenzymes cleave only the terminal 1–4 bond. The pectic enzyme complexes of the soft-rot bacteria (*Erwinia* spp.) have been extensively studied and are best understood (Collmer and Keen 1986). Although much is known of their regulation in culture, little is understood of their exact role as disease factors. Production of some pectic enzymes increases in the presence of cell-wall fragments and is repressed by glucose. They are among the first enzymes to be produced in

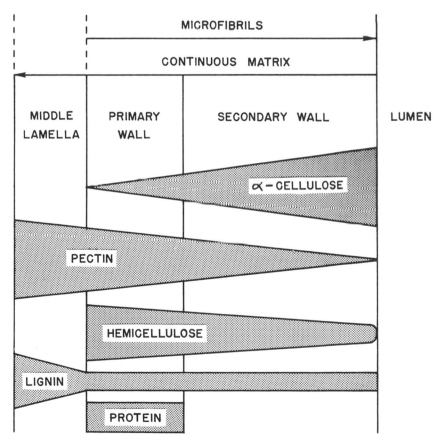

FIGURE 7.7 Distribution in the plant cell wall of the major wall components. Redrawn from Bateman and Basham (1976), with permission from Springer-Verlag, New York.

diseased tissues, but their production does not correlate with host susceptibility or tolerance.

Most fungi and some bacteria are able to synthesize one or more pectin-degrading enzymes. These enzymes undoubtedly assist many microorganisms in their colonization of plant tissues. Not only do pectic materials provide some nutritional needs for pathogens, but their removal and the associated cell-wall maceration also assist in penetration of the host. Host cells are usually quickly killed when macerated by pectic enzymes. Exactly how this occurs is not understood, but maceration renders the cells more osmotically fragile and may account for its lethal effect. However, calcium bridges between carboxyls of adjacent pectic chains limit degradation, as does the presence of phenolic materials and the presence of indoleacetic acid.

In addition to this resistance, there are many host-produced inhibitors to maceration, which are effective against pectic enzymes. In some host/pathogen combinations, higher production of inhibitors is correlated with disease resistance. They are produced constitutively, are found ionically bound to host cell walls, and protect cell walls against some pectic enzymes.

Some pectic enzymes elicit host synthesis of phytoalexins. As enzymatic degradation begins, heat-stable elicitors are released from plant cell walls. The elicitors are pectic-fragments of certain size released from plant cell walls (Collmer and Keen 1986).

Cellulose is a major structural constituent of plant cell walls in the form of microfibrils (Fig. 7.7). Cellulose is a linear polymer of β-1,4-linked D-glucose units (Fig. 4.8) and may have a molecular weight of more than a million. The molecules align themselves within the microfibrils to form crystalline regions known as micelles. Outside the micelles the cellulose molecules are not oriented in any particular way in what is known as the amorphous region. Spaces within and between the microfibrils may be filled with pectins and hemicellulose (Fig. 7.7). Hemicelluloses are mixtures of water-insoluble polysaccharides of simple sugar components and act as a cementing agent to bind the cell-wall components into a cohesive, single structure. There are many hemicellulases, but their role in pathogenesis is not well known. It is believed that degradation of hemicelluloses exposes the embedded cellulose microfibrils and lignin to enzymatic degradation.

Because cellulose is the most abundant polysaccharide in plant tissues, it represents a valuable potential source of carbon for fungi. Many fungi produce a series of enzymes, collectively known as cellulase. The C_1 cellulase causes a loosening of the cellulose chains within the microfibrils, perhaps by cleaving the cross-linkages between the cellulose chains and converting it to a soluble form. This loosening allows greater accessibility to the chain-splitting C_x cellulases. These cellulases, sometimes called β-1,4-glucanases, cleave the β-1,4 linkages of the cellulose chains, hydrolyzing the soluble cellulose to monosaccharides. A third enzyme, cellobiase, may also be present and hydrolyzes the disaccharide cellobiose to glucose.

Lignin, too, is a major component of plants and is second only to cellulose as the most abundant organic polymer (Fig. 7.7). Lignin is, thus, a potentially important food base for parasitic and saprophytic fungi. Lignin is found primarily within the cell wall of plants and especially in xylem cells. The structures of lignin, though, are complex and vary with age and species of plant. In general, lignins contain three aromatic alcohols (Fig. 7.8). The aromatic alcohols in lignin or other aromatic base

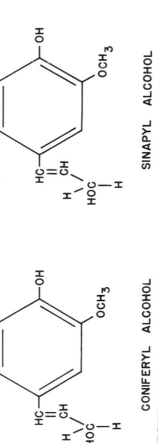

CONIFERYL ALCOHOL   SINAPYL ALCOHOL   p-COUMARYL ALCOHOL

FIGURE 7.8 The structure of three common phenolic subunits found in lignin.

221

units form a framework that replaces pectic substances as tissues get older and ramifies within and between cell walls.

Although lignin is a widespread potential source of carbon for fungi, its complex structures make it resistant to all but a few hundred white-decaying fungus species. In addition to these fungi, a few brown-decay Basidiomycetes and a few soft-decay Ascomycetes and Deuteromycetes can also degrade lignin. It is obvious, though, that these fungi are very successful, as evidenced by the enormous amounts of lignin that are degraded in nature annually. One reason why lignin is more resistant to enzymatic degradation than some other plant substances is that, although the lignin precursors are formed enzymatically, construction of the lignin polymer itself is a result of an autocatalytic chain reaction between these precursors. Lignin degradation requires a great deal of energy input to sustain the process, and wood-decaying fungi cannot grow on lignin as their sole carbon source.

7.5.5   Organic Materials Within the Cytoplasm of Host Cells

For parasitic fungi, host plasmalemma represents another important host cell component that must be dealt with. Lipases are produced by some fungi, but it is not known if they play a role in pathogenesis. Necrotrophic fungi cause breakdown of the plasmalemma coincident with host cell death. Obligate parasites, such as rusts, are unable to penetrate the plasmalemma that invaginates in response to an invading haustorium.

Starch is the main reserve carbohydrate produced by green plants. It is a glucose polymer and exists as amylose, a linear molecule of α-(1,4) glucosidic linkages; amylopectin, a highly branched molecule linked by α-(1,6) bonds to the main chain.

Pathogens degrade starch through the action of three enzymes-α-amylase, β-amylase, and isoamylase. Both α- and β-amylase attack amylose and amylopectin, α-amylase attacks α-(1,4) linkages at random, finally leaving only maltose, glucose, and the α-(1,6) branch points. β-amylase degrades from the nonreducing end by hydrolysing α-(1,4) linkages to form maltose (a disaccharide). β-amylase will degrade down the chain until α-(1,6) linkages are reached. It cannot degrade these high molecular-weight fractions called dextrins. Isoamylase attacks α-(1,6) linkages of amylopectin and produces shorter chains, which are then attacked by α- and β-amylases.

Proteins. Proteins are readily degraded by most pathogenic fungi into simple amino acids that are absorbed and can be used for synthesizing their own proteins. Proteinases break proteins down into polypeptides, which peptidases break down into smaller peptides and amino acids.

Lipids. Lipids are degraded by fungi, but little is known of the specific enzymes involved. Lipases liberate fatty acids from the lipid molecule, and these are probably taken up and used by the pathogen. Neutral lipids serve an important storage function in fungi, and are often abundant in hyphae. Lipids store large amounts of energy and carbon building blocks in relatively small volumes and are drawn upon during rapid growth and during sporulation.

7.5.6   Alteration of Growth Induced by Hormones

Growth and development in trees is regulated by several groups of organic compounds known as growth regulators. In healthy trees, growth regulators make sure that different aspects in plant growth such as cell elongation, differentiation, activation, dormancy, and senescence occur in proper sequence. These substances include auxins, gibberellins, cytokinins, ethylene, and other growth promotors or inhibitors.

Auxins include some naturally occurring substances such as indoleacetic acid (IAA) (Fig. 7.9), phenylacetic acid, and synthetic auxins such as α-naphthalene acetic acid, indole butyric acid, 2,4-D, and 2,4,5-T.

Indoleacetic acid is produced continuously in growing plant tissues and is enzymatically destroyed almost as fast in older tissues. Generally it is formed from the amino acid tryptophan, which is primarily found in young meristems such as shoots, leaves, and fruits. It is required for cell elongation and differentiation as well as promoting the synthesis of enzymes and structural proteins. It promotes adventitious root development. Easy-to-root woody species such as willow and poplar have preformed adventitious root buds that remain dormant unless stimulated by an auxin. Auxin translocation is not through the normal pathways in xylem or phloem but rather from cell to cell through living protoplasm. It moves slowly and always in a basal direction.

Some of the synthetic auxins such as 2,4-D and 2,4,5-T have had wide commercial applications as herbicides. Exactly how these substances kill certain weeds is not understood, although they appear to have an effect on coordinated growth.

Plant pathogens, including fungi, bacteria, mycoplasmas, nematodes, parasitic seed plants, and viruses, have a major influence in their hosts on auxin production, concentration, and metabolism. Many pathogens can also synthesize auxin, which results in serious abnormal plant growth. This growth may benefit the invading pathogen be-

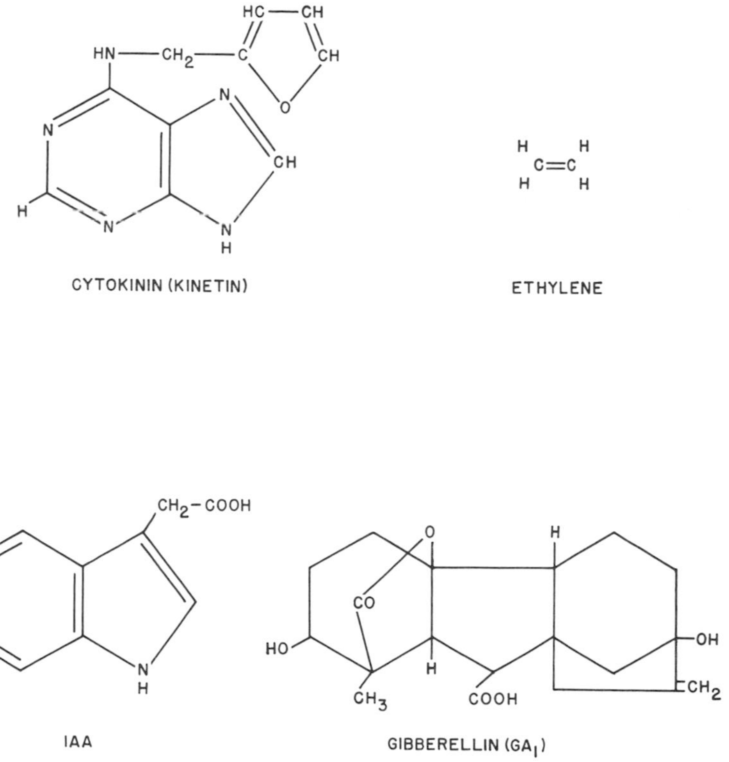

FIGURE 7.9   Chemical structures of some important growth regulators.

cause the increased plasticity of host cell walls makes them more susceptible to enzymatic action. High auxin levels have also been associated with increased respiration rates in diseased tissues.

Gibberellins are a group of chemically similar substances having growth regulator activity (Fig. 7.9). They are chemically different from auxins and cause some distinct growth patterns in plants. They are produced by plants and some pathogens. Gibberellins stimulate the mobilization of nutritional elements in seeds. These are secreted by the embryo and activate the cells in the aleurone layer to secrete hydrolytic enzymes, such as α-amylase, to break down and solubilize the reserve starch. In germinating seeds, gibberellins also enhance cell elongation so that the radicle can push through the seed coat. Gibberellins can substitute for low-temperature, long-day, or red-light requirements of dormant plants.

The gibberellins enhance elongation of stems, especially of monocots and dicots. Some species in the Pinaceae, however, show little response to gibberellins. In many dwarf mutants of plant species, application of gibberellins will reverse the effects of dwarfism. The mechanism of its action may be to activate genes that have been turned off. Little has been studied on the effects of disease on gibberellins, but application of gibberellins to virus-infected plants will overcome some of the symptoms.

Cytokinins are substances that are derivatives of adenine and that have potent effects on cell growth and differentiation of plants, more specifically, they promote cell division (Fig. 7.9). Cytokinins are present in green plants, mosses, some algae, diatoms, and some bacteria and fungi. They enhance the formation of chlorophyll. They direct the flow of metabolites toward the site of infection by pathogens and are thought to have some contributory role during pathogenesis. They delay senescence and are associated with the formation of green islands in many rust-diseased leaves. These areas contain high levels of cytokinins, probably produced by the fungus, and are surrounded by starch-rich areas that retain their green color even though the remainder of the leaf is yellow and senescent. Cytokinins may also contribute to the formation of mycorrhizae.

In dicots, cytokinins will promote lateral bud development. The bacterium *Corynebacterium fascians,* the fungus *Exobasidium* spp., and the parasitic flowering plant *Arceuthobium* sp. cause witches' brooms to form in trees. Witches' brooms form following production of multiple lateral buds that grow into a compact mass of branches. The prolific amount of twig and branch growth that is induced to grow will produce a diversion of nutrients into the broomed tissues, and the rest of the tree will eventually decline. The loss of growth and the defects introduced because of the heavy, pitch-soaked branches that support the brooms are two reasons why the dwarf mistletoes are considered by many forest pathologists to be among the most serious forest tree pathogens in North America.

Ethylene is a volatile hormone that stimulates fruit ripening, senescence, and abscission of leaves (Fig. 7.9). It is produced by green plants, a few bacteria, and several fungi, especially those that are normally found in soil. With these characteristics, it may play an important role in encouraging the diseases caused by soilborne organisms. Some symptoms of virus infection have also been associated with ethylene production.

Ethylene has been shown to play a role in leaf epinasty resulting from flooding. Under anaerobic conditions, ethylene synthesis is largely inhibited, but the small amount that is produced is trapped in the cortical cells and causes them to synthesize cellulase. This enzyme degrades cortex cell walls forming air-filled tissue. More ethylene is then produced, which causes leaf epinasty.

7.5.7 Other Growth Effects

Some volatile fungal metabolites have biological activity (Hutchinson 1973). Carbon dioxide is fixed by some fungal hyphae and at low concentrations commonly stimulates growth. At higher concentrations it is inhibitory to many fungi, especially soil fungi. Many fusaria, however, can tolerate high carbon dioxide levels.

Hydrogen cyanide is a normal fungal metabolite that may be liberated as free HCN. It can delay spore germination or reduce growth rates and cause some inhibition of germination of seeds. A recovery of these effects, however, suggests that some organisms have a mechanism to overcome inhibition of any one pathway by HCN.

Some bacteria and fungi can tolerate large concentrations of compounds of arsenic, selenium, and tellurium and methylate them to form a volatile compound. This technique has been used to provide a delicate test for traces of arsenic. In the past, arsenic was added to wallpaper paste or wall plaster to inhibit insects. Fungi would thrive if conditions were damp and metabolize the arsenic to trimethyl arsine which was inhaled and did cause poisoning. It produced a garliclike odor.

There are many other odorous materials produced by fungi. Truffles and other edible fungi produce pleasant odors. Phenyl crotonaldehyde from the stinkhorn, *Ithyphallus impudicus,* smells like carrion and serves to attract flies, which assist in spore dissemination (Hutchinson 1973). There are also many odorous carbonyl compounds, esters, and other materials in extracts from *Ceratocystis, Endoconidiophora,* the sap-staining fungi and some species of *Fomes.*

7.5.8 Genetic Information Passed into the Host

The crown-gall bacterium *Agrobacterium tumefaciens* causes the formation of hard swellings or knoblike growths on stems or on roots of trees. These are most prevalent near the soil line but may also be found higher on the stem or even on branches. Small trees may be stunted or killed, but large trees, even though heavily infected, appear to function normally. The pathogen is found in the soil and enters through wounds.

The mechanism of gall formation by this bacterium has attracted much interest among scientists, and a great deal is known of specific steps in the genesis of crown-gall tumor cells. Lippincott and Lippincott (1976) have divided this sequence into three major periods. During the conditioning period, the tissue is wounded, and this wound may accompany the introduction of the bacterium. If the bacterium is not present, the wounded cells will form callus, the wound will heal, and no tumor will form. If the bacterium is present, wound size has a direct bearing on the size and rate of appearance of the resulting tumor. In the early stages of the plant cell's response to wounding, the cells become receptive to infection by the crown-gall bacteria. These become attached to the plant cell wall and release a large bacterial plasmid into the cell. This was formerly known as the tumor-inducing principle and has subsequently been determined to be a circular extrachromosomal molecule of DNA. In the second period, that of transformation, a fragment of this molecule with the genes that encode for tumor production is inserted into the genetic code of the infected host plant cell.

In the development stage, the inserted genes direct the host cell to produce abnormal amounts of cytokinins, auxins, and gibberellinlike substances. At this stage the bacterium may no longer be present. The tissues within the young tumor then become organized, the tumor grows in size and will develop some internal differentiated tissues, and secondary growth centers will be induced to form.

Agrobacterium tumefaciens has proven itself to be an excellent vector for the genetic transformation of plants in the dicotyledons. Except for the Lilales and Arales, its ability to infect monocotyledons is limited. Restrictions to infection appear to involve the lack of bacterium-host binding (Rao et al. 1982) and/or the inability of the plant to produce the required chemical(s) to activate the virulence genes on the tumor-inducing plasmid (Stachel et al. 1985, Usami et al. 1987).

7.5.9   Host Selectivity as Related to Recognition Factors

Pathogens frequently find themselves in the presence of plants that are not their hosts. It is a rare instance in which the pathogen finds itself on a plant that happens to be a susceptible host. Cell-wall carbohydrates, with their myriad of differing wall structures and side chains, represent a major barrier to infection. Albersheim et al. (1969) have presented evidence that cell-wall polysaccharides play a significant role in the recognition by pathogens of their susceptible hosts. The specific enzyme(s) able to cleave a single cell-wall linkage might be induced or repressed in response to proper stimuli. Sugars and sugar-containing-molecules involved in these linkages produce specific recognitions to these enzymes. An analogy would be that of the specific blood group recognition phenomenon between A, B, and O blood groups, which is based solely on differing terminal residues of glycoproteins on the blood-cell surface (Albersheim et al. 1969).

Enzymes degrading a polysaccharide are dramatically affected by minute alterations in the structure of the polysaccharide. Cell-wall polysaccharides vary greatly from one part of a plant to another and also through time. A slight change in cell-wall composition can render a susceptible plant resistant to a specific pathogen. In many instances this resistance is genetically controlled in the host. The ability to overcome this resistance is determined by a virulence gene in the pathogen. Many pathogenic races are conditioned by a single gene for virulence.

The variety of variation in plant cell-wall polysaccharide structures is large, which means that early pathogenesis could release a vast array of enzyme effectors or repressors that either induce or stop synthesis of specific enzymes produced by probing pathogens.

7.5.10   Interference with Normal Water, Nutrient, or Metabolite Movement

Dysfunction in the Flow of Food.   Pathogenesis has an effect on both short- and long-distance translocation of food. Most is known about the movement of carbon. Obligate parasites seem to have the greatest effect. There is a more rapid movement of carbon toward the pustules of obligate parasites and an increased transport of carbon into tissues affected by biotrophic organisms.

The simplest explanation is that this movement results from a concentration gradient from host to parasite caused by removal or fixation of substrates by the parasite. The conversion to polyols and trehalose in many obligate parasites may be a way by which a gradient is maintained. Host carbon may also be sequestered into starch by the host. Some obligate parasites are associated with formation of green islands around fruiting lesions. These green islands contain high concentrations of starch, which are depleted during sporulation.

This concentration effect may also be under some degree of hormonal control. Translocation in rusted plants has been mimicked by application of kinetin. Increase in hormones has been well documented for diseases caused by obligate parasites and biotrophs.

Many pathogens, especially viruses and mycoplasmas, cause damage to the phloem, which results in the inability of infected plants to translocate carbohydrates. The deleterious effect this has on plant growth and crop yield is a major factor in the economic losses that result.

Less immediate effects may be a reduction in root systems, which will reduce vigor or yield or predispose them to attack by secondary parasites. Other indirect effects of altered patterns of translocation may be important in the movement of fertilizers or systemic fungicides.

Dysfunction in the Water System. Maintenance of cell permeability is a primary characteristic of living cells. Cells must regulate all kinds and quantities of substances they take in and release. This regulation requires an expenditure of energy. Dysfunction of cell permeability is often one of the first effects of pathogenesis on living cells. Diseased or injured cells are less able to maintain proper electrolyte balance and usually show significant increases in electrolyte losses. Toxins and other pathogen-produced metabolic products may injure cell membranes. An increase in conductivity of the electrolytes leached from tissues bathed in distilled water is sometimes used as a measure of disease-related injury.

Living plants require relatively large amounts of water for photosynthesis and transpiration. There is a continuum of water within the plant between the root cells in contact with the soil solution, through the vessels or tracheids of the xylem, and the liquid-cell wall-air interface of leaves. As each molecule of transpired water is lost by the leaves, it is replaced by a molecule of water taken up by the roots. This movement results in a water column that is usually under tension, especially during sunny days. An individual tree may transpire more than 300 l/day of water. The large amount of water that evaporates from the leaves cools the leaves and prevents heat damage to the cells and tissues. Any disruption that stops this flow of water can have lethal and often dramatic results.

An extensive root system is critical for plant well-being, especially during drought. Diseases that reduce root growth and development may have a significant harmful effect on later plant health. The transpiration rate also has a great influence on the plant's water status but in a slightly different sense from that previously mentioned. Pathogens may influence the transpirational behavior of infected plants. For example, rust infection in the early stages will partially inhibit stomatal opening and will decrease transpiration. Soon, however, after uredinia mature, the epidermis and cuticle are damaged by the erupting sori, and the transpiration rate increases. Some fungal toxins may promote transpiration by causing abnormal stomatal opening. In the cases of Fusarium and Verticillium wilts, there is no alteration in transpirational behavior. With these two pathogens and others, resistance to water flow in the stem was found to be much more important, with resistance as much as 200 times the resistance of healthy stems (Threlfall 1959). In twigs with late stages of oak wilt, resistance was found to approach infinity (Gregory 1971).

Vascular wilt diseases are caused by pathogen invasion of xylem vessels. They tend to be more devastating in ring-porous trees because water movement occurs in the current vessels. Exactly how vascular wilt pathogens initiate the wilting syndrome is not completely understood, although several interrelated processes are involved, and all may be important to some degree.

Pathogens that enter the sapstream are faced with a unique set of environmental conditions. They must grow and reproduce in a dilute and constantly varying nutrient solution that contains some inorganic nutrients but is low in amino acids and sugars (Dimond 1970). Some pathogens supplement their diet by degrading cell-wall polysaccharides, which can free many additional nutrients, especially sugars and amino acids, into the sapstream. This release may increase the viscosity of tracheal fluid and contribute to reduced transpiration. Large molecular weight-degraded cell-wall components tend to become entrapped within the conductive elements (Hodgson et al. 1949) and reduce water flow through the system. Smaller molecules move up the sapstream and accumulate in leaf margins, producing marginal drying (Hess and Strobel 1969). Pectolytic and cellulolytic enzymes are themselves of fairly high molecular weight and may reduce transpirational flow when they become entrapped among the rest of the clutter accumulated at perforation plates (Dimond 1970). After the vascular tissue is invaded, the fungi or bacteria may partially fill the cross-sectional area of vessels that may also contribute an increased resistance to the flow of water.

An important distinctive feature of the vascular wilts is that spores may be produced in the vascular elements. These become dislodged and are transported upward in the transpiration system, causing rapid spread of the pathogen within the host. This cycle is significant in the Angiosperms because it allows the spores to migrate great distances quickly within the sapstream. In the Gymnosperms, water movement depends on tracheids. These tracheids are short and are connected by pits that restrict movement of large spores and mycelial fragments. Spores become lodged at the pits and germinate, and the hyphae grow through the pits and sporulate on the other side. This time-consuming process must be repeated each time as it moves from tracheid to tracheid; consequently, the growing season may not be long enough to allow the pathogen to invade enough of the host to cause significant and nonreversible injury. The chances are then good that it will become buried in secondary xylem as next year's growth resumes.

Another cause of vascular disruption is the formation of air embolisms in the transpirational water (Van Alfen 1989). This water is under extreme tension during hot, sunny days when transpiration is at a maximum. Fungus sporulation and growth may cause gas bubbles to form; they quickly enlarge and block water flow.

The pathogen may also initiate a defense reaction by the host. In this process, which consists of two parts, the plasma membrane of living parenchyma cells adjacent to vessels balloons out through the simple pits in the walls connecting the two, and these occlude the vessels. The parenchyma cells may then also cause the deposition of phenolics, which presumably are for the purpose of killing the invading pathogen. Toxins may also affect the water-pumping ability of the parenchyma cells.

7.6 PLANT RESISTANCE

Invading pathogens usually elicit a resistance reaction from the host which may or may not be susccessful in repelling the invader. Not all these responses occur all the time, and some seem to be more effective than others. Some induced defense mechanisms include histological defenses such as formation of periderm, abscission layers, tyloses, and gums, swelling of cell walls, and sheathing of invading hyphae. Some important chemical defenses include production of phenols and phytoalexins.

7.6.1 Periderm Formation

Periderm is a protective tissue of secondary origin that replaces the epidermis in stems and roots that have continued secondary growth (Esau 1966). The periderm forms after surfaces are exposed following abscission or formation of wounds resulting from growth and expansion of secondary tissues. It may also result from mechanical wounding or invasion by parasites. Periderm of the latter type forms after injured cortical cells dedifferentiate and form cork cambium. Cork cells are nonliving at maturity and are arranged compactly like tiny building blocks. They have heavily suberized cell walls, which may be quite thick. The cork layer provides excellent mechanical and chemical protection to the underlying host cells from invading pathogens. Periderm may also form around lesions or infection loci. The flow of nutrients and water into the infected area is prevented and deprives the pathogen of nourishment. The periderm may form a scab beneath the lesion which continues to grow and eventually slough away the infection. Figure 7.10 shows periderm formation in eastern larch by the eastern dwarf mistletoe. One reason why this host is infrequently found to be

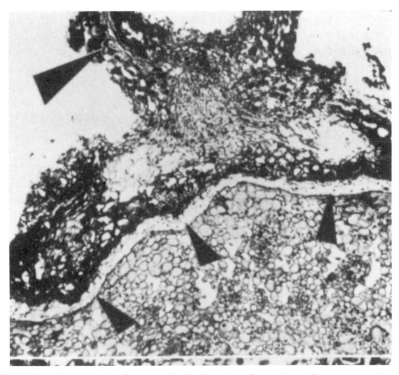

FIGURE 7.10   Spur shoot of eastern larch showing formation of wound periderm (small arrows) around the endophytic system of dwarf mistletoe. Large arrow marks a basal cup of the mistletoe. From Tainter (1970). Courtesy of F. H. Tainter.

infected is that the invading pathogen, in this case a parasitic flowering plant, is sloughed off and killed before it can reproduce. The effectiveness of these cork layers depends on the speed at which the host forms them following infection. There is evidence that infection of plants by parasitic flowering plants, fungi, bacteria, and some viruses and nematodes is largely prevented by periderm formation.

7.6.2 Abscission Layer

An abscission layer forms when a layer of host cells break down and cause separation of a plant part from the main body of the plant. Leaf abscission is a normal process that occurs seasonally in many woody dicotyledons. Premature leaf abscission is a common symptom of many vascular wilt diseases. Portions of leaves of stone fruit trees surrounding infection loci will frequently fall away following attack by pathogens. This reaction may represent one step beyond formation of the zone of periderm surrounding the infected area. With leaf diseases, abscission of infection loci may be a highly effective way for the host to limit spread of the invading pathogen. Leaf abscission following systemic infection by a vascular wilt pathogen has little chance of reducing spread but often merely indicates that the host is one step closer to death.

7.6.3 Tyloses

Tyloses are outgrowths from a ray or longitudinal parenchyma cell through a pit cavity in a vessel wall, which partially or completely blocks the lumen of the vessel. Some woody species such as the oaks and black locust produce abundant tyloses during the growing season. Tyloses are also induced to form during pathogenesis. They are not easily penetrated by pathogens and form an effective barrier to spread. Speed of their formation is very important. With some vascular wilt pathogens of trees, they do not form quickly enough to reduce the spread of pathogens.

7.6.4 Gum

Gum is a nontechnical term often applied to breakdown products of plant cell carbohydrate (Esau 1966). Gums are common in many species, especially in stone fruit trees. They may be deposited in host cells surrounding the infection and tend to form an impenetrable barrier to invasion. Gums also are deposited around tyloses and assist in helping to stop the advance of vascular pathogens.

7.6.5 Reactions of Host Cell Wall

Swelling of the cell wall and sheathing of invading hyphae are two reactions sometimes seen in response to fungal pathogens. Thickening of the cell wall is sometimes accompanied by deposition of phenolic or gummy materials, which probably assist in increasing its resistance. Sheathing of hyphae is infrequent with many fungi but is common with haustoria of some of rust fungi. Figure 4.4b shows an early state of penetration in which a haustorium has been enveloped by host cell-wall material as it invaded the host cytoplasm. In this particular host–pathogen combination, however, the rust can outgrow the sheath and establish an apparently effective parasitic relationship.

7.6.6 Hypersensitivity

With many host -parasite combinations, early stages of infection will initiate what is known as a necrotic or hypersensitive type of reaction. This reaction is especially common with fungi, which are obligate parasites, and with viruses and nematodes. The physiology of the reaction is not well understood, but the net effect is a rapid death of host cells in the immediate vicinity of the invading pathogen. Host protoplasm becomes darkened, and toxic materials are likely to be produced. This defense tends to isolate the invading pathogen and renders its attack unsuccessful. Hypersensitivity is a valuable disease resistance feature that has been used extensively by plant breeders. Its usefulness in forest pest management has less potential because it is often conditioned by a single gene. Continued exposure between potential pathogens and long-lived tree crops markedly increase the likelihood that this resistance will be overcome.

7.6.7 Phytoalexin Production

Phytoalexins are low molecular-weight antimicrobial compounds that accumulate in plant tissues exposed to biotic or abiotic stresses (Ebel 1986). They show a range of structural complexity but are predominantly phenylpropanoids, isoprenoids, and acetylenes. Most known phytoalexins have been identified in dicotyledons and only a few in monocotyledons or gymnosperms. It is not known if phytoalexin production is universal among all higher plants. Most is known about phytoalexins produced in the plant families Leguminosae and Solanaceae, both of which are important as producers of agricultural crops.

In general, phytoalexins are toxic to fungi, bacteria, higher-plant cells, and also animal cells. Effective dosages within plant tissues may range from 10^{-5} to 10^{-4} M, but levels of the soybean phytoalexin may increase from undetectable amounts to greater than 10% of the dry weight of infected tissue within 24–48 hours.

Because most phytoalexins are absent in healthy plants, it appears that resistance to disease depends upon the speed of synthesis and accumulation following challenge by a plant pathogen. Although it has been suggested that phytoalexins function by causing disruption of membrane integrity, their mode of action is not known. Phytoalexins may be a consequence of cell death rather than a cause of it (Johal and Rahe 1988). Phytoalexin production has a potent effect on invading microorganisms, but it is just one of several inducible defense mechanisms in plants. Other inducible mechanisms expressed at the infection site include reinforcement of the plant cell wall through synthesis of ligninlike materials, the deposition of callose, the accumulation of hydroxyproline-rich glycoproteins, and the induced synthesis of β-glucanase and chitinase. Because chitin is not a component of plant cell walls, the production of chitinase is likely to have a role in defense against pathogens. The role of β-glucanase is less well understood.

An intriguing aspect of phytoalexins is that the same phytoalexins formed in response to host–pathogen interaction can also be produced by treatment of plants with chemical and physical stimuli. Because there is a close correlation between phytoalexin accumulation and a large number of factors that cause formation of phytoalexins, there may be a common route in plant cells through which these factors act.

It also has been suggested that since phytoalexin accumulation is often associated with the rapid death of plant cells during the hypersensitive response, the exogenous elicitors of phytoalexin formation may be merely acting as potent phytoalexins that cause events similar to those occurring in dying plant cells in the hypersensitive response. These dying plant cells then release either endogenous elicitors that activate phytoalexin synthesis in adjacent cells or endogenous elicitor-active fragments from plant cell walls during invasion by the pathogen.

The term *elicitor* has been broadly used in reference to compounds that induce a typical defense reaction, including synthesis of phytoalexins. There are many types of elicitors, including complex carbohydrates from fungal and plant cell walls, enzymes, polypeptides, and some abiotic elicitors such as heavy-metal salts, detergents, cold, and UV light.

The best characterized elicitor from fungal cell walls is that of a hepta β-glucoside fragment isolated from the cell wall of *Phytophthora megasperma* f. sp. *glycinea* (Fig. 7.11). Plant cell walls may be a source of endogenous elicitors that are released following exposure to cell wall-degrading enzymes. The exact mode of action of elicitors has not been determined. Some evidence suggests that the primary interaction of elicitors could be located at the cell plasmalemma.

Because phytoalexins are toxic compounds synthesized by plants in response to microbial infection, potential pathogens are faced with a choice of either preventing their expression or developing tolerance to the phytoalexin. Phytoalexin degradation can be a phytoalexin tolerance mechanism and a pathogenicity determinant (Van Etten et al. 1989). The amount of disease resulting may be the result of a balance between the relative rates of phytoalexin synthesis by the plant and phytoalexin detoxification by the pathogen. *Nectria haematococca,* a pathogen of pea, synthesizes the enzyme pisatin demethylase, which is a cytochrome P450 similar in function to a similar enzyme in mammalian liver, which serves to detoxify xenobiotic compounds. In pea it encodes pisatin demethylase and demethylates the pea phytoalexin pisatin. The cytochrome P450 gene was introduced by transformation into *Cochliobolus heterostrophus,* a leaf pathogen of maize (Leong and Holden 1989). The nontransformed parent strain was unable to attack leaves of the nonhost pea, but the transformant could attack. Thus the host range was artificially extended, and role of the P450 gene was proven.

Tolerance to phytoalexins is most common in the higher fungi, which are primarily

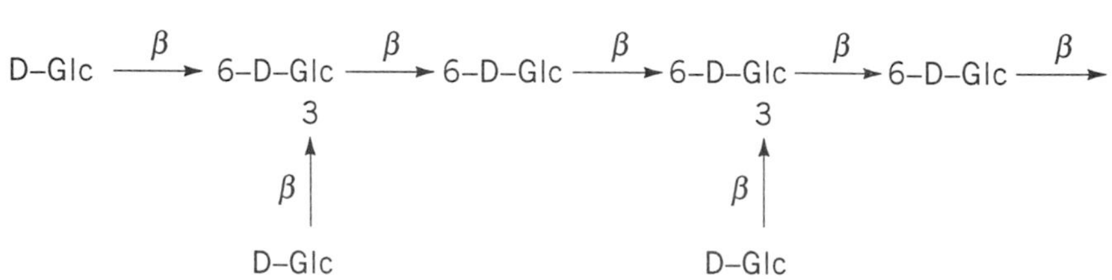

FIGURE 7.11   Structure of a biologically active hepta β-glucoside elicitor fragment isolated from the *Phytophthora megasperma* f. sp. *glycinea* cell wall. From Sharp et al. (1984). Courtesy of J. K. Sharp, M. McNeil, and P. Albersheim.

necrotrophic. Oomycetes and bacteria do not commonly rely on phytoalexin tolerance for pathogenicity.

7.6.8 Dysfunction in Photosynthesis and Respiration

A rise in respiration is characteristic for most plant diseases. Possible exceptions are for some plants systemically infected with virus. Both obligate and facultative parasites cause an increase in respiration as do mechanical and chemical injury.

Respiration begins to increase shortly after inoculation and continues to rise through the appearance of symptoms. It reaches a maximum during sporulation and then quickly declines to preinfection levels or lower. In resistant varieties respiration tends to rise more quickly and then also decline more quickly after it reaches a maximum.

Although glycolysis may increase somewhat during pathogenesis, increased respiration is usually accompanied by increased activity in the pentose pathway and the accumulation of phenolic compounds. During glycolysis in healthy plants, the availability of adenosine diphosphate (ADP) governs the rate of this portion of the respiratory cycle. Pathogenesis is accompanied by a shift from the glycolytic pathway to the pentose pathway. Because the pentose pathway is not linked to oxidative phosphorylation, this shift, in effect, uncouples the braking effect of the relative scarcity of ADP, and an abundance of pentose pathway intermediates and products are produced. The Krebs cycle tends to be less active in diseased plants, and there may also be much more fermentation.

The effect of the shift toward the pentose pathway is to produce less ATP and, hence, make a less efficient use of energy. On the other hand, the pentose pathway is the main source of phenolic compounds that play important roles in the resistance of plants to infection.

Rates of apparent photosynthesis generally decrease in infected leaves. This increase is often accompanied by varying degrees of chlorosis. When infected tissues remain green, the decrease in photosynthesis does not seem to be a result of damage to the photosynthetic apparatus. The decrease is greatest at the time of heavy sporulation, just when high rates might be beneficial to both the host and the pathogen.

LITERATURE CITED

Albersheim, P., T. M. Jones, and P. D. English. 1969. Biochemistry of the cell wall in relation to infective process. *Annu. Rev. Phytopath.* 7:171–194.

Bassett, C., M. Buchanan, R. T. Gallagher, and R. Hodges. 1970. A toxic difuranthroquinone from *Dothistroma pini.Chem. and Indust.* 52:1659–1660.

Bassett, C., R. T. Sherwood, J. A. Kepler, and P. B. Hamilton. 1967. Production and biological activity of fommanosin, a toxin sesquiterpene metabolite of *Fomes annosus*. *Phytopathology* 57:1046–1052.

Bateman, D. F., and H. G. Basham. 1976. Degradation of call walls and membranes by microbial enzymes, pp. 316–355. In R. Heitefuss and P. H. Williams (eds.), *Encyclopedia of Plant Pathology, New Series Vol. 4, Physiological Plant Pathology,* Springer-Verlag, Berlin. 890 pp.

Bell, A. A., and M. H. Wheeler. 1986. Biosynthesis and functions of fungal melanins. *Annu. Rev. Phytopath.* 24:411–451.

Bruck, R. I., and P. D. Manion. 1980. Mammatoxin assay for genetic and environmental predisposition of aspen to cankering by *Hypoxylon mammatum*.*Pl. Dis.* 64:306–308.

Casinovi, C. G. 1970. Chemistry of the terpenoid phytotoxins, pp. 105–125. In R. K. S. Wood, A. Ballio, and A. Graniti (eds.), *Phytotoxins in Plant Diseases*. Academic Press, New York. 530 pp.

Collmer, A., and N. T. Keen. 1986. The role of pectic enzymes in plant pathogenesis. *Annu. Rev. Phytopath.* 24:383–409.

Daly, J. M. 1981. Mechanisms of action, pp. 331–394. In R. D. Durbin (ed.), *Toxins in Plant Disease*. Academic Press, New York. 515 pp.

Dimond, A. E. 1947. Symptoms of Dutch elm disease reproduced by toxins of *Graphium ulmi* in culture. *Phytopathology* 37:7 (abst.).

Dimond, A. E. 1970. Biophysics and biochemistry of the vascular wilt syndrome. *Annu. Rev. Phytopath.* 8:301–322.

Dimond, A. E., and P. E. Waggoner. 1953. On the nature and role of vivotoxins in plant disease. *Phytopathology* 43:229–235.

Dimond, A. E., G. H. Plumb, E. M. Stoddard, and J. G. Horsfall. 1949. *An Evaluation of Chemotherapy and Vector Control by Insecticide for Combatting Dutch Elm Disease*. Conn. Agric. Exp. Sta. Bull. 531. 69 pp.

Ebel, J. 1986. Phytoalexin synthesis: The biochemical analysis of the induction process. *Annu. Rev. Phytopath.* 24:235–264.

Esau, K. 1966. *Anatomy of Seed Plants*. John Wiley & Sons, New York. 376 pp.

Gilchrist, D. G. 1983. Molecular modes of action, pp. 81–136. In J. M. Daly and B. J. Deverall (eds.), *Toxins and Plant Pathogenesis*. Academic Press, New York. 181 pp.

Gregory, G. F. 1971. Correlation of isolability of the oak wilt pathogen with leaf wilt and vascular water flow resistance. *Phytopathology* 61:1003–1005.

Griffin, D. H. 1981. *Fungal Physiology*. John Wiley & Sons, Inc., New York. 383 pp.

Griffin, D. H., and P. D. Manion. 1985. Host–pathogen interactions as measured by bioassay of metabolites produced by *Hypoxylon mammatum* with its host *Populus tremuloides*.*Phytopathology* 75:674–678.

Hess, W. H., and G. A. Strobel. 1969. Ultrastructural investigations of tomato stems treated with the toxin glycopeptide of *Corynebacterium sepedonicum*.*Phytopathology* 59:12.

Hodgson, R., W. H. Peterson, and A. J. Riker. 1949. The toxicity of polysaccharides and other large molecules to tomato cuttings. *Phytopathology* 39:47–62.

Hubbes, M. 1964. The invasion site of *Hypoxylon pruinatum* on *Populus tremuloides*.*Phytopathology* 54:896.

Hutchinson, S. A. 1973. Biological activities of volatile fungal metabolites. *Annu. Rev. Phytopath.* 11:223–246.

Johal, G. S., and J. F. Rahe. 1988. Glyphosate hypersensitivity and phytoalexin accumulation in the incompatible bean anthracnose host–parasite interaction. *Phys. Mol. Pl. Path.* 32:267–281.

Leong, S. A., and D. W. Holden. 1989. Molecular genetic approaches to the study of fungal pathogenesis. *Annu. Rev. Phytopath.* 27:463–481.

Lippincott, J. A., and B. B. Lippincott. 1976. Morphogenic determinants as exemplified by the crown-gall disease, pp. 356–388. In R. Heitefuss and P. H. Williams (eds.), *Encyclopedia of Plant Pathology, New Series Vol. 4, Physiological Plant Pathology*. Springer-Verlag, Berlin. 890 pp.

McWain, P., and G. F. Gregory. 1972. A neutral mannan from *Ceratocystis fagacearum* culture filtrate. *Phytochemistry* 11:2609–2612.

Owens, L. 1969. Toxins in plant disease: Structure and mode of action. *Science* 165:18–25.

Rao, S. 1982. *Agrobacterium* adherence involves the pectic portion of the host cell wall and is sensitive to the degree of pectin methylation. *Physiol. Plant Path.* 56:374–380.

Rudolph, K. 1976. Non-specific toxins, pp. 270–315. In R. Heitefuss and P. H. Williams (eds.), *Encyclopedia of Plant Pathology, New Series Vol. 4, Physiological Plant Pathology.* Springer-Verlag, Berlin. 890 pp.

Russo, P. S., F. D. Blum, J. D. Ipsen, Y. J. Abul-Haij, and W. G. Miller. 1981. The solubility and surface activity of the *Ceratocystis ulmi* toxin cerato-ulmin. *Physiol. Plant Path.* 19:113–126.

Scheffer, R. P. 1976. Host-specific toxins in relation to pathogenesis and disease resistance, pp. 247–269. In R. Heitefuss and P. H. Williams (eds.), *Encyclopedia of Plant Physiology, New Series Vol. 4, Physiological Plant Pathology.* Springer-Verlag, Berlin. 890 pp.

Scheffer, R. P. 1983. Toxins as chemical determinants of plant disease, pp. 1–40. In J. M. Daly and B. J. Deverall (eds.), *Toxins and Plant Pathogenesis.* Academic Press, New York. 181 pp.

Scheffer, R. J., and D. M. Elgersma. 1981. Detection of a phytotoxic glycopeptide produced by *Ophiostoma ulmi* in elm by enzyme-linked immunospecific assay (ELISA). *Physiol. Plant Path.* 18:27–32.

Schipper, A. L., Jr. 1978. A *Hypoxylon mammatum* pathotoxin responsible for canker formation in quaking aspen. *Phytopathology* 68:866–872.

Shain, L., and R. A. Franich. 1981. Induction of Dothistroma blight symptoms with dothistromin. *Physiol. Plant Path.* 19:49–55.

Sharp, J. K., M. McNeil, and P. Albersheim. 1984. The primary structures of one elicitor-active and seven elicitor-inactive hexa (B-D-glucopyranosyl)-D-glucitols isolated from the mycelial walls of *Phytophthora megasperma* f. sp. *glycinea.J. Biol. Chem.* 259:11321–11336.

Stachel, S. 1985. Identification of the signal molecules produced by wound plant cells that activate T-DNA transfer in *Agrobacterium tumefaciens.Nature* 318:624–629.

Stermer, B. A., R. P. Scheffer, and J. H. Hart. 1984. Isolation of toxins of *Hypoxylon mammatum* and demonstration of some toxin effects on selected clones of *Populus tremuloides.Phytopathology* 74:654–658.

Stevenson, K. J., and S. Takai. 1981. Structural studies on cerato-ulmin—A wilting toxin of Dutch elm disease fungus, *Ceratocystis ulmi*, pp. 178–194. In *Proc. Dutch Elm Disease Symposium and Workshop, Oct. 5–9,* Winnipeg, Manitoba. 517 pp.

Stevenson, K. J., M. Kutryk, and S. Takai. 1979. Characterization of cerato-ulmin, a wilting toxin of the Dutch elm disease fungus. *Proc. Can. Phytopathological Soc.* 46:70.

Strobel, G. A. 1982. Phytotoxins. *Annu. Rev. Biochem.* 51:309–333.

Strobel, G. A., N. Van Alfen, K. D. Hapner, M. McNeil, and P. Albersheim. 1978. Some phytotoxic glycopeptides from *Ceratocystis ulmi*, the Dutch elm disease pathogen. *Biochem. Biophys. Acta* 538:60–75.

Svircev, A. M., R. S. Jeng, and M. Hubbes. 1988. Detection of cerato-ulmin on aggressive isolates of *Ophiostoma ulmi* by immunocytochemistry and scanning electron microscopy. *Phytopathology* 78:322–327.

Tainter, F. H., and S. W. Fraedrich. 1986. Compartmentalization of *Ceratocystis fagacearum* in turkey oaks in South Carolina. *Phytopathology* 76:698–701.

Takai, S. 1974. Pathogenicity and cerato-ulmin production in *Ceratocystis ulmi*. *Nature* 252:124–126.

Threlfall, R. J. 1959. Physiological studies on the *Verticillium* wilt disease of tomato. *Ann. Appl. Biol.* 47:57–77.

Usami, S. 1987. Absence in monocotyledonous plants of the diffusible plant factors inducing

T-DNA circularization and vir gene expression in *Agrobacterium.Mol. Gen. Genet.* 209:221–226.

Van Alfen, N. K. 1989. Reassessment of plant wilt toxins. *Annu. Rev. Phytopath.* 27:533–550.

Van Alfen, N. K., and N. C. Turner. 1975. Influence of a *Ceratocystis ulmi* toxin on water relations of elm *(Ulmus americana)*. *Plant Phys.* 55:312–316.

Van Etten, H. D., D. E. Matthews, and P. S. Williams. 1989. Phytoalexin detoxification: importance for pathogenicity and practical implications. *Annu. Rev. Phytopath.* 27:143–164.

Wheeler, H., and H. H. Luke. 1963. Microbial toxins in plant disease. *Annu. Rev. Microbiol.* 17:223–242.

White, F. G. 1955. Toxin production by the oak wilt fungus. *Endoconidiophora fagacearum. Am. J. Bot.* 42:759–764.

8

EPIDEMIOLOGY*

Synopsis: Epidemiology is the ecology of disease and its practical application in disease management. We will be concerned with the quantitative aspects of disease at the level of the individual, the population, and the ecosystem. Understanding how the disease at each level changes with time and in space is critical to managed efforts to control its spread.

8.1 THE SCOPE OF EPIDEMIOLOGY

Etiology is the study of cause, in which the application of Koch's "Rules of Proof of Pathogenicity" results in a verdict of guilt or innocence for suspected pathogens. Identification of the pathogen begins rather than ends the study of plant pathology. Next, the nature and function of those factors that influence disease development, given a known host and pathogen, must be understood. This body of knowledge is the concern of epidemiology and is a prerequisite for disease management.

Epidemiology was defined early as the "ecology of disease" (Whetzel 1926) but was primarily autoecological and limited to the study of the effects of secondary factors (climatic, edaphic, and biotic environmental factors) on disease development. More recently (Zadoks 1974), epidemiology was broadened to include synecological concepts, including populations of suscepts, pathogens, and holistic environments. In this latter sense, epidemiology includes not only the roles of secondary environmental factors but also those of suscept and pathogen, cultural practices, and control strategies as they influence the epidemic. Conceived as such, epidemiology is the ecology of disease and a forerunner of disease management.

The human medical terms *epidemic* (Greek for "on the people") and *epidemiology* are adopted and most used in contemporary plant pathology in place of the more etomologically correct terms *epiphytotic* (Greek for "on the plants") and *epiphytology* (Snell and Dick 1971). Unfortunately, within plant pathology, the terms *epidemic* and especially *endemic* (Greek for "in the people") are used ambiguously and are often misused. Too often endemic is used to refer to a static or small amount of disease; used correctly endemic refers to a native or indigenous (as opposed to exotic) pathogen or disease. The mistake arises because many endemic (indigenous) diseases exist

* Written by R. A. Schmidt, School of Forest Resources and Conservation, University of Florida, Gainesville.

at a static, low level. Unfortunately, endemic disease can increase rapidly with subtle changes in the host, pathogen, environment, or cultural practices. Such is the case with fusiform rust, an endemic disease of southern pines. This disease, caused by an indigenous pathogen of indigenous suscepts, was rare prior to 1900 but since 1940 has become an important obstacle to the management of southern pine (Dinus 1974, Griggs and Schmidt 1977). Such epidemics which increase for more than one year are called polyetic epidemics (Zadoks 1974). The term *epidemic* is often misused to refer to a rapid and widespread increase of disease (Snell and Dick 1971). Despite their traditional misuse, endemic and epidemic are not opposites. Epidemic refers to the quantitative progress of disease irrespective of the initial or final amount or the rapidity of change (Van der Plank 1963). Whether the amount of disease is great or small, increasing, decreasing, or static, it is an epidemic.

Epidemics are characterized as the amount of disease relative to time (disease progress curves) or the amount of disease relative to space (disease gradient curves). Thus an epidemic may be conceived as the quantitative development of the amount of disease relative to time and space as influenced by environmental, host, and pathogen factors as well as cultural practices including control. Epidemiology is the study of such epidemics and is a cornerstone for the development of control strategies and pest management.

8.2 INCREASE AND SPREAD OF PLANT DISEASE

8.2.1 Measuring the Epidemic

To study the temporal and spatial relations of epidemics, it is necessary to measure disease, time, and distance. Disease is measured most easily as incidence but most precisely as severity (James 1974). Incidence refers to the number of diseased units relative to the number of such units examined and is often expressed as a percentage, for example,

$$\frac{\text{number of leaves, roots, or stems diseased}}{\text{number of leaves, roots, or stems examined}} \times 100$$

Severity refers to the amount of disease per unit examined and is often expressed as a percentage, for example,

$$\frac{\text{area of leaf, root, or stem diseased}}{\text{area of leaf, root, or stem examined}} \times 100$$

Severity may also be recorded as the number of lesions per unit, for example, number of cankers per infected tree or number of lesions per square centimeter of leaf surface. Indirect severity estimates can include inoculum concentration measurements, for example, number of spores or nematodes per gram of soil. A numerical index is commonly used to facilitate measurement and recording of severity data.

Incidence and severity are correlated for some pathogen systems. For example, if 60% of the trees in a lodgepole pine stand have dwarf mistletoe, we can estimate the average dwarf mistletoe rating (DMR) as a measure of severity. However, the relationships between incidence and severity are not known for many forest pathogen systems.

Time and distance are measured in appropriate units. Days and weeks are commonly used for phenomena that progress rapidly (e.g., local lesion diseases of hardwood foliage), while years are appropriate for diseases of perennial tissues (e.g., cankers of hardwoods or conifers).

8.2.2 The Temporal Epidemic

The development of an epidemic is demonstrated in Figure 8.1 for a typical local lesion foliar pathosystem producing secondary cycles. Following the deposition of inoculum on the leaf surface of the suscept (inoculation), spore germination and penetration of the stomata (infection site or court) occur. The suscept is infected, but disease is not visible until lesions (symptoms) appear at the end of day **3**. The interval from penetration until symptoms appear is the incubation period—3 days in this example. Spores form in the lesions at the end of day **7**. The interval from penetration to sporulation is the latent period—7 days in this example. The period during which spores are produced in a lesion is the infectious period—2 days in this example. Thus at the end of the seventh day, during the eighth and ninth days and multiples thereof, new inoculum initiates secondary cycles. Infectious and especially latent periods are important features of epidemics since the shorter the latent period and the longer the infectious period, the more rapid the increase of inoculum and, given favorable environmental conditions, disease.

A quantitative example is shown in the disease progress curve of Figure 8.2a, where number of lesions and proportion of tissue diseased are given as a function of time. If 1,000 spores are deposited on the leaf surface and 1% successfully penetrate the stomata, 10 lesions will result and be visible at the end of day 3. At the end of day 7, the lesions will produce spores for secondary cycles. In our example, each lesion produces two new (daughter) lesions at the end of each latent period such that cumulative lesion numbers are 10, 30 (20 new plus 10 original), 90, 270, and 810. Assuming a lesion occupies 0.001% of the total host tissue, the proportions of diseased

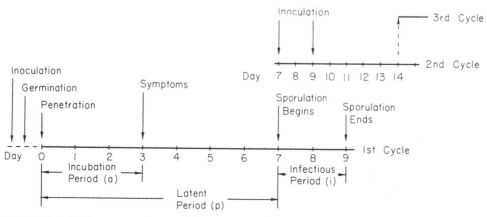

FIGURE 8.1 Sequence of events in an epidemic in which secondary cycles occur. Adapted from Van der Plank (1963). Courtesy of Academic Press and J. E. Van der Plank.

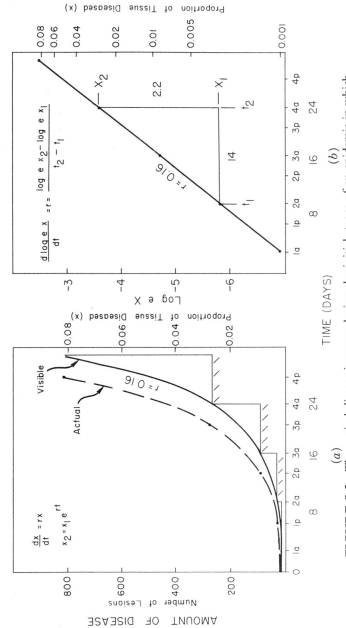

FIGURE 8.2 Theoretical disease increase during the initial stages of an epidemic in which secondary cycles occur. (a) Number of lesions and proportion of tissue diseased as a function of time. (b) $\log_e$ of the proportion of tissue diseased as a function of time.

tissues are 0.001, 0.003, 0.009, 0.027, and 0.081 or 0.1, 0.3, 0.9, 2.7, and 8.1%, respectively. Initially disease increases slowly because of limited inoculum, but as inoculum becomes more available, disease increases rapidly. This rapid rate of increase is associated with pathosystems having abundant secondary cycles and abundant susceptible host tissue and are called "compound interest diseases" (Van der Plank 1963). Note that the curve for visible disease lags behind the curve for actual disease by the length of the incubation period. If the latent period is shorter or longer than 7 days, the rate of disease increase is more and less rapid, respectively. Actual epidemics are more complex, and discrete stepwise increases are masked as initial inoculum arrives on more than one occasion and normal variability in the length of the latent and infectious periods occurs in response to environmental factors (e.g., temperature).

By transforming the amount of disease x to $\log e\ x$, a linear relationship results (Fig. 8.2b), and the infection rate r can be calculated as the slope of the regression line. In this manner the early stages of an epidemic can be quantified by reference to the infection rate. In our example, $r = 0.16$ units per unit per day (i.e., lesions per lesion per day).

This exponential increase is maintained (given favorable environmental conditions) until the amount of healthy susceptible host tissue becomes a limiting factor, at which time the rate of disease increase slows. Thus for epidemics that involve a high proportion of host tissue, disease progress curves are typically sigmoid (S-shaped) and have a lag, exponential, and stall phase (Fig. 8.3a). The infection rate may be calculated by transforming the amount of disease x to the log of $x\ ^c\log e[x/(1-x)]$. This results in a linear disease progress curve (Fig. 8.3b) and r is the slope of the regression line.

Infection rates (change in amount of disease per unit time) are useful to compare epidemics and to assess the effects of disease control measures. The equations for the derivations and calculation of r are given in Box 8.1.

Some plant diseases, including forest tree diseases, increase at less than exponential rates, even when environmental conditions are favorable. Inoculum may be a limiting factor (e.g., when secondary cycles are absent), or the availability of susceptible host tissue may be limiting from the beginning (e.g., when fresh wounds are required for infection). These so-called "simple interest diseases" increase slowly and are exemplified by root diseases where infection is dependent on inoculum in the soil coming into contact with growing roots, which provide the infection sites. In this case, disease progress curves may be linearized by transforming amount of disease x to $\log e$ $[1/(1-x)]$, and the infection rate r is computed as the slope of the regression line (Van der Plank 1963). These epidemics (often mistakenly referred to as endemics) are typical of some, even many, forest tree diseases.

Examples of disease progress curves for several forest tree disease epidemics are in Figure 8.4. Figures 8.4a and b show the effects of sanitation (removal of diseased elms) on epidemics of Dutch elm disease. In both instances the average infection rates were greater during those periods (Fig. 8.4a) and those locations (Fig. 8.4b) when sanitation was not practiced. Figure 8.4c shows oak wilt epidemics in West Virginia and Pennsylvania. The r value for the epidemic in West Virginia is 3.2 times as great as that of the epidemic in Pennsylvania. Figure 8.4d shows the effect of genetic disease resistance in slash pine to *Cronartium quercuum* f. sp. *fusiforme* (Uredinales) on epidemics of southern fusiform rust. Although the r values are not calculated, the mitigating effect of resistant R, relative to intermediate I and susceptible S varieties is demonstrated. For multiple-lesion-type diseases, such as fusiform rust galls on pine,

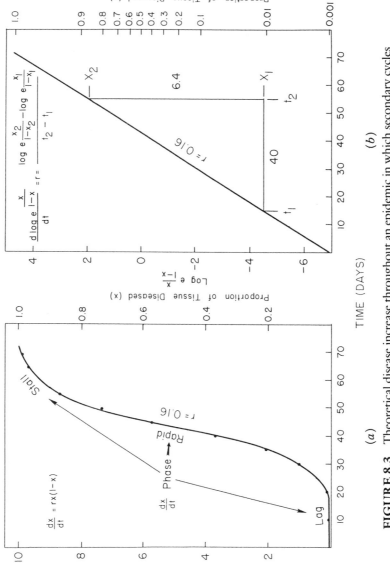

FIGURE 8.3 Theoretical disease increase throughout an epidemic in which secondary cycles occur. (a) Number of lesions and proportion of tissue diseased as a function of time. (b) $\log_e [x/(1-x)]$ as a function of time where x is the proportion of tissue diseased.

BOX 8.1
CALCULATION OF THE INFECTION RATE r*

When lesions, through the production of inoculum, generate new lesions, the amount of disease in the future depends on the present amount of disease. Thus the change in the amount of disease dx with the change in time dt is proportional to the amount of disease x at that time, that is,

$$\frac{dx}{dt} = rx \tag{1}$$

where r is the constant of proportionality or the infection rate. This differential equation when integrated gives

$$x = e^{rt} \tag{2}$$

where e is a constant (2.718), the base of the natural log. Graphically the curve depicted by this exponential function is

The infection rate r determines the exact shape of the curve, and disease at some future time x_2 can be calculated knowing the present amount of disease x_1 and the infection rate r.

$$x_2 = x_1 e^{rt} \tag{3}$$

Likewise, knowing x_1 and x_2, the infection rate can be calculated. The calculation of r is facilitated by linearizing the equation. Thus x is transformed to the natural log of x such that

$$\ln x_2 = \ln x_1 + rt$$

and $\tag{4}$

$$r = \frac{\ln x_2 - \ln x_1}{t_2 - t_1}$$

Graphically this is seen as

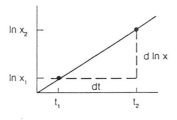

and the infection rate r is the slope of the regression line.

In this model of disease increase, the only variable that conditions the amount of disease in the future is the present amount of disease, and disease would increase exponentially forever. However, exponential increase occurs only early in the epidemic when susceptible tissue is abundant. As the epidemic progresses, the proportion of nondiseased tissue $(1 - x)$ limits the epidemic such that

$$\frac{dx}{dt} = rx\,(1 - x) \tag{5}$$

Here $1 - x$ is the amount of healthy susceptible tissue and is called "the correction factor" because as x increases, $1-x$ decreases and the epidemic slows. Conversely, when x is small, $1-x$ approaches 1 and the epidemic increases exponentially. Equation 5 is similar to the differential form of the logistic model, which describes the growth of many biological phenomena. The equation can be integrated to solve for x, and when the accumulated amount of disease is graphed as a function of time, the typical sigmoid (S-shaped) curve results.

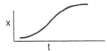

Sequential increments of disease are shown by the typical bell-shaped curve.

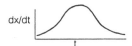

So disease increases slowly at first when inoculum is limited, more rapidly as inoculum increases, and finally slowly once again as the amount of healthy susceptible tissue decreases and becomes a limiting factor.

The infection rate r is calculated as before by transforming x to $\ln\,[x/(1-x)]$ such that

$$r = \frac{\ln \dfrac{x_2}{1-x_2} - \ln \dfrac{x_1}{1-x_1}}{t_2 - t_1} \tag{6}$$

Graphically

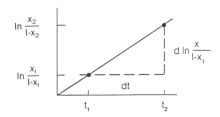

The infection rate r is the slope of the regression line and represents the average infection rate for the epidemic. As such, r is very useful to compare epidemics and to devise disease-control strategies.

Several points need to be made in regard to these preliminary disease models.

1. The infection rate includes all suscept, pathogen, and environmental factors that affect disease development. This is an advantage in that it allows a useful average index of the epidemic but a disadvantage in that there is no means of driving the model with these important variables.

2. The infection rate is assumed to be a constant but in actual epidemics r varies.

3. These models form a basis for understanding some disease epidemics, especially those for local lesion pathogens with abundant inoculum from secondary cycles and abundant susceptible tissue. These models are not necessarily appropriate for other type of diseases. Unfortunately, many forest tree diseases fall into the latter category.

* Adapted from Van der Plank, (1963). Courtesy of Academic Press.

when data are given as a percentage of trees infected, a multiple infection transformation (Gregory 1948) can be used to determine the number of lesions. Figure 8.4e gives data on the early stage of an epidemic of dwarf mistletoe (Hawksworth and Graham 1963). Again, data are given as a percentage of trees infected, but they are not for a single epidemic. The disease progress curve is a composite of data from many stands of various ages. This approach is common to forest pathology because data from long-term permanent plots are often unavailable.

8.2.3   The Spatial Epidemic

Plant disease is distributed both horizontally and vertically in space. Strictly speaking, only pathogens spread (Chapter 5), but if we understand the process we can refer to the occurrence of disease in new locations as spread of disease. Disease increase implies disease spread, if only within one plant. Spread is measured in distance and most often characterized as short-distance (local) or long-distance spread. Spread is also characterized (Dimond and Horsfall 1960) as one-, two-, or three-dimensional, that is, distributed horizontally along a line, horizontally within a plane, and horizontally as well as vertically, respectively. Examples are a bird carrying a mistletoe seed, mycelium "fanning out" at one level in the soil, and spores disseminated by horizontal and turbulent wind.

Disease commonly spreads horizontally in the following way. An initial lesion develops from a source of inoculum of distant origin (alloinfection). As new inoculum is produced by this initial lesion, nearby plant tissues become diseased (autoinfection) and a focus (a local concentration of lesions) develops. New disease occurrences are most frequent near these foci, but newly diseased plants can occur at relatively long distances from the focal source of inoculum. This distance is dependent on many

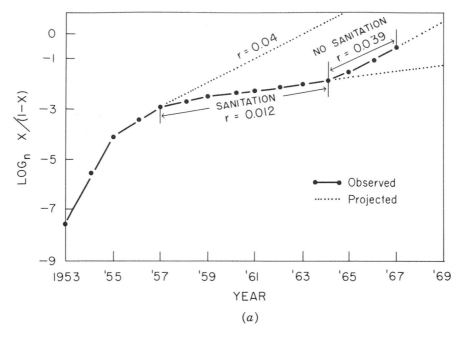

FIGURE 8.4 Disease increase curves for forest tree disease epidemics. (a) Rate of increase (r = per unit per month) of Dutch elm disease with and without sanitation. Redrawn from Berger (1977) and Miller et al. (1969). Reproduced, with permission, from the *Annual Review of Phytopathology,* Vol. 15, 1977, by Annual Reviews Inc. (b) Rate of increase (r = per unit per month) of Dutch elm disease with (Fredericton) and without (all others) sanitation. Redrawn from Berger et al. (1976). (c) Rate of increase (r = per unit per year, based on 15-m radius infection center) of oak wilt in Pennsylvania and West Virginia. Redrawn from Merrill (1967). (d) Rate of increase (r = per unit per year, based on percentage of trees infected) of fusiform rust on resistant R, intermediately resistant I, and susceptible S slash pine families. Redrawn from Griggs and Schmidt (1976). (e) Increase of dwarf mistletoe on ponderosa pine stands (age 5–28) within 9 m of infected residual stands. Redrawn from Hawksworth and Graham (1963). Reprinted from *Journal of Forestry,* Vol. 61, No. 8, published by the Society of American Foresters, 5400 Grosvenor Lane, Bethesda, MD. 20814-2198. Not for further reproduction.

factors, especially the means of dissemination. For example, windborne inoculum typically can travel long distances to produce distant infections.

Spread in the vicinity of foci or sources of abundant inoculum is characterized by disease gradient curves which show the amount of disease as a function of distance from the source (Gregory 1968). Disease gradients vary depending on the disease, especially the means of spread. Typically, for local-lesion-type diseases for which there is abundant wind-disseminated inoculum, susceptible tissue, and favorable environment (three-dimensional spread), disease decreases exponentially with increasing distance from the source of inoculum. A rule of thumb is that disease x_1 decreases inversely as the square of the distance d from the source x_0, that is, $x_1 = x_0/d^2$ (Fig. 8.5a). I illustrate this specific relationship using a $\log_{10}$ transformation for both amount of disease and distance (Fig. 8.5b), a linear regression (disease gradient) with a slope of -2 results. Slopes of transformed disease gradient curves are variable depending on several factors (Gregory 1968) but are useful to characterize epidemics.

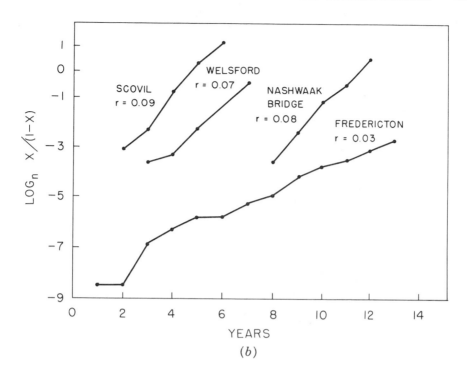

(*b*)

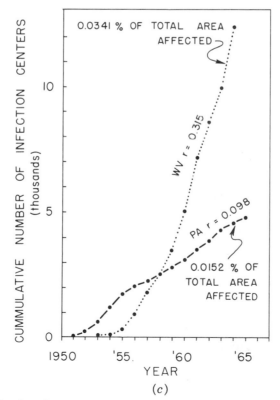

(*c*)

FIGURE 8.4   (*Continued*)

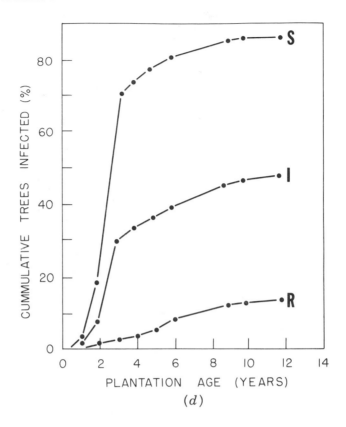

(d)

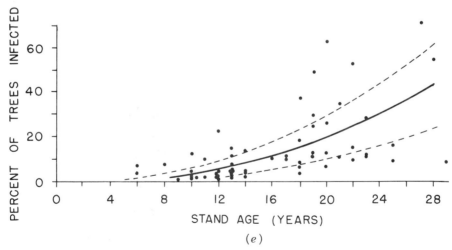

(e)

FIGURE 8.4   *(Continued)*

Disease gradient curves can aid control strategies based on spread of disease (e.g., quarantines) and isolation or eradication zones.

Untransformed disease gradient curves for several forest tree diseases appear in Figure 8.6. Figure 8.6a shows the disease gradient for the percentage of trees diseased and the percentage of infection sites attacked by the insect vector in Dutch elm dis-

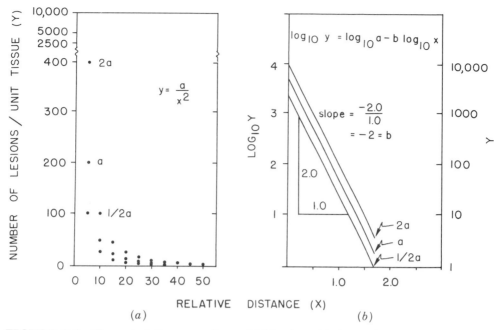

FIGURE 8.5 Theoretical disease gradients. (a) Number of lesions/unit tissue as a function of distance from the source of inoculum. (b) $\log_{10}$ of the number of lesions per unit tissue as a function of $\log_{10}$ of the distance from the source of inoculum.

ease. These data are from different studies, but suggest logically that the beetle is found at much farther distances from the source than is the disease. Also, the unproportionally high incidence ($> 90\%$) of disease near the source likely reflects root graft spread in addition to local spread associated with the vector. The general shape of the disease gradient curves for oak wilt in Figure 8.6b and white pine blister rust cankers in Figure 8.6c are similar to one another and to those for Dutch elm disease. However, the gradient for dwarf mistletoe (Fig. 8.6d) is not as steep initially. A comparison of spread in open and dense stands is also provided in the latter example. Gradients reach an asymtote, and "background contamination" from another source becomes a factor. This point is reached at approximately 107, 18, and 14 m for Dutch elm, oak wilt, and white pine blister rust, respectively, and represents the end of significant spread for each epidemic.

Small amounts of inoculum may be carried long distances by wind, water, insects, and humans (Chapter 5). The location of the resulting disease is scattered and is therefore less predictable. These occurrences are important as potential new foci from which disease may increase and spread. Some plant disease control measures are aimed primarily at preventing these new foci (e.g., quarantines).

The maximum distance of spread is often not the most important factor to consider in control programs because most disease, and therefore damage, occurs close to the source of inoculum. The individual tree that becomes diseased at a distant point may be of little consequence, especially if this tree does not serve as a new focus for new infections.

Definitive patterns of spread occur for many forest tree diseases. For example, oak wilt, Dutch elm disease, annosum root disease, and white pine blister rust have patterns of the sort just described. That is, after initial infection occurs in oak wilt and

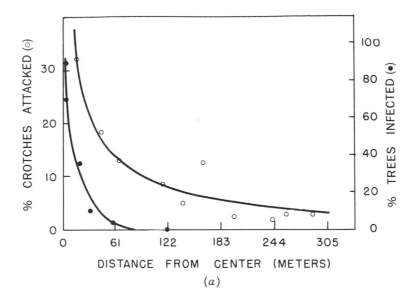

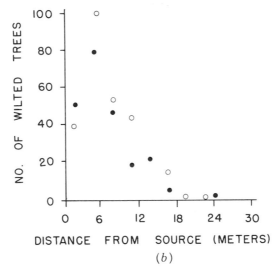

FIGURE 8.6   Disease gradients for forest tree disease epidemics. (a) Percent of trees infected with Dutch elm disease at distances from an isolated infected tree, and percentage of elm twig crotches fed on by *Scolytus multistriatus* at distances from a beetle breeding center. Redrawn from Zentmyer et al. (1944) and Wadley and Wolfenbarger (1944). (b) Number of oaks wilted at distances from an isolated infected oak. Redrawn from Boyce (1957) and Jones (1971). (c) Number of white pine blister rust cankers per million needles at distances from infected *Ribes*. Redrawn from Buchanan and Kimmey (1938). (d) Proportion of lodgepole pine in open and dense stands infected with dwarf mistletoe at distances from infected stands. Redrawn from Hawksworth (1958). Reprinted from *Journal of Forestry,* Vol. 56, No. 6, published by the Society of American Foresters, 5400 Grosvenor Lane, Bethesda Lane, Md. 20814-2198. Not for further reproduction.

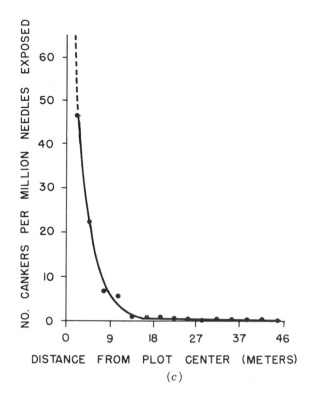

(c)

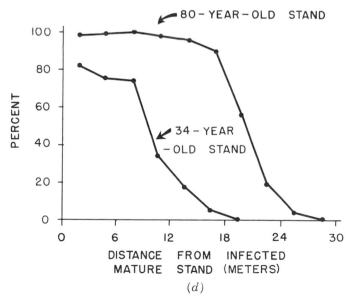

(d)

FIGURE 8.6   *(Continued)*

Dutch elm disease, adjacent trees become diseased due to root grafts and insect feeding. Subsequently, long distance spread occurs due to beetle feeding. A similar pattern develops with annosum root disease where local spread is primarily through root contact, and long distant spread is the result of wind-disseminated inoculum. White pine blister rust typically enters new (distant) areas via aeciospore infection on *Ribes* where disease increases with secondary cycles of urediniospores. Following pine infection by basidiospores, aeciospores from sporulating galls provide additional inoculum. These foci provide inoculum for additional long-distance spread of the disease. It is also possible that the long- versus short-distance spread of aeciospores and basidiospores could be reversed with white pine blister rust; such is the case in the fusiform rust pathosystem.

8.3 THE ROLE OF ENVIRONMENTAL FACTORS AND CULTURAL PRACTICES IN THE DEVELOPMENT OF THE EPIDEMIC

Environmental factors and cultural practices influence disease development and play important roles in the epidemic. Environmental factors may of themselves be responsible for plant disease (Chapter 3), but here these factors are discussed as variables that regulate the epidemic given the presence of a susceptible host and a biotic pathogen.

Environmental factors are classified as climatic, edaphic, or biotic, although the placement of specific factors is sometimes arbitrary. Climatic factors are those common to the atmosphere, for example, radiation (ultraviolet, visible light, and heat) pressure, wind, temperature, precipitation, water vapor, CO_2, air pollutants, and various airborne particles. Edaphic factors are those common to the soil, for example, texture, pH, nutrition, temperature, moisture, and oxygen. Biotic factors, excluding the pathogen and suscept, may occur in the air or soil environment or in and on the plant (e.g., antagonistic, competitive, beneficial, predisposing, wounding, vector organisms, and plant exudates).

Cultural practices, including site preparation, cultivation, fertilization, burning, thinning, species selection, resistant genotypes, and harvesting systems, can greatly affect disease development (Carlson and Main 1976). Perennial plants and forest trees in particular are vulnerable to silvicultural practices that alter delicately balanced natural ecosystems and destroy natural defense profiles (Schmidt 1978). This is not an argument against practicing forestry but rather an argument for good forestry practices.

As shown in Figure 8.7, numerous environmental factors and cultural practices can affect a multitude of phenomena in the pathosystem, and they may do so differentially. For example, optimal temperature for the suscept, pathogens, and disease may be quite different. Some common phenomena affected in the host include susceptibility and amount of plant tissue, defense reactions, and exudates; those in the pathogen include sporulation, dissemination, germination, and penetration; those in the disease interaction include colonization, incubation, latent, and infectious periods.

Usually, for any one disease, there is a limited number of critical interactions. Detailed knowledge of all interactions is normally not necessary to develop disease control strategies. It is the task of the epidemiologist to identify these critical interactions rather than to gain a comprehensive understanding of all possible interactions.

Environmental factors regulate epidemics by influencing the occurrence or the rate of processes. Spore germination is often critically affected by temperature and mois-

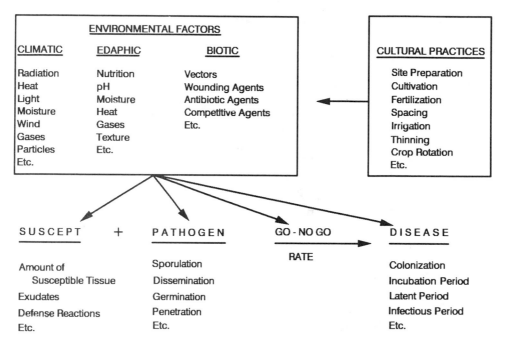

FIGURE 8.7 Environmental factors and cultural practices as they affect the suscept, pathogen and disease development.

ture. Temperatures at intermediate levels act as a rheostat and determine the rate of spore germination, while temperature beyond the maximum and minimum, or the presence or absence of moisture act as an on or off switch to allow or prevent germination.

Another important aspect of the epidemic is time. The disease square is immersed in time so that the timing of events is critical. It is not enough for inoculum to be abundant; it must be abundant when environmental factors are favorable and when susceptible tissue is available. Thus occurrence of disease is probablistic, involving a multitude of happenings and their probabilities of occurrence.

In Part Two under the topic of epidemiology in the Disease Profiles, important environmental interactions, including site and cultural factors, are discussed for each disease.

8.3.1   Levels of the Environment

Environmental factors, especially climate, affect forest tree diseases at several levels. Macroclimate, stand (crop), climate, and microclimate significantly affect temporal and spatial development of disease. The prefixes *macro* and *micro* are overall relative, but within the context of forest pathology they can be rather specifically defined.

Macroclimate is the climate of a region, determined by large-scale dynamic atmospheric processes, and measured by standard U.S. Weather Bureau Stations, and recently measured via weather satellites. Macroclimate is important to disease development directly because it affects dissemination of pathogens and indirectly because it conditions stand climate.

Stand climate is the climate associated with a stand of trees, determined by the interaction of stand vegetation with macroclimate, and typically measured by instruments such as hygrothermographs (in the case of temperature and moisture) in standard weather shelters. The forest canopy significantly alters and ameliorates both solar and terrestrial radiation, wind, and precipitation (Reifsnyder and Lull 1965), all of which greatly affect temperature and moisture factors immediately above, within, and below the crown (Geiger 1965). Stand climate is important with respect to plant growth, dissemination of inoculum, and, indirectly, how it conditions microclimate.

Most important to plant pathogens and disease development is microclimate. *Microclimate* is the climate in the biotope (Dansereau 1957) of the pathogen, that is, at the level where the interaction between the pathogen and the suscept occur (Schmidt and Wood 1969). For example, the microclimates common to foliar pathogens include the temperature and moisture of the foliosphere and folioplane, while those of a root pathogen include the rhizosphere and rhizoplane. A pathogen may be exposed to several unique microclimates as, for example, *Heterobasidion annosum* (Aphyllophorales), which occupies stump top, root, wound, and needle habitats. Microclimate is the product of the interaction of vegetation with surrounding climate and is measured with miniature probes (e.g., thermisters in the case of temperature) placed in the biotope (e.g., on or in a leaf). Microclimate is all important to plant disease because it conditions phenomena at the infection court and at fruitification. Spore production, release, germination, and penetration are regulated by microclimate. The biotope is the arena where the game is played, and the occurrence and rates of processes are controlled by the microenvironment.

In as much as macroclimate and stand climate influence and are estimates of microclimate, they are often used as substitutes. The validity of this practice depends on how accurately microclimate reflects or can be predicted from macroclimate and/or stand climate.

8.4   GEOPHYTOPATHOLOGY

Geophytopathology (Weltzien 1972) refers to the geographic distribution of plant disease incidence and severity. Clearly, diseases occur only where the appropriate suscepts and pathogens coexist. Regional distributions of diseases are the result. Just as important, there exist pathogen and disease ecotypes, that is, pathogens and disease that require unique environmental factors. So, even in areas where the pathogen and suscept coexist, disease may not develop in the absence of the critical environmental factors. For example, *Cronartium ribicola* (Uredinales), which causes white pine blister rust, is a cool-wet weather ecotype, and blister rust incidence and severity are greatest in white pine areas where summers are relatively cool and wet. Accordingly, rust incidence and severity were highest in the Lake states: in the northern latitudes, at high elevations, in frost pockets, and in stand openings having a negative radiation balance (Van Arsdel 1961). These areas provided prolonged cool-wet periods for sporulation on *Ribes* and infection of pine. In other areas the suscept escapes disease because climate is limiting.

There are numerous examples of these climate-induced disease distributions as well as those conditioned by vectors, edaphic factors, competitive organisms, and alternate hosts. The latter is demonstrated in Figures 8.8a–c, which show the geographic pat-

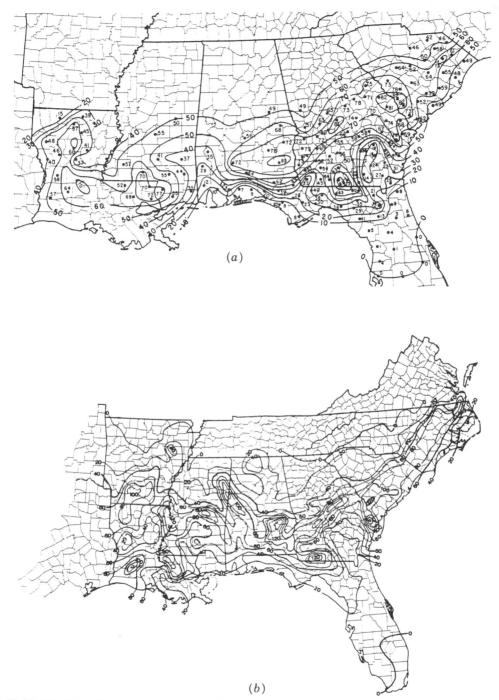

(a)

(b)

FIGURE 8.8 Geographic patterns of fusiform rust incidence and oak abundance in the southeast. (a) Percent of trees infected in 8- to 12-year-old slash pine plantations. (b) Volume (cubic feet per acre) of water oak on commercial forest lands. From Squillace et al. (1978). (c) Regression of percent of infected slash pines and volume of water oaks per acre (Squillace et al. 1978).

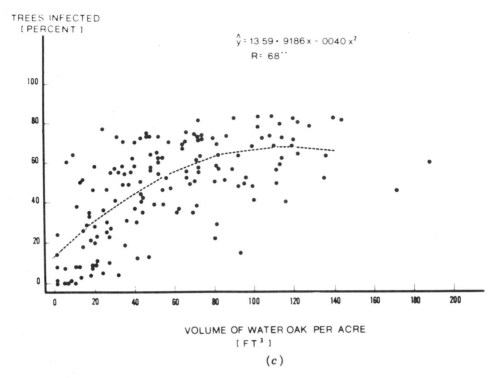

FIGURE 8.8   *(Continued)*

tern of fusiform rust incidence on slash pine and the abundance of water oak (an alternate host). There is a similarity between the two distribution patterns and, while factors other than oak abundance influence rust incidence, oak distribution accounts for a major portion of the variability in the distribution of rust incidence (Fig. 8.8c).

8.5   QUANTITATIVE ANALYSIS OF EPIDEMICS

Epidemics are complex biological systems that can be quantified. As such, mathematical models are extremely useful to understand the structure, function, and control of epidemics. Mathematical models organize and communicate knowledge, discipline research, and aid in decision making (Kranz 1974). Of the many kinds of mathematical models available in plant pathology, only three will be present here, namely, differential equation, regression equation, and simulation models.

Differential equation models are useful to characterize rate functions. Their utility for analysis of plant disease epidemics was demonstrated (Van der Plank 1963) relative to the calculation of infection rates (Box 8.1). However useful, differential equation models are limited because they do not readily accommodate the large number of independent variables contained in plant disease epidemics.

Multiple regression models overcome partially this deficiency and allow numerous independent variables. These models are essentially data fitting routines, but are useful to identify and rank the important factors that influence epidemics and to construct predictive equations of epidemics.

Multiple regression equations are of the form $y = b_0 + b_1 x_1 + b_2 x_2 + \cdots + b_n x_n$, where y is the dependent variable (e.g., amount of disease), $x_1, x_2, \ldots, x_n$ are the independent variables (e.g., amount of inoculum) or climatic variables, and b_0-b_n are constants—b_0 is the initial condition or the y intercept and b_1-b_n are the partial regression coefficients.

Regression equations, both simple and multiple, are common to pathology (Butt and Royle 1974). For example, in forest pathology the following multiple regression equations are used to describe the geographic distribution of fusiform rust on slash pine in North Florida (Schmidt et al. 1974), the role of site factors in this distribution (Hollis and Schmidt 1977), and the association of oak and temperature with southwide distribution of fusiform rust on slash pine (Squillace et al. 1978), respectively.

1. % Rust = $-2065.42 + 18.42$ (longitude) + 18.77 (latitude)
2. % Rust = $-40.41 + 0.48$ (oak leaf surface area) + 0.19 (amount of susceptible pine shoot tissue)
3. % Rust = $-7056.5 + 0.50$ (volume of laurel oak) $- 0.002$ (volume of water oak)^2 + 0.06 (volume of laurel oak) + 0.00002 (volume of laurel oak)^2 + 206.99 (average April–May temperature) $- 1.51$ (average April-May temperature)^2.

The disadvantages of regression equations are that they accommodate with difficulty or not at the intercorrelation of independent variables so common in holistic natural environments, they can accommodate only a limited number of variables, and they are posteriori models derived from fitting existing data, and, while they are useful to analyze and predict epidemics, their reliability depends on the fundamental nature (cause and effect relationships) of the variables. Regression models can be easily misinterpreted or in error when used in a new situation if the dependent variables are not truly of a causal nature.

With the development of computers and system analysis languages, it became possible to develop simulation analyses of plant disease (Waggoner 1974, Zadoks 1971). Simulation models provide for a logical, sequential, real-time accounting of the amount of disease as affected by the many influencing factors. Figure 8.9, which coincides with Figure 8.1, illustrates a partial and hypothetical simulation of an epidemic. This simulator is an accounting of the number of infection sites of the host that are occupied by the pathogen (Zadoks 1971). Infection sites are accumulated over time; they are the output variables, and their numbers can be obtained at any time during the epidemic. Infection sites (perhaps stomata) move from vacant (healthy) to occupied (diseased); the latter are characterized as latent (not yet sporulating), sporulating, and removed (no longer sporulating nor infectious). Numbers of infection sites are regulated by rate functions (occupation, sporulation, and removal), which, in turn, are conditioned by the effects of environmental factors and cultural and control practices on host and pathogen phenomena. The initial number of infection sites and initial amount of inoculum are original input variables (initial conditions). Following initialization, the number of spores able to germinate and penetrate, as conditioned by the duration of the dew period, is entered when the functional relationship among these variables is known. Similarly, the effect of a protective fun-

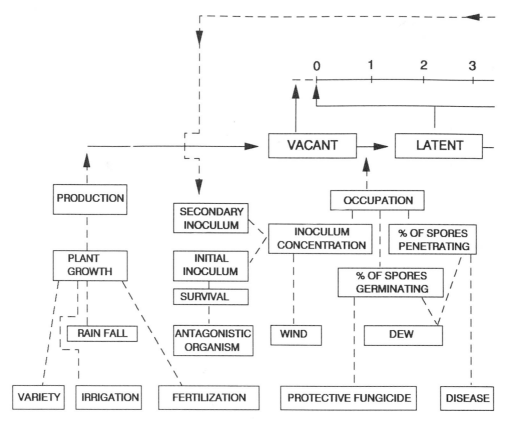

FIGURE 8.9 A hypothetical diagram of a simulation model of an epidemic in which secondary cyles occur. Examples of action and reaction pathways for cultural practices, disease control, environmental factors, and pathogen and host phenomena are shown relative to rates and numbers of infection sites (Adapted from Zadocs 1971).

gicide or of disease resistance on disease development is entered into the simulator at the appropriate place. The production of new infection sites (plant growth) is conditioned by cultural practice (irrigation, fertilization, etc.), and secondary inoculum are input variables generated by the simulator as the epidemic progresses.

The great advantage of a simulator is its flexibility to accept varied and large numbers of inputs. This strength becomes a disadvantage in that quantitative data on the functional relationship among variables is often not available. Building a simulator is time-consuming and expensive if data are not available. In practice, simulators function as posteriori models and thus become late warning systems if used for forecasts. Even so, simulators are a splendid means of organizing data and discerning gaps in available data so that simulations can become an important tool for directing research.

Simulation models for *Armillaria* and annosum root diseases and for dwarf mistletoe spread and intensification (Strand and Roth 1976) are available as submodels of the Forest Vegetation Simulator (FVS, formerly known as Prognosis), which is used by the USDA Forest Service. Submodels linking diseases and important factors can also be incorporated (e.g., root decay and bark beetles).

A diagram of the FVS submodel for annosum root disease is in Figure 8.10. The

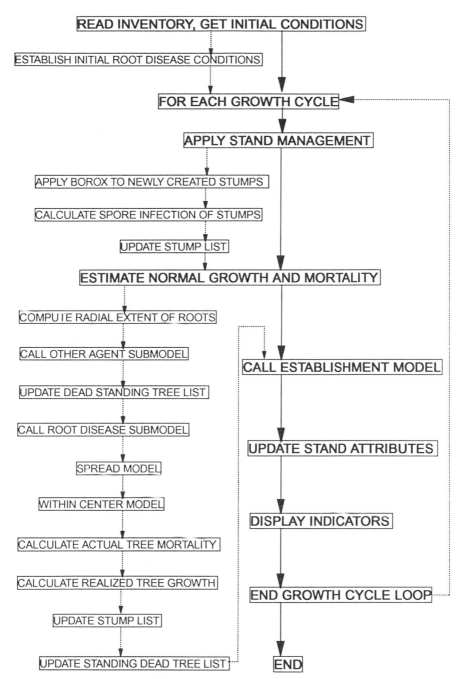

FIGURE 8.10   A diagram of the FVS annosum root disease submodel. Solid arrows refer to FVS subroutines; the dashed arrows describe program flow when the root disease model is invoked. Adapted from McNamee et al. (1991).

annosum model simulates the development of inoculum in the form of infected stumps, the most important source of inoculum. New infections resulting from spores and management actions such as stump removal or borax application affect the amount of inoculum. The root disease model then simulates the extent of tree roots and the inoculation of uninfected trees. As the disease progresses in these trees, their growth is reduced, their susceptibility to other agents especially bark beetles increases, and their probability of dying increases. After calculating these variables, the stump list is updated to reflect increased inoculum from dead trees. The FVS then computes stand growth and structure information from the disease-affected and disease-free trees.

This example is an extreme simplification of a very complex model, but several key points should be made. First, the annosum model operates at a much larger scale than a model of the status of infection sites such as that illustrated in Figure 8.9. This scale is probably appropriate for epidemics such as root diseases where disease cycles occur over many years in a crop that may not mature for more than 100 years. Secondly, forest managers are not concerned with the detailed quantitative aspects of the epidemic, but their need is to understand the effects of root disease on stand growth and structure. Simulation models can meet this need. Yet, developing a simulation model requires understanding the quantitative aspects of the epidemic. The rate of stump colonization, the longevity of inoculum (infectious period), and the probability of root infection at various distances from an infected stump (disease gradient) are just a few of the many characteristics of the epidemic used in simulation. Even if simulation models are not initially directly useful for forest managers, they provide researchers with insight into the epidemic. Researchers often find models useful in testing hypotheses. The exercise of building a simulation model identifies aspects of the epidemic that are not well characterized and serves to focus research.

8.6 DISEASE FORECASTING AND HAZARD EVALUATION

Forecasting the occurrence and amount of plant disease is an applied aspect of epidemiology that has great utility for disease control. If the regulatory roles of environmental factors are understood, accurate and meaningful disease forecasts are possible. Most plant disease forecasts are based on weather, but amount of inoculum, presence of vectors, infection rates, time of initial disease, plant growth, and other factors are most useful (Miller 1959, Squillace et al. 1978). Such forecasts may be emperical or experimental, early or late-warning, and positive or negative. Initially, forecasts were based on experience and emperical information; later forecasts were based on experimentally determined cause-and-effect relationships. Late-warning forecasts use measurements of variables as they occur (their effect on disease has occurred), and symptoms and inoculum appear following one incubation and latent period, respectively. Control, if necessary, is primarily directed at destroying the new inoculum and protecting new susceptible tissue. Early-warning forecasts use 3−, 5−, and 30−day synoptic weather forecasts (Bourke 1970) to predict infection before it occurs. Such forecasts provide ample time for control but are only as accurate as the weather predictions. Positive disease forecasts predict that a significant amount of disease will occur and are useful to ensure that control measures are initiated. Negative forecasts predict that little or no disease will occur and are useful to prevent unnecessary controls. Negative forecasts have become more important as the concern for environmental pollution by pesticides increases. Recent innovations in the measurement and prediction of climate (both macro- and microclimate) parameters; data acquisition, storage, and processing; computers; and system analysis techniques have significantly aided disease forecasting (Krause and Massie 1975).

It is not practical to forecast all diseases, and the following criteria are necessary for a useful forecast:

1. an economically important, widely grown crop,
2. a destructive disease,
3. a disease of sporadic occurrence (diseases that are always abundant, rare or punctual require no prediction), and
4. biologically effective, economically sound, and environmentally safe control measures.

Of course, it is a prerequisite that the data on the independent variables be available and their role in the epidemic be understood.

Unfortunately, most forest tree diseases do not meet these requirements, completely or in part. Perhaps criterion 4 is most often not satisfied. Nurseries and seed orchards offer the most promise to use disease forecasts (Davis and Snow 1968, Foster and Kruger 1961), but even here common nursery diseases such as damping-off or fusiform rust occur so regularly that control measures are routinely applied without regard for forecasting.

However, another type of disease forecast, namely, site hazard evaluation, is especially useful in forests. Forest sites (or practices) can be classified (usually low, intermediate, or high disease hazard) relative to the probability of occurrence; using known relationships among climatic, edaphic, biotic, and site factors; silvicultural practices; and disease incidence and severity. Once disease hazard evaluations are available, control strategies are forthcoming.

The nature and utility of a disease-hazard evaluation is evident for white pine blister rust in the Lake states. Here, four blister-rust hazard zones (Fig. 8.11) are based on favorable or unfavorable climate for disease development, as previously described. General recommendations are for little, minor, modified, and augmented control on low, intermediate, and high disease-hazard zones. Specific control recommendations appear in Part Two (Chapter 15) with white pine blister rust.

Not all hazard evaluations are based on climate: edaphic and biotic factors and cultural practices also provide a basis of delineating disease-hazard zones. The details of other useful disease-hazard evaluations are provided in Part Two for example, annosum root disease (Froelich et al. 1977, Morris and Frazier 1966, Chapter 13), littleleaf (Campbell et al. 1953, Chapter 11), Scleroderris canker (Dorworth 1972, Chapter 16), fusiform rust (Squillace et al. 1978, Chapter 15), and dwarf mistletoe (Hawksworth 1961, Chapter 19).

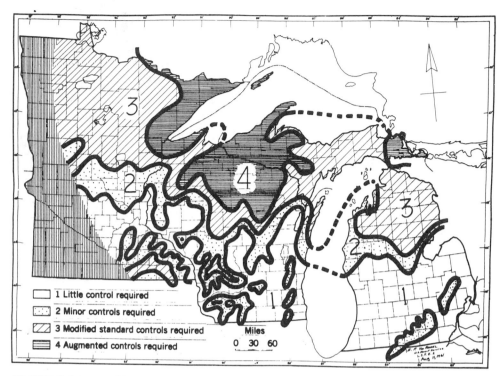

FIGURE 8.11 White pine blister rust hazard zones and control strategies in the Lake states. From Van Arsdel (1961).

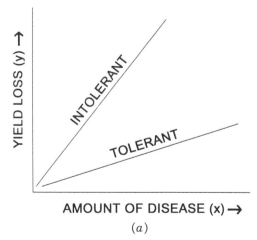

(a)

FIGURE 8.12 Yield loss due to the epidemic. (a) Hypothetical relation of amount of disease and yield loss showing tolerant and intolerant plants. (b) Relation of cull percentage to tree diameter for tree populations with different percentages of basal wounds; S = scarlet oak, Y = yellow-poplar. Redrawn from Hepting and Hedgcock (1937). (c) Relation of percentage mortality at age 10 to percentage stem cankers at age 5 for fusiform rust of slash and loblolly pine. Note that loblolly pine is more tolerant than slash pine (Wells and Dinus 1978). Reprinted from the *Journal of Forestry,* Vol. 76, No.1. Published by the Society of American Foresters, 5400 Grosvenor Lane, Bethesda, MD 20814-2198. Not for further publication.

8.7   YIELD LOSSES DUE TO THE EPIDEMIC

In forests, direct economic yield losses occur as a result of mortality, growth reduction, unmerchantability (cull), or product grade reduction. Indirect losses occur as delayed regeneration, reduced stocking, changes in species composition, deterioration of site quality, reduced management opportunities, and pest control costs. Yield loss is more important to the forest manager than the amount of disease. Diseases that cause limited yield loss are considered of little importance even though the incidence and severity may be high (e.g., some hardwood foliage diseases). Yield loss (aside from grade reduction) is commonly measured as board feet, cubic feet, or metric volume and is a function of the amount of disease as shown in Figure 8.12a. The growth of the host and the mode of pathogenesis greatly influence the nature of the

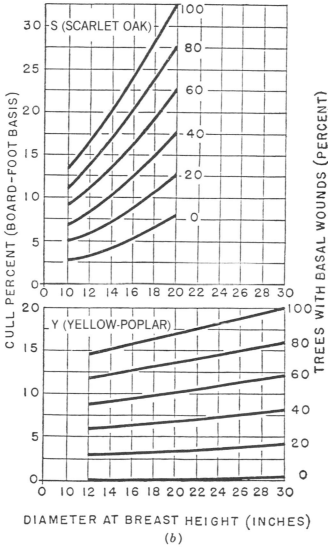

FIGURE 8.12   *(Continued)*

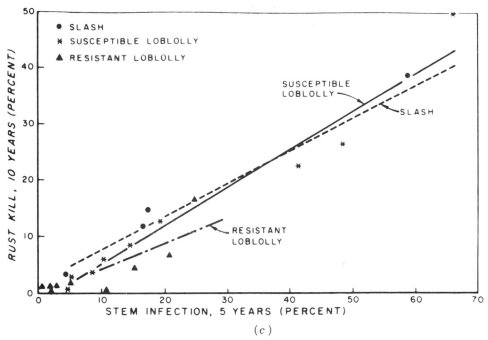

FIGURE 8.12   *(Continued)*

amount of disease—yield loss relationship. Also, site quality, the growth stage, and the age of the plant when infected influence yield loss.

The importance of determining the relationship between the amount of disease and the amount of yield loss lies in its utility for determining the amount of yield loss in volume estimates for timber sales; determining control strategies (i.e., is control necessary and what is the benefit–cost ratio?); and aiding production, utilization, and marketing decisions.

In order to develop quantitative models, information on yield loss and amount of disease (usually severity, but sometimes incidence data are appropriate) are obtained from various conditions of site, age, species, and so on. In annual crops (James 1974), both critical- and multiple-point models are used depending on the nature of yield accumulation. Critical-point models give yield loss as a function of the amount of disease at one growth stage of the plant and are useful for diseases of plants in which yield accumulation occurs in a short period. Multiple-point models characterize yield loss as a function of the amount of disease at several stages of plant growth and acknowledge that the rate of disease development and the nature of yield accumulation affect the amount of yield loss. For example, defoliation of hardwoods early in the season causes more growth loss than the same amount of defoliation later in the growing season. Critical-point models are most common in forest pathology, but multiple-point models would appear more appropriate because of the long period of yield accumulation. However, in some instances where yield (seed) is more rapidly accumulated (e.g., cone rust), critical-point models would seem appropriate.

Cull studies of losses due to wood-decay fungi were among the earliest of yield-loss models in plant pathology. Those studies quantified volume loss associated with

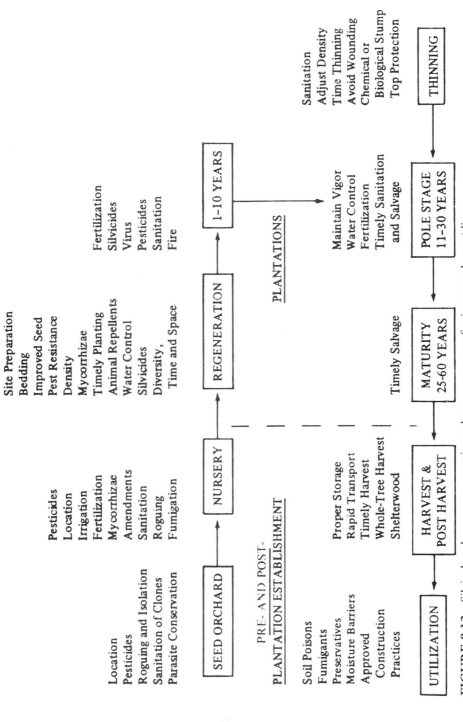

FIGURE 8.13 Silvicultural management units and some prospects for integrated pest (diseases and insects) management in intensively managed slash pines. From Schmidt and Wilkinson (1981).

TABLE 8.1 Hypothetical Life Table of Some Slash Pine Diseases

Age Interval When Disease Is Prevalent and Damaging	Disease	Pathogen(s)
Nursery		
0–3 months	Damping-off	*Fusarium oxysporum* *Pythium ultimum* *Rhizoctonia solani* *Macrophomina phaseolina* *Phytophthora* spp.
3 months–3 years	Seedling root disease	*Macrophomina phaseolina* and other damping-off fungi
1–5years	Needle blights	*Coleosporium* spp. *Hypoderma* spp. *Mycosphaerella dearnessii*
1–10 years	Fusiform rust	*Cronartium quercuum* f. sp. *fusiforme*
10–20 years	Pitch canker	*Fusarium subglutinans*
Thinning—rotation	Annosum root disease	*Heterobasidion annosum*
Seed Production Orchard		
15 years—rotation	Cone rust	*Cronartium strobilinum*
20 years—rotation	Root decay	*Phaeolus schweinitzii* *Armillaria (Clitocybe) ta-bescens*
40 years—rotation	Heartwood decay	*Phellinus pini*
Harvest—utilization	Blue stain of logs and lumber	*Ophiostoma* and *Ceratocystis* spp.
Post harvest	Decay of wood in use	*Meruliporia incrassata* *Gloeophyllum sepiarium* and many others

Rotation age is 20–30 years for pulpwood and 60–80 years for lumber.
Thinning age is 10–15 years.
Source: From Schmidt (1978).

varying amount of sporophores, swollen knots, wounds, and so on and acknowledged effects of site, species, and age or size of tree. An example of these data is shown in Figure 8.12b. Also shown in Figure 8.12c is a critical-point yield loss model of fusiform rust, which predicts future mortality based on percentages of stems infected at age 5 years.

Some species or varieties of trees are tolerant of disease, that is, they suffer less yield loss than other species with the same amount of disease. A general example of tolerance is seen in Figure 8.12a and specifically in Figure 8.12c where loblolly pine suffers less mortality due to stem-breakage than does slash pine when both have the same disease incidence (percent of stems with rust galls). Figure 8.12b, which gives board feet yield losses relative to age and percent of trees with basal wounds, also shows that scarlet oak is less tolerant than yellow-poplar to stem decay.

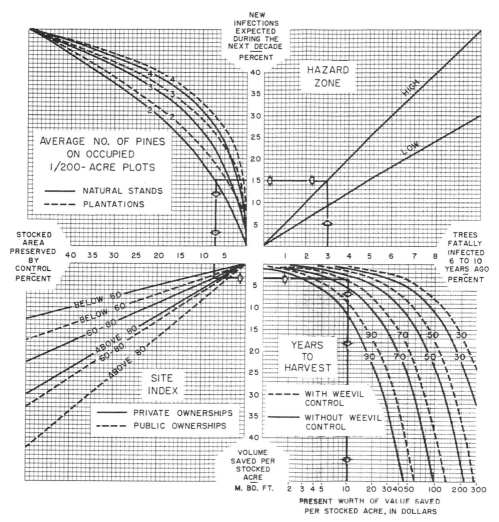

FIGURE 8.14 The financial value (present worth of value saved in dollars per stocked acre of eastern white pine) of blister rust control (*Ribes* eradication) relative to rust hazard zone, stocking levels of natural and planted stands, site index, ownership, years to harvest, and white pine weevil control. From Marty (1966).

Another useful approach for characterizing yield loss is through the development of stand life tables (Waters 1969). These tables provide an accounting of the cause and amount of the loss relative to stand development. Such life tables can be constructed for one disease, all important diseases, all pests, or all impacts. A composite of life tables that details losses can be constructed for management units (Fig. 8.13) or larger areas, and strategies for integrated pest management and multiple use can be optimized. Baxter in his text (Baxter 1943) and elsewhere (Baxter 1967) provides excellent examples of the life table concept. Table 8.1 is a partial, hypothetical life table of slash pine diseases, but no quanitative yield-loss estimates are included. Disease management options may also be developed for silvicultural management units (Fig. 8.14), since specific diseases are associated with specific management units.

Eventually, yield-loss studies must lead to economic considerations. In other words, what is the value of the crop and what is the financial benefit of control? Economic analyses include the uncertainties of the market place and often require inputs from an economist familiar with appropriate models (Carlson and Main 1976).

Figure 8.14 shows the financial value as present worth of value saved in dollars per stocked acre of eastern white pine for blister-rust control (via *Ribes* eradication) relative to rust hazard zones, stocking levels of natural and planted stands, site index, ownership, years to harvest, and white pine weevil control. Other economic analyses of control are discussed later for dwarf mistletoe (Flora 1966), Dutch elm disease (Cannon and Worley 1976), and white pine blister rust (King et al. 1960). With such models forest managers can decide if disease control is economically feasible. Of course, not all control decisions are based solely on financial considerations, but such information should be available to aid control decisions.

LITERATURE CITED

Baxter, D. V. 1943. *Pathology in Forest Practice*. John Wiley & Sons, New York. 618 pp.

Baxter, D. V. 1967. *Disease in Forest Plantations: Thief of Time*. Cranbrook Inst. Sci. Bull. 51. 251 pp.

Berger, R. D. 1977. Application of epidemiological principles to achieve plant disease control. *Annu. Rev. Phytopath.* 15:165–183.

Berger, R. D., G. A. Van Sickle, and T. E. Sterner. 1976. Sanitation: A practical protection against Dutch elm disease in Fredericton, New Brunswick. *Pl. Dis. Reptr.* 60:336–338.

Bourke, P. M. A. 1970. Use of weather information in the prediction of plant disease epiphytotics. *Annu. Rev. Phytopath.* 8:345–370.

Boyce, J. S., Jr. 1957. Oak wilt spread and damage in the southern Appalachians. *J. For.* 55:499–505.

Buchanan, T. S., and J. W. Kimmey. 1938. Initial tests of the distance of spread to and intensity of infection on *Pinus monticola* by *Cronartium ribicola* from *Ribes lacustre* and *R. viscosissimum*. *J. Agric. Res.* 56:9–30.

Butt, D. J., and D. J. Royle. 1974. Multiple regression analysis in the epidemiology of plant disease, pp. 78–114. In J. Kranz (ed.), *Epidemics of Plant Diseases: Mathematical Analysis and Modeling*. Springer-Verlag, New York. 170 pp.

Campbell, W. A., O. L. Copeland, Jr., and G. H. Hepting. 1953. *Managing Shortleaf Pine in Littleleaf Disease Areas*. USDA For. Serv., SE For. Expt. Sta. Pap. 25. 12 pp.

Cannon, W. N., Jr., and D. P. Worley. 1976. *Dutch Elm Disease Control: Performance and Costs*. USDA For. Serv., Res. Pap. NE-345. 7 pp.

Carlson, G. A., and C. E. Main. 1976. Economics of disease-loss management. *Annu. Rev. Phytopath.* 14:381–403.

Dansereau, P. M. 1957. *Biogeography: An Oecological Perspective*. Ronald Press Co., New York. 394 pp.

Davis, R. T., and G. A. Snow. 1968. Weather systems related to fusiform rust infection. *Pl. Dis.Reptr.* 52:419–422.

Dimond, A. E., and J. G. Horsfall. 1960. Inoculum and the diseased population, pp. 1–22. In

J. G. Horsfall and A.E. Dimond (eds.), *Plant Pathology: An Advanced Treatise,* Vol. 3. Academic Press, New York. 675 pp.

Dinus, R. J. 1974. Knowledge about natural ecosystems as a guide to disease control in managed forests, pp. 184–190. In Symposium on impacts of disease epidemics on natural plant ecosystems. *Proc. Am. Phytopath. Soc.* 1:170–199.

Dorworth, C. E. 1972. Epidemiology of *Scleroderris lagerbergii* in central Ontario. *Can. J. Bot.* 50:751–765.

Flora, D. F. 1966. *Economic Guides for Ponderosa Pine Dwarf Mistletoe in Young Stands of the Pacific Northwest.* USDA For. Serv., Res. Rep. PNW 29. 16 pp.

Foster, A. A., and D. W. Kruger. 1961. *Protection of Pine Seed Orchards and Nurseries from Fusiform Rust by Timing Ferbam Sprays to Coincide with Infection Periods.* Ga. For. Res. Pap. 1. 4 pp.

Froelich, R. C., E. G. Kuhlman, C. S. Hodges, M. J. Weiss, and J. D. Nichols. 1977. *Fomes Annosus Root Rot in the South. Guidelines for Prevention.* USDA For. Serv. 17 pp.

Geiger, R. 1965. *The Climate Near the Ground.* Harvard University Press, Cambridge, Mass. 611 pp.

Gregory, P. H. 1948. The multiple-infection transformation. *Ann. Appl. Biol.* 35:412–417.

Gregory, P. H. 1968. Interpreting plant disease gradients. *Annu. Rev. Phytopath.* 6:189–212.

Griggs, M. M., and R. A. Schmidt. 1977. Increase and spread of fusiform rust, pp. 25–31. In R. J. Dinus and R. A. Schmidt (eds.), *Management of Fusiform Rust in Southern Pines.* Southern Forest Disease and Insect Research Council meeting Dec. 7–8, 1976 at Gainesville, Florida (Proceedings). 163 pp.

Hawksworth, F. G. 1958. Rate of spread and intensification of dwarf mistletoe in young lodgepole pine stands. *J. For.* 56:404–407.

Hawksworth, F. G. 1961. *Dwarf Mistletoe of Ponderosa Pine in the Southwest.* USDA For. Serv., Tech. Bull. 1246. 112 pp.

Hawksworth, F. G., and D. P. Graham. 1963. Spread and intensification of dwarf mistletoe in lodgepole pine reproduction. *J. For.* 61:587–591.

Hepting, G. H., and G. G. Hedgecock. 1937. *Decay in Merchantable Oak, Yellow-Poplar, and Basswood in the Applachian Region.* USDA, Tech Bull. 570. 29 pp.

Hollis, C. A., and R. A. Schmidt. 1977. Site factors related to fusiform rust incidence in north Florida slash pine plantations. *For. Sci.* 23:69–77.

James, W. C. 1974. Assessment of plant disease losses. *Annu. Rev. Phytopath.* 12:27–48.

Jones, T. W. 1971. *An Appraisal of Oak Wilt Control Programs in Pennsylvania and West Virginia.* USDA For. Serv., Res. Pap. NE-204. 15 pp.

King, D. B., C. H. Stoltenberg, and R. J. Marty. 1960. *The Economics of White Pine Blister Rust Control in the Lake States.* USDA For. Serv. 39 pp.

Kranz, J. 1974. The role and scope of mathematical analysis and modeling in epidemiology, pp. 8–54. In J. Kranz (ed.), *Ecological Studies 13. Epidemics of Plant Disease: Mathematical Analysis and Modeling.* Springer-Verlag, New York. 170 pp.

Krause, R. A., and L. B. Massie. 1975. Predictive systems: Modern approaches to disease control. *Annu. Rev. Phytopath.* 13:31–47.

Marty, R. 1966. *Economic Guides for Blister Rust Control in the East.* USDA For. Serv., Res. Pap. NE-45. 14 pp.

McNamee, P. J., W. A. Kurz, C. J. Daniel, D. C. E. Robinson, and M. G. Deering. 1991. Description of the annosus root disease model. Prepared by ESSA Ltd., Vancouver, B. C. for USDA Forest Service, San Francisco, CA. 70 pp.

Merrill, W. 1967. The oak wilt epidemic in Pennsylvania and West Virginia: An analysis. *Phytopathology* 57:1206–1210.

Miller, N. C., S. B. Silverborg, and R. J. Campana. 1969. Dutch elm disease: Relation of spread and intensification to control by sanitation in Syracuse, New York. *Pl. Dis. Reptr.* 53:551–555.

Miller, P. R. 1959. Plant disease forecasting, pp. 557–565. In *Plant Pathology. Problems and Progress 1908–1958.* University Wisconsin Press. Madison. 588 pp.

Morris, C. L., and D. N. Frazier. 1966. Development of a hazard rating for *Fomes annosus* in Virginia. *Pl. Dis. Reptr.* 50:510–511.

Reifsnyder, W. E., and N. W. Lull. 1965. *Radiant Energy in Relation to Forests.* USDA For. Serv., Tech. Bull. 1344. 111 pp.

Schmidt, R. A. 1978. Diseases in forest ecosystems; The importance of functional diversity, pp. 287–315. In J. G. Horsfall and E. B. Cowling (eds.), *Plant Disease: an Advanced Treatise,* Vol. 2. Academic Press, New York. 436 pp.

Schmidt, R. A., R. E. Goddard, and C. A. Hollis. 1974. *Incidence and Distribution of Fusiform Rust in Slash Pine Plantations in Florida and Georgia.* Florida Agric. Expt. Sta. Tech. Bull. No. 763. 21 pp.

Schmidt, R. A., and R. C. Wilkinson. 1981. Prospects for integrated pest management in slash pine ecosystems, pp. 610–615. In T. Kommedahl (ed.) *Ninth International Congress of Plant Protection.* Aug. 1979, Washington, DC., Symposium Proc. Vol. II. Integrat. Plant Protect. Agric. Crops and Forest Trees.

Schmidt, R. A., and F. A. Wood. 1969. Temperature and relative humidity regimes in the pine stump habitat of *Fomes annosus. Can. J. Bot.* 47:141–154.

Snell, W. H., and E. A. Dick. 1971. *A Glossary of Mycology.* Harvard University Press, Cambridge, Mass. 181 pp.

Squillace, A. E., R. J. Dinus, C. A. Hollis, and R. A. Schmidt. 1978. *Relation of Oak Abundance, Seed Source, and Temperature to Geographic Patterns of Fusiform Rust Incidence.* USDA For. Serv., Res. Pap. SE-186. 20 pp.

Strand, M. A., and L. F. Roth. 1976. Simulation model for spread and intensification of western dwarf mistletoe in thinned stands of ponderosa pine saplings. *Phytopathology* 66:888–895.

Van Arsdel, E. P. 1961. *Growing White Pine in the Lake States to Avoid Blister Rust.* USDA For. Serv., Lake States For. Expt. Sta. Pap. No. 92. 11 pp.

Van Arsdel, E. P. 1964. *Growing White Pine to Avoid Blister Rust—New Information for 1964.* USDA For. Serv. Res. Note LS-42. 4 pp.

Van der Plank, J. E. 1963. *Plant Diseases: Epidemics and Control.* Academic Press, New York. 349 pp.

Van Sickle, G. A., and T. E. Sterner. 1976. Sanitation: A practical protection against Dutch elm disease in Fredericton, New Brunswick. *Pl. Dis. Reptr.* 60:336–338.

Wadley, F. M., and D. O. Wolfenbarger. 1944. Regression of insect density on distance from center of dispersion as shown by a study of the smaller European elm bark beetle. *J. Agric. Res.* 69:299–308.

Waggoner, P. E. 1974. Simulation of epidemics, pp. 137–160. In J. Kranz (ed.), *Epidemics of Plant Diseases: Mathematical Analysis and Modeling.* Springer-Verlag, New York. 170 pp.

Waters, W. E. 1969. The life table approach to analysis of insect impact. *J. For.* 67:300–304.

Wells, O. O., and R. J. Dinus. 1978. Early infection as a predictor of mortality associated with fusiform rust of southern pines. *J. For.* 76:8–12.

Weltzien, H. C. 1972. Geophytopathology. *Annu. Rev. Phytopath.* 10:277–298.

Whetzel, H. H. 1926. The terminology of plant pathology. *Proc. Intern. Congr. Pl. Sci.,* Ithaca, 2:1204–1215.

Zadoks, J. C. 1971. Systems analysis and the dynamics of epidemics. *Phytopathology* 61:600–610.

Zadoks, J. C. 1974. The role of epidemiology in modern phytopathology. *Phytopathology* 64:918–923.

Zentmyer, G. A., P. P. Wallace, and J. G. Horsfall. 1944. Distance as a dosage factor in the spread of Dutch elm disease. *Phytopathology* 34:1025–1033.

9

PRINCIPLES OF DISEASE MANAGEMENT

Synopsis: Detecting pests is the first step in pest management. Sound management considers pest impacts on timber and other resources, forest management objectives, and economic and social results of management alternatives. Although specific actions can reduce pest impact, incorporating pest management principles into stand management activities will help minimize pest problems over a rotation.

9.1 INTRODUCTION

Disease management, and pest management in general, is a topic of great interest to forest managers, especially when a pest outbreak is occurring or is imminent. Pest managers are often viewed like plumbers—called when there is a problem and ignored when there is none. Many potential pest problems, however, can be minimized or prevented if pest managers are involved in planning the silvicultural manipulations in a stand. Incorporating pest management concepts into the realm of forest management can provide ecologically sound strategies for managing the host crop and potential pests. This is the core of integrated pest management (IPM). Although ecologically sound silvicultural practices, or IPM, can reduce the potential for pest outbreaks, occasionally direct pest management activities will be needed. The effects of these management activities, and indeed all management actions, must be considered in relation to the insect, weed, and wildlife populations of forests. In this chapter we will discuss general principles and illustrate them with examples drawn from pathology. These principles also apply to agents other than those that cause disease.

Disease situations in forest crops are particularly difficult to treat because of the low value of the crop. Agricultural crops, conversely, are of much greater value, and disease control costs need to be carried for only one growing season. Of the crops listed in Table 9.1, wheat is of low value and usually will not justify even a single pest control treatment during the growing season. The only exception might be a chemical spray applied by aircraft to save the crop from total devastation. Wheat is an extensively managed crop, and a single individual may manage several thousands of hectares. Tomatoes are of much higher value, and the manager can afford to expend a great deal of effort to ensure a good crop. The intensity of labor required is such that the full-time services of one or two people are required for each hectare or so of production.

These relatively lucrative crops may be contrasted with those of forest lands. A

TABLE 9.1 Relative Values of Some Agricultural Crops

Crop	Value ($/ha/yr)
Wheat	250
Soybeans	325
Spinach	875–1250
Tomato	12,500

typical hardwood stand in the piedmont containing 15 Mbf/ha may have a stumpage value of $38/Mbf at harvest time. This would amount to a total value per hectare of $570. Because such a stand may have required 100 years to reach maturity, this stand produces an income of $5.70/ha/yr. We can easily understand why private landowners in the piedmont of the southeastern United States do not like to grow hardwoods.

The situation for pine culture is only a little better. A reasonably productive pine stand might be expected to yield 50 Mbf/ha at maturity, with a value of $190/Mbf, or a total value of $9,500/ha. Because a reasonable rotation length of 60 years might be required to achieve this yield, income would be approximately $157.50/ha/yr. Although this amount is considerably more than that produced by hardwoods, it is still much less than what is produced by wheat, which produced such a small yield that income could pay only for the cost of land preparation, seeding, and perhaps some fertilization. We can easily appreciate why disease control, or pest control in general, in forests is severely limited by cost.

Disease controls generally fall into one of three general categories: legal, cultural, or direct.

9.1.1 Legal Procedures

The simplest examples of legal procedures are the state and federal seed certification programs. Even though they are designed primarily to ensure that the customer receives a quality product with known viability, a major side benefit is that the seeds are also free of pests.

More restrictive are plant quarantines. These restrictions limit the movement of plant materials either into, out of, or within the county. They may include movement of soil, nursery stock, Christmas trees, and even logs. The need for quarantines has been evidenced by our experience with white pine blister rust, chestnut blight, and Dutch elm disease. Another legal procedure is the enactment of laws restricting the growing of alternate hosts, such as black currents and gooseberries for control of white pine blister rust.

These laws may require the active participation of the land manager, but they are not control methods used in an operational strategy. In extensively managed and noncommercial forests, these laws may constitute the only defense against diseases.

9.1.2 Cultural Control

Cultural control represents one major step upward in intensiveness and includes all manipulations of stand growth designed to maintain overall health and vigor of trees.

The philosophy here is to prevent diseases from occurring rather than stopping them after they have started.

Cultural controls may include a strategy of avoidance in which we discriminate against species susceptible to diseases that are known to occur in an area or we use genetically resistant stock. Knowledge of the degree of risk to pathogens is often useful. This has led to the development of risk or hazard maps that indicate the relative degree of risk to many pathogens. In the southeast, maps of varying degrees of specificity have been developed for annosum root disease, littleleaf disease (Fig. 9.1), and fusiform rust. For some pathogens, the most notable example being dwarf mistletoe, it has been impossible to predict risk, but once infection occurs, it is relatively simple to project losses based on site index, stocking levels, and other stand parameters.

Another major component of cultural controls includes sanitation. Sanitation attempts to deny a pest population the opportunity to persist. It may be ineffective against aggressive or virulent pest species, large populations, or distant sources of infection. Removal of diseased elm for Dutch elm disease control and burning to control brown spot disease of longleaf pine are two examples.

Sanitation cutting may be combined with salvage to make the salvage more economically attractive, but the latter generally involves removal of trees that have already been killed or damaged. For most product needs, hardwoods must be salvaged before death as significant deterioration may have already occurred. Most conifers must be salvaged soon after death, although some deterioration following invasion by blue stain is not nearly so critical because end uses are usually for rough construction where appearance is less important.

The feasibility of salvage depends upon market values of the potential products and accessibility of the damaged timber. The affected volume may be so small so as

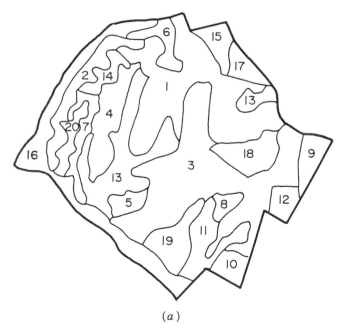

(a)

FIGURE 9.1 Compartment map showing (a) stand boundaries and (b) relative risk to littleleaf disease. From Oak and Tainter (1988).

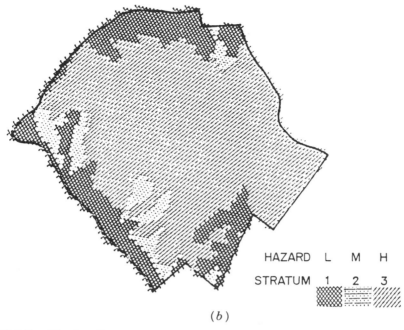

HAZARD   L   M   H

STRATUM   1   2   3

(b)

FIGURE 9.1   (Continued)

to not be worth salvaging. Healthy, higher-valued timber may be included in the harvest sale to make salvage more lucrative. A large epidemic or loss could saturate the market and reduce prices. This happened following Hurricane Hugo in 1989. Much timber was simply left to decay because of depressed markets and the inability to extract and handle large volumes of logs.

9.1.3   Direct Control

Direct controls tend to act directly on the pathogen and involve the use of chemical fungicides, prescribed fire, and physical methods. Fungicides are generally not used much in forests, being limited to intensively managed areas such as Christmas tree plantations, nurseries, seed orchards, and urban and suburban situations. Prescribed fire is more widely used as it is much less expensive and is very effective for some pathogens. Physical methods tend to be expensive and are used only in high-value situations.

Disease management is a component of the total forest management system (Fig. 9.2). The pest management subsystem is invoked with detection or recognition of a pest problem. Pest detection has two parts: determining that a plant has a pest problem and diagnosing or identifying the problem. This process of disease diagnosis for most major tree pathogens is outlined in the Disease Profiles in Part Two. In the forest, however, most trees have several pests affecting them. Foresters manage stands and thus are concerned about the importance of pests in a stand. This chapter discusses manipulation of the pest at the stand level. Then, knowing where the pest occurs, the damage may be appraised. Foresters are concerned not only with what has happened in a stand but also with what will happen during the time remaining until

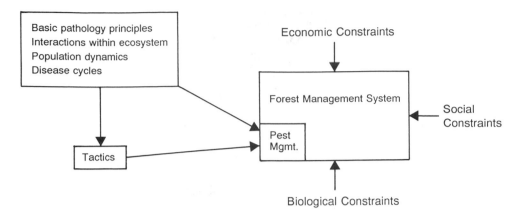

FIGURE 9.2   Integrating disease management into the forest management system.

a stand is harvested. This assessment often involves some type of impact projection. Armed with this information, and considering economics and other resource values, a forester may choose to apply some type of management action. More often than not, however, the forester chooses to do nothing and permits the disease to run its course. The no-action alternative is often employed when effective control measures are unavailable, when economics do not justify the cost of management, or when the proposed action harms other resource values.

In order to consider dealing with a pest, it must be present or pose a threat of occurring in the management unit. Determining whether a pest is present (distribution) and the proportion of the forest affected (incidence) is a common goal of pest detection efforts. We will discuss them in Section 9.2. Estimating the severity of the pest and the extent of damage will be discussed in Section 9.3. Then, some general principles of pest control will be discussed in Section 9.4. In Section 9.5 we will discuss integrated pest management. In Section 9.6 we will illustrate how a pest management system might be organized and implemented.

9.2   DISEASE DETECTION

The methods used to detect a given pest depend upon the characteristics of both the pest and the management unit. The extent to which the symptoms of the disease and the signs of the pest are readily visible influences the ease of detection. Foliar or stem pests are easily seen, while more serious but less evident root diseases are more difficult to detect. Detecting pests on roots often requires that roots be excavated, increasing the labor needed for detection. Culturing may be necessary to confirm the presence and identity of pathogens that produce indistinct symptoms or those easily confused with others. Immunochemical techniques (based on antibodies) or molecular techniques based on enzymes or nucleic acids may prove useful in pest identification.

The size of the management unit, the intensity of management, and the crop value also have roles in disease detection. In a small, intensively managed unit, it may be feasible to inspect each tree for symptoms and signs of disease. The value of the vegetation on the unit also contributes to the feasibility of pest detection. Each highly valued elm in a community can be examined for symptoms of Dutch elm disease at 2-week intervals during the entire growing season. In contrast, remote drainages with low-valued timber species may be examined at 10- or 20-year intervals, and then primarily for measuring timber volume.

The distribution pattern of the disease also influences how we look for it. A disease that occurs randomly in the forest can often be detected during the course of stand examination or during inventories. Pests that occur in a certain habitat type or that are associated with roads or other disturbances may not be adequately detected. The anticipated incidence of the disease must also be considered in designing detection programs. Describing the incidence of a widespread parasite like the lodgepole pine dwarf mistletoe requires less intensive efforts than for a pathogen of very low incidence such as tomentosus root disease.

Pest detection efforts can be divided into two categories: informal and formal surveys. *Formal surveys* are activities intended to gather information about pest distribution and incidence. *Informal surveys* provide information, primarily about distribution, gathered during other activities. Informal surveys may not always be recognized as such, but they often play a very important role in pest detection. Alert foresters often see something wrong and send a sample to a clinic or diagnostic laboratory or contact a pest specialist. These contacts often lead to the discovery of a disease problem. Opportunistic in nature, informal surveys are extremely important additions to the efforts of pest-management specialists.

Formal surveys can be described by how they are executed. Ground surveys are done by crews on foot. These surveys may be casual walk-throughs, a special extension to the normal stand inventory procedure, or they may be specifically designed to gather information about a particular pest. Information collected during the USDA Forest Service Stage 2 stand inventory is considered a ground survey. Ground surveys usually provide information about a particular stand, although data may be combined for forest-level estimates. Time often constrains the aggregation of these data for a large area because, by the time the entire area is surveyed, the disease conditions have changed in the first stands examined.

Roadside surveys can provide pest information when road systems traverse a significant amount of the forest type of interest. The disease must be distinguishable from a moving vehicle. The large areas that can be sampled permit estimation of pest incidence within districts or forests. Roadside surveys have been done for western dwarf mistletoes and for black stain root disease, a problem associated with road construction and other human disturbance (Merrill et al. 1985).

Pest incidence on larger areas or in inaccessible areas can be estimated using aerial surveys, if the disease causes symptoms visible from the air. Satellite imagery has not yet proven useful for detecting pests, but advancing technology in remote sensing offers potential. Thus most aerial surveys are done from aircraft. Data are collected by direct observation or through the use of aerial photography or other remote sensing. Oak wilt and Dutch elm disease cause yellowing, wilting branches. These symptoms can be seen from the air. The difference between symptomatic and unaffected foliage may be enhanced on color infrared aerial photography. Other diseases cause aggregated mortality and, consequently, openings in the crown canopy. The dwarf mistletoes, *Arceuthobium pusillum* in black spruce stands and *A. americanum* in jack pine stands, are easily detected by virtue of their mortality centers and the large witches' brooms in trees bordering the center (Baker et al. 1992, Meyer and French 1966). Root diseases may also be detected from the air, but the identity of the pathogen must be determined in a ground survey (Williams and Leaphart 1978). Aerial and roadside surveys often provide results with large margins of error because each tree is not directly examined. Ground surveys in part of the sampled area can improve the accuracy of roadside or aerial surveys.

Regardless of the type of survey, the expertise of those collecting the information determines its value. Field crews must be well trained to identify the pests they are likely to encounter. Detecting diseases requires appropriate sampling methods. For example, using aboveground symptoms and signs will underestimate root disease incidence. Root examination provides a more accurate picture. Excavating roots will greatly increase the time and cost of a survey, but it must be done if determining root disease incidence is a survey objective.

In designing surveys, carefully consider the scope of the detection effort. A little can be learned about a large area, or a great deal can be learned about a small area. To know a great deal about a large area is often impossible due to time and budget constraints.

Pest surveys are done by many different organizations. Informal surveys are done by anyone who visits the forest. Foresters, loggers, wildlife managers, and even the public on recreational visits can detect a problem in the forest. These informal discoveries often alert pest managers to a need for more formal surveys. The formal surveys are usually done by pest management groups. Forest Pest Management (FPM), in some regions called Forest Health (FH), is a USDA Forest Service group in State and Private Forestry. They are responsible for assessing forest conditions on federal lands. They also provide financial and technical assistance to the state forester for assessing pest conditions on state, county, and private lands. As a result, many states have a pest-management specialist who works with state foresters and the public. In Canada, the Canadian Forestry Service has a similar group called the Forest Insect and Disease Survey (FIDS), which is responsible for monitoring pest conditions on public lands. Both FIDS and FH produce Annual Regional Forest Condition Reports, which summarize the status of pests in the region.

9.3   DAMAGE APPRAISAL

The distribution of a pest may be of use in determining whether more detailed surveys are warranted. However, knowing that a pest is present does not permit a forester to answer questions such as, "How much wood volume will be lost in the next 10 years?" Or, "Which stands are or will be experiencing the most serious losses?" Clearly, foresters need some measure of pest impact on stand value. A pathogen is not a disease, and a diseased tree is not always damaged. For example, the presence of *Hypoxylon mammatum*, the fungus that causes Hypoxylon canker, does not always indicate a disease problem in an aspen stand. In fact, we might question the value of *Hypoxylon* and other canker fungi as thinning agents that reduce the density of aspen stands. Many pathogens play important and beneficial roles in decomposition and nutrient recycling in a stand while causing little damage. And the presence of diseased trees does not mean a loss has occurred or will occur. In many situations, diseased trees can be salvaged during intermediate stand entries. In others, the diseased trees will have only slight damage at rotation. Or diseased trees may be of greater value than disease-free trees. These issues must be considered during damage appraisal.

Once foresters know the distribution of a pest problem, they can determine the severity of the problem, or "how bad" it is. If the problem is not serious today, foresters must consider the likelihood of the situation deteriorating during the remainder of the rotation. Future projections often require knowledge of the amount of pest (disease severity) on a tree. Well-established procedures for assessing disease severity exist only for a few pests. The 6-class system for rating dwarf mistletoe severity (DMR) is perhaps most familiar to foresters (Hawksworth 1977). And further, we understand the relationship between current and future severity for only a few of the most serious disease problems such as fusiform rust, dwarf mistletoes, and root diseases. As foresters seek to improve their ability to project future stand conditions under different management and pest scenarios, the need for improved pest severity rating is becoming clear.

Incidence, the proportion of trees infested, damaged, or killed, may be useful in situations where a pest is just entering a new area or where the unit of management is the individual tree. In urban situations or in campgrounds where individual trees have value and are treated, incidence can be used to determine tree removal and tree-planting needs. Incidence is of less value in forest stands, unless the pest affects the value of all trees in the stand equally. For a few pests, most notably dwarf mistletoes, incidence has been related to severity and is much more useful for quantifying damage.

When a pest kills trees rapidly, incidence may be a good indicator of damage. However, when the pest causes a slower demise, foresters need some measure of severity or the amount of disease present. Campground managers face a need to determine severity of decay in campground trees. Decay in a tree may ultimately lead to its failure. The incidence of decay is often high. In some trees, however, the decay is not extensive enough to place the tree at a great risk of immediate failure. The campground manager must know the extent of decay to decide which trees should be removed and which can remain to provide a pleasant environment with trees, while exposing users to a minimal risk of injury from tree failure.

Expressing pest damage in terms of the stand volume affected is often the most useful and the most difficult approach. Such information permits the forest manager to consider the timber volume lost and the volume that prompt salvage operations may recover. Furthermore, converting volume loss figures to dollar values permits economic analysis of management options. Keep in mind that economics may not be the sole justification for action. The benefit of protecting tree health in a particularly popular state park campground may justify exorbitant costs.

This discussion follows the practice of most foresters and considers pest impact only on timber volume. Forests can no longer be thought of as timber factories; they are much more. Forests produce recreational, wildlife, and watershed values. In many regions these other products are more important than timber. We recognize these values, but quantifying them is very difficult. And although pests may damage trees or even kill them, their entire value may not be lost. A damaged tree may be harvested and all or part of its value as a wood product recovered. In some regions the public values, even prefers, insect- and disease-killed trees for use as firewood. Dead or top-killed trees often benefit wildlife. Foresters are beginning to manage for such habitats. In many areas of the western United States, great efforts are made to create snags for wildlife. A disease such as a rust, which top-kills, may increase the value of a stand for these wildlife. Damaged trees may occur in areas where the value cannot be recovered. Trees killed by pests in national parks, wilderness areas, and other reserved forests are not harvested. Is this a loss? And how do we consider the chance that pests will spread from these "reservoirs" to adjacent lands managed for other purposes? And what about pest impacts on scenic quality? Quantifying the damage and benefits to nontimber resources and appraising their value are among the emerging issues in forest management challenging foresters and pest managers.

Understanding the impact of pests in stands is a difficult task. Foresters' need to know what will happen in the future further complicates the issue. Projecting the likely stand condition at future points in time allows foresters to consider management strategies and the timing of those activities. Incorporating pest projection tools into growth and yield models permits examining stand management alternatives. The forest vegetation simulator (FVS, formerly known as Prognosis), the model used by the USDA Forest Service in the western United States, has submodels dealing with dwarf mistletoes and root diseases. Routines are being developed to simulate the interaction between bark beetles and root diseases (Stage et al. 1990). Such tools will permit a more thorough consideration of management alternatives.

9.4 DISEASE MANAGEMENT STRATEGIES

There are six categories of disease management actions: avoidance, exclusion, eradication, protection, resistance, and therapy. These measures attempt to disrupt the disease process by changing the suscept, the pathogen, the physical environment, or the biological environment needed for disease. Measures from one or more of these categories may reduce or control a disease. Conversely, some actions may fall into more than one of the categories. Successful pest management strategies often employ multiple strategies and are almost always based on sound silvicultural practices for the host species.

9.4.1 Avoidance

Management practices can often be modified to avoid a pest. Avoidance may be as simple as planting in areas where the disease is not present or where the environment does not favor disease development. Many nurseries are located near prisons to capitalize on low-cost, "captive" labor, and, most often, without regard to the site conditions. Consequently, nurseries must try to produce a very high-value crop of seedlings on ground that would not produce trees. Poor soils are responsible for many nursery pest problems. Poorly drained soils contribute to diseases caused by *Phytophthora* spp. In western states, many nurseries are in valleys, well below the natural range of trees. Spring comes earlier to these lower elevations. Seedlings must be lifted while they are still dormant. Because the forests at higher elevation are still snow-covered, seedlings are stored in refrigerated buildings. Often, fungi damage seedlings while in storage, creating a whole new pest problem. Locating the nursery in a less convenient site, but one more like where the trees would naturally grow, would greatly help to reduce nursery pest problems.

Hazard-rating systems can aid foresters in identifying sites where pests are or may

become severe. Recognizing potential problems, foresters may avoid planting susceptible species on these sites, they may use measures to counteract the disease such as
resistant varieties, or they may plan for increased losses. Environmental factors are
most often used to rate hazard. For rust diseases, hazard is often related to the frequency with which weather-permited infection occurs. In the Ozark plateau of Arkansas, successful infections by aeciospores of comandra blister rust are dependent on
violent winds generated by the passage of cold frontal systems in spring and the precipitation and higher relative humidity that results from the precipitation (Dolezal
and Tainter 1979). In a given year the frontal systems may not be frequent or strong
enough to ensure adequate secondary infection of comandra plants, ultimately resulting in a "low" rust year. Soil characteristics have been used to estimate hazard to
annosum root disease (Baker et al. 1993, Froelich et al. 1966) and littleleaf disease
(Campbell and Copeland 1954).

Timing management activities to avoid having susceptible trees or tissues during
pest infection periods can reduce pest problems. Many nurseries avoid pest problems
by sowing certain species at specified times. In slash and loblolly pine stands, thinning
during the summer, when stump surfaces are too hot to be colonized by germinating
spores reduces infection by *Heterobasidion annosum*.

9.4.2 Exclusion

If a pest is absent from an area, potential problems can be minimized by keeping it
out. Quarantines are laws that prohibit the transport of materials that may harbor
pests. Quarantines could have kept the white pine blister rust out of the United States,
but once established here, quarantines between states were of little use. Fungi do not
respect the artificial barriers posed by state lines. These same fungi, however, often
have no means of crossing geologic barriers, such as oceans or mountain ranges, unless we help them. People have been responsible for carrying pests over natural barriers. Transport of seeds, seedlings, green forest products, and other plant materials
introduced our most devastating forest diseases. There is always a risk of introducing
a pest on plant materials. Because of this risk, a biological viewpoint would suggest
that countries should not trade in plant materials. However, countries must import
plant products they cannot produce. For economic and political reasons a country
may import products it can produce. Trade will always occur, providing some risk
that new pests will become established. When that risk is considered unacceptable,
quarantines may be enacted. A country may encourage quarantines not only to protect itself or others from a potential pest problem but also to improve the market
for a substitute product that it produces. Again, the political forces may overpower
biological issues in choosing a course of action.

Other measures, less stringent than quarantines, may be implemented to reduce
the risk of unknown pests. Shipments of plants into the United States are often held
in warehouses for 7–28 days and examined by U.S. Customs Agricultural Pest Inspec-
tors for pests before the shipment is released. Monitoring these shipments is a gargan-
tuan effort considering the large volume of imported plant materials. If a pest is detected, the plants may be destroyed, or the materials may be treated to eliminate the
pest. In some situations a shipment is treated before entering a country. At one time
it was suggested that oak logs from the United States be fumigated with methyl
bromide before shipping to prevent introducing the oak wilt fungus into Europe.

Unless there is some assurance that all shipments are effectively treated, there is some potential for introducing a pest. A ship waiting in port for its cargo to be treated is losing money. Storms or other situations may disrupt the log fumigation effort. It would be very easy to resolve the delays by loading untreated logs onto the ship. For these and other reasons, log fumigation was found to be a waste of money and time, and so is not done.

Starting with pest-free plants will help to keep them free of pests. This also may be considered a means to avoid a pest. Inspection and certification programs are often used with agricultural crops to ensure that lots of seed and plants are pest-free. A phytosanitary certificate warrants that the plant was grown in an area where a specified disease is absent. Phytosanitary certification has been used to establish that a particular shipment of oak logs came from a county that does not have oak wilt or for shipments into Canada of logs originally from the United States. Fumigating these logs is unnecessary. Certification programs can be effective with good inspection and if shipments are not confused. However, there are examples of failed certification programs in agriculture.

Straddling the line between exclusion and eradication are biological controls. Some biological controls function by taking over a substrate, preventing colonization by (excluding) pests. *Phanerochaete gigantea* is a saprophytic decay fungus that often colonizes pine stumps. If *Phanerochaete* colonizes the stump before *Heterobasidion annosum*, *Heterobasidion* will not colonize the stump and infect trees that contact the root system. Other biological controls involve organisms that parasitize pests. In some years these organisms may effectively maintain pests at a low level. In other years they are ineffective. It is unlikely that they will eliminate their host, for where would they live? Much attention has been given to *Tuberculina maxima*, a fungus that parasitizes rusts. Although it may be prevalent in some years, this fungus has not effectively controlled any rust.

9.4.3 Eradication

Eradication is the total removal of the disease from a site. Either the pathogen, or, if the pathogen is an obligate parasite, its host may be eradicated. If a pest has no suitable host to infest or has no means to get to one, it will die. This is one of the principles supporting the use of crop rotation. Even though it is one of nature's solutions, rotating species often conflicts with human management objectives and is not done.

Extreme efforts to eradicate a pest may not be feasible, let alone possible. Pests often occur over extensive areas and may even be integral parts of natural ecosystems. Pests often have such a tremendous reproductive potential that they cannot be eliminated. Attempts to eradicate the chestnut blight and white pine blister rust fungi failed because of the ability of these pests to reproduce rapidly, although white pine blister rust can be effectively eliminated by pruning over a period of years. Pests with a limited reproductive ability, however, may be eradicated from a stand. The eastern dwarf mistletoe should be eradicated from regenerating black spruce stands because even low levels of dwarf mistletoe will lead to unacceptable mortality at rotation. This parasite discharges its seeds to a maximum of 17 m only once each year. Removing the infected host trees and the trees at least 17 m around them will eliminate the dwarf mistletoe. Clear-cutting, and burning if slash is available, or using mechanical equipment or herbicides to kill residuals are consistent with the silvicultural needs of black spruce and can eradicate the dwarf mistletoe.

Foresters most often attempt eradication in high-value areas such as nurseries and seed orchards. On these sites, for example, removing root debris from the soil and then fumigating the soil with methyl bromide can eliminate root diseases. In nurseries only the seed beds are usually treated. The fungi often invade the treated beds from the untreated areas between beds, contributing to the need for retreatment.

9.4.4   Protection

Protection is the application of some substance or technique to protect a plant or prevent the plant from becoming infected. Applications of fungicides, nematicides, or insecticides (to control vectors) come first to mind, but mycorrhizae and fertilizers also can protect plants from pests. The extensive area of forests, the relatively low value of the forest crop, and the real or perceived effects on nontarget resources often preclude the use of chemicals. The high value of areas such as nurseries or Christmas tree plantations often justifies chemical use.

Some pests occur on or in seeds of many plants, and trees are no exception. These pests often attack the seedling as it emerges from the seed. Seedlings are protected by treating the seed coat with a fungicide. A dye is usually added to the treatment to identify treated seeds. Because the fungicides are applied to the seed coat, seed treatment is most effective against fungi that occur on the seed coat. Fungi in or beneath the seed coat may escape and cause problems.

Cultural practices also serve to protect plants from pests by changing the environment to discourage the pest or disrupt its development. Thinning, which permits air drainage, and pruning the lower branches, which removes the susceptible tissues in humid areas near the ground, can reduce the incidence of white pine blister rust (Hungerford et al. 1982, Hunt 1982). Keep in mind that cultural treatments that discourage one pest may favor another. Pruning lower branches and increasing the spacing between white pines to reduce white pine blister rust often increases damage by white pine weevils.

9.4.5   Resistance

To most of us, genetic resistance means immunity, where a plant has inherited a trait that prevents the plant from becoming infected. However, other inherited traits may be useful in pest management. A trait similar to the slow rusting in wheat occurs in western white pines (Hoff 1984). Other traits permit the plant to become infected but not damaged. Such plants are considered tolerant to the pest. Many eastern white pines are tolerant to ozone. Finally, some plants produce morphological barriers that allow them to avoid a pest. A ponderosa pine with drooping needles is "resistant" to dwarf mistletoes because the seeds slide off the needles and are removed from the host (Roth 1964).

Regardless of how it is manifested, genetic resistance can be an important tool for foresters in long-term pest management. There are, however, several caveats. First, there must be genetic resistance in the population of host trees. If that resistance cannot be found in this country, geneticists often go to the origin of the pest to seek resistance there. Where the pest is native, we often find that its host plant has developed some resistance to the pest. If resistance is found in another closely related species, selective breeding can often incorporate the genes into the new host. Desirable traits (e.g., tree form and branch retention) must be retained in the breeding program. Resistant trees must then be tested repeatedly to ensure that the resistance is effective. After a tree has proven resistant, it must then be propagated to provide stock for outplanting. Selection, breeding, and propagating resistant trees can take many years. Pests, too, have genetic variability. If the only suscept is one resistant to most of the pest population, only those pests that can attack the resistant tree will survive. After enough generations, the pest population will contain enough pests able to attack the trees' resistance, and losses will occur. Then, a new source of resistance must be added to the suscept population. Foresters must continue to monitor the pest's ability to attack resistant trees and prepare to use new sources of resistance. As you can see, using genetic resistance requires an ongoing and expensive commitment for research. Manion (1981) discusses breeding for resistance in greater detail.

Considerable effort, and success, has been made to select trees resistant to rust diseases. Resistant varieties are available for white pine blister rust, fusiform rust, and jack pine stem rusts (Bingham 1983). Considerable progress had been made to find resistance in aspens to Hypoxylon canker (Anderson et al. 1990). This research has involved the selection and testing of superior genotypes, the use of molecular genetic markers to assist tree improvement efforts, the development of methods for introducing foreign genes into aspen, and the propagation of these genotypes by tissue culture. Much of this resistance may be attributed to resistance to wounding insects in addition to resistance to infection, resistance to canker development, and spatial resistance that is influenced by stand density factors (Ostry and Anderson 1990). Having resistance does not permit a total disregard for the diseases. If this happens, the resistance can break down, and one pest management tool will be lost. Rather, resistance should be considered as only one part of an integrated program for reducing pest losses.

9.4.6 Therapy

Some pests can be controlled therapeutically, that is, after they have infected a host. The public, and many foresters, relate to injection because of familiarity with the technique: A person gets sick, and he or she goes to the doctor and gets a shot. And so it should be for trees. Unlike people, trees have no circulatory system. We have a limited understanding of the vascular system of a tree and the transport of chemicals within it. Yet, we must rely on this system for translocating the pesticides to the site where they will combat the pest. Consequently, our ability to deliver chemicals effectively within a tree inhibits our use of therapy.

Injecting elms with fungicides to eliminate the Dutch elm disease fungus is perhaps one of the more successful uses of therapy in forestry (Stennes and French 1987). Injection is effective only for trees with less than 10% wilt, and only when feeding beetles introduced the pathogen. Proper injection is expensive and time-consuming and thus can be considered only for very highly valued trees. Therapeutic injection would be of little use in forests. Therapeutic injection is being tested for curing or preventing infection of ornamental white oaks with oak wilt. Injections of tetracycline antibiotics can often cure diseases caused by mycoplasmas (Raju and Wells 1986).

Heat therapy is extensively used in ornamental plant production to free plants of viruses before vegetative propagation. When heated, the cells in a tissue culture of the plant reproduce more rapidly than the virus, and the plant essentially outgrows the virus. Meristematic tissues are subcultured and transferred to new culture media before the virus can invade those newly formed cells. Poplar clones have been freed from viruses in this manner. As vegetative propagation increases the likelihood of virus transmission, these kinds of techniques will be used increasingly.

Removing a diseased branch can often free trees from disease. If detected early enough, Dutch elm disease can be removed from infected elms (Baker and French 1985). Similarly, pruning can remove white pine blister rust and other rust fungi from pines before the pathogen reaches the main stem (Lehrer 1982). Pruning has been used in campgrounds to free pines from dwarf mistletoes and to increase their vigor and longevity even if the parasite cannot be eliminated (Scharpf et al. 1987).

9.5   INTEGRATED FOREST PROTECTION

For many years foresters have managed forests considering only the response of the trees. When a pest became evident, pest managers were called in to solve the problem. This approach often led to another pest problem. Foresters have come to realize that forests must be considered as an ecosystem and that manipulations to one part of that ecosystem affect the other parts. This ecosystem concept is the foundation of IPM. Pests have long influenced forest management. We now recognize that forest management can affect pest populations (Baker 1988). We are beginning to understand how insects, diseases, and environmental stresses interact to cause pest problems and how forest managers can prevent these outbreaks (Schowalter and Filip 1993). The emphasis on prevention to protect our forests requires ecologically sound forest management. There is still a long way to go; pests are still often considered singly and separate from forest-management actions. Only when forest protection is truly integrated into forest management can we reach the goal of pest prevention.

The second major contribution of IPM links pest control with economics. From this comes the concept of a threshold damage level. That is, damage may be present, but at some level the loss becomes unacceptable (Fig. 9.3). We commonly think of the loss in terms of timber volume, but as discussed earlier, values of the forest for other uses may also be diminished. The economic injury level is that level at which the value of the injury exceeds the cost of preventing it. At this level, investments in pest control are economically justified. To prevent this damage from occurring, the management action must be applied before the incidence or severity reaches the damage threshold. This approach implies that pest levels be monitored and that pest incidence or severity can be related to damage. For many forest pests these relationships are poorly understood.

The threshold concept suggests that some damage is tolerated. It may in fact be unavoidable. Complete control of a disease is not always justifiable, if even possible, in forest situations. In other situations "zero tolerance" must be strived for. For example, 5% decay in a wood utility pole is too much when it occurs at the ground line and you are clinging to the top 95% decay-free part of the pole, well above the ground! A thorough examination of pest impact must also consider broad-scale social effects. The impact of the late blight of potato on the Irish people is well known. Although the dollar value of the chestnuts lost to chestnut blight is substantial, we often over-

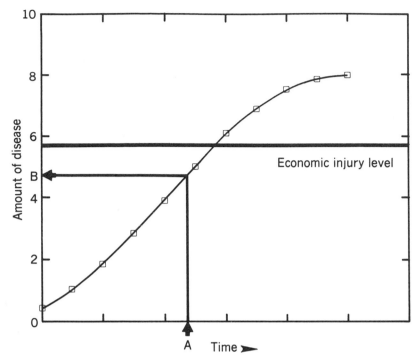

FIGURE 9.3   A disease control threshold diagram showing (A) time and (B) amount of disease at which control must be applied to keep disease level below the economic injury level.

look the impact of this disease on the populations of Appalachia. No doubt that the poverty there would be less if chestnuts attained commercial size.

The USDA Forest Service Forest Pest Management, in some regions called Forest Health, often provides biological evaluations of pest situations. These evaluations consider the incidence and severity of the pest and the damage that has occurred and is likely to occur in some future timeframe. Then, considering economic criteria and other management objectives, they suggest management alternatives. Note that pest managers do not choose the management action; this is done by the forest manager. The pest manager's role is to provide information about alternatives and the likely effects of those alternatives on the forest ecosystem and on the ability of that system to meet management objectives.

9.6   DEVELOPMENT OF A PEST MANAGEMENT SYSTEM

The development of a tree disease management protocol should be integrated into a larger pest management system that not only includes insects but in some cases weed control and fire control as well. Increasingly, as stated previously, the interrelatedness of these factors must be taken into account in order for the pest management system to be effective.

Pest management is almost always a small part of the overall management program and in many respects is just another decision-making process. Decisions regarding pest management should be dealt with just like any other potential treatment, such as thinning, fertilization, and site preparation. A wise pest management decision involves a series of steps:

1. define the problem,
2. quantify the pest's impact,
3. identify control options,
4. perform cost–benefit analysis, and
5. make a decision.

9.6.1   Define the Problem

This step usually involves a statement by the landowner of his or her objectives for that forest and a recognition of the potential pest complexes pertinent to that system. What is defined as a pest depends on the objectives prescribed by the land manager. Most forest land is owned by individuals who have no sort of management goal. They may be holding the land as speculation, or they may simply be too detached from it. For these people there are not pest problems because they have no specific management goal. Conversely, on most public and industry lands, ownership objectives are clearly defined. As a result, pest problems are usually, but not always, clearly defined, and preventive and control measures are understood.

The nature, extent, and strategy of control is largely related to the intensiveness of management as well as to the identity of the pest and purposes of management. There are eight general levels of management intensity (Fig. 9.4). Note that these examples of forest management types are listed in order of increasing intensity of management.

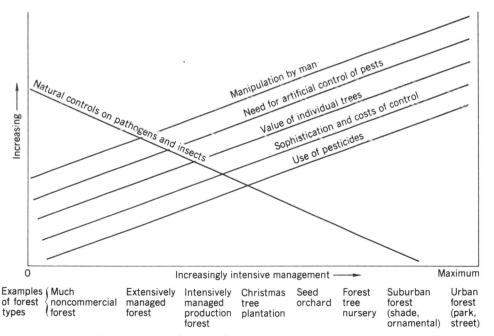

FIGURE 9.4   Characteristics of various forest management situations in regard to pest control practices. From National Academy of Sciences (1975).

Noncommercial Forest. Approximately one third of all the nation's forest land is in this category. Included are lands that have been withdrawn from timber production either because of adverse site conditions or because of their uniqueness. Pest control measures are usually minimal and are only justified when a potentially serious pest could spill over onto commercial lands. The existence of these lands has immeasurable educational value with respect to biology, ecological processes, and a holistic approach to pest-control decision making.

Extensively Managed Forest. This category includes approximately one half of the national forest area and much private land. Sites are variable but tend to be poor or small in area such as small woodlots. Pest control tends to be ignored but owners may participate in the National Tree Farm Program. The Forest Pest Control Act of 1947 was established to assist in *pest control* on a cost-sharing basis.

Intensively Managed Production Forest. In most of the forests to the left of "intensively managed forests," diseases are simply accepted and tolerated. This may not be passive control, though. Acceptance of injury is the single most widely used strategy on forest lands today. These forests tend to be comprised of relatively small production units that are, as a result, more likely to receive the appropriate treatment. Intensive pest control usually implies increased expenditures, which are warranted only where soil capability and climate allow high productivity. The land manager can make such expenditures only when land costs are high, competition for raw wood supplies is great, and harvest and manufacturing operations can be carried out efficiently. Another limitation is the availability of investment capital. At the present, only forests in the Pacific Northwest and in the southeastern United States fit into the category of intensively managed production forests. Little of the early cost of pest control is recoverable in less than two to three decades. In the meantime the young stand is exposed to many other natural hazards. These forests may be plantations but are not necessarily plantations. A plantation can be established and then never managed. But more intensive practices can be made either on plantations or natural stands. Often the latter respond favorably with minimal input.

For pest control the operational units must be small enough and accessible enough for close surveillance. High monetary values in stand establishment impel the manager to respond promptly to threats of loss. The levels of damage that are economically acceptable are directly determined by the investment costs.

Pest damage is of two general kinds. Damage may be in the form of occasional, partial, or brief delays in establishment or in the form of fractional loss of yearly wood or fiber production. An example of the latter form of loss was the needle cast epidemic caused by *Lophodermella cerina* and *Hypoderma lethale* in 1971. In the southeastern United States pines in an area over 22 million ha were affected. Causes for the epidemic are not understood, and it has not recurred.

Conversely, diseases that prevent full stocking cause persistent defects or continuously retard growth demand control because they result in large cumulative losses. Examples are fusiform rust and dwarf mistletoe.

Some disease problems can be avoided by not planting susceptible species, but there may be few opportunities because there are few species to choose from. Longleaf pine may be a good replacement for slash and loblolly pines on high-risk fusiform rust sites, but it is not suited for short rotations.

Development of genetic resistance will be a major factor in the near future. Over one half of industrial forests already are replanted using genetically improved stock. As management intensifies, some diseases will increase in occurrence because of the greater amount of succulent tissues produced by the faster growing trees. Root decays will also increase as a result of root damage to seedlings during planting and to older trees during thinning. The tendency toward even-aged stands comprised of clonal material also ensures greater risk because of more uniform growth and genetic homogeneity. Heart decays in shorter rotations will tend to decrease in importance, but some of these volume savings will be offset by losses resulting from butt decays. An interesting problem experienced by managers on some southern national forest lands is to enhance the formation of heart decay by *Phellinus pini* in relatively young trees. In this case a major management objective is to increase nesting and feeding habitats for the red-cockaded woodpecker.

Although these forests tend to receive some form of intensive management, the relatively low value of products dictates that direct control of disease by chemical or mechanical means is restricted to two situations: before or during stand establishment, and for intermittent outbreaks that threaten excessive damage.

Christmas Tree Plantations. Christmas tree culture is important from a disease control standpoint because of the fact that each individual tree receives a high degree of attention and the fact that Christmas trees are usually species that are planted out of range and, hence, are susceptible to many pests. Most Christmas trees are systematically harvested and replanted as a crop. This management involves large annual investments in shaping, fertilization, vegetation control, and disease and insect management. Foliar diseases are of major concern because a tree with discolored or missing foliage is of little value. Some root decays are also of concern because tree mortality may occur before they can be harvested. Foliar sprays, up to a dozen or more, may be applied each year to control pests. Unlike forest products such as sawtimber, in which defects can be tolerated, and pulpwood, where more chemicals can be added during processing to alleviate excessive blue stain, Christmas trees have a very low acceptance level, which is based entirely on appearance. A tree with insect injury or excessive needle cast has little market. Although off-color needles can be dyed green just before harvest, there is no treatment that will keep the needles from being prematurely cast.

Seed Orchards. Seed orchards have increased dramatically in acreage since 1974 when there were 3,705 ha in all states. Since 1974, most states in which forestry is an important industry and virtually all forest industries that manage their own or client's lands plant a large proportion of their holdings in superior stock.

Southern pine management, which includes 40% of all commercial forest acreage in the United States, relies on artificial regeneration of genetically improved planting stock. In 1980 there were approximately 845 ha of clonal loblolly pine seed orchards in the North Carolina State Tree Improvement Cooperative. In 1983, 68,500 bushels of cones were produced; they yielded 49 tons of genetically improved seeds. This was enough seed to grow 784 million seedlings and to regenerate 530,000 ha of forested land. Genetically improved stock was used in 82% of plantings on industry lands. Of seedlings grown in state nurseries, 60% were of genetically improved stock. This improved stock resulted in a very high value of the seed crop, $440/kg or more.

Desirable traits usually include fast growth balanced by desired product quality, good form, and, frequently, pest resistance. For example, slash and loblolly pine seedlings are routinely planted on sites that are at high risk for fusiform rust. In the southeastern states breeding for rust resistance is just as important as any other trait, perhaps more so. A tree that may be very good in all traits except rust resistance will be worthless if it is girdled by rust.

Diseases are quite numerous but are of variable importance, depending on the tree species, the diseases unique to that area, and the cultural treatment of the seed orchards, which are very intensively managed, much as with traditional agriculture. The orchards are mowed, fertilized, subsoiled, pruned, and otherwise maintained. There is a single purpose, not multiple use, and that is to produce genetically improved planting stock. For this reason seed orchards usually are small areas of extremely valuable forest property.

Seed orchards should not be confused with seed production areas that are phenotypically superior stands. During their formation, poor phenotypes, with poor growth rate, height, or form class, for example, are rogued out. They are treated in various ways to produce large quantities of seed. Seed-production areas are usually an intermediate step until the full-fledged seed orchard is up to full production.

Seed and cone insects are undoubtedly of much more importance than are diseases in influencing seed production in orchards. In many instances, insect damage is the major factor determining the economic feasibility of a seed orchard. In some untreated southern pine orchards, losses may be as great as 90%. Developing flowers, seeds, and cones are very rich in nutrients, particularly amino acids, and can cause insect populations to flourish. Seed and cone insects have caused annual losses to southern forest enterprises alone in excess of $25 million.

Forest Tree Nurseries. Forest tree nurseries have some very significant disease problems but, interestingly, few insect pests. In 1974 there were approximately 3,600 ha of forest tree nurseries in the United States. This area has increased dramatically since then. Although many tree seedlings can be produced per square meter of nursery bed, treatments are many, varied, and expensive. Seed beds must be fumigated, carefully prepared, inoculated with mycorrhizae, planted at the proper time, irrigated, fertilized, mulched, weeded, sprayed for foliage diseases, root pruned, and lifted when seedlings have grown to the proper size. Fumigation costs alone may exceed $1,500/ha and, unless carefully done with proper weather and soil conditions, may be partially or totally ineffective.

Because of the short growing cycle and intensive treatments, insects are seldom able to build up large enough populations to cause damage. Fungi, though, can cause heavy losses, primarily resulting from damping-off, and some root decays and foliage diseases. Damping-off can cause quick and spectacular losses and little can be done to stop an epidemic once it has started.

Surburban and Urban Forests. The suburban and urban forests are difficult to define. Although the suburban forest may have some value for products, as cover for wildlife, or for protection for watersheds, the major value of both forests is in their aesthetics. Here, an individual tree can have a considerable value, and it need not necessarily be one under the shade of which George Washington once slept. Individual trees add considerable value to the landscape and can be worth thousands of dollars apiece. Of course, they may require many years to grow and attain their present stature, and in the urban forest time is especially valuable.

As a result, the special needs of the suburban and urban forest have created a whole cadre of tree specialists, professionals and nonprofessionals, who cater to nearly every whim of the property owner, and the problems are many. Species are often planted off-site and often the site is a harsh one. Arborists, tree surgeons, and others have devised numerous treatments for root feeding, for foliar and injection feeding, and for prevention of desiccation by watering, mulching, covering, and spraying with wax emulsions or plastics. Often trees are treated even though they may not need it or the actual damage they were suffering was minimal.

The Dutch elm disease did the most harm to the suburban and urban forest in this century. More than 40% of the planted elms have been killed, representing a loss of 400,000 trees/yr at a cost of over $100 million.

9.6.2 Quantify the Pest's Impact

Quantification may have to start with identification of the pest of concern. Determining impact may be the most difficult problem of all. Usually determining impact involves translating some measure of the pathogen into an anticipated amount of loss or damage. Loss may be death of the tree, growth loss, or loss of a particular product. Very often the relationship between the amount of inoculum detected and resulting injury is speculative. It may not be a linear cause-and-effect relationship. For fusiform rust determining this relationship would involve finding the connection among the number of branch or stem galls at age 5, the predicted morality at age 7, and the expected volume yield at age 16. This example would be for pulpwood production. If sawtimber was the desired goal, a somewhat greater level of mortality, which would lead to lower stocking, might be acceptable. A large number of nonlethal stem galls, however, could cause reduced product value.

9.6.3 Identify Control Options

Identification of management options is the second most difficult step. The manager must know how much control is necessary and what it costs. Could the damage have been prevented? Which management tactics are most suitable? Should they be biological, social, or economical? With the application of any management tactic, we must define the impact on the rest of the ecosystem. Will the tactic be permanent or temporary? Does it have a residual effect? Will it allow for the increase of another pest? Planting different species may be one recommendation, but the new species may also have more pests.

9.6.4 Perform Cost–Benefit Analysis

Once information is available on disease impacts, control costs, and the effect of the level of control on impact, then the costs of control can be compared with the reduction in losses (Fig. 9.5).

Control costs can be short- or long-term. Short-term costs can be compounded out to the end of the rotation. Long-term costs are very difficult to quantify, especially in a monetary sense, because they usually relate to the environment. Benefits may also be variable. Timber values depend on present prices for pulpwood, sawtimber, or fuel. These prices are not stable. Not only are there local variations, but these values are also difficult to predict. Real values may be even more difficult to estimate. Water-

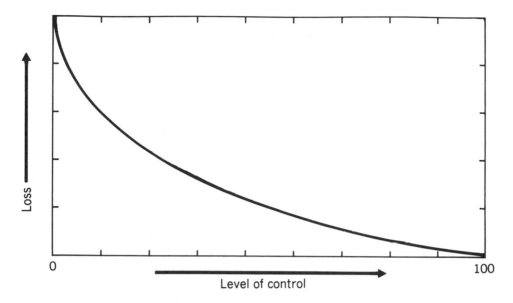

FIGURE 9.5   Losses to be expected following various levels of control.

shed protection and erosion control certainly are important as are aesthetics and the preservation of a wildlife habitat.

The following simple, but very real, example illustrates how fusiform rust and various other factors affect the cash flow (and internal rate of return, or IRR) of a slash pine plantation.*

Management Item	Year	Cash Flow ($ Acre)	IRR (%)
No Rust			
site prep. and plant	1	− 120.00	
annual mgmt. costs (fire control, taxes)	(1–26)	− 1.00	
harvest (get 45 cords at $15/cord at 0.5%/yr real price increase)	26	+ 764.00	7.26
With Rust (identical stand but with 50% rust infection at age 11)			
harvest (get 29 cords at $15/cord at 0.5%/yr real price increase)	26	+ 493.00	5.33

* The stand has 900 stems/acre, site index of 60 (base year 25).

How does rust infection affect your return on the investment? The IRR for the rust-free stand is 7.26%. In the stand with rust, the IRR is 5.33%. If an alternative rate of return is 6.00%, then the latter stand is not effective in paying a suitable rate of return and should probably be terminated.

An assumption was made that the ideal rotation age was 26 years. If, however, each of the stands was harvested at age 21, maybe the IRR would increase to 6.00%. Consider the following example.

Management Item	Year	Cash Flow ($ Acre)	IRR (%)
No Rust harvest—36 cords	21	+600.00	7.91
With Rust harvest—24 cords	21	+400.00	5.67

Although the IRR did not reach 6.00%, it did improve somewhat in the stand with rust. At this point in time, harvesting the rust-free stand at age 21 would also yield a better IRR than harvesting it at age 26.

Other options that could be explored include letting the stands grow to sawtimber size, capturing mortality by salvage cutting, or harvesting earlier. Refined growth and yield models are available for most commercial timber species. As pest information is added to these models, very powerful tools that can greatly assist land managers in making the best decision possible are yielded.

9.6.5 Make a Decision

A decision is finally made regarding the implementation of a disease-management program. It may be based entirely on the results of the benefits versus the cost or, as sometimes happens on public lands, on a political or public relations basis. This decision may be easy, but often it is very difficult because of a lack of information.

Once the decision is made to implement a disease-control program, a strategy is developed based upon the best information available (Fig. 9.6). What is the best way to apply the treatment? Should helicopters be used? Will a full-time salvage crew be needed? After the treatment is applied, a survey must be made to see if there is the appropriate response. If not, the remaining treatments must be modified to increase the likelihood of successful control.

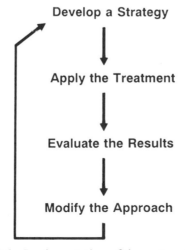

FIGURE 9.6 Implementation of the pest control program.

LITERATURE CITED

Anderson, N. A., D. W. French, G. R. Furnier, W. P. Hackett, and C. A. Mohn. 1990. A summary of aspen genetic improvement research at the University of Minnesota, pp. 231–235. In R. Adams (ed.), *Aspen Symposium '89, Proceedings*. USDA For Serv., Gen. Tech. Rept. NC-140. 348 pp.

Baker, F. A. 1988. The influence of forest management on pathogens. *Northwest Environ. J.* 4:229–246.

Baker, F. A., and D. W. French. 1985. Economic effectiveness of operational therapeutic pruning for control of Dutch elm disease. *J. Arboric.* 11:247–249.

Baker, F. A., M. Slivitsky, and K. Knowles. 1992. Impact of dwarf mistletoe on jack pine forests in Manitoba. *Pl. Dis.* 76:1256–1259.

Baker, F. A., D. L. Verbyla, C. S. Hodges, Jr., and E. W. Ross. 1993. Classification and regression tree analysis for assessing hazard of pine mortality caused by *Heterobasidion annosum*. *Pl. Dis.* 77:136–139.

Bingham, R. T. 1983. *Blister Rust Resistant Western White Pine for the Inland Empire: The Story of the First 25 Years of the Research and Development Program*. USDA For. Serv., Gen. Tech. Rept. INT-146. 45 pp.

Campbell, W. A., and O. L. Copeland. 1954. *Littleleaf Disease of Shortleaf and Loblolly Pines*. USDA Circ. 940. 41 pp.

Dolezal, W. E., and F. H. Tainter. 1979. Phenology of comandra blister rust in Arkansas. *Phytopathology* 69:41–44.

Froelich, R. C., T. R. Dell, and C. H. Walkinshaw. 1966. Soil factors associated with *Fomes annosus* in the Gulf states. *For. Sci.* 12:356–361.

Hawksworth, F. C. 1977. *The 6–Class Dwarf Mistletoe Rating System*. USDA For. Serv., Gen. Tech. Rep. RM-48. 7 pp.

Hoff, R. J. 1984. *Resistance to Cronartium ribicola in Pinus monticola: Higher Survival of Infected Trees*. USDA For. Serv., Res. Note INT-343. 6 pp.

Hungerford, R. D., R. E. Williams, and M. A. Marsden. 1982. *Thinning and Pruning Western White Pine: A Potential for Reducing Mortality Due to Blister Rust*. USDA For. Serv., Res. Note INT-322. 7 pp.

Hunt, R. S. 1982. White pine blister rust control in British Columbia. 1. The possibilities of control by branch removal. *For. Chron.* 58:136–138.

Lehrer, G. F. 1982. Pathological pruning: A useful tool in white pine blister rust control. *Pl. Dis.* 66:1138–1139.

Manion, P. D. 1981. *Tree Disease Concepts*. Prentice-Hall. Englewood Cliffs, New Jersey. 399 pp.

Meyer, M. P., and D. W. French. 1966. Forest disease spread. *Photogramm. Eng.* 32:812–814.

Merrill, L. M., F. G. Hawksworth, and D. W. Johnson. 1985. Evaluation of a roadside survey procedure for dwarf mistletoe on ponderosa pine in Colorado. *Pl. Dis.* 69:572–573.

National Academy of Sciences. 1975. *Pest Control: An Assessment of Present and Alternative Technologies*. Vol. IV. Printing and Publishing Office, Washington, D.C. 170 pp.

Oak, S. W., and F. H. Tainter. 1988. *How to Identify and Control Littleleaf Disease*. USDA For. Serv., Prot. Rept. R8–PR12. 15 pp.

Ostry, M. E., and N. A. Anderson. 1990. Disease resistance in a wild system: Hypoxylon canker of aspen, pp. 237–241. In R. D. Adams (ed.), *Aspen Symposium '89, Proceedings*. USDA For Serv., Gen. Tech. Rept. NC-140. 348 pp.

Raju, B. C., and J. M. Wells. 1986. Diseases caused by fastidious xylem-limited bacteria and strategies for management. *Pl. Dis.* 70:182–186.

Roth, L. F. 1966. Foliar habit of ponderosa pine as a heritable basis for resistance to dwarf mistletoe, pp. 221–228. In H. D. Gerhold, E. J. Schreiner, R. E. McDermott, and J. A. Winieski (eds.). *Breeding Pest-Resistant Trees, Proceedings of a NATO and NSF Advanced Study Institute on Genetic Improvement for Disease and Insect Resistance of Forest Trees, University Park, PA., 1964.* Pergammon Press, Oxford. 505 pp.

Scharpf, R. F., R. S. Smith, and D. Vogler. 1987. *Pruning Dwarf Mistletoe Brooms Reduces Stress on Jeffrey Pines, Cleveland National Forest, California.* USDA For. Serv., Res. Pap. PSW-196. 7 pp.

Schowalter, T. D., and G. M. Filip. 1993. *Beetle–Pathogen Interactions in Conifer Forests.* Academic Press, San Diego. 252 pp.

Stage, A. R., C. G. Shaw, III, M. A. Marsden, J. W. Byler, D. L. Renner, B. B. Eav, P. J. McNamee, G. D. Sutherland, and T. M. Webb. 1990. *User's Manual for Western Root Disease Model.* USDA For. Serv., Gen. Tech. Rep. INT-267. 49 pp.

Stennes, M. A., and D. W. French. 1987. Distribution and retention of thiabendazole hypophosphite and carbendazin phosphate injected into mature American elms. *Phytopathology* 77:707–712.

Williams, R. E., and C. D. Leaphart. 1978. A system using aerial photography to estimate area of root disease centers in forests. *Can. J. For. Res.* 8:214–219.

PART TWO

BIOLOGY AND MANAGEMENT
OF FOREST DISEASES

Part Two contains detailed descriptions of representative tree diseases of the North American continent. Each disease is selected on the basis of regional importance and historical or current impact, its usefulness to demonstrate disease principles of management, the differences between symptoms and causal agent, or priorities in tree disease research, particularly during the last decade. Each disease will be presented according to the following profile:

1. Importance
2. Suscepts
3. Distribution (map or figure as appropriate)
4. Causal Agent(s)
5. Biology
6. Epidemiology
7. Diagnosis: symptoms; signs (figures)
8. Control Strategy (management)
9. Other Selected Diseases (optional)
10. Selected References

In cases where a given disease substantiates a particularly important principle or well illustrates a unique feature that students should be encouraged to remember, a boxed format will be used for emphasis.

10

ROOT PATHOLOGY

10.1 BIOLOGY OF ROOT DISEASES

Plant roots have been subject to invasion by fungi and bacteria ever since plants adapted to land colonization during the Devonian Period (ca. 300 million years ago). As plant roots evolved, they were in turn invaded by fungi. During the ensuing millennia, very complex relationships were established among roots, parasites, saprophytes, and the soil. Symbioses that imparted special survival advantages to each of the partners formed. Because of the intense competition in the soil environment, relationships that did not adjust to changing conditions did not survive.

When agriculture began to develop a few thousands of years ago, a dramatic shift in the root environment occurred. This shift caused an imbalance in the natural competitiveness of soil organisms. Natural biological control of root pathogens was diminished, and epidemic root diseases began to take their toll. Today, root diseases are still responsible for major losses of most agricultural crops, and a great expenditure of money, resources, and energy are spent to minimize them.

This does not mean to imply that a pristine forest is not affected by root diseases. Root diseases undoubtedly took their toll in the geologic past as weather changes or climate fluctuations caused stress, which weakened trees and increased their susceptibility to colonization by root pathogens. Root pathogens, thus, tended to be scavengers, attacking those trees that were in less than a perfect state of health. They were always present to some extent but were likely never dominant in a given population. The loss in old growth caused by root disease was largely as mortality (Childs and Shea 1967) (Fig. 10.1). While root diseases cause some mortality in younger growth (these are usually forests that have been under some level of intensive management), a far greater volume loss occurs in the form of lost growth in trees that remain alive.

If we examine the entire spectrum of forestry activities today, there is an apparent close correlation between increased intensity of management and losses due to root pathogens. In older western North American forests, this relationship may result from

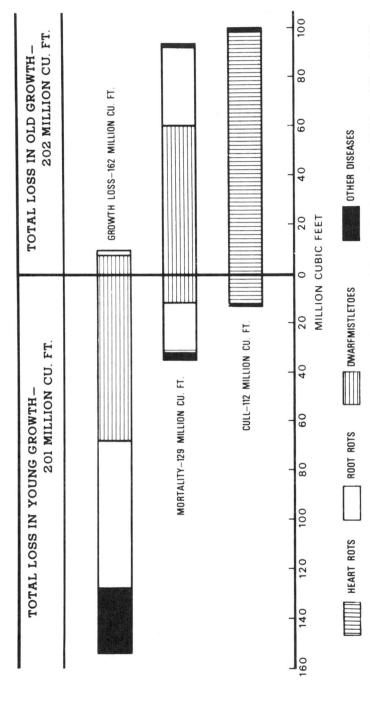

FIGURE 10.1 Principal forms and causes of disease loss in Washington and Oregon. Redrawn from Childs and Shea (1967).

disturbances such as logging, windthrow, fire, or fire control. Monoculture and con-taminated soil exacerbate root problems in forest tree nurseries. In fertilized and well-cared-for seed orchards, root and soil compaction may lead to losses.

10.2 DIAGNOSTIC FEATURES

Root pathogens produce either of two basic symptoms that may be useful for diagno-sis. Both, however, can be confused with attack by other pathogens or may simply result from adverse abiotic factors.

Premature mortality is the most striking result of root attack and can occur in very young seedlings, very old trees, or any age class in between. Overnight decline and death of 60–80% of the seedlings in a conifer nursery may be rather easy to diagnose but will nonetheless still have a traumatic effect on the manager of that nursery when he or she views the damage the next day. Death of pole-sized trees affected with annosum root disease or littleleaf disease, likewise, produces dramatic and unequivo-cal indications that there is a serious tree health problem. Windthrow of root-decayed older specimen trees in residential or recreation areas also have a unique quality that assists in diagnosis.

Growth loss and predisposition to other pests, however, is another feature of root disease that may be much more difficult to diagnose. Often, loss may not be easily apparent until secondary organisms invade and produce characteristic symptoms, or very tedious and detailed examination is made of the affected trees or stand. Even then the exact role of the suspected root pathogen may not be determined with any degree of certainty.

10.3 EVOLUTION OF ROOT-DISEASE CONCEPTS

10.3.1 Physical Edaphic Influences

Soil is a complex mechanical system. It consists largely of mineral matter such as sand, silt, and clay but may also include liberal amounts of organic matter. The organic matter and clay may cement the mineral particles together to form lumps or clods of varying sizes.

Soil is formed by the physical weathering of rocks to cause their eventual disinte-gration into smaller pieces. Physical disintegration is accompanied by chemical de-composition, largely through the actions of exposure to air and water.

At any given time, a particular soil will contain some amount of water. The water will be a solution of nutrient elements that, with the exception of carbon and oxygen, are derived from the breakdown of mineral and organic matter in the soil. The soil solution has certain pH characteristics depending particularly on the parent material and nutrient concentration that, by themselves, have a large influence on plant growth. Plant roots tend to be rather leaky and add sugars and amino acids to the soil solution. The rich variety and abundance of mineral and organic nutrients ensures an attractive environment for growth of soil microbes. The buffering capacity of soil to extreme fluctuations of temperature and moisture provide additional support for an active microbial population. Fungi, bacteria, and actinomycetes, in turn, aid in the formation of soil aggregates by insoluble gummy secretions.

When a root dies or a piece of dead organic material falls to the soil surface, an intense coordinated attack begins by the primary consumers (Fig. 10.2). Bacteria and fungi simultaneously begin to degrade it. Larger consumers such as snails, insects, or earthworms may eat the entire piece; smaller consumers nibble away parts. All the primary consumers use the energy stored in the plant material. Secondary consumers

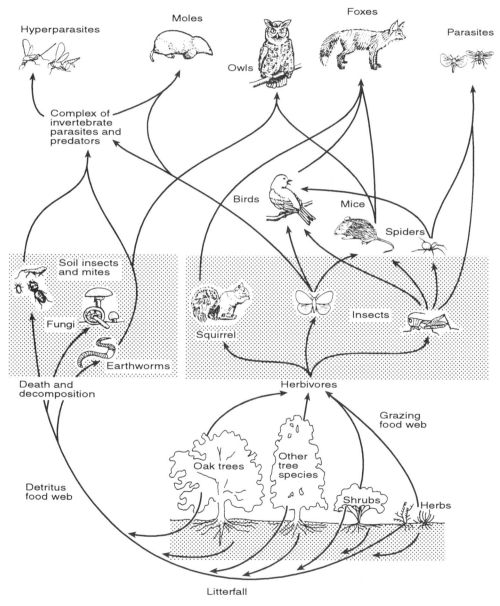

FIGURE 10.2   Energy flow through the trophic web of an oak woodland. Drawn by Valerie Mortensen.

devour the primary consumers. The secondary consumers could include carnivorous animals, which feed on the snails, insects, or earthworms, or fungi and bacteria, which feed on the dead bodies of the primary consumers, including bacteria and fungi. Still further up the food chain are the tertiary consumers, which prey on the carnivores and fecal material and dead bodies of the secondary consumers. Fungi and bacteria are also active as tertiary consumers.

Because of their rapid reproductive traits, soil microbes are normally present in soils in extremely high numbers. Bacteria, for example, may range from a few hundred million to several billion individual cells in each gram of soil. Actinomycetes are also abundant, reaching levels of about one tenth the number of bacteria. The abundance of fungi is more difficult to estimate as one individual may produce hyphae of considerable length. Weights of fungal biomass of 568–5,682 kg/ha are not unusual. A large portion of fungal biomass is in a beneficial and economically important form known as mycorrhizae, which are discussed in Chapter 12.

Roots of higher plants grow and die, not only changing soil properties and structure but also leaving behind organic debris and exudates that greatly influence populations of primary consumers, including potential root pathogens.

The zone of soil in the immediate vicinity of the growing root tip is known as the rhizosphere. This term was first used by L. Hiltner in 1904 to describe this zone of soil and to emphasize the special influence it had on soil microflora. While the growing root tip does slough off some dead organic material, its major influence is via the chemical exudates that leak from the dividing and elongating cells in the root tip. The ratio of bacteria, actinomycetes, and fungi in the rhizosphere may be 10 to 20 times greater than that in surrounding soil. Some of this increased population may include potential pathogens.

Root exudates have a significant effect on some root pathogens. The stimulatory effect on zoospores of *Phytophthora cinnamomi* (Zentmyer 1980) is well known (Fig. 10.3). Not only are zoospores attracted by concentration gradients of root exudates, but they also have the capacity to induce germination of spores of some pathogenic fungi.

10.3.2 Microbial Ecology

Although soilborne fungi and bacteria are often well dispersed, they tend to have larger populations in the upper horizons. Some species are prevalent in certain vertical zones. Dispersal may be assisted by movement of soil by water and by the activity of soil animals.

A major role of soil microorganisms is the decomposition of organic residues. A large volume and variety of dead biological material finds its way into the soil. This is broken down and is resynthesized as microbial boiomass or is excreted as metabolic waste. Both of these are in turn broken down by other microorganisms.

The time required for decomposition varies depending on the composition of the organic matter. Sugars and other water-soluble materials are degraded very quickly, usually within days. Proteins are hydrolyzed and also disappear within a few days or so. Simple polysaccharides such as starch take a little longer. Decomposition of complex polysaccharides such as cellulose may take months. Lignin may take years. Suberin and cutin are degraded very slowly and often remain after the lignin and cellulose have disappeared. These form much of the organic portion of humus.

FIGURE 10.3   Within a few minutes after placing a shortleaf pine seedling in a suspension of actively swimming zoospores of *Phytophthora cinnamomi,* the spores have been attracted to the region of elongation, encysted, and germinated. Courtesy of S. W. Fraedrich, USDA Forest Service.

The breakdown of complex plant and animal residues is carried out by a wide variety of soil flora and fauna. Larger materials may be chewed into smaller pieces by insects and passed along as fecal material. Further breakdown by fungi and bacteria is done by extracellular enzymes that are released into the substrate. These enzymes in turn release water-soluble organic fragments that diffuse away from the site and are absorbed by microorganisms in the immediate vicinity. Availability of these fragments may cause intense competition between microorganisms. Much of the carbon is released into the air as carbon dioxide.

The fate of nitrogen depends largely on the form of the nitrogen released and the soil type. In acid soils, nitrogen is released as ammonia and is quickly absorbed by roots or microorganisms or is complexed by organic matter. Forested systems may be extremely efficient and recycle nearly all the nitrogen in this manner. In other soils, nitrate, which is more susceptible to leaching, may be formed.

Denitrification is another form of nitrogen loss. Some microorganisms can decompose the nitrate and release free nitrogen as a gas. Nitrogen loss from both of these systems can be significant.

10.3.3 Biological Control

Even though the vast majority of soilborne microorganisms are beneficial in that they actively contribute to the recycling of nutrients, only a small proportion can injure crop plants. The purpose of biological control is to take advantage of the antagonistic qualities of some soil microorganisms and use these qualities to inhibit growth or reproduction of the pathogen.

The major strategies employed in the management of the associated microbiota for the purpose of reducing disease are generally of two major types.

1. The inoculation of the soil with selected competitive or antagonistic microorganisms. This strategy has been generally disappointing because of the buffering capacity of the resident microflora. It may be more suitable for agricultural soils or forest tree nurseries where we can get a temporary seasonal advantage.
2. The alteration of soil conditions by the appropriate management so as to enhance natural biological control by the already existing soil microflora.

Cook (1977) has presented a good review of the various aspects involved in management of the soil.

1. *Reducing inoculum density of the pathogen.* The aim is to either destroy propagules of pathogens or prevent their formation. Crop rotation may starve the pathogen or weaken it, and it is eventually consumed by other microbiota. Addition of organic or sugar amendments may temporarily cause an increase in competitors that excrete chemicals that cause lysis of propagules following their germination. Favoring growth of nematode-trapping fungi has been considered, but practical results obtained thus far in the field have been disappointing.

2. *Replacement of a pathogen in plant refuse.* This aspect is effective for those pathogens that must survive in plant residues for extended periods and use the residues both as a refuge and as a food base. It illustrates the "possession principle" of plant pathogens. The key to success in active possession is a continued slow metabolism on part of the pathogen.

The root pathogen *Armillaria* spp. is a classic example. Low-valued, or scrub, hardwoods are often killed with herbicides or by other means to assist in preparing the site for other species. The dead tree roots are quickly invaded by rhizomorphs of *Armillaria*. Once this positional advantage is gained, the fungus can withstand attack by competitors and remain alive for years. A decade, or more, later the pathogen still has sufficient inoculum potential to quickly invade roots of trees that were subsequently planted and cause significant mortality.

Conditions that need to be changed to weaken the hold of the pathogen include:

a. Producing a physical environment unfavorable to metabolism of the pathogen but not to potential colonists. *Armillaria* is controlled in Africa by girdling trunks of standing trees before the harvest. This procedure encourages invasion of the roots by saprophytes in advance of the pathogen. Extraction of stumps of ponderosa pine in Oregon causes drying of the wood, which reduces the inoculum potential of *Armillaria*.

b. Causing the lack of an essential nutrient normally supplied by the soil solution.

c. Imposing stress by a sublethal dose of a fungicide or fumigant. Stumps of citrus and forest trees are treated, weakening *Armillaria* and permitting the fumigant-tolerant *Trichoderma* to invade the wood.

3. *The suppression of germination and growth of pathogens.* This aspect takes advantage of the so-called widespread fungistasis, which is a very active area of interest at the present. Inhibitor substances present in soil include ethylene and ammonia, and there are likely others. Many propagules also contain powerful self-inhibitors that are deactivated by volatiles in the soil, sugars, and amino acids or by alternate periods of wetting and drying. As a result, propagules either fail to germinate or germinate at the wrong time, and thus fail to colonize an available substrate successfully.

For example, *Rhizina undulata* will not invade fire sites on alkaline soils due to heat-resistant bacteria such as *Bacillus subtilis*. These bacteria have been artificially applied to acidic soils following burning to inhibit invasion by this root pathogen. Natural control of *Heterobasidion annosum* was achieved by use of prescribed fire 9 months before thinning. The fire encouraged growth of *Lespedeza* and *Pensacola* grasses, and the soil fungi associated with these were antagonistic to *H. annosum*.

4. *Protection of an infection court involves any measure that slows or prevents infection by a pathogen.* It may involve production of an antibiotic or germination suppressant. A good example is the natural protection offered by ectomycorrhizal fungi. Under the proper conditions, root tips are actively sought by pioneer fungal colonists. Whoever gets there first wins. If the ectomycorrhizal fungus arrives first and becomes established, it can offer good mechanical and chemical protection against pathogens.

H. annosum is a devastating root pathogen of recently thinned southern yellow pine stands growing on high-risk sites. The infection court in this case is the freshly cut stump surface of the trees removed during thinning. During the sequence of events favorable to *H. annosum*, spores of this fungus land on the stump surface and germinate, and the hyphae colonize the stump and eventually grow into the roots of adjacent trees and kill them. It is, however, a poor competitor. If the stumps are inoculated with spores of *Phanerochaete gigantea*, the latter germinate, and the resulting mycelium grows faster and soon out-competes *H. annosum* for a food base. Urea has also been applied to stump surfaces for control of *H. annosum*, but it enhances growth of *Trichoderma viride*, which is antagonistic toward *H. annosum*.

5. *Stimulation of a resistance response in a potential host.* There are many instances in agricultural crops where a near pathogen is inoculated to the crops. This inoculation initiates a defense reaction that protects the plant against the more destructive pathogen.

These general management tools have enjoyed considerable success when applied to agricultural crop systems. A few of them are appropriate for forest systems, especially in intensive management applications such as seed orchards and nurseries.

Organic amendments are very useful, and few soilborne diseases cannot be controlled by their use. In Australia the root pathogen *P. cinnamomi* causes great losses in avocado orchards. The root disease can be largely eliminated if organic matter is added to the soil in the orchards. The rapid population buildup of bacteria and fungi form a pathogen-suppressive soil with increased germination and lysis of pathogen propagules.

Nitrogen fertilizers have the general effect of introducing a nitrogen burst into a system where most nitrogen is already tied up within the bodies of living microbes. Saprophytes take advantage faster than pathogens and will often replace them because of their increased competitiveness. In Oregon *Phellinus weirii* has been controlled by adding urea to the soil. The urea encourages growth of *Trichoderma*. Willow stumps have been treated with ammonium sulphamate and then inoculated with *Trametes versicolor*. This decay fungus rapidly decayed stump and root tissues before *Armillaria* could colonize.

Treatments with soil fumigants or other pesticides at sublethal dosages are not designed to eradicate the pathogen, which coincidentally might be very difficult to do, but are aimed at reducing its vigor to such an extent that competitors will find it much easier to invade the same substrate. Fumigation of stumps of ponderosa pine by methyl bromide debilitates *Armillaria,* and it is replaced by *T. viride*.

Tillage is effective, especially in nurseries. It provides temporary improvement in aeration and accelerates the drying of soil and a redistribution of substrates. During the process, pathogens are predisposed to antagonistic effects.

A cover crop may be used in conjunction with tillage. In tree nurseries a leguminous agricultural crop species is often planted between tree crops when the nursery is fallow. Nutrient reserves of the tree pathogens from the previous crop are depleted. When the cover crop is tilled in preparation for the next tree-planting cycle, resulting decay yields a burst of saprophytic microbial activity that produces additional beneficial antagonistic effects.

10.4   CLASSIFICATION OF ROOT-INFECTING AND SOIL-INHABITING ORGANISMS, CONCENTRATING ON FUNGI

Soil contains a rich and diverse population of flora and fauna. As far as number of individuals are concerned, bacteria and actinomycetes account for the largest portion of the flora. Bergey's Manual (Breed et al. 1957) lists more than 50 genera that have soil as their natural habitat, which includes approximately 1,600 species.

Soil is an extremely heterogeneous system. Within that system, bacteria tend to exist in colonies rather than as individual, isolated cells. The colonies tend to be closely attached to soil particles. The degree of clumping is greatly affected by the physical condition of the soil. The different species of bacteria may have great antagonistic influences on their neighbors. Populations of bacteria can respond very quickly to changing composition of the soil environment.

The various genera associated with nitrogen fixation and oxidation of ammonia and nitrite numerically comprise a negligible fraction of the soil bacterial population (Burges 1965) but play a vital role in these processes as well as sulfur oxidation.

Actinomycetes resemble fungi in that they are filamentous, but they are unicellular and, thus, are more closely related to the bacteria. They are of great practical importance and are responsible for the "earthy" odor of soil. They are undoubtedly of great importance in the decomposition of organic debris in the soil.

Most unicellular freshwater green or blue-green algae and diatoms can grow on moist soil. Although algae can use sunlight to synthesize organic molecules needed for growth, they can also grow as heterotrophs. For this reason, although they are more prevalent near the soil surface, they are not restricted to the surface and can be found at great depths. Algae have the ability to maintain their viability for long periods under extremely adverse conditions. They certainly contribute to the organic content of the soil.

In considering fungi in soil, Burges (1965) lists 200 Phycomycetes, 32 Ascomycetes, and 385 Deuteromycetes that occur in soil. To this list must be added a growing number of Basidiomycetes, especially important for their role in forming ectomycorrhizae.

10.5   ECOLOGY OF THE RHIZOSPHERE

Roots serve several vital functions that are important for the growth and well-being of trees. They serve as anchorage, holding the tree in place to a particular plot of soil and protecting it against windthrow. Roots serve as an underground absorption and conduction link between the aerial portion of the tree and the nutrient solutions in the soil, absorbing these nutrients and water from the soil and translocating them into the aerial portions. Roots may also serve as storage organs for nutrients, including photosynthates accumulated during photosynthesis. The abundance of these stored materials may be attractive to potential pathogens. Conversely, small amounts of stored reserves indicate a host that is not well able to withstand the attack of either root or aerial pests.

The first root that originates in the embryo is called the primary root. This root may become a taproot, which grows directly downward, giving rise to lateral roots. Many pines and oaks produce deep taproots. Other tree species, such as spruces and poplars, produce spreading root systems that tend to not penetrate as deeply as do taproot systems.

The lateral, or branch, roots form specialized roots with two quite different functions. Long roots grow outward and penetrate uncolonized soil. From these long roots arise so-called short, or feeder, roots that tend to be much shorter in length. These roots further divide as they grow and fill in the newly colonized soil volume. They tend to become mycorrhizal.

Root growth is a continuous process whose rate is influenced by soil moisture and temperature. Cell divisions within the apical meristem produce the root cap, which protects the meristem and aids in penetration of the root through the soil, and three primary meristems, which differentiate into the epidermis, cortex, or vascular cylinder, respectively. The epidermal cells of nonmycorrhizal roots develop root hairs, which increase the absorbing surface of the root. The cortex is composed of large cells with large intercellular spaces. The inner cortex contains a layer of compactly arranged cells known as the endodermis containing the Casparian strip. All the substances that are taken up by the root must pass through the Casparian strip before they can subsequently pass into the vascular cylinder for transport to other parts of the tree.

Based on their structure, the outer portion of roots tends to be easily penetrated by pathogens. The areas of active growth also tend to be nutrient-rich. Because the elongating cells are leaky, nutrients become concentrated in the surrounding soil, which, along with sloughed cell-wall debris, serves as a potential nutrient base for soilborne microorganisms, some of which may be pathogenic.

Growth of invaders in the outer cortex is also facilitated by the large intercellular spaces and the leaky nature of the cortical cells. The endodermis, however, poses a formidable barrier to many pathogens because of the Casparian strip that is impregnated with suberin and sometimes lignin. Ingress of pathogens past this point is facilitated by lateral roots, which orginate from deep within the root. As the lateral root pushes its way outward through the cortex, it digests cortical cells in its path, producing an opening in the Casparian strip. This creates a zone of entry that some pathogens use to good advantage.

All soil is characterized by an intense competition between microflora and fauna for nutrients released by the weathering of parent material and organic debris cast off by living plants and other microflora and fauna and scavanging their bodies after they die. The zone of soil immediately adjacent to roots has been called the rhizosphere. Rhizosphere activity is especially intense because of the leakage of organic materials and sloughing phenomena noted earlier, providing a rich booster shot of unexpected nutrients for those microorganisms able to claim them first. Most important root pathogenic fungi have poor competitive ability as saprophytes in the soil. They survive mainly as resistant spores or microsclerotia. Root exudates may break their dormancy and induce germination (Smith 1969) or stimulate germ-tube growth (Agnihortri and Vaartaja 1967). Tree-root exudates released into the rhizosphere probably favor the pathogen and, are thus, indirectly deleterious to the host. In some agricultural crops a growing body of evidence suggests that some root exudates may contain toxic substances that directly affect pathogens or indirectly affect them by stimulating antagonistic microorganisms. Additional influences occasioned by changes in pH, oxygen tension, and moisture in the rhizosphere may also have a deleterious effect.

10.6 ROOT PATHOLOGY CHARACTERIZATION

There are basically two types of root pathogenic fungi, and these are identified based on that portion of the root system that is attacked. Most of the previous discourse on biology of root diseases set the stage for those pathogens that attack the feeder roots of trees. These pathogens attack either in the zone of elongation of trees of any age or the young, succulent root tips of seedlings that have not yet hardened off. These feeder root diseases are described in Chapter 11.

There is another major group of root pathogens that has an entirely different mode of attack, and this group produces losses that are just as important as those that occur only in feeder roots. Members of this latter group attack entire root systems, especially those that have produced secondary growth, decaying not only the xylem of roots but often decaying wood of the butt and lower bole. These diseases are described in Chapter 13.

LITERATURE CITED

Agnihortri, V. P., and O. Vaartaja. 1967. Root exudates from red pine seedlings and their effects on *Pythium ultimum. Can. J. Bot.* 45:1031–1040.

Breed, R. S., E. G. D. Murray, and N. R. Smith. 1957. *Bergey's Manual of Determinative Bacteriology.* Bailliere, Tindall and Cox Ltd., London. 1094 pp.

Burges, A. 1965. The soil microflora—Its nature and biology, pp. 21–32. In K. F. Baker and W. C. Snyder (eds.), *Ecology of Soil-Borne Plant Pathogens—Prelude to Biological Control.* University of California Press, Berkeley, Los Angeles. 571 pp.

Childs, T. W., and K. R. Shea. 1967. *Annual Losses from Disease in Pacific Northwest Forests.* USDA For. Serv., Res. Bull. PNW-20. 19 pp.

Cook, R. J. 1977. Management of the associated microbiota, pp. 145–166. In J. G. Horsfall and E. B. Cowling (eds.), *Plant Disease: An Advanced Treatise. Vol. 1.* Academic Press, New York. 674 pp.

Smith, W. H. 1969. Germination of *Macrophomina phaseoli* sclerotia as affected by *Pinus lambertiana* root exudate. *Can. J. Microbiol.* 15:1387–1391.

Zentmyer, G. A. 1980. *Phytophthora cinnamomi and the Diseases it Causes.* Monograph No. 10, American Phytopathological Soc., St. Paul., MN. 96 pp.

11

FEEDER ROOT DISEASES

11.1 PREGERMINATION (SEED) DISEASE

Forest tree seeds which are mistreated in some way during collection or storage may exhibit reduced viability due to pathogeneses that are unrelated to potential root diseases. Although diseases of stored seed are not generally caused by root pathogens, this small, but important, subject is placed at the beginning of this chapter as an introduction to root pathology.

Fungi in Stored Seed

Most forest tree seed is stored under conditions unfavorable for seed-inhabiting fungi, but if not so maintained, viability of the seed will be reduced. Tree seed, at least that of some species, is not readily available every year; thus, seed storage is necessary to ensure annual supplies for nurseries. As long as seed is stored at a low temperature (0–5°C) and preferably at a moisture content of less than 12%, any fungi present in or on the seed will not be able to develop sufficiently to reduce seed germination. Seed such as yellow birch is completely nonviable after 18 months at 20–24°C, while at 2–4°C in airtight containers it survives for at least 4 years (Clausen 1985).

The percentage germination of spruce pine seed decreased with increased moisture content of the seed and increased storage temperature as follows (Barnett and McLemore 1967).

Moisture Content	Percent Germination After 1 Year		
	−18°C	−14°C	+1°C
6	95	89	80
9	94	80	77
12	90	83	60
15	88	61	40

With seeds of other plants such as wheat, fungi that invade these seeds can cause complete loss of viability if the moisture content is 14% or higher. In some seeds the lower limit for fungus development may be less than 14%. Very likely tree seed is similar in this respect, and we can assume that moisture contents above 12–15% may be favorable for the invasion and deterioration of seed. Low temperatures, below freezing, would limit or completely inhibit the development of fungi.

The fungi, commonly called storage fungi, that can invade and kill seed are species in the genera *Aspergillus* and *Penicillium*. These species often are xerophytic in nature and able to develop normally at lower moisture contents than most kinds of fungi. In wheat seeds and the seeds of other agricultural crops, these storage fungi invade the seed after it is in storage. The fungi that occur on and in seed in the field are different species, referred to as field fungi. The common genera of fungi that have been isolated from tree seed are *Alternaria*, *Aspergillus*, *Cephalosporium*, *Chaetomium*, *Cladosporium*, *Gliocladium*, *Penicillium*, *Aureobasidium*, and *Trichoderma* (Timonin 1964). The kinds and numbers of fungi vary among tree species. The presence or absence of some combinations of fungi may affect the early survival of seedlings. Damping-off fungi such as species in the genera *Pythium*, *Phytophthora*, and *Rhizoctonia* were not isolated from jack pine, lodgepole pine, and white spruce; however, damping-off symptoms occurred when these seeds were surface sterilized (with sodium hypochlorite) and planted in sterilized soil, thus suggesting that the saprophytic fungi in the seed could attack seeds and seedlings when competing organisms were not present (Timonin 1964). It is unlikely that surface treatments to supposedly sterilize the outer surface of the seed can eliminate fungi within the seed. Seeds artificially pierced to resemble damage to the seed coat were susceptible to the fungi present on the seed (Gibson 1957).

A seed pathogen that does not quite fit into either of the preceding general groups is *Fusarium subglutinans* (Hyphomycetes), the causal agent of pitch canker (Dwinell et al. 1985) (Fig. 11.1). Although this pathogen causes significant losses because of stem malformation, growth suppression, and tree mortality, it can also cause unacceptable loss of seed and cones in production seed orchards and loss of seedlings in infested seed beds and subsequent outplanting. It is, thus, an active pathogen that does not necessarily depend on poor storage conditions in order to infect seed.

LITERATURE CITED

Barnett, J. P., and B. F. McLemore. 1967. Improving storage of spruce pine seed. *Tree Planters' Notes* 18:2.

Clausen, K. E. 1965. *Yellow and Paper Birch Seeds Germinate Well after 4 Years in Storage.* USDA For. Serv., Note LS-69. 2 pp.

Dwinell, L., J. B. Barrows-Broaddus, and E. G. Kuhlman. 1985. Pitch canker: A disease complex. *Pl. Dis.* 69:270–276.

Gibson, I. A. S. 1957. Saprophytic fungi as destroyers of germinating pine seeds. *East African Agri. J.* 22:203–206.

Sutherland, J. R., T. Miller, and R. Salinas Q. 1987. *Cone and Seed Diseases of North American Conifers.* North Amer. For. Comm. Pub. No. 1, Victoria, B. C. 77 pp.

Timonin, M. I. 1964. Interaction of seed-coat microflora and soil microorganisms and its effects on pre- and post-emergence of some conifer seedlings. *Can. J. Microbiol.* 10:17–22.

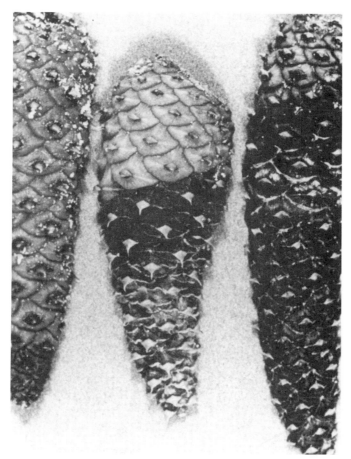

FIGURE 11.1 Necrosis of loblolly pine cones caused by *Fusarium subglutinans*. From Sutherland et al. (1987).

11.2 POSTGERMINATION (ROOT) DISEASES

The first root of a seed plant develops from the root promeristem of the embryo. This forms the taproot, which, with its various lateral roots, soon forms the root system. Root systems that do not have a taproot are diffuse, or spreading, root systems. Roots are mainly concerned with anchorage, absorption, and conduction and to a lesser degree with storage.

The young root is a simple structure with no separation into nodes and internodes. In cross section a clear separation into three distinct tissues is possible: the epidermis, the cortex, and the vascular system (Fig. 11.2). A rootcap may be present only if the root tips have not formed a mycorrhizal union with symbiotic fungi. Likewise, root hairs are present only when the root tips are nonmycorrhizal. Even without root hairs, the epidermis seems to be specialized for absorption. A thin cuticle may be present on the outer epidermal surface. Nevertheless, the root epidermis offers little resistance to potential pathogens.

Immediately inside the epidermis is the cortex. This is composed of parenchyma

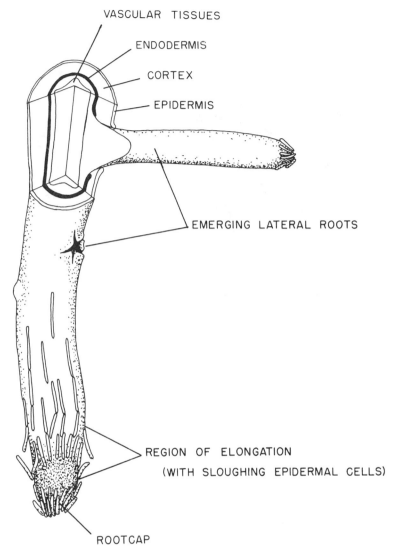

FIGURE 11.2   Morphology of a young conifer root.

cells, usually with conspicuous intercellular spaces. Cortical cells usually do not contain chloroplasts but often do contain large amounts of starch. The vascular system is found in the central axis of the root, which is separated from the surrounding cortex by a suberized bandlike structure known as the Casparian strip. The Casparian strip apparently has some regulatory function in the movement of materials from the cortex into the vascular system and perhaps vice versa. The vascular system consists of xylem and phloem elements, which are responsible for conduction into and out of the root tip.

As stated earlier, the root tips of young germinating seedlings and the feeder roots of older trees are very susceptible to attack by pathogens. The region of elongation, which is located just behind the root tip, is comprised of cells that are rapidly increasing in length. They are loosely joined to each other and because the cell walls are not completely formed and matured, they leak components of the cell protoplasm. Sugars, nucleic acids, amino acids, and other organic materials seep into the immediate soil environment and have a profound influence on microorganisms. In addition, the generally loosely packed nature of the epidermal and cortical cells allows easy entrance to pathogens that are in the immediate vicinity.

After several weeks the roots of a young seedling will harden off as they produce secondary growth. When this happens, the root tips become resistant to attack by some feeder root pathogens, often referred to as the damping-off fungi. In woody species the secondary tissues that form include a cylindrical vascular cambium, increased amounts of thick-walled sclerenchyma cells in the phloem, and a periderm, which is a barklike structure. Cork may also form. Usually many cells have lignified walls. Roots with secondary structure are generally much more resistant to damping-off fungi, but for a variety of reasons they may be attractive to other root pathogens. This attraction is especially noticeable where lateral roots form. Lateral roots are formed at the periphery of the vascular cylinder. As the young lateral roost grow, they push aside cortical cells and emerge at the surface of the epidermis. Although the growth of lateral roots is important in creating new absorbing surfaces for the tree, it also creates new infection courts where each root emerges from the parent root.

DISEASE PROFILE

Damping-Off

Importance: Damping-off is the most prevalent nursery disease and a serious constraint to tree seedling production, particularly of conifers, in regions typified by cool, wet spring weather.

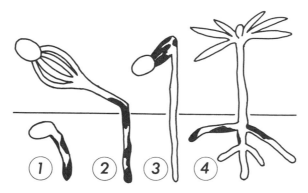

An illustration of the four types of damping-off.

Suscepts: Most conifers (except cedars, junipers, and cypress) and some hardwoods (particularly sweetgum, black locust, and yellow-poplar) are susceptible. No species of either group, however, is known to be completely immune.

Distribution: Damping-off is cosmopolitan to most soils but is especially prevalent in those under cultivation.

Causal Agents: These soil-inhabiting fungi are primarily of the genera *Fusarium* (Hyphomycetes), *Pythium* and *Phytophthora* (Peronosporales), and *Rhizoctonia* (Agonomycetes).

Biology: By direct penetration, the damping-off fungi cause rapid decay of seed via the developing radicle or of the succulent roots and stem tissue of emergent seedlings; seedling resistance ultimately develops as suberization of tissue occurs.

Epidemiology: Damping-off is influenced primarily by soil temperature-moisture interactions as a function of particular soils and/or critical weather conditions at the time of seeding; the various edaphic optima for the pathogens and suscept are generalized and follow.

Diagnosis: Generally, four basic types of damping-off occur during seedling development: (1) preemergence damping-off involves decay of the radicle and seed and is seen only as germination failure because death has prevented emergence; (2) postemergence damping-off is visible in emergent seedlings as necrosis of root and stem tissue at or near the ground line, thus producing a weakening effect that causes their toppling over and imminent death; (3) postemergence damping-off may occur also as top damping-off, wherein persistent seed coats with attached fungi account for cotyledonary infection, particularly under conditions of crowding and poor air circulation; or (4) postemergence damping-off may occur in time as seedling root decay or late-damping off, in which the resistance of tissue suberization, as marked by secondary root development, is overcome by growth conditions that stress the seedlings and favor the damping-off fungi.

Control Strategy: Assuming that the nursery soil is a well-drained loam, damping-off is prevented normally by good cultural practice, namely, (1) by seeding after soil temperature is above 16°C, (2) adjusting soil pH to 6.0 or below, (3) avoiding dense stands, (4) delaying nitrogen fertilization until 6 weeks after seeding, and (5) mulching lightly only to conserve moisture. In problem beds, chemical protection or eradication may be warranted by employing, respectively, fungicides as seed treatments and soil drenches or all-purpose fumigants in preplant sterilization of the beds.

SELECTED REFERENCES

Cordell, C. E., R. L. Anderson, W. H. Hoffard, T. D. Landis, R. S. Smith, Jr., and H. V. Toko (Tech. Coords.). 1989. *Forest Nursery Pests.* USDA For. Serv., Agri. Handbook No. 680. 184 pp.

Hodges, C. S., Jr. 1962. *Diseases in Southeastern Forest Nurseries and Their Control.* USDA For. Serv., SE For. Exp. Sta. Pap. 142. 16 pp.

Vaartaja, O. 1964. Chemical control of seed beds to control nursery diseases. *Bot. Rev.* 30:1–91.

11.3 DAMPING-OFF

Damping-off, which may be the most important pathological problem in nurseries, is a disease involving rapid decay of the newly germinated seeds or of the roots and stems of succulent seedlings. The term *damping-off* is derived from the fact that it often develops during damp weather. The seedlings of most plants are susceptible to damping-off; those with stems and roots that remain succulent for some time are likely to be more susceptible than those whose roots soon become hard and woody. For this reason, conifer seedlings are susceptible for a longer period of time than the seedlings of most other plants.

Economic Importance

Damping-off occurs commonly throughout the world wherever seedlings grow, in greenhouses and outside in nurseries and natural areas. Damping-off has been at times a major problem in the production of sugar beets, tomatoes, other agricultural plants, and many species of ornamental flowering plants; it is one of the chief obstacles in the way of raising coniferous seedlings. In 1964, 129 nurseries in the United States produced almost 740 million seedlings and 81% of these were species of pine (Abbott and Eliason 1968). Most conifers are very susceptible to damping-off; cedars and cypress are resistant. Susceptibility to damping-off seems to be inversely correlated with degree of shade tolerance, the pines being intolerant of shade and highly susceptible. In contrast the white cedars are shade-tolerant and resistant to damping-off. Among the deciduous species, sweetgum, black locust, and yellow-poplar are susceptible. Although losses would be more serious in nurseries if not controlled, damping-off fungi take their toll in natural stands and when seeds are sown directly. In a directly seeded logged area in western Colorado, damping-off accounted for almost 15% loss in lodgepole pine and Englemann spruce seedlings (Ronco 1967).

Symptoms

Generally, four types of damping-off are recognized.

1. *Germination or preemergence damping-off.* The seedlings are attacked immediately after germination, before they emerge from the soil, and the only outward symptoms of the disease is the nonappearance of the seedlings. Unless the percentage germination of the seeds has been determined previously, this type of damping-off might be confused with lack of plants due to nonviable seeds.

2. *Normal damping-off.* The roots and stems of the young seedlings are attacked. If the stems are rotted at the ground line, the seedlings fall over and either dry out or are completely rotted. Sometimes the roots are killed up to within a centimeter of the soil surface, and the trees die slowly, but remain standing.

3. *Late damping-off.* Late damping-off occurs after stems and some of the roots have begun to become woody. The small, new roots of seedlings are susceptible to rot by these fungi as long as they remain tender. Because the roots become woody gradually, there is no sharp division between normal and late damping-off. Usually late-damping off is that which appears after secondary roots have developed. The smaller roots of trees may be attacked by these soil-inhabiting fungi at any time in the life of the tree.

4. *Top damping-off.* Sometimes the cotyledons of a seedling are held in the seed coat for some time after the plant emerges from the soil. Fungi may attack and kill these cotyledons, causing top damping-off, but this type of the disease is relatively rare and unimportant.

Cause and Development

Damping-off is caused by fungi that usually live saprophytically on organic matter and inorganic salts in the soils. The most important of these fungi are species of *Fusarium, Phytophthora, Pythium,* and *Rhizoctonia.* Species of *Fusarium* are most commonly associated with damping-off in southern nurseries. Although most tree species can be attacked by several species of fungi, *Pythium sylvaticum* has been important on sweetgum seedlings (Filer 1967) while black locust is highly susceptible to *Pythium ultimum* (Wright 1957). Species of *Cylindrocladium* and other genera may cause damping-off, but because of their ability to attack seedlings more than 1 year old, these fungi are classed with the root-decaying organisms.

Damping-off fungi penetrate directly through the cell walls of the root epidermis and grow intracellularly. They digest the cell contents, kill the cells, and decompose the cell walls to a certain extent. Most damping-off due to *Pythium debaryanum* occurs within 20 days of seedling emergence because of the barrier presented by the suberized endodermis. Breaks in this endodermis caused by such things as emergence of secondary roots may allow fungi to colonize parenchyma tissue inside the endodermis (Hock and Klarman 1967).

As a general rule of thumb, conifer seedlings are susceptible to damping-off only during the first 4 weeks of growth or so after seed germination (Fig. 11.3). Most cultural control strategies are aimed at minimizing this period of susceptibility.

Fusarium oxysporum, which can cause damping-off, also causes a hypocotyl rot of sugar pine, white fir, and grand fir. These hosts develop resistance to the fungus 3 weeks after emergence. Roots were not infected (Brownell and Schneider 1983). Terminal shoots of loblolly and slash pine are killed by *Diplodia gossypina* and *F. subglutinans* (Rowan 1982).

Epidemiology

Sclerotia of *Rhizoctonia* may be carried by running water, by the wind, and by dirt adhering to plant parts when the seedlings are transplanted. Basidiospores of *Thana-*

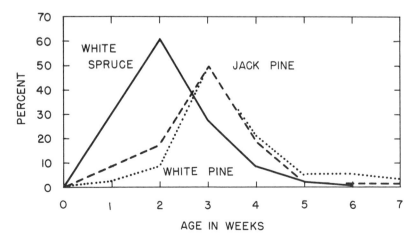

FIGURE 11.3   Rate of damping-off in several conifer species. Redrawn from Hansen et al. (1923).

tephorus species (perfect stage of *Rhizoctonia*) are wind-disseminated. *Pythium* and *Phytophthora* species form oospores and zoospores in sporangia, the first of which may be carried by the wind and by dirt adhering to plants or machinery. Sporangia may be splashed about by rain and carried by wind. Zoospores are able to swim short distances and, thus, spread the fungus locally. These fungi, at least some forms of them, are present in most soils.

Temperature, moisture, light, soil composition, pH, stand density, nutrition, and time of sowing are major factors in determining extent of damping-off (Table 11.1).

1. *Temperature*. Temperatures that allow seedling growth usually favor the growth of these fungi, although temperatures low enough to retard the germination of seeds may increase damping-off because of the longer period of exposure. Different strains of the fungi have different optimum temperatures for growth; one strain of *Pythium* has an optimum of 8°C, while another has an optimum of 25°C.

2. *Moisture*. Damping-off may occur in a relatively dry soil; some strains of *Rhizoctonia* cause damping-off when the soil contains only 30% of its water-holding capacity. However, it is likely to be much more prevalent when the soil is moist and the atmospheric humidity high. Wet soils favor species of *Phytophthora* and *Pythium* while species of *Fusarium* are more common in drier soils.

3. *Light*. Many tree species, especially conifers, require high light intensity for both seed germination and good seedling growth. Long periods of cloudy weather, thick mulches, and high seedling density often result in spindly, weak seedlings that are more susceptible to attack by fungi.

4. *Soil composition*. Soils that contain a large amount of organic matter are favorable for the growth of these fungi, and crops planted in such soils are likely to be more heavily attacked than those in sandy soils. Oat, rye, and clover green manure caused an increase in post emergence damping-off (Wall 1984).

5. *pH*. The fungi that cause damping-off grow best in a somewhat acid soil, but as the concentration of acid is increased, their growth is inhibited. Coniferous seedlings are able to withstand more acid soils than the seedlings of most other plants, and they will grow in soils too acid to permit the growth of most of these fungi.

6. *Stand density*. Dense stands favor damping-off.

7. *Nutrition*. In general, proper fertilization will result in a more vigorous healthy

TABLE 11.1 Various Edaphic Optima for Genera of Fungi Causing Damping-off and Their Major Conifer Suscept

Genus	Temperature (°C)	Moisture[a]	pH
Pythium	about 16	about FC	6.0–7.5
Fusarium	about 21	<70% FC	4.0–6.0
Rhizoctonia	24	<70% FC	4.0–6.0
Pinus	21–27		5.5

[a]FC = field capacity.

seedling that is more resistant to attack by damping-off fungi than a weak seedling. However, high-nitrogen fertilization of seedlings up to 6 weeks of age often results in fast growing, spindly seedlings which are more susceptible to fungus attack.

8. *Time of sowing*. In some nurseries and in some years, fall sowing results in less damping-off than spring sowing; sometimes the opposite is true, and it varies considerably from year to year. The same applies to early and late spring sowing. The seeds should be sown at a time when the weather favors the rapid growth of seedlings and inhibits the growth of fungi. If the weather is constant, no definite time can be said to be best; the timing must be determined for each nursery.

Control Strategy

The various means of controlling damping-off fungi can be grouped under soil sterilization, seed treatment, and improved cultural practices. In southern forest tree nurseries, with few exceptions, adjustment of cultural practices will minimize losses. Because of rapid growth, the seedlings are susceptible for a very short time. The various factors that promote damping-off are discussed in the previous section on Epidemiology. These factors can usually be adjusted to some degree to lessen the damaging effects of damping off. The rest of this section will concentrate on chemical controls, although it will discuss other controls as appropriate.

Treating the soil with Thiram, Captan, or pentachloronitrobenzene (PCNB) as soil drenches affords temporary protection but is of doubtful value after the disease appears. In nurseries where damping-off is a problem, seed treatment with Thiram will reduce losses from the fungi, but the chemical alone can cause 5–15% loss. Many nurseries in the southern United States, in order to control root rots, fumigate their seed beds with methyl bromide, which will control damping-off. Fungi such as *Fusarium oxysporum* can be carried in seeds, resulting in significant preemergence damping-off. Disease-free seeds would solve this problem (Graham and Linderman 1983). Cultural practices include selecting a well-drained site, planting after soil temperature is above 16°C, maintaining soil pH at less than 6, avoiding dense stands, keeping nitrogen level in soil low until seedlings are about 6 weeks old, turning under cover crops at least 2 months before seeding, and using the minimum of mulch to conserve moisture (Hodges 1962).

In areas other than the southern United States, methyl bromide is used in nurseries where root rots are a problem. Otherwise, damping-off must be controlled by some combination of soil treatment, cultural practices, and seed treatment. The latter is probably least effective simply because tree species such as the pines are susceptible for reasonably long periods, and the effectiveness of the chemical applied to the seeds is greatly reduced before the seedlings have developed resistance to the fungi. Seeds of pine and other conifers germinate in such a way that their seed coats are lifted out of the soil, thus removing the fungicide. Many fungicides have been tested as seed protectants, and the results vary with the prevailing weather, the kinds of fungi present in the seed beds, and the amount of chemical applied to the seeds. Even the more successful seed treatments will not provide satisfactory control if the conditions for damping-off are unusually favorable. If the fungicide is phytotoxic, only limited quantities can be added to the seeds, but less phytotoxic materials can be applied in larger doses. To add to the holding power of the seed, methyl cellulose can be used as a sticker. The methyl cellulose apparently reduces phytotoxicity but also may slow the germination process. The more successful seed treatment chemicals have been Thiram, Captan, Zineb, and Chloranil. Dichlone and several mercury compounds are excessively phytotoxic. Thiram added to seeds at 12.5% of the weight of the seeds along with methyl cellulose solution (in water, 4%) at 25% of the weight of seed protected red pine against damping-off (Berbee et al. 1953). To protect seed from predation by birds, anthroquinone can be used.

Soil treatments are more commonly used than seed treatments, although the two methods can be combined. Soil treatments would include acidification of the soil, fumigation, and application of fungicides. Another possibility would be the addition of competing microflora to hold damping-off fungi in check. Although soil acidification has been used commonly in the past, the unsatisfactory results in many nurseries have encouraged the use of more effective techniques. Sulphuric acid (0.09–0.24 g/cm^2), aluminum sulphate (0.1–0.3 g/cm^2), and other acids were used. The aluminum sulphate can be used after planting.

Soil fumigation involving several materials has been successfully used for controlling damping-off and has the added advantage of eliminating nematodes and weed seeds (Fig. 11.4). Although the costs of fumigation are higher than the costs of other control measures, they can be justified at least in part by the fact that weeds are controlled as well. Disadvantages of soil fumigants are the high cost, the temporary nature of control leaving the seed beds open to reinvasion, the elimination of beneficial microorganisms including mycorrhizae and other species that compete with damping-off fungi, and the hazard of handling fumigants, some of which are very toxic to

FIGURE 11.4   Application of methyl bromide to a forest tree nursery.

people (Vaartaja 1967). Fumigation may delay formation of mycorrhizae for several weeks and alter the species mix of the fungi involved (Danielson and Davey 1969). If fumigants are used, it is important to treat the surrounding areas and, of course, contaminated soil, and the like must not be introduced to the treated area.

Formaldehyde is one of the older fumigants. Mylone, Vapam, methyl bromide in various formulations, and chloropicrin are the now more commonly used fumigants. Methyl bromide and chloropicrin, which are sometimes used in combination, are the most toxic and thus the most effective. Mylone, when dry, is an inert material, but it decomposes in soil to produce such compounds as isothiocyanates, formaldehyde, hydrogen sulphide, and monomethylamide. Methyl bromide and chloropicrin can be injected into the soil to a depth of about 20 cm and then covered within the hour to contain the gases within the soil long enough to treat the soil effectively, especially the upper layer of soil. Vapam, which is used at the rate of 472 l/ha under plastic, can be used at 943 l/ha with a 1.3-cm water seal. This is not the case with methyl bromide, which needs a plastic tarp left in place for 24–48 hours. A 10-cm foam cover has been used, but it has not been tested extensively (Braud and Esphahani 1971). Seeds can be planted 48 hours after the tarp is removed. Soil to be treated with any of these chemicals must be well cultivated and optimally at a temperature of at least 21°C. It can be done at 10°C, but is very slow. Although methyl bromide has been applied to the soil surface under a tarp, this method is less effective, more dangerous to the applicator, and more difficult to apply. The cost may be above $2,000/ha, and this is not much higher in cost than what it would be for Vapam or Mylone. In southern nurseries, the seed beds are treated prior to planting a new crop of seeds; the same is true of northern nurseries, although because of the slower growth rate, the time between treatments may be 3 years or more rather than 1 or 2 years, as it is for the southern nurseries. There have been cases where, because of cold temperatures, improper cultivation, or other factors, the methyl bromide is slow in leaving the soil. The result is damage to susceptible plants. Spring fumigation with methyl bromide has resulted in stunted seedlings; in contrast, fall fumigation is not likely to result in damage. The seedlings are larger and thus may reduce by a year the time required to produce plantable seedlings. One year less in the nursery could mean a saving of more than $12,500/ha.

A major concern regarding the use of soil fumigants such as methyl bromide is their environmental risk and health risk to the people who manufacture and apply them. These materials will lose their registration for these applications in the near future. Because of this threat, practitioners are actively researching alternative means to minimize losses due to damping-off. Although other chemicals are likely to continue to be an important part of an integrated approach to damping-off in the nursery, control by altering cultural practices is likely to be stressed in future nursery management practices.

Several chemicals less volatile than the soil fumigants have provided varying degrees of protection. Successful treatment of the soil with these fungicides depends on the fungi involved and the environment. Each nursery needs to learn by trial and error which fungicide is most effective for its conditions. Copper and zinc compounds are potential fungicides for damping-off, especially when used together. Some species of *Phytophthora* are sensitive to copper, but copper compounds can injure seeds, especially in acid soils. Organic compounds such as Thiram at 165–275 kg/ha, and at rates as high as 1,100–2,750 kg/ha, have controlled damping-off. Other possibilities include Maneb, Captan (mixed in soil at 500 ppm), and Vancide 51 (used at rates as high as 2,200 kg/ha).

Fungicides such as Captan, Thiram, and Zineb have been more effective when applied at intervals of 7–10 days (Vaartaja 1964). Possibly the first application would involve a higher concentration than each of the subsequent applications. Foliage sprays are sometimes necessary, although they are far less effective than soil applications. Foliar applications of Bordeaux mixture, Captan, Ferbam, Maneb, and Zineb have been used to protect against top damping-off, especially a problem with tree species such as locust. The foliar sprays protect against other fungus diseases as well as damping off. Certain of the fungicides discussed here can be systemic and thus protect the seedlings, at least for a brief period of time, against other pathogens. Combinations of fungicides have been more effective in some nurseries, one chemical controlling one species and the other chemical another fungus. This is not always the case, however; and some combinations have resulted in more damping-off, possibly because of detrimental effects on soil microorganisms that serve to inhibit damping-off fungi.

Antibiotics, of which there are many, have been used with minor degrees of success. Among the more common ones are actidione, chloromycetin, endomycin, griseofulvin, neomycin, streptomycin, and terramycin. Probably far more important are the naturally occurring fungi, such as species in the genera *Trichoderma*, *Gliocladium*, *Penicillium*, and *Chaetomium* and species of the bacterial genus *Streptomyces*. These organisms growing actively can produce more antibiotic than the preceding proprietary materials and are actually assisted by many of the chemical treatments used in nurseries. Because of their tolerance and rapid growth, they can quickly colonize partially sterilized soil and play an important role in controlling damping-off fungi.

Effectiveness of soil fungicides is conditioned by many factors (Vaartaja 1964). For example, Vancide 51 was effective at 21°C but not at 28°C. Captan, on the other hand, was more effective at 28°C than 21°C. Increasing the pH of the soil improved the effectiveness of fungicides while adding organic material decreased their efficiency. Fungi can survive better in organic matter, especially as sclerotia. Poor distribution of fungicides in soils can have a great deal to do with uneven results, and certain fungicides are more easily leached from the soil. The composition of the soil helps control the rate of leaching, and soils can inactivate fungicides. Chemicals such as Mylone become more active when placed in soil, but other materials, especially most antibiotics, are inactivated rather rapidly. Methods of application can affect the results, and some fungicides should be placed on or near the surface, while others need to be mixed into the soil. The best technique depends, in part, on where the damping-off fungi are most active (near the surface or in the root area) and the rate of leaching.

The use of cover crops between planting cycles to reduce disease losses has had variable results. The theory is that cover crops increase the populations of antagonistic microorganisms and that this increase will cause disease levels to decrease. Unfortunately, some cover crops can be colonized by pathogens before being turned under and not all the individuals will be killed by fumigation. Hansen et al. (1990) found population densities of *Fusarium* and *Pythium* to be lower in bare, fallow plots than in those under grass or legume cover crops.

LITERATURE CITED

Abbott, H. G., and E. J. Eliason. 1968. Forest tree nursery practices in the United States. *J. For.* 66:704, 706, 708–711.

Berbee, J. G., F. Berbee, and W. J. Brener. 1953. The prevention of damping off of coniferous seedlings by pelleting seed. *Phytopathology* 43:466 (abst.).

Braud, H. J., and M. Esphahani. 1971. *Foam Cover for Soil Fumigation.* Amer. Soc. Agr. Eng., Pap. No. 71–149. 19 pp.

Brownell, K. H., and R. W. Schneider. 1983. *Fusarium hypocotyl* rot of sugar pine in California forest nurseries. *Pl. Dis.* 67:105–107.

Danielson, R. M., and C. B. Davey. 1969. Microbial recolonization of a fumigated nursery soil. *For. Sci.* 15:368–380.

Filer, T. H. 1967. Damping off of sweetgum by *Pythium sylvaticum. Phytopathology* 57:1284.

Graham, J. H., and R. G. Linderman. 1983. Pathogenic seedborne *Fusarium oxysporum* from Douglas-fir. *Pl. Dis.* 67:323–325.

Hansen, E. M., D. D. Myrold, and P. B. Hamm. 1990. Effects of soil fumigation and cover crops on potential pathogens, microbial activity, nitrogen availability, and seedling quality in conifer nurseries. *Phytopathology* 80:698–704.

Hansen, T. S., W. H. Kentz, G. H. Wiggin, and E. C. Stakman. 1923. *A Study of the Damping-off Disease of Coniferous Seedlings.* Univ. Minn. Agri. Exp. Sta., Tech. Bull. 15. 35 pp.

Hock, W. K., and W. L. Klarman. 1967. The function of the endodermis in resistance of Virginia pine seedlings of damping off. *For. Sci.* 13:108–112.

Hodges, C. S. 1962. *Diseases of Southeastern Forest Nurseries and Their Control.* USDA For. Serv., SE For. Exp. Sta. Pap. 142. 16 pp.

Ronco, F. 1967. *Lessons from Artificial Regeneration Studies in a Beetle-Killed Spruce Stand in Western Colorado.* USDA For. Serv., Res. Note RM 90. 8 pp.

Rowan, S. J. 1982. Tip dieback in southern pine nurseries. *Pl. Dis.* 66:258–259.

Vaartaja, O. 1964. Chemical treatment of seed beds to control nursery diseases. *Bot. Rev.* 30:1–91.

Vaartaja, O. 1967. Reinfestation of sterilized nursery seed beds by fungi. *Can. J. Microbiol.* 13:771–776.

Wall, R. E. 1984. Effects of recently incorporated organic amendments on damping-off of conifer seedlings. *Pl. Dis.* 68:57–58.

Wright, E. 1957. Influence of temperature and moisture on damping off of American Siberian elm, black locust, desertwillow. *Phytopathology* 47:658–662.

DISEASE PROFILE

Black Root Disease

Importance: Black root disease, a synergistic interaction of fungi and possibly nematodes on southern pines, is potentially the most serious summer-season nursery disease of the region. A similar problem in western nurseries, the charcoal root disease, is not truly synonymous because it is caused solely by the primary pathogen of the black root disease complex.

Suscepts: All of the commercially important southern pines are susceptible. Only spruce pine, sweetgum, cedars, and cypresses are known to be sufficiently resistant for safe planting in infested soil.

Distribution: Since 1936, black root disease has caused economic losses in at least 10 southern nurseries as shown. Its severity in the first year of production in nurseries

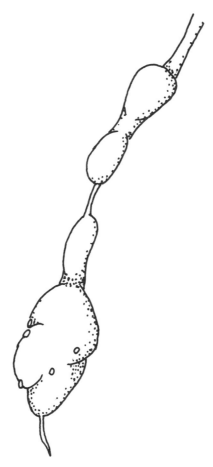

Root swellings.

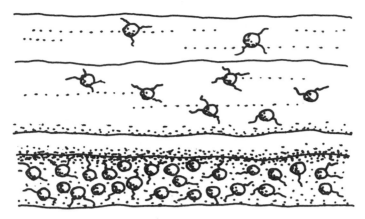

Sclerotia on root.

established on newly cleared forest land suggests a more cosmopolitan hazard than is indicated.

Cause: Infection is initiated by *Macrophomina phaseolina* (= *Sclerotium bataticola*) (Coelomycetes), and further aggravated by species of *Fusarium* (Hyphomycetes), either *F. oxysporum* or *F. solani.* Also, certain parasitic nematodes are sometimes involved.

Biology: As soil inhabitants, *M. phaseolina* and *Fusarium* spp. overwinter and persist even longer by means of sclerotia and chlamydospores, respectively. Sporulation of *M. phaseolina* is rare while that of the fusarial species becomes quite profuse on dead seedlings during wet weather. The primary pathogen, *M. phaseolina,* kills succulent roots, penetrates the phellem of suberized roots, and causes drastic proliferation of adjacent phelloderm without colonizing it. The phelloderm, which is not normally meristematic, presumably is activated by translocated indoleacetic acid, perhaps produced by *M. phaseolina* but not by the fusarial associates. The latter, as secondary pathogens, intensify the later stages of root decline.

Epidemiology: As observed on slash pine, black root disease is maximized by high soil temperature (32–38°C) and nitrogen fertilization. Mortality is rare while soil moisture is adequate but becomes pronounced as irrigation is withheld in late-season preparations for lifting.

Diagnosis: By mid-summer, root examination will reveal distinctive reddish-black swellings, ranging from localized to diffuse on the taproot and larger laterals. Rootlets emanating from such enlargements usually succumb, and the root system degenerates to a tap root stub. Aerial symptoms, namely chlorosis and stunting, may or may not appear in correlation with root symptomatology. Sign detection requires slide preparation and/or tissue isolation to reveal the black microsclerotia of *M. phaseolina* and the hyaline macro- and microconidia of *Fusarium* spp.

Control Strategy: Temporary eradication is successfully practiced by annual fumigation of problem soils with methyl bromide at 48.8 g/m². With the loss of methyl bromide, cultural treatments will become increasingly important. Established infection can be reduced by adding organic matter, irrigating frequently, and maintaining adequate mulch. Fortunately, *M. phaseolina* may be more of an oppportunistic, weak pathogen, and variations of these treatments may be effective.

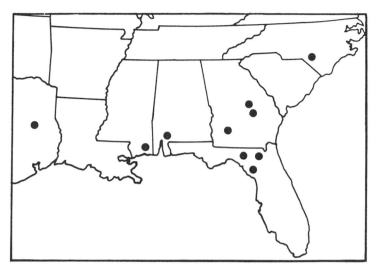

Distribution of affected nurseries in southeastern United States.

SELECTED REFERENCES

Hodges, C. S., Jr. 1962. Black root rot of pine seedlings. *Phytopathology* 52:210–219.
Rowan, S. J. 1971. Soil fertilization, fumigation, and temperature affect severity of black root rot of slash
    pine. *Phytopathology* 61:184–187.

11.4 BLACK ROOT DISEASE

Also known as charcoal root disease, black root disease is caused by a fungus, *Macrophomina phaseoli (Sclerotium bataticola)*. Other fungi such as species of *Fusarium* and nematodes may increase intensity of the disease. In the southern United States, slash and loblolly pines are especially susceptible (Hodges 1962) (Fig. 11.5). In the western states, sugar pine seedlings are susceptible, as well as white fir, Douglas-fir, and ponderosa pine (Smith 1966, Smith and Krugman 1967).

As the name of the disease implies, the tap root and deeper laterals become enlarged, blackened, and roughened. At first, small reddish-brown areas form on the larger roots, gradually enlarge, and become darker, eventually covering the entire tap root. Few lateral roots survive. Tiny black microsclerotia, which remain on or in dead root and lower stem tissues, are produced. These germinate when new roots grow close to them, thus initiating new infections. The microsclerotia are the major means of survival and also are important in dispersal of infected soil within the nursery. No conidial state is associated with trees.

The genus *Macrophomina* has a long host list. It is responsible for a root and stem disease of over 300 plant species (Smith 1969, Seymour and Cordell 1979, Short et al. 1978). The most frequent Macrophomina disease is referred to as black root disease or charcoal root disease, but it also causes damping-off, stem disease, and loss of plant structural integrity (Joye 1990). This fungus causes serious infections of a num-

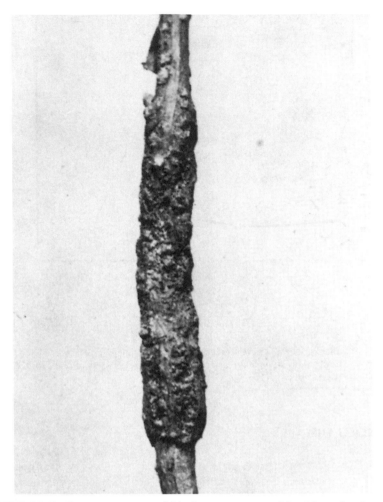

FIGURE 11.5   Root symptoms of black root rot caused by *Macrophomina phaseoli*. From Hodges (1962).

ber of economically important plant species, including all major southern pine species (Seymour and Cordell 1979, Smalley and Scheer 1963), giant sequoia (Smith and Bega 1964), other conifers (Smith 1964), many hardwoods (Rowan et al. 1972), soybean (Short and Wyllie 1978), corn, sorghum, snapbean, peanut (Smith 1969), sunflower (Jimenez-Diaz et al. 1983, Gulya et al. 1991), cotton (Watanabe et al. 1970), and others. It is also a significant pathogen of *Euphorbia lathyris,* the gopher plant, which is currently being investigated in southern Arizona as a source of fuel and animal feed (Young and Alcorn 1982, 1984) and *Salicornia figelovii,* a plant native to the salt marshes of Mexico, which is also being studied as a source of forage and cooking oil (Stanghellini et al. 1992). A related species, *M. phaseolina* has been found to be such a destructive pathogen of *Hydrilla verticillia,* the aquatic plant that is a troublesome pest in waterways in Florida and other tropical areas, that it has been suggested as a possible biocontrol agent for this annoying plant (Joye 1990).

Macrophomina disease is favored by high soil temperatures and moisture stress. It has been an important problem in nurseries in Florida (Seymour and Cordell 1979, Barnard et al. 1985, Seymour 1969, Smalley and Scheer 1963) and California (Smith and Bega 1964, Smith 1964, Smith and Krugman 1967). It has also caused significant problems in forest nurseries in Georgia, Alabama, North Carolina, and Texas (Hodges 1962). Large numbers of pine seedlings have been destroyed by charcoal root disease. In the mid 1960s, more than 50% of the pine seedlings were killed in three important forest nurseries in Florida; in 1976, 20% of Florida's nursery-grown pine seedlings were killed (Seymour and Cordell 1979). Losses of soybean from *Macrophomina* stem and root rot are also significant in the United States every year (Meyer et al. 1973).

Current methods for detection and identification of *Macrophomina* are labor-intensive, time-consuming, and require special equipment, skill, and knowledge of fungal morphology. Presently, the only method to diagnose this disease definitively is to culture the fungus from diseased tissue and identify it by its morphology.

Fumigation with methyl bromide (33% chloropicrin) effectively eliminates the fungus, and both Mylone and Vapam reduce its population to a low level (Foster 1961). Mortality in white fir has been lowered to 50% of what occurred in untreated seedlings using a fumigant consisting of 61% methyl bromide, 31% trichloronitromethane, and 8% propargyle bromide at a rate of 225 kg/ha. A fumigant consisting of 57% methyl bromide and 43% trichloronitromethane at 365 kg/ha was even more effective and reduced the percent of infection of white fir and ponderosa pine roots at the time of lifting from 100% for untreated seedlings to 25% (Smith and Krugman 1967). The amount of root disease was reduced significantly when soil was fumigated with DD-menes or drenched with benomyl. However, the benomyl also reduced mycorrhizae, while soil fumigation did not.

LITERATURE CITED

Barnard, F. L., G. M. Blakeslee, J. T. English, S. W. Oak, and R. L. Anderson. 1985. Pathogenic fungi associated with sand pine root disease in Florida. *Pl. Dis.* 69:196–199.

Foster, A. A. 1961. *Control of Black Root Rot of Pine Seedlings by Fumigation in the Nursery.* Georgia For. Res. Council, Rept. 8. 5 pp.

Gulya, T. J., D. M. Woods, R. Bell, and M. K. Mancl. 1991. Diseases of sunflower in California. *Pl. Dis.* 75:572–574.

Hodges, C. S. 1962. Black root rot of pine seedlings. *Phytopathology* 52:210–219.

Jimenez-Diaz, R. M., M. A. Blanco-Lopez, and W. E. Sackston. 1983. Incidence and distribution of charcoal rot of sunflower caused by *Macrophomina phaseolina* in Spain. *Pl. Dis.* 67:1033–1036.

Joye, G. F. 1990. Biocontrol of *Hydrilla verticillata* with the endemic fungus *Macrophomina phaseolina*. *Pl. Dis.* 74:1035–1036.

Meyer, W. A., J. B. Sinclair, and M. N. Khare. 1973. Biology of *Macrophomina phaseoli* in soil studied with selective media. *Phytopathology* 63:613–620.

Rowan, S. J., T. H. Filer, and W. R. Phelps. 1972. *Nursery Diseases of Southern Hardwoods.* USDA For. Serv., For. Pest Leaflet 137. 7 pp.

Seymour, C. P. 1969. Charcoal rot of nursery-grown pines in Florida. *Phytopathology* 59:89–92.

Seymour, C. P., and C. E. Cordell. 1979. Control of charcoal root rot with methyl bromide in forest nurseries. *So. J. Appl. For.* 3:104–108.

Short, G. E., and T. D. Wyllie. 1978. Inoculum potential of *Macrophomina phaseolina*. *Phytopathology* 68:742–746.

Short, G. E., T. D. Wyllie, and V. D. Ammon. 1978. Quantitative enumeration of *Macrophomina phaseolina* in soybean tissues. *Phytopathology* 68:736–741.

Smalley, G. W., and R. L. Scheer. 1963. Black root rot in Florida sandhills. *Pl. Dis. Reptr.* 47:669–671.

Smith, R. S., Jr. 1964. Effect of diurnal temperature fluctuations on linear growth rate of *Macrophomina phaseoli* in culture. *Phytopathology* 54:849–852.

Smith, R. S., Jr., and R. V. Bega. 1964. *Macrophomina phaseoli* in the forest tree nurseries of California. *Pl. Dis. Reptr.* 48:206.

Smith, R. S. 1966. Effect of diurnal temperature fluctuations on the charcoal root disease of *Pinus lambertiana*. *Phytopathology* 56:61–64.

Smith, R. S., and S. L. Krugman. 1967. Control of the charcoal root disease of white fir by fall soil fumigation. *Pl. Dis. Reptr.* 51:671–674.

Smith, W. H. 1969. Comparison of mycelial and sclerotial inoculum of *Macrophomina phaseoli* in the mortality of pine seedlings under varying soil conditions. *Phytopathology* 59:379–752.

Stanghellini, M. E., J. D. Mihail, S. L. Rasmussen, and B. C. Turner. 1992. *Macrophomina phaseolina:* A soilborne pathogen of *Salicornia bigelovii* in a marine habitat. *Pl. Dis.* 76:751–752.

Watanabe, T., R. S. Smith, Jr., and W. C. Snyder. 1970. Populations of *Macrophomina phaseoli* in soil as affected by fumigation and cropping. *Phytopathology* 60:1717–1719.

Young, D. J., and S. M. Alcorn. 1982. Soilborne pathogens of *Euphorbia lathyris: Macrophomina phaseolina, Pythium aphanidermatum,* and *Rhizoctonia solani. Pl. Dis.* 66:236–238.

Young, D. J., and S. M. Alcorn. 1984. Latent infection of *Euphorbia lathyris* and weeds by *Macrophomina phaseolina* and propagule populations in Arizona field soils. *Pl. Dis.* 68:587–589.

DISEASE PROFILE

Cylindrocladium Root Disease

Importance: Known since the 1930s, Cylindrocladium root disease more recently has caused severe losses of conifers and hardwoods in forest nurseries throughout the eastern, southern, and Lake states regions. Because of its broad geographic and host range potential, control is costly and difficult to achieve.

Suscepts: The most serious losses have occurred on red and white pines, black and white spruces, yellow-poplar, and black walnut. Additional suscepts include Austrian, Mugo, and Scots pines; blue and Norway spruces; balsam and Fraser firs; Douglas-fir; azalea; and rhododendron. Apart from nursery culture, some peach replant losses are attributed to Cylindrocladium root disease.

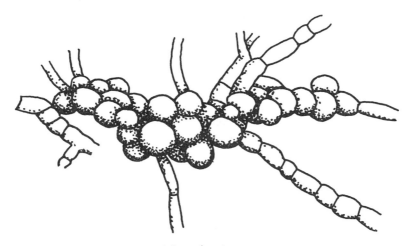

Microsclerotium.

Distribution: As shown by the number of reported nursery infestations, the disease range currently occurs in 17 states in the United States and only Quebec in Canada.

Causal Agents: Of 18 world species in the soil-inhabiting fungal genus *Cylindrocladium* (Hyphomycetes), only *C. floridanum* and *C. scoparium* are responsible for seedling diseases in forest nurseries. Some mycologists recognize these as one species, *C. scoparium. Cylindrocladium crotalariae,* isolated from peanut soils in Georgia, is virulent on yellow-poplar and could become a problem.

Biology: The causal species overwinter as microsclerotia in infected tissue and soil. As roots grow and reach this sclerotial inoculum, germination is stimulated, and mycelial penetration of intact root tissue occurs. During hot, humid weather, foliage can be infected from airborne conidia originating from previously established aerial symptoms. Any infected tissue may support the formation of sclerotia, which incorporate with the soil through deterioration or tillage.

Epidemiology: To date, the epidemiology is not well defined.

Diagnosis: Although root rot is most typical, other symptoms may occur—damping off and needle blight—ranging from fascicle necrosis to total reddening, as in red pine, to distal yellowing and wilt in other conifers. Other symptoms that might occur are leaf blight, necrotic leaf spots in hardwoods, and stem cankers, particularly in white pine following

MN(6) CT(I)
WI(5) NJ(I)
MI(3) PA(I)
NC(2) DE(I)
TN(2) MD(I)
SC(2) VA(I)
AL(I) WV(I)
MS(I) KY(2)
       ME(I)

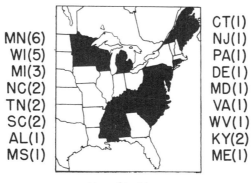

Map of incidence.

the needle blight stage. In conifers, root tip infection progresses inward to the root crown; the necrotic cortex slips easily in advanced stages of rot. Roots of infected hardwood seedlings develop pronounced blackening and longitudinal cracking of the cortex.

Control Strategy: Preplant fumigation has been done with 340 kg/ha of 67% methyl bromide and 33% chloropicrin under plastic tarping to a depth of 30 cm. These chemicals are extremely hazardous. After their use is curtailed, it is possible that cultural treatments will be developed to control this very destructive root disease. At the time of printing, however, there is no other effective control; infested nurseries are abandoned.

SELECTED REFERENCES

Cordell, C. E., and D. D. Skilling. 1975. Cylindrocladium root rot, pp. 23–26. In G. W. Peterson and R. S. Smith, Jr., *Forest Nursery Diseases in the United States.* USDA For. Serv., Agric. Handbook 470. 125 pp.

Thies, W. G., and R. F. Patton. 1970. The biology of *Cylindrocladium scoparium* in Wisconsin forest tree nurseries. *Phytopathology* 60:1662–1668.

11.5   CYLINDROCLADIUM ROOT DISEASE

Several *Cylindrocladium* species are highly significant plant pathogens causing serious infections in a large number of plant hosts. In an extensive literature review, Cox (1954) referred to at least 37 hosts for *C. scoparium* including conifers (hemlock, pine, spruce, and fir), fruits and berries (apple, plum, cherry, apricot, strawberry, and raspberry), forage legumes (clover and alfalfa), ornamentals (*Eucalyptus* and rose), and tea. Diseases associated with this fungus include root decay, damping-off, stem cankers, shoot wilt, leaf spot, and seedling blight.

Since 1954 numerous reports of *Cylindrocladium* disease on other economically important hosts include root disease of tulip-poplar (Kelman and Gooding 1965), black walnut (Filer 1968), sweetgum (Cordell and Rowan 1975, Kuhlman 1980), cherrybark oak (Smyly and Filer 1977), and peanut (Bell and Sobers 1966); decline of peach (Sobers and Seymour 1967); severe blights of rhododendron, lilac, and sweet william cuttings (Horst and Hoitink 1968), *Ilex* (Gill et al. 1971), azalea (Timonin and Self 1955, Horst and Hoitink 1968, Cox 1969); scurf of sugar beets (Lentz 1955), and others.

Cylindrocladium damping-off and root disease are important problems of seedlings and cuttings grown in greenhouses and nursery transplant beds. Nursery conditions such as overhead irrigation and high temperature, humidity, and seedling density exacerbate the disease and enhance chances for its development (Barnard 1984). As early as 1917, *Cylindrocladium* was reported as a serious pathogen of greenhouse roses (Massey 1917, 1921). Later, Cox (1951; 1953a,b) expressed concern over the threat of crown canker of rose as well as the incidence of damping-off and root decay in seedbeds of conifers in Delaware. At one time, the Delaware State Department of Forestry had abandoned production of red pine and was considering abandoning white pine as a result of *Cylindrocladium* root disease (Cox 1954). Large losses of nursery plants including forest trees and ornamentals have been reported since that time. In 1955 several nurseries in North Carolina, Virginia, and Tennessee reported heavy seedling losses of both yellow-poplar and pine (Kelman and Gooding 1965). In 1962 losses ranging from 10 to 90% in seedling and transplant beds of pine and spruce were reported in 12 forest and ornamental tree nurseries in Minnesota, Wisconsin, and Michigan as a result of root or needle blight from *C. scoparium* (Anderson et al. 1962). *Cylindrocladium* blight was described by Hodges (1962) as the most important seedling disease of white pine. Although not definitively associated with diseased trees, *C. scoparium* was also isolated at two Quebec provincial nurseries in 1965 (Sutherland 1967). In 1968 extensive mortality resulted when 70% of 250,000 yellow-poplar seedlings were infected by *C. floridanum* in a North Carolina state forest tree nursery (Cordell et al. 1969). That same year large losses were reported in azalea, rhododendron, lilac, and sweet william cuttings in Ohio and Florida (Horst and Hoitink 1968, Cox 1969) and many of the pine and all the yellow-poplar and walnut seedlings were killed at the Parsons Forest Nursery at Parsons, West Virginia (Cordell et al. 1969).

Since 1968 *Cylindrocladium* has sporadically continued to plague both forest and ornamental nurseries with heavy losses (Barnard 1984, Cordell et al. 1971, Filer 1970, Hunter and Barnett 1976, Smyly and Filer 1977). *Cylindrocladium* has been described as one of the most virulent pathogens in forest nurseries (Cordell et al. 1969).

There are three species of *Cylindrocladium* that may cause root disease of tree species: *C. floridanum, C. crotalariae,* and *C. scoparium* (Fig. 11.6). Species characteristics, as illustrated, are marked by the shape of the stalked vesicle and by the number of conidial septations. Recently, mycologists have recognized *C. scoparium* and *C. floridanum* as one species.

Cylindrocladium floridanum, reported as a cause of root disease of peach trees in Georgia, is apparently the same species that causes root disease of a wide range of tree species including white pine, Fraser fir, and yellow-poplar in the southeastern United States and black spruce, red pine, white pine, and other conifers in the northern

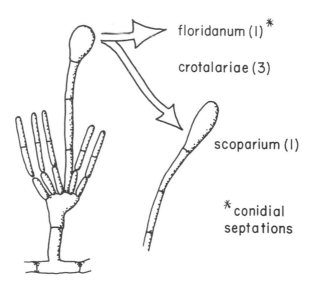

FIGURE 11.6 Illustration of species characteristic of *Cylindrocladium* including the shape of the stalked vesicle and the number of conidial septations.

FIGURE 11.7   Seedlings of black walnut with symptoms of *Cylindrocladium* root disease.

United States (Anderson et al. 1962, Morrison and French 1969). *Cylindrocladium floridanum* was found causing root decay of *Pinus* spp. and yellow-poplar seedlings in New Zealand (Boesewinkel 1974). In the greenhouse, the fungus has killed all conifers tested except northern white cedar. *Cylindrocladium scoparium* causes root disease of azaleas and other hosts.

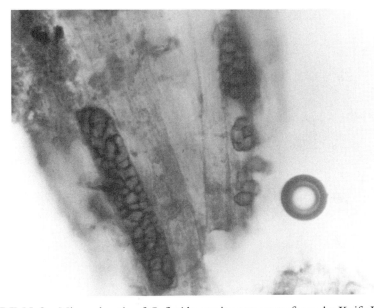

FIGURE 11.8   Microsclerotia of *C. floridanum* in grass roots from the Knife River Nursery, Minnesota, 14 years after it had been abandoned for production of black spruce seedlings.

Cylindrocladium floridanum attacks the roots resulting in the needles turning yellow and the current year's growth drooping (Fig. 11.7). In advanced stages a major part of the root system is destroyed. The fungus invades the cortex of the roots and forms microsclerotia (Fig. 11.8). *Cylindrocladium floridanum* also causes stem cankers and a needle blight. Both pine and spruce are infected after 2–4 days' incubation in a saturated atmosphere. The fungus produces conidia on the needles. Microsclerotia as well as conidia could be responsible for spread of the disease.

Control Strategy

Soil fumigation with methyl bromide or equivalent materials is the only means of successfully controlling the root-rot phase of the disease (Hodges 1962). Seed and transplant beds, as well as the walkways between beds, should be treated. Soil temperatures for fumigation should be between 10 and 27°C, soil moisture should be near field capacity, and the soil should be disked to a depth of 15–20 cm. Cover crops should be turned under at least 2 months prior to fumigation. On most soils, 284 kg/ha of actual methyl bromide is sufficient; it should be injected into the soil 15–20 cm, and the area should be covered within 1 hour with a 2-mil-thick plastic tarp, which can be removed 24–40 hours after treatment. The tarp can be reused, and small holes are not a problem. Planting can start 48 hours after the cover is lifted. Where soil temperatures or the weather after application is cooler, more time will be needed for the fumigant to vaporize and dissapate. Planting too soon may damage or even kill the seedlings. Protection will normally last for at least 2 years. Fumigation also provides control of weeds, which reduces the cost of treatment.

Vapam or VPM injected into the soil at the rate of 530 l/ha and covered with plastic for 48 hours is comparable in effectiveness with methyl bromide and costs about the same. Planting, however, must be delayed for 2 weeks after the plastic cover is removed, and the soil should be disked at least once during this time to facilitate escape of the chemical. A 1.3-cm water seal can be used in place of the plastic tarp, but the volume of chemical must then be increased from 530 to 1,060 l/ha. Mylone, another soil fumigant, as an 85% wettable powder can be used at 200–341 kg/ha. This chemical can be applied with a fertilizer spreader and then should be disked to a depth of 15–20 cm. No plastic cover is needed, but planting must be delayed 2–3 weeks. The cost is similar to the preceding fumigants but is less effective. The needle blight phase can be controlled using Ferbam at 2.3 kg/ha. In the southern states, stem cankers caused by *C. floridanum* are controlled by increasing the volume of spray containing Ferbam to 3,180–4,240 l/ha. The losses of azaleas and rhododendrons have been controlled in the greenhouse with the methyl ester of 1-(butylcarbamoyl)-2-benzimidazole carbamic acid (benomyl). Even at optimum conditions for disease (24–27°C) and high moisture, 100% control was obtained with this fungicide (Horst and Hoitink 1968).

Recovering *Cylindrocladium* from soil samples is difficult and very time-consuming. Several "baiting" techniques have been described. These consist of growing a susceptible plant in soil samples under conditions suitable for the germination of microsclerotia and then culturing the fungus from the plant on agar media. In order for this technique to be successful, the microsclerotia present in the soil must germinate and infect the plant. Current techniques involving this principle are alfalfa trapping (Bugbee and Anderson 1963, Thies and Patton 1966, Cordell et al. 1971, Hodges and May 1972), geranium leaf baiting (Hunter et al. 1980), and azalea leaf trapping (Linderman 1972).

Quantification of microsclerotia in soil samples can be accomplished by the method of Thies and Patton (1970) in which microsclerotia are separated from the soil by washing and wet-sieving and grown on a selective medium for counting. The results are expressed in propagules per gram of oven dry weight of sample. Another technique for estimating numbers of infectious units in soil is the quantitative alfalfa assay of Menge and French (1976). In their assay, a certain number of 1-g soil samples are planted with alfalfa and incubated. The inoculum potential of *Cylindrocladium* is calculated by the percentage of samples that contain alfalfa with visible signs of disease.

Altering cultural practices may help limit the losses caused by species of *Cylindrocladium;* however, the fungus is very persistent and survived in one nursery for 14 years after the nursery was abandoned. Although the fungus was not recovered from areas outside the nursery itself, it succeeded in some way to invade newly opened nursery beds before they could produce their first crop. Infection potentials could be measured using alfalfa seedlings, and it was found that potentials were high in June, low in August, and high again in October. Cover crops such as oats, wheat, and rye increased infection potentials, but soybeans and buckwheat did not, even though the latter plants could be infected. Corn did not increase potentials above those existing in fallow soil. Infection potentials were higher in a loamy sand soil than in a Waukegan silt loam. Fresh cover crop residues, more so than partially decayed debris, enhanced germination of *C. floridanum*. Remains of clover plants, which could be blown about, harbored viable sclerotia. Obviously, suitable hosts should be avoided as cover crops and any crop should be plowed under well before planting time.

LITERATURE CITED

Anderson, N., D. W. French, and D. P. Taylor. 1962. Cylindrocladium root rot of conifers in Minnesota. *For. Sci.* 8:378–383.

Barnard, E. L. 1984. Occurrence, impact, and fungicidal control of girdling stem cankers caused by *Cylindrocladium scoparium* on eucalyptus seedlings in a south Florida nursery. *Pl. Dis.* 66:471–474.

Bell, D. K., and E. K. Sobers. 1966. A peg, pod, and root necrosis of peanuts caused by a species of *Calonectria. Phytopathology* 56:1361–1364.

Boesewinkel, H. J. 1974. *Cylindrocladium floridanum,* a new recording for New Zealand. *Pl. Dis. Reptr.* 58:705–707.

Bugbee, W. M., and N. A. Anderson. 1963. Host range and distribution of *Cylindrocladium scoparium* in the north-central states. *Pl. Dis. Reptr.* 47:512–515.

Cordell, C. E., and S. J. Rowan. 1975. *Cylindrocladium scoparium* infection in a natural sweetgum stand. *Pl. Dis. Reptr.* 59:775–776.

Cordell, C. E., C. E. Affeltranger, and A. H. Maxwell. 1969. *Cylindrocladium* Root Rot—Damaging Yellow-Poplar Seedlings in a North Carolina Nursery. USDA For. Serv., S.E. Area Rept. 69–1–31. 12 p.

Cordell, C. E., A. S. Juttner, and W. J. Stambaugh. 1971. *Cylindrocladium floridanum* causes severe mortality of seedling yellow-poplar in a North Carolina nursery. *Pl. Dis. Reptr.* 55:700–701.

Cox, R. S. 1951. A preliminary report on etiological and control studies of damping-off and root rot in the conifer seedbed in Delaware. *Pl. Dis. Reptr.* 35:374–378.

Cox, R. S. 1953a. Crown canker on a field planting of multiflora rose cuttings. *Pl. Dis. Reptr.* 37:447.

Cox, R. S. 1953b. Etiology and control of a serious complex of diseases of conifer seedlings. *Phytopathology* 43:469 (abst.).

Cox, R. S. 1954. *Cylindrocladium scoparium* on conifer seedlings. Univ. Del. Agric. Exp. Stn. Tech. Bull. 301. 40 pp.

Cox, R. S. 1969. *Cylindrocladium scoparium* on azalea in south Florida. *Pl. Dis. Reptr.* 53:139.

Filer, T. H., Jr. 1968. *Cylindrocladium* root rot of black walnut. *Phytopathology* 58:728 (abst.).

Filer, T. H., Jr. 1970. Virulence of three *Cylindrocladium* species to yellow-poplar seedlings. *Pl. Dis. Reptr.* 54:320–322.

Gill, D. L., S. A. Alfieri, Jr., and E. K. Sobers. 1971. A new leaf disease of *Ilex* spp. caused by *Cylindrocladium avesiculatum* sp. nov. *Phytopathology* 61:58–60.

Hodges, C. S. 1962. *Diseases in Southeastern Forest Nurseries and Their Control.* USDA For. Serv., S.E. For. Exp. Stn. Pap. 142. 16 pp.

Hodges, C. S., and L. C. May. 1972. A root disease of pine, *Araucaria* and *Eucalyptus* in Brazil caused by a new species of *Cylindrocladium*. *Phytopathology* 62:898–901.

Horst, R. K., and H. A. Hoitink. 1968. Occurrence of *Cylindrocladium* blights on nursery crops and control with fungicide 1991 on azalea. *Pl. Dis. Reptr.* 52:615–617.

Hunter, B. B., and H. L. Barnett. 1976. Production of microsclerotia by species of *Cylindrocladium*. *Phytopathology* 66:777–780.

Hunter, B. B., M. A. Sylvester, and J. Balling. 1980. A rapid method for identifying and recovering species of *Cylindrocladium* from soil *via* geranium leaf baiting. *Proc. PA Acad. Sci.* 54:157–160.

Kelman, A., and G. V. Gooding. 1965. A root and stem rot of yellow-poplar caused by *Cylindrocladium scoparium*. *Pl. Dis. Reptr.* 49:797–801.

Kuhlman, E. G. 1980. *Cylindrocladium* root rots of sweetgum seedlings in southern forest tree nurseries. *Pl. Dis.* 64:1079–1080.

Lentz, P. L. 1955. A scurf of sugar beets caused by *Cylindrocladium scoparium*. *Pl. Dis. Reptr.* 39:654–655.

Linderman, R. G. 1972. Isolation of *Cylindrocladium* from soil of infected azalea stems with azalea leaf traps. *Phytopathology* 62:736–739.

Massey, L. M. 1917. The crown canker disease of rose. *Phytopathology* 7:408–417.

Massey, L. M. 1921. Experimental data on losses due to crown canker of rose. *Phytopathology* 11:125–134.

Menge, J. A., and D. W. French. 1976. Determining inoculum potentials of *Cylindrocladium floridanum* in cropped and chemically-treated soils by a quantitative assay. *Phytopathology* 66:862–867.

Morrison, R. H., and D. W. French. 1969. Taxonomy of *Cylindrocladium floridanum* and *C. scoparium*. *Mycologia* 61:957–966.

Smyly, W. B., and T. H. Filer, Jr. 1977. *Cylindrocladium scoparium* associated with root rot and mortality of cherrybark oak seedlings. *Pl. Dis. Reptr.* 61:577–579.

Sobers, E. K., and C. P. Seymour. 1967. *Cylindrocladium floridanum* sp. n. associated with decline of peach trees in Florida. *Phytopathology* 57:389–393.

Sutherland, J. R. 1967. Occurrence of *Cylindrocladium scoparium* Morg. in Quebec forest nurseries, Canada. *Dept. For. and Rural Dev. Bi-m. Res. Notes* 23:4–5.

Thies, W. G., and R. F. Patton. 1966. Spot-plate technique for the bioassay of *Cylindrocladium scoparium*. *Phytopathology* 56:1116–1117.

Thies, W. G., and R. F. Patton. 1970. An evaluation of propagules of *Cylindrocladium scoparium* in soil by direct isolation. *Phytopathology* 60:599–601.

Timonin, M. I., and R. L. Self. 1955. *Cylindrocladium scoparium* Morgan on azaleas and other ornamentals. *Pl. Dis. Reptr.* 39:860–863.

DISEASE PROFILE

Nematode-Caused Diseases in Nurseries and Plantations

Importance: Nematodes thrive under the cultural practices and continuous cultivation found in many forest nurseries but tend to not be a problem when rotations are less than 1 year and fumigation is practiced between crops. They may become serious when abandoned farm land is reforested. As forest management intensifies, nematodes are likely to increase in importance.

Nematodes and necrotic root tip.

Littleleaf symptoms.

Suscepts: Virtually all plant species are susceptible, including most forest trees, with orchard species being particularly susceptible. Many hardwoods and pines, especially in southern nurseries, have been injured.

Distribution: Nematodes are present in virtually all soils, being more abundant in warmer climates. All need free water in the soil in order to function but may survive long periods in dry soil. Their location within soil is influenced largely by feeder root concentration.

Causal Agents: Nine genera of nematodes attack tree roots. These include *endoparasitic* species that enter, feed, and reproduce within tree roots and *ectoparasitic* species that either remain outside plant roots where they feed on the epidermal or cortical cells or, on occasion, may penetrate partially or wholly into cortical tissue.

Biology: Some species overwinter as eggs, as second-stage juveniles, or as adults. Both larvae and adults move to and feed on plant roots. The sedentary forms eventually enter the root and feed on only a few modified host cells. The free-ranging forms probe on root epidermis and move around to a limited extent in the soil. Parasitic nematodes feed with a hollow spear or stylet, which is used to pierce the plant cell wall, withdrawing cellular contents and perhaps injecting toxic materials. A single nematode usually causes little injury to the plant even though it feeds on hundreds or thousands of cells. Injury to the plant results when it is attacked by thousands of nematodes. Nematodes frequently predispose trees to attack by other pathogens. They also cause wounds on roots, which serve as infection courts for root pathogens such as *Fusarium* spp.

Epidemiology: Nematodes can actively move in the soil only a few centimeters, at most, meters a year. They are carried in soil, infected plants, or drainage water for long distances. An average generation takes about 30 days, and each adult female may produce up to 500 eggs. Populations can build up more quickly in sandy soils because of the warmer soil temperatures. A cold winter with little snow cover may greatly reduce the size of the nematode population.

Diagnosis: The root-knot nematode (*Meloidogyne* spp.) causes formation of prominent galls on the roots. The pine-cystoid nematode (*Meloidodera* spp.), root-lesion nematode (*Pratylenchus* spp.), dagger nematode (*Xiphinema* spp.), lance nematode (*Hoplolaimus* spp.), sheath nematode (*Hemicycliophora* spp.), ring nematode (*Macroposthonia* spp.), pin nematode (*Paratylenchus* spp.), and sting nematode (*Belonolaimus* spp.) cause less specific injuries such as stunting, chlorosis, and mortality.

Control Strategy: Preplant fumigation for damping-off or *Cylindrocladium* control (e.g., 340 kg/ha of 67% methyl bromide and 33% chloropicrin under plastic tarping to a depth of

30 cm) also controls nematodes. Neither of these chemicals is likely to be available for use in the near future.

SELECTED REFERENCES

Ruehle, J. L. 1969. Forest nematology—A new field of biological research. *J. For.* 67:316–320.
Sutherland, J. R. 1977. Corky root disease of Douglas-fir seedlings: Pathogenicity of the nematode *Xiphenema bakeri* alone and in combination with the fungus *Cylindrocarpon destructans*. *Can. J. For. Res.* 7:41–46.

DISEASE PROFILE

Littleleaf Disease

Importance: This root disease complex has seriously curtailed the sustained management of some 2,400,000 ha of the shortleaf pine type in the southeastern United States.

Suscepts: Hosts are primarily shortleaf pine; however, the host range of the littleleaf pathogen encompasses more than 900 species of plants, including many evergreen and deciduous trees, either planted or native to the temperate regions of the world.

Distribution: Littleleaf disease occupies the contiguous commercial range of shortleaf pine, and to a lesser extent loblolly pine, east of the Mississippi River, an eight-state area largely comprising the piedmont plateau. Shortleaf pine west of the Mississippi River is not affected.

Causal Agents: Littleleaf disease is caused by *Phytophthora cinnamomi* (Peronosporales).

Biology: Host root tips are penetrated by zoospores, and unsuberized tissue is quickly killed; modes of saprophytic survival are not fully known.

Epidemiology: The disease develops as a function of nitrogen deficiency in the tree and is associated with reduction in absorptive capacity of the rootlets due to predisposing soil factors, specifically poor aeration, low fertility, multiple attacks by *P. cinnamomi* and periodic moisture stress. If sustained, littleleaf disease prevents root rejuvenation and tree recovery.

Diagnosis: Littleleaf is typified by general crown decline (thinning, dwarfing, and chlorosis of the foliage) and by associated drastic reduction in shoot and diameter growth; soil samples aid in identification of littleleaf sites and in laboratory detection of the causal fungus.

Control Strategy: Losses from established littleleaf can be offset by the returns from salvage cuts in affected stands; future management of shortleaf pine should avoid littleleaf soils as judged by site risk classification.

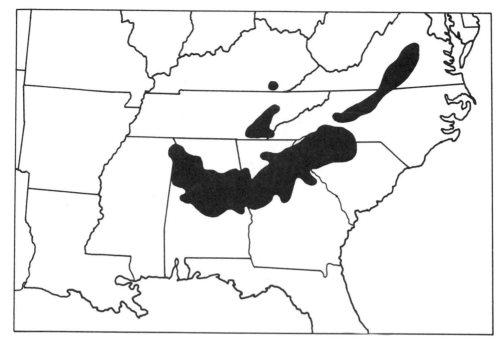

Disease incidence.

SELECTED REFERENCES

Oak, S. W., and F. H. Tainter. 1988. *How to Identify and Control Littleleaf Disease.* USDA For. Serv., Prot. Rept. R8–PR12. 14 pp.

11.6 LITTLELEAF DISEASE

Importance

Littleleaf is a root disease complex involving a soil-inhabiting fungus, namely *Phytophthora cinnamomi,* which affects shortleaf pine and to a lesser extent loblolly pine as both become predisposed by unfavorable soil conditions (Fig. 11.9). Because of the interesting sequence of events that led to our present understanding of this disease, a more thorough discussion than usual will follow. A minimum of scientific references occur within the text. These references are listed at the end of the chapter. Littleleaf disease represents a classic example of how man's abuse of the land and natural forested ecosystems has directly led to the development of a disease that was insignificant before European settlement.

Colonization of the southern piedmont of the southeastern United States was especially rapid after 1790, the decade in which the cotton gin was invented (Fig. 11.10). This device deseeded cotton inexpensively. The resulting high world demand for cotton encouraged widespread clearing of steep slopes that should have probably remained forested. Repeated cropping, of cotton in particular, with little or no atten-

FIGURE 11.9   In the foreground are a healthy shortleaf pine on the left and a typical littleleaf tree on the right. From Zak (1957).

tion to proper soil management and erosion control deteriorated the soil so badly that yields progressively declined, and farming was no longer profitable. The origin and ecology of the disease is primarily related to the widespread abandonment of this farmland in the southern states following the Civil War, the depression of the 1880s, and the economic depression of 1928–1929. Much of this idle land, by then reduced to subsoil from sheet and gully erosion (Fig. 11.11), reverted to natural successions or was colonized primarily by shortleaf and other southern pines. Some 20–50 years later, littleleaf disease was found to occur in greatest severity on these infertile old field sites.

This ecological progression to littleleaf was first detected shortly after the turn of the century in Tuscaloosa and Tallapoosa Counties, Alabama. By 1940 littleleaf oc-

DISTRIBUTION OF POPULATION 1740, 1770, 1810

ONE DOT EQUALS 200 PEOPLE

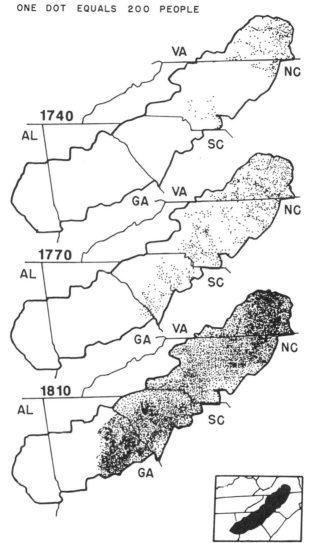

FIGURE 11.10   Changes in patterns of population on the southern Piedmont from 1740 to 1810. Redrawn from Trimble (1974). Courtesy of the Soil and Conservation Society.

curred widely in Alabama, South Carolina, and Georgia, where, ultimately, losses were most severe. Because shortleaf pine is commercially important in 12 states within a botanical range that covers 22 states from New Jersey to Texas and Oklahoma, the threat of spread and catastrophic losses was very real. During the next decade, littleleaf received priority funding and commanded almost exclusively the research attention of regional pathologists. By 1950 extensive surveys had established that the disease oc-

FIGURE 11.11 Severe gully erosion resulting from poor agricultural practices and subsequent land abandonment. Courtesy of Department of Forest Resources, Clemson University, Clemson, S.C.

curred within the contiguous commercial range of shortleaf pine east of the Mississippi River, an eight-state area that includes the piedmont plateau (Alabama through Virginia), the Appalachian plateau (eastern Kentucky and Tennessee), and north-central Mississippi in the Atlantic coastal plain (Fig. 11.12). There was, however, no evidence of spread beyond the area.

Throughout the disease range, relatively large areas remained free of littleleaf. Spotty distribution of the disease was correlated with stand age, poorly drained soils, and degree of erosion. Symptoms were not apparent until stands were 20–50 years old. On soils with poor internal drainage, the incidence of littleleaf was 17–31% compared with 6–15% on well-drained soils. On slightly to moderately eroded sites, littleleaf incidence was 4–9% compared with 14–53% on severely eroded land. Three factors (stand age, soil internal drainage, and degree of erosion) interacted with one or more pathogenic fungi to cause littleleaf disease.

Once established, the disease intensified within affected stands very rapidly. From observation plots averaging 1, 12, and 33% littleleaf, the disease increased to 49, 75, and 86%, respectively, within 16 years. In such areas, as shortleaf pine lost dominance in the stand, it was replaced by other pines or by hardwoods.

In the comprehensive disease survey of 1952 (Hepting and Jemison 1958), littleleaf ranked eighth nationally in accounting for sawtimber loss with an annual growth impact of 146 million board feet or 0.8% of the total disease impact. The disease ultimately affected some 6,000,000 ha or about 35% of the commercial shortleaf pine area. It reached such important proportions on 2,400,000 ha that it was regarded as the most serious limitation to the sustained management of shortleaf

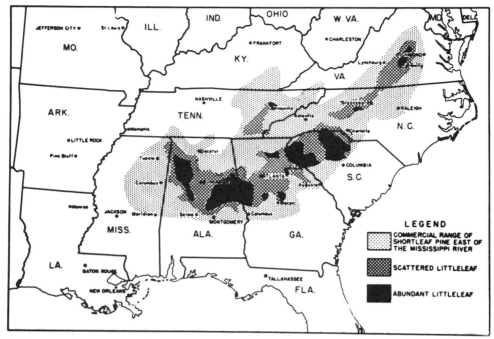

FIGURE 11.12   The distribution of littleleaf disease of shortleaf and loblolly pines. From Campbell et al. (1953).

pine in much of the Appalachian piedmont and in the upper coastal plain of Alabama. Translated to economic impact, the annual stumpage loss from littleleaf for the southern states as a whole was estimated at $5,000,000 (in 1958 dollars) on the basis of salvage returns being offset by growth reduction and understocking.

Littleleaf is a complicated disease and is a classic example of the many interacting factors generally operating in soilborne diseases. That this complex problem was resolved in a practical way is largely a tribute to the painstaking work of the research personnel of the USDA Forest Service.

Suscepts and Distribution

Phytophthora cinnamomi was first reported in the United States on rhododendron. By 1954 the host list totaled 109 species of woody perennials and annual plants. More recently in Australia, the fungus was isolated from the root zones of 87 species of native forest trees of which 72 species were diseased; altogether these species encompassed 31 plant genera in 16 families. Release of these findings expanded the host range to approximately 400 species of plants representing a wide distribution in both the northern and southern hemispheres.

The pre-littleleaf history of *P. cinnamomi* in the southeastern United States presents an interesting account of disease as a factor in the evolution of forest composition. There is intriguing evidence that *P. cinnamomi* was responsible for the epidemic recession of the American chestnut and associated Ozark chinkapin from the lower piedmont and Atlantic coast portions of their common natural range during the period

FIGURE 11.13   Dieback of jarrah *(Eucalyptus marginata)*. From Podger (1972).

1850–1875. The early distribution map of *P. cinnamomi* isolations from surviving chestnut and chinkapin is strikingly similar to the littleleaf disease range plotted nearly a century later. Thus it appears that the pathogen was well established prior to the succession of soil deterioration, land abandonment, and suscept revegetation, which led to littleleaf development. During the interim years after 1912, chestnut survivors and in particular the untouched elite stands in the Appalachians were destroyed by yet another fungus, the introduced chestnut blight pathogen, *Cryphonectria parasitica*.

Through the years, the disease capacity of *P. cinnamomi* has severely limited the economic productivity of a number of hosts in addition to shortleaf pine. Spectacular damage has occurred from avocado root disease in California and pineapple root disease in Hawaii and Queensland. On forest trees in the United States, it has caused losses in nurseries and in young plantations of Lawson cypress, sand pine, and Fraser fir; in Louisiana, *P. cinnamomi* and *Pythium* species have been associated with the decline of natural stands of loblolly pine.

In the southern hemisphere, the epidemic development of *P. cinnamomi* in exotic plantations of shortleaf and Monterey pines in New Zealand was the harbinger of succeeding and more severe devastation of entire forest communities in Australia. The pine disease in New Zealand displayed essentially the same features as those of littleleaf in the United States; in both locales, seriousness of the disease was determined largely by edaphic factors. Jarrah *(Eucalyptus marginata)* dieback in Western Australia, known since 1921, was ultimately linked to root pathogenesis by *P. cinnamomi* in soils characterized by poor moisture-storage capacity (Fig. 11.13). To date, of the approximately 1,400,000 ha of jarrah and plant associates in state forest reserves, some 80,000–100,000 ha have been destroyed in Australia and Victoria. Although these losses are most economically critical with regard to members of *Eucalyptus* and *Monocalyptus*, Australia's major timber species, total damage is compounded by the fact that more than 100 species of the native flora are affected. Growth in the highly susceptible *Banksia grandis* understory enables *P. cinnamomi* to spread within the jarrah forest. The pathogen is dispersed in water ponding on concrete laterites 10–100 cm below the soil surface. Severe infection occurs in vertical woody roots of jarrah-penetrating channels in the concreted laterite at a depth of 1 m. In a thorough review of the gravity of this situation, which culminated with the prognosis that jarrah faces commercial extinction, it was concluded that *P. cinnamomi* is an introduced pathogen. Certainly, the omnivorous pathogenicity of *P. cinnamomi* in Australia's forests poses a threat of extreme magnitude in marked contrast with other epidemics, namely chestnut blight, Dutch elm disease, and white pine blister rust, each caused by a pathogen limited to one or two host genera.

Cause

The search for the cause of littleleaf touched upon all conceivable possibilities, primarily because first attempts at isolations from deteriorating roots by conventional methods yielded inconclusive results. Nevertheless, as a consequence of this work, the duplication of brown patch symptoms in roots inoculated with *Wolfiporia cocos (Poria cocos)* and *Torula marginata* raised important questions about the food storage capacity of diseased trees. *Brown patch* is described as premature formation of root bark that occurs normally in all southern pines but excessively, and at the expense of cortical storage tissue, in the more defective root systems of littleleaf trees.

Some potential causes were discounted completely after thorough investigation. No evidence for virus involvement developed after long-term observation of successful bark-patch and approach grafts of tissue from diseased to healthy trees. In addition, cleft-grafting of littleleaf scions to healthy seedlings showed complete symptom reversal, some within a year. Evaluation of weather station data, particularly that comparing precipitation within and beyond the littleleaf belt, showed only a negligible deficit of 15 cm of rainfall for the 18-year period leading to the littleleaf outbreak. Ultimately, however, littleleaf symptoms were duplicated, in part, by the 5-year influence of artificially induced drought. Under prolonged conditions of moisture stress, shortleaf pine showed reductions in growth and in needle length and retention but reduced nitrogen assimilation, and foliage yellowing so characteristic of littleleaf did not occur.

Various aspects of tree nutrition and translocation provided additional clues in deciphering the littleleaf phenomenon. Food reserves of root but not stem tissues of affected trees were found to be only 50% of normal. Brown patch in roots is often a secondary, but not always a consistent, result in trees with advanced littleleaf. The differences in carbohydrate content between roots of diseased trees presented a pattern of deterioration suggestive of pathogenic action rather than a uniform decline that could be attributed solely to unfavorable soil conditions. Further explanation of impeded carbohydrate synthesis in diseased trees was derived indirectly from amendment of littleleaf soils with a variety of inorganic salts and organic materials. The response of littleleaf trees was such that Roth et al. (1948) concluded that "littleleaf seems to be associated with failure of the trees to absorb sufficient nitrogen, even where soil nitrogen is present in amounts regarded as adequate for the development of normal shortleaf pine." They suggested the strong probability that deficient nitrogen absorption was due to faulty mycorrhizae or killing of fine roots by soil fungi. Evidence for the latter was soon forthcoming.

The breakthrough in identifying the primary pathogen of littleleaf came when Campbell (1948) applied the differential baiting method of Tucker (1931) in isolating *Phytophthora cinnamomi* (Peronosporales) from rootlets of diseased trees at several locations. However, less than 2% of diseased rootlets yielded the pathogen, especially in clay soils typical of littleleaf sites. Campbell soon adapted the method to isolate *P. cinnamomi* directly from soil sampled in the root zone of littleleaf trees. In so doing, he obtained isolation frequencies as high as 75%. This method uses apples as bait. These are inoculated with suspect soil, sealed, and incubated at 15–27°C for 5–10 days. The fungus, if present, rapidly and selectively colonizes the apple tissue causing a firm, dry decay. Bits of this corky apple tissue are then transferred to cornmeal agar for pure culture confirmation. Isolation and detection procedures have been much refined by the development of commercial serological kits using enzyme-linked immunosorbent assay (ELISA) to detect *P. cinnamomi* in soil and plant samples. Unfortunately, this technique was not available to the earlier scientists attempting to unravel the littleleaf disease complex.

Subsequent apple–soil assay of shortleaf pine stands near Athens, Georgia, yielded *P. cinnamomi* in approximately 5% of the soil samples taken under healthy trees on disease-free sites; 15% of the soil samples taken under healthy trees on littleleaf sites; and 42% of those from under littleleaf-diseased trees. As the soil-assay area expanded, the fungus was not only isolated commonly from littleleaf soils within the disease range (e.g., in South Carolina it was present in 91% of the 154 plots sampled), but it was also detected in the soils from 31 of 61 plots broadly established outside the littleleaf area. Later the apple technique was used to confirm the ubiquity of *P. cinnamomi* in soils associated with littleleaf in Kentucky and Tennessee, with healthy pines in Virginia and Maryland at locations near or well beyond the littleleaf range, and with a number of woody hosts at numerous locations in North Carolina, including 28 of 78 stands of healthy shortleaf and loblolly pines sampled in the coastal plain. Although the baiting method qualitatively substantiated the constant-association link between *P. cinnamomi* and the littleleaf disease, a quantitative method employing soil dilution and a selective medium was soon developed for propagule counts and soil population assay.

Second-stage proof of littleleaf causation (i.e., regular isolation of the pathogen from necrotic rootlets) has been adequately demonstrated by the abundance of *P. cinnamomi,* as just indicated, in association with high rootlet mortality. The host–pathogen interaction and resultant disease is most pronounced in an edaphic environment characterized by wet soils, either as a function of the water-holding capacity of the soil itself or by structural features, such as clay pans, that impede downward percolation of water. Early failures and later difficulties in isolating *P. cinnamomi* directly from roots are readily explained by the locus and ephemeral nature of root pathogenesis. The fungus attacks and quickly kills only the succulent root tips of the pine host wherein saprophytic dominance and survival of the pathogen, as a typical soil-inhabitant, sensu Garrett (1956), is short-lived due to the contaminating and suppressing effect of competitive saprophytic colonization by soil microorganisms. Hence, the difficulty of direct isolation and the success of selective enhancement by the use of baits and specialized media.

Critical fulfillment of Koch's postulates in demonstrating pathogenicity of *P. cinnamomi* on shortleaf pine has been satisfied in greenhouse studies; however, field inoculations of trees 20 years of age and older have yielded inconclusive results.

Biology

The disease diagram of littleleaf disease of shortleaf pine is shown in Figure 11.14. The genus *Phytophthora,* commonly referred to as a Phycomycete or water mold, has been mycologically reassigned to the class Oomycetes, order Peronosporales, and family Pythiaceae. The genus contains nearly 70 described species of worldwide distribution and is unique among facultative parasites in that all species are pathogens of higher plants; hence, the nomenclature, which translated from the Greek means plant destroyer.

All structures of the fungus are hyaline so that stained slide preparations are necessary for detailed microscopic examination. In gross morphology, the colony of *P. cinnamomi* on agar presents a petaloid or camelloid pattern of growth. Microscopically, its vegetative morphology is marked by coenocytic, sparingly branched, thin-walled hyphae containing abundant fat globules; some lateral and terminal branches give rise to globose to pyriform chlamydospores that form singly or in clusters of 3–10 (Fig. 11.15).

Asexual reproduction occurs in the presence of free water with the development of stalked, terminally borne, lemon-shaped sporangia (Fig. 11.16); with maturity, the thin-walled sporangium becomes filled with an indefinite number of sporangiospores (8–40). After rupture of the sporangial wall, the sporangiospores become motile (zoospores). The reniform (kidney-shaped) zoospores are propelled by submicroscopic biflagella emanating from the concave side. The zoospores, upon swimming to a suitable substrate such as suscept root tips, make contact, attach firmly by en-

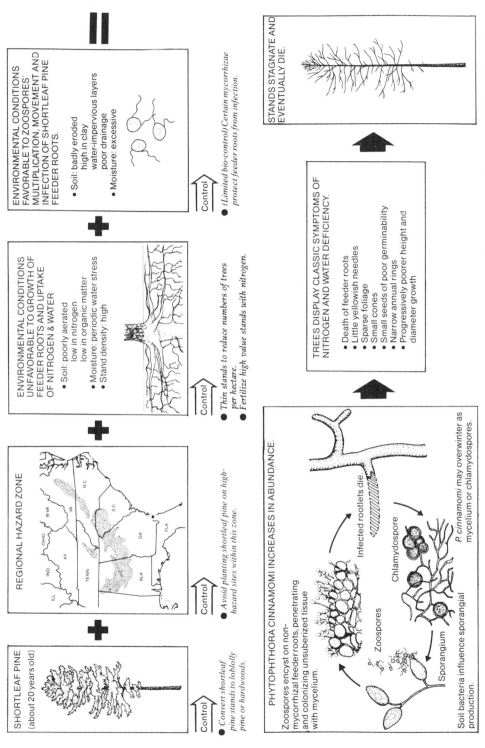

FIGURE 11.14 Disease diagram of littleleaf of shortleaf pine induced by a combination of environmental factors and *Phytophthora cinnamomi*. Drawn by Valerie Mortensen.

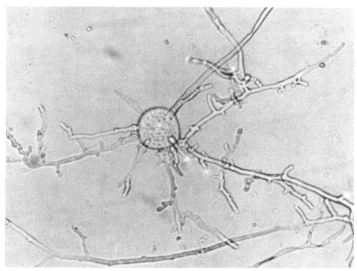

FIGURE 11.15 Germinating chlamydospore of *Phytophthora cinnamomi*. From Mircetich et al. (1968).

cystment, and germinate, thus establishing a new vegetative colony. Emptied sporangia are often succeeded by sympodial proliferation of additional sporangia.

Sexual reproduction occurs through the genetic recombination of heterothallic, bisexual mating types; pairing of compatible isolates results in the formation of amphygynous antheridia and oogonia, the male and female gametangia, respectively (Fig. 11.17). Within 4–6 days after fertilization, the oogonia transform into thick-walled, spherical oospores. Nuclear ontogeny and meiosis have been determined in but few species of *Phytophthora*. In *Phytophthora cactorum* a single nucleus from the

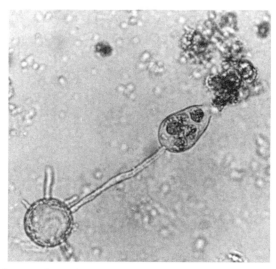

FIGURE 11.16 Sporangium of *Phytophthora cinnamomi*. Note discharge of zoospores. From Mircetich et al. (1968).

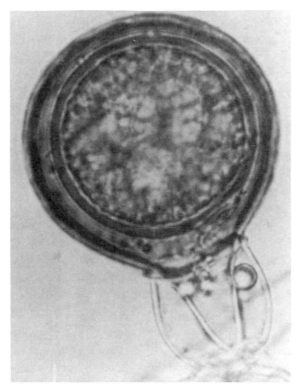

FIGURE 11.17 Amphigynous antheridium and oogonium with oospore of *Phytophthora cinnamomi*. From Zentmyer (1980). Courtesy of G. A. Zentmyer.

antheridium migrates into the oogonium and unites with one of many oogonial nuclei; the others disintegrate. Meiosis occurs both in the antheridium and oogonium which, if generically representative, suggests that Oomycetes are diploid.

Cultural Physiology. The unique requirements of *P. cinnamomi* for some microbial principle in soil leachate in stimulating sporangial production has been further elaborated. Marx and Haasis (1965) induced aseptic sporangial formation with the metabolic diffusates of two species of *Pseudomonas;* species of *Chromobacterium* were similarly active. The role of the bacteria in this process is not known, but there are suggestions that the bacteria produce a stimulatory metabolite or deplete nutrients.

The germination capacity of the various spore stages of *P. cinnamomi* has important implications in the survival and pathogenicity of the fungus. Chlamydospore germination requires exogenous nutrients, pH 3–9, and temperatures between 9–12°C and 33–36°C with an optimum of 18°C. The amount of chlamydospore germination is more or less directly influenced by nutrient availability; likewise, the type of germination in response to optimal and suboptimal nutrition, respectively, ranges from multiple germ-tube production (8–16/spore) to much-reduced germ-tube emergence (1–4/spore) and some spores with but a single short germ tube bearing a sporangium. Sporangial germination in the genus *Phytophthora* is typically bimodal and rather unique among the fungi. The bimodality refers to direct and indirect germination as influenced primarily by temperature and moisture. Indirect germination, or zoospore formation, requires free water and brief chilling; otherwise, the sporangia germinate to form mycelium directly in situ. Oospores of most species of *Phytophthora* germinate slowly and erratically for reasons not understood but are attributed to constitutive dormancy. It has been noted, however, that oospores of homothallic species usually germinate more quickly and in greater abundance than those of heterothallic species. Because oospores are essentially dual mechanisms of survival and genetic adaptability in relation to pathogenicity, elucidation of the factors attendant to their germination is critically needed.

The biology of *Phytophthora* zoospores has been the subject of several reviews. Zoospores, after emerging from the sporangium into a water environment, pass successively through phases of motility, encystment, and germination.

Several aspects of zoospore behavior should also be noted with regard to littleleaf development. Zoospores of some 11 species of *Phytophthora* are attracted nonspecifically to plant roots, while those of *P. cinnamomi*, the singular exception to date, are attracted specifically to avocado. Because the stimulus is partly contained in constituents of root exudates, as demonstrated by zoospore aggregation at the zone of root elongation from whence exudation primarily emanates, the motility response is termed *chemotaxis*. The gradient of exudate attraction from the root may be very steep due to use by rhizosphere flora and dilution in surrounding soil. In soil leachate in petri dishes, zoospores of *P. cinnamomi* fail to reach suscept root pieces placed at distances greater than 8 cm. The substances that induce chemotaxis are primarily sugars and amino acids, which, respectively, are the main energy sources for fungi and bacteria in the rhizosphere. Zoospores of *P. cinnamomi* respond to gradients of ethanol, accumulating in proportion to source concentrations. This poses significant implications in that ethanol production by plant roots during brief periods of fermentive metabolism, as induced by soil saturation, may function in zoospore attraction. Environmental influences upon suscept physiology and, in turn, the quantity and quality of root exudates such as the example cited merit greater attention in resolving the mechanisms of zoospore attraction, which in itself, has yet to be demonstrated in soil. Essentially, anaerobic conditions are necessary in the rhizosphere before pine root tips are predisposed to attack by *P. cinnamomi* (Fig. 11.18).

Upon accumulating at the root, zoospores encyst and germinate rapidly, within 30–60 minutes, if *P. cinnamomi* on avocado roots is any indication of the speed of the process on other suscepts. Germ tubes of the fungus also grow toward the region of root elongation and respond to root exudates and their constituent amino acids. Such growth orientation to a chemical stimulus is called *chemotropism*.

Electrotaxis is yet another aspect of zoospore trophism that has been studied. Zoospores respond differently to various intensities of current (0–5 μA). These responses were similar for the seven species of *Phytophthora* studied (including *P. cinnamomi*) and were not altered by the addition of organic acids, sugars, antibiotics, metabolic inhibitors, or surfactants. Both motile and encysted zoospores carry a net negative charge. The cationic and anionic exchange properties of the surfaces of actively growing roots are sufficient to attract and hold negatively charged zoospores by positive electrostatic forces. Root pathogenesis by soilborne species of the Pythiaceae is unique and highly efficient in that zoospore inoculum is actively disseminated and trophically responsive to suscept roots. Zoospore motility and, as a consequence, root infection are primarily dependent upon continuous water films for passage of inocu-

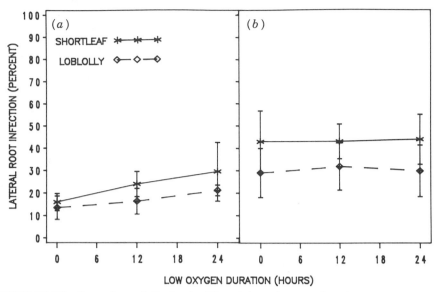

FIGURE 11.18   Lateral root infection by *Phytophthora cinnamomi* at a low oxygen concentration of 0.25-0.50 mg/l. (a) and (b) represent two experiments. Note that loblolly was infected less often than shortleaf pine. From Fraedrich and Tainter (1989). Courtesy of S. W. Fraedrich.

lum from source to suscept. As this soil water linkage becomes discontinuous through drying, however, hyphal growth and direct penetration may constitute the mode of infection, especially in soils lacking a certain complement of bacteria as shown in Figure 11.19. By the latter mode, intracellular ramification of the fungus in the cortical cells of aseptic shortleaf and loblolly pine roots is more extensive and marked by prolific vesicle development. In contrast, when zoospores are the inoculum, the resulting infection exhibits limited intracellular hyphal development and no vesicles. By either mode, the net effect in terms of root necrosis is the same—the fungus destroys the rootlet back to suberized tissue and essentially negates its function in nutrient absorption. Littleleaf expression then is governed by the magnitude and persistence of rootlet infection in whole root systems. The rootlet mortality differential in 37-year-old shortleaf pine between unaffected and littleleaf trees averaged 18 and 34%, respectively. Under comparable conditions of soil and tree age, rootlet mortality in healthy loblolly pine was only 6% (Fig. 11.20). These differences and the greater susceptibility of shortleaf pine to *P. cinnamomi* may be explained in part by the rooting characteristics of the two species. The feeder roots of shortleaf pine are most abundant in the upper few centimeters of soil, while loblolly pine roots, although fewer in number and larger in size, are more deeply distributed and show greater ability to penetrate heavy soils.

   In addition to the bacterial influence upon sporangial formation and mode of infection by *P. cinnamomi*, other soil organisms can influence root pathogenesis. Plant-parasitic nematodes, the sheath *(Hemicycliophora vidua)* and spiral *(Helicotylenchus dihystera* and *H. erythrinae)* nematodes in particular, were twice as abundant in diseased as in healthy stands of shortleaf pine. Although species such as these may contribute to the overall root damage in the littleleaf complex, the possibility has not been further examined. Naturally occurring mycorrhizae also deter *P. cinnamomi* on shortleaf

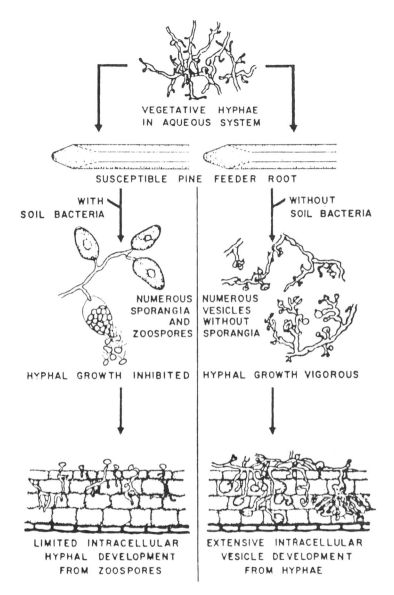

VEGETATIVE HYPHAE
IN AQUEOUS SYSTEM

SUSCEPTIBLE PINE FEEDER ROOT

WITH
SOIL BACTERIA

WITHOUT
SOIL BACTERIA

NUMEROUS
SPORANGIA
AND
ZOOSPORES

NUMEROUS
VESICLES
WITHOUT
SPORANGIA

HYPHAL GROWTH INHIBITED

HYPHAL GROWTH VIGOROUS

LIMITED INTRACELLULAR
HYPHAL DEVELOPMENT
FROM ZOOSPORES

EXTENSIVE INTRACELLULAR
VESICLE DEVELOPMENT
FROM HYPHAE

FIGURE 11.19 The influence of bacteria from a forest soil on the mode of infection of aseptic shortleaf and loblolly pine roots by *Phytophthora cinnamomi*. From Marx and Bryan (1969).

pine, physically and chemically. Mycorrhizae of shortleaf pine were completely resistant to zoospore infection of *P. cinnamomi*, whereas nonmycorrhizal roots were completely susceptible. A series of classic experiments on mycorrhizal protection of host roots against soilborne pathogens is discussed in more detail in Chapter 12.

Saprophytism and Survival. Members of the genus *Phytophthora* are generally classified as root-infecting fungi, sensu Garrett (1956), because they lack one or more of

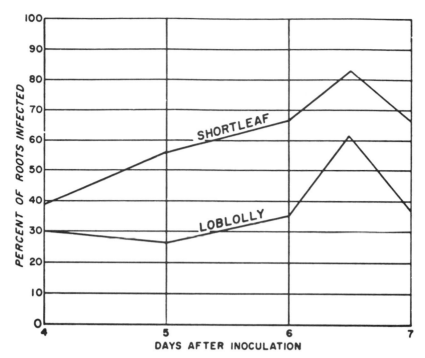

FIGURE 11.20   Relative susceptibility of shortleaf and loblolly pine roots to infection by
Phytophthora cinnamomi. From Zak and Campbell (1958). Reprinted from *Forest Science*, Vol.
No. 4, published by the Society of American Foresters, 5400 Grosvenor Lane, Bethesda, MD
20814-2198. Not for further reproduction.

the criteria for competitive saprophytism. This deficiency may be related largely to
their sensitivity to moisture conditions, because all *Phytophthora* structures except
chlamydospores and oospores die under dry conditions. *Phytophthora cinnamomi*,
however, may be an exception to these generalizations. In the absence of a host, *P.
cinnamomi* can persist for up to 6 years in soil maintained at 20°C and approximately
20% moisture content. Additional tests for saprophytism showed moderate mycelial
growth through nonsterile soil and appreciable colonization of dead organic matter,
particularly under conditions of high soil moisture. Conversely, *P. cinnamomi* may be
a poor competitive soil saprophyte based on failure of the fungus to colonize dead
Douglas-fir tissue in natural soil amended to 50% or less by volume with an alfalfa–
sand–fungus mixture. The fungus did survive, however, for up to 19 months in natural agricultural and forest soils that were artificially infested with the fungus and
stored under field conditions. These apparent contradictions regarding the saprophytic status of *P. cinnamomi* may simply be a function of differences in experimental
materials and physical conditions.

Researchers generally agree that long-term survival of *P. cinnamomi* in soil and
host tissue is by means of chlamydospores. The chlamydospore is a primary unit of
survival for *Phytophthora* spp. in soil and with the heterothallic species, which includes
P. cinnamomi, and may play a more important role than oospores. Nevertheless, the
evidence to date on chlamydospore survival of *P. cinnamomi* is only strongly circumstantial at best, while oospore survival of the fungus awaits evaluation of any sort.

Epidemiology

Because various aspects of littleleaf epidemiology have already been discussed and documented, a brief statement on the causal complex at this point seems more appropriate in review. In summary, littleleaf symptoms arise as a result of nitrogen deficiency in the tree associated with reduction in absorptive capacity of the rootlets due to primary attack of *P. cinnamomi;* other soil factors, specifically poor aeration, low fertility, and periodic moisture stress, are also injurious to the rootlets and, if sustained, prevent root rejuvenation and tree recovery.

The influence of the edaphic environment upon the suscept, the pathogen, and the host–pathogen interaction is the primary determinant of disease expression. Excessive soil moisture and its consequence, poor aeration, are particularly important when temperature is favorable. Essentially, high moisture favors motility of the pathogen, while poor aeration is debilitating to the suscept and may even influence the composition of root exudates in functional chemotaxis. In the long term, shallow, poorly drained soils of heavy texture or impervious, poorly aerated soils low in fertility and depleted of organic matter favor the disease primarily by reducing host vigor and ability to regenerate normal roots following local infection of short duration.

Pathogenicity of *P. cinnamomi* seems remarkably uniform in view of its wide geographic distribution, extensive host range, and capacity for genetic adaptation. Although isolates vary in comparative pathogenicity to a number of seedling tree species, the inherent susceptibility of most temperate zone hosts has been more a function of environmental regulation. The status of Douglas-fir in this regard provides an interesting case study. The discovery of *P. cinnamomi* in the Pacific Northwest at Salem, Oregon, in 1951 and recognition that Douglas-fir is a highly susceptible host prompted research evaluation of what then appeared to be an impending threat to the commercial forests of the region and, in particular, the more valuable Douglas-fir component. Preliminary and representative soil-survey isolations in Oregon showed that the fungus was absent from forested areas but present in some nurseries. Subsequently, the fungus was introduced three times in 5 years into the soil of cutover sites and into the root zones of a sapling fir thicket, but it did not attack the fir nor did it survive the growing season in any year or location. Therefore, it was concluded that *P. cinnamomi* is poorly adapted to the soil-site conditions of the study. It must be cautioned, however, that these results may not apply to canopied sites in a more moderate environment. This qualifier was taken into account in later studies of the effects of regulated soil moisture–temperature regimes upon the infection potential of *P. cinnamomi* on seedling Douglas-fir. A temperature threshold of 16°C was found to be necessary for infection. This translates to the following field situation. In southern-exposure soils, summer temperatures are favorable, but moisture is continuously below field capacity and too dry for infection. Conversely, in northern-exposure soils, moisture is adequate, but temperatures are too low. Thus the prognosis that the requirements for pathogenesis of *P. cinnamomi* do not occur concurrently in the Douglas-fir region, and, therefore, disease development of any consequence is unlikely.

Diagnosis

Primary symptoms of littleleaf are hidden from view, occurring as rootlet mortality. As root loss progresses and without regeneration, the secondary symptoms of crown

FIGURE 11.21 A branch of shortleaf pine from (a) a littleleaf diseased tree and (b) a healthy tree. From Hepting et al. (1945).

decline become visible. Initially, only a slight yellowing and dwarfing of new foliage may be evident. At this stage the symptoms are typical of those arising from nutrient or water deficiencies. Within a few years, however, the symptoms advance to the diagnostic littleleaf condition (Fig. 11.21). The crown is thin, shoot growth is stunted, and the branches, with only terminal tufts of needles to support, often assume an ascending habit. The sparse foliage that remains becomes chlorotic and may be reduced in size from a normal 8–13 cm to 1.3 cm in length. As the foliage declines, the crown dieback advances, and radial growth is drastically reduced (Fig. 11.22).

FIGURE 11.22 Increment cores from (left) healthy shortleaf pines and (right) littleleaf-affected trees showing reduced growth in recent years. From Hepting et al. (1945).

Final symptoms may be evidenced by abnormally heavy production of smaller cones that persist and contain mostly abortive seeds. As the crown dies, sprouts of normal vigor and foliage color develop profusely along the lower part of the stem. Ordinarily, afflicted trees succumb within 6 years after the onset of visible symptoms; however, some may die as quickly as 1 year, and others may languish in gradual decline for 12 or more years. During the past two decades, these stressed trees are often prematurely killed by bark beetles, especially the southern pine beetle.

Control Strategy

Perhaps a consideration of all the principles of disease control, as viewed within the context of what is known about *P. cinnamomi* and the diseases it causes, might serve as a useful approach to the courses of action that have been taken.

Exclusion and Eradication. In most circumstances, the application of both of these principles seems inappropriate to current knowledge and status of the problem. Exclusion is presently employed primarily in disease-free jarrah forests in Australia where controlled-access logging and sanitation of logging equipment offers hope for preventing introduction of the fungus. In the southeastern United States and in southwestern Australia, the fungus is already so ubiquitous in the soil that its presence must be assumed and counteracted by other measures. Eradication of the pathogen at its source, namely soil and host tissue, is impractical for the same reason and also for obvious difficulties to be encountered in root-system extraction and application of chemical eradicants. The latter approach has been attempted in Australia in small-scale tests without success. In the nursery, where the economics of intensive culture permit and often necessitate direct control of root disease, chemical eradicants did not give practical control of *P. cinnamomi* in Fraser fir nursery beds, but they were effective in Eucalyptus transplant beds in Australia. Subsequently, however, metalaxyl has been effectively employed as a prophylactic fungicide against *P. cinnamomi* root disease of Fraser fir.

Therapy. The response of affected trees to nutrient amendment by soil fertilization is the only successful example of therapy to date. In the United States the application of 5–10–5 commercial fertilizer at the rate of 2273 kg/ha, together with 454 kg of ammonium sulfate, prevents symptom development in healthy trees and improves the condition of littleleaf trees only in the early stages of disease development. The fertilizer should be broadcast over the soil in the spring with applications repeated every 4 years. The limited benefits thus attained are attributed to improved nitrogen assimilation, which offsets the imbalance of full crown demands upon a declining root system under the attack of *P. cinnamomi*.

In New Zealand, where littleleaf soils are more characteristically phosphorus-deficient clays, the aerial application of superphosphate (568 kg/ha) produced spectacular and long-lasting response in Monterey pine stands with littleleaf. The control has been attributed to a complex series of changes that directly promote host recovery (crown improvement and closure, root rejuvenation and mycorrhizal gains, and improved soil aeration as a function of reduced water-logging due to the increased transpiration and closure of recovered crowns) and ultimately limitation of the pathogen through lowered soil moisture and possibly mycorrhizal protection. These successes, especially the latter, merit further study and application in view of increasing aerial fertilization in forest practice.

Avoidance. Obviously, the forest manager cannot avoid what has already happened; but he/she can manipulate forests of the future by recognizing high-risk littleleaf sites and treating them accordingly. The potential for littleleaf in shortleaf pine stands in the Appalachian piedmont region can now be judged on site using a simple soil-rating scale (Fig. 11.23), which is based on four readily measurable soil characteristics, or even by site index classification, which is inversely correlated with the soil–littleleaf incidence relationship. Note that a high-risk site would score low (0–50) on a soil-rating scale of 100, the composite of erosion (40), subsoil consistency (32), depth to the zone of greatly reduced permeability (15), and subsoil mottling (13); likewise, high risk is indicated by low-site indices. Thus upon identifying high-risk sites, the forester is advised to convert the composition of the stand to less susceptible species. In practice, this generally means favoring loblolly pine by planting or by natural regeneration.

Soil rehabilitation is an alternative that can ultimately lead to avoidance of the pathogen but at the sacrifice of time in fallow or reduced productivity for long-term gains in disease prevention. Silviculturally converting the site to soil-building species of hardwoods can renew the productivity of littleleaf sites, while providing forest cover. Many of these species, such as sweetgum or yellow-poplar, are considered to be "low value" on such poor sites. Their value as a forest product (although reduced), their other ecological values, and the value of site rehabilitation often justifies the site conversion.

Protection. To date the only evidence of functional inactivation of *P. cinnamomi* at the site of suscept penetration is by mycorrhizal protection. Demonstration of this additional role of mycorrhizae has aroused much speculation regarding its occurrence

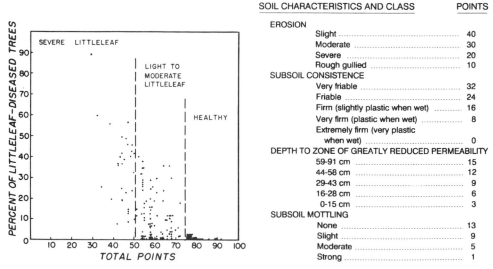

FIGURE 11.23 A soil rating scale for risk rating a site for littleleaf. The higher the index, the lower the risk to littleleaf. From Campbell et al. (1953).

as a natural phenomenon. The protection efficiency of mycorrhizae may be a function of their frequency on the root system. High mortality levels in mycorrhizal seedlings of two races of sand pine, the Choctawhatchee race in particular, were attributed to invasion of nonmycorrhizal roots by *P. cinnamomi;* the ectomycorrhizae of these plants were not penetrated by the fungus. These kinds of results seem to bode ill for the achievement of field control by manipulation of biological balances. Nevertheless, mycorrhizae furnish many benefits that in toto increase the host's capacity for growth and endurance under adverse physical, chemical, and biological conditions; progress toward their manipulation is one of the most exciting areas of current research in applied forest pathology.

Disease Resistance. Once the role of *P. cinnamomi* in the littleleaf complex was understood, field observations and greenhouse inoculation trials indicated that individual shortleaf pines exhibited natural resistance and probably were not disease escapes. This apparent gene pool of resistance was immediately tapped by means of controlled pollinations of selected field trees, from which scion material was also grafted to seedling root stock. Ultimately, the grafted trees were established in a seed orchard that has facilitated breeding experiments and subsequent seed procurement for progeny tests. As seeds from these sources became available, a laboratory method was developed for mass screening of seedling progeny for resistance to the pathogen alone and whereby resistant lines could be identified for further tests under field conditions.

Presently, shortleaf pine from open-pollinated parents in seven states are being tested for long-term growth response on severe littleleaf sites in Georgia, South Carolina, and Virginia. Evaluation of these plantings after 16 years showed that shortleaf pine can be successfully selected for adaptability to littleleaf sites. More recently, height measurements of 34 progenies from controlled pollinations of seed-orchard clones have shown, after 6 years of growth on a littleleaf site, that the fastest growing

FIGURE 11.24   Loblolly pine with moderately severe symptoms of littleleaf disease.

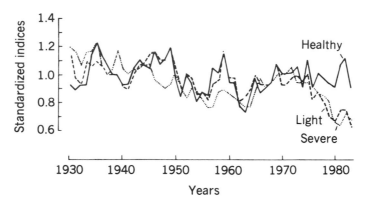

FIGURE 11.25 Radial growth chronologies of healthy and littleleaf-affected loblolly pines with light and severe crown symptoms. From Jacobi et al. (1988).

progenies are derived from a single clone. Implications are that their high vigor is related to resistance to littleleaf. These reports hold promise that a source of shortleaf pine tailored for productive growth on high-hazard sites is developing if supply can ultimately meet the demand. Tissue culture is a possibility to help meet that demand.

There is considerable within-species variation of resistance to attack by *P. cinnamomi* in seed lots of shortleaf, loblolly, loblolly × shortleaf hybrids, and Virginia pines. In a pine-callus tissue system, cellular responses to infection were easily segre-

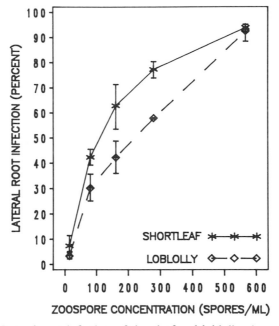

FIGURE 11.26 Lateral root infection of shortleaf and loblolly pine roots by *Phytophthora cinnamomi*. Note that only at zoospore concentrations between 100 and 300/ml are there relative differences in susceptibility. From Fraedrich et al. (1989). Courtesy of A. E. Miller and S. W. Fraedrich.

gated into either susceptible or resistant reactions. In resistant reactions there were fewer hyphal penetrations, a greater accumulation of electron-dense materials, and morphological changes of the host cell wall. It was not a hypersensitive reaction, however. Tissue culture of resistant clones had sparse overgrowth of hyphae, and this provided a convenient means of visibly assessing degree of resistance.

Loblolly pine has often been planted or favored over shortleaf on high-hazard littleleaf sites. Experience has shown that loblolly pine may survive littleleaf symptoms somewhat better than does shortleaf pine and grow into the small- to medium-sized sawlog class, thus producing a much more valuable product (Fig. 11.24). Loblolly pine probably suffers from the same complex of factors as that which produces littleleaf disease in shortleaf pine (Fig. 11.25). The point system and the soils-series risk-prediction systems developed for shortleaf pine have been successfully used to predict risk to loblolly pine. Lateral roots of loblolly pine are less susceptible to infection by *P. cinnamomi* than are roots of shortleaf pine (Fig. 11.26). At zoospore concentrations greater than 280/ml, individual infections tended to coalesce, and these between-species differences were not detectable.

Salvage. Salvage cuts, although not a control per se, have been prescribed to minimize losses from littleleaf. This is difficult to achieve in reality, however, as the southern pine beetle finds and kills littleleaf trees before they may show visible symptoms. Vigorous salvage or control treatments for southern pine beetle often preempt littleleaf control efforts.

SELECTED REFERENCES

Allen, R. N., and F. J. Newhook. 1973. Chemotaxis of zoospores of *Phytophthora cinnamomi* to ethanol in capillaries of soil pore dimensions. *Trans. Brit. Mycol. Soc.* 61:287–302.

Anderson, E. J. 1951. The *Phytophthora cinnamomi* problem in pineapple fields of Hawaii. *Phytopathology* 41:1–2 (abst.).

Batini, F. E., and E. R. Hopkins. 1972. *Phytophthora cinnamomi* Rands– A root pathogen of the jarrah forest. *Aust. For.* 36:57–68.

Benson, D. M. 1991. Detection of *Phytophthora cinnamomi* in azalea with commercial serological assay kits. *Pl. Dis.* 75:478–482.

Bertus, A. L. 1974. Control of *Phytophthora cinnamomi* root rot with soil drenches of sodium *p*-(dimethylamino) benzenediozo-sulfonate and 5-ethoxy-3-(trichloromethyl)-1,2,4,-thiadiazole. *Pl. Dis. Reptr.* 58:437–438.

Blackwell, E. 1943. The life history of *Phytophthora cactorum. Trans. Brit. Mycol. Soc.* 26:71–89.

Brasier, C. M. 1971. Induction of sexual reproduction in single A2 isolates of *Phytophthora* species by *Trichoderma viride. Natura New Biol.* 251:283.

Bruck, R. I., and C. M. Kenerley. 1983. Effects of metalaxyl on *Phytophthora cinnamomi* root rot of *Abies fraseri. Pl. Dis.* 67:688–690.

Bryan, W. C. 1965. *Testing Shortleaf Pine Seedlings for Resistance to Infection by Phytophthora cinnamomi.* USDA For. Serv., Res. Note SE-50. 4 pp.

Bryan, W. C. 1973. *Height Growth of Shortleaf Pine Progenies from Trees Selected for Resistance to Littleleaf Disease.* USDA For. Serv., Res. Note SE-185. 7 pp.

Campbell, W. A. 1948. *Phytophthora cinnamomi* associated with the roots of littleleaf-diseased shortleaf pine. *Pl. Dis. Reptr.* 32:472.

Campbell, W. A. 1949a. A method of isolating *Phytophthora cinnamomi* directly from soil. *Pl. Dis. Reptr.* 33:134–135.

Campbell, W. A. 1949b. Relative abundance of *Phytophthora cinnamomi* in the root zones of healthy and littleleaf-diseased shortleaf pine. *Phytopathology* 39:752–753.

Campbell, W. A. 1951. The occurrence of *Phytophthora cinnamomi* in the soil under pine stands in the southeast. *Phytopathology* 41:742–746.

Campbell, W. A. 1961. Littleleaf disease of shortleaf pine: Present status and future needs, pp. 1529–1532. In *Recent Advances in Botany.* University of Toronto Press, Toronto. 766 pp.

Campbell, W. A., and O. L. Copeland, Jr. 1954. *Littleleaf Disease of Shortleaf and Loblolly Pines.* USDA Circ. 940. 41 pp.

Campbell, W. A., O. L. Copeland, Jr., and G. H. Hepting. 1953. *Managing Shortleaf Pine in Littleleaf Disease Areas.* USDA For. Serv., SE Sta. Pap. No. 25. 12 pp.

Campbell, W. A., and A. F. Verrall. 1963. *Phytophthora cinnamomi* associated with Lawson cypress mortality in Louisiana. *Pl. Dis. Reptr.* 47:808.

Campbell, W. A., G. V. Gooding, Jr., and F. A. Haasis. 1963. The occurrence of *Phytophthora cinnamomi* in Kentucky, North Carolina, Tennessee, and Virginia. *Pl. Dis. Reptr.* 47:924–926.

Copeland, O. L., Jr. 1949. Some relations between soils and the littleleaf disease of pine. *J. For.* 47:566–567.

Copeland, O. L., Jr. 1952. Root mortality of shortleaf and loblolly pine in relation to soils and littleleaf disease. *J. For.* 50:21–25.

Copeland, O. L., Jr. 1954. *Estimating the Littleleaf Hazard in South Carolina Piedmont Shortleaf Pine Stands Based on Site Index.* USDA For. Serv., S.E. For. Exp. Sta. Res. Note 57. 1 p.

Copeland, O. L., Jr. 1955. The effects of an artificially induced drought on shortleaf pine. *J. For.* 53:262–264.

Copeland, O. L., Jr., and R. G. McAlpine. 1955. The interrelations of littleleaf, site index, soil, and ground cover in Piedmont shortleaf pine stands. *Ecology* 36:635–641.

Crandall, B. S., G. F. Gravatt, and M. M. Ryan. 1945. Root disease of *Castanea* species and some coniferous and broadleaf nursery stocks, caused by *Phytophthora cinnamomi*. *Phytopathology* 35:162–180.

Erwin, D. C., G. A. Zentmyer, J. Galindo, and J. S. Niederhauser. 1963. Variation in the genus *Phytophthora*. *Annu. Rev. Phytopath.* 1:375–396.

Fraedrich, S. W., and F. H. Tainter. 1989. Effect of dissolved oxygen concentration on the relative susceptibility of shortleaf and loblolly pine root tips to *Phytophthora cinnamomi*. *Phytopathology* 79:1114–1118.

Fraedrich, S. W., F. H. Tainter, and A. E. Miller. 1989. Zoospore inoculum density of *Phytophthora cinnamomi* and the infection of lateral root tips of shortleaf and loblolly pine. *Phytopathology* 79:1109–1113.

Gallegly, M. E. 1970. Genetical aspects of pathogenic and saprophytic behavior of the Phycomycetes with special reference to *Phytophthora*, pp. 50–54. In T. A. Toussoun, R. V. Bega, and P. E. Nelson (eds.), *Root Diseases and Soil-borne Pathogens.* University of California Press, Berkeley. 252 pp.

Gallindo, J., and G. A. Zentmyer. 1964. Mating types in *Phytophthora cinnamomi*. *Phytopathology* 54:238–239.

Garrett, S. D. 1956. *Biology of Root-infecting Fungi.* Cambridge University Press, London. 292 pp.

Grand, L. F., and N. A. Lapp. 1974. *Phytophthora cinnamomi* root rot of Fraser fir in North Carolina. *Pl. Dis. Reptr.* 58:318–320.

Haasis, F. A., R. R. Nelson, and D. H. Marx. 1964. Morphological and physiological characteristics of mating types of *Phytophthora cinnamomi*. *Phytopathology* 54:1146–1151.

Hansen, T. S., W. H. Kentz, G. H. Wiggin, and E. C. Stakman. 1923. *A Study of the Damping-off Disease of Coniferous Seedlings.* Univ. Minn. Agri. Exp. Sta., Tech. Bull. 15. 35 pp.

Hendrix, F. F., Jr., and E. G. Kuhlman. 1965. Factors affecting direct recovery of *Phytophthora cinnamomi* from soil. *Phytopathology* 55:1183–1187.

Hepting, G. H. 1945. Reserve food storage in shortleaf pine in relation to littleleaf disease. *Phytopathology* 35:106–119.

Hepting, G. H., and G. M. Jemison. 1958. Forest protection, pp. 185–220. In *Timber Resources for America's Future.* USDA For. Serv., For. Res. Rept. 14. 713 pp.

Hepting, G. H., and F. J. Newhook. 1962. A pine disease in New Zealand resembling littleleaf. *Pl. Dis. Reptr.* 46:570–575.

Hepting, G. H., T. S. Buchanan, and L. W. R. Jackson. 1945. *Littleleaf Disease of Pine.* USDA Circ. No. 716. 15 pp.

Hickman, C. J. 1970. Biology of *Phytophthora* zoospores. *Phytopathology* 60:1128–1135.

Hickman, C. J., and H. H. Ho. 1966. Behavior of zoospores in plant-pathogenic Phycomycetes. *Annu. Rev. Phytopath.* 4:195–220.

Jackson, L. W. R. 1945. Root defects and fungi associated with the littleleaf disease of southern pines. *Phytopathology* 35:91–105.

Jackson, L. W. R., and G. H. Hepting. 1964. Rough bark formation and food reserves in pine roots. *For. Sci.* 10:174–179.

Jacobi, J. C., F. H. Tainter, and S. W. Oak. 1988. The effect of drought on growth decline of loblolly pine on littleleaf sites. *Pl. Dis.* 72:294–297.

Jang, J. C., and F. H. Tainter. 1990. Hyphal growth of *Phytophthora cinnamomi* on pine callus tissue. *Pl. Cell Repts.* 8:741–744.

Jang, J. C., and F. H. Tainter. 1990. Cellular responses of pine callus to infection by *Phytophthora cinnamomi*. *Phytopathology* 80:1347–1352.

Jones, S. M. 1988. Old-growth forests within the piedmont of South Carolina. *Nat. Areas J.* 8:31–37.

Khew, K. L., and G. A. Zentmyer. 1974. Electrotactic response of zoospores of seven species of *Phytophthora*. *Phytopathology* 64:500–507.

Kuhlman, E. G. 1964. Survival and pathogenicity of *Phytophthora cinnamomi* in several western Oregon soils. *For. Sci.* 10:151–158.

Kuhlman, E. G., and F. F. Hendrix, Jr. 1963. Phytophthora root rot of Fraser fir. *Pl. Dis. Reptr.* 47:552–553.

Lorio, P. L., Jr. 1966. *Phytophthora cinnamomi* and *Pythium* species associated with loblolly pine decline in Louisiana. *Pl. Dis. Reptr.* 50:596–597.

Manning, W. J., and D. F. Crossan. 1966. Evidence for variation in degree of pathogenicity of isolates of *Phytophthora cinnamomi* to broadleaf and coniferous evergreens. *Pl. Dis. Reptr.* 50:647–649.

Marx, D. H. 1969. Antagonism of mycorrhizal fungi to root pathogenic fungi and soil bacteria. *Phytopathology* 59:153–163.

Marx, D. H. 1969. Production, identification, and biological activity of antibiotics, produced by *Leucopaxillus cerealis* var. *piceina*. *Phytopathology* 59:411–417.

Marx, D. H., and W. C. Bryan. 1969. Effect of soil bacteria on the mode of infection of pine roots by *Phytophthora cinnamomi*. *Phytopathology* 59:614–619.

Marx, D. H., and W. C. Bryan. 1970. The influence of soil bacteria on the mode of infection of pine roots by *Phytophthora cinnamomi*, pp. 171–172. In T. A. Toussoun, R. V. Bega, and P. E. Nelson (eds.), *Root Diseases and Soil-Borne Pathogens*. University of California Press, Berkeley. 252 pp.

Marx, D. H., and C. B. Davey. 1967. Ectotrophic mycorrhiaze as deterrents to pathogenic root infections. *Nature* 213:1139.

Marx, D. H., and C. B. Davey. 1969. Resistance of aseptically formed mycorrhizae to infection by *Phytophthora cinnamomi*. *Phytopathology* 59:549–558.

Marx, D. H., and C. B. Davey. 1969. Resistance of naturally occurring mycorrhizae to infections by *Phytophthora cinnamomi*. *Phytopathology* 59:559–565.

Marx, D. M., and F. A. Haasis. 1965. Induction of aseptic sporangial formation in *Phytophthora cinnamomi* by metabolic diffusates of soil micro-organisms. *Nature* 206:673–674.

McCain, A. H., O. V. Holtzmann, and E. E. Trujillo. 1967. Concentration of *Phytophthora cinnamomi* chlamydospores by soil sieving. *Phytopathology* 57:1134–1135.

McQuilken, W. E. 1940. The natural establishment of pine in abandoned fields in the Piedmont plateau region. *Ecology* 21:135–147.

Mircetich, S. M., and G. A. Zentmyer. 1970. Germination of chlamydospores of *Phytophthora*, pp. 112–115. In T. A. Toussoun, R. V. Bega, and P. E. Nelson (eds.), *Root Diseases and Soil-Borne Pathogens*. University of California Press, Berkeley. 252 pp.

Mircetich, S. M., G. A. Zentmyer, and J. B. Kendrick, Jr. 1968. Physiology of germination of chlamydospores of *Phytophthora cinnamomi*. *Phytopathology* 58:666–671.

Newhook, F. J. 1970. *Phytophthora cinnamomi* in New Zealand, pp. 173–176. In T. A. Toussoun, R. V. Bega, and P. E. Nelson (eds.), *Root Diseases and Soil-Borne Pathogens*. University of California Press, Berkeley. 252 pp.

Newhook, F. J., and F. D. Podger. 1972. The role of *Phytophthora cinnamomi* in Australian and New Zealand forests. *Annu. Rev. Phytopath.* 10:299–326.

Oak, S. W., and F. H. Tainter. 1988. Risk prediction of loblolly pine decline on littleleaf disease sites in South Carolina. *Pl. Dis.* 72:289–293.

Oxenham, B. L. 1957. Diseases of the pineapple. *Queensland Agr. J.* 83:13–26.

Podger, F. D. 1972. *Phytophthora cinnamomi*, a cause of lethal disease in indigenous plant communities in Western Australia. *Phytopathology* 62:972–981.

Podger, F. D., R. F. Doepel, and G. A. Zentmyer. 1965. Association of *Phytophthora cinnamomi* with a disease of *Eucalyptus marginata* forest in Western Australia. *Pl. Dis. Reptr.* 49:943–947.

Pratt, B. H., and W. A. Heather. 1972. Method for rapid differentiation of *Phytophthora cinnamomi* from other *Phytophthora* species isolated from soil by lupin baiting. *Trans. Brit. Mycol. Soc.* 59:87–96.

Pratt, B. H., and W. A. Heather. 1973. The origin and distribution of *Phytophthora cinnamomi* Rands in Australian native plant communities and the significance of its association with particular plant species. *Aust. J. Biol. Sci.* 26:559–573.

Pratt, B. H., J. H. Sedgley, W. A. Heather, and C. J. Shepherd. 1972. Oospore production in *Phytophthora cinnamomi* in the presence of *Trichoderma konigii*. *Aust. J. Biol. Sci.* 25:861–863.

Ross, E. W., and D. H. Marx. 1972. Susceptibility of sand pine to *Phytophthora cinnamomi*. *Phytopathology* 62:1197–1200.

Roth, E. R. 1954. Spread and intensification of the littleleaf disease of pine. *J. For.* 52:592–596.

Roth, E. R. 1960. Plots demonstrate 16 years' loss from littleleaf. *J. For.* 58:322–323.

Roth, E. R., E. R. Toole, and G. H. Hepting. 1948. Nutritional aspects of the littleleaf disease of pine. *J. For.* 46:578–587.

Roth, L. F. 1963. *Phytophthora cinnamomi* root rot of Douglas-fir. *Phytopathology* 53:1128–1131.

Roth, L. F., and E. G. Kuhlman. 1963. Field tests of the capacity of *Phytophthora* root rot to damage Douglas-fir. *J. For.* 61:199–205.

Roth, L. F., and E. G. Kuhlman. 1966. *Phytophthora cinnamomi*, an unlikely threat to Douglas-fir forestry. *For. Sci.* 12:147–159.

Ruehle, J. L. 1962. Plant-parasitic nematodes associated with shortleaf pine showing symptoms of littleleaf. *Pl. Dis. Reptr.* 46:710–711.

Ruehle, J. L., and W. A. Campbell. 1971. *Adaptability of Geographic Selections of Shortleaf Pine to Littleleaf Sites.* USDA For. Serv., Res. Paper SE-87. 8 pp.

Sansome, E. 1961. Meiosis in the oogonium and antheridium of *Pythium debaryanum* Hesse. *Nature* 191:827–828.

Schmitthenner, A. F. 1970. Significance of populations of *Pythium* and *Phytophthora* in soil, pp. 25–27. In T. A. Toussoun, R. V. Bega, and P. E. Nelson (eds.), *Root Diseases and Soil-Borne Pathogens.* University of California Press, Berkeley. 252 pp.

Shea, S. R., B. Shearer, J. Tippett, and P. M. Deegan. 1984. A new perspective on jarrah dieback. *For. Focus* 31:3–11.

Siggers, P. V., and K. D. Doak. 1940. *The Littleleaf Disease of Shortleaf Pine.* USDA For. Serv., So. For. Exp. Sta. Occas. Paper 95. 5 pp.

Thorn, W. A., and G. A. Zentmyer. 1954. Hosts of *Phytophthora cinnamomi* Rands. *Pl. Dis. Reptr.* 38:47–52.

Torgeson, D. C. 1954. Root rot of Lawson cypress and other ornamentals caused by *Phytophthora cinnamomi*. *Contrib. Boyce Thompson Inst.* 17:359–373.

Trimble, S. W. 1974. *Man-Induced Soil Erosion on the Southern Piedmont 1700–1970.* Soil Conservation Service of America, Ankeny, IA.

Tucker, C. M. 1931. *Taxonomy of the Genus Phytophthora* DeBary. Mo. Agr. Exp. Sta. Res. Bull. 153. 208 pp.

Waterhouse, G. M. 1956. *The Genus Phytophthora.* Commonwealth Mycol. Inst. Misc. Publ. 12. Kew, Surrey. 120 pp.

White, R. P. 1930. Two *Phytophthora* diseases of rhododendron. *Phytopathology* 20:131 (abst.).

Woods, F. A. 1953. Disease as a factor in the evolution of forest composition. *J. For.* 51:871–873.

Zak, B. 1949. The search for virus as cause of littleleaf. *For. Farmer* 8(9):8,10.

Zak, B. 1957. *Littleleaf of Pine.* USDA For. Serv., Forest Pest Leafl. 20. 4 pp.

Zak, B. 1961. *Aeration and Other Soil Factors Affecting Southern Pines as Related to Littleleaf Disease.* USDA For. Serv., Tech. Bull. 1248. 30 pp.

Zak, B., and W. A. Campbell. 1958. Susceptibility of southern pines and other species to the littleleaf pathogen in liquid culture. *For. Sci.* 4:156–161.

Zentmyer, G. A. 1961. Chemotaxis of zoospores for root exudates. *Science* 133:1595–1596.

Zentmeyer, G. A. 1965. Bacterial stimulation of sporangial production in *Phytophthora cinnamomi*. *Science* 150:1178–1179.

Zentmyer, G. A. 1970. Tactic responses of zoospores of *Phytophthora*, pp. 109–111. In T. A. Toussoun, R. V. Bega, and P. E. Nelson (eds.), *Root Diseases and Soil-Borne Pathogens*. University of California Press, Berkeley. 252 pp.

Zentmyer, G. A. 1980. *Phytophthora cinnamomi* and the Diseases It Causes. Monograph No. 10, Amer. Phytopath. Soc., St. Paul. 96 pp.

Zentmyer, G. A., and D. C. Erwin. 1970. Development and reproduction of *Phytophthora*. *Phytopathology* 60:1120–1127.

Zentmyer, G. A., and S. M. Mircetich. 1966. Saprophytism and persistence in soil by *Phytophthora cinnamomi*. *Phytopathology* 66:710–712.

Zentmyer, G. A., A. O. Paulus, and R. M. Burns. 1967. *Avocado Root Rot*. Calif. Agr. Exp. Sta. Circ. 511. 16 pp.

DISEASE PROFILE

Phytophthora Root Disease of Port Orford Cedar

Importance: A serious problem since its discovery (1938) in Oregon's Willamette Valley, this disease has destroyed Port Orford cedar in nurseries, landscape and windbreak plantings, and forests within and local to its natural range.

Suscepts: Pathogenicity is confined to species of *Chamaecyparis* with Port Orford cedar being primarily affected. Asiatic species have a degree of resistance that suggests possible origin of the fungus; however, the disease is not known today in either Europe or Asia.

Distribution: This disease is found in California, Oregon, Washington, and British Columbia.

Causal Agents: This disease is caused by *Phytophthora lateralis* (Peronosporales).

Biology: The pathogen lies dormant during the normally warm, dry summer, grows very slowly in the fall and winter, but multiplies rapidly in the spring by sporulating from infected roots and cast foliage. The latter contain resistant spores that germinate, after winter chilling, to colonize and sporulate on newly accumulated cedar litter. Soil leachates stimulate the formation of chlamydospores and zoosporangia. Zoospores are carried passively in rain splash or surface water to new infection sites. As a result, multiple rootlet infections spread to the root collar where death from girdling ensues within weeks in small trees and 2–4 years in the largest individuals. Aerial infection occurs when lowermost branches contact infested litter or soil during wet, windy weather. Sporangia develop from newly infected foliage after 36–48 hours of wetting at 10–20°C and, because they rarely detach, zoospore release and rain-splash dispersal accounts for secondary spread and upward progression of foliage infection. Inward spread averages 5 cm per month and results in stem cankering, girdling, and distal blight and dieback. Foliage infection stops at the end of the rainy season in late spring, whereupon the fungus dies, forming resistant chlamydospores and oospores in the drying foliage.

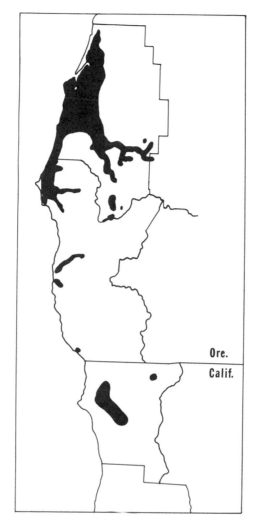

Disease incidence.

Epidemiology: The rapidity by which *P. lateralis* traversed the native cedar range within a span of 2 years is circumstantially explained by inoculum transport via soil and water movement as occasioned by road construction and maintenance; logging operations; cattle grazing; game movement, especially of elk; and water courses. Local upslope spread is attributed to intraspecific root grafting that provides a continuous pathway of infection.

Diagnosis: Foliage symptoms are localized and progress slowly from the lower crown upward and outward in an irregular triangle of necrosis, whereas root pathogenesis is typified by uniform foliage decline. Crown-fading reaches a lusterless green stage and progresses through bronzing and yellowing to the light tan of foliage death, quite distinct from the reddish-brown trees killed by fire, beetles, or other causes. Infected roots show a cinnamon-brown discoloration of the inner bark, which contrasts sharply with the pinkish white color of healthy tissue. The discoloration also marks the progress of the fungus upward in the lower stem, usually in a spiral pattern and to an extent coincident with crown decline. Because Port Orford cedar is equally susceptible to *P. cinnamomi* and the symptoms of the two root diseases are identical, diagnostic distinction should be made, particularly in nursery and ornamental plantings where both fungi are likely to occur. Species separation is best made in the laboratory from tissue isolations from the apex of inner bark discolorations.

Control Strategy: Many believe that the disease is now beyond control, even under conditions where economic feasibility is justified. Wherever cedar can grow within the known disease range, there are no edaphic or climatic conditions that appear to limit the disease. No evidence of genetic resistance has been found in the native cedar population, and even though Asiatic species and Alaska cedar are somewhat resistant, they are not suitable for replacement. Exclusion of the pathogen from unaffected natural stands has been prescribed, but it would require a rigid policy of nonentry of people, animals, and equipment.

SELECTED REFERENCES

Trione, E. J. 1959. The pathology of *Phytophthora lateralis* on native *Chamaecyparis lawsoniana*. *Phytopathology* 49:306–310.

Roth, L. E., H. H. Bynum, and E. E. Nelson. 1972. *Phytophthora Root Rot of Port-Orford-Cedar*. USDA For. Serv., For. Pest Leaflet 131. 7 pp.

DISEASE PROFILE

Senna Seymeria, Root Parasite of Southern Pines

Importance: This unique disease, known only since 1969, is the only one of its kind involving root pathogenesis of southern pines by a parasitic seed plant, with up to 50% mortality in Georgia and Florida. Although the disease is not widespread, the severity of localized epidemics and the southwide distribution of the parasite warn of a disease potential that could intensify.

Suscepts: Loblolly and slash pine are naturally infected, and longleaf pine was susceptible in field inoculations.

Distribution: The pathogen occurs in the Atlantic and Gulf coastal plains of eight states from Virginia to Louisiana. However, the disease is limited to the area of first observation near Panama City, Florida, and to ten counties in southeast Georgia.

Causal Agents: Commonly called senna seymeria or black senna, *Seymeria cassioides* (Scrophulariaceae), is an annual figwort in a family that includes some 500 species of green-root parasites.

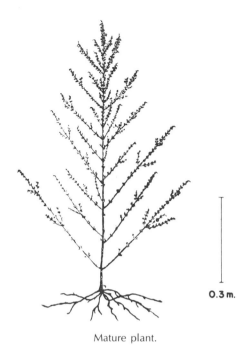

0.3 m.

Mature plant.

Biology: Senna seeds germinate in early spring when air temperature approximates 20°C. Senna germlings remain in the cotyledon stage until penetration of suscept roots is made; failing to make such contact, seedlings die within 4–8 weeks. The slender, smooth, and pale light tan senna roots intertwine with those of pine, forming many secondary lateral holdfasts. The haustoria are vascular extensions that become sheathed vessels in penetrating and ramifying throughout the host root cortex and xylem. As parasitism becomes well established, the annual shoots emerge, grow rapidly (7 cm/wk), and attain maximum height of 0.8 m at maturity. Flowering ordinarily begins in late August and continues into October; the annual plants die soon after seed ripening and dispersal in late October and early November.

Epidemiology: Abundant inoculum is produced as nearly microscopic seeds, about 75 per capsule, which measures only 0.25 cm in diameter. Because a single large plant may bear 2,000 capsules, and parasite populations on epidemic sites may range from 1,250 to 5,000 plants/ha, inoculum availability of 188–750 million seeds/ha may occur. While the small seeds do not have wings or other appendages, they are probably carried by the wind for short distances. Secondary dispersal may occur by water transport. Infection is limited to drier, bare-soil microsites. Seeds do not survive well under standing water and do not settle through or germinate under surface litter.

Diagnosis: Best detection is by recognizing the senna plant growing in close proximity to the pine host, which most likely will exhibit crown decline symptoms. Senna plants are bushy with scalelike foliage that is light green until early summer, then often turning to purple. From a distance, the plants resemble seedlings of eastern redcedar. The hundreds of bright yellow, five-petaled flowers about 1.3 cm in diameter on each plant are distinctive. The blackened remains of dead plants, which often retain some empty seed capsules, can be recognized without difficulty through the winter.

Control Strategy: Prescribed burning may destroy seeds on infested areas in site preparation for pine regeneration. Selective herbicides would seem to have potential but none have yet been recommended. Defoliation by larvae of the buckeye butterfly *(Precis orithya evarete)* may account for some degree of biological control.

SELECTED REFERENCES

Grelen, H. E., and W. F. Mann, Jr. 1973. Distribution of senna seymeria *(Seymeria cassioides)* a root parasite on southern pines. *Econ. Bot.* 27:339–342.

Fitzgerald, C. H., M. Reines, S. Terrell, and P. P. Kormanik. 1975. Early root development and anatomical parasite-host relationships of black senna with slash pine. *For. Sci.* 21:239–242.

12

FEEDER ROOT SYMBIOSES: MYCORRHIZAE

Not all fungi are detrimental to the plants with which they live. Those fungi that, together with a plant root, form structures termed mycorrhizae are beneficial to the plant. In some instances, this association is essential to the survival and subsequent growth of the host plant.

12.1 DEFINITION

The term *mycorrhiza* (plural, mycorrhizae) is derived from two Greek words, *mycos* meaning fungus and *rhizome* meaning root and is applied to structures resulting from the association of the mycelium of certain fungi with the small roots of a higher plant. This original definition, however, is such that a parasitic association between the mycelium of a fungus and a higher plant root could be included in its broadest interpretation. Most researchers in this field recognize mycorrhiza as a beneficial association between a fungus and a plant root rather than a detrimental one and consider mycorrhiza to function as a mutualistic, symbiotic biotophy between a fungus and a higher plant.

12.2 HISTORY

Vittadini in 1842 first observed a fungal mantle around tree roots and proposed that it nourished the roots. Mycorrhizae were first described by Theodore Hartig on coniferous trees, but he did not investigate their function. Frank, also a German, published in 1885 the results of his observations on the relation of mycorrhizae to the growth of trees and to the growth of fungi in the forests and coined the term *mycorrhiza*. Melin in Sweden, Bjorkman in Poland, Harley in Great Britain, and Hatch and Doak in the United States pioneered research on mycorrhizae of forest trees. Considerable progress has been made in understanding these fungus–root relationships and in the practical application of this knowledge to maximize yields in forest and agricultural systems.

12.3   ECOSYSTEMATIC FUNCTIONS

Mycorrhizae provide strong physiological and ecological benefits to both plant and fungus partners. A strong interdependence necessary for survival has evolved. Many forest plantations have failed, for example, because of lack of proper fungal symbionts. Conversely, most fungal symbionts (mycobionts) depend on the living plant host to such an extent that they grow poorly, if at all, in pure culture. The appearance and subsequent development of land plants may have been enhanced by the ability of fungi and higher plants to develop mutualistic symbiotic relationships. Although life started some 3.5 to 4.2 billion years ago, the first four fifths of life history was typified by slow, conservative, simple evolution of life forms. The rapid establishment of complex forms of plants from bacteria through algae is difficult to understand based on the previous slow events. However, during the Cambrian period, about 1.7 billion years ago, atmospheric oxygen had reached a high enough concentration to allow colonization of surface waters by algae, which in turn provided screening from ultraviolet radiation. Concurrently, more efficient aerobic respiration fostered the development of more complex cellular organizations in the form of floating colonies. These colonies served as oxygen oases for primitive plants and animals, which later merged to form primitive Oomycetes.

The rather "sudden" appearance of land plants 4 million years ago was possibly a natural evolution of the mutualistic symbiosis between the two partners, algae and Oomycetes. Neither was fully equipped to exploit the land environment: Algae were unable to extract essential nutrients efficiently, and fungi were unable to manufacture carbohydrates. The two life forms collaborated to colonize dry land and survive the major hazards—desiccation and starvation. Today, of 200 extant plant families, only 14 are regularly nonmycorrhizal. Marsh and aquatic plants lack mycorrhizae when growing in water but need them when growing on land. Other modern plant groups lacking mycorrhizae are the Amaranthaceae, Chenopodiaceae, and Cruciferae, which are successful only when growing in disturbed soils. They cannot compete with later mycorrhizal plants.

There is growing evidence that mycorrhizal fungi occurring in forests are successional and reflect the age of their tree hosts as well as host species. Early-stage fungi, which are abundant in nurseries, are not competitive after outplanting into nutrient-poor soils but are highly competitive in fertile nursery soils that have been fumigated. An example is *Thelephora terrestris*. Fungal diversity also increases as trees age. After canopy closure, understory herbaceous vegetation declines, and more lignin and polyphenols become incorporated into the litter. This change in resource quality strongly influences the potential mycobionts that may form mycorrhizae. Early successional mycobionts tend to be "r" strategists, produce small fruit bodies, and grow rapidly on simple media containing low levels of sugars. Later-stage successional mycobionts are difficult to culture and require large amounts of sugars and complex mixtures of vitamins, which they obtain from their tree host. These attributes, along with their large fruit bodies and persistent mycelial structures, are characteristic of "K" strategists.

What was once viewed as a simple symbiotic relationship is, thus, now being realized as an extremely complex interaction between many fungal taxa and higher plants.

12.4 ECONOMIC IMPORTANCE

The failure of coniferous plantations in many different countries including Puerto Rico, Costa Rica, Phillipines, and Java and some prairie sites in the central United States has been ascribed to the absence of proper mycorrhizal fungi. The introduction of humus from forest stands or pure cultures of mycorrhizal fungi has enabled trees to grow in these areas. In Florida, direct seeding of conifers on reclaimed swamp lands failed unless humus with mycorrhizal fungi was added with the seeds. The seeds of many commercially grown orchids will germinate readily only when the appropriate mycorrhizal fungi are present or when certain organic food supplements are added to the soil.

12.5 HOST PLANTS

Mycorrhizae occur more or less commonly on an estimated 95% of all flowering plants, including most forest tree species. They are most important to plants in the families *Orchidaceae* (Fig. 12.1), the orchids, *Ericaceae,* the heath plants, and conifers.

FIGURE 12.1 Lady's slipper, *Cypripedium acaule.*

Different groups of plants form different types of mycorrhizae, and the same is true for fungi. These groups are discussed in more detail in the next section.

12.6   MORPHOLOGY

The principle roots involved in formation of mycorrhizae are naturally ephemeral (i.e., last only one season), ordinarily do not exhibit secondary growth, and have no root cap on which root hairs, when present, arise from epidermal cells. These roots are termed *feeder roots*. In conifers, feeder roots comprise 90–95% of the total root system. The remaining roots, 5–10% for conifers, are permanent in structure, exhibit secondary growth, have a root cap, and bear root hairs that arise from the second or third layer of cortical cells. These roots seldom form mycorrhizae. Thus conifer roots are classified into either long and short (feeder) roots, the latter being involved princi-pally in mycorrhizae formation.

The general morphology of the root may be greatly altered in some instances in a mycorrhizal association. Normally, the roots are simple, unbranched, seldom more than 2.5 cm in length, and most often a pallid white color. In conifers, upon infection by mycorrhizal fungi, the roots (now properly called mycorrhizae) become noticeably swollen, in many cases (Pinaceae) branch dichotomously, and may become colored due to the presence of fungus tissue. Angiosperm mycorrhizae do not often exhibit the increase in volume or degree of branching found in conifer mycorrhizae.

12.7   CLASSIFICATION

Mycorrhizae are classified into three major types based on the physical relationship of the fungus and the root cells. The terminology of these types has undergone changes. The older terminology is placed in parentheses after the newer terms. The three major types of mycorrhizae are as follows:

12.7.1   *Endomycorrhiza* (endotrophic mycorrhiza)

Members of this highly artificial group of mycorrhizae are related only because host cells are penetrated by the fungus (Fig. 12.2). The mycelium is principally intracellu-lar (within cells) in the cortical cells of the host root. There is usually no fungus mantle or sheath produced. The hyphae within the cells grow for a period of time and then disintegrate or are digested by the host.

The two major types of endomycorrhizae are based on whether the mycobiont has septate or nonseptate hyphae.

1. *Nonseptate hyphae* form vesicular–arbuscular (VA) endomycorrhiza. This group is the most widely distributed and occurs everywhere in all terrestrial habitats. The Pteridophyta, Cupressaceae, Taxodiaceae, and most monocots and dicots have VA endomycorrhizae. The Salicaceae are VA in youth but become ectomycorrhizal with age.

All mycobionts involved in this group are species of the Endogonaceae (Mucor-ales, Zygomycetes). None form true spores, but some form large, dark-colored structures that technically are chlamydospores. These are produced in root tissues,

FIGURE 12.2   Growth habit of endomycorrhizae.

on the root surface, or free in the soil. They generally function as overwintering propagules and are the source of inoculum the following growing season. No VA fungi have yet been observed to form zygospores, but some genera form azygo-spores, while others form chlamydospores.

VA mycorrhizae are characterized by intracellular (within cells) hyphae, which produce haustoria that frequently branch many times within the host cells, thus the term *arbuscules*. The hyphae are large in comparison with other types of mycor-rhizal fungi, and large swollen vesicles are formed both inside and outside of the host tissues. A loose weft of mycelium may occur around the outside of the root. The hyphae within the cells are subsequently digested by the host, and the fungus is controlled in this way. The VA fungi tend to be host nonspecific, although some host–fungus combinations are more efficient. The VA fungi do tend to be specific to soil conditions and are sensitive to pH, soil structure, temperature, and the like.

2. *Septate hyphae* form mycorrhizae of two types: the orchidaceous and the erica-ceous.

The *orchidaceous type* is confined to the Orchidales and is one of the most com-plex of symbiotic interactions. Because the occurrence of orchids is limited, this type is inconsequential in most temperate and boreal ecosystems, but it is im-portant in the tropics. All the mycobionts are members of the form-genus *Rhizocto-nia*. Some perfect states that have been found are in the genera *Ceratobasidium*, *Sebacina*, and *Tulasnella*.

Orchidaceous mycorrhizae form clumps of intracellular hyphae within cells of the orchid root (Fig. 12.3). The host cell contents are digested as needed.

The *ericaceous type* are of arbutoid or ericoid types. The arbutoid mycorrhizae occur in a number of genera of Ericaceae, including *Arbutus, Arctostaphylos, Azalea, Gaul-*

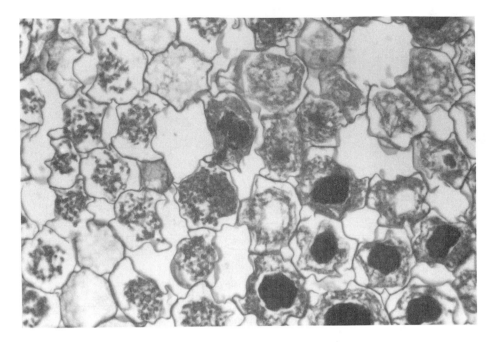

FIGURE 12.3 Orchidaceous endomycorrhizae showing active endomycorrhizae and digested hyphae.

theria, Leucothoe, and *Vaccinium.* They have a fungal sheath and intracellular penetration. Of the few fungal partners identified thus far, *Amanita, Cortinarius,* and *Boletus* have been determined to be arbutoid symbionts, chiefly through observation of mycelial connections between their fruit bodies with mycorrhizal roots. Ericoid mycorrhizae also occur on the fine roots of ericaceous plants. Strands or wefts of mycelium occur close to the root from which hyphae ramify within root cells to form a knotlike swelling. *Pezizella ericae* (Discomycetes) is the only fungus that has been both identified and used to synthesize the ericoid mycorrhiza.

12.7.2 *Ectomycorrhiza* (ectotrophic mycorrhiza)

Ectomycorrhizae are most common on forest trees of temperate regions especially on Pinaceae, Betulaceae, Fagaceae, Tiliaceae, and Salicaceae with some also found in the Ericaceae, Leguminosae, Myrtaceae, and Rosaceae. The feeder roots are surrounded by a fungus mantle, which may be up to 60 μm thick and equal to as much as 40% of the whole organ. Intercellular (between cells) penetration by the fungus hyphae is abundant in external cortical tissues, at times completely isolating cortical cells from each other. Intercellular hyphae with this degree of development form the Hartig net (Fig. 12.4). Intracellular penetration is relatively rare. Permanent or long roots have intercellular hyphae, except at the tips, but no fungus mantle. Ectomycorrhizae are usually short-lived, lasting from a few months to a maximum of 3 years.

Fungi that form ectomycorrhizae are species in the Agaricales (Boletaceae, Tricholomataceae, Amanitaceae, Cortinariaceae, Paxillaceae, Gomphidiaceae), Gastromycetes, and occasionally Deuteromycetes, Ascomycetes, and Phycomycetes. Some species of fungi are specific for certain tree species. *Suillus grevillei* and *S. cavipes*

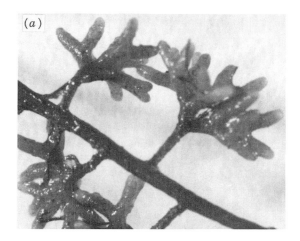

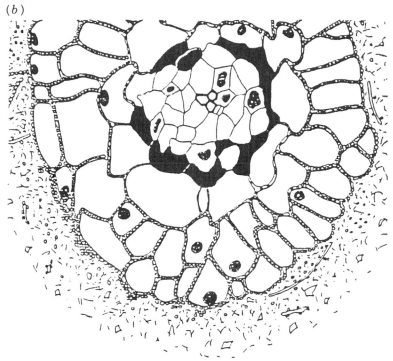

FIGURE 12.4   Ectomycorrhizae: (a) external view and (b) Hartig net.

apparently form mycorrhizae only with species of *Larix*. In contrast, *Pisolithus tinctorius* forms mycorrhizae with 14 species of *Pinus* and *Thelephora terrestris* with 21 species. *Cenococcum graniforme* is perhaps the most prolific mycorrhizal fungus and forms mycorrhizae with at least 154 species and varieties of woody plants. On southern pines, *Laccaria laccata, C. graniforme, Suillus brevipes,* and *Leucopaxillus cerealis* var. *piceina* form mycorrhizae. On Virginia pine, species of *Amanita, Boletus, Paxillus, Russula* and *Rhizopogon* form mycorrhizae. Many different fungi may form mycorrhizae on the same tree, and as many as seven different fungi have been found on roots of a single pine seedling.

12.7.3 *Ectendomycorrhiza* (ectendotrophic mycorrhiza)

The ectendomycorrhizae are intermediate between the preceding two types. They are characterized by having intracellular hyphae in the cortical cells and a Hartig net. A fungus mantle may or may not be present. This type of mycorrhiza was considered relatively rare because the fungal mantle can be sparse and thin. Recent studies have indicated that ectendomycorrhizae are common on nursery stock of many species. Ectendomycorrhizae usually lack root hairs and are usually not swollen. Microscopic verification may be necessary to detect Hartig net formation or intracellular fungus growth. This mycorrhiza has been found on species of *Pinus, Larix, Abies, Picea* as well as several species of hardwoods. Fungi shown to form ectendomycorrhizae are species of Ascomycotina.

12.8 FUNCTION

Nutrition

Mycorrhizal root systems benefit their respective hosts by increasing the capacity of the roots to absorb nutrients from the soil. This is accomplished in three ways.

1. The root-absorbing surface is markedly increased.
2. Mycelium radiating from the fungus mantle or the mycorrhizal root is able to penetrate farther into soil than the root hairs of nonmycorrhizal roots.
3. Selective absorption of advantage to the host may occur through mycorrhizal roots.

A discussion of each of these three ways of improving host nutrition follows.

12.8.1 Mycorrhizae Increase Absorptive Surface Area

Mycorrhizae can greatly increase the absorptive surface area of roots. Measurements indicated that in some instances total root surface was increased 30 times more than an uninfected root. In six subalpine spruce communities, the Af horizon had 0–8,800 mycorrhizal root tips/100 cm^3 (10 cm × 10 cm × 10 cm) soil, the An horizon had 3,600–16,00 tips/100 cm^3 soil, and the B horizon had 30–1,650 tips/100 cm^3 soil. A young stand of beech was estimated to have 2,600 kg/ha of mycorrhizal root tips. In a 450-year-old stand of Douglas-fir, the top 10 cm of soil contained 5,150–5,420 kg/ha, the dry weight of which was equivalent to 11% of the total root biomass.

Cenococcum graniforme, which produces black ectomycorrhizae, forms black sclerotia, which coincidently allow its abundance to be easily measured. In a spruce stand this fungus produced 41 billion sclerotia in the top 1 cm. In the upper 10 cm, there were 148–209 kg/ha dry weight of sclerotia.

12.8.2 Structures of Mycorrhizae Are Important

The structures of mycorrhizae are important in the nutrient cycling process because they can extend farther and faster than can roots. Not only are mycorrhizal root tips larger, but the fungal mycelium that grows into the soil effectively extends the root system. A single hypha of *Cenococcum* spp. was traced more than 2 m and had more than 120 lateral branches or fusions with other hyphae. As many as 200–2,000 hyphae may emerge from a single mycorrhiza, and a single cubic centimeter of soil can contain as much as 4 m of hyphae. The outgrowing hyphae are well adapted for substrate exploitation. They produce extracellular auxins, vitamins, cytokinins, enzymes, and so on, and influence root tissue and ion uptake.

The stimulation or inhibition of soil microorganisms is well known. Presence of the mycobiont alters rhizosphere chemistry. It uses root exudates and produces it own exudates, some of which can deactivate soil phytotoxins that can injure roots. Mycobionts do not fix nitrogen, but some do stimulate nitrogen-fixing bacteria such as *Azotobacter*. The host can have a profound influence on the mycobiont as well. For example, *Lactarius obscuratus* is an associate of alder. If alder is absent, the fungus does not fruit.

Mycorrhizae have an important influence on water relations in that mycorrhizal seedlings resist drought. Mycorrhizal seedlings can grow in solutions of much higher osmotic concentration than that which plasmolyzes root hairs of nonmycorrhizal root. This feature allows symbionts such as *Pisolithus tinctorius* to be used when replanting on acid soils of spoil banks. A dramatic illustration of their importance in water relations was seen in a case where several spruce shoots were discarded and left laying on the ground. One shoot was colonized by a mycobiont from the soil. This shoot remained green for 8 months, while the others soon died.

12.8.3 Mycorrhizal Roots Selectively Absorb Nutrients

Mycorrhizal roots can selectively absorb nutrients, providing further benefits to the host tree. Radiotracer studies have shown mycorrhizae to be important in cycling of macronutrients and several micronutrients, especially phosphorus, which is accumulated and passed steadily to the host. Mycorrhizal roots also respire more. Much of this soil respiration in early energy flow studies was erroneously attributed to soil bacteria. Respiration of mycorrhizal roots influences the rhizosphere or mycorrhizosphere. This influence is becoming increasingly appreciated. Yet only recently have mycorrhizae been considered in nutrient-uptake studies of plants. Their high respiration indicates active ion uptake and sporocarps contain greater concentrations of calcium, potassium, nitrogen, sodium, phosphorus, and zinc than do pine needles. Sporocarps are usually eaten within a few weeks, thus providing a concentrated nutrient source for decomposers and consumers.

Mycorrhizae also play a role in carbon cycling. Labeled carbon moves readily from host to mycobiont, usually in the form of sugars. It can also be transferred to other adjacent green plants and achlorophyllous plants via the mycobionts. In fact, the current concepts of shade tolerance may need revision because the "shade-tolerant" understory plants may actually be epiparsitizing overstory plants through the mycobionts.

12.9 PROTECTION

In addition to absorption of nutrients, ectomycorrhizae may protect roots from invasion by pathogenic and saprophytic microorganisms. This protection has been proven experimentally. Ectomycorrhizae protect roots in the following ways.

1. Mycorrhizal fungi use carbohydrates, amino acids, and other compounds produced by the root that would normally attract root pathogens.
2. They provide a physical barrier to pathogens by the formation of the mantle and Hartig net.
3. They produce and secrete antibiotics that are lethal to pathogens.
4. They influence the rhizosphere to the extent that more favorable microorganisms are favored over pathogens.
5. They stimulate host cells to produce compounds inhibitory to pathogens.

12.10   FORMATION

12.10.1   VA Endomycorrhizae

Inoculum consists of infested litter or humus. For inoculation of nurseries, pot-cultured inoculum may be used. Hyphae of VA mycorrhizal fungi enter the young root through the epidermis behind the meristematic region or through root hairs. Entrance is usually by direct penetration via an appressorium. Hyphae colonize the root cortex but do not penetrate the endodermis, stele, or meristematic regions. Hyphae may be intra- or intercellular, but in the latter instance the development is not as extensive as the intercellular hyphae in ectomycorrhizae.

12.10.2   Ectomycorrhizae

Infection of the host by ectomycorrhizal fungi begins in the spring when tree growth begins. Inoculum consists of active mycorrhizae, spores, resting structures, mycelium in the soil, and occasionally rhizomorphs. Long roots are infected first, and the short feeding roots are infected before they emerge from the cortex. The roots may change to those characteristic of mycorrhizal roots before being invaded by the fungus. The number of short roots is about double on infected as compared to uninfected plants, and the presence of the fungus delays abortion or loss of short roots. The amount of mycorrhizal root development is related to deficiencies in nitrogen, phosphorus, and possibly potassium. If pine seedlings are well supplied with nutrients, few mycorrhizae develop. Thus plants in a fertile soil normally have fewer mycorrhizae than those in an infertile soil.

12.11   MANAGEMENT

Foresters are faced with an unprecedented challenge to rapidly and successfully reforest harvested lands. In the field, management is unnecessary if there is adequate inoculum. Factors that can influence inoculum availability include how long the site has been without the particular tree host, whether the site been severely disturbed, what the inoculum potential of the site is, what consideration should be given to the ecology of the host and fungus, and whether ecto- or endomycorrhizae are involved.

Ectomycorrhizae are strongly associated with the organic layers in soil. During dry seasons, ectomycorrhizae are very active in soil wood, which can make up as much as

15% of the soil volume. Rotten wood on the soil surface is also exploited by mycorrhizal fungi and their associated plants. Some plants, such as hemlock seedlings, are often associated with "nurse" trees, that is, dead logs laying on the soil surface. Although the hemlock seedlings may live several years without mycorrhizal fungi, their growth is minimal until they become mycorrhizal. To ensure mycorrhizal activity, foresters have learned to leave coarse woody materials on the site. Declining timber revenues in rural communities in the Pacific Northwest have resulted in an increase in harvesting of nontimber products such as mushrooms. Washington and California and national forests closely regulate mushroom collecting, largely to prevent destruction of forest habitat. There is some fear that excessive mushroom picking could limit mycorrhizal fungi.

In the nursery the manager has the opportunity to manipiulate fungal associates before outplanting. When establishing new nurseries where mycorrhizal populations are absent or low, these fungi or humus containing them can be added to the nursery beds. It should be remembered that detrimental fungi may unknowingly be added as well, and caution should be exercised.

Such nursery practices as soil fumigation can limit the proper development of mycorrhizae, and it has been observed that the type of soil fumigant and soil are factors. In pure culture, high concentrations of some insecticides and nematicides inhibit fungal growth and ectomycorrhizal development. Fumigation with methyl bromide resulted in an 18% loss in acceptable seedlings due to reduction in endomycorrhizae.

In general, many fungicides decrease ectomycorrhizal development (Table 12.1).

TABLE 12.1 Fungicides That Decrease Ectomycorrhizal Development

Active Ingredient	Trade Name
banrot	Banrot
triadimefon	Bayleton
benodanil	Benodanil[a]
chlorothalonil	Bravo[a]
captan	Captan[a]
chloroneb	Chloroneb
tridiazol	Etridiazol
fenaminosulf	Lesan
maneb	Maneb
mancozeb	Mansate
olpisan	Olpisan
quintozene	PCNB
folpet	Phaltan
sulfuric acid	sulfuric acid
thiram	Thiram
zinc white	zinc oxide
zineb	Zineb[a]
ziram	Ziram

[a] Fungicides that had no effect at low application rates tended to decrease ectomycorrhizae at higher rates.

Source: Castellano and Molina (1990).

Fungicides applied to control root diseases have had variable effects on mycorrhizal development. Benodanil prevented growth of *Pisolithus tinctorius* and *Thelephora terrestris* on agar medium. Captan and PCNB were less inhibitory to these fungi. Metalaxyl, on the other hand, may stimulate ectomycorrhizae. In southern nurseries the systemic fungicide Bayleton is used to control fusiform rust. Unfortunately, Bayleton inhibits ectomycorrhizal formation by *P. tinctorius*. Ferbam was formerly used, but it required more applications to be effective and hence was more costly to apply.

Some herbicides decrease ectomycorrhizal development. Bifenox stimulates early formation of mycorrhizae on loblolly and ponderosa pines; napropamide stimulated mycorrhizae on Douglas-fir. Some herbicides that had no effect at low application rates tended to decrease ectomycorrhizae at higher rates.

Fertilizers can also have an effect. Although foresters previously thought that fertilizers would reduce mycorrhizae, a range of sodium nitrate applications made 40 days after planting red oak increased ectomycorrhizae. With VAM present, six of eight species of trees had greater heights and diameters when fertilized than comparable seedlings without VAM. Foliar application of fertilizers may favor tree growth and stimulate ectomycorrhizae. Sugar maple and black walnut did not benefit from fertilizer applications with or without mycorrhizae.

Other cultural manipulations can have profound influences of development of mycorrhizae. Burning after harvesting may decrease mycorrhizae of the next generation of trees. Cover crops can aid in increasing inoculum of vesicular-arbuscular fungi in nurseries.

Artificial inoculation in the nursery of pine seedlings destined to be planted on harsh sites, such as strip mine banks and spoil banks, has improved seedling survival and growth. It is now possible to produce tree seedlings in nurseries or greenhouses with mycorrhizal root systems suited to each tree species and region where they will be planted. This practice does not always result in larger seedlings, such as with containerized Sitka and white spruce, but will provide trees that may outperform those without mycorrhizae. Many fungi are potential candidates for producing mycorrhizae, and only through experimentation and experience will the best combinations be determined. Certain fungi, especially *P. tinctorius,* have received much attention in the southeastern states and in other warm areas of the world. Ultimately, we will need to use many fungi to achieve maximum survival and growth of trees. For example, 50 fungi form ectomycorrhizae on western hemlock, whereas the genus *Alnus* may be very specialized in its fungus associates. *Pisolithus tinctorius* forms mycorrhizae at temperatures (up to 34°C) much higher than what most other mycorrhizal fungi can develop at. Because high temperature is one of the limiting factors in the establishment of seedlings on these adverse sites, it is possible that if *P. tinctorius* is introduced and forms mycorrhizae with pine seedlings that are to be planted on such sites, survival will be greatly increased. For the future, we should be able to produce seedlings and transplants with the right combination of fungi for that tree species and the site where they are to be planted.

The ability to use artificial inoculation effectively in nurseries depends on the specific objectives of the nursery supervisor. In the Pacific Northwest, increased stem caliper or leader growth in the nursery and/or field may be highly desirable, while increased outplanting survival may be of more importance in the southern states. Protection against pathogens may be important in a nursery in the Lake states. There is no one fungal species or ecotype that will meet all these objectives. While a great deal of knowledge has been acquired in the practical manipulation of mycosymbionts in the nursery, future success will likely be measured by our ability to produce tree seedlings with the optimal combination of fungi for that particular tree species and the specific site on which they will be planted.

Scorne not the least.
He that high growth on Cedars did bestow;
Gave also lowly mushrumpes leave to grow.

—Robert Southwell, sixteenth century poet

LITERATURE CITED

Beckford, P. R., R. E. Adams, and D. W. Smith. 1980. Effects of nitrogen fertilization on growth and ectomycorrhizal formation of red oak. *For. Sci.* 26:529–536.

Castellano, M. A., and R. Molina. 1990. Mycorrhizae, pp. 101–167. In T. D. Landis, R. W. Tinus, S. E. MacDonald, and J. P. Barnett (eds.), *The Container Tree Nursery Manual,* Vol. 5, USDA For. Serv., Agric. Handbook 674. 171 pp.

Daniels, B. A. 1980. Factors affecting spore germination of the vesicular-arbuscular mycorrhizal fungus, *Glomus epigaeus. Mycologia* 72:457–471.

Dighton, J., and P. A. Mason. 1985. Mycorrhizal dynamics during forest tree development, pp. 117–139. In D. Moore, L. A. Casselton, D. A. Wood, and J. C. Frankland (eds.), *Developmental Biology of Higher Fungi.* Cambridge University Press, Cambridge.

Dixon, R. K., H. E. Garrett, G. S. Cox, P. S. Johnson, and I. L. Sander. 1981. Container- and nursery-grown black oak seedlings inoculated with *Pisolithus tinctorius:* Growth and ectomycorrhizal development following outplanting on an Ozark clear-cut. *Can. J. For. Res.* 11:492–496.

Englander, L. 1982. Endomycorrhizae by septate fungi, pp. 11–13. In N. C. Schenk (ed.), *Methods and Principles of Mycorrhizal Research.* APS Press, St. Paul., MN. 244 pp.

Gerdemann, J. W. 1965. Vesicular-arbuscular mycorrhizae formed on maize and tuliptree by *Endogone fasciculata. Mycologia* 57:562–577.

Gerdemann, J. W. 1968. Vesicular-arbuscular mycorrhizae and plant growth, pp. 387–418. In J. G. Horsfall, K. F. Baker, and D. C. Hildebrand (eds.), *Annu. Rev. Plant Path.,* Vol. 6. Annu. Reviews, Inc., Palo Alto, CA.

Gerdemann, J. W., and J. M. Trappe. 1974. *The Endogonaceae in the Pacific Northwest.* Mycologia Memoir No. 5. 76 pp.

Graham, J. H., and R. G. Linderman. 1981. Inoculation of containerized Douglas-fir with the ectomycorrhizal fungus *Cenococcum geophilum. For. Sci.* 27:27–31.

Hacskaylo, E. 1971. *Mycorrhizae.* Proc. 1st N. Am. Conf. of Mycorrhizae. USDA For. Serv., Misc. Pub. 1189. 255 pp.

Hacskaylo, E. 1973. The Torrey Symposium on current aspects of fungal development. IV. Dependence of mycorrhizal fungi on hosts. *Bull. Torrey Bot. Club* 100:217–223.

Hacskaylo, E., and A. G. Snow. 1959. *Relation of Soil Nutrients and Light to Prevalance of Mycorrhizae on Pine Seedlings.* USDA For. Serv., N. E. For. Exp. Sta. Paper No. 125. 13 pp.

Harley, J. L. 1969. *The Biology of Mycorrhiza.* 2nd ed. Leonard Hill Ltd., London. 334 pp.

Harvey, A. E., M. F. Jurgensen, and M. J. Larsen. 1986. Residues, beneficial microbes, diseases and soil management in cool, east slope Rocky Mountain lodgepole pine ecosystems. Presented at workshop on management of small-stem stands of lodgepole pine, Fairmont Hot Springs, Mont. June 30–July 2, 1986.

Harvey, A. E., M. J. Larsen, and M. F. Jurgensen. 1980. Partial cut harvesting and ectomycorr-hizae: Early effects in Douglas-fir-larch forests of western Montana. *Can. J. For. Res.* 10:436–440.

Hatch, A. B. 1935. *The Physical Bases of Mycotrophy in Pinus.* The Black Rock Forest Bull. 6:1–168.

Hatch, A. B., and K. D. Doak. 1933. Mycorrhizae and other features of the root systems of *Pinus. J. Arnold Arbor.* 14:85–89.

Kenerly, C. M., R. I. Bruck, and L. F. Grand. 1984. Effects of metalaxyl on growth and ectomycorrhizae of Fraser fir seedlings. *Phytopathology* 68:32–35.

Kormanik, P. P., W. C. Bryan, and R. C. Schultz. 1980. Increasing endomycorrhizal fungus inoculum in forest nursery soil with cover crops. *So. J. Appl. For.* 4:151–153.

Kropp, B. R., and J. M. Trappe. 1982. Ectomycorrhizal fungi of *Tsuga heterophylla. Mycologia* 74:479–488.

Marks, G. C., and T. T. Kozlowski (eds.). 1973. *Ectomycorrhizae, Their Ecology and Physiology.* Academic Press, New York. 444 pp.

Marx, D. H. 1969. The influence of ectotrophic mycorrhizal fungi on the resistance of pine roots to pathogenic infections. I. Antagonism of mycorrhizal fungi to root pathogenic fungi and soil bacteria. *Phytopathology* 59:153–163.

Marx, D. H. 1969. The influence of ectotrophic mycorrhizal fungi on the resistance of pine roots to pathogenic infections. II. Production, identification, and biological activity of anti-biotics produced by *Leucopaxillus cerealis* var. *piceina. Phytopathology* 59:411–417.

Marx, D. H. 1981. Variability in ectomycorrhizal development and growth among isolates of *Pisolithus tinctorius* as affected by source, age, and reisolation. *Can. J. For. Res.* 11:168–174.

Marx, D. H., and C. B. Davey. 1969. The influence of mycorrhizal fungi on the resistance of pine roots to pathogenic infections. III. Resistance of aseptically formed mycorrhizae to infection by *Phytophthora cinnamomi* Rands. *Phytopathology* 59:549–558.

Marx, D. H., and C. B. Davey. 1969. The influence of mycorrhizal fungi on the resistance of pine roots to pathogenic infections. IV. Resistance of naturally occurring mycorrhizae to infections by *Phytophthora cinnamomi* Rands. *Phytopathology* 59:559–565.

Marx, D. H., and S. J. Rowan. 1981. Fungicides influence growth and development of specific ectomycorrhizae on loblolly pine seedlings. *For. Sci.* 27:167–176.

Marx, D. H., W. C. Bryan, and C. B. Davey. 1970. Influence of temperature on aseptic synthe-sis of ectomycorrhizae by *Thelephora terrestris* and *Pisolithus tinctorius* on loblolly pine. *For. Sci.* 16:424–431.

Marx, D. H., J. G. Mexal, and W. G. Morris. 1979. Inoculation of nursery seedbeds with *Pisolithus tinctorius* spores mixed with hydromulch increases ectomycorrhizae and growth of loblolly pines. *So. J. Appl. For.* 3:175–178.

Maser, C., and J. M. Trappe (eds.). 1984. *The Seen and Unseen World of the Fallen Tree.* USDA For Serv., Gen. Tech. Rept. PNW-164. 56 pp.

Mason, P. A., J. Wilson, F. T. Last, and C. Walker. 1983. The concept of succession in relation to the spread of sheathing mycorrhizal fungi on inoculated tree seedlings growing in unster-ile soils. *Plant and Soil* 71:247–256.

Melin, E. 1953. Physiology of mycorrhizal relations in plants. *Annu. Rev. Pl. Path.* 4:325–346.

Mikola, P. 1973. Application of mycorrhizal symbiosis in forestry practice, pp. 383–412. In G. C. Marks and T. T. Kozlowski (eds.), *Ectomycorrhizae, Their Ecology and Physiology.* Aca-demic Press, New York. 444 pp.

Molina, R. 1981. Ectomycorrhizal specificity in the genus *Alnus. Can. J. Bot.* 59:325–334.

Molina, R., T. O'Dell, D. Luoma, M. Amaranthus, M. Castellano, and K. Russell. 1993. *Biology, Ecology, and Social Aspects of Wild Edible Mushrooms in the Forests of the Pacific North-*

west: A Preface to Managing Commercial Harvest. USDA For. Serv., Gen. Tech. Rept. PNW-GTR-309. 42 pp.

Perry, D. A., R. Molina, and M. P. Amaranthus. 1987. Mycorrhizae, mycorrhizospheres, and reforestation: Current knowledge and research needs. *Can. J. For. Res.* 17:929–940.

Pirozynski, K. A., and D. W. Malloch. 1975. The origin of land plants: A matter of mycotrophism. *BioSystems* 6:153–164.

Riffle, J. W. 1973. Pure culture synthesis of ectomycorrhizae on *Pinus ponderosa* with species of *Amanita, Suillus* and *Lactarius. For. Sci.* 19:242–250.

Riffle, J. W. 1980. Growth and endomycorrhizal development of broadleaf seedlings in fumigated nursery soil. *For. Sci.* 26:403–413.

Ruehle, J. L. 1980. *Inoculation of Containerized Loblolly Pine Seedlings with Basidiospores of Pisolithus tinctorius.* USDA For. Serv., Res. Note SE-291. 4 pp.

Schramm, J. R. 1966. Plant colonization studies on black wastes from anthracite mining in Pennsylvania. *Trans. Am. Phil. Soc.* 56:1–194.

Schultz, R. C., P. P. Kormanik, and W. C. Bryan. 1981. Effects of fertilization and vesicular-arbuscular mycorrhizal inoculation on growth of hardwood seedlings. *Soil Sci. Amer. J.* 45:961–965.

Shaw, C. G., R. Molina, and J. Walden. 1982. Development of ectomycorrhizae following inoculation of containerized Sitka and white spruce seedlings. *Can. J. For. Res.* 12:191–195.

Shoulders, E. 1972. *Mycorrhizal Inoculation Influences Survival, Growth, and Chemical Composition of Slash Pine Seedlings.* USDA For. Serv., Res. Pap. SO-74. 12 pp.

Sinclair, W. A. 1974. Development of mycorrhizae in a Douglas-fir nursery: I. Seasonal characteristics. *For. Sci.* 20:51–56.

Sinclair, W. A. 1974. Development of ectomycorrhizae in a Douglas-fir nursery. II. Influence of soil fumigation, fertilization and cropping history. *For. Sci.* 20:57–63.

Trappe, J. M. 1962. Fungus associates of ectotrophic mycorrhizae. *Bot. Rev.* 28:538–606.

Trappe, J. M. 1983. Effects of the herbicides Bifenox, DCPA, and napropamide on mycorrhiza development of ponderosa pine and Douglas-fir seedlings in six western nurseries. *For. Sci.* 29:464–468.

Trappe, J. M., and R. D. Fogel. 1977. Ecosystematic functions of mycorrhizae, pp. 205–214. In *The Belowground Ecosystem: A Synthesis of Plant-Associated Processes.* Range Sci. Dept., Sci. Ser. No. 26, Colorado State Univ., Fort Collins.

Trappe, J. M., and R. F. Strand. 1969. Mycorrhizal deficiency in a Douglas-fir region nursery. *For. Sci.* 15:381–389.

Vozzo, J. Z., and E. Hacskaylo. 1971. Inoculation of *Pinus caribaea* with ectomycorrhizal fungi in Puerto Rico. *For. Sci.* 17:239–245.

Zak, B. 1964. Role of mycorrhizae on root diseases. *Annu. Rev. Phytopath.* 2:377–392.

Zak, B. 1969. *Mycorrhizae. 4. Characterization and Identification of Douglas-Fir Mycorrhizae.* Proc. First North Amer. Conf. on Mycorrhizae. USDA For. Ser., Misc. Pub. 1189. 255 pp.

DISEASE PROFILE

Ectomycorrhizae

Importance: Not all fungi are detrimental to the plants with which they live. Ectomycorrhizal fungi increase the total number of feeder roots, thus increasing root surface areas. They capture nutrients from the soil and help reduce root respiration, which contributes to root longevity. They have a fungus mantle that provides a barrier against entry by pathogens, and some produce antibiotics that are antagonistic to some root pathogens.

Suscepts: Trees that form ectomycorrhizae include most temperate and boreal commercial forest species and 70% of the species planted in the tropics. Important host families include Pinaceae, Fagaceae, Betulaceae, Juglandaceae, Myrtaceae, and Dipterocarpaceae and comprise over 2,000 species of woody plants.

Distribution: Worldwide in occurrence, ectomycorrhizae are often important on infertile tropical soils.

Causal Agents: The fungi that form ectomycorrhizae are primarily in the Basidiomycotina (*Amanita, Boletus, Hebeloma, Laccaria, Lactarius, Pisolithus, Rhizopogon, Russula, Scleroderma, Suillus, Tricholoma*) and the Ascomycotina (*Cenococcum* and *Tuber*).

Biology: Ectomycorrhizae produce a characteristic fungal sheath or mantle that envelopes the short, feeder roots. Hyphae penetrate the cortical region, forming a structure called the Hartig net between the cortical cells. As the young root tips are colonized, they become swollen and often become bifurcately branched. As they mature, ectomycorrhize

Macro view of ectomycorrhizal roots.

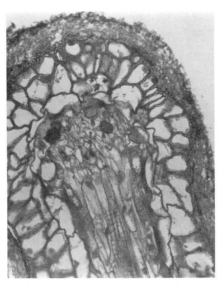

Section of ectomycorrhizal root tip.

branch several times in regular or irregular patterns. Ectomycorrhizal feeder roots do not have root hairs.

Epidemiology: While many sporocarps of ectomycorrhizal fungi produce large quantities of airborne spores, artificial inoculations are usually employed to ensure that the most beneficial fungus inoculum will be used.

Diagnosis: Most ectomycorrhizal root tips may be easily identified from their swollen tips covered with a fungus mantle and characteristic branching habit. The internal structure of the intercellular hyphal growth between cortical cells forming the Hartig net is characteristic when sections are viewed under the light microscope. Mantle color, habit, and morphology of the branching pattern are diagnostic for a few host/fungus ectomycorrhizal associations.

Control Strategy: Historically, soil inoculum was used to infest nursery soils and even new planting areas for exotic tree species. Spores or macerated fruiting bodies are frequently used today and are applied in the nursery shortly after seeding. Mycelial inoculum has the advantage of being produced in pure culture and, hence, may be more carefully tailored for a specific need. The inoculum must be leached with water before applying to remove excess nutrients. *Pisolithus tinctorius* was the first ectomycorrhizal fungus to be used widely in forest tree and container nurseries.

SELECTED REFERENCES

Castellano, M. A., and R. Molina. 1990. Mycorrhizae, pp. 101–171. In T. D. Landis, R. W. Tinus, S. E. McDonald, and J. P. Barnett (eds). *The Container Tree Nursery Manual*, Vol. 5. USDA For. Serv., Agric. Handbook 674. 171 pp.

DISEASE PROFILE

Endomycorrhizae

Importance: Endomycorrhizal fungi explore great volumes of soil far beyond the plant root zone and, thus, return water and nutrients, especially phosphorus. They reduce exudations from plant roots and thus reduce pathogen activity.

Suscepts: Endomycorrhizal fungi form associations with agricultural field crop plants such as legumes and cereals; temperate forest trees, especially maple, sweetgum, cedar, and redwood; tropical timber trees; and horticultural and ornamental crop plants. Some tree species such as alder, willow, and eucalypts form both endo- and ectomycorrhizal associations.

Distribution: Its widespread occurrence includes tropic and temperate areas.

Causal Agents: The fungi with nonseptate hyphae that form endomycorrhizae are in the Zygomycotina (*Acaulospora, Endogone, Entrophospora, Gigaspora, Glomus, Sclerocystis,* and *Scutellospora.* The fungi with septate hyphae that form endomycorrhizae are species of *Rhizoctonia* or a Discomycete, *Pezizella ericae.* Ectendomycorrhizae are in the Ascomycotina *(Phialophora, Chloridium).*

Biology: Of the three subgroups of endomycorrhizae, the vesicular–arbuscular type are the most common. Fungal hyphae enter the root cells but cause no noticeable structural

Vesicles.

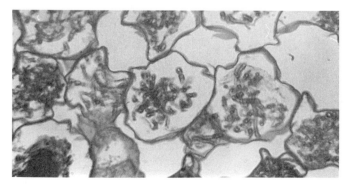

Active and digested hyphae.

changes in the root's exterior. The name vesicular-arbuscular comes from the structures that are formed within root cortical cells: vesicles, which are spherical shaped, and arbuscules, which are branched. Arbuscules are short-lived structures that serve mainly for nutrient exchange between host and fungus. Vesicles are usually filled with lipids and serve as storage organs and as reproductive structures.

Epidemiology: The VA mycorrhizae fungi form large chlamydospores in the soil, which are moved primarily by small animals and insects. Because as much as 20–50% of the total amount of carbon fixed by plants may be used by VA fungi, excessive cropping of the host plant can be deleterious to the fungal population levels.

Diagnosis: To assess presence of VA mycorrhizae, plant roots must be stained and observed under a light microscope. The balloon-shaped vesicles and finely branched arbuscules are distinctive and can be used to estimate the degree of root colonization.

Control Strategy: The major strategy is one of increasing their incidence in certain cropping situations. Some soils supporting agricultural or horticultural crops have been artificially inoculated with vesicles of desired VA fungi. With forest tree species, artificial inoculations have been done mainly in production of containerized seedlings. Because VA mycorrhizal fungi cannot be grown on synthetic media, inoculation consists of either using infested soil or, preferably, spores concentrated from soil/root growth medium.

SELECTED REFERENCES

Castellano, M. A., and R. Molina. 1990. Mycorrhizae, pp. 101–171. In T. D. Landis, R. W. Tinus, S. E. McDonald, and J. P. Barnett (eds.), *The Container Tree Nursery Manual,* Vol. 5. USDA For. Serv., Agric. Handbook 674. 171 pp.

13

ROOT SYSTEM DECAY

13.1 INTRODUCTION

Almost every tree species is subject to one or more of the many pathogens that can attack its roots. The fungi that attack tree roots can be conveniently divided into two groups:

1. Those that affect the small feeder roots, such as *Phytophthora cinnamomi,* were discussed in Chapter 11.
2. Those that attack the larger roots and have the capacity to decay wood will be discussed in this chapter. An example is *Heterobasidion annosum.*

Trees planted outside their normal range may be more susceptible to root diseases, as well as the other kinds of diseases. Root diseases can increase in severity over a period of years when successive crops of the same species are grown in one area. The decline in production can be so gradual that losses often are not recognized. Fumigation or total eradication of microorganisms in nursery beds will sometimes indicate the extent of these losses. There is increasing evidence, though, that root diseases are a significant problem in plantations and forest stands and are far more devastating than formerly believed. Root diseases can become more severe when the fungi involved are allowed to increase their numbers on a previous crop. In general, root diseases are more of a problem in warm, moist climates and are more common in plantations or in natural forests where unnatural stand compositions exist. Man's attempt to change natural plant succession can lead to problems with root decays as well as other diseases.

The total extent of root diseases is not known because of the difficulties involved in studying root systems of trees. Root diseases often are complexes involving more than one kind of microorganism, and all are affected by the environment. The resultant disease is due to the interaction of many kinds of microorganisms, several environmental factors, and the host.

Although *Heterobasidion annosum, Armillaria* spp., *Phellinus weirii, Inonotus tomentosus,* and other related fungi can cause heart decay, our main concern with these decay fungi is their propensity for attacking young, vigorously developing trees and killing them. In one sense, the so-called heart-decaying fungi may be nothing more than saprophytes; but each of the previously listed fungi can also be a parasite. As a group, root diseases rank high on the list of serious pathogens with which foresters must contend.

DISEASE PROFILE

Annosum Root Disease

Importance: This disease is one of the most economically important diseases of conifers in the north temperate zone, especially where intensive thinning and monoculture are practiced.

Suscepts: Some 150 species of trees, including a few hardwoods but conifers are the major hosts, particularly species of *Abies, Juniperus, Larix, Picea, Pinus, Pseudotsuga,* and *Tsuga.*

Distribution: This disease is essentially endemic with suscepts in the north temperate zone of North America, Europe, and Asia; in the southern hemisphere it is found in Australia on hoop pine *(Araucaria cunninghamii)* and in the South American Andes Mountains on planted exotics.

Causal Agent: Annosum root disease is caused by *Heterobasidion annosum (Fomes annosus)* (Aphyllophorales). The imperfect stage is *Spiniger meineckellus (Oedocephalum lineatum)* (Hyphomycetes).

Biology: The fungus decays woody root systems and then advances to the root collar where it may surface to the cambium and kill by girdling as in hard pines, or it may progress more slowly through roots to the stem and cause butt decay as in soft pines. The fungus is known to colonize stem wounds above the root-collar zone and has been found up to 10 m aboveground in western redcedar.

Epidemiology: The epidemic potential of *H. annosum* is primarily associated with suscept thinnings, particularly of pure stands, and the wind deposition of basidiospores onto the surface of freshly cut 2- to 4-week-old stumps; saprophytic colonization of the stump body progresses downward and outward through the roots (2 m/yr in slash pine) where, by root contact, the fungus may penetrate the living roots of adjacent trees. The fungus may persist in stump systems of southern pines for 5–10 years, and in other species as long as 25–50 years. In addition to colonized stumps, basidiospores may also be

Sporocarps. From Mook and Eno (1961).

inoculum. Basidiospores can move through the soil and penetrate live roots. Basidiospore inoculum may be more significant than currently recognized. Disease development is favored by certain edaphic factors, such as deep sands, in the eastern United States, and by alkaline, ex-agricultural soils, in Great Britain.

Diagnosis: Progressive root system decay, seen as a light-yellowish stringy decay, may lead to crown thinning and mortality or windthrow of live stems before aerial symptoms become evident. Basidiocarp signs show a grey-brown upper surface and a white poroid undersurface. They are found resupinate on the underside of decayed roots and slash or applanate under the duff at the base of stumps and infected trees or inside stumps in arid areas.

Control Strategy: Management alternatives include: (1) avoidance (e.g., Virginia soil hazard rating or summer thinning from central Georgia southward); (2) exclusion to prevent further introductions into the south temperate zone; (3) eradication, specifically stump extraction; (4) protection of stumps by chemical (Borax or TIM-BOR) and biological *[Phanerochaete gigantea (Peniophora gigantea)]* means are operationally prescribed by USFS and British Forestry Commission, respectively; (5) resistance (e.g., administer some tests of progeny from seed-orchard selections); and (6) therapy (e.g., trenching to sever root-contact connections ineffective).

SELECTED REFERENCES

Hadfield, J. S., D. J. Goheen, G. M. Filip, L. L. Schmitt, and R. D. Harvey. 1986. *Root Diseases in Oregon and Washington Conifers*. USDA For. Serv., R6–FPM-250–86. 27 pp.

Hodges, C. S., Jr. 1974. Cost of treating stumps to prevent infection by *Fomes annosus*. J. For. 72:402–404.

Morris, C. L., and D. H. Frazier. 1966. Development of a hazard rating for *Fomes annosus* in Virginia. *Pl. Dis. Reptr.* 50:510–511.

Otrosina, W. J., and R. F. Scharpf (tech. coords.). 1989. Proceedings of the Symposium on Research and Management of Annosus Root Disease *(Heterobasidion annosum)* in Western North America. *April 18–21, 1989. Monterey, CA.* USDA For. Serv., Gen. Tech. Rep. PSW-116. 177 pp.

13.2   ANNOSUM ROOT DISEASE

Importance

Plantations tend to have greater damage than natural stands. They have close uniform spacing, are thinned frequently, and have a continuity of root systems of the same species that favors disease spread. In contrast, natural stands are often mixed with various conifers and nonsusceptible species of hardwoods with irregular spacing and age distribution that results in less continuous root systems and less chance for spread.

Economic losses from *H. annosum* are due to mortality and decreased growth. Mortality and related growth loss on high-risk sites have resulted in serious losses in northern Europe and in the southeastern United States. On the highest-hazard sites, losses in the first 5–7 years after thinning can be expected to exceed 25–62 trees/ha. The monetary loss from this mortality was approximately double the cost of applying control measures in 1966. Even if mortality due to *H. annosum* stops, the economic losses continue because of decreased growth of infected trees and understocking of the residual stand. The difference in diameter growth between healthy and diseased loblolly pine trees was 19% over a 5-year period. Slash pines with more than 50% of their root systems colonized by *H. annosum* showed both reduced height and diameter growth.

The amount of the loss has been greater in thinned stands, especially in plantations established on former cropland. In a survey of thinned stands of pine from New England to Texas, in some plots as many as 30% of the residual trees were either dead or dying because of *H. annosum*. Loblolly pines appeared most susceptible, with 59% of stands having some infected trees and almost 3% of the trees dead or dying. The incidence of the disease and mortality was less in slash pine. Red pine seemed to be the most resistant with less than 1% of the trees dead or dying. In Tennessee *H. annosum* caused losses in 96% of the loblolly pine plantations surveyed; average volume loss was 0.25 cord/ha/yr and mortality increased steadily until 7 years after thinning, after which losses declined rapidly. In loblolly pine stands in Virginia, maximum losses for 5 years after thinning amounted to approximately 1 cord/ha/yr. Losses peaked at 7 years and declined rapidly after the ninth year following thinning. In young western hemlocks in western Washington, *H. annosum* was found in 0–57% of the stands prior to thinning. *Heterobasidium annosum* was present in 50–90% of the stumps 2–8 months after thinning. *Heterobasidium annosum* has been found in 28 of the 58 counties in California and was first recorded in the state as early as 1909. At the Institute of Forest Genetics in Placerville, California, 26 species of pines, three varieties, and one hybrid have been killed by the fungus. In one California state forest, 60 infection centers involving 800 dead trees were found on five sections of land and infection centers ranged in size from 1 to 62 trees. The primary concern about this disease is not necessarily the losses occurring now but the potential hazard that *H. annosum* root disease poses to the extensive plantations of the future.

It has been realized recently that, although the losses due to mortality described previously are sizeable, a more significant source of loss may be in the form of lost growth on sites of lower risk. Loblolly pines stressed by *H. annosum* were preferred by the southern pine beetle, and other bark beetles often attack trees with annosum root disease. These kills are usually attributed to bark beetles, further underestimating the damage caused by disease. This predisposition may be as important as *H. annosum*-caused mortality, particularly in western forests.

Annosum root disease is probably increasing in importance in western forests because stands have been entered several times and because of the increasing component of shade-tolerant true firs in these stands. Increasing emphasis on uneven-aged silvicultural systems and restrictions on the use of prescribed fire will further increase damage caused by *H. annosum*.

Suscepts

The fungus *Heterobasidion annosum* (Fr.:Fr.) Bref. [formerly *Fomes annosus* (Fr.) Karst.] causes a root and butt decay of a wide range of woody plants throughout the world. The common name is annosum root disease. The reported host range includes 100 species of angiosperms and 126 species of gymnosperms, although it is much more important on the latter. Species of true fir, juniper, larch, spruce, pine, Douglas-fir, and hemlock are the major hosts. It is a serious problem on eastern redcedar. In the United States annosum root disease has traditionally caused the greatest losses in southern pines, although more recent surveys indicate that losses in western North American forests may be much greater than formerly believed.

Distribution

Originally described by Robert Hartig (1894) during the 1870s as causing the "death circle" in conifer stands, *H. annosum* now has been reported from most north temperate regions and some tropical and subtropical areas. This fungus is present in most forested areas of the United States and Canada and has caused mortality in the eastern states from southern Ontario to Florida and in western states from British Columbia to southern California. The fungus seems to be limited in its northward movement by weather. Survival of the fungus is less common under cool, wet conditions, and damage is markedly less severe under these climatic conditions.

The first reports of mortality in southern pines caused by *H. annosum* came from South Carolina and Georgia in 1954. In 1961 a survey was conducted to determine the extent and severity of the disease in the eastern and southern pine forests. The incidence of annosum root disease was higher in the coastal states from Virginia to Texas than in the northeastern states. The survey found that annosum root disease occurred in 59% of the planted loblolly pine and 44% of the planted slash pine plantations. Natural stands of slash pine had only 8% incidence. In some plantations 30% or more of the trees were dead or dying. The high incidence and mortality were of concern because these are the two most important timber species in the southern United States.

We know considerably less about the distribution of annosum root disease in the vast western forests. It is extremely common in stands with true firs and will likely increase its distribution and importance as true firs begin to dominate stands as a result of fire control and partial cutting.

Cause

As the common name indicates, this root disease is caused by *Heterobasidion annosum* (Basidiomycotina), a fungus species that can be spread, probably for many kilometers, by means of wind-disseminated basidiospores.

Biology and Epidemiology

The life history of *H. annosum* is shown in the disease diagram in Figure 13.1. *Heterobasidion annosum* also produces asexual spores (*Spiniger* spp.) (*Oedocephalum* stage), which may be spread by wind or other agents. The role of the conidia is not clear, but they are useful in detecting the fungus. When freshly cut pine disks are exposed in stands and incubated for 8–12 days, conidiophores form on the disk, indicating the presence of *H. annosum*. The *Spiniger* state also forms on infected tissues incubated in warm, moist conditions.

Infection commonly occurs in stump surfaces but can occur to a lesser extent in roots. A second major form of spread is by growth of mycelium across root contacts between healthy and infected trees. The establishment of new infection centers is by spores, and enlargement of infection centers is by growth of mycelium across root contacts.

Basidiospores are produced in basidiocarps and wind disseminated long distances. Conidia are produced during periods of high humidity and are also wind disseminated. Conidia are rarely observed on diseased trees or stumps and probably do not play a major role in dispersal. However, as mentioned earlier, the conidia are important in determining whether tissues are infected. The production of spores slows or stops during periods of hot, dry weather and stops during freezing weather. In Europe, Canada, and the northern United States, the period of greatest spore production occurs during late summer and fall. In the southeastern United States, greatest spore production occurs during the fall, winter, and spring with few if any spores produced during the summer. In Georgia, spores of *H. annosum* were not detected from June 1 to October 15. Trapping stations in the southern United States collected few spores when the temperature was above 32°C or mean temperature over 21°C.

The freshly cut stump surface is highly selective for colonization by *H. annosum*. The stump surface of loblolly pine remains susceptible to infection for up to 2 weeks after felling. Eastern white pine stumps are highly susceptible to invasion for only 1–3 days after cutting. Freshly cut stumps of western species remain susceptible for at least 4 weeks, though susceptibility decreases within 2 weeks after cutting. During February in Georgia, loblolly pine stumps are susceptible for as long as 14 days after

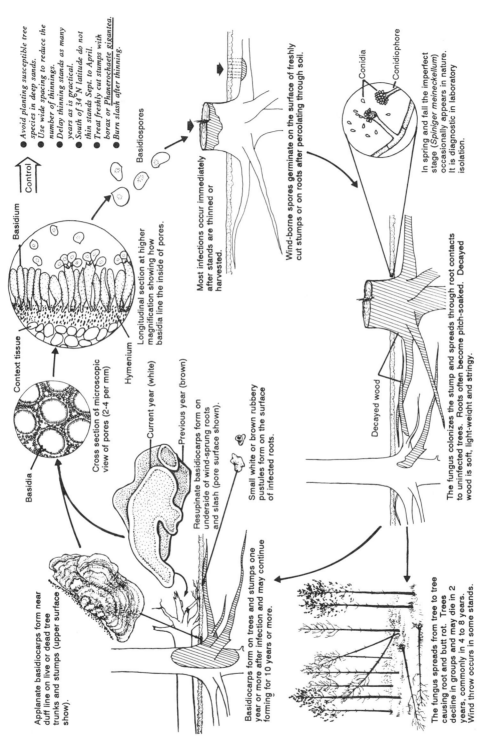

Avoid planting susceptible tree
species in deep sands.
Use wide spacing to reduce the
number of thinnings.
Delay thinning stands as many
years as is practical.
South of 34°N latitude do not
thin stands Sept. to April.
Treat freshly cut stumps with
borax or *Phanerochaete gigantea*.
Burn slash after thinning.

Control

Basidiospores

Basidium

Context tissue

Basidia

Cross section of microscopic
view of pores (2-4 per mm).

Hymenium

Longitudinal section at higher
magnification showing how
basidia line the inside of pores.

Most infections occur immediately
after stands are thinned or
harvested.

Wind-borne spores germinate on the surface of freshly
cut stumps or on roots after percolating through soil.

Conidia

Conidiophore

In spring and fall the imperfect
stage (*Spiniger meineckellum*)
occasionally appears in nature.
It is diagnostic in laboratory
isolation.

Current year (white)

Previous year (brown)

Resupinate basidiocarps form on
underside of wind-sprung roots
and slash (pore surface shown).

Small white or brown rubbery
pustules form on the surface
of infected roots.

Applanate basidiocarps form near
duff line on live or dead tree
trunks and stumps (upper surface
show).

Decayed wood

The fungus colonizes the stump and spreads through root contacts
to uninfected trees. Roots often become pitch-soaked. Decayed
wood is soft, light-weight and stringy.

Basidiocarps form on trees and stumps one
year or more after infection and may continue
forming for 10 years or more.

The fungus spreads from tree to tree
causing root and butt rot. Trees
decline in groups and may die in 2
years, commonly in 4 to 8 years.
Wind throw occurs in some stands.

FIGURE 13.1 The disease diagram of annosum root disease of Pinaceae spp. caused by *Heterobasidion annosum*. Drawn by Valerie Mortensen.

cutting. Further infection is limited by an increase in the incidence of *Phanerochaete* (*Peniophora*) *gigantea* and other microorganisms. The largest amount of stump infection generally corresponds to periods of greatest spore production by *H. annosum* (and lower spore production by competing fungi). Colonization of slash pine stumps in Georgia was greatest during a period from mid-October to mid-December. Higher populations of competing microorganisms like *P. gigantea* in the winter and spring excluded *H. annosum* despite the fact that spore concentrations were similar to those in the fall.

Following stump surface colonization the mycelium moves down into the stump and roots at a rate of about 1 m/yr. Vertical penetration in ponderosa pine stumps averaged 13–16 cm/month during summer months. The rate of development of the fungus in the roots of a living tree and mortality of the host varies, but usually 4–10 years are required from the time of stump invasion to appearance of symptoms in adjacent trees. Death of surrounding trees has occurred, however, within 2 years of when a stand was thinned. Based on artificial inoculations of 13-year-old trees, *H. annosum* grew an average of 16 cm in loblolly pine, 21 cm in slash pine, and 5.8 cm in longleaf pine in 6 months time. *Heterobasidion annosum* grew more rapidly in roots of suppressed trees than in roots of dominant trees. The fungus can survive in a stump for many years but is not likely to exist very long in soil. Once established in a stand, *H. annosum* remains indefinitely. It has been spread to new areas with infested fence posts.

The fungus infects adjacent healthy trees by the spread of mycelium across contacts between infected and healthy roots. The fungal mycelium is unable to grow through field soil. Rishbeth (1950, 1951a,b) observed that in pine the mycelium passes directly from the bark of diseased roots to that of living roots. In roots 3 cm in diameter and smaller, the mycelium invades the wood soon after colonizing the bark. The fungus has been observed on bark of roots far ahead of infection in the wood. This can dramatically increase the rate of spread. In southern pines, the number of infected pines in a disease center increases for the first 4–8 years following thinning and decreases thereafter, usually stabilizing 10 years after thinning. This behavior seems to be less common in western forests. The infected tree dies after it is girdled at the root collar, windthrown, or attacked by bark beetles or secondary pathogens such as *Armillaria*. *Heterobasidion annosum* can also enter through wounds made for injecting some herbicides such as picloram-2,4-D and can move from wounds on the stem into base of the tree. Wounds for injection of monosodium acid methanearsonate were not infected.

Direct root infection by spores is another possible method of spread. Rishbeth (1950, 1951a,b) placed samples of field soil on freshly cut stumps, and infection resulted, indicating that there was an infectious agent in the soil. Evidence from infestations in the western United States also suggest a role for spore infection. A relatively low level of 44 conidia per inoculation point was sufficient to cause infection of non-wounded roots. Wounding of intact roots greatly increased the number of infections obtained.

For root infection to occur, fungal spores must move down into the soil and remain viable. Conidia were found to move 17.5 cm deep in a sand column with as little as 1.24 ml of water. In field soil the amounts of conidia and basidiospores that reach the lower soil layers are greatly reduced by the filtering effects of the upper soil layers, decreasing the chances for direct root infection. Downward movement of spores was greater in high-hazard soils than low-hazard soils. This is generally due to the higher sand content of the high-hazard soils, which more closely approximates the ideal conditions in the sand column. The spores must also remain viable for a long enough period to come in contact with the root. Kuhlman (1969a) reported that conidia remained viable for up to 10 months in field soil. Spores of *H. annosum* have been recovered from infected and noninfected stands. Overall, direct root infections appear to be limited to less than 10% of the roots because of the low numbers of spores that reach the roots. If the amount of direct root infection was greater, there would probably be more infection in nonthinned stands. In forests in the western United States, however, some undisturbed stands appear to be virtually 100% infected.

Other possible methods of spread include insects and burrowing animals. Conidia are produced abundantly in insect galleries on infected trees, and it seems possible that insects could carry conidia to healthy trees. Southern pine beetle *(Dendroctonus frontalis)* is the most likely candidate for transmission in southern pines because it is commonly associated with trees stressed by *H. annosum*. Conidia have also been produced in tunnels of burrowing animals, yet no experimental evidence has been given to support transmission by insects or burrowing animals.

The incidence and severity of the disease are related to a number of site factors and management practices. Losses increase with time since thinning. A southwide survey of *H. annosum* found that disease severity was directly related to the number of thinnings in the stand and the proportion of sand in the soil. In England the most severely diseased stands were on sandy, alkaline soils. The greater rate of spread in these soils was attributed to their lower populations of competing microorganisms. In the coastal plain of the Gulf states, soils of healthy plantations had lower soil pH, higher organic matter, higher silt content, and more grass cover than soils of severely damaged plantations. Stands with coarser-textured A horizons and lighter soils had more root disease. Trees on slopes apparently are more susceptible. Induced drought conditions resulted in more rapid penetration of roots by *H. annosum*. The incidence of *H. annosum* was greater on sites with less organic matter, higher pH, more sand or clay, and less grass. One plot with a high incidence of *H. annosum* root disease had 0.5% organic material, 6.1 pH, 4% clay, 88% sand, and no grass. In contrast, stands free of root disease had 2.2% organic matter, 4.8 pH, 8% clay, 48% sand, and 90% grass cover. Less root disease occurs in stands that have been burned, possibly because fire stimulates microflora less favorable for *H. annosum*. In Great Britain, species such as Douglas-fir are killed when young but become more resistant with age; in California the trees do not seem to increase in resistance with age.

Forest pathologists recognize two physiological races of *H. annosum*. One, referred to as the "P-type," causes disease on pines throughout North America. The other race is called the "S-type," and it causes disease in spruces, true firs, western hemlock, and redcedar. In North America the S-type has been found only in the western states. Although these two races may be found on hosts in the other group (e.g., P-type may be found in fir stands), they cause disease in their respective host groups only. Forest managers may use this host specificity to avoid root-disease problems by favoring species that are not a host for the *H. annosum* strain present.

Diagnosis

Annosum root disease produces irregularly shaped sporophores or conks at the base of trees (Fig. 13.2). The conks are variable in size from less than 2.5 cm to several centimeters across and are frequently located below the surface of the litter layer. They are difficult to locate unless the litter layer is removed. The undersurface of the conk is chalky-white with pore openings visible to the unaided eye and has a thin sterile margin around the outer edge. The upper surface is variable in color ranging from tan to reddish brown, becoming darker with age. Another diagnostic feature is that the conk is difficult to tear. The fruiting bodies seldom last more than a year in the southeastern United States because they are attacked by fungi and insects, but they may be produced for up to 10 years following death of the tree. The conks are not always present on infected trees, and their presence on slash does not necessarily indicate that the living trees are infected. One can be very suspicious that *H. annosum* is involved if trees are dying around a stump with annosum conks or decay in it (Fig. 13.3). If a tree is infected, it is likely to have a thin, unhealthy appearing crown; however, some trees die so quickly that no crown thinning is evident. In the drier forests of the western United States, conks are rarely formed at the base of trees. Most often, conks are produced inside well-decayed stumps and on the underside of exposed roots of infected, windthrown trees.

Aerial symptoms of annosum root disease can also be confused with other decay organisms. The symptoms are decline and death (Fig. 13.4). Severely affected trees may have thin, light green to yellow foliage with short needles tufted at the ends of branches. However, some trees die so quickly that crown symptoms do not appear. This commonly happens on high-risk sites. Infected roots may have a small reddish or brownish discoloration column. Later, they may become resin soaked and brownish red in color with soil adhering to them. The roots eventually become a white stringy mass of decayed tissue (Fig. 13.5). *Heterobasidion annosum* can cause heart decay without killing the tree. Usually groups of trees are infected, and infection

FIGURE 13.2   Heterobasidion annosum fruiting at the base of windthrown subalpine fir.

FIGURE 13.3   Dead ponderosa pine near stumps implicate *Heterobasidion annosum* as a potential cause of mortality.

FIGURE 13.4   A plantation of slash pines breaking up from annosum root disease following thinning. From Powers and Hodges (1970).

FIGURE 13.5 The typical stringy decay produced by *Heterobasidion annosum* in a white pine. From Powers and Hodges (1970).

centers are irregular patches in the stand that increase in size as more trees become infected. When infected trees are windthrown or the trees are felled and the major roots examined, the typical soft stringy decay is evident. In western conifers, decayed wood often separates along annual rings, and small pits (1–2 mm) may be present on one side of the delaminated wood. Infected trees are susceptible to invasion by bark beetles.

Control Strategy

In stands, especially plantations, where factors favor *H. annosum,* thinning should be delayed as long as possible, and the stand should be thinned only once. Cutting even a single living tree in a stand might allow the establishment of *H. annosum,* which, once present, can continue to cause increasing losses during subsequent years. In stands that have been thinned and root disease has not resulted in losses, control measures would not be necessary. To eliminate the danger posed by precommercial thinning in southern pines, a stocking density of 1,112 stems/ha has been recommended. This density results in 3–m spacing between stems; however, in young plantations the expense of vegetation control may be excessive.

Because most new infection centers are formed by colonization of freshly cut stumps after thinning, stump protection can keep annosum root disease out of uninfested stands. Once the disease becomes established in a stand, eradication is not economically feasible. Stumps are protected immediately following cutting to prevent colonization by *H. annosum.* Stumps of loblolly pine need to be protected for 2 weeks following thinning to escape colonization by *H. annosum.*

The two protectants most widely recommended are borax and *P. gigantea.* Other protectants tested include urea, creosote, and sodium nitrate. In tests, borax provided the most effective and consistent protection against stump surface colonization, even though it adds to the cost of thinning stands. Borax is applied in granular form, from a container with holes in the lid like a salt shaker (Fig. 13.6). A dye is often added for ease in evaluating the treatment. A light, uniform covering of the stump is adequate for protection. For an average thinning in a southern pine stand, 5.8 kg/ha of borax is needed. After application, the borax penetrates into the stump to a depth of 2.5–5 cm and remains at toxic levels for up to 2 years. The cost of treating stumps has been estimated at $7.41/ha, compared with losses of up to $106/ha in the southeastern United States. It is important to recognize sites where damage will be substantial, such as deep, well-drained sandy soils where cost of treating stumps would be justified. Treatment of stumps on most piedmont and poorly drained coastal plain soils would be of doubtful value. Although a variety of compounds have been tested, the general consensus is that either urea or borax is most effective, and the latter compound is easier to apply in the field. Creosote was recommended originally but is not

FIGURE 13.6   Granular borax is applied to fresh stump "salt-shaker" style to control *Heterobasidion annosum*. From Hodges (1974a). Reprinted from *Journal of Forestry,* Vol. 72, No. 7, published by the Society of American Foresters, 5400 Grosvenor Lane, Bethesda, Md. 20814-2198. Not for further reproduction.

as effective. Even though stumps of some tree species remain susceptible for up to 4 weeks, stumps should be treated immediately after the tree is felled. Thorough coverage is important. The dry borax (technical grade sodium tetraborate decahydrate at 0.1 kg/m^2) applied with a hand shaker is faster, easier to apply, and cheaper than liquid urea, even though the latter results in better coverage. Whether resin inundation of a stump surface will prevent infection is not known, but in the spring both slash and longleaf pine stumps are completely covered with resin in 5–10 minutes. In the forests of the western United States, the value of treating stumps in infested stands is questionable. For example, many true-fir stands are riddled with annosum root disease. Applying borax to the stump surface does little to inhibit the fungus already in the root systems. At the present time the granular formulation of borax (sodium tetraborate decahydrate), and the water-soluble formulation of borax sold under the trade name of TIM-BOR, are no longer commercially available. A repackaged formulation of granular borax has been registered recently and should now be available under the trade name of Sporax.

Colonization of stumps by fungi other than *H. annosum* can limit or prevent invasion by the latter species. The stumps of loblolly pine are not susceptible to *H. annosum* 12 days after cutting, apparently because of the wood-decaying fungus *Phanerochaete gigantea*. This fungus quickly grows into lateral roots and prevents entrance by *H. annosum*. Of all the fungi that colonize pine stumps, it is the most vigorous competitor of *H. annosum*. The antagonistic action of *P. gigantea* probably results from hyphal interference having a short-range antibiotic effect. *Phanerochaete gigantea* can replace *H. annosum* in roots colonized after felling, but it generally failed to do so in roots invaded before felling. *Phanerochaete gigantea* is recommended instead of borax for use during the second thinning on infected sites because it decomposes stump tissues and helps prevent further spread of the disease.

The fungus is mixed with water, making a spore suspension to spray on the stump surface. It is commercially available. One culture plate makes enough spore suspension to treat about 500 stumps 25 cm in diameter. To coat the stump surface, 2–4 ml are required. The spore suspension may be applied to the stump surface with a plastisqueeze bottle or hand garden sprayer.

The application of *P. gigantea* takes longer and is somewhat more difficult than borax. A fresh spore suspension of *P. gigantea* must be made daily. It also costs about 10–12 cents more per cord than borax. On moderate- to high-hazard sites, it is economically beneficial to treat stumps with borax instead of *P. gigantea;* but on low-hazard sites, the advantage is only slight.

Several other treatments have also shown some promise in preventing *H. annosum* from colonizing stumps. Loblolly pine stumps were totally protected from *H. annosum* for 8 months after treatment with oidia of *P. gigantea* mixed in SAE 30 motor oil. Presumably, the stumps can be inoculated with *P. gigantea* as the trees are felled by using the oil mixture as bar oil in the chain saw. *Trichoderma viride* also inhibits *H. annosum* but is only half as effective as *P. gigantea*. In the future *P. gigantea* or a solution of borax may be used to treat tree stumps in combination with shearing by a mechanical harvester.

Because root disease caused by *H. annosum* is primarily a problem in thinned stands, practices that reduce the number of thinnings decrease the opportunity for the fungus to invade the stand. Planting trees at wider spacings increases the time before the first thinning and decreases the potential for root contact. However, wider spacing may be undesirable because the possibility of damage from fusiform rust *(Cronartium quercuum* f. sp. *fusiforme)* increases with the width of spacing.

Another control measure is summer thinning. For stands below 34° N latitude, annosum root disease can be controlled by thinning during the summer (April to August). In southern Georgia and similar climatic zones, infection of stumps is unlikely from March until September because the temperatures are too high for the fungus. If the mean daily air temperature is above 21°C, which may mean that stump temperatures are over 38°C, the spores cannot infect stumps, and *H. annosum* cannot establish infection centers. Soil temperatures, however, are not as high, and root infection could occur during the summer. In slash pine plantations in the sand hills of South Carolina, mortality occurred when trees were inoculated during summer months. In milder climatic zones, stumps are susceptible to infection all year, while in still colder climates infection is probably limited to the summer months. Greatest deposition of spores was during winter in southern California and in fall in central Sierra Nevada. Most spores were deposited at night. In Oregon and Washington more spores were deposited in southern than in northern sites. The most spores were deposited in fall and spring; the fewest, in winter and summer. Above 34° N latitude stumps must be protected year round. The low rate of stump infection during the summer is due to thermal inactivation of spores and mycelium of the fungus and low rates of spore production. Basidiospores are rapidly inactivated in full sunlight with temperatures above 15°C. Temperatures of 40°C killed all actively growing mycelium in 2 hours. During May, June, and July, temperatures 0.6 cm below the surface of the stump reached 40°C for 2 hours or more on 50% of the days.

Planting resistant or tolerant species is another recommended control. Results indicate that longleaf pine is less susceptible to annosum root decay and should be favored on high-hazard sites over the more susceptible slash and loblolly pines. However, even on heavily infected southern pine sites, the losses due to *H. annosum* were less than 6%. In contrast, losses in pine plantations in England may be 50% by age 10.

In the Pacific Northwest, mountain hemlock, western hemlock, white fir, grand fir, and Pacific silver fir are highly susceptible. Ponderosa pine, lodgepole pine, California red fir, subalpine fir, and noble fir are intermediate in susceptibility when compared with Douglas-fir, sugar pine, western white pine, western red cedar, incense cedar, Port-Orford cedar, western larch, Englemann spruce, and Sitka spruce.

Prescribed burning is frequently practiced in stands of southern pine to reduce hardwood competition. Mortality and infection by annosum root disease were significantly reduced by the use of prescribed fire before and after thinning. The beneficial effects of the fire were greatest where disease was most serious. The exact reason why fire reduces infection is not known. Burning increases soil temperatures and also increases populations of *Trichoderma* sp., a competitor of *H. annosum*.

With large potential losses, foresters in Great Britain use more intense methods of control. On the highest-hazard sites, the stumps are removed to reduce the source of inoculum. Studies have shown that destumping reduced losses by *H. annosum* from 60 to 20% of the Scots pine at 18 years of age. Destumping may be necessary in temperate areas like England, where stumps decompose slowly. Rishbeth (1951c) recovered *H. annosum* from pine stumps up to 44 years after cutting. But pine stumps in the southeastern United States decompose in as little as 8 years. This decomposition suggests the fungus has nowhere to persist.

Other control measures used on an experimental basis are the application of the fungicides sulfur and methyl bromide. Spread of *H. annosum* was reduced by very heavy application of sulfur (2,272 kg/ha). The application significantly reduced the soil pH for a period of 3 years. The lower soil pH probably limited the spread of the fungus. The soil fumigant methyl bromide was used to stop the spread of infection centers. The fumigation line formed a barrier to the spread of annosum root disease between infected and healthy trees. These methods of control seem cost prohibitive. Forest crops, because of their relatively low value and long rotation times, cannot remain profitable with the large monetary investment required with these control measures.

A particularly useful management tool involves forecasting incidence and severity of annosum root disease in a soil risk rating system. Soil risk rating determines if the use of control measures such as stump treatment are necessary. It is based on the fact that *H. annosum* inoculum and favorable conditions for infection are generally present, except during the summer in the southern United States. The spread of the disease after stump infection largely depends on soil type. Ross (1973) reported that the Aulander, North Carolina, site had higher initial stump colonization than the Bainbridge, Georgia, site, but the latter site had greater losses 5 years after thinning. The Aulander site had a high water table, which reduced disease potential.

A hazard rating system was developed in Virginia to predict the rate of spread of *H. annosum* using soil texture, depth of A horizon, and height of the water table. The soil is classified as high hazard if a sandy A horizon extends deeper than 0.3 meter. A high-hazard rating means that significant mortality will occur unless control measures are taken. If high-hazard soils are underlain by a clay horizon at 0.25 m or less, the soil is classified as intermediate or low hazard. If the soil water table is high enough to produce flooding for 2 months of the year or soil mottling, the hazard is reduced regardless of the depth of the A horizon. In general, clays and clay loams are low-hazard soils, silt and silt loams are intermediate-hazard soils, and sandy loams and sands are high-hazard soils.

There are other methods to determine soil hazard rating. In many areas of the southern United States, the soil hazard rating can be identified using soil series maps. Low-hazard soils with poor internal drainage can be identified using indicator plants such as gallberry *(Ilex glabra)* and pitcher plant *(Sarracenia purpurea)*. Froelich et al. (1966) developed equations that predict soil hazard rating based on the soil pH and the amount of sand, clay, and organic matter. These equations have not been thoroughly tested, and they use soil parameters that are not easily measured in the field.

The method of forecasting developed by Morris and Frazier (1966) has been quite effective and reliable in practice. However, the hazard rating system was developed for use in southern pine stands, and other species like white pine do not seem to fit the system. In one study, loblolly pine stands classified as high hazard had an average infection incidence four times greater than low-hazard sites. A major advantage of risk rating is that it can easily be applied in the field using only a soil auger or shovel. Based on the three soil characteristics of texture, depth of the A horizon, and height of the water table, the forest manager can determine if the site needs special management considerations.

REFERENCES

Alexander, S. A., J. M. Skelly, and C. L. Morris. 1975. Edaphic factors associated with incidence and severity of disease caused by *Fomes annosus* in loblolly pine plantations in Virginia. *Phytopathology* 65:585–591.

Alexander, S. A., J. M. Skelly, and C. L. Morris. 1981. Effects of *Heterobasidion annosum* on radial growth in southern pine beetle-infested loblolly pine. *Phytopathology* 71:479–481.

Applegate, H. W. 1971. Annosus root rot mortality in once-thinned loblolly pine plantations in Tennessee. *Pl. Dis. Reptr.* 55:625–627.

Artman, J. D., and E. L. Sharpe. 1971. An inoculation test using *Peniophora gigantea* on stumps of eastern white pine. *Pl. Dis. Reptr.* 55:834–836.

Artman, J. D., and W. J. Stambaugh. 1970. A practical approach to the application of *Peniophora gigantea* for control of *Fomes annosus*. *Pl. Dis. Reptr.* 54:799–802.

Bega, R. V. 1962. Tree killing by *Fomes annosus* in a genetics arboretum. *Pl. Dis. Reptr.* 46:107–110.

Bega, R. V. 1963. Symposium on root diseases of forest trees: *Fomes annosus*. *Phytopathology* 53:1120–1123.

Bega, R. V., and R. S. Smith. 1966. Distribution of *Fomes annosus* in natural forests of California. *Pl. Dis. Reptr.* 50:832–836.

Bradford, B., S. A. Alexander, and J. M. Skelly. 1978. Determination of growth loss of *Pinus taeda* L. caused by *Heterobasidion annosum* (Fr.) Bref. *Eur. J. For. Path.* 8:129–134.

Cobb, F. W., and R. A. Schmidt. 1964. Duration of susceptibility of eastern white pine stumps to *Fomes annosus*. *Phytopathology* 54:1216–1218.

Driver, C. H., and J. H. Ginns. 1964. The effects of climate on the occurrence of annosus root rot in thinned slash pine plantations. *Pl. Dis. Reptr.* 48:509–511.

Driver, C. H., and J. H. Ginns. 1968. Practical control of *Fomes annosus* in intensively managed young-growth western hemlock stands. *Pl. Dis. Reptr.* 52:370–372.

Driver, C. H., and J. H. Ginns. 1969. Ecology of slash pine stumps: Fungal colonization and infection by *Fomes annosus*. *For. Sci.* 15:2–10.

Fago, C. E. 1969. *Operational Aspects of Chemical Stump Treatment for Fomes annosus* Protection on Boggs Mountain State Forest. Calif. State For. Notes 38. 4 pp.

Froelich, R. C., T. R. Dell, and C. H. Walkinshaw. 1966. Soil factors associated with *Fomes annosus* in the Gulf States. *For. Sci.* 12:356–361.

Froelich, R. C., and T. R. Dell. 1967. Prescribed fire as a possible control for *Fomes annosus*. *Phytopathology* 57:811.

Froelich, R. C., and J. D. Nicholson. 1973. Spread of *Fomes annosus* reduced by heavy application of sulfur. *For. Sci.* 19:75–76.

Froelich, R. C., C. S. Hodges, and S. S. Sackett. 1978. Prescribed burning reduces severity of annosus root rot in the south. *For. Sci.* 24:93–100.

Goheen, D., C. Schmitt, E. M. Goheen, and S. Frankel. 1986. Effects of management activities and dominant species type on root disease-caused mortality in two Oregon forests. In *Proc. Western International Forest Disease Work Conference, Juneau, Alaska, Sept. 9–12, 1986*.

Gooding, G. V., C. S. Hodges, and E. W. Ross. 1966. Effect of temperature on growth and survival of *Fomes annosus*. *For. Sci.* 12:325–333.

Greig, B. J. 1984. Management of East England pine plantations affected by *Heterobasidion annosum* root rot. *Eur. J. For. Path.* 14:393–397.

Hadfield, J. S., D. J. Goheen, G. M. Filip, L. L. Schmitt, and R. D. Harvey. 1986. *Root Diseases in Oregon and Washington Conifers*. USDA For. Ser., R6–FPM-250-86. 27 pp.

Hartig, R. 1894. *Textbook of the Diseases of Trees.* London: Geo. Newnes, Ltd. 331 pp.

Hendrix, F. F., and E. G. Kuhlman. 1964. Root infection of *Pinus elliottii* by *Fomes annosus.* *Nature* 201:55–56.

Hodges, C. S. 1969a. Relative susceptibility of loblolly, longleaf and slash pine roots to infection by *Fomes annosus. Phytopathology* 59:1031 (abst.).

Hodges, C. S. 1969b. Modes of infection and spread of *Fomes annosus. Annu. Rev. Phytopath.* 7:247–266.

Hodges, C. S. 1974a. Cost of treating stumps to prevent infection by *Fomes annosus. J. For.* 72:402–404.

Hodges, C. S. 1974b. Relative susceptibility of slash, loblolly and longleaf pines to infection by *Fomes annosus,* pp. 86–92. In E. G. Kuhlman (ed.), *Proc. Fourth Internat. Conf. on Fomes annosus,* Athens, Georgia, Sept. 17–22, 1973.

Hodges, C. S. 1974c. *Symptomology and Spread of Fomes annosus* in Southern Pine Plantations. USDA For. Serv., Res. Pap. SE-114. 10 pp.

Houston, D. R. 1975. *Soil Fumigation to Control Spread of Fomes annosus:* Results of Field Trials. USDA For. Serv., Res. Pap. NE-327. 4 pp.

Kuhlman, E. G. 1969a. Survival of *Fomes annosus* spores in soil. *Phytopathology* 59:198–201.

Kuhlman, E. G. 1969b. Number of conidia necessary for stump root infection by *Fomes annosus. Phytopathology* 59:1168–1169.

Kuhlman, E. G., and F. F. Hendrix. 1964. Infection, growth rate, and competitive ability of *Fomes annosus* in inoculated *Pinus echinata* stumps. *Phytopathology* 54:556–561.

Kuhlman, E. G., and E. W. Ross. 1970. Regeneration of pine on *Fomes annosus* infested sites in the southeastern United States, pp. 71–76. In C. S. Hodges, J. Rishbeth, and A. Yde-Anderson (eds.), *Proc. Third Internat. Conf. on Fomes annosus,* Aarhus, Denmark, July 29–Aug.3, 1968.

Kuhlman, E. G., and R. C. Froelich. 1976. *Minimizing Loses to Fomes annosus in the Southeastern United States.* USDA For. Serv., Res. Pap. SE-151. 16 pp.

Laird, P. P., and M. Newton. 1973. Contrasting effects of two herbicides on invasion by *Fomes annosus* in tree-injector wounds on western hemlock. *Pl. Dis. Reptr.* 57:94–96.

Miller, T., and A. Kelman. 1966. Growth of *Fomes annosus* in roots of suppressed and dominant loblolly pines. *For. Sci.* 12:225–233.

Mook, P. V., and H. G. Eno. 1961. *Fomes annosus—What Is It and How To Recognize It.* USDA For. Serv., Northeast. For. Exp. Sta. Pap. No. 146. 33 pp.

Morris, C. L. 1970. Volume losses from *Fomes annosus* in loblolly pine in Virginia. *J. For.* 68:283–284.

Morris, C. L., and D. H. Frazier. 1966. Development of a hazard rating for *Fomes annosus* in Virginia. *Pl. Dis. Reptr.* 50:510–511.

Powers, H. R., and C. S. Hodges. 1970. *Annosus Root Rot of Eastern Pines.* USDA For. Serv., For. Pest Leaf. No. 76. 8 pp.

Powers, H. R., and A. F. Verrall. 1962. A closer look at *Fomes annosus. For. Farmer* 21:8–9, 16–17.

Rishbeth, J. 1950. Observations on the biology of *Fomes annosus,* with particular reference to East Anglian pine plantations. I. The outbreaks of disease and ecological status of the fungus. *Ann. Bot., N.S.* 14:365–383.

Rishbeth, J. 1951a. Observations on the biology of *Fomes annosus,* with particular reference to East Anglian pine plantations. II. Spore production, stump infection, and saprophytic activity in stumps. *Ann. Bot., N.S.* 15:1–21.

Rishbeth, J. 1951b. Observations on the biology of *Fomes annosus,* with particular reference to

East Anglian pine plantations. III. Natural and experimental infection of pines, and some factors affecting severity of the disease. *Ann. Bot., N.S.* 15:221 246.

Rishbeth, J. 1951c. Butt rot by *Fomes annosus* Fr. in East Anglian conifer plantations and its relation to tree killing. *Forestry* 24:114–120.

Rishbeth, J. 1975. Stump inoculation: A biological control of *Fomes annosus*, pp. 158–162. In G. W. Breuhl (ed.), *Biology and Control of Soil-Borne Pathogens*. The American Phytopathological Society, St. Paul, Minnesota.

Ross, E. W. 1968. Duration of stump susceptibility of loblolly pine to infection by *Fomes annosus*. *For. Sci.* 14:206–211.

Ross, E. W. 1973. *Fomes annosus* in the Southeastern United States: Relation of Environmental and Biotic Factors to Stump Colonization and Losses in the Residual Stand. USDA For. Serv., Tech. Bull. 1459. 26 pp.

Ross, E. W., and C. H. Hodges. 1981. *Control of Heterobasidion annosum* Colonization in Mechanically Sheared Slash Pine Stumps Treated with *Peniophora gigantea*. USDA For. Serv., Res. Pap. SE-229. 3 pp.

Russell, K. W., J. H. Thompson, J. L. Stewart, and C. H. Driver. 1973. *Evaluation of Chemicals to Control Infection of Stumps by Fomes annosus* in Precommercially Thinned Western Hemlock Stands. Wash. Dept. Nat. Res. Rep. 33. 16 pp.

Shaw, C. G., III. 1989. Is *Heterobasidion annosum* poorly adapted to incite disease in cool, wet environments?, pp. 101–104. In W. J. Otrosina and R. F. Scharpf (eds.), *Proc. Symp. on Research and Management of Annosus Root Disease (Heterobasidion annosum) in Western North America*. USDA For. Serv., Gen. Tech. Rept. PSW-116.

Sinclair, W. A. 1964. *Root- and Butt-Rot of Conifers Caused by Fomes annosus,* with Special Reference to Inoculum Dispersal and Control of the Disease in New York. Cornell Agr. Exp. Sta. Mem. 391. 54 pp.

Smith, R. S. 1970. Borax to control *Fomes annosus* infection of white fir stumps. *Pl. Dis. Reptr.* 54:872–875.

Tegethoff, A. C. 1973. Known distribution of *Fomes annosus* in the intermountain region. *Pl. Dis. Reptr.* 57:407–410.

Towers, B., and W. J. Stambaugh. 1968. The influence of induced soil moisture stress upon *Fomes annosus* root rot of loblolly pine. *Phytopathology* 58:296–272.

Weiss, M. J., P. H. Peacher, J. L. Knighten, and C. E. Affeltranger. 1978. *Annosus Root Rot Stump Treatment: A Pilot Project*. USDA For. Serv., For. Rep. SA-FR1. 21 pp.

DISEASE PROFILE

Laminated Root Disease

Importance: Known since 1929, this destructive root disease of Pacific Northwest conifers is becoming an increasingly significant limitation in the management of second-growth Douglas-fir. In western Oregon and Washington alone, the disease already accounts for annual losses of 2,977 m³ in terms of mortality and growth reduction of Douglas-fir at age 25–125 years.

Suscepts: Douglas-fir, Pacific silver fir, lowland white fir, and mountain hemlock are the most consistently affected species. When these suscepts are found in stands with western larch, alpine fir, western white and lodgepole pines, and Sitka and Engelmann spruces, the latter are also attacked. Ponderosa pine and western redcedar are rarely infected.

Distribution: The disease is common from sea level to the upper elevations of commercial forests, on a variety of soils and sites from central British Columbia to southern Oregon and eastward into northern Idaho.

Causal Agents: Laminated root disease is caused by *Phellinus (Poria) weirii* (Aphyllophorales). The fungus has two resupinate forms, namely annual, primarily on Douglas-fir in the Pacific Northwest, and perennial on western redcedar in the northern Rocky Mountains. The latter is pathogenic to both suscepts, whereas the former is so only on fir.

Biology: Root infection occurs when healthy roots grow into contact with decayed root systems of the previous stand. The fungus grows ectotrophically well in advance of root-wood infection and penetrates intact bark of roots as large as 6 cm in diameter whether they are part of living trees or stumps of trees cut as much as 12 months earlier.

Laminated decay. From Thies (1984). Reprinted from *Journal of Forestry*, Vol. 82, No. 6, published by the Society of American Foresters, 5400 Grosvenor Lane, Bethesda, Md. 20814-2198. Not for further reproduction.

Epidemiology: Apparent limitations in aerial spread of the pathogen are compensated by its saprophytic persistence in woody debris and roots, ranging in the latter from 11 years in 2-cm diameter material to 50 years or more in larger roots. Clonal mapping substantiates holdover infections and suggests that the fungus may be more widespread than is indicated by aboveground symptoms. Active centers double their infections every 15 years or at an annual rate of spread of 32 cm.

Diagnosis: A radial progression from dead snags and windthrows to marginal living trees with thin crowns is typical of established infection centers. Incipient decay in butts and main roots is reddish brown to brown; advanced decay is typically laminate in that the wood tends to separate along the annual rings. Basidiocarps are relatively inconspicuous occurring as poroid, cinnamon-brown crusts on the underside of the butt portion of windthrown or root-sprung trees.

Control Strategy: Management alternatives include: (1) conversion to resistant species of trees but the choices (viz., pines and western redcedar) are poorly adapted to Douglas-fir sites; (2) mechanical stump removal, which is limited in application by expense and topography; (3) biological control as predicated on the microbial ecology of red alder and its indirect effects upon *P. weirii*. Improved markets for red alder make it an economic alternative. The evidence to date suggests that if alder prevails on a site long enough, *Phellinus* might well be eradicated.

SELECTED REFERENCE

Thies, W. G. 1984. Laminated root rot - The quest for control. *J. For.* 82:345–356.

13.3 LAMINATED ROOT DISEASE

13.3.1 Importance

The fact that the fungus responsible for laminated root disease parasitizes Douglas-fir, 20–40 years old, makes this a potentially dangerous pathogen. The disease yearly reduces forest productivity by about 4.4 million m^3 with 0.9 million m^3 of loss in western Oregon and western Washington alone (Fig. 13.7). In British Columbia in one stand of Douglas-fir, 58% of the expected volume was lost during a 15–yr period. Laminated root disease probably results in less than a 5% loss in the average stand of Douglas-fir, especially of old growth. This loss is expected to increase, though, because management of second-growth stands encourages buildup of inoculum. After two or three rotations of a highly susceptible host like Douglas-fir on an infested site, losses of 50–90% of the predicted harvest volume should be expected.

13.3.2 Suscepts

Most conifer species in the northwestern United States are susceptible. The most important hosts are Douglas-fir and western hemlock. Pacific silver fir, mountain hemlock, and grand fir are very susceptible. Subalpine fir, western larch, Engelmann spruce, sitka spruce, and western hemlock are intermediately susceptible. Disease inci-

FIGURE 13.7   A laminated root disease infection center in an 80-year-old stand of Douglas-fir. From Childs and Nelson (1971).

dence is significantly higher in western hemlock when this species is mixed with Douglas-fir. Lodgepole, ponderosa, and western white pines, incense-cedar, and western redcedar are either tolerant or resistant.

13.3.3   Diagnosis

Crown symptoms include reduced shoot length and a rounding of the crown as a result. In later stages of the disease, the foliage turns yellow and a "distress" crop of small cones is produced.

Laminated root disease occurs in centers or patches of affected trees ranging in size from only one or two trees to areas a hectare or larger (Fig. 13.8). Symptoms may be evident in plantations 15–20 years old, but by the time stands reach 40 years, obvious mortality centers will be present. Larger infection centers will have fallen trees or leaning trees, usually in a random pattern (Fig. 13.9).

The root collar of living infected trees may be covered with a sheath of superficial

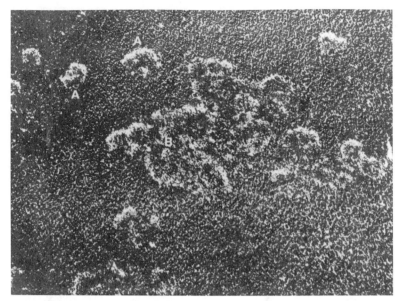

FIGURE 13.8 Aerial photograph showing (A) single and (B) multiple infection centers caused by *Phellinus weirii*. From Nelson and Hartman (1975). Reprinted from *Journal of Forestry*, Vol. 73, No. 3, published by the Society of American Foresters, 5400 Grosvenor Lane, Bethesda, Md. 20814-2198. Not for further reproduction.

FIGURE 13.9 "Root-thrown" Douglas-fir resulting from attack by *Phellinus weirii*. Note stubs of decayed roots. From Childs and Nelson (1971).

mycelium ranging in color from grey-white to tawny to light purple. Examination with a hand lens will reveal scattered reddish-brown, wiry setal hyphae. A brown, crustlike mycelial growth over this sheath is very typical of this fungus.

When diseased trees are felled, a red-brown stain is initially very evident on the stump surface as irregular or crescent-shaped stains. This stain may extend 2–4 m up the bole (Fig. 13.10). As the decay progresses, small pits form, and the wood tends to separate in layers along annual rings, eventually producing sheets of yellowish, pitted wood—hence the name laminated root rot (Fig. 13.11). Resupinate basidio-carps, which are not always present, have a brown pore surface surrounded by sterile white margins. They may form on the underside of fallen trees, on uprooted stumps, and sometimes on the boles of dead, standing trees. They tend to be rare.

13.3.4 Cause

Phellinus weirii (Murrill) R. L. Gilbertson causes this disease, which is often referred to as yellow laminated root disease or formerly yellow ring rot.

13.3.5 Biology and Epidemiology

Most infections result from mycelium in roots and stumps already present in the area, having survived from the previous generation of trees. This fungus can survive for at

FIGURE 13.10 Cross section of base of Douglas-fir showing incipient and advanced decay caused by *Phellinus weirii*. From Childs and Nelson (1971).

FIGURE 13.11 Stump of western redcedar showing advanced decay of the heartwood caused by *Phellinus weirii*. From Boyce (1961), reproduced with permission of McGraw-Hill, Inc.

least 50 years in old Douglas-fir stumps, and perhaps much longer. Host tissues supporting inoculum must be large enough to allow the fungus to survive and remain infective until host roots of the next rotation trees contact it. Because most spread is by root contact, the diseased trees usually occur in patches or infection centers. In one stand, for example, decay caused by *P. weirii* was visible on the surfaces of 62% of the stumps within 8 m of killed trees, 20% of the stumps at 8–13 m, and on 4% of the stumps beyond 13 m. In forest soils *P. weirii* always lives in host wood and never grows more than a few centimeters from infected wood. The fungus, however, can easily grow from infected root residues to healthy roots that grow into contact with infected wood and apparently can penetrate roots directly within 12 months.

Although it may be assumed that new infection centers are the result of wind-blown basidiospores, it has not been possible to infect Douglas-fir stumps with spores. Inoculation with mycelium has resulted in penetration exceeding 30 cm in 1 year in the sapwood.

The fungus is restricted to the heartwood region of resistant trees. These trees replace roots killed by the fungus with adventitious roots. Susceptible trees may be killed within 3 years; no adventitious roots are formed, and the fungus is not stopped short of the root collar. Resistant trees can serve as sources of inoculum.

13.3.6   Control Strategy

Laminated root disease is probably more severe now than it was formerly, and it is likely to increase in importance as successive rotations are produced on infested sites. In unmanaged, old-growth stands, the larger infected trees would windthrow. In falling, the stump and some larger support roots would be largely wrenched from the soil. This removes much inoculum from the soil and opens up remaining pieces to invasion by antagonistic competitors. *Phellinus weirii* would decay the smaller inoculum remaining relatively quickly; this infected root material would serve as inoculum for a shorter time. Thus, before host species reoccupied the site, much inoculum was removed or rendered ineffective.

Successful management depends on knowing where the disease is located. Although laminated root disease is widely distributed in many of the Douglas-fir stands in Oregon and Washington, not every hectare or stand is affected. Specific management strategies involve stump removal, high-nitrogen fertilization, chemical agents, biological agents, or species manipulation.

Infection centers should be located during harvesting and when fresh stumps are available. The diseased stumps should be extracted or pushed out. Bulldozers with solid blades tend to move too much soil and leave large holes. More recent work with brush blades or log forks shows that less soil is mixed and smaller holes are left. With both of these, much of the soil adhering to roots falls back into the hole. A vibrating stump puller is a new concept that does even better at separating soil from roots as the stump is pulled free. For a level site with a high site index, stump removal is presently probably the most feasible control strategy. It is, however, expensive.

A possible alternative to stump removal is the use of fumigants such as chloropicrin, allyl alcohol, Vapam, or Vorlex. The most effective dosages and proper application techniques have not yet been developed, and it has not yet proven to be a cost-effective treatment. In addition to treating stumps with fumigants to remove *P. weirii*, a second approach has been to use fumigants to treat living diseased trees.

Before the infested site is reforested, it may be heavily fertilized with nitrogen. Field tests with application of urea have reduced survival of *P. weirii*. Reasons for the reduced survival are unknown, but it is suspected that nitrogen causes populations of the antagonistic organism *Trichoderma* spp. to increase. Studies are underway to determine if nitrogen applied to infested stumps will encourage soilborne organisms to invade and displace *P. weirii*. Initial studies indicate that competing microorganisms such as *Streptomyces* spp. may act against *P. weirii*. Also, a grain-positive bacterium similar to *Bacillus cereus* inhibits *P. weirii*. Biological control agents such as *Trichoderma viride* can be introduced into stumps, thereby reducing inoculum potential. These agents are not presently available for use by foresters.

Burning is not likely to reduce infection. Root raking to a depth of 60 cm will eliminate or drastically reduce the amount of root disease in the next generation of trees. Burning is not necessary except to clear the area of refuse so that all the area can be planted.

If the preceding treatments are not feasible, then a more resistant species should be planted in an area, extending 30 m beyond known diseased trees. Douglas-fir should not be replanted within 30 m of an infested stump. The purpose of planting resistant tree species on *P. weirii*-infested sites is to reduce inoculum to acceptable limits. Planting alder results in shorter survival time for *P. weirii,* adds fertility, and results in production of phenolic acids inhibitory to *P. weirii.* Evidence to date suggests that two rotations of alder, 60–80 total years, are required before replanting Douglas-fir. The influence of red alder as a biological control agent is based on several bits of evidence that have been accumulated: (1) from paired plot surveys, the incidence of *P. weirii* infection was 6.3% in pure Douglas-fir but only 1.2% in mixed alder-fir and then only where Douglas-fir occurred in groups without alder nearby; (2) alder is resistant to *Phellinus* as a function of root phenols that, upon release, may also reduce the pathogen's longevity in buried inoculum; (3) in addition, red alder roots are typically nodulated (Actinomycetales) and through N-fixation maintain high nitrate levels that result in a differential increase in populations of organisms that actively compete with, inhibit, or parasitize *P. weirii,* which in turn cannot use nitrate nitrogen; (4) under these conditions, zone line formation and associated persistence of the fungus are reduced through the net effects of microbial suppression. The hypothetical potential of these phenomena is a decrease in viable *Phellinus* in the soil.

Resistant Douglas-fir has not been found. Western red cedar is resistant and, if feasible in other respects, could be planted in place of Douglas-fir. In young stands, symptoms may not be evident. Spacing should not be attempted if small centers of dead and dying trees are common (10 or more per hectare). Delay thinning in suspect stands until symptoms are evident.

LITERATURE CITED

Arnold, R. 1981. Nothing can stump this pulling machine. *West. Cons. J.* 38:38–42.

Bloomberg, W. J., P. M. Cumberbirch, and G. W. Wallis. 1980. *A Ground Survey Method for Estimating Loss Caused by Phellinus weirii Root Rot.* II. Can. For. Serv., Pac. For. Res. Cent. Inf. Rep. BC-R-4. 44 pp.

Bloomberg, W. J. 1983a. *A Ground Survey Method for Estimating Loss Caused by Phellinus weirii Root Rot.* III. Can. For. Serv., Pac. For. Res. Cent. Inf. Rep. BC-R-7. 25 pp.

Bloomberg, W. J. 1983b. *A Ground Survey Method for Estimating Loss Caused by Phellinus weirii Root Rot.* IV. Can. For. Serv., Pac. For. Res. Cent. Inf. Rep. BC-R-8. 16 pp.

Boyce, J. S. 1961. *Forest Pathology,* 3rd ed. McGraw-Hill, New York. 572 pp.

Childs, T. W. 1963. *Poria weirii* root rot. *Phytopathology* 53:1124–1127.

Childs, T. W., and E. E. Nelson. 1971. *Laminated Root Rot of Douglas-Fir.* USDA For. Serv., For. Pest Leaflet 48. 7 pp.

Childs, T. W., and K. R. Shea. 1967. *Annual Losses from Disease in Pacific Northwest Forests.* USDA For. Serv., Res. Bull. PNW 20. 19 pp.

Hutchins, S. H., and C. Li. 1981. Relative capacities of filamentous and non-filamentous bacteria from two forest soils to inhibit *Phellinus weirii* in culture. *Northwest Sci.* 55:219–224.

Nelson, E. E. 1971. *Invasion of Freshly Cut Douglas-Fir Stumps by Poria weirii.* USDA For. Serv., Res. Note PNW-144. 5 pp.

Nelson, E. E. 1975. Survival of *Poria weirii* in wood buried in urea-amended forest soil. *Phytopathology* 65:501–502.

Nelson, E. E. 1976. Effect of urea on *Poria weirii* and soil microbes in an artificial system. *Soil Biol. Biochem.* 8:51–53.

Nelson, E. E., and T. H. Hartman. 1975. Estimating spread of *Poria weirii* in a high-elevation, mixed conifer stand. *J. For.* 73:141–142.

Nelson, E. E., and W. G. Thies. 1981. Chemical and biological means of reducing laminated root rot inoculum, pp. 71–73. In *Proc. 29th Annual West Int. For. Dis. Work Conf.*, Vernon, B. C. 146 pp.

Nelson, E. E., N. E. Martin, and R. E. Williams. 1981. *Laminated Root Rot of Western Conifers.* USDA For. Serv., For. Insect and Disease Leafl. 159. 6 pp.

Rose, S. L., C-Y. Li, and A. S. Hutchins. 1980. A streptomycete antagonist to *Phellinus weirii*, *Fomes annosus* and *Phytophthora cinnamomi*. *Can. J. Microbiol.* 26:583–587.

Roth, L. F., L. Rolph, and S. Cooley. 1980. Identifying infected ponderosa pine stumps to reduce cost of controlling *Armillaria* root rot. *J. For.* 78:145–152.

Thies, W. G. 1984. Laminated root rot—The quest for control. *J. For.* 82:345–356.

Thies, W. G., and E. E. Nelson. 1982. Control of *Phellinus weirii* in Douglas-fir stumps by the fumigants chloropicrin, allyl alcohol, Vapam or Vorlex. *Can. J. For. Res.* 12:528–532.

Wallis, G. W. 1976. *Phellinus (Poria) weirii Root Rot. Detection and Management Proposals in Douglas-Fir Stands.* Can. For. Serv., Tech. Rept. 12. 16 pp.

Wallis, G. W., and G. Reynolds. 1962. Inoculation of Douglas-fir roots with *Poria weirii*. *Can. J. Bot.* 40:637–645.

DISEASE PROFILE

Armillaria Root Disease

Importance: First described by Hartig in 1873, Armillaria root disease is now known to exist as a disease of opportunity on a worldwide scale and wherever woody vegetation is common.

Suscepts: Some 600+ species of woody plants representing 280 genera support this disease. In addition, the fungus has been reported on strawberries, potatoes, cactus, dahlia, and other nonwoody hosts and is essentially regarded as omnivorous on woody perennials and in particular angiosperms.

Distribution: This disease is found in both temperate zones and the tropics.

Causal Agents: *Armillaria* spp. (Agaricales) are pathogens causing Armillaria root disease. The genus *Armillaria* comprises numerous species with varying and distinct pathogenicities. The better-known species with a circumboreal distribution include: *A. mellea*, *A. gallica*, and *A. ostoyae*. *A. mellea* and *A. ostoyae* have high pathogenicity.

Sporocarps of *Armillaria*.

Biology: The fungus sporulates sexually, generally in the fall, with the production of clustered, honey-colored, annulated mushrooms. Although colonization of stump surfaces by windborne basidiospores must account for a great portion of its saprophytic establishment, there is no experimental evidence to support this. Local spread from either stumps or dead trees is facilitated by radial extension of subterranean rhizomorphs to maxima of about 18 m. Rhizomorph biology has been intensively studied; however, the infection process seems limited by poorly understood host predisposition factors and variance in pathogenicity of species and clonal sources.

Epidemiology: The aerial phase of establishment via windborne transport of basidiospores in colonization of wounds and/or woody substrates such as stumps is strongly implied. Once decay of underground substrates is established, the fungus progresses outward to healthy roots of adjacent trees where it may infect by way of contact or by the production and growth of rhizomorphs.

Diagnosis: Affected trees show crown decline and growth reduction as correlated with the extent of root decay. Excavation of the root collar will reveal decayed roots with rhizomorphs. Debarking of the root collar at the juncture of root decay will expose white mycelial fans. Resin exuding from diseased areas often causes soil to adhere to the root. After the tree dies and the bark loosens a bit, subcortical rhizomorphs develop as a rather extensive and interconnected netlike growth. Clustered mushrooms develop at the base of and from roots on infected trees and stumps, but they are present only as diagnostic aids for a few weeks in the fall.

Control Strategy: Management objectives determine the control strategy. Scattered infections may actually improve stand quality. Small openings may improve forage for wildlife. Conversely, even a few diseased trees in a recreation area may be unacceptable. Stump removal or root raking has been used in western forests. On high-risk sites, consideration should be given to replanting with less-susceptible species. Maintain stands in a vigorous condition. Simulation using the Western Root Disease Model can identify management schemes that minimize disease impact.

SELECTED REFERENCES

Shaw, C. G., III, and G. A. Kile. 1991. *Armillaria Root Disease*. USDA For. Serv., Agric. Handbook No. 691. 233 pp.

13.4 ARMILLARIA ROOT DISEASE

13.4.1 Importance

Armillaria spp. are among the more common fungi in our forests (Fig. 13.12). The root disease caused by some species of *Armillaria* has been one of the most prominent killers of deciduous and coniferous trees in natural forest stands and plantations. In North America there is a major contrast between pathogenic relationships. In eastern deciduous forests *Armillaria* is predominantly a secondary pathogen on stressed trees whereas in western coniferous forests the fungus is often an aggressive pathogen (Fig. 13.13). Viruslike particles have been extracted from haploid and diploid isolates of *A. ostoyae* but were absent in isolates from hardwood species.

Attacks of *Armillaria* are not often of an epidemic nature, although during and after periods of drought, or other stress, they occasionally may appear to be so, due to the predisposition of the trees because of insufficient moisture. Trees of all ages, from seedlings to overmature individuals, are attacked, but the damage is likely to be more severe on older or weakened trees. In many instances, invasion is of a secondary nature, but apparently when trees are moved outside their normal geographic range, *Armillaria* can act as a primary parasite. This is particularly true of fruit trees in California, some of which were originally from the Mediterranean region.

FIGURE 13.12   Sporocarps of *Armillaria mellea*.

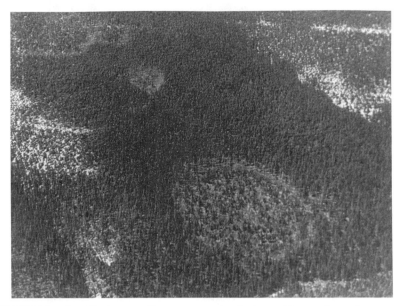

FIGURE 13.13 Armillaria root disease center in virgin coniferous forest in western North America. The lowermost center covers about 8 ha. From Shaw and Kile (1991).

When red pine, and probably other pine species, are planted in unsuitable sites, such as in light soils in areas where there is a high rate of evaporation and insufficient rainfall, they will be affected by *Armillaria* root disease (Fig. 13.14). Also, when pine are established on sites where hardwoods such as oak had been growing, the large amount of inoculum present on the root systems of the original trees can result in infection and mortality in the young pine. In Minnesota, losses in red pine have been as high as 45%. In Alberta, Canada, losses in lodgepole pine were as high as 15%.

13.4.2 Suscepts

Armillaria root disease has been found on hundreds of woody plant species; most forest, shade, and fruit trees are attacked by it. In addition to trees and shrubs, *Armillaria* spp. have been reported on strawberries, potatoes, cacti, and other plant species, but so far as is known, it seldom causes much damage on any but its tree hosts.

13.4.3 Distribution

An account of the fungus and of the disease caused by it was published by Robert Hartig in 1873, and it has been studied since that time by many others in many parts of the world. Prior to the 1970s, *Armillaria mellea* was viewed as a single variable or polymorphic species with an extremely wide distribution in both temperate and tropical regions. It was subsequently determined that *Armillaria* contains several intersterile groups known as biological species. The genus now contains about 55 species, of which several have restricted geographical distribution or host associations.

FIGURE 13.14   Young red pine planted on former scrub hardwood site. Roots of declined tree on right have contacted oak roots colonized by *Armillaria*, which has subsequently invaded and killed the pine roots and has recently girdled the root crown. Courtesy of D. W. French, University of Minnesota, St. Paul.

13.4.4   Cause

Armillaria root disease is caused by species of *Armillaria* (Basidiomycotina). Based on their nucleic acid composition, Anderson and Ulbrich (1979) and others have placed some European or North American taxa into distinct biological species (BS): rDNA BS 1, *A. ostoyae;* BS 2, *A. gemina;* BS 3, *A. borealis;* BS 4, *A. sinapina;* BS 5, *A. calvescens, A. gallica, A. cepistipes;* BS 6, *A. mellea.* All biological species except for BS 4, were considered to be natural groupings, with BS 2 and 3 derived from the more widely distributed BS 1.

13.4.5   Biology and Epidemiology

In eastern deciduous forests, *Armillaria* acts as a secondary pathogen. It attacks the root systems of trees, and the success of the fungus depends in large part on the condition of the host as well as the amount and kind of inoculum. For the most part, *Armillaria* probably grows as a saprophyte and is beneficial as a digester of downed and dead timber. It can also cause heart decay, a white decay, and not necessarily kill the infected trees. Initial infection can result from basidiospores, which are wind-disseminated, but undoubtedly the fungus is so well distributed that it already exists in many stands in the form of mycelium or as rhizomorphs. Shigo and Tippett (1981) observed that *Armillaria* spread into dying sapwood beneath and beyond the area of killed cambium of several northeastern conifer and hardwood species, but it did not spread radially outward into new wood that formed.

Trees that have been weakened by such stresses as insect defoliation, drought, or air pollution may become colonized by rhizomorphs growing from roots of dead colonized trees or by mycelium from quiescent lesions (Fig. 13.15). The fungus can spread between trees as rhizomorphs and mycelium grows between roots in close contact. Infection of young trees is also probably aided by some form of predisposition such as drought. In Minnesota *Armillaria* infected trees that were girdled but did not infect any comparable trees that were not girdled. The tops of the girdled trees appeared healthy, even after *Armillaria* had become established in their root systems. It seemed that the fungus moved into these trees after they had started to die. If stress is abated and tree vigor is restored, colonization does not continue. Aggressive root diseases form infection centers by progressive colonization and mortality of adjacent trees; in eastern forests *Armillaria* does not form such centers.

In the southeastern United States, *Armillaria* seems to be prevalent primarily as a colonizer of recently dead trees, both deciduous and coniferous, and does not parasitize living trees regardless of their state of vigor. It is frequently found, however, in decaying dead roots of declining oaks in urban settings. These trees frequently

FIGURE 13.15   Decline of white oak, resulting from stress increased when the site was converted from a woodland to a housing development and subsequent invasion by *Armillaria mellea*.

FIGURE 13.16 Decline and windthrown white oak shade tree. Its root system was extensively decayed by *Armillaria*.

windthrow (Fig. 13.16). The role that *Armillaria* may have played in their premature death is not known. At any rate, the behavior of the fungus in eastern forests suggests that it depends heavily on host stress for fulfilling its pathogenic role.

In conifer forests of western North American, *Armillaria* occurs commonly as a butt decayer in old trees and a decayer of dead and downed trees in coastal forests. In the drier interior region, however, *Armillaria* can attack, colonize, and kill apparently healthy trees of all ages (Fig. 13.17). This produces enlarging infection centers that can cover several hectares. Infection centers are frequently associated with harvesting operations but are also common in unmanaged virgin forests. Many former pine stands have been converted to more disease-susceptible spruce and fir species as the result of past cutting practices and an 85–year-old policy of fire suppression.

In Japan, in larch plantations, *Armillaria* occurred in patches and sporadically throughout the area. The disease became evident 1 year after planting, reached a maximum level in 3 or 4 years, and then ceased. It has occurred in one plantation 30 years old. In Minnesota, maximum development occurred 4 years after the plantation was established with some additional mortality in subsequent years. How important predisposition is has not been established. This phenomenon of a period of several years after a harvest or thinning before the appearance of *Armillaria*, followed by a period of increasing disease incidence, and then a period of declining disease incidence may be due to inoculum dynamics. *Armillaria* species are aggressive saprophytes. When trees are cut, *Armillaria* rapidly colonizes the stumps of infected trees. From these stumps, the fungus sends out rhizomorphs, which colonize other dead material. If the rhizomorphs contact a healthy root and the fungus has enough energy available in its "food base," it can overcome the host resistance and infect the

FIGURE 13.17   Armillaria infection center in pole-sized ponderosa pine, showing disease progression through the stand. From Shaw and Kile (1991).

tree. Host stress weakens the tree's resistance response, making it easier for *Armillaria* to infect the tree. The cyclical nature of disease incidence may be due to the quality of the food base. Initially, *Armillaria* must colonize stumps, which may take several years. Then, with the stump serving as a high-quality food base, *Armillaria* can infect adjacent trees. As the stump is decayed, the food base declines in quality, and fewer trees become infected. In plantations, the killed trees have much smaller stumps than did the previous generation of larger trees, and these stumps do not serve as an effective foodbase. Thus, the incidence of *Armillaria* usually peaks within 10 years, and the disease is of minor importance after that time.

Defoliation and other stresses can result in changes within the host, which allow secondary organisms such as *Armillaria* to invade and kill that tree. Increased levels of glucose favor this fungus, and glucose levels increase in roots of defoliated trees. Increased levels of glucose enable *Armillaria* to grow in the presence of phenolic compounds that otherwise would inhibit the fungus. Gallic acid is inhibitory to *Armillaria,* but inhibition is reduced when more glucose is available. Defoliation also results in increases in nitrogen, thus favoring *Armillaria*. Defoliation can alter the activity of bark enzymes, which serve as a defensive mechanism. In Wisconsin plantations, from 18–34% of the red and white pines artificially inoculated became infected; 45% of the potted trees became infected. Apparently, rhizomorphs are the means by which the fungus can invade its host, and they must form and come in contact with roots for infection to result. Rhizomorph development varies with isolates and apparently is stimulated by compounds such as ethanol, extracts of red alder, and other microorganisms including *Aureobasidium pullulans*. These compounds enable *Armillaria* to grow in the presence of inhibitory substances such as phenolics. In healthy tissues, fungal enzymes cannot oxidize phenolics, and the fungus is confined to wounded and necrotic tissues. In trees that are stressed, oxidative enzymes of the fungus can injure adjacent host cells, allowing colonization to proceed. Different isolates vary greatly in their ability to oxidize phenolics. There is little if any correlation between virulence and ability of strains to produce rhizomorphs on agar. There are 10 biological species of what was *Armillaria mellea* in North America. *Armillaria ostoyae* is pathogenic on conifers, and *A. mellea* is pathogenic mostly on hardwoods. Others are found primarily on stressed trees or dead wood.

Diagnosis

The fungus forms mats of mycelium called mycelial fans in the inner bark and between the bark and sapwood of the infected roots, and mycelium may extend to a height of several meters in the phloem and cambium of the trunk (Fig. 13.18). These fans of mycelium, which glow in the dark, are an unmistakable sign of the fungus. The mycelium does not grow as abundantly under the bark of some hardwoods as in pines, but often in hardwoods a network of rhizomorphs is found between the wood and the bark (Fig. 13.19). These rhizomorphs are reddish-brown to black and may be somewhat round or flattened, depending on whether they are growing beneath the bark or arc outside of roots. Rhizomorphs may also form in culture under certain conditions (Fig. 13.20). Resin is exuded by the infected roots of the pines and some other conifers and solidifies in the soil, causing masses of soil to adhere to the roots (Fig. 13.21). The sporocarps are mushrooms, which have a central stem and ring (which sometimes disappears quickly) and white spores (normally very evident on the caps of the mushrooms and surrounding vegetation) (Fig. 13.12). The cap and stem

FIGURE 13.18 Mycelial fan of *Armillaria* at base of red pine with bark and soil removed. Courtesy of P. Wargo, USDA Forest Service.

FIGURE 13.19   Rhizomorphs of *Armillaria* on a root of a windthrown black walnut.

usually are honey colored, although the color varies considerably. In the northern United States, sporocarps are produced in the fall in clumps around the stumps of infected trees and above decayed roots; 500 mushrooms of this species were found in an area 3.7 m² around stumps in a hardwood stand in central Minnesota, which indicates the prolific fruiting of the fungus.

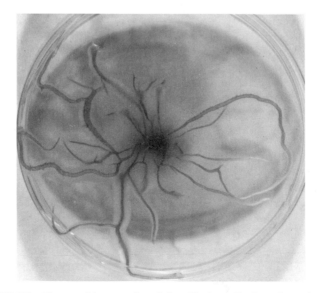

FIGURE 13.20   Young rhizomorphs of *Armillaria mellea* in culture, bottom view.

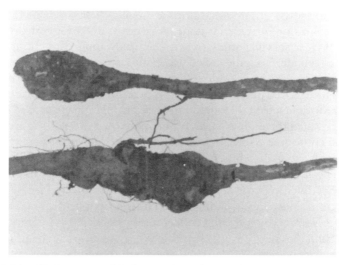

FIGURE 13.21   Pitch-encrusted lesions on lateral root of ponderosa pine infected by *Armillaria*. From Shaw (1980).

The crown of an affected tree may die either gradually, one limb at a time, or rather suddenly, depending on the extent of injury to the roots and the abundance of water. If there is sufficient water, the crown of a tree may remain green and apparently alive for a year or two after practically all the roots have been killed by the fungus, but it is likely to die suddenly when the water supply becomes insufficient. This is the reason that periods of drought accentuate the damage done by the fungus. Infected ponderosa pine have rounded crowns. Bark beetles frequently attack coniferous trees after they have been weakened by root disease, and the death of the trees may be ascribed to the beetles, whose presence often is obvious, rather than to the fungus, the presence of which is not always easy to detect. Mature lodgepole pine infected with *Armillaria* were infested by endemic population levels of the mountain pine beetle with greater frequency than uninfected trees. Naturally, it sometimes is very difficult to find the exact cause of death of a tree that has been attacked by root-decaying fungi or bark beetles and subjected to drought or other unfavorable growing conditions. Young pines affected by *Armillaria* turn color to an off-green and stop growing, and the entire tree seems to die uniformly. Reduced leaders and poor color indicate the presence of root disease.

Control Strategy

Control strategies vary with region and amount of loss caused by *Armillaria* root disease. The cost of control must be balanced by the value of the crop saved. Because of the low crop value, low-cost controls through silvicultural modifications should be given first priority. In western states, the best time to deal with *Armillaria* root disease is at the time of harvest. Diseased trees need to be identified. Older stumps are more likely to become infected than recent stumps. Black-and-white enlargements of color positive aerial photos have been used to record locations of diseased stumps and were found to be superior to freehand sketches or tracings of projected images of color photos. Resolution was accurate for stumps as small as 9 cm in diameter. Small dead trees near the stumps or in the root zone of larger stumps suggest presence of the fungus. In a 10–year survival study, seven chemicals applied to the root collars of small-diametered ponderosa pine failed to reduce mortality caused by *Armillaria*. Slash burning will not reduce root decay. In a high-risk site, probably the most feasible approach is to remove (push) infected stumps and with a brush blade or root rake disrupt roots and stumps from a zone 20 m wide around infected trees. If root decay is generally distributed through the stand (i.e., more than 20% of the area), it may be best to rake the entire area to a depth of 60 cm. Removed stumps are not hazardous to residual trees. Exercise care in selecting a site for a new plantation. If the site is of high risk, expensive site preparation costs are probable. During thinning, diseased trees should be removed. Those trees adjacent to known infected trees also probably should be removed. However, thinning and fertilization may predispose Douglas-fir to infection by *Armillaria* by lowering concentrations of defensive compounds in root bark and increasing the energy available to the fungus to degrade them. Less susceptible species such as ponderosa pine and western larch should be favored. Most susceptible species are the true firs, Douglas-fir, and western hemlock.

In the northcentral states, *Armillaria* root disease can cause substantial losses in red pine planted on sites originally occupied by oaks and to a lesser degree other species such as aspen. This might be due to stumps of these species serving as excellent food bases for *Armillaria*. Species other than red pine should be considered for planting on such sites. Newly planted trees seem to fare well after planting and then die. When the pine reach 2 m in height, no further mortality occurs, apparently because the source of inoculum in the decaying oak roots has been exhausted. On poor sites that are marginal for red pine, it is advisable to plant a species more tolerant of the site such as jack pine.

REFERENCES

Anderson, J. B., and R. C. Ulbrich. 1979. Biological species of *Armillaria mellea* in North America. *Mycologia* 71:402–414.

Arno, S. F. 1980. Forest fire history in the northern Rockies. *J. For.* 78:460–465.

Baranyay, J. A., and G. R. Stevenson. 1964. Mortality caused by Armillaria root rot, Peridermium rusts, and other destructive agents in lodgepole pine regeneration. *For. Chron.* 40:350–361.

Christensen, C. M., and A. C. Hodson. 1954. Artificially induced senescence of forest trees. *J. For.* 52:126–129.

Entry, J. A., K. Cromack, Jr., R. G. Kelsey, and N. E. Martin. 1991. Response of Douglas-fir to infection by *Armillaria ostoyae* after thinning or thinning plus fertilization. *Phytopathology* 81:682–689.

Filip, G. M., and D. J. Goheen. 1984. Root diseases cause severe mortality in white and grand fir stands of the Pacific Northwest. *For. Sci.* 30:138–142.

Filip, G. M., and L. F. Roth. 1987. *Seven Chemicals Fail to Protect Ponderosa Pine from Armillaria Root Disease in Central Washington.* USDA For. Serv., Res. Note PNW-RN-460. 8 pp.

Martin, N. E., and R. E. Williams. 1986. *Using Aerial Photos to Fingerprint a Stand for Root Disease Research.* USDA For. Serv., Res. Note INT-360. 3 pp.

Morrison, D. J. 1981. *Armillaria Root Disease. A Guide to Disease Diagnosis, Development and Management in British Columbia.* Can. For. Serv., Pac. For. Res. Cent., Victoria, B.C. 15 pp.

Ono, K. 1965. *Armillaria Root Rot in Plantations of Hokkaido.* Bull. of the Gov't Forest Exp. Sta. No. 179. 62 pp.

Patton, R. F., and A. J. Riker. 1959. Artificial inoculations of pine and spruce trees with *Armillaria mellea. Phytopathology* 49:615–622.

Pentland, B. D. 1965. Stimulation of rhizomorph development of *Armillaria mellea* by *Aureobasidium pullulans* in artificial culture. *Can. J. Micro.* 11:345–350.

Raabe, R. D. 1967. Variation in pathogenicity and virulence in *Armillaria mellea. Phytopathology* 57:73–75.

Reaves, J. L., T. C. Allen, C. G. Shaw, III, W. V. Dashek, and J. E. Mayfield. 1988. Occurrence of viruslike particles in isolates of *Armillaria. J. Ultra. Molec. Struct. Res.* 98:217–221.

Rehill, P. S. 1968. Stimulation of *Armillaria mellea* rhizomorphs with alder extracts. *Can. Dep. Forest. Bi-mo. Res. Notes* 24:34.

Roth, L. F., L. Rolph, and S. Cooley. 1980. Identifying infected ponderosa pine stumps to reduce cost of controlling *Armillaria* root rot. *J. For.* 78:145–148, 151.

Shaw, C. G. 1980. Characteristics of *Armillaria mellea* on pine root systems in expanding centers of root rot. *Northwest Sci.* 54:137–145.

Shaw, C. G., and G. A. Kile. 1991. *Armillaria Root Disease.* USDA For. Serv., Agric. Handbook No. 691. 233 pp.

Shigo, A. L., and J. T. Tippett. 1981. *Compartmentalization of Decay Wood Associated with Armillaria mellea in Several Tree Species.* USDA For. Serv., Res. Pap. NE-488. 20 pp.

Snider, P. J. 1959. Stages of development in rhizomorphic thalli of *Armillaria mellea. Mycologia* 51:693–707.

Tkacz, B. M., and R. F. Schmitz. 1986. *Association of an Endemic Mountain Pine Beetle Population with Lodgepole Pine Infected with Armillaria Root Disease in Utah.* USDA For. Serv., Res. Note INT-353. 7 pp.

Wargo, P. M. 1981. Defoliation and secondary-action organism attack: With emphasis on *Armillaria mellea. J. Arboric.* 7:64–69.

Wargo, P. M., and C. G. Shaw. 1985. Armillaria root rot: The puzzle is being solved. *Pl. Dis.* 69:826–832.

Weinhold, A. R. 1963. Rhizomorph production by *Armillaria mellea* induced by ethanol and related compounds. *Science* 142:1065–1066.

DISEASE PROFILE

Clitocybe Root Disease

Importance: Regarded as the southern counterpart of some *Armillaria* diseases, Clitocybe root disease is destructive on fruit, forest, shade, and ornamental trees in the southeastern United States. In the absence of fruiting bodies or cultural studies, the symptoms produced by Clitocybe are so similar to these produced by *Armillaria* spp. that the two diseases may be readily confused. Known since 1901, the disease had devastating effects in orchards of apple, cherry, and peach and later of grapevines. In Florida and Georgia, Clitocybe root disease has caused up to 25% mortality in some plantations of sand pine.

Suscepts: The host list comprises 213 species of plants belonging to 137 genera and 59 families. Of these, the Casuarinaceae, Rosaceae, Leguminosae, Rutaceae, and Myrtaceae are the families most frequently affected. Conifers as well as broad-leaved or hardwood trees are attacked. Many exotic tree species that have been planted in Florida have been affected.

Distribution: Although fruiting bodies have been collected as far north as Michigan and New York, damage has been recorded only from eastern Texas, Oklahoma, Missouri, southern Illinois and Indiana, West Virginia, and Virginia southward to Florida and in southern Europe.

Causal Agents: Armillaria (Clitocybe) tabescens (Agaricales) is the source of Clitocybe root disease.

Biology: The role of basidiospores is unknown. Rhizomorphs are produced in culture but not in nature. Basal lesions may coalesce to girdle the tree and cause stem swelling above

Mycelial mats. From Rhoades (1956).

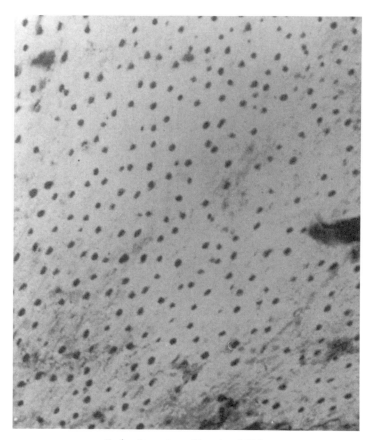

Perforations. From Rhoades (1956).

the infected area. Sand pines may produce resin exudations around diseased roots. Girdling mats of white or cream-colored mycelium form beneath the bark. Once the root collar has been invaded, wilting and death follow quickly. The fungus is favored by higher temperatures. In autumn, clusters of mushrooms develop at the base of diseased trees.

Epidemiology: Armillaria tabescens persists for many years in dead roots. New host roots growing in close proximity are subject to attack. It is considered to be an opportunistic fungus that depends on weakened hosts for infection. In some highly susceptible species, however, it appears to be an aggressive killer.

Diagnosis: The clusters of persistent mushrooms that form at the base of diseased trees are an important indicator of infection. They resemble those of *Armillaria* species but lack an annular ring. Mushrooms may occur infrequently on dry sites or in dry years. Visible elliptical perforations are sometimes present on the mycelial mats that form in the cambial region. These are diagnostic on sand pine. Diseased roots also may be covered with a crusty mass of hardened resin and soil particles and may display crisscrossing black ridges of fungal tissue. Resin soaking of the wood beneath the sunken canker at ground line is typical.

Control Strategy: Avoid wounding when transplanting. Do not plant susceptible species in

Ridges. From Rhoades (1956).

areas where infected stumps or roots are known to occur. Partially infected plants may be saved by exposing the diseased root collar area to drying for a time. In Florida, infections tend to be more prevalent in soils that are acidic and droughty. The Ocala race of sand pine apparently is resistant.

SELECTED REFERENCES

Rhoads, A. S. 1956. The occurrence and destructiveness of Clitocybe root rot of woody plants in Florida. *Lloydia* 19:193–240.
Ross, E. W. 1970. Sand pine root rot—Pathogen: *Clitocybe tabescens. J. For.* 68:156–158.

DISEASE PROFILE

Tomentosus Root Disease

Importance: An older name for this disease is red root and butt rot. In Canada it is known as stand opening disease. In older stands of spruce, dying and dead trees occur in patches that increase slowly in size 20–30 years after initial attack. In young plantations, seedlings are killed if planted near infected roots of stumps from the preceeding stand. Some southern pine species may be attacked if planted off-site.

Suscepts: Many conifers, especially white, blue, and Engelmann spruces and red, western white, and slash pines.

Distribution: Considered to be a northern temperate disease, its success in southern U. S. forests may be reduced somewhat by more rapid decay and, consequently, shorter survival in infected roots.

Causal Agents: Tomentosus root disease is caused by *Inonotus (Polyporus) tomentosus* (Aphyllophorales) and its variety *circinatus*.

Biology: Small, stalked basidiocarps are produced in summer, usually growing from decayed roots. Infection can be by basidiospores and also by mycelium. In slash pine, infection may occur through basal cankers caused by the fusiform rust fungus. In western white pine, infection is associated with fire scars. Declining roots of weakened trees may be predisposed to infection, in this case by mycelium. Mycelium can also enter healthy roots through wounds created by root weevils (*Hylobius* sp.). Survival of *I. tomentosus* in spruce roots for at least 20–30 years suggests that it has a good survival mechanism, especially in northern climes. Its longevity in southern pine roots may be much shorter.

Epidemiology: In existing stands, infections initiated by basidiospores are enhanced by wounds caused by poor logging practices or root weevils. Although the fungus can colonize sapwood, it grows slowly, only about 4 cm/yr. Nevertheless, it can withstand

Windthrown diseased trees. From Whitney (1988).

host defenses and competing microorganisms and can survive for at least 30 years after death of the tree. It is this trait that is responsible for losses when a new stand is established on the site. Young trees are infected following root contact.

Diagnosis: Declined crowns and premature death of trees in patches are suggestive of attack by *I. tomentosus*. The basidiocarps are distinctive. Advanced decay is of the white pocket type.

Control Strategy: If the previous stand was comprised of species that can harbor the pathogen for many years, regeneration with susceptible seedlings should be delayed. Unfortunately, 20–30 years or more may be an unacceptable delay. Stump removal might be an alternative. In these cases, immune species should be favored. With white spruce, high risk of disease is associated with relatively infertile soils with low water-holding capacity or underlain with rock or hardpan. Similarly, sand pine is at highest risk if a zone of reduced permeability lies close to the surface. Use of prescribed fire after clear-cutting has not reduced the incidence of decay in stumps.

SELECTED REFERENCES

Barnard, E. L., and W. N. Dixon. 1983. *Insects and Disease: Important Problems of Florida's Forest and Shade Tree Resources.* Fla. Dept. Agri. and Consumer Serv., Bull. No. 196–A. 120 pp.

Lewis, K., D. J. Morrison, and E. M. Hansen. 1992. Spread of *Inonotus tomentosus* from infection centres in spruce forests in British Columbia. *Can. J. For. Res.* 22:68–72.

Myren, D. T., and R. F. Patton. 1971. Establishment and spread of *Polyporus tomentosus* in pine and spruce plantations in Wisconsin. *Can. J. Bot.* 49:1033–1040.

13.5 TOMENTOSUS ROOT DISEASE

13.5.1 Importance

We know of tomentosus root disease as an important problem primarily of spruces. In Canada this disease causes mortality centers in white spruce stands and, hence, is also known as stand opening disease. Tomentosus root disease is a minor problem on many other North American conifers. Its true importance, however, may be masked by other more easily detected root diseases and by bark beetles, which attack trees weakened by tomentosus root disease.

In some situations, tomentosus root disease can causes serious losses. For example, Myren and Patton (1971) report that the area of ten mortality centers in a white spruce plantation increased by 14% annually. In white spruce stands 70–110 years old, root diseases accounted for volume losses averaging 28%, with an additional 12% of the volume in trees with "poor condition and frequently with heavy root rot."

13.5.2 Suscepts

The spruces are the most important hosts. Other hosts include the pines, larches, Douglas-fir, hemlocks, and western redcedar.

13.5.3 Distribution

The causal pathogen is found throughout North America, Europe, Russia, and India.

13.5.4 Cause

Tomentosus root disease is caused by the basidiomycete *Inonotus tomentosus* (Fr.:Fr.) S. Teng. This fungus was known as *Polyporus tomentosus*. A similar fungus was known as *P. circinatus* because it had curved setae in its spore-producing layer rather than straight setae.

13.5.5 Biology

Much of what is known of this disease comes from Canada, where tomentosus root disease was recognized as a serious problem in white spruce stands.

In the western United States, tomentosus root disease can lead to stand devastation. Small groups of trees may be killed and/or windthrown. These trees are then attacked by the spruce beetle *(Dendroctonus rufipennis),* which prefers to breed in weakened or dying trees. Using the substrate provided by the root-diseased trees, the spruce beetle population increases. When the spruce beetle population is large enough, the beetles can attack and kill even vigorous trees and over several years eliminate the larger spruce in a stand. Such a situation is well documented on the Dixie National Forest in southern Utah. Here, spruce beetle has damaged the forests for many years, but only recently was the true role of *I. tomentosus* recognized.

13.5.6 Epidemiology

Infection can occur by means of basidiospores invading through deep wounds in roots and by spreading along root grafts and root contacts. Feeder roots may be a major entry court for *I. tomentosus.* There is no evidence that stump surfaces are colonized by basidiospores. Once the stand is cut, however, *I. tomentosus* can survive for at least 30 years in stumps of infected trees. When roots of regeneration contact this inoculum, they too become infected.

Basidiospores are produced on mushroomlike fruiting bodies with a central stalk (Fig. 13.22). Basidiocarps are produced in August or September in wet years but may be absent in dry years.

FIGURE 13.22   Basidiocarp of *Inonotus tomentosus.*

13.5.7 Diagnosis

Roots infected with *I. tomentosus* show a pink to brownish stain 12 months after infection (Fig. 13.23). Spread is most rapid in larger roots, but smaller roots can also become infected. Eighteen months after infection, a white pocket rot is evident (Fig. 13.24). With time, this decay may spread 1–2 m aboveground. Decay is often most severe in heartwood and thus may be quite difficult to detect by exposing roots. Although detection of the fungus is time consuming, the best way to find it is to extract two cores at right angles to each other from the base of the tree.

Trees may be infected with no outward symptoms for many years. Tomentosus root disease causes a characteristic "unthriftiness" of trees: leader growth is reduced and foliage is chlorotic. Trees are windthrown, often in patches or centers. Diseased trees may be attacked by bark beetles, which are often erroneously credited for the kill.

13.5.8 Control Strategy

Management of tomentosus root disease does not have a long history, so tried-and-true methods are lacking. Mortality centers should be identified, and their locations should be recorded on maps that are stored in a permanent file for future use. If exposing the stand to windthrow is not a concern, trees in mortality centers should be harvested, along with all merchantable trees within 9–15 m. Harvesting around the mortality center will not stop the progress of the fungus; rather, it is designed to salvage imminent mortality.

Where windthrow is a concern, light thinnings at 10- to 20-yr intervals will give

FIGURE 13.23 Spruce root infected with *Inonotus tomentosus,* showing discoloration of early decay.

FIGURE 13.24   White pocket decay caused by *Inonotus tomentosus*.

other trees an opportunity to grow and provide volume for removals without exposing the stand excessively. During these entries, salvage recently killed and symptomatic trees.

Where possible, consider converting the site, or at least the root disease centers and some area beyond, to more resistant or tolerant species such as pines. In southern Utah, trembling aspen colonize root disease centers naturally. Allowing aspen to occupy the site for one rotation should permit the inoculum to decay and the fungus to die.

Removing the large stumps that serve as inoculum for this fungus is being evaluated. Stump removal is done by a crawler tractor. This process is expensive (>$600/ ha), perturbs site and especially soil conditions, and thus may not be an option on some sites.

REFERENCES

Etheridge, D. E. 1956. Decay in subalpine spruce on the Rocky Mountain Forest Reserve in Alberta. *Can. J. Bot.* 34:805–816.

Hinds, T. E., F. G. Hawksworth, and R. W. Davidson. 1965. Beetle killed Engelmann spruce: Its deterioration in Colorado. *J. For.* 63:536–542

Lachance, D. 1978. The effect of decay on growth rate in a white spruce plantation. *For. Chron.* 54:20–23.

Lewis, K., D. J. Morrison, and E. M. Hansen. 1992. Spread of *Inonotus tomentosus* from infection centres in spruce forests in British Columbia. *Can. J. For. Res.* 22:68–72.

Mielke, J. L. 1950. Rate of deterioration of beetle-killed Engelmann spruce. *J. For.* 48:882–888.

Myren, D. T., and R. F. Patton. 1971. Establishment and spread of *Polyporus tomentosus* in pine and spruce plantations in Wisconsin. *Can. J. Bot.* 49:1033–1040.

Tkacz, B. M., and F. A. Baker. 1991. Survival of *Inonotus tomentosus* in spruce stumps after logging. *Pl. Dis.* 75:788–790.

Whitney, R. D. 1961. Root wounds and associated root rots of white spruce. *For. Chron.* 37:401–411.

Whitney, R. D. 1962. Studies in forest pathology: XXIV. *Polyporus tomentosus* Fr as a major factor in stand-opening disease of white spruce. *Can. J. Bot.* 40:1631–1658.

Whitney, R. D. 1963. Artificial infection of small spruce roots with *Polyporus tomentosus*. *Phytopathology* 55:441–443.

Whitney, R. D. 1966. Susceptibility of white spruce to *Polyporus tomentosus* in healthy and diseased stands. *Can. J. Bot.* 44:1711–1716.

Whitney, R. D. 1967. Comparative susceptibility of large and small spruce roots to *Polyporus tomentosus*. *Can. J. Bot.* 45:2227–2229.

Whitney, R. D. 1973. Root rot losses in upland spruce at Candle Lake, Saskatchewan. *For. Chron.* 49:176–179.

Whitney, R. D. 1988. *For Practical Use in the Field: A Forester's Guide to Identification and Reduction of Major Root Rots in Ontario.* Can. For. Serv., Great Lakes For. Center. 35 pp.

DISEASE PROFILE

Schweinitzii Root and Butt Disease

Importance: A major disease of older conifers, this disease, also known as red-brown butt rot, causes decay of the roots and lower stem. Resulting strength loss predisposes trees to windthrow and breakage.

Suscepts: In the United States, Douglas-fir, pines, and spruces are common hosts.

Distribution: It is widespread throughout Asia, Europe, and North America, attacking many conifers, but it is rare on hardwoods.

Causal Agents: *Phaeolus (Polyporus) schweinitzii* (Aphyllophorales) causes schweinitzii root and butt disease. The fungus is also known as the velvet top fungus, and is recognized by annual basidiocarps, which develop in late summer and fall on the butts of trees or grow up from a decayed root. On the tree, the basidiocarp resembles a thin bracket. On the

Sporocarp. Courtesy of C. M. Christensen, University of Minnesota, St. Paul.

ground, it is circular in form, sunken in the center, and with a short, thick stalk. The upper surface is velvety when fresh and is comprised of concentric zones of reddish-brown and a light yellow-brown margin. The pore surface is dirty green when fresh and turns red-brown when bruised. The pores are large and irregular in outline.

Biology: Infection occurs through mechanical butt wounds or fire scars and perhaps through root contact with diseased trees. Roots subjected to periodic flooding, drought, or soil compaction may be predisposed to infection. Roots previously invaded by *Armillaria* have also been suggested as possible avenues of infection.

Epidemiology: Considered to be mainly a disease of mature and senescing trees, *P. schweinitzii* may also infect roots or basal wounds of young trees. Its presence is first known when the tree breaks or is windthrown. Younger trees growing off-site seem to be more susceptible to infection.

Diagnosis: The presence of basidiocarps, often growing up from decayed roots, is a good indicator that a substantial amount of decay is likely present. Incipient decay appears as a light-yellowish to pale reddish-brown color. As the decay advances, the color intensifies, and the wood rapidly loses strength. In an advanced stage, the wood breaks up into large cubes and is easily crushed. Decayed wood has a pungent, turpentinelike odor. Decay is confined to the roots and lower 1–2 m of the butt log but may rarely extend much higher. In the old-growth forests of the Pacific Northwest, decay caused by *P. schweinitzii* was a major cause of cull, especially on Douglas-fir. Actual losses were greater, however, because decayed trees were predisposed to windthrow and more than just the decayed portion was lost.

Control Strategy: Harvesting of old-growth stands has eliminated much of the actual losses due to *P. schweinitzii*. Where old growth is to be maintained, ground fires hot enough to cause basal wounds should be avoided, as should movement of heavy equipment or livestock, which might cause soil compaction and root injury. In recreation areas, high-risk trees should be recognized and removed before they are windthrown. Younger stands should be suited for the site on which they are to be established. Eastern white pine and red pine are particularly susceptible when on shallow, poorly drained soils.

SELECTED REFERENCES

Barrett, D. K. 1985. Basıdiospores of *Phaeolus schweinitzii:* a source of soil infestation. *Eur. J. For. Path.* 15:417–425.

Barrett, D. K., and B. J. W. Greig. 1985. The occurrence of *Phaeolus schweinitzii* in the soils of Sitka spruce plantations with broadleaved and non-woodland histories. *Eur. J. For. Path.* 15:412–417.

Blakeslee, G. M., and S. W. Oak. 1980. Residual naval stores stumps as reservoirs of inoculum for infection of slash pines by *Phaeolus schweinitzii. Pl. Dis.* 64:167.

Sinclair, W. A., H. H. Lyon, and W. T. Johnson. 1987. *Diseases of Trees and Shrubs.* Cornell University Press, Ithaca, New York. 574 pp.

DISEASE PROFILE

Rhizina Root Disease

Importance: This root disease is caused by a cosmopolitan fungus that parasitises seedlings planted shortly after a fire. In 1880 the disease was discovered on maritime pine in France and there named *la maladie du ronde.* In years since, group dying of conifer regeneration of various ages on burned and usually cut sites has been reported. Known in North America since 1915, the disease has developed occasionally in nurseries, mostly in young plantations, and rarely in old-growth western hemlock in Washington and red spruce in Vermont.

Suscepts: The host range may well include most conifers.

Distribution: The causal fungus occurs across Canada, but only in British Columbia has it been associated with significant mortality of Douglas-fir. In the United States it is present

Sporocarps. From Gremmen (1958).

across the northern tier of states above the 38° latitude; but the disease has occurred only in Washington, Oregon, Idaho, Montana, Minnesota, New York, Vermont, and Maryland.

Causal Agents: Rhizina root diseases is caused by *Rhizina undulata* (= *R. inflata*) (Pezizales).

Biology: Ascospores are wind-dispersed in the fall from apothecia that develop from infected roots. Subsequent precipitation washes the spores into the soil where they lie dormant for as long as 2 years. Surface fire breaks dormancy and also influences the exudation of conifer roots, which also stimulates the spores. The roots may be so weakened by heat that the fungus is ideally positioned for immediate colonization. In the soil proper, however, mycelial growth from germinated spores is limited and saprophytic establishment is governed by the availability of conifer roots of fire-killed and recently cut (1–2 yr) trees.

Epidemiology: Infection of regeneration develops from the periphery of burn sites where heating of underlying soil to sublethal temperatures of 38–45°C stimulates ascospore germination. Within this concentric activation zone, the presence of green roots from recently killed or cut conifers promotes active saprophytic colonization and vegetative increase by the fungus. Seedling suscepts planted in this zone during the first few years after the fire are likely to become infected, whereas natural regeneration often escapes this fate through delayed root contact with the inoculum, which, in itself, begins to dissipate with time. Recurrence of fire is not necessary for radial progression of tree mortality (about 1 m/yr) to continue via coalescent root systems; however, most infection centers stabilize after a few years and infection inexplicably abates.

Diagnosis: Detection is best accommodated by recognizing the specificity of the disease to burned areas as mortality of peripheral seedling–saplings, and the chestnut-brown, white-margined, crustlike masses of apothecia, which measure 2 × 6 cm across. These annual fructifications appear in July–September, darken with age, and may persist in a blackened, partially decayed condition through the winter. The apothecia are connected by buff-yellow mycelial strands to underlying infected roots, which are matted together with white mycelium and are more or less resinous. The asci have eight hyaline, fusiform spores with peculiar tapering ends.

Control Strategy: The impact of the disease in this country to date is minor; consequently, bans on fire use in site preparation are not warranted. If preplant examination of burned sites reveals abundant apothecia of the fungus, consideration might be given to delayed planting or, better still, favoring direct seeding on heavily infested patches, as prescribed in Europe.

SELECTED REFERENCES

Gremmen, J. 1958. A dieback of pine-species in the Netherlands and its probable cause. *Horte Mededeling Nr.* 35:201–208.

Gremmen, J. 1971. *Rhizina undulata.* A review of research in the Netherlands. *Eur. J. For. Path.* 1:1–6.

Thies, W. G., K. W. Russell, and L. C. Weir. 1977. Distribution and damage appraisal of *Rhizina undulata* in western Oregon and Washington. *Pl. Dis. Reptr.* 61:859–862.

Thompson, J. H., and T. A. Tattar. 1973. *Rhizina undulata* associated with disease of 80–year-old red spruce in Vermont. *Pl. Dis. Reptr.* 57:394–396.

14

FOLIAGE PATHOLOGY

14.1 HARDWOODS

The leaf is highly variable, both in structure and function. The leaf is usually in a flattened form and is specialized as a photosynthetic unit. Like the root, the leaf consists of a dermal system, a vascular system, and a ground tissue system. The epidermis consists of a compact arrangement of epidermal cells, a cuticle, and stomata. Stomata may occur on both sides or on only one side (usually the lower) of the leaf. In the broad leaves of dicotyledons, the stomata are scattered.

The structure of leaves is related, to a great extent, to the availability of water in the habitat in which the tree is growing. The epidermal cells are compactly arranged and covered with the cuticle, both of which help reduce water loss. Stomata permit gas exchange with the atmosphere but close during dry periods to reduce water loss. They may also be sunken in depressions or surrounded with epidermal hairs to reduce transpirational water loss further.

The majority of the ground or internal tissue of a leaf is composed of the mesophyll, which contains an abundance of chloroplasts. Large intercellular spaces assist in movement of gases between the mesophyll cells and the exterior environment. The mesophyll may be differentiated into palisade parenchyma and spongy parenchyma. The spongy parenchyma consists of large cells of irregular shapes with large intercellular spaces, which allow gas flow and exchange within the leaf. The palisade parenchyma consists of cells elongated perpendicularly to the external leaf surface and helps provide structural support for the leaf.

The vascular system of a leaf occupies a close spatial relationship with the mesophyll, forming veins that are part of an interconnected system through the leaf petiole to the vascular system of the stem. The vascular system supplies the leaf with a steady supply of water to replace that lost to transpiration and organic and inorganic nutrients which may be needed for photosynthesis and synthesis of organic molecules necessary for growth.

As we saw in Chapter 7, the leaf may seem like a rather delicate structure, but it can present a formidable barrier to attacking pathogens because of a variety of structural and biochemical defenses. Because of the concentration of simple photosynthetic products such as sugars and starches, though, it is a prime target for many pathogens, many of which either have devised strategies to enter the leaf through stomates or directly through the cuticle and epidermis.

Leaves, however, have one other defense mechanism not available to most other plant tissues. This defense is the ability to separate the leaf from the stem—the process of *abscission*. This process occurs normally each season but may also form prematurely to isolate and cast off an invading pathogen. Many leaf pathogens have responded to abscission by developing elaborate mechanisms to survive it and still produce a large quantity of inoculum to attack the emerging foliage in a subsequent growth season.

In the following section, we will examine several of the more important leaf pathogens of trees and elucidate the strategies that each has used to ensure success. A series of Disease Profiles illustrates the wide variety of leaf diseases with which foresters may have to contend.

REFERENCE

Esau, K. 1966. *Anatomy of Seed Plants*. John Wiley & Sons, Inc. New York. 376 pp.

14.1.1   Anthracnose of Sycamore, Oak, and Other Hardwoods

Importance

Hardwood, or broadleaf, anthracnoses cause a variety of disease symptoms on their respective hosts. Some foliar pathogens cause single to scattered discrete leaf lesions, which may have an almost undetectable effect on the host. Other more aggressive pathogens not only kill leaves but also can attack and, in some cases, kill twigs and branches as well. An example of the latter is sycamore anthracnose, which is shown in the disease diagram in Figure 14.1 and briefly discussed in the Disease Profiles. Although most hardwood tree species are affected to some degree by anthracnoses, anthracnose of sycamore is one of the most important diseases of that species, causing defoliation, twig dieback, and cankers. Severe and repeated defoliations may weaken and predispose trees to secondary agents (e.g., borers and winter injury) and result in mortality. A primary impact of anthracnose is on shade and ornamental trees because of the unsightly effects of premature defoliation and twig dieback. Species of oaks, especially white oaks, are significantly damaged by anthracnose fungi.

Suscepts

American, Arizona, and California sycamore are susceptible. London plane tree is somewhat resistant, and oriental or European plane tree is most resistant to anthracnose fungi. Red, black, and white oaks are susceptible, especially the latter two.

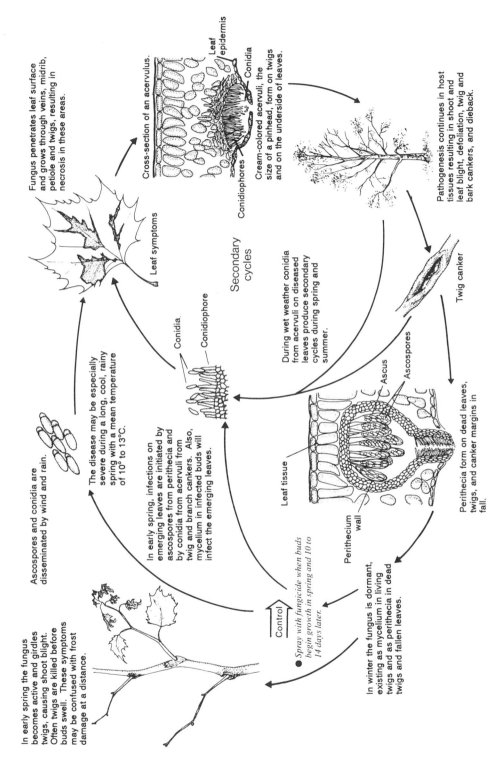

Leaf epidermis

Conidia

Cross-section of an acervulus.

Conidiophores

Cream-colored acervuli; the size of a pinhead, form on twigs and on the underside of leaves.

Fungus penetrates leaf surface and grows through veins, midrib, petiole and twigs, resulting in necrosis in these areas.

Leaf symptoms

Pathogenesis continues in host tissues resulting in shoot and leaf blight, defoliation, twig and bark cankers, and dieback.

Secondary cycles

Ascospores and conidia are disseminated by wind and rain.

The disease may be especially severe during a long, cool, rainy spring with a mean temperature of 10° to 13°C.

Conidia.

Conidiophore

During wet weather conidia from acervuli on diseased leaves produce secondary cycles during spring and summer.

Twig canker

Ascospores

Ascus

In early spring the fungus becomes active and girdles twigs, causing shoot blight. Often twigs are killed before buds swell. These symptoms may be confused with frost damage at a distance.

In early spring, infections on emerging leaves are initiated by ascospores from perithecia and by conidia from acervuli from twig and branch cankers. Also, mycelium in infected buds will infect the emerging leaves.

Leaf tissue

Perithecium wall

Perithecia form on dead leaves, twigs, and canker margins in fall.

Control

• Spray with fungicide when buds begin growth in spring and 10 to 14 days later.

In winter the fungus is dormant, existing as mycelium in living twigs and as perithecia in dead twigs and fallen leaves.

FIGURE 14.1 Disease diagram of anthracnose of sycamore (*Platanus* spp.) caused by *Apiognomonia veneta* (*Gnomonia platani*). Drawn by Valerie Mortensen.

Distribution

Anthracnose is widespread wherever susceptible host species and favorable climate occur. Seasonal weather fluctuations may cause anthracnose to be locally more abundant in a given year.

Cause

Historically the fungus *Gnomonia venata* and its imperfect stage *Gloeosporium nervisequum* were recognized as the pathogens of anthracnose of sycamore and oak. More recently, the pathogen on sycamore was recognized as *Gnomonia platani* and its imperfect stage *Gloeosporium platani*. It is now *Apiognomonia veneta* (Pyrenomycetes) and its imperfect stage is *Discula platani* (Coleomycetes). The pathogen on oak is now *Apiognomonia errabunda (Gnomonia quercina),* and its imperfect stage is *Discula umbrinella (Gloeosporium quercinum)*. Cross-inoculations of oak and sycamore with these two fungi were unsuccessful.

Biology

Sycamore anthracnose is initiated in the spring, primarily by conidia produced from overwintering mycelium in infected twigs remaining on the tree and perhaps from conidia and ascospores produced on fallen debris. Spores germinate on and penetrate the leaf surface, and grow through the vein and midrib into the petiole and twig. The fungus mycelium may girdle the twig causing a dieback, or a canker may be formed and girdle the twig in subsequent years. A secondary bud may form below the girdled area, producing a new twig at a noticeable angle from the parent twig. Recurring heavy infections tend to produce a crown full of branches, many with a rather zig-zag or angular appearance (Fig. 14.2). The fungus overwinters as mycelium in infected twigs and in sporocarps in dead debris. Secondary cycles occur throughout the summer, during moist periods, as conidia produced in acervuli on infected leaves and twigs reinfect leaves.

Epidemiology

The relative amount of injury that the fungus can inflict on sycamore largely depends on the amount and frequency of favorable moisture periods in early spring as spores are released and infect emerging leaves. While the first cycle may not be severe, if cool-wet weather continues, the diseased leaves will produce large quantities of conidia in a series of secondary cycles. These conditions often set the stage for an exponential buildup of both inoculum and disease. Most anthracnoses are cool-wet weather diseases.

Shoot blight is severe when temperatures are below 10°C; moderate to slight if temperatures are 13–16°C; and slight to nil when temperatures are above 16°C. Rain does not affect the shoot blight stage, but the leaf blight stage. Secondary cycles on leaves are favored by moist conditions, especially rain. Inocula are wind- and rain-disseminated.

FIGURE 14.2   Sycamore anthracnose, which has defoliated this tree in mid-June.

Diagnosis

Symptoms are distinguished as leaf, bud, shoot, and twig blight and often defoliation and cankers, the latter especially on sycamore. The most obvious symptom is shoot blight, which occurs in the spring and is initiated by the girdling action of overwintered mycelium in the shoot. The necrotic and blackened leaves, buds, and shoots resemble frost damage and, when severe, can result in complete crown defoliation. Early defoliation may be followed by a second flush of leaves. Leaf blight, the result of spores infecting local or discrete areas of leaves, is typified by various-sized necrotic areas (round to irregularly shaped) primarily along the veins and midrib. Entire leaves may be killed and cast or become deformed as areas of infected leaves expand differentially. Signs of the pathogen are acervuli (often cream-colored) occurring along veins and midrib of infected leaves, on diseased twigs, and on cankers. Black perithecia with two-celled, hyaline ascospores occur on dead and fallen leaves and twigs.

Control Strategy

In forests no control is practiced; in plantations recommendations are to provide wide spacing to allow air movement and sunlight, which reduces atmospheric and leaf moisture thereby inhibiting secondary disease cycles, and to avoid planting pure stands. In shade, ornamental, and nursery trees the disease is controlled with one application of an eradicant and protective fungicide in the spring during bud break. A second application after 14 days is suggested if prolonged cool weather occurs following the initial application. Forecasting of weather favorable or unfavorable for shoot blight (i.e., mean daily temperature during leaf emergence) provides a basis for spray schedules such that fungicides need to be used only in those years when temperature conditions are favorable for disease development. Pruning infected twigs is recommended, but removing fallen debris appears of little value because the disease is initiated by mycelium and spores in infected tissues on the tree. Fertilization that promotes shoot growth can help restore a healthy crown. London plane tree and especially Oriental or European plane (sycamore) and hybrids are disease-resistant and should be favored in plantings. Black oak appear more resistant than white oak and should be favored when applicable.

Other Anthracnose Diseases of Hardwoods

There are numerous species of anthracnose fungi, which attack a variety of hardwood suscepts. Most pathogens are species of *Apiognomonia* or the imperfect genus *Discula*. However, there are other genera, for example, *Glomerella, Elsinoe,* and *Marssonina*. Suscepts include ash, barberry, basswood, birch, catalpa, dogwood, elm, hickory, horsechestnut, maple, privet, snowberry, yellow-poplar, and walnut. Although not much is known about most of these diseases, the biology, epidemiology, diagnoses, and control are perhaps similar to those for anthracnose of sycamore and oak.

REFERENCES

Berry, F.H., and W. Lautz. 1972. *Anthracnose of Eastern Hardwoods*. USDA For. Serv., For. Pest Leafl.133. 6 pp.

Sinclair, W.A., H.H. Lyon, and W.T. Johnson. 1987. *Diseases of Trees and Shrubs*. Cornell University Press. Ithaca and London. 574 pp.

DISEASE PROFILE

Anthracnose of Sycamore, Oak, and Other Hardwoods

Importance: Anthracnose is the most important foliage disease of sycamore, resulting in defoliation, twig dieback, and cankers. Severe and repeated occurrences may weaken and predispose trees to secondary agents (e.g., borers and winter injury) and result in mortality.

Suscepts: American, Arizona, and California sycamore are susceptible. London plane tree is somewhat resistant, and the oriental or European plane tree is resistant. Oaks are significantly injured.

Distribution: Anthracnose is widespread in the United States and is present wherever susceptible hosts and favorable climate occur.

Causal Agents: The fungus *Apiognomonia (Gnomonia) veneta* (Diaporthales) and its imperfect stage *Discula platani* (Coelomycetes) are the pathogens on sycamore. The pathogen on oak is *A. errabunda,* and its imperfect stage is *D. umbrinella.*

Biology: Anthracnose is initiated in the spring, primarily by conidia produced from overwintering mycelium in infected twigs remaining on the tree. Infections penetrate the leaf surface and grow into the petiole and twig. The twig may be girdled causing a dieback, or a canker may form and girdle the twig in subsequent years. The fungus overwinters as mycelium in infected twigs and in sporocarps in dead debris. Secondary cycles occur throughout the summer during moist periods.

Epidemiology: Anthracnose is a cool-wet weather disease. The amount of shoot blight of sycamore is severe when temperatures for a 2–week period following leaf emergence are below 10°C; moderate to slight if temperatures are 13–16°C; and slight when temperatures are above 16°C. Rain does not affect the shoot blight stage, but the leaf blight stage and, thus, secondary cycles, are favored by moisture. Inocula are wind- and rain-disseminated.

Diagnosis: Shoot blight occurs in the spring and is initiated by the girdling action of overwintered mycelium in the shoot. The necrotic and blackened leaves, buds, and shoots resemble frost damage and can result in complete crown defoliation. Early defoliation may be followed by a second flush of leaves. Leaf blight, the result of spores infecting local areas of leaves, is typified by various-sized necrotic areas primarily along the veins and midrib. Entire leaves may be killed and cast or become deformed as infected areas expand differentially. Acervuli (often cream-colored) occur on infested tissues. Black perithecia with two-celled, hyaline ascospores occur on dead and fallen tissues.

Control Strategy: In forests no control is practiced. In plantations wide spacing reduces atmospheric and leaf moisture thereby inhibiting secondary disease cycles. Pure stands should be avoided. In shade, ornamental, and nursery trees the disease is controlled with one application of a protective fungicide during bud break followed by a second application after 14 days. Pruning infected twigs is recommended, but removing fallen debris appears of little value because the disease is initiated by mycelium and spores in infected tissues on the tree. Fertilization that promotes shoot growth can help to restore a healthy crown. London plane tree and especially Oriental or European plane (sycamore) and hybrids are disease resistant and should be favored in plantings. Black oak appear more resistant than white oak and should be favored when applicable.

SELECTED REFERENCES

Berry, F. H., and W. Lautz. 1972. *Anthracnose of Eastern Hardwoods.* USDA For. Serv., For. Pest Leafl. No. 133. 6 pp.
Sinclair, W. A., and W. T. Johnson. 1968. *Anthracnose Disease of Trees and Shrubs.* N.Y. State Coll. Agric., Cornell Tree Pest Leafl. A-2. 7 pp.

14.1.2 Powdery Mildews of Oak and Other Hardwoods

Importance

Powdery mildews are of little importance in the forest, and it is doubtful if they ever cause any appreciable damage to forest trees in this continent. Powdery mildew of cultivated grapes threatened to eradicate this crop in France until control measures were found and it is still one of the most important diseases of grapes in California. Mildew of cereals is also a very important disease in Europe, and it causes occasional losses here. Apple mildew causes losses in some areas. Most plants grown in greenhouses, such as roses, are attacked by these fungi, and this is one of the serious troubles with which florists have to contend. Occasionally mildew has been a problem on yellow-poplar in nurseries.

Suscepts

The fungi that cause powdery mildew have been found on about 1300 species of plants; they are common on most broad-leaved trees and shrubs, on cereals, and on ornamental plants and flowers.

Distribution

Fungi causing powdery mildews are of historic interest because they were the first fungi that were proven by the Tulasne brothers in France, about a hundred years ago, to produce more than one kind of spore. During the present century, methods to control the diseases that these fungi cause have been studied thoroughly in many countries. Powdery mildew is found throughout the world wherever the hosts grow.

Cause

The powdery mildews with the aid of a simple key can be identified to genus and with the aid of a host index identified to species. The fungi that cause powdery mildews are in the Ascomycotina (Erysiphaceae), and the genera are classified according to the following key:

One Ascus	*Several Asci*
Simple appendages	Simple appendages
Sphaerotheca	*Erysiphe*
Branched appendages	Branched appendages
Podosphaera	*Microsphaera*
	Tips of appendages curled
	Uncinula
	Swollen bases on appendages
	Phyllactinia

Biology

The disease diagram in Figure 14.3 illustrates the life history of a typical powdery mildew. A spore falls on a leaf and germinates, the mycelium ramifies over the surface of the leaf, and haustoria penetrate into the epidermal or subepidermal cells and absorb nutrients. Simple chains of oval conidia are produced on upright conidiophores soon after infection; these spores are wind-disseminated during the growing season, and each spore can initiate a new infection. Cleistothecia (Fig. 14.4) form on the surface of the leaf in the fall, ascospores develop in these and remain there until spring, when the walls of the cleistothecia rupture and the ascospores are shot out of the asci thus causing early infection in spring. All the powdery mildew fungi are obligate parasites. Some species are divided into numerous physiologic forms that are morphologically similar but differ from each other in the particular species of host that they are able to parasitize.

Epidemiology

A high relative humidity and a fairly high temperature favor infection; these conditions often occur in greenhouses and in the understory in forests where the light intensity is low and the air does not circulate much, hence the prevalence of mildew in such situations. A film of moisture over the leaf surface is not required for spore germination and invasion by these fungi.

Diagnosis

Infected areas of plants are covered with superficial mycelium, which usually is abundant enough to appear as white or tan patches; when cleistothecia are present, the infected leaves are covered with what looks like small black specks that can be seen more easily with a hand lens.

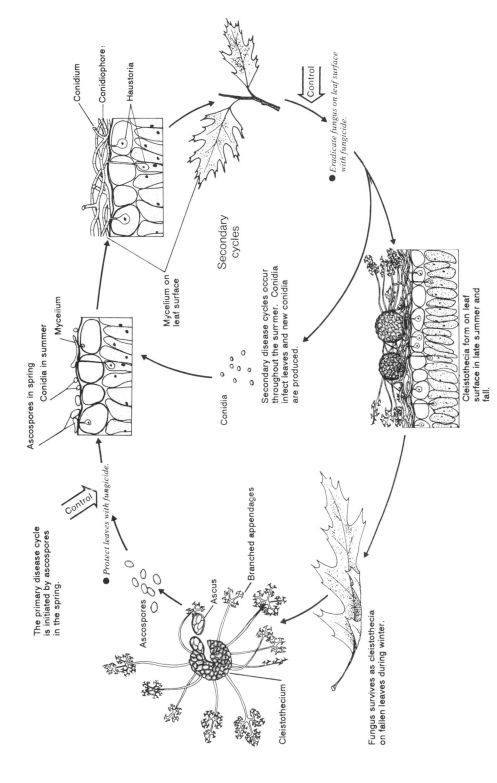

The primary disease cycle is initiated by ascospores in the spring.

Control

● Protect leaves with fungicide.

Ascospores in spring
Conidia in summer

Mycelium

Conidium
Conidiophore
Haustoria

Mycelium on leaf surface

Secondary cycles

Conidia

Secondary disease cycles occur throughout the summer. Conidia infect leaves and new conidia are produced.

Control

● Eradicate fungus on leaf surface with fungicide.

Cleistothecia form on leaf surface in late summer and fall.

Ascospores

Ascus

Branched appendages

Cleistothecium

Fungus survives as cleistothecia on fallen leaves during winter.

FIGURE 14.3 Disease diagram of powdery mildew on *Quercus* spp. caused by *Microsphaera penicillata*. Drawn by Valerie Mortensen.

453

FIGURE 14.4   Cleistothecium of *Microsphaera* spp. Note the branched appendages and multiple asci.

Control Strategy

The mycelium of the powdery mildew fungi is superficial; hence, they can be controlled by applying fungicides such as sulfur dust. This is one of the few fungus diseases that can be successfully controlled after it once is established.

REFERENCE

Anonymous. 1960. *Index of Plant Diseases in the United States*. USDA Agric. Handbook 165. 531 pp.

DISEASE PROFILE

Powdery Mildews of Oak and Other Hardwoods

Importance: Powdery mildews are among the most abundant and obvious foliage diseases of forest trees. Although omnipresent, they cause little significant injury except occasionally on young sprouts and nursery seedlings.

Suscepts: Powdery mildews occur on more than 1,300 species of plants, with 40 genera of hardwood trees and shrubs affected by these diseases; within the genera *Quercus* at least 30 species are suscepts. Conifers are not suscepts.

Distribution: Powdery mildews are ubiquitous throughout the world.

Causal Agents: There are seven genera of fungi (*Erysiphe, Phyllactinia, Microsphaera, Podosphaera, Sphaerotheca, Cystotheca,* and *Uncinula,* Pyrenomycetes, Erysiphales) that cause powdery mildews, and representatives of each genus occur on forest trees and shrubs. Some species of these obligate parasites, though, occur on many suscept genera. For example, *Microsphaera penicillata* occurs on 16 suscept genera and *Phyllactinia guttata* occurs on 26 genera. *Microsphaera penicillata* and *Cystotheca (Sphaerotheca) lanestris* are the most common on oak. *Oidium* (Hyphomycetes) is the imperfect form genus.

Biology: The disease is initiated in the spring by wind-disseminated ascospores ejected from overwintered cleistothecia on fallen leaves or by conidia produced by overwintered mycelium. Spores germinate on and penetrate the leaf surface and then form haustoria in the epidermal and subepidermal cells. The haustoria absorb nutrients from the suscept and sustain the development of the extracellular and superficial mycelium, which covers the leaf. Conidia formed on simple conidiophores provide inoculum for secondary cycles. In the fall cleistothecia form on the leaf surface and ascospores are discharged the following spring.

Epidemiology: Temperature and moisture relations differ greatly among species of the pathogen. Optimum temperature for germination and growth is 21°C, slightly lower than for many other pathogens. However, minimum and maximum temperatures of 0 and

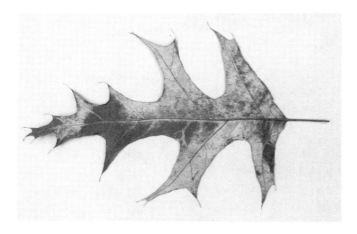

36°C, respectively, are reported. Free moisture on leaf surfaces inhibits germination and mycelial growth, and spore germination in many species occurs at less than saturation humidities. Germination and appressorial formation, including host colonization, are favored by low light intensities and darkness. These conditions—a moist atmosphere, dry leaf surfaces, and low light intensity—exist in the understory of dense stands where powdery mildews thrive.

Diagnosis: Leaf surfaces above colonized cells are blistered, and young leaves may become curled and distorted. Similar symptoms appear on colonized shoots, flowers, or fruits. Powdery mildews derive their name from their signs and are easily identified by the white powdery mycelium, conidiophores, and conidia, which occur on the plant surface with black cleistothecia visible among the mycelia. Identification of the genera is easily accomplished by reference to appendage morphology and the number of asci.

Control Strategy: No control is necessary in forests, although wide spacing will allow air movement, reduce atmospheric moisture, and increase light, which, in turn, will inhibit powdery mildew development. In situations where the fungus overwinters on fallen leaves only, collecting and burning leaves will reduce primary inoculum. On nursery and ornamental trees fungicide dusts and sprays are effective.

SELECTED REFERENCES

Gardner, M. W., C. E. Yarwood, and T. Duafala. 1972. Oak mildews. *Pl. Dis. Reptr.* 46:313–317.

Hepting, G. H. 1971. *Diseases of Forest and Shade Trees of the United States.* USDA For. Serv., Agric. Handbook No. 386. 658 pp.

Schnathorst, W. C. 1965. Environmental relationships in the powdery mildews. *Annu. Rev. Phytopath.* 3:343–366.

Yarwood, E. C. 1957. Powdery mildews. *Bot. Rev.* 23:235–300.

DISEASE PROFILE

Sooty Molds

Importance: Most sooty mold fungi are superficial on leaf surfaces and cause little or no damage to trees. Leaves blackened by sooty mold fungi are unsightly but, more importantly, indicate associated insect problems that may retard tree growth and kill young trees.

Suscepts: Many hardwood species (e.g., basswood, catalpa, elm, fig, holly, magnolia, and yellow-poplar) and conifers are hosts (not suscepts) for this disease. These saprophytic fungi are not host-specific.

Distribution: Sooty molds occur extensively throughout North America.

Sooty mold on beech leaf.

Causal Agents: A very diverse group of fungi is associated with sooty mold, and its
taxonomy is not well determined. Most are ascomycetes or imperfect fungi, some with
pleomorphic conidial forms. Historically, species of the genus *Capnodium* (Dothideales)
(e.g., *C. elongatum* and *C. tuba*) appear as sooty molds. *Leptoxyphium axillatum*
(Hyphomycetes) is a sooty mold of catalpa. *Fumago vagans,* a sooty mold on leaves of
various trees and on bark of sugar maple associated with sapsucker injury, is on many
species a mixture of two fungi, *Aureobasidium (Pullularia) pullulans* (Hyphomycetes), and
Cladosporium herbarum (Hyphomycetes). It has no connection to *Capnodium.* Sooty
molds are often common on conifers. To mention only a few, species of the genus
Dimerosporium (Meliola) and *Adelopus* occur on grand fir; *Dimeriella,* on Douglas-fir;
and *Capnodium,* on eastern white pine.

Biology: Most sooty mold fungi are saprophytes that grow superficially on nutrients on plant
surfaces. Commonly, these fungi are associated with excrement ("honeydew") of aphids
and scale insects. These insects extract the nitrogen in the tree's sap. They must process
very large volumes of sap because the concentration of nitrogen is very low. The sugar
content of the sap passes through their bodies virtually untouched and emerges in a
concentrated form, as honeydew. Ants often protect colonies of aphids and scale insects
from predators and use the excreted honeydew as a source of carbohydrates. Sooty mold
fungi typically have brown or black mycelium and spores. There is, of course, no
infection biology of the sooty molds because they are not parasites. They are pathogens
when they screen sunlight from the leaf tissues. Biologically, these fungi are most
interesting as they are a component of the abundant yet subtle microflora and fauna
which inhabit plant surfaces.

Epidemiology: Sooty molds appear to be most abundant in mild climates, but little is known
about their epidemiology.

Diagnosis: No symptoms are associated with sooty mold disease unless chlorosis results
from lack of sunlight and photosynthesis. The disease is easily recognized by the black
sooty appearance of the vegetative and reproductive signs of the pathogen on the leaf or
other plant surfaces. These signs are ascocarps and imperfect sporulation structures such

as pycnidia and/or naked conidiophores and conidia. Of course, the aphids or scale insects associated with this disease are often obvious when not covered by the sooty mold fungus.

Control Strategy: No control is practiced in the forest. Insecticides are useful against the aphid or scale insects when ornamental or nursery trees require treatment.

SELECTED REFERENCE

Hughes, S. J. 1976. Sooty molds. *Mycologia* 68:693–820.

DISEASE PROFILE

Tar Spots of Maples

Importance: Tar spots are obvious and fairly common on maples. When severe, this disease can cause defoliation and is unsightly on shade, ornamental, and nursery trees. However, most often tar spots cause no significant damage and are of little economic importance.

Suscepts: Large tar spot is common on red and silver maples and also occurs on Norway, sycamore, sugar, and bigleaf maples. Small tar spot (speckled tar spot) occurs on bigleaf

Tar spot on maple leaf.

maple and also on mountain, silver, striped, Rocky Mountain (dwarf), Norway, red, and sugar maples and boxelder.

Distribution: Tar spots occur in both eastern and western North America. Large and small tar spot are apparently more abundant in the eastern and western United States, respectively.

Causal Agents: Rhytisma acerinum and *R. punctatum* (Rhytismatales) are the pathogens of large and small tar spots, respectively. The imperfect stages are recognized as *Melasmia acerina* and *M. punctata* (Coelomycetes), respectively.

Biology: Infection of new foliage occurs in the spring when ascospores, released from apothecia on fallen leaves, germinate on the upper surface of leaves. The mycelium enters the stomata and grows within the epidermal and mesophyll cells. Thick, shiny, black stromata (tar spots) are formed on the upper leaf surface during the summer. Apothecia develop in these stromata during the fall; following defoliation, asci and ascospores form during the winter and early spring. Ascospores are forcibly ejected from the ascus when apothecia rupture during wet spring weather. This inoculum is disseminated by wind to new foliage. Conidia are produced on the stromata during the summer, but their function is unknown.

Epidemiology: Primary inoculum is released from overwintering ascocarps during moist weather in the spring and is wind-disseminated to suscept leaves. Secondary cycles are unknown although spermatia occur.

Diagnosis: Symptoms that occur in the late spring and summer (6–8 weeks following inoculation) are yellow-green leaf spots. Premature defoliation can occur if tar spot is severe. This disease is easily recognized by the signs (tar spots) that occur on the upper surface of infected leaves. These shiny, black stromata are approximately 0.5–1.0 cm in diameter or larger when coalesced in large tar spot and 0.2–0.3 cm for small tar spot. Mature stromata split along radial lines and appear wrinkled. Ascospores are needle-shaped (filiform) and quite long.

Control Strategy: No control is practiced in forests. For shade, ornamental, and nursery trees, burning off fallen leaves is an effective means of control. Sanitation in the northern part of the United States is effective because all inoculum is contained on the fallen leaves. Also, the fungus is an obligate parasite and apparently does not produce conidia for secondary cycles. If necessary, protective foliar fungicides may be applied to ornamental or nursery trees in the spring as the leaves emerge and expand.

Other Tar and Ink Spots: Rhytisma arbuti, R. liriodendri, and *R. salicinum* cause tar spot on madroño, yellow-poplar, and willow, respectively. A disease of aspen leaves known as ink spot is caused by several species of fungi in the genus *Ciborina.*

SELECTED REFERENCES

Jones, S. G. 1925. Life history and cytology of *Rhytisma acerinum* (Pres.) Fries. *Ann. Bot.* 39:41–75.
Woo, J. Y., and A. D. Partridge. 1969. The life history and cytology of *Rhytisma punctatum* on bigleaf maple. *Mycologia* 61:1085–1095.

DISEASE PROFILE

Leaf Blister and Related Diseases

Importance: These diseases do not cause economically important losses in forest trees. Leaf blister of oak sometimes causes defoliation on many different species of oak in eastern and southern United States. Oaks may be 50–85% defoliated by midsummer, and this lack of leaves makes them unsightly as shade trees and undoubtedly reduces their vigor.

Suscepts: Oaks, plum, wild and domestic cherry, aspen, and alder are susceptible.

Distribution: Found throughout the United States, except for the arid southwestern region, leaf blisters are more common and severe in the southeastern states where the pathogen can remain active throughout the winter.

Causal Agents: More than 50 species of *Taphrina* (Taphrinales) are significant pathogens of trees or treelike shrubs that may be used for ornamental plantings.

Biology: New infections begin in the spring about the time the buds start to swell. Germ tubes from germinating ascospores enter through the stomata of enlarging leaves. The mycelium is entirely intercellular; no haustoria enter the cells.

Epidemiology: Asci, containing usually eight round ascospores, arise from the intercellular mycelium, and are borne in a palisade layer all over the infected areas. Usually this is on the concave surface of a blister. The ascospores may bud, like yeasts, in the ascus producing as many as 64 spores per ascus. These spores are discharged and carried by the wind to the bud scales of susceptible plants. They remain there over winter and germinate about the time that buds expand. Infection is favored by low temperature and a high relative humidity. Asci are produced shortly after full symptom development, arising from just beneath the cuticle, which they rupture as they grow. At maturity, they are quite exposed. Secondary infections during the remainder of the season are apparently not common.

Diagnosis: Infection causes the size and number of host cells to increase, resulting in overgrowth and often in considerable distortion of the infected parts. *Taphrina caerulescens* and *T. populina* cause leaf blisters of oak and poplar, respectively. On the

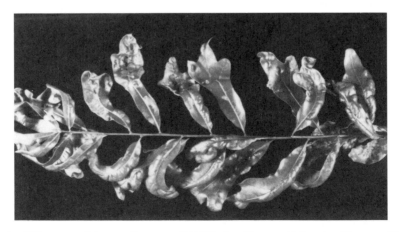

Leaf blister on oak leaves. Courtesy of W. Witcher, Clemson University, Clemson, SC.

latter host, infections by *Taphrina* may be confused with those of *Melampsora* rust. Plums infected with *T. communis* are elongated, stoneless, and much larger than healthy plums. *Taphrina wiesneri* causes leaf curl and witches' brooms on twigs of cherry and apricot. *Taphrina johansonii* and *T. robinsoniana* cause deformation of catkins on aspen and alder, respectively.

Control Strategy: Control measures for these diseases on forest trees are not warranted. On shade trees, a fungicidal spray may be applied just as the buds start to swell. Applications after buds have opened and leaves emerged are not effective.

SELECTED REFERENCES

Sinclair, W. A., H. H. Lyon, and W. T. Johnson. 1987. *Diseases of Trees and Shrubs.* Cornell University Press, Ithaca, NY. 575 pp.

von Arx, J. A., J. P. van der Walt, and N. V. D. M. Liebenberg. 1982. The classification of *Taphrina* and other fungi with yeast-like cultural states. *Mycologia* 74:285–296.

DISEASE PROFILE

Marssonina Leaf Spot of Poplars

Importance: Fungi of the genus *Marssonina* cause serious leaf and shoot diseases of many poplar species and hybrids. Infection can result in premature defoliation and growth reduction. Repeated attacks cause branch dieback and predispose trees to other pathogens and to winter injury. Seedlings in nurseries may be killed.

Marssonina leaf spot.

Suscepts: Eastern cottonwood, black cottonwood, balsam poplar, quaking aspen, and various Euroamerican hybrid poplars are susceptible.

Distribution: Found worldwide, this disease is especially prevalent where poplars are intensively managed. Seedlings in nurseries and plantation trees are most severely affected. The disease was first described in this country in 1889, but only in the past few decades has it caused severe losses. It has caused great economic loss in Europe since its discovery there in 1958.

Causal Agents: *Marssonina tremulae, M. castagnei,* and *M. populi* (Coelomycetes). The perfect states are *Drepanopeziza tremulae, D. populi-albae,* and *D. populorum,* respectively (Helotiales). Because the perfect state has not been reported in North America, the names of the conidial forms are commonly used here.

Biology: In North America primary infection results in spring soon after rain-dispersed conidia are deposited on leaves. In Europe ascospores also initiate primary infections. Acervuli then develop, and conidia produced by them initiate repeating cycles that continue as long as wet weather continues. Numerous infections cause leaves to turn yellow and then brown, and if severe, they may coalesce to form angular, necrotic blotches.

Epidemiology: New infections in the spring are caused by conidia and possibly ascospores that are produced in infected fallen leaves and in shoot lesions on the tree. They are carried by wind-splashed rain to developing leaves and succulent shoots where infection takes place. The secondary infections, which often become epidemic, can occur into the fall. The disease moves upward into the crown and can cause defoliation of susceptible trees long before normal leaf fall.

Diagnosis: Leaf spots are first visible as small discrete reddish-brown to purple discolorations on both leaf surfaces and on leaf petioles. On leaves, lesions soon darken and are circular, about 1 mm across. When the light-colored macroconidia are released from the middle of the lesion, it produces a characteristic ringlike structure. On veins and petioles, the spots may be elongated. Infected leaves remain flattened (unlike rusted leaves that become rumpled) and are shed prematurely. Two types of conidia are produced: macroconidia and microconidia, both in fruiting structures that are indistinguishable from each other. Macroconidia are hyaline, oval, divided unequally into two cells, and range in size from $11-12 \times 3.5-7$ μm. Microconidia are hyaline, one-celled, elliptical, and are $3.3-5.5 \times 1.2-1.8$ μm in size.

Control Strategy: Fungicidal sprays may be effective, especially in nurseries. Removal and disposal of infected leaves may reduce inoculum levels in nurseries and plantations. The best strategy is to plant resistant or tolerant poplar clones.

SELECTED REFERENCES

Harniss, R. O., and D. L. Nelson. 1984. *A Severe Epidemic of Marssonina* Leaf Blight on Quaking Aspen in Northern Utah. USDA For. Serv., Res. Note INT-339. 6 pp.

Jokela, J. J., J. D. Paxton, and E. J. Zegar. 1976. Marssonina leaf spot and rust on eastern cottonwood. *Pl. Dis. Reptr.* 60:1020–1024.

Ostry, M. E., and H. S. McNabb, Jr. 1986. *Populus* Species and Hybrid Clones Resistant to *Melampsora, Marssonina,* and *Septoria.* USDA For. Serv., Res. Pap. NC-272. 7 pp.

DISEASE PROFILE

Dogwood Anthracnose

Importance: Dogwoods in at least 22 states are affected, as well as in British Columbia, Canada. The disease increased dramatically in the southeastern United States from 0.2 million ha affected in 1988 to 5.1 million ha in 1992. In some areas, particularly in woodland understories at high elevations, flowering dogwood has been eliminated from some sites.

Suscepts: The major host in the eastern United States is flowering dogwood *(Cornus florida)*. Pacific dogwood *(C. nuttallii)* is affected in the northwest. *Cornus kousa* is regarded as more resistant than is *C. florida*.

Distribution: Discovered in the late 1970s in New York and southern New England and in the Pacific Northwest, the dogwood anthracnose pathogen is known as far south as northern Alabama in the mountains and foothills of the eastern United States and also occurs in an isolated population of *C. nuttallii* in Idaho.

Causal Agents: A newly described species of fungus, *Discula destructiva* (Coelomycetes), is responsible for dogwood anthracnose. Its origin is unknown, and it may have been introduced. A complex of closely related species may be involved, however, as at least two distinct types (Type 1 and Type 2) have thus far been detected. Type 2 is probably a different, as yet undescribed, species of *Discula*.

Biology: Understanding disease biology has been hampered by the difficulty of reproducing the disease under controlled conditions. In nature the disease usually develops first on leaves as a typical anthracnose. Infections then progress down the petioles into shoots in winter and eventually produce cankers. Heavy leaf infection and twig dieback induce proliferation of epicormic shoots to form on the main stem and on large branches.

Diseased stem sprouts.

Epicormic shoots are very prone to infection and permit entry of the pathogen into the main stem. Infected leaves hang on trees or abort. A toxin may kill branches.

Epidemiology: Large numbers of conidia are produced on acervular conidiomata formed on necrotic leaves or bark tissues during wet weather. Natural infection apparently occurs in flowers in early spring. Infection is favored by cool temperatures and high relative humidity. Most isolates of the Type 1 strain are infected with a mycovirus, but what effect it has, if any, is unknown.

Diagnosis: Leaf infections produces tan spots that develop purple margins. Large necrotic blotches may also appear; leaves sometimes abscise. Tiny cankers will develop on twigs. Heavy epicormic branching is evident in older infections, and lower branches are killed. Removing bark at the base of dead epicormic branches often reveals brown, elliptical annual cankers. Bracts may be infected if rainy conditions prevail during flowering.

Control Strategy: Dogwoods that which receive good cultural care can withstand anthracnose. This care includes mulching, watering (without wetting the foliage), and fertilizing. The disease must be detected before extensive dieback begins. Pruning of epicormic branches will prevent development of trunk cankers. Reduce mechanical injury. High-hazard sites are close to water, on north-facing slopes, and above 1,000 m elevation (in eastern states) where clouds may persist. Open stands in full sunlight are at much lower risk to infection. Fungicide sprays protect plants when environmental conditions are favorable for disease development.

SELECTED REFERENCES

Daughtrey, M. L., C. R. Hibben, and G. W. Hudler. 1988. Cause and control of dogwood anthracnose in northeastern United States. *J. Arboric.* 14:55.

Hibben, C. R., and M. L. Daughtrey. 1988. Dogwood anthracnose in northeastern United States. *Pl. Dis.* 72:199–203.

McElreath, S. D., J.-M. Yao, P. S. Coker, and F. H. Tainter. 1994. Double-stranded RNA in isolates of *Discula destructiva* from the eastern United States. *Curr. Microbio.* 29:57–60.

DISEASE PROFILE

Air Pollution Effects on Hardwoods

Importance: Air pollutants cause significant injury to hardwood forest, shade, and ornamental trees. Exposure to large dosages results in leaf necrosis, defoliation, and mortality. Growth loss and predisposition of injured trees to secondary factors can accompany air pollution injury.

Suscepts: Hardwoods vary considerably in their response to various air pollutants. For example, white oak is sensitive to several pollutants while dogwood tolerates these pollutants. Other trees such as sycamore are sensitive to some but tolerant of other pollutants.

Interveinal necrosis. From Skelly, Undated.

Distribution: Air pollution injury is directly related to the source of contamination and is normally restricted to and most severe in the immediate vicinity of the pollution source, for example, in industrial areas or near electrical generating plants. Injury can occur on sites distant to the source when pollutants are disseminated there by wind.

Causal Agents: More than a dozen air pollutants are recognized as toxic to plants. These pollutants include gaseous, aerosol, and particulate matter derived from a variety of primary or secondary sources. Major pollutants from primary sources are sulfur dioxide (SO_2) and fluorine-containing compounds such as hydrogen fluoride (HF) and silicon tetrafluoride (SiF_4). Important pollutants from secondary photochemical reactions include ozone (O_3) and peroxyacetyl-nitrate (CH_3–$COONO_2$). The latter, referred to as PAN, derives chiefly from motor vehicles and is a common constituent along with ozone in urban smog.

Biology: Air pollutants are toxic to plants at very low concentrations, some at less than 1 ppm or even 1 ppb. Generally, air pollutants enter leaves through stomata and once inside may be toxic to cell components or react to form toxic substances. Inhibition of mitochrondria activity, lipid synthesis and membrane permeability, and photosynthesis are possible modes of pathogenesis.

Epidemiology: Environmental factors that affect plant growth such as light, temperature, and relative humidity condition plant response and injury before, during, and after exposure to air pollutants. Dosage (concentration × duration of exposure) of air pollutants is the critical factor conditioning plant injury and is affected by concentration at the source, height of the source aboveground, horizontal wind speed, and vertical mixing of air (turbulence).

Diagnosis: Symptoms, which are restricted to leaves, may be classified as chronic (tissue injury but no necrosis) resulting from low dosages or plant tolerance and acute (tissue necrosis) resulting from large dosages or very sensitive plants.

SO_2—Chronic, Tissue between veins becomes chlorotic; Acute, Ivory, tan, red-brown necrotic tissue between green veins.

Fluoride—Chronic, Marginal chlorosis; Acute, Marginal necrosis. O_3—Chronic, Metallic flecking or pigmented stippling of the upper leaf surface; Acute, Necrotic flecking or stippling.

PAN—Chronic, Glazing, silvering, or bronzing of undersurface; Acute, Interveinal necrosis and defoliation.

Foliage symptoms are similar to those of other physiological disorders such as temperature-, moisture-, or nutrition-induced diseases and also some biotic problems such as virus, mite, aphid, or leafhopper injury. Therefore, diagnosis may require knowledge of pollution sources, weather patterns, site and vegetation histories, and chemical analysis in addition to symptomology.

Control Strategy: Elimination or modification of the source of pollution is achieved by air quality standards and appropriate laws. Injury is minimized by the use of resistant species and varieties; selection and breeding is especially attractive as a control because pathogenic variability is not a factor. The ability to forecast meteorological conditions when inversions will occur or periods when plants are most susceptible can be used to temporarily curtail emissions or change to low-sulfur fossil fuels.

SELECTED REFERENCES

Loomis, R. C., and W. H. Padgett. 1975. *Air Pollution and Trees in the East.* USDA For. Serv., State and Private Forestry. 28 pp.

Skelly, J. M., ed. Undated. *Diagnosing Injury to Eastern Forest Trees.* Nat. Acid Precip. Assess. Program. 122 pp.

14.2   CONIFERS

In the first portion of this chapter, we examined the role of major leaf pathogens in hardwood tree species. Next, we will examine the biology and characteristics of needle-infecting fungi, indigenous problems and potential threats.

The leaves of gymnosperms (i.e., conifers) are less variable in structure than those of the angiosperms and are highly independent of the environment. Except for *Ginkgo, Larix,* and *Taxodium,* most gymnosperms are evergreen. Conifer needles originate singly or most commonly in groups of two to several. Depending on the number in the group, the cross-sectional shape may vary from circular to triangular. Conifer needles have a thick-walled epidermis, a thick cuticle, and deeply sunken stomata. The stomata are limited to rows between sclerified fibrous tissues. The mesophyll is of only one form and is not differentiated into palisade and spongy parenchyma as in the dicotyledons. The vascular system usually is found in one or two bundles in the central portion of the needle. Resin canals may be present in the needle, ranging from none in *Taxus* to one in *Sequoia* and *Tsuga* and two in *Pinus* to many in *Araucaria.* Except for the deciduous genera, most conifers retain their needles for several years. Although conifer needles tend to have a structure that is rather resistant to invasion by pathogens, the more successful pathogens can take advantage of the needles being retained for more than one growing season in order to complete their life cycle. As with the hardwood foliage pathogens, those pathogens that are most successful on conifers often take advantage of certain cultural management practices which lessen natural host resistance.

REFERENCE

Esau, K. 1966. *Anatomy of Seed Plants.* John Wiley & Sons, New York. 376 pp.

DISEASE PROFILE

Brown Spot

Importance: The most serious disease of grass-stage longleaf pine and the major limitation, along with competing vegetation, in its regeneration is brown spot.

Suscepts: Longleaf pine is the major host; however, an additional 25 species of pine, 10 native to the southeast and Scots and red pine in the Lake states, are known to be susceptible.

Distribution: Brown spot occurs throughout the original range of longleaf pine, which extends along the Atlantic coast from North Carolina to Florida and the Gulf coast to Texas and inland to Arkansas, Tennessee, and Ohio; fringe areas include Delaware, Arizona, Idaho, Oregon, Wisconsin, Minnesota, Iowa, Kansas, Missouri, Kentucky, and Manitoba.

Causal Agents: The fungus pathogen is *Mycosphaerella dearnessii (Scirrhia acicola)* (Loculoascomycetes). The imperfect stage is *Lecanosticta acicola* (Coelomycetes).

Biology: Conidia (two- to four-celled, yellow-olive) from acervuli on dead fallen needles are rain-splashed to a height of approximately 80 cm (i.e., the grass-stage zone of longleaf pine). Foliage penetration is through stomates. Ascospores are discharged from stromatic pseudothecia and wind-dispersed for long-distance transport. Northern and southern races differ in spore germination optima (20 and 25°C, respectively), in cultural characteristics, and in pathogenicity between regional suscepts. In the south, brown spot on longleaf pine is mainly a disease of the grass stage. Once the longleaf pine leaves the grass stage and begins height growth, brown spot is no longer of concern. In northern states, especially on Christmas trees and in nurseries, brown spot can cause serious economic losses.

Epidemiology: In southern Mississippi both types of inoculum are available throughout the year to initiate infection during wet periods. Conidia are released during all rains in some proportion to rainfall duration with maximum sporulation from May through August and at temperatures as low as 2–3°C. Ascospores are discharged during rains, dews, and fogs

Brown spot symptoms. Courtesy of D. W. French, University of Minnesota, St. Paul, MN.

within a temperature range of 4–27°C. In Wisconsin, conidia do not mature until late May, and peak sporulation follows in July and September associated with seasonal rains.

Diagnosis: Symptoms develop as yellow spots that progress to brown bands and then total or distal necrosis. Defoliation begins toward the close of the growing season. Fruiting occurs as subepidermal gray spots that are flush with the needle when dry but protrude slightly when wet. In culture the fungus grows slowly, producing masses of conidia in a dark, olive-green gelatinous matrix, mostly above the agar surface.

Control Strategy: In the nursery, protectant fungicidal sprays may be applied at 10- to 30-day intervals as governed by frequency of rain. Root-dip treatment with Benomyl has given excellent disease control following outplanting. In grass-stage plantations, prescribed winter burning is recommended when more than one third of the foliage was infected the previous fall. Resistance to brown spot infection is heritable (0.57), but inherent fast height growth is not the major mechanism of resistance. Selections and hybrid progeny provide a 15–20% gain in resistance over natural populations. Early initiation of height growth, as favored by seedling quality, weed competition, and nutrition, reduces the effects of brown spot needle blight.

SELECTED REFERENCES

Siggers, P. V. 1944. *The Brown Spot Needle Blight of Pine Seedlings.* USDA, Tech. Bull. 870. 36 pp.
Skilling, D. D., and T. H. Nichols. 1974. *Brown Spot Needle Disease—Biology and Control in Scotch Pine Plantations.* USDA For. Serv., Res. Paper NC-109. 19 pp.

14.2.1   Brown Spot

Importance

Since colonial days, longleaf pine has been one of the major forest species in the southeastern United States. During the first half of the twentieth century, the area of land in longleaf pine decreased rapidly, longleaf being replaced by other species of pines, chiefly loblolly and slash, and by various hardwoods such as scrub oak and sweetgum. Chapman (1932a) attributed the failure of longleaf pine to regenerate naturally to the control of fire. Fire serves several useful purposes in management of this species, one of which is the effective control of brown spot disease. This disease has been a major factor in the reduction of the longleaf pine type and its control by use of prescribed burning was essential for this trend to be halted. Within the past three decades, the application of the knowledge of the factors responsible for this decline to forest management has resulted in its halt, and the area has even begun to increase. Brown spot, or brown spot needle blight as it is also called, causes its greatest losses on the seedlings of longleaf pine and is a major problem in longleaf pine stands established either naturally or artificially. Defoliation in 3 successive years will often kill the seedlings. Where defoliation is not sufficient to kill the seedlings, they are weakened to such an extent that adverse environmental factors will kill them. Brown spot may increase greatly the length of time during which the seedlings remain in the grass stage and thus increase the danger of their suppression by competing vegetation. This grass stage under favorable condition lasts usually from 3 to 5 years, but under adverse conditions, such as may occur when the plants are defoliated wholly or partially year after year, the grass stage may continue for 8–10 years.

Some injury may result from partial defoliation of older trees. Longleaf trees are quite resistant to attack after they pass the seedling stage, but occasionally the current-year needles of loblolly pines of pole size and larger may be damaged severely. This fall and winter loss of foliage reduces tree growth during the following year, but it has never been observed to cause death of the tree. Loss to Christmas tree growers due to brown spot is due to defoliation or color change of needles.

Partial defoliation affects terminal growth. Destruction of 10% of the foliage of seedlings resulted in a 50% reduction in growth the following year, while a 30% loss of foliage virtually stopped terminal growth. In older trees, the effect was much less severe. When brown spot was controlled with fungicides, a tremendous increase in growth was observed in the sprayed seedlings. Ninety percent of the sprayed seedlings were considered as vigorous, whereas only 73% of the unsprayed seedlings were in this category. In a similar test, Siggers (1932) determined that 2-year-old sprayed seedlings had a mean stem diameter of 1.6 cm as compared to a diameter of 1.0 cm in the unsprayed plots. Two years later, the sprayed 4-year-old seedlings were 27.4 cm tall, while the unsprayed seedlings were only 4.6 cm tall. Wakeley (1970) reported that, after 30 years, many trees which as seedlings were lightly infected, produced averages of 1.7–2.4 times as much pulp wood as an equal number of trees that were moderately to heavily infected as seedlings. Brown spot has been a major factor in the failure of longleaf to regenerate itself on sites where it was clearcut.

Suscepts

Siggers (1944) tested 24 species of pines and found all to be susceptible to brown spot. These pines included 10 species native to the southeastern United States and 14 exotic species that had been introduced there. The 10 native pine species tested by Siggers were slash, shortleaf, spruce, longleaf, pitch, pond, sonderegger, white, loblolly, and Virginia.

Brown spot is primarily a disease of longleaf seedlings, and it causes its greatest damage to this species. Boyce (1952) reported brown spot as causing extensive needle dieback on loblolly pine, and Luttrell (1949) reported moderate to severe injury on ponderosa pine growing in Georgia. Nicholls and Hudler (1972) reported that brown spot resulted in serious defoliation of red pine in Wisconsin. Laut et al. (1966) reported severe defoliation of jack pine and lodgepole pine in Canada. Defoliation of Christmas tree plantings of Scots pine in Wisconsin is economically important. In Pennsylvania, eastern white pine is a rare host of brown spot.

Juvenile longleaf pines vary in their susceptibility to brown spot. Verrall (1934) observed that certain individual seedlings showed the "bar spot" symptoms, which he associated with resistance. These bar spots occurred naturally on all longleaf plants after they passed the seedling stage and on certain individual seedlings. This resistance was associated with a high content of resin in the needles of these resistant seedlings and in the needles of older trees. Verrall concluded that the resin was toxic to the fungus, resulted in a change in moisture conditions in the lesion, or served as a mechanical barrier to the fungus. This bar spot was most common on needles of longleaf and slash pines, both of which are well known for their high resin content. Siggers, in 1937, found a longleaf seedling that appeared to be resistant. This was planted near Alexandria, Lousiana, and progeny from this tree were resistant to brown spot. Snyder and Derr (1972) compared longleaf seedlings from five locations throughout the South and found that seedlings from southwestern Alabama were more resistant to brown spot than those from elsewhere. Derr and Melder (1970) reported that seedlings resulting from cross pollination of two resistant parents were much more resistant to brown spot than seedlings produced from open-pollinated seed. Prey and Morse (1971) noted that short needle varieties of Scots pine were more susceptible than varieties with longer needles.

Distribution

Brown spot needle blight was first observed and collected near Aiken, South Carolina, by H. W. Ravenal in 1876 (Fig. 14.5), and it has been recognized as a serious disease at least since 1919.

FIGURE 14.5   Distribution of brown spot. Redrawn from Cordell et al. (1989).

Up to 1915, no concerted effort at fire protection had been made in large longleaf areas. Prior to then, the longleaf pine sites had been burned periodically for many years, ignited either by lightning or man. Both the pre-Columbian residents and the early European settlers used fire regularly, and it was rare that the intervals between fires exceeded 3 or 4 years.

After widespread fire control began to be practiced about 1915, the evidence of the dependence of the longleaf pine on periodic fires began to build up rapidly. In the absence of fire, longleaf pine often did not regenerate itself but was gradually replaced by other species of pines, chiefly loblolly and slash pines, and various hardwoods. After fire control became more widespread and more effective, brown spot, which had been limited by the periodic burning, became much more severe. This increase in the severity of brown spot was one of several factors that has resulted in the failure of the longleaf to regenerate. Currently, brown spot is undoubtedly the major disease of longleaf pine in the southeastern states.

Early reports identify brown spot in the United States from Pennsylvania west to Kansas and south to Florida and Texas and in Oregon and Idaho and it was especially prevalent in the coastal and Gulf states from North Carolina to Texas. The northern limits of the disease were not well defined but generally were thought to coincide with the northern limits of loblolly and shortleaf pines. The disease was later confirmed as causing serious losses on ponderosa pine in Missouri, and Laut et al. (1966) even reported that it may sometimes cause serious losses in Canada. Brown spot is important on Scots pine Christmas trees in Wisconsin, and on red pine in Wisconsin.

Cause

The pathogen was first described under the conidial stage of the fungus [*Cryptosporium acicolum* (Melanconiaceae)]. Later it was named *Lecanosticta acicola*. The perfect stage of the pathogen was first described as *Oligostroma acicola* (Dothidiales). Siggers (1939) performed a laboratory comparison of single ascospore and single conidial cultures and shifted the fungus from the genus *Oligostroma* to the genus *Scirrhia*. Until recently the most commonly used name has been *Scirrhia acicola*. The correct current name is *Mycosphaeralla dearnessii* (Cooke & Harkn.) Lindan and the name for the imperfect stage is *Lecanosticta acicola* (Thum.) Syd. and Petrak.

C. W. Edgerton, in Louisiana in 1923, was the first to prove the causal relationship of the pathogen to brown spot by successfully inoculating seedling pines with conidial suspensions. E. C. Tims repeated the work of Edgerton in several tests in the years following, but neither of these workers published his results. Hedgcock (1929) produced the first published report of successful inoculations. Although he was unable to obtain infection with conidial suspensions of the fungus, he did get infection with a macerated agar culture suspension. Subsequent recent inoculations have been successful. In all these reports as well as others, moisture was a very important factor influencing disease development.

Biology and Epidemiology

The development of brown spot in the South is continuous throughout the year (Fig. 14.6). During the winter the pathogen becomes less active, but during warm periods, even in midwinter, conidia and ascospores may be produced and infection may occur.

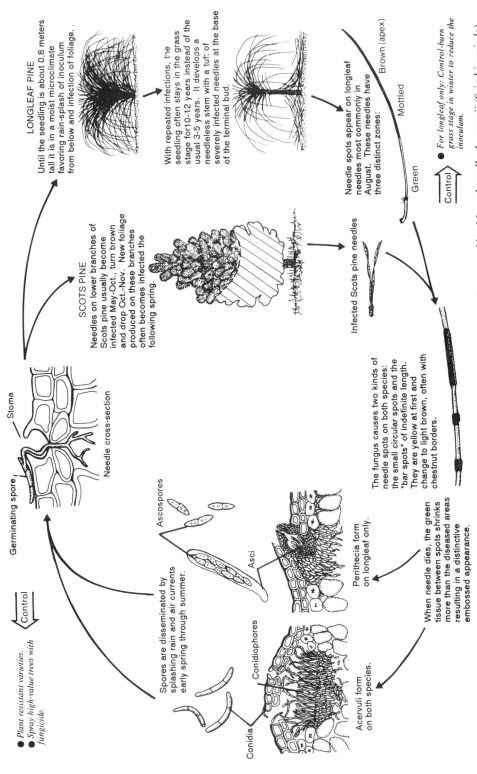

FIGURE 14.6 Disease diagram of brown spot needle blight of longleaf pine and Scots pine caused by *Mycosphaerella dearnessii* (*Scirrhia acicola*). Drawn by Valerie Mortensen.

Germinating spore

Stoma

Needle cross-section

LONGLEAF PINE

Until the seedling is about 0.8 meters tall it is in a moist microclimate favoring rain-splash of inoculum from below and infection of foliage.

With repeated infections, the seedling often stays in the grass stage for 10-12 years instead of the usual 3-5 years. It develops a needleless stem with a tuft of severely infected needles at the base of the terminal bud.

Needle spots appear on longleaf needles most commonly in August. These needles have three distinct zones:

Brown (apex)

Mottled

Green

Control

- *For longleaf only: Control-burn grass stage in winter to reduce the inoculum.*

SCOTS PINE

Needles on lower branches of Scots pine usually become infected May-Oct., turn brown and drop Oct.-Nov. New foliage produced on these branches often becomes infected the following spring.

Infected Scots pine needles

The fungus causes two kinds of needle spots on both species: the small circular spots and the "bar spots" of indefinite length. They are yellow at first and change to light brown, often with chestnut borders.

When needle dies, the green tissue between spots shrinks more than the diseased areas resulting in a distinctive embossed appearance.

Ascospores

Asci

Peritheci form on longleaf only.

Conidia

Conidiophores

Acervuli form on both species.

Spores are disseminated by splashing rain and air currents early spring through summer.

Control

- *Plant resistant varieties.*
- *Spray high-value trees with fungicide.*

472

In Wisconsin sporulation occurs in two peaks, one in early July and the other in late September.

Primary infection in the spring is either by conidia or ascospores, which are produced on dead needles or dead portions of living needles. These needles may be still on the plant or lying on the ground near the plant. The inoculum usually lands near the tips of needles, and the first symptoms appear there. Ascospores are abundant in late winter and early spring and increase from early spring and peak at the end of August. They are forcibly discharged during rains, dews, and fogs and are airborne to the living needles of the host plant. Windblown ascospores may be responsible for all the primary infection that occurs on the foliage of trees above 0.5 m from the ground and some of the infection on seedlings near the ground. No one has reported producing the disease with ascospores as inoculum, so the role in dissemination attributed to ascospores is based on circumstantial evidence. Brown spot has been observed to spread very slowly into large burned areas. This slow rate of spread would indicate that windborne ascospores probably play only a minor role in dissemination. In Wisconsin, Prey and Morse (1971) found only the imperfect stage, suggesting that brown spot may become serious without producing ascospores.

Conidia are produced in large numbers in the spring and may serve as primary inoculum. The conidia, which are curved, have one to three septations, are light brown and are extruded from the stromata in a water-soluble mucoid matrix and adhere in black masses on the surface of the lesion until they are washed away by rain or dew. Thus their spread is due almost entirely to splashing or windblown rain, and they would generally be restricted to less than 0.5 m above the ground level. Because the conidia are exuded in a sticky mass, they can adhere to the feet of cattle or hogs. The rate of spread from outside into a burned area may be increased greatly if cattle graze in the two areas.

The germination of the two spore types and the succeeding infection probably occurs in the same manner. The spores germinate by means of a germ tube that penetrates into the mesophyll of the needle through the stomata. Snow (1961) made an impression of a pine needle with the germinated spore and hypha entering a stomatal cavity. Germ tubes from the germinating conidia usually grew appressed to the needle surface of Scots pine and followed contours of the epidermis. Penetration occurred through stomata after germ-tube growth seemed to be directed specifically toward individual stomata (Fig. 14.8). Siggers (1950) noted that, under certain conditions, the multicelled conidia may break up into their component cells and each cell functions as a one-celled spore (endospore). He doubted that this occurs in nature. Crosby (1966) observed these one-celled spores but found that they represented the cellular contents of individual, enlarged cells of the conidium, which were released by the spontaneous rupturing of the spore walls. The released contents were spheroidal to elliptical and thick walled. Conidia produced one or two of these spores per cell, and a four-celled conidium might produce four to eight spores.

Light may enhance infection of longleaf pine by stimulating the opening of the stomata. In the absence of light, the fungus functions primarily as a wound parasite. Temperatures above 35°C during the day and 27°C at night are inhibitory. Needles are less susceptible as they elongate and mature. Tissue of mature needles proximal to the stem is more susceptible to infection than either medial or distal tissue of the same needles. The hyphae within the leaf are initially intercellular. Within a few days after infection, the hyphae penetrate the thin walls of the mesophyll cells, and the cells collapse and die. The cell lumen then quickly becomes filled with masses of hyphae. The hyphae are restricted to the mesophyll parenchyma; the vascular bundles are not affected and apparently continue to function normally, since the invaded tissue distal to the lesion remains green thoughout the season or until killed by new infections. Within 3–14 days after the initial appearance of symptoms, the conidial stromata are formed, and mature conidia appear on the surface of the lesions. These same stromata may continue to produce conidia for an indefinite period. Perithecial stromata form much more slowly and then only after a major portion of the needle is killed.

Infections that occur during spring and summer initially are localized. During cool weather, however, the mycelium within the lesions may invade the living tissue between the lesions causing its death. This rapid invasion, which occurs primarily in the cool weather of fall and spring, may result in the death of more tissue than is killed by the original spotting. The fungus lives as a parasite only for a very short time. After the cell walls are penetrated and the cells killed, it may continue to live for many months in the dead needle tissue as a saprophyte.

Numerous secondary cycles of infection may occur, normally beginning during May and continuing until cool weather or late fall, resulting in a recession of activity of the pathogen. Even during the winter, however, secondary infection may occur when several consecutive warm days permit conidial formation and infection.

The secondary inoculum is thought to consist almost entirely of conidia. Although several workers have observed some ascospores throughout the year, their numbers were very small in comparison to the number of conidia except in early spring. Moreover, a minimum period of 2–3 months after the tissue dies is required for ascospore production. Therefore, while it is theoretically possible for ascospores to function as secondary inoculum, they are certainly relatively unimportant in this capacity.

Secondary spread in larger longleaf pines is of little consequence. The ascospores that are windborne up into the needles may produce lesions of the bar spot type. In loblolly pine, the primary infection may be due to windborne ascospores, and the secondary infection, to conidia that are washed down onto lower needles from lesions higher in the tree.

The primary environmental factor affecting the prevalence of brown spot on longleaf pine seedlings is rainfall. The dissemination of conidia is dependent on splashing or windblown rain. Ordinary summer showers are not sufficient to obtain very much spore discharge, and 24–48 or more consecutive hours of rainy or wet weather are necessary to obtain abundant spore discharge and infection. The conidia will not withstand drying; therefore, the period of continuous wet weather must be long enough for the conidia to be produced, to be disseminated to the infection court, and to germinate and produce infection.

Temperature is relatively less important than rainfall, since *Mycosphaerella dearnessii* is active over a wide range. The optimum temperature for development of the fungus is 25°C, the maximum is near 35°C, and the minimum is between 5 and 19°C. This wide range indicates that temperature would probably not often limit disease development in the longleaf pine area except in the winter. Kais (1972) found that northern and southern isolates had different temperature optima of 20 and 25°C, respectively.

There is a close negative correlation between ground cover after seedling emergence and disease severity. In plots with a normal ground cover of weeds, pine needles, and other debris, pine seedlings showed an average of 24% defoliation in two consecutive seasons. In corresponding plots that had been mechanically denuded of cover, the trees showed 43 and 68% defoliation, respectively, for the 2 years. The difference was attributed to the excessive erosion and resulting poorer growth on the denuded areas. In addition, however, the debris probably reduced the amount of splashing and thus reduced dissemination of the conidia.

Diagnosis

The first symptoms of brown spot on pine needles are small chlorotic flecks approximately 0.5 mm in diameter with necrotic centers (Fig. 14.7), and they appear in April. Three days later they are approximately 1.5 mm in diameter, and spore masses may be found. The margin of the spot may be a slightly darker brown and in the cooler weather of late fall, purplish in color. The needle tissue between the spots remains green, since girdling of the needle does not result in death of the portion of the needle distal to the girdle. By May secondary inoculum is produced. Tip dieback is usually first apparent in June. Eventually, the coalescing of numerous spots may result in the death of the needle tips so that the needles are divided into three distinct zones—a green basal zone, a spotted central zone, and a dead tip. In the cool weather of late fall and early spring, the mycelium of the fungus rapidly invades the green tissue between the spots so that most or all the needle may be killed. This invasion of green tissue probably results in the destruction of more tissue than the original infection.

As the tips of the needles are killed, they bend outward and downward until, by midwinter, the dead needles drooping around the stem resemble a tussock of dead grass surmounted by the erect-standing spotted needles. As new needles are devel-

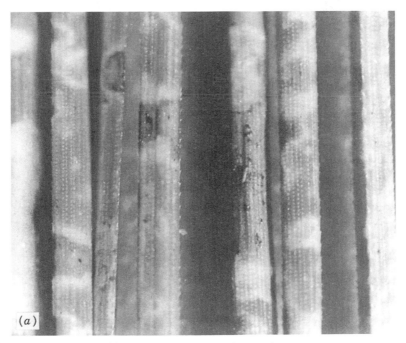

FIGURE 14.7 (a) Typical brown spot lesions on longleaf pine. From Cordell et al. (1989). (b) Typical brown spot lesions on Scots pine. From Phelps and Kais (1975).

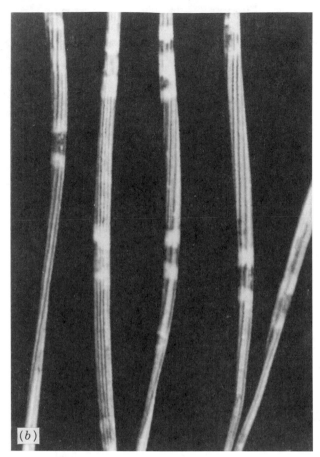

FIGURE 14.7   *(Continued)*

oped, each crop may be attacked successively. This results in stunting and, if serious defoliation occurs for 3 or more successive years, many seedlings die.

The browning caused by brown spot is most pronounced in the fall, whereas the destruction of needles by Hypoderma needle blight occurs primarily in the spring. Snow (1961) pointed out that the age of needles is a factor in susceptibility and that infection occurred only on immature longleaf pine needles 8–15 cm long. Old loblolly pine needles were more susceptible than young needles. Newly emerging needles of Scots pine appeared to be initially resistant to brown spot.

Verrall (1934) described a special type of spot that regularly occurs on older longleaf and slash saplings and on certain individual seedlings. This spot, which was designated a bar spot, is brown, and is enclosed on each side by a pale yellow band 1–3 mm broad.

On older loblolly pines, brown spot often results in a dieback of the needles so that the trees look as if they have been burned over. This dieback occurs in early fall, and the burned appearance remains until new growth in the spring covers the dead needles. Trees of any age may be attacked.

Within 3 days after the lesions appear on the needles, conidia begin to appear in the necrotic areas. The stromata, which are composed of thick-walled brown cells, originate within the mesophyll and usually within the substomatal cavities. On the surface toward the leaf epidermis, conidiophores and conidia form in large numbers. The conidia push out from the palisades of conidiophores, eventually rupturing the overlying hypodermis and epidermis. The conidial stromata develop throughout the year but are much more profuse during the summer.

The linear black perithecial stromata develop on dead needles and dead portions of living needles primarily during late winter and early spring, although they may develop in smaller numbers throughout the year. They vary in length from 0.3 to 2.5 mm and in width from 0.3 to 0.5 mm. The ascospores are hyaline to light brown, two-celled, with the upper cell in the ascus larger than the lower cell. Each cell usually has two distinct oil globules. The ascospores vary in size from 15–19 by 7–9 μm.

The conidial stromata develop within 3–14 days after a lesion appears, whereas perithecia do not appear until 2–3 months after the tissue is killed. Ascospores are produced in greatest numbers in early spring and conidia, from midspring to late fall, but both may occur throughout the year. Kais (1971) found that the largest numbers of conidia are discharged from May through August.

In northern regions, infection occurs June to August, and the spots appear in August and September. The needles are cast in October and November, usually involving the lower half of the tree. Defoliation is often more severe on the north side of trees. In Wisconsin the fungus spores may be carried by wind.

Germ tubes from the germinating conidia usually grew appressed to the needle surface of Scots pine and followed contours of the epidermis. Penetration occurred through stomata after germ-tube growth seemed to be directed specifically toward individual stomata (Fig. 14.8).

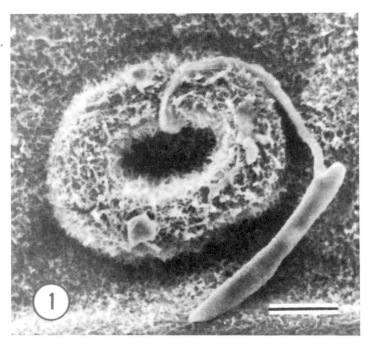

FIGURE 14.8   Entry of germ tube into a stoma of a Scots pine needle. From Patton and Spear (1978). Courtesy of R. F. Patton and R. N. Spear (1978).

Control Strategy

Exclusion. Brown spot is generally distributed over the entire southeastern United States where longleaf pine occurs. Therefore, the principle of exclusion can be used only in preventing reinfestation of large burned areas. Demmon (1935) suggested that the pasturing of livestock on these areas will result in more rapid spread of the pathogen back into a burned area because the sticky conidia will adhere to the feet of cattle and be spread more rapidly than would occur naturally. The complete exclusion of cattle would thus tend to delay the reappearance of the disease in these areas and would increase the benefits obtained from the use of controlled fires.

Eradication. Chapman (1926) was the first to suggest the use of fire to encourage natural regeneration. His suggestion created considerable controversy among southern foresters, who had recently adopted fire control. However, they soon realized the value of fire when used properly.

Controlled burning serves several useful purposes, as well as aiding in the control of brown spot (Fig. 14.9). Demmon (1935) listed some of these additional advantages of controlled burning as follows:

1. the removal of excessive combustible litter which reduces the hazard of uncontrolled fires;
2. the control and elimination of less fire-hardy competing vegetation;
3. the preparation of an adequate seedbed in the winter before a seed crop occurs;
4. the improvement of pastures by stimulating early growth of grass; and
5. the removal of dense ground cover for game management.

FIGURE 14.9   Prescribed burn in longleaf pine regeneration area. From Williston et al. (1980).

Demmon also stressed the potential danger from the use of fire by inexperienced persons, indicating that the beneficial effects could be offset very quickly if the fire were not completely controlled at all times.

Fire is recommended to prepare the seedbed in the winter before a seed crop is expected the following fall. Where the litter on the ground is heavy, the seed will not reach the mineral soil and will not germinate. An early fall fire just before seeds are shed is not recommended, according to Demmon (1935), because the seeds falling on completely cleared ground would be too easily found and eaten by birds and rodents.

Siggers (1934) suggested the following program of controlled burning in longleaf areas (Fig. 14.10). Seedlings are very sensitive to fire for the first year after they emerge, so fire should not be used in the first winter. From the second year until they begin rapid stem elongation, the seedlings are very resistant to fire. Siggers suggested a single controlled winter fire after the second season of growth. Then controlled burning should be repeated at 3-year intervals until the stems begin to elongate, after which trees not only become resistant to the disease and more susceptible to fire injury, but also grow up out of reach of the splashing conidia.

A longleaf seedling under good growing conditions normally will remain in the grass stage from 3 to 5 years, but under adverse conditions it may not make much growth in height for 8–10 years. Thus one to three, or rarely four, controlled fires may be used in young longleaf pine stands before they begin rapid terminal growth.

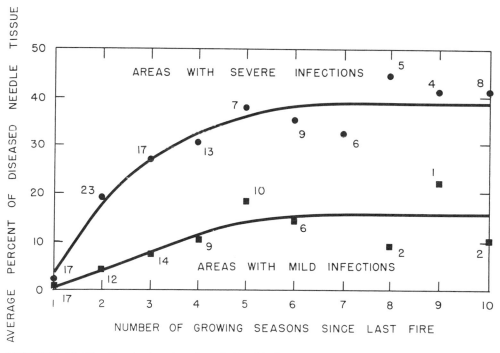

FIGURE 14.10 Graph showing the relationship between the infection due to brown spot and the number of growing seasons elapsed since fire. Numbers represent the number of sample points in each average. From Siggers (1934). Reprinted from *Journal of Forestry*, Vol. 32, No. 5, Published by the Society of American Foresters, 5400 Grosvenor Lane, Bethesda, MD 20814-2198. Not for further reproduction.

This increase in height occurs only after the roots and stem diameter have made considerable growth. The growth largely depends on carbohydrate accumulation, which in turn depends on the retention of needles from year to year.

Maple (1976) suggested that prescribed burns are advisable when the mean infection rate of crop seedlings reaches 20%. To reduce the risk of excessive seedling mortality, stands should be burned during the cool dormant season, as soon after a 1 cm (or more) rain as the fire will burn. Steady northerly winds are best, with wind speeds of 5–7 mph within the stand. Air temperature several degrees above freezing and relative humidity between 35 and 50% will ensure maximal fire coverage and minimal injury to the crop seedlings.

Annual and biennial fires applied around May 1 are more beneficial to the growth of young longleaf pines than March 1 fires. More grass-stage seedlings survived, began height growth, and grew taller. The May fires also controlled brown spot more effectively.

After burning, mortality increased with percentage of infection but usually declined with increased height, root collar diameter, and time since overstory removal (Fig. 14.11). Survival of heavily infected seedlings was best when root collar diameter was between 0.8 and 1.8 cm in the grass stage or larger than 3.8 cm in the height-growth stage.

Siggers (1932, 1934) and others stressed the importance of excluding fire during the first winter, since a fire at that time will kill a high percentage of the seedlings. Then, too frequent fires, especially annual winter fires, tend to destroy many living needles and prevent needle carryover from season to season, thus causing the same type of annual defoliation that would result from the disease if the fire were not used. After terminal elongation begins, fire is no longer needed for brown-spot control, and its use must be dictated by other possible benefits.

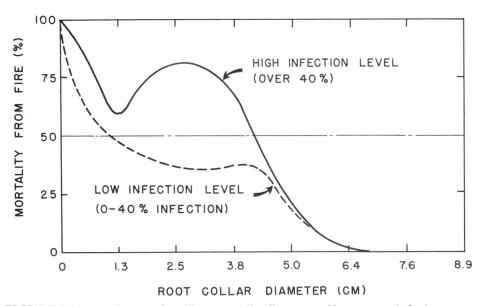

FIGURE 14.11   Influence of seedling root collar diameter and brown-spot infection on mortality from fire. Redrawn from Maple (1975).

The following additional general conclusions concerning the use of controlled fires for brown-spot control were derived from a study of the works of various authors.

1. Winter fires were more effective than fires during the growing season and caused much less damage to the young pines.
2. The use of fires on small areas is of questionable value because the disease may spread very quickly back into the area from adjacent unburned areas.
3. Annual fires are not beneficial and often result in the eventual death of the seedlings.

Protection. Fungicides effectively control brown spot, but except for use in the nursery, Siggers (1934) considered the cost to be too high to be economically justified. Hedgcock (1929) obtained excellent control with a 4–4–50 Bordeaux mixture applied at 14-day intervals beginning when new growth started in the spring and continuing until the needles had all reached maturity. Wolf and Barbour (1941) felt that spraying could not be justified even in the nursery, because applications would be required throughout the year. Siggers (1944) obtained effective control and greatly increased growth of seedlings with two applications of Bordeaux mixture per year, one in the spring and one in the fall, for the first two years. Derr (1957) reported effective control in planted longleaf pine by spraying with Ferbam in May and November of the second year and with Bordeaux mixture in May of the third and fourth years. This great reduction in the number of applications results in such a reduction in cost that the practice might, under certain conditions, be economically justified even on field plantings. McGrath et al. (1972) found that brown spot on Scots pine could be controlled in Wisconsin with two applications of Bordeaux mixture (8-8-100) in June and July. Chlorothalonil was effective if applied when needles were half grown and again 3 or 4 weeks later.

Rapid height growth of longleaf pine seedlings is presently an important management goal. It can be achieved by controlling brown spot and weed competition and by increasing soil fertility. Barnett and Kais (1986) dip-treated seedlings in a benomyl/clay mix (10% a.i. benomyl) and planted them in scalped rows that had been treated with 4 or 8% nitric acid. The combined use of benomyl and nitric acid resulted in the most rapid height growth. Longleaf seedlings stored in a benomyl–clay slurry for 3 weeks had a 60% better survival over seedlings that received only a clay slurry root treatment. Kais et al. (1986) recommended a root treatment of 5% a.i. benomyl/kaolin mixture for seedlings planted throughout the natural range of longleaf pine. Seedlings inoculated in the nursery with the ectomycorrhizal fungus *Pisolithus tinctorius* had higher survival rates than nontreated seedlings. This procedure would tend to ensure a vigorous seedling that would grow rapidly through the brown-spot susceptible grass stage.

Cultivating to remove grasses and similar vegetation around longleaf seedlings is not recommended. Siggers (1932) found that the mechanical removal of this vegetation for 2 successive years resulted in a great increase in disease severity.

Siggers (1944) reported that the use of commercial fertilizers increased the growth rate of longleaf seedlings and thus indirectly offset partially the detrimental effects of the disease. He concluded, however, that the benefit obtained certainly did not justify the expenditure for the fertilizer. Derr (1957) reported that fertilization had no effect

FIGURE 14.12   Longleaf pine seedlings during fourth grow season. Tree at center was selected as resistant to brown spot; those in foreground are heavily infected. From Snyder and
Derr (1972).

on the severity of brown spot but merely that fertilization reduced seedling survival.
The fertilizer in his tests was placed too close to the roots of the seedlings, however,
and the reduced survival was due to root injury.

Resistance.   Brown spot normally causes severe injury only on longleaf pine seedlings
with occasional severe defoliation in other species. Managing other species of pines,
especially slash, loblolly, and shortleaf pines, on sites where they are adapted, will tend
to reduce losses from this disease. Longleaf pine sites frequently have considerable
populations of scrub oaks, however, so that fusiform rust hazard may be high. Thus
using slash or loblolly pines on these sites would merely shift the major problem from
brown spot to fusiform rust. Where fusiform rust is not a serious hazard or where
shortleaf pine is adapted, the shift of species will result in effective brown-spot
control.

Varietal resistance in longleaf pine was considered in the 1930s. Verrall (1934)
considered resistance to be apparent in individual seedlings in nurseries and plantations because of the bar spot symptoms. Based on a tree selected by Dr. Siggers in
1937, it became apparent that resistance did exist in longleaf pine. Derr (1963) tested
the progeny from the single resistant seedling selected by Siggers in 1937 and found
them to be much more resistant than average seedlings. Resistance was correlated
with early height growth, and control-pollinated progenies expressed higher levels of
resistance than did open pollinated ones. Seed sources from the western extremity of
the range were generally more heavily infected than sources from the central part of

the longleaf pine range when tested in southern Mississippi. Tests with exposed and protected progeny indicated that inherent rapid height growth was not a major mechanism of resistance and that southwestern Alabama was the best source for both height growth and brown-spot resistance. Short-needled Scots pine varieties such as Spanish and French Green are more susceptible than the long-needled varieties.

In reviewing results of progeny tests of 540 parents with different disease histories and from several geographic locations, it was found that heritability of brown-spot resistance was 0.57, and that of height was 0.52, at age 3 years (Fig. 14.12). Infection in progeny of the best 10% of parents averaged 48%, compared to a population average of 63%. Offspring from parents selected 30 years earlier from a heavily infected planting averaged 55% taller and had about 10% less brown spot infection than those from parents with an unknown history. Resistant families delayed the onset of dieback 5–7 weeks and had lower levels of maximum needle dieback.

SELECTED REFERENCES

Barnett, J. P., and A. G. Kais. 1986. *Longleaf Pine Seedling Storability and Resistance to Brown-Spot Disease Improved by Adding Benomyl to the Packing Medium.* USDA For. Serv., Gen. Tech. Rept. SE-42. pp. 222–224.

Boyce, J. S., Jr. 1952. A needle blight of loblolly pine caused by the brown-spot fungus. *J. For.* 50:686–687.

Bruce, D. 1951. Fire, site, and longleaf height growth. *J. For.* 49:25–28.

Chapman, H. H. 1926. Factors determining natural reproduction of longleaf pine on cut over lands in La Salle Parish, Louisiana. *Yale Univ. School of For. Bull.* 16:1–44.

Chapman, H. H. 1932a. Some further relations of fire to longleaf pine. *J. For.* 30:602:604.

Chapman, H. H. 1932b. Is the longleaf pine a climax? *Ecology* 13:328–334.

Cordell, C. C., R. L. Anderson, W. H. Hoffard, T. D. Landis, R. S. Smith, Jr., and H. V. Toko. 1989. *Forest Nursery Pests.* USDA For. Serv., Agric. Handbook No. 680. 184 pp.

Crosby, E. S. 1966. Endospores in *Scirrhia acicola. Phytopathology* 56:720.

Demmon, E. L. 1935. The silvicultural aspects of the forest fire problem in the longleaf pine region. *J. For.* 33:323–331.

Derr, H. J. 1957. Effect of site treatment, fertilization, and brown spot control on planted longleaf pine. *J. For.* 55:364–367.

Derr, H. J. 1963. *Brown-Spot Resistance Among F₁ Progeny of a Single, Resistant Longleaf Parent.* Proc. Forest Genetics Workshop, Macon, GA. pp. 16–17.

Derr, H. J. 1966. Longleaf × slash hybrids at age 7: survival, growth, and disease susceptibility. *J. For.* 64:236–239.

Derr, H. J., and T. W. Melder. 1970. Brown spot resistance in longleaf pine. *For. Sci.* 16:204–209.

Griggs, M. M., and R. A. Schmidt. 1986. *Disease Progress of Scirrhia acicola* in Single and Mixed Family Plantings of Resistant and Susceptible Longleaf Pine. USDA For. Serv., Gen. Tech. Rept. WO-50. pp. 5–10.

Hedgcock, G. G. 1929. *Septoria acicola* and the brown spot disease of pine. *Phytopathology* 19:993–999.

Henry, B. W. 1954. Sporulation of the brown spot fungus on longleaf pine needles. *Phytopathology* 44:385–386.

Henry, B. W., and O. O. Wells. 1967. *Variation in Brown-Spot Infection of Longleaf Pine from Several Geographic Sources.* USDA For. Serv., Res. Note SO-52. 4 pp.

Kais, A. G. 1971. Dispersal of *Schirrhia acicola* spores in southern Mississippi. *Pl. Dis. Reptr.* 55:309–311.

Kais, A. G. 1972. Variation between southern and northern isolates of *Scirrhia acicola.* (abst.) *Phytopathology* 62:768.

Kais, A. G. 1975. Environmental factors affecting brown-spot infection on longleaf pine. *Phytopathology* 65:1389–1392.

Kais, A. G. 1977. Influence of needle age and inoculum spore density on susceptibility of longleaf pine to *Scirrhia acicola. Phytopathology* 67:686–688.

Kais, A. G. 1978. Pruning of longleaf pine seedlings in nurseries promotes brown-spot needle blight. *Tree Planters' Notes* 29(1):3–4.

Kais, A. G., C. E. Cordell, and C. E. Affeltranger. 1986. Benomyl root treatment controls brown-spot disease on longleaf pine in the southern United States. *For. Sci.* 32:506–511.

Kais, A. G., R. C. Hare, and J. P. Barnett. 1984. *Nitric Acid and Benomyl Stimulate Rapid Height Growth of Longleaf Pine.* USDA For. Serv., Res. Note SO-307. 4 pp.

Laut, J. G., B. C. Sutton, and J. J. Lawrence. 1966. Brown spot needle blight in Canada. *Pl. Dis. Reptr.* 50:208.

Lightle, P. C. 1969. *Brown Spot Needle Blight of Longleaf Pine.* USDA For. Serv., For. Pest Leafl. 44 (Rev.). 7 pp.

Luttrell, E. S. 1949. *Scirrhia acicola, Phaeocryptopus pinastri,* and *Lophodermium pinastri* associated with the decline of ponderosa pine in Missouri. *Pl. Dis. Reptr.* 33:397–401.

Maple, W. R. 1969. Shaded longleaf seedlings can survive prescribed burns. *For. Farmer* 29:3, 13.

Maple, W. R. 1975. *Mortality of Longleaf Pine Seedlings Following a Winter Burn Against Brown-Spot Blight.* USDA For. Serv., Res. Note SO-195. 3 pp.

Maple, W. R. 1976. How to estimate longleaf seedling mortality before control burns. *J. For.* 74:517–518.

McGrath, W. T., A. J. Prey, and F. S. Morse. 1972. Brown spot blight of Scotch pine Christmas trees in Wisconsin: Sporulation and control. *Pl. Dis. Reptr.* 56:99–102.

Nicholls, T. H., and G. W. Hudler. 1972. Red pine—A new host for brown spot *(Scirrhia acicola). Pl. Dis. Reptr.* 56:712–713.

Parris, G. K. 1967. Field infection of loblolly pine seedlings in Mississippi with naturally produced inoculum of *Scirrhia acicola. Pl. Dis. Reptr.* 52:552–556.

Patton, R. F., and R. N. Spear. 1978. Scanning electron microscopy of infection of Scotch pine needles by *Scirrhia acicola. Phytopathology* 68:1700–1704.

Phelps, W. R., and A. G. Kais. 1975. *Brown Spot Needle Blight.* USDA For. Serv., SE-FPM Dis. Bull. 1 pp.

Prey, A. J., and F. S. Morse. 1971. Brown spot needle blight of Scotch pine Christmas trees in Wisconsin. *Pl. Dis. Reptr.* 55:648–649.

Siggers, P. V. 1932. The brown-spot needle blight of longleaf pine seedlings. *J. For.* 30:579–593.

Siggers, P. V. 1934. Observations on the influence of fire on the brown-spot needle blight of longleaf pine seedlings. *J. For.* 32:556–562.

Siggers, P. V. 1939. *Scirrhia acicola* (Dearn.) n. comb., the perfect stage of the fungus causing the brown needle of pines. *Phytopathology* 29:1076–1077.

Siggers, P. V. 1944. *The Brown Spot Needle Blight of Pine Seedlings.* USDA, Tech. Bull. 870. 36 pp.

Siggers, P. V. 1950. Possible mechanism of variation in the imperfect stage of *Scirrhia acicola*. *Phytopathology* 40:726–728.

Skilling, D. D., and T. H. Nicholls. 1974. *Brown Spot Needle Disease Biology and Control in Scotch Pine Plantations.* USDA For. Serv., Res. Paper NC-109. 19 pp.

Snow, G. A. 1961. Artificial inoculation of longleaf pine with *Scirrhia acicola*. *Phytopathology* 51:186–188.

Snyder, E. B., and H. J. Derr. 1972. Breeding longleaf pines for resistance to brown spot needle blight. *Phytopathology* 62:325–329.

Stanosz, G. 1990. Premature needle drop and symptoms associated with brown spot needle blight on *Pinus strobus* in northcentral Pennsylvania. (abst.) *Phytopathology* 80:124.

Verrall, A. F. 1934. The resistance of saplings and certain seedlings of *Pinus palustris* to *Septoria acicola*. *Phytopathology* 24:1262–1264.

Verrall, A. F. 1936. The dissemination of *Septoria acicola* and the effect of grass fires on it in pine needles. *Phytopathology* 26:1021–1024.

Wakeley, P. C. 1968. Rust susceptibility in longleaf pine associated with brown-spot resistance and early commencement of height growth. *For. Sci.* 14:323–324.

Wakeley, P. C. 1970. Thirty-year effects of uncontrolled brown spot on planted longleaf pine. *For. Sci.* 16:197–202.

Wells, O. O., and P. C. Wakeley. 1970. Variation in longleaf pine from several geographic sources. *For. Sci.* 16:28–42.

Williston, H. L., T. J. Rogers, and R. L. Anderson. 1980. *Forest Management Practices to Prevent Insect and Disease Damage to Southern Pine.* USDA For. Serv., Rept. SA-FR 9. 8 pp.

Wolf, F. A. and W. J. Barbour. 1941. Brown spot needle disease of pines. *Phytopathology* 31:61–74.

DISEASE PROFILE

Elytroderma Needle Blight

Importance: This exclusively perennial needle cast, noted since 1940, periodically causes such severe growth reduction and mortality as to necessitate salvage in some areas.

Suscepts: Ponderosa and Jeffrey pines are the major hosts; jack, lodgepole, and shortleaf pines are also susceptible.

Distribution: The disease ranges from British Colombia south into California and east inclusive of Arizona and New Mexico into western Montana. Isolated locales include Lake Timagami, Ontario, and Gilpin County, Colorado.

Causal Agent: The causal fungus is *Elytroderma deformans* (Rhytismatales). The imperfect stage produces black pycnidia but has not been named or fully described.

Biology: Ascospores mature in late summer and early fall in hysterothecia on infected needles and are discharged upon wetting. They are wind-dispersed to new foliage. The

Witches' brooms. From Childs (1968).

fungus colonizes the phloem of needles and twigs. The needles do not fade until the following spring and ascospores develop to repeat the cycle months later. *Elytroderma deformans* is unique among the needle-cast fungi in its ability to colonize new needles as they form. Brooming and shoot deformation are common symptoms, and progression of damage is not solely dependent upon spore inoculum. The fungus colonizes twigs at a rate two to three times greater in the upper crown than in lower branches where it averages 10–13 cm annually. The combined effect of premature defoliation of 1-year-old foliage and perennial branch dieback leads to drastic growth reduction and, very often, tree death.

Epidemiology: The disease is favored by high stand density. Low temperature apparently promotes growth of the fungus because the disease is most severe at higher elevations, and infected trees moved from higher to lower elevations have recovered.

Diagnosis: Trees of all ages can be attacked. Symptoms are most conspicuous in spring when branch terminals assume a "red-flag" appearance, which fades through summer as partially obscured by new growth. Once infected, twigs repeatedly flag each spring and gradually curve upward, ultimately to display globose, dense witches' brooms on the more vigorous branches. The inner bark of such branches and even stems of young trees will show distinctive brown necrotic lesions. In severe cases, affected trees may be killed in a few years, but, more often, the pathogen gradually invades additional crown, weakening the tree and predisposing it to attack by bark beetles and other secondary agents. The first signs of the fungus appear in May and June as pycnidial aggregates, small 1-mm blisters that are concolorous with the blighted foliage. During wet weather, rod-shaped conidia (0.4–0.6 × 4–6μm) are exuded in tendrils that have a granular appearance. Immediately following, the hysterothecia of the perfect stage begin to develop as brownish lines 1–2 mm long. At maturity, the hysterothecia are black and shiny, averaging 10 mm in length; asci are fusiform-clavate (30–45 × 140–240 μm), bearing eight, hyaline, two-celled, fusiform-shaped ascospores (6–8 × 90–118 μm).

Control Strategy: Fungicidal protection is economically and biologically impractical. Damage can be reduced by maintaining thrifty stands and salvaging threatened mature trees before they die. Dense spacing should be avoided in young stands, and crop-tree selection should be based on lack of infection or, if infected, no flags within 1–2 m of the main leader. Salvage in mature stands is guided by half or more of the crown flagged or killed.

SELECTED REFERENCES

Childs, T. W., K. R. Shea, and J. L. Stewart. 1971. *Elytroderma Disease of Ponderosa Pine.* USDA For. Serv., For. Pest Leafl. 42. 6 pp.

Lightle, P. C. 1954. The pathology of *Elytroderma deformans* on ponderosa pine. *Phytopathology* 44:557–569.

DISEASE PROFILE

Lophodermium Needle Cast

Importance: This century-old problem of pine in Europe has caused serious nursery losses in the United States only since 1966. Affected seedlings are often shipped before symptoms become evident, only to blight after outplanting. This mode of spread, followed by subsequent local intensification, has been most critical in the Christmas tree industry.

Suscepts: Some 26 species and varieties of pines can be affected, but greatest injury has occurred on red pine and the short-needled varieties of Scots pine.

Distribution: Strains of the fungus are cosmopolitan in the North Temperate Zone. In the United States the disease has been reported from 17 states; in Canada it is known to occur in British Columbia, Nova Scotia, and Ontario.

Hysterothecia on needles. Courtesy of D. W. French, University of Minnesota, St. Paul, MN.

Causal Agents: Lophodermium pinastri (Rhytismatales) is the perfect stage. The function of the pycnidial state, *Leptostroma pinastri* (Coelomycetes), is unknown.

Biology: Recent recognition of biotypes differing in their pathogenicity explains the long-standing contradiction between European and North American views of the fungus as parasite versus saprophyte, respectively, and the likelihood that introduction of the pathogenic strain accounts for the recent disease outbreaks in the United States and Canada. Presymptom detection in new areas could be masked by the ubiquitous fruiting of the saprophytic strain so common in pine litter. Foliage infection in the north-central states occurs from August through September as a function of inoculum maturation and rain periodicity. The hysterothecia absorb water, open, and discharge ascospores that are carried by wind to healthy trees. Successful germination, penetration, and colonization are not expressed until May or June of the following spring when foliage first spots, then yellows and turns brown, and defoliation ensues. Hysterothecia begin development in these needles by June or July and are mature by August to repeat the cycle.

Epidemiology: In the north-central region, the disease can reach epidemic levels yearly in nurseries because of high seedling density and favorable moisture ensured by irrigation. In plantations it usually takes 2–3 years to become severe due to a more exposed microclimate and fluctuations in seasonal rains. Rain provides sufficient moisture to open the hysterothecia, but significant sporulation and infection have been noted also during dry weather.

Diagnosis: Brown spots with yellow margins develop in spring on 1–year-old needles, which turn brown and drop through June and July when hysterothecia appear, and mature in August. In possible litter mixture with the saprophytic form, the pathogen can be distinguished by absence of black-line partitions between fruiting, larger hysterothecia (1–2 vs 0.8–1.0 mm), and hooked paraphysis tips.

Control Strategy: Cultural measures for avoidance and eradication can be sufficient if generally and thoroughly administered. Periodic inspection of the crop for early detection of the disease is vital, especially in the nursery where infection builds so rapidly. Once disease is evident, fungicidal protection during the spore-dispersal period (July–September) has given 90% control. Resistant long-needled Scottish and Austrian varieties of Scots pine should be favored over the susceptible, short-needled Spanish and French sources. Red pine is uniformly susceptible but only as nursery seedlings.

SELECTED REFERENCES

Minter, D. W. 1981. *Lophodermium on Pines.* Comm. Agr. Bur., U. K., Mycological paper 147. 54 pp.
Nichols, T. H., and D. D. Skilling. 1974. *Control of Lophodermium Needlecast Disease in Nurseries and Christmas Tree Plantations.* USDA For. Serv., Res. Paper NC-110. 11 pp.

DISEASE PROFILE

Diplodia Tip Blight

Importance: Death of immature pine shoots typifies this usually nonlethal disease, which, on planted exotics, can seriously reduce their windbreak and ornamental values. Native pines may be affected in plantations but never in natural stands. The disease has damaged extensive plantings of radiata pine and other exotics in Australia, New Zealand, and South Africa. In South Africa and neighboring Swaziland, this pathogen is also responsible for a killing root stain disease of stressed loblolly and slash pines.

Suscepts: The host range includes 33 species of pine, but the disease is most frequently encountered on Austrian pine and is commonly seen on ponderosa, red, Scots, and mugo pines.

Hysterothecia on needles. From Skelly, Undated.

Distribution: The disease spans the globe in the temperate zones between 40° and 50° north and south latitudes, respectively, but is not found in Asia. In the United States it is known to occur in a contiguous 30-state area delineated by North Dakota, Maine, South Carolina, and Oklahoma; it also is found in California and Hawaii.

Causal Agents: *Sphaeropsis sapinea* (*Diplodia pinea*) (Coelomycetes) is the cause. The perfect stage, if any, is unknown.

Biology: Pycnidiospores are released and dispersed, primarily during wet weather, from before bud-break and through the growing season. Infection of pine is on developing shoots and foliage via stomates and on umbos of second-year cones. In Nebraska, maximal susceptibility starts when buds open in late April and extends until mid June; previous years' needles are not affected. Although wounds are not necessary for entry, those created by hail and spittle bugs have favored infection. Symptoms, which can appear as early as 2–3 weeks after inoculation, involve only new growth. After several years, affected branches may be killed back to the main stem by unexplained modes of penetration; evidence is lacking for growth of the fungus into older tissues from primary infections. Immature pycnidia can appear within weeks following symptoms; however, maturation does not occur until late summer.

Epidemiology: Although pines of all ages are susceptible, injury is more severe in plantings as they approach age 30. This age corresponds with the onset of self-thinning in red pine plantations, and the fungus may be a stress-sensitive contributor to density-dependent mortality. Occasional infection of younger trees is attributed in part to local high inoculum levels. Where the fungus is already established, inoculum is abundant the year after an abnormally wet growing season. Profuse pycnidial production on second-year and older seed cones is a major source of inoculum, which may go undetected for years by occurring on trees that have not yet expressed tip blight. Pine infection is optimal under conditions of free moisture and a temperature of 24°C for 4 hours for maximum spore germination to occur. Tip blight generally increases with time.

Diagnosis: The most conspicuous symptom is stunting and browning of new shoots and needles, usually beginning in the lower crown. Successive attacks result in dieback of branches and tops, but progression to tree death is rare. The small black ostiolate pycnidia appear on killed shoots and stunted needles and usually concentrate under the sheath of the latter. The spores, measuring 35–40 × 16–18 μm, are hyaline when formed but become brown and are rarely septate. Tip blight sometimes displays a "shepherd's crook" conformation, making it indistinguishable from low-temperature injury due to late spring frosts; pycnidia then are definitive in diagnosis.

Control Strategy: Fungicidal protection of new shoots during the critical 2-week period following bud break is the principal focus of control. Two applications of protectant or systemic fungicide at 1-week intervals has proven effective. Chemical or mechanical eradication of inoculum sources on seed cones is impractical; likewise, winter pruning of infected shoots/branches may have cosmetic appeal, but the biological gains are nil. Avoidance of wounds immediately before and during the growing season also applies to shearing of Christmas trees in disease areas.

SELECTED REFERENCES

Peterson, G. W. 1981. *Pine and Juniper Diseases in the Great Plains.* USDA For. Serv., Gen. Tech. Rpt. RM-86. 47 pp.

DISEASE PROFILE

Dothistroma Needle Blight

Importance: Exotic pines in both hemispheres are threatened. In the United States, ornamental and Christmas tree plantings, especially those of Austrian pine, have been damaged. In the southern hemisphere, the disease is linked with extensive forest planting of radiata pine.

Suscepts: The world list includes over 30 species of two-, three-, and five-needled pines, of which 20 are known hosts in North America. Austrian, radiata, and ponderosa pines are especially susceptible planted as exotics.

Distribution: The disease occurs in 22 states of the United States, including southeastern Alaska, and four Canadian provinces.

Causal Agents: Only the asexual stage *Dothistroma septospora* (Coelomycetes) is recognized and has received the most attention since the ascomycetous stage is so rarely found (e.g., only in Alaska, California, Oregon, and British Columbia in North America).

Biology: Conidia from rain-gelatinized acervuli on diseased foliage are rain-splash released and dispersed locally from May to October in the central United States. Old pine foliage is immediately susceptible and remains so through the growing season, whereas new foliage does not become susceptible until about mid July. Red band symptoms appear first in early fall and are attributed, in part, to production of the toxin dothistromin. The blight phase and eruption of the epidermis by acervular stromata soon follow, but effective sporulation does not occur until the next spring.

Epidemiology: Disease impacts are primarily limited to planted stands. Nursery losses have seldom been seen in North America; however, fully half of the origin of *D. septospora* in Christmas tree plantings in Nebraska was explained by movement of infected planting stock. Because of latency up to 6 months, detection is not always possible. Rain-splash dispersal of conidia within infected tree crowns has been verified. One possible mode for

Dothistroma needle blight. Courtesy of D. W. French, University of Minnesota, St. Paul, MN.

wider dispersal is the release of spores into low clouds, which could transport spores long distances. Temperature optima for infection vary considerably, ranging from 18°C in New Zealand to 24°C in Nebraska. Under these respective optima, susceptible pine stands have been subjected to explosive rates of blight increase, as high as 4 (Van der Plank's r value) within just a few years.

Diagnosis: Foliage symptoms first consist of dark-green bands and yellow/tan spots that soon turn brown to reddish brown. The reddish coloration is more intense on pines in west coast states where red band disease is common nomenclature for the blight. Distal necrosis ensues, sometimes within 2–3 weeks, and premature, but extremely time-variable, defoliation follows. Infection is typically most severe in the lower crown but can progress to total crown involvement. Subepidermal black acervular stromata (ca. 0.5 × 1.0 mm) begin to develop in the fall and mature in the spring. Conidia are cylindrical to slightly curved, usually three-septate (range 1–5), and hyaline. The perithecial stage being so scarce has little diagnostic value.

Control Strategy: Various epiphytotics in extensive plantings of radiata pine abroad opened the search for control. Early success was attained by use of copper fungicides. Besides fungicidal protection, the future holds promise for selection from blight-resistant provenances in major suscepts. Such clones have been identified in Austrian, radiata, and ponderosa pines.

Other Related Diseases: *Dothistroma* blight and brown-spot diseases on common hosts are so similar that diagnostic separation comes down to conidial differences, (i.e., those of *D. septospora* are hyaline in contrast to olive coloration of the other). The fact that Scots pine is resistant to *D. septospora* and highly susceptible to brown spot is useful for both diagnosis and species selection purposes.

SELECTED REFERENCES

Gibson, I. A. S. 1972. Dothistroma blight of *Pinus radiata. Annu. Rev. Phytopath.* 10:51–72.
Peterson, G. W. 1981. *Pine and Juniper Diseases in the Great Plains.* USDA For. Serv., Gen. Tech. Rpt. RM-86. 47 pp.

DISEASE PROFILE

Rhabdocline Needle Cast of Douglas-fir

Importance: *Rhabdocline* needle cast can severely retard and devalue young plantations of Christmas trees. It is caused by a fungus endemic in native stands of Douglas-fir. Beyond this origin the disease developed after introduction of the host. Plantation success has been most seriously curtailed in western Europe.

Suscepts: Douglas-fir in its various growth forms is the only suscept of consequence.

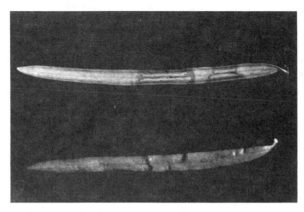

Light colored fruiting bodies on needles. From Bega (1979).

Distribution: Injury was first noted in Scotland (1914) and then Germany (1926) and is now found throughout western Europe, including Scandinavia. In the United States this disease has been sporadically and locally epidemic in plantings in the northeast since the 1930s. Now, it is nearly cosmopolitan wherever Douglas-fir is grown exotically in the United States.

Causal Agents: Rhabdocline needle cast is caused by *Rhabdocline pseudotsugae* (Rhytismatales), with subspecies *epiphylla,* characterized by a subepidermal apothecium. Forms having ascal tips with pores are of another species, *weirii,* with subspecies *oblonga,* and *obovata.* Asexual stages are not known, but *Rhabdogloeum pseudotsugae* (Coelomycetes) is suggested as one in association with *R. weirii.*

Biology: Most of the literature is based on the monotype *R. pseudotsugae,* the only form known in Europe. With recognition in 1969 of additional taxa, however, the pathogenicity and distribution of each has been a matter of concern. In general, ascospores are wind-dispersed to new foliage as it emerges during late May and early June. Wet weather optimizes spore germination and direct penetration when sustained for 3 days at approximately 13°C. Latency is marked by a growth optimum of 10°C and intercellular massing of mycelium in the meosphyll underlying the developing yellow spots on the foliage in autumn, progressing to purplish-brown mottling by spring. Sporulation follows and peaks after bud break, at which time the affected 1-year-old foliage is cast.

Epidemiology: Relative humidity of 100% at 13°C for 3 days is optimum for ascospore release and foliage infection, resulting in 94% infection; lesser times and higher day–night temperatures are much less favorable. Differences in the various taxa of *Rhabdocline* may have significant epidemiological consequences. For example, *R. pseudotsugae* fruits in British Columbia from late May to mid July, while that of *R. weirii* appears in mid June to early August. Depending on when Douglas-fir flushes in the region, it may escape the bulk of inoculum from one or the other. This could be offset, however, by late season, secondary flushes of "Lammas" shoots found susceptible to *R. weirii* in New York state. Spread of *Rhabdocline* in plantations in New York state was estimated at 30–76 m per year downwind; at these rates and over 4 years of climatic extremes, disease incidence reached 55–75%.

Diagnosis: First symptoms appear in autumn as slight yellowing in scattered spots. By late winter, infected foliage is easily distinguished by the sequence of purple-brown mottling, rupturing of the lower epidermis by maturing apothecia, and shedding. The apothecium (0.5–10 mm long) could be mistaken for a hysterothecium, but close inspection will show a raised flap of epidermis and partial exposure of the orange-brown hymenium. The asci are clavate, measuring 80–130 × 15–22 μm. Those of *R. weirii* have an apical pore that stains blue (for iodine) in Melzer's reagent; those of *R. pseudotsugae* do not take the stain. The hyaline, one-celled ascospores are bone-shaped with a size range of 13–22 × 4–9 μm for all taxa; spore septation occurs after discharge when one of the cells darkens to germinate.

Control Strategy: Losses in Christmas trees can be minimized by early detection and timely application of a fungicide, which must first be applied at bud break and repeated at 2- to 3-week intervals until shoots and foliage are fully grown. Two years of fungicidal protection should restore moderately infected trees to full foliage and marketability. Most plantations contain some trees with expressed resistance that should be favored whenever possible and used as seed sources.

SELECTED REFERENCES

Bega, R. V. 1979. *Diseases of Pacific Coast Conifers*. USDA For. Serv., Agric. Handbook No.521. 206 pp.

Parker, A. K. and J. Reid. 1969. The genus *Rhabdocline* Syd. *Can. J. Bot.* 47:1533–1545.

DISEASE PROFILE

Swiss Needle Cast of Douglas-fir

Importance: Swiss needle cast can severely affect young plantations of Douglas-fir, especially when grown for Christmas trees. This disease is caused by a fungus endemic in and innocuous to native stands of Douglas-fir. Beyond this origin, the disease developed in the wake of exotic introduction of the host at a number of locations in both temperate zones. Plantation success has been seriously curtailed in southern Germany. The limiting effects of this disease epitomizes the dangers encountered in attempting to establish a tree species beyond its natural range.

Suscepts: Douglas-fir in its various growth forms is the only suscept of consequence.

Distribution: Swiss needle cast was discovered in Switzerland in 1925, hence the common name. This disease now occurs to varying degrees throughout western Europe, including Scandinavia. In addition, it has also been reported from Japan and New Zealand. In the United States it has occasionally become epidemic in Christmas tree plantings since its recognition in the northeast in the 1930s. Now, it occurs with Douglas-fir wherever it is grown exotically in the United States.

Causal Agents: Swiss needle cast is caused by *Phaeocryptopus (Adelopus) gaumannii* (Dothideales), as distinguished by stomatal perithecia. Only the sexual stage is known.

Small, black fruiting bodies on needles. Courtesy of D. R. Bergdahl, University of Vermont, Burlington, VT.

Biology: In general, ascospores are wind-dispersed to infect new foliage as it emerges during late May and early June. Wet weather may favor *P. gaumannii* as suggested by accounts of disease prevalence in regions typified by cool, wet springs. *P. gaumannii* is systemic and gradual in its development, with slight apical yellowing of the foliage in the fall and subsequent petiolar advance over the next several years. Shedding of these needles is time-variable, therefore, sporulation may occur once in the spring or recur up to three times on persistent needles.

Epidemiology: The importance of moisture for ascospore release and foliage infection has been derived from field observations.

Diagnosis: First symptoms in autumn are barely discernable as slight yellowing at needle tips. Yellow-green discoloration and casting of infected foliage is time-variable and may span 3 years during which ventral, sootlike bands of stomatal perithecia develop in successive winters. First fruiting sometimes appears as early as November on still green, healthy-appearing needles. The perithecium is small, about 0.1 mm in diameter, and globose with a conical base that fills the substomatal cavity of the needle. Each contains about 15 clavate asci (30 × 13 μm) bearing eight hyaline, unequally elliptical, uniseptate ascospores, measuring 14 × 4 μm.

Control Strategy: Suppression of this disease is an imperative in the culture of Douglas-fir Christmas trees. Losses can be minimized by early detection and timely application of a fungicide. Time of application(s) is especially critical. First application can be delayed until new shoots are about 1–5 cm long. A second spray 2–3 weeks later and even a third if rainfall is high are recommended. Two years of fungicidal protection should restore moderately infected trees to full foliage and marketability; heavily infected plantings may take longer. Resistance in Douglas-fir is widely reported. European experience notes that the mountain form of Douglas-fir is susceptible to *P. gaumannii.* Trees that appear resistant should be favored whenever possible and used as seed sources.

SELECTED REFERENCE

Skilling, D. D., and H. L. Morton. 1983. *How to Identify and Control Rhabdocline and Swiss Needlecasts of Douglas-Fir.* USDA For. Serv., Leafl. HT-59. 1 p.

DISEASE PROFILE

Air Pollution Effects on Conifers

Importance: The first significant and well-documented example of air pollution injury to conifers was on the West Coast and began in 1953 in the San Bernardino National Forest, 144 km east of Los Angeles. Known initially as chlorotic decline, it was later found to be due to interaction of components of smog. A similar disease in the eastern United States was known as emergence tip burn.

Suscepts: In western forests, ponderosa, Coulter, sugar, and Jeffrey pines were most susceptible. These species were stressed by the air pollutants and then attacked by root-decay fungi and bark beetles. Sugar pine was relatively tolerant, and giant sequoia was completely resistant. In the East, eastern white pine was the species first studied, but symptoms have subsequently been observed on many coniferous species.

Distribution: Stringent auto emission reductions in California have reduced the impacts of air pollution. In the East there is increasing evidence that air pollution symptoms are widespread.

Causal Agents: In the San Bernardino National Forest, ozone, nitrogen oxides, and peroxyacetyl nitrate, all forming directly or indirectly from automobile emissions, were the major causes of symptom development and tree predisposition. In the East, symptoms of injury were first observed on trees in the vicinity of a coal-fired electricity-generating plant, a pulp mill, an iron smelter, and some other industries. Causal pollutants included sulfur dioxide, nitrogen dioxide, and ozone.

Biology: Sulfur dioxide and ozone enter the needles through stomates, destroy the chlorophyll, and impair metabolic processes. Nitrogen oxides react with water on or within the leaf surfaces to produce toxic compounds that cause tissue injury.

Epidemiology: In the Los Angeles basin, automobile emissions, as smog, began to concentrate in the early morning as traffic increased. Prevailing westerly cool marine breezes carried the smog eastward and upward into the San Bernardino Mountains where

Needle-tip chlorosis. From Skelly, Undated.

it was prevented from passing over the mountains by a persistent temperature inversion line overlain by subtropical warm air. This produced ozone concentrations in excess of 0.4 ppm. Needle symptoms were later duplicated by exposure to synthetic ozone at 0.4 ppm, 8 hr/day for 2–3 weeks. In the East, injurious concentrations of sulfur dioxide are found mainly in the vicinity of major pollution sources. Nitrogen oxides are rarely found in concentrations sufficient to cause visible injury to vegetation. Ozone presently occurs in concentrations sufficient to cause visible injury in most of eastern North America.

Diagnosis: Sulfur dioxide causes tip necrosis or banding of needles and premature defoliation. Ozone produces chlorotic mottling, with older needles being more sensitive. Nitrogen oxides produce lesions on needle surfaces and often cause needle tip burn. Symptoms appear first on older needles.

Control Strategy: Reduction of harmful automobile emissions is a major goal of federal and some state legislation. A secondary goal is that of reducing industrial emissions. The sheer magnitude of the combined emissions, however, is likely to make the process of overall reduction a difficult and expensive one. Many tree species and varieties are resistant to specific pollutants, and some resistance is being incorporated into breeding programs.

SELECTED REFERENCES

Malhotra, S. S., and R. A. Blauel. 1980. *Diagnosis of Air Pollutant and Natural Stress Symptoms on Forest Vegetation in Western Canada.* Can. For. Serv., Info. Rept. NOR-X-228. 84 pp.

Skelly, J. M. (ed.). Undated. *Diagnosing Injury to Eastern Forest Trees.* Nat. Acid Precip. Assess. Program. 122 pp.

Wood, F. A. 1968. Sources of plant-pathogenic air pollutants. *Phytopathology* 58:1075–1084.

15

STEM, FOLIAGE, AND CONE RUSTS

15.1 BIOLOGY OF TREE RUSTS

Among the thousands of fungi that cause rust diseases, several are responsible for major losses in forest trees. The rust fungi are unique in many respects. Many of the species of rust have five spore stages. Although the majority of the rust fungi are *autoecious,* which means they complete their life cycle on one host, the important rust diseases of trees are caused by fungi that are *heteroecious,* which means they complete their life cycles on two entirely unrelated hosts. The species that invade the main stem are more destructive and thus more important than those that develop only in needle or leaf tissues. Normally the rust diseases are of greatest consequence to young trees, especially in nurseries and in recently established plantations.

15.2 PINE STEM RUSTS

498

DISEASE PROFILE

White Pine Blister Rust

Importance: White pine blister rust is an internationally important disease, introduced into the eastern United States (New York) about 1906 on diseased pine nursery stock from Germany, with a similar introduction from France occurring on the west coast (British Columbia, Canada) about 1910. This disease presently limits the planting and management of white pine in areas of high-rust incidence.

Suscepts: Nearly all species of white pines are susceptible. Naturally infected white pines in the United States include eastern, western, sugar, white bark, and limber pines. The disease has not been found in native stands of foxtail, bristlecone, or Mexican pines. Several foreign species, *Pinus cembra* and *P. peuce,* are resistant. Nut or pinyon pine together with its varieties *P. parryone, P. edulis* and *P. monophylla* and an oriental white pine, *P. aramandi,* are immune. Most, if not all native species of currant or gooseberry (*Ribes* spp.) are susceptible. About one half of these, some 40 species of *Ribes,* occur within the white pine range.

Distribution: White pine blister rust is widespread through the ranges of eastern, western, and sugar pines, except perhaps for the southern extremities of their ranges.

Causal Agents: White pine blister rust is caused by the fungus *Cronartium ribicola* (Uredinales).

Biology: Cronartium ribicola is a typical heteroecious, macrocyclic rust with pycnial and aecial stages on pine galls and uredinial and telial stages on *Ribes* leaves. Pycnia occur in late spring or early summer 2–3 years after infection and are followed 1 year later by

Aecial gall. Courtesy of D. W. French, University of Minnesota, St. Paul, MN.

aecia, also in the spring. Aeciospores infect *Ribes* leaves through the stomata, and within 2 weeks uredinia form on the undersurface. Urediniospores can reinfect *Ribes* throughout the summer, producing secondary cycles. Telia grow from the uredinial pustules in several weeks to months. Mature teliospores germinate in moist conditions via a promycelium producing four basidiospores, which infect pine needles through the stomata in the fall. Mycelium grows through the needles and via the inner bark into the branch or stem where galls are formed in 12–18 months. Pycnia form on these galls. In both pine and *Ribes*, the mycelium grows intercellulary, and haustoria penetrate host cells.

Epidemiology: White pine blister rust is a cool, wet-weather disease. Telia require 97–100% relative humidity and 10–18°C for germination and production of basidiospores, which, in turn, require moisture on needle surfaces for germination and subsequent infection of pine. Despite the steep infection gradients from *Ribes* to pine and the climatic limitations, *C. ribicola* has spread rapidly. The fungus *Tuberculina maxima* invades and kills rust-infected cortical tissues destroying *C. ribicola*.

Diagnosis: The diagnostic spindle-shaped gall that occurs on branches and stems is preceeded by needle lesions (often yellow), but these are neither obvious nor always present. Blister rust galls are initially visible as yellow-orange discolored areas in the bark at the base of infected needles. Girdling occurs as aecia erupt through the cambium and bark, producing blisters that allow desiccation and death of host tissues. Flagging or dieback of individual cankered branches often occurs. Signs include honey to brown pycnial droplets and cream-colored aecia and aeciospores produced abundantly on cankers. On *Ribes*, yellow to orange uredinial pustules and short brown telial columns are obvious.

Control Strategy: Eradication of *Ribes* (mechanically, or with 2,4-D or 2,4,5-T), formerly used as a control, has been discontinued in many areas. In high-hazard areas, sanitation (eradication) has not been effective. In low-hazard areas with few *Ribes* and unfavorable climate, eradication is not necessary. However, in intermediate-hazard areas, eradication is biologically and economically feasible. Natural resistance to *C. ribicola* exists, and selection and breeding for disease resistance and propagation of desirable genotypes holds promise. Pruning of branch cankers is feasible to prevent stem cankers in high-value trees. Treatment of diseased trees with *T. maxima* may provide control leads for the future.

SELECTED REFERENCES

Anderson, R. L. 1973. *A Summary of White Pine Blister Rust Research in the Lake States*. USDA For. Serv., Gen. Tech. Rept. NC-6. 12 pp.
VanArsdel, E. P. 1961. *Growing White Pine in the Lake States to Avoid Blister Rust*. USDA For. Serv., Lake States For. Expt. Sta. Pap. 92. 11 pp.

15.2.1   White Pine Blister Rust

Importance

White pine blister rust is one of the most important diseases of the white pines. It has changed white pine management wherever it has been introduced. The rust severely affected timber management in both eastern and western North America. Eastern white, western white, and sugar pines are very important timber species in their respective regions where they are valued for their good growth characteristics and their fine-textured, lightweight woods with excellent dimensional stability and machining qualities. Trees of all ages are killed, but the rust is particularly damaging in young stands, preventing them from reaching a merchantable age (Fig. 15.1).

In the Lake states, it has been locally destructive, and in the East it has not caused large losses, but in the Inland Empire it is one of the most important factors with which foresters must contend. The financial success of forestry in that region depends to a large degree on white pine. If blister rust were allowed to spread in that region, it would kill a large proportion of white pines. The spread of blister rust would mean the abandonment of selective logging in these areas so far as white pine is concerned and doubtless would make it impossible for private companies to practice good management. By 1952, 30 years after the introduction of white pine blister rust, annual losses amounted to 90,000,000 bd ft of saw timber and 75,000,000 bd ft of younger trees. In Europe this disease was one of the chief factors that caused the failure of rather extensive plantations of white pine. Black currants are an important cultivated crop in Europe, and the rust has caused some loss on this host.

FIGURE 15.1 Eastern white pine with multiple blister rust infections. Courtesy of D. W. French, University of Minnesota, St. Paul, Minn.

Suscepts

The white pines (those with five needles in a fascicle) and species of the genus *Ribes* (Grossulariaceae) (currants and gooseberries) are hosts of the fungus that causes this disease. The very susceptible pines are western white pine, sugar pine, limber pine, and white-barked pine. The latter two are not commercially important timber trees. The moderately susceptible pines are eastern white pine and foxtail pine. Eastern white pine is an important forest tree in the Lake states and northeastern forest regions, western white pine and sugar pine are important along the west coast, and western white pine is important in the Inland Empire and in the northern Rocky Mountains. The ranges of the western species of five-needled pines are geographically separate from eastern white pine.

The resistant species are Swiss stone pine and *Pinus peuce,* both endemic to Europe and Asia. Those immune are nut or pinon pine, and its varieties, *P. cembroides, P. parryane, P. edulis, P. monophylla,* and *P. aramandi,* an Oriental white pine. All our native wild Ribes are susceptible; *Ribes nigrum,* European black currant, is very susceptible; *R. alpinum,* an ornamental shrub, is immune (although female plants are susceptible). In the white pine regions of Idaho, *R. petiolare* and *R. inerme* are very susceptible; *R. viscosissimum* is moderately susceptible, and *R. lacustre* is the least susceptible of those four common species. Cultivated currant varieties such as Viking and Red Dutch are resistant and can be used in place of the susceptible varieties such as *R. nigrum.*

Geographic Distribution

It is theorized that the causal agent of white pine blister rust was endemic to Siberia where it coevolved with *P. cembra.* It was brought to Europe by plant collectors between 1750 and 1850. It became epidemic there about 1880 on *P. strobus,* which had been introduced to Europe about 1705 and was widely planted. Specimens on pines and on Ribes were collected in the Baltic region in 1854, and by 1880 it was generally spread over Europe. Klebahn in 1880 determined that the disease on pines and that on currants was caused by the same fungus in different stages of its life cycle. The fungus was brought into China about 1900 and Japan in 1905. It was first recognized in North America in New York state in 1906 but may have been in the northeastern states as early as 1898 when it may have arrived on diseased nursery stock from Germany. White pine blister rust was found in Vancouver, British Columbia, in 1921 and was traced to shipments of infected nursery stock from France in 1910. Thus the pathogen was introduced to both the east and west coasts of North America from Europe. In the United States, white pine blister rust was initially epidemic along a spreading front, but it is now endemic over most of the northern ranges of native five-needled pines (Fig. 1.4). Recent reports indicate that the rust is continuing to spread south along the edges of the pathogen range in the Sierra Nevada and in New Mexico. In parts of the range where the climate is not conducive to the rust, it may be epidemic during favorable weather patterns.

Cause

White pine blister rust is caused by *Cronartium ribicola* J. C. Fish. (Uredinales). *Cronartium ribicola* is an obligate parasite on five-needled pines, but it also requires an alternate stage on *Ribes* spp. to complete its life cycle.

Biology

Beginning with the deposition of a viable basidiospore on a susceptible pine needle in the fall, the germ tubes penetrate the needle, apparently through the stomata, and cause a spot. These spots are evident 4–10 weeks after infection. The fungus grows through the vascular bundles of the needle down to the stems or branch, where an orange bark discoloration appears within 12–18 months of initial infection (Fig. 15.2). The margin of this discoloration continues to expand as the fungus colonizes new tissue by passive growth between the cells. The intercellular hyphae derive nutrients through haustoria, which penetrate host cells. The fungus will continue to grow within the bark through intercellular spaces, expanding the canker, until either the fungus or the tree dies. A typical 2-year-old canker will have a yellowish border and spindle-shaped swelling (Fig. 15.3). As the canker ages, it builds up callus ridges that may crack, resulting in a hardened resin flow.

Pycnia (spermagonia) are produced 2–4 years after infection, in late spring or early summer, appearing as yellow-brown blisters on the bark, and pycniospores ooze out

FIGURE 15.2   Eastern white pine seedling with rust infection. The rust infected the lower needle and has nearly girdled the stem. Courtesy of D. W. French, University of Minnesota, St. Paul.

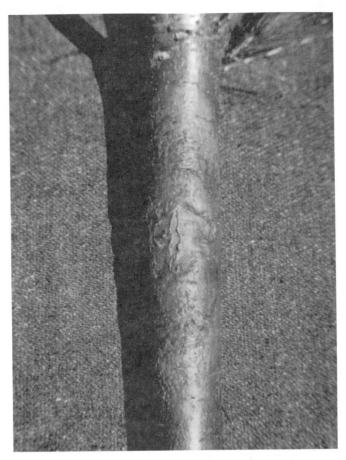

FIGURE 15.3   Spindle-shaped swelling resulting from a young infection on stem of white pine. Courtesy of D. W. French, University of Minnesota, St. Paul.

in a sticky yellow fluid for a short time. Then the blisters and fluid turn hard and black and persist in this form for a long time. Aecia are produced 3–6 years after infection, in spring, appearing as yellow blisters. Where aecia form, the fungus hyphae exert mechanical force and possibly possess pectinase activity. Each "blister" is an aecium covered with a thin membranous peridium containing aeciospores. The peridium breaks, exposing the powdery mass of spores, and the spores are blown by the wind to the leaves of *Ribes* plants. The empty ruptured aecia persist on white pine bark for a long time. On the underside of the *Ribes* leaves, the basidiospores germinate, and the germ tubes enter host cells through the stomata. This mycelium is only local in extent and produces urediniospores 7–18 days after infection. Neither aeciospores or urediniospores can infect pine.

The urediniospores are called repeating spores because they infect other *Ribes* plants. Seven generations of urediniospores may be formed in one season. In a few weeks the same mycelium that gave rise to urediniospores produces telial columns, consisting of teliospores (Fig. 15.4). The teliospores germinate from July to October and produce basidiospores, four per teliospore, which are blown to the needles of

FIGURE 15.4 Telial stage on leaf of currant. From Anderson et al. (1980).

pine, completing the life cycle. The fungus can produce secondary or even tertiary basidiospores in the event that the original spores land on an unsuitable host. This process, called repetitive germination, is possible only if the temperature and moisture conditions remain favorable. The entire life cycle, with emphasis on spore stages, is summarized in Table 15.1 and illustrated in Fig. 15.5. The injury to *Ribes* is minor, although heavy infection can cause premature defoliation.

The mycelium is intercellular in both pine and *Ribes* hosts, and haustoria penetrate into the cells. The fungus stimulates the growth of cells in the bark of pines, the mycelium pushes the cells apart, and when aecia are produced in quantity all around the stem, the bark is disrupted to such an extent that the tree dies of girdling.

TABLE 15.1 Summary of the Life Cycle of *Cronartium ribicola*

Kind of Spores	Host on Which Formed	Season Production	Maximum Viability	Agent of Dissemination	Distance Dissemination
Pycniospore	Pine	Late spring, early summer	Not known	Insects, rain	Few centimeters by rain, farther by insects
Aeciospore	Pine	Spring	9 months, may overwinter	Wind	Many kilometers (up to 560 km)
Urediniospore	*Ribes*	Late spring, summer	9 months, may overwinter	Wind	A kilometer or more
Teliospore	*Ribes*	Summer, fall	90 days	None	Not disseminated
Basidiospore	*Ribes*	Summer, fall	30 hours	Wind	Few meters; few kilometers

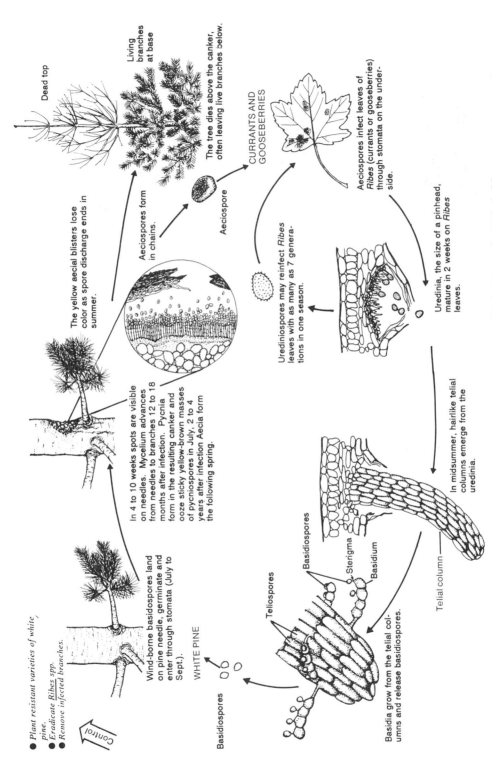

Dead top

Living
branches
at base

The tree dies above the canker,
often leaving live branches below.

Aeciospores form
in chains.

Aeciospore

CURRANTS AND
GOOSEBERRIES

Aeciospores infect leaves of
Ribes (currants or gooseberries)
through stomata on the underside.

The yellow aecial blisters lose
color as spore discharge ends in
summer.

Urediniospores may reinfect *Ribes*
leaves with as many as 7 generations in one season.

Uredinia, the size of a pinhead,
mature in 2 weeks on *Ribes*
leaves.

In 4 to 10 weeks spots are visible
on needles. Mycelium advances
from needles to branches 12 to 18
months after infection. Pycnia
form in the resulting canker and
ooze sticky yellow-brown masses
of pycniospores in July, 2 to 4
years after infection Aecia form
the following spring.

In midsummer, hairlike telial columns emerge from the
uredinia.

Telial column

Teliospores

Basidiospores

Sterigma

Basidium

Basidia grow from the telial columns and release basidiospores.

Wind-borne basidospores land
on pine needle, germinate and
enter through stomata (July to
Sept.).

WHITE PINE

Basidiospores

● *Plant resistant varieties of white pine.*
● *Eradicate Ribes spp.*
● *Remove infected branches.*

Control

506

FIGURE 15.5 Disease diagram of white pine blister rust caused by *Cronartium ribicola.* Drawn by Valerie Mortensen.

Epidemiology

The multiplicity of hosts and spore types provides many opportunities for environmental influence. The most important part of the life cycle from the standpoint of control is the infection of the pine host by basidiospores. Basidiospores need moisture for germination, and their viability decreases directly with increasing dryness of the air and length of time exposed to dry air.

Two weeks of cool temperatures are required before teliospores will germinate. Their germination is inhibited by daytime temperatures >35°C. Telia formed at <20°C are more fertile than those produced at higher temperatures. The degree of *Ribes* leaf maturity may also influence teliospore production. Actual germination of teliospores, formation of basidiospores, and infection by these spores requires approximately 48 hours of a saturated atmosphere and a temperature of less than 28°C. Basidiospores are probably released at night following rain.

Germination of aeciospores, production of urediniospores, and germination of urediniospores also require cool conditions. Aeciospores germinate from 8 to 24°C (optimum at 16°C). Urediniospores are formed with daytime temperatures from 16 to 28°C and nighttime temperatures from 2 to 20°C. Urediniospore germination occurs from 16 to 28°C (optimum at 20°C) and requires free water.

Climate has a major impact on the intensity of white pine blister rust. In California and Oregon it was determined that rust hazard on sugar pine depended on conditions in summer and fall, which allowed maintenance and intensification of rust on *Ribes*. Prevailing moist winds from the Pacific coast helped spread the disease rapidly to the north and east. However, spread to the south was much slower because prevailing northerly winds occur primarily in dry years, and thus the rust could not complete its life cycle.

Basidiospores are only 10–12 nm in diameter. They do not easily settle out of air currents and are subject to long-distance transport. They are also very delicate and have short viability. They were initially thought to be responsible only for short distance local spread from *Ribes* to pine. Uredinia, aecia, and transport of infected plants by humans, were held responsible for any movement of the disease over a few hundred meters. However, Van Arsdel (1965) found that, in the Lake states, basidiospores released into favorable night breezes could be spread considerable distances. Topography, vegetation patterns, and air-cooling patterns influence the movement of air masses at night (Fig. 15.6). If the conditions in these air masses are moist and cool enough, basidiospores can be carried several kilometers and remain viable. Balloons and smoke were used to demonstrate the potential movement of spores. These techniques showed that updrafts from bogs may carry basidiospores to trees on slopes some distance from the *Ribes* plants, which serve as the source of inoculum. Thus where the macroclimate is favorable for survival of basidiospores, infection of pines cannot be easily managed. In the Inland Empire, macroclimate is usually favorable for infection of pines, but in the Lake states and the northeastern United States, macroclimate is variable enough to permit the mapping of hazard zones (Fig. 8.11). Incidence of infection is so related to environment that forested regions can be divided into zones ranging from minimal disease to high incidence.

Where the macroclimate is not favorable, infections of pines can occur only a short distance from the site of basidiospore production. Local effects, such as slope, slope position, and stand opening size can temper this distance somewhat by modifying

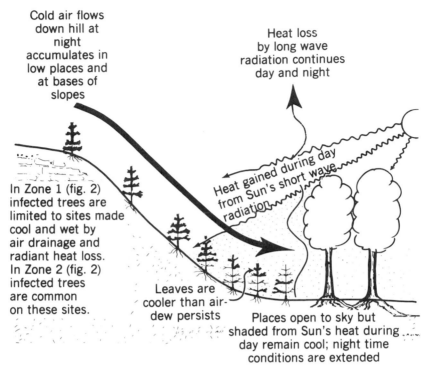

FIGURE 15.6 Two processes (air drainage and local radiant heat loss) that result in local cool-moist areas favorable to rust. From Van Arsdel (1961).

moisture and temperature conditions. Under a tree canopy, microclimate conditions will be less favorable because nighttime temperatures will not be as cool, and dew will form on the overstory rather than on young trees near the forest floor. In a small clearing, cool air concentrates and dew forms. Air movements in clear-cut areas resulted in 50 times as much infection in the middle of the opening as compared to the edges, even though the alternate hosts were uniformly present.

Stand characteristics are also important because they affect the alternate host. *Ribes* are widespread, and different species are common components of both moist and dry pine habitats, particularly in western North America. Under fully stocked, mature stands, *Ribes* will decline and produce less seed. But seeds may be retained in the organic litter layer for decades without losing viability. Silvicultural practices that expose stored seed and remove the overstory encourage abundant *Ribes* regeneration. Practices that maintain forest cover on a site discourage growth of *Ribes*. Burning releases stored seed. The timing of fires for site preparation can determine whether *Ribes* will be released or destroyed. However, along streams and ridges, no amount of overstory manipulation will eliminate *Ribes*.

Thinning from below usually increases *Ribes,* but crown thinnings and selection thinnings do not favor *Ribes* germination and development. Pruning does not increase *Ribes* populations if the slash is not burned.

Ribes species differ in susceptibility to white pine blister rust, but even the least susceptible can produce appreciable inoculum. Cultivated black currant and gooseberries are highly susceptible, as are many wild species.

Pine susceptibility varies greatly by needle and tree age as well as by species and genotype. Young pine foliage is generally more susceptible than old needles. Susceptibility of eastern white pine needles appears to increase with age. On sugar pine, the second-and third-season needles are most susceptible. The fourth-season needles are slightly less susceptible, and current-season needles are significantly less susceptible.

Vigorous hosts are more susceptible, thus, 4-year-old seedlings are more susceptible than grafts of 4-year-old trees. Comparison of grafts of different aged trees indicates that the needles and bark of older trees are more resistant. Bole canker growth rate decreases with tree age or size, and most fatal bole infections develop from branch cankers. If the branch is shaded to death before the canker reaches the bole, a fatal bole canker will not develop.

As white pine blister rust has become endemic in stands, it has also become a part of other pest interactions. Kulhavy et al. (1984) proposed a decline sequence in western white pine stands beginning with invasion by blister rust followed by root diseases and finally bark beetle attack.

Diagnosis

Infected small branches of pine become swollen and orange brown; large stems frequently become ridged when the mycelium has been present for some years. Cankers form on the stems; the bark in the cankered area breaks, and resin flows down the stem and hardens in masses providing a characteristic symptom. Branches girdled by the fungus will have dead, brown needles; the dead branch is called a flag and can be easily spotted. In May and June, blisters filled with yellow-orange spores will appear on the infected branches. The infections on *Ribes* leaves result in the formation of spots sometimes so numerous that they cover the underside of the leaf. Orange masses of urediniospores are produced on these spots in early summer, and later the characteristic telial columns, which appear as brownish hairs on the under surface of the leaves, are formed.

Control Strategy

Various cultural practices, chemicals, and biological agents have been proposed or used to manage white pine blister rust. With changes in disease distribution, knowledge of disease relationships, and public awareness, control theory for white pine blister rust has gone through several distinct stages. After the failure of early attempts to eradicate *Cronartium ribicola* from North America, the prevailing control method until the 1950s was eradication of *Ribes* (Fig. 15.7). Although *Ribes* eradication continued into the 1960s, during the period from the late 1950s to the early 1960s, antibiotics received much attention. From the 1960s until the present, a more integrated approach including resistance, site hazard recognition, and silvicultural practices has been employed.

Ever since the introduction of white pine blister rust to the United States and until recently, *Ribes* eradication has been the primary control measure for the disease. *Ribes* eradication was based on the fact that basidiospores usually live less than 10 minutes. In Europe the eradication of *Ribes* species in the vicinity of imported eastern white pine was never widely accepted because the *Ribes* were more valued than the nonna-

FIGURE 15.7   A *Ribes* eradication crew applying chemical herbicides. From Pack (1933).

tive pine. For eastern white pine in the United States, removal of the top and root crown of all currant and gooseberry within 200 to 300 m and cultivated black currant within 1 km of pine was recommended.

In the western United States, *Ribes* was more abundant than in the east so hand eradication was not as feasible. Even with the use of chemicals such as 5% sodium fluoride in 4% aqueous sodium hydroxide and chemical silvicides such as 2,4-D and 2,4,5-T, *Ribes* eradication was labor-intensive and required constant attention to achieve limited success.

Unfortunately, the effectiveness of eliminating *Ribes* was not adequately evaluated; thus the success of such programs is not actually known. Economists have shown that *Ribes* eradication is feasible, returning $4 for each $1 invested, but their study was based on the questionable premise that reduction of the alternate host population to low levels prevented infection of the pine. This premise might be true if eradication programs eliminated all the *Ribes* plants, but this is not the case. Currant and gooseberry plants are missed, and a great deal of inoculum can be generated by a few small plants. In areas with a recent fire history, pulling *Ribes* plants actually resulted in more plants. The use of certain chemical silvicides has been curtailed making it even more impractical to eliminate these alternate hosts from extensive regions where pine existed.

By the 1960s, the economics of *Ribes* eradication were finally questioned in both the eastern and western United States. In the west, eradication alone did not reduce infection to an acceptable level, and in the Lake states, eradication was ineffective in areas with a favorable microclimate. In 1968 the *Ribes* eradication program was discontinued on national forests in the northern Rocky Mountain region. Presently, eradication is useful only in localities where white pine is valued for more than its timber (e.g., in recreation areas).

Chemical treatment of *C. ribicola* infections on pine was initially seen as the solution to all blister rust problems. Acti-dione (cycloheximide) and another antibiotic, phytoactin, were claimed effective on branch cankers and somewhat effective on trunk cankers. These two compounds are antibiotics produced by one of the actinomycetes, *Streptomyces griseus*. These materials were applied in oil or water to excised cankers, to the basal portion of trees with cankers, and to foliage by aerial applications. The USDA Forest Service treated thousands of hectares of western white pine with phytoactin and cycloheximide before they determined it was ineffective and discontinued treatment in 1965. Although it appeared at first that the antibiotics were translocated away from the point of application, the fungus survived the treatment and later continued normal development. Apparently these first attempts at a direct chemical control of blister rust have not succeeded, but for the future this remains as a possibility.

The most promising control measure for the future is the development of resistant white pine. In early work with eastern white pine, vegetatively produced trees from original resistant selections were resistant, while seedlings from these same trees were quite susceptible. This difference was probably due in part to the differences in age of needles. Many mechanisms of resistance to white pine blister rust have subsequently been identified in five-needled pines, including prevention of needle lesions, reduced frequency of needle lesions, premature shedding of infected needles, a fungicidal reaction in the short shoot, a reaction that eliminates established bark infections, and the ability to remain alive when infected. When 18 species of white pine were compared on these factors, Eurasian species had a high level of resistance, mostly due to prevention of needle lesions. Eastern white pine was the most susceptible species, but sugar pine was also very susceptible. Western white pine of resistant parentage was used in this particular study, but it was still moderately susceptible.

Resistance mechanisms of the major pine species have been individually explored. Breeding programs for resistant white pine were initiated by the USDA Forest Service in 1950. The first stage was selection of phenotypically resistant trees from natural stands and analysis of their genetic combining ability for resistance traits. Western white pine has two main resistance responses: early shedding of infected needles and failure to develop stem cankers despite infected needle retention. Field tests have shown high gain in resistance in the F_1 generation and additional gain in the F_2 generation. Much of the inbred resistance is monogenic and, thus, might be easily overcome by pathogen population shifts toward a more virulent race, so geneticists are trying to incorporate all three vertical resistance responses plus additional horizontal resistance mechanisms into breeding programs. In newer seed orchards, breeding is being manipulated to combine more genes and thus stabilize the resistance. Use of resistant seed in areas of low to moderate infection hazard has been suggested to maintain a genetic breadth between highly resistant western white pine and highly virulent pathogen populations and to reduce the demand for the limited supply of highly resistant seeds. Low-and moderate-level resistance can be used effectively in combination with other control methods to minimize losses and maintain genetic diversity.

Whereas western white pine has multiple-factor resistance with simple additive effects, resistant sugar pine has a single-gene, simply inherited, nearly absolute resistance mechanism. This major gene resistance factor causes a hypersensitive reaction in secondary needles and bark, but not in primary needles. This major resistance gene response can be detected in seedlings as young as 8 weeks and in tissue culture embryos. Unfortunately, a race of *C. ribicola* virulent to the major gene resistance factor has been confirmed. Fortunately, there is also multigene "slow-rusting" resistance, which appears to have good potential for development and improvement.

Shifts in the virulence of the *C. ribicola* population have been a concern in all white pine types. The population is variable, as is demonstrated by segregation of lesion types and differences in epidemiological fitness traits of single aeciospore cultures. In addition to the aforementioned virulent strain on sugar pine, a strain has been reported on western white pine in the Cascade Mountains that has different virulence characteristics than the typical wild-type inoculum from Oregon. This strain requires a longer incubation time and causes increased rust intensity, reduced shedding-of-the-needle response, reduced proportion of seedlings with retarded canker appearance, increased proportion of seedlings with stunted leaders, reduced period from inoculation to mortality, and reduced proportion of trees healthy 2 years after inoculation.

Eastern white pine breeding strategies have included cloning of resistant individuals, use of resistant European pines in breeding programs, and exploitation of low-level resistance in low-risk areas. Planting of resistant eastern white pine is recommended in an integrated management scheme, but other control strategies seem to be more important in eastern United States.

Cronartium ribicola, like all fungi, has the ability to vary, and thus new more-virulent strains may attack the resistant trees. Only time will tell how effective resistant white pine will be in reducing losses to blister rust.

Site hazard rating has become one very useful management tool, particularly for eastern white pine. The purpose of hazard rating is to escape as much blister rust as possible by site selection and stand manipulation. Hazard zones can be delineated on the basis of macroclimate patterns. Individual site hazard ratings depend on the physical site and stand characteristics that affect air movement patterns, such as stand structure and canopy closure, slope, slope position, aspect, and topography.

In the Lake states, four hazard zones are identified on the basis of macroclimate. In the low-hazard zone, losses will be less than 5% with no additional control. In the moderate-hazard zones, species other than white pine should be planted on high-hazard microsites, and where white pine is planted, *Ribes* in the stands and in air drainages into the stands should be eradicated. In high-hazard zones, pine should be planted only under overstories or where *Ribes* has been eradicated. In the very high-hazard zone, pine should be planted in large blocks to warrant the costs of blister rust control. To minimize losses in this zone, an overstory cover should be maintained, resistant stock should be planted, *Ribes* should be eradicated, and barrier strips should be established to reduce air transport of aeciospores. The closed canopy keeps the air dry below the canopy, and the more susceptible and cankered branches are killed because of the shade. White pine should not be planted in small openings. Openings in crown canopy with a diameter less than the height of the surrounding trees are cool, wet, and ideal for blister rust infection.

Charlton (1963) derived a probability equation based on climatic factors and used it to delineate three hazard zones for the 14 eastern states. He considered cool temperatures, moisture or high humidity, air movement, and duration of favorable conditions as they were affected by weather, topography, and vegetation. His equation correlated quite well with actual incidence of white pine blister rust.

Hazard areas have also been identified in the southern Appalachians. The known high-hazard zones are in parts of Virginia and Kentucky. On the highest-hazard sites in these zones, planting eastern white pine is not recommended. Where white pine is planted, *Ribes* control and pathological pruning are incorporated into management practices.

On western white pine, the only hazard rating appears to be in conjunction with resistance programs. Blister rust is a more recent arrival in the sugar pine region, so fewer programs have been developed.

Most infections occur within 2 m of the ground, and in well-stocked stands, as well as in plantations, pruning is a feasible control measure. It is labor-intensive but effective. Infected branches should be removed allowing 15 cm of what appears to be healthy tissue between canker and main stem. However, for pruning to be economical, all the lower branches should be removed, thus ensuring more complete elimination of the rust and improving wood quality. Pruning is particularly useful on moderate hazard sites, but it may be beneficial in any of the zones. Plantations and ornamental white pine should be surveyed in May when infected branches are most apparent, and these branches should be removed. Another practice that can be considered is manipulation of stand density to cause early natural pruning of lower branches. Infection can also be carved out of the main stem, but this is feasible only for trees with very high values. Rodents such as squirrels and porcupines often feed on the bark surrounding cankers, and this has sometimes eradicated that infection (Fig. 15.8).

In both western and eastern United States a fungus, *Tuberculina maxima,* occurs

FIGURE 15.8 Western white pine with rodent-feeding activity around blister rust canker. Courtesy of D. W. French, University of Minnesota, St. Paul.

on blister rust cankers and is responsible for the death of the fungus (Fig. 15.9). It causes a disease called purple mold. Approximately 67% of the lethal-type cankers were inactivated by *Tuberculina maxima*, and aecial production was 18% of normal in 1965. Cankers are susceptible to infection only while producing aecia or pycnia, and sporulation varies with the year thus restricting *T. maxima*. *T. maxima* invades and enzymatically degrades rust-invaded cortical tissues of western white pine thus destroying *C. ribicola*. There is no evidence that *T. maxima* is parasitic on the rust fungus; thus it is not a hyperparasite. *Tuberculina maxima* is prevented from invading healthy pine tissue by the middle lammella.

Tuberculina maxima has several characteristics of a good biological control agent, including inability to invade nondiseased host tissue, stability of infectivity and pathogenicity, a short incubation period and asexual life cycle, and ease of field inoculation. However, the presence of *T. maxima* on declining cankers is probably an indication of low canker vigor rather than the cause of inactivation, and *T. maxima* often dies before invasion reaches the extent of *C. ribicola* infection.

Diseases forecasts are valuable on annual crops or fruit crops to estimate and predict epidemic disease intensity so that control can be efficiently used. Forecasts may

FIGURE 15.9   Eastern white pine with mature aecia and adjacent infection by *Tuberculina maxima*. Courtesy of D. R. Bergdahl, University of Vermont, Burlington.

be based on the amount of initial inoculum, the conditions affecting secondary inoculum, or both. For a forecasting system to be effective, it must be reliable, the disease must be important but sporadic, the crop must be important, control methods must be available, and communication to crop growers must be sufficient.

Unfortunately, white pine blister rust on five-needled pines does not readily fit the purpose of nor criteria for disease forecasting. The final timber crop from a forest depends on many ecosystem interactions (including disease) over many seasons of growth, so a forecast for a particular season may not give a reliable estimate of impact on the final crop. *Cronartium ribicola* has become a naturalized stand component over most of the ranges of white pines, although the occurrence of white pine blister rust in a stand may be both important and sporadic. White pines are certainly important timber species, but the economics of timber management are marginal under the current market conditions. Most blister rust control methods are long-term practices aimed at prevention of unacceptable levels of loss rather than postinfection treatment.

Although disease forecasting is not as applicable to white pine blister rust as to other diseases, a computer simulation model based on western white pine has been developed to study the behavior of rust epidemics. This model is not a forecasting system but rather a tool to combine all the facts known about white pine blister rust and analyze different control strategies.

The model is organized into five separate submodels: *Ribes* density, *Ribes* infection, pine target, pine infection, and stand infection. *Ribes* density is calculated as a function of light intensity or can be set at a given value to simulate eradication practices. The *Ribes* infection submodel uses spore characteristics and climate conditions to predict the amount of inoculum available to infect pine. The pine target submodel calculates the available infection surface. The pine infection submodel tallies the total infection on a simulated host and calculates the annual amount of new infection based on the available inoculum, past tree height, and eight input variables. The stand infection submodel expands the per tree annual infection to a stand basis.

The simulation is based on an average tree in a stand with specified characteristics. The average infection levels are calculated based on the characteristics of an average tree, and these characteristics are extrapolated to yield stand values. This value is assumed to give a reasonable estimate since most white pine stands are even-aged. Host resistance factors, control measures, and pathogen epidemiological fitness traits can be specified and environmental affects can be manipulated. The model is currently being verified.

SELECTED REFERENCES

Anderson, R. L., J. D. Artman, C. Doggett, and C. E. Cordell. 1980. *How to Control White Pine Blister Rust in the Southern Applachian Mountains.* USDA For. Serv., For. Bull. SA-FB/ P23. 1 pp.

Anderson, R. L. 1973. *A Summary of White Pine Blister Rust Research in the Lake States.* USDA For. Serv., Gen. Tech. Rep. NC-6. 12 pp.

Bingham, R. T. 1968. Breeding blister rust resistant western white pine. IV. Mixed-pollen crosses for appraisal of general combining ability. *Silvae Genetica* 17:133–138.

Bingham, R. T., R. J. Hoff, and G. I. McDonald. 1973. *Breeding Blister Resistant Western White Pine. VI. First Results from Field Testing of Resistant Planting Stock.* USDA For. Serv., Res. Note INT-179. 12 pp.

Bingham, R. T., A. E. Squillace, and J. W. Duffield. 1953. Breeding blister rust-resistant western white pine. *J. For.* 51:163–168.

Buchanan, T. S., and J. W. Kimmey. 1938. Initial tests of the spread to and intensity of infection on *Pinus monticola* by *Cronartium ribicola* from *Ribes LaCustre* and *R. viscosissimum. J. Agri. Res.* 56:9–30.

Charlton, J. W. 1963. *Relating Climate to Eastern White Pine Blister Rust Infection Hazard.* USDA For. Serv., Eastern Region. Upper Darby, PA. 38 pp.

Dainer, A. M., and R. L. Mott. 1982. Major gene resistance to blister rust in *Pinus lambertiana* is expressed in tissue culture. (abst.) *Phytopathology* 72:978–979.

Goddard, R. E., G. I. McDonald, and R. J. Steinhoff. 1985. *Measurement of Field Resistance, Rust Hazard, and Deployment of Blister Rust-Resistant Western White Pine.* USDA For. Serv., Res. Paper INT-358. 8 pp.

Harvey, A. E. 1972. Influence of host dormancy and temperature on teliospore induction by *Cronartium ribicola. For. Sci.* 18:321–323.

Hoff, R. J., and G. I. McDonald. 1971. Resistance to *Cronartium ribicola* in *Pinus monticola:* short shoot fungicidal reaction. *Can. J. Bot.* 49:1235–1239.

Hoff, R. J., R. T. Bingham, and G. I. McDonald. 1980. Relative blister rust resistance of white pines. *Eur. J. For. Pathol.* 10:307–316.

Hoff, R. J., G. I. McDonald, and R. T. Bingham. 1973. *Resistance to Cronartium ribicola in Pinus monticola: Structure and Gain of Resistance in the Second Generation.* USDA For. Serv., Res. Note INT-178. 8 pp.

Ketcham, D. E., C. A. Wellner, and S. S. Evans. 1968. Western white pine management programs realigned on northern Rocky Mountain national forests. *J. For.* 66:329–332.

Kimmey, J. W. 1969. Inactivation of lethal-type blister-rust cankers on western white pine. *J. For.* 67:296–299.

King, D. B. 1958. *Incidence of White Pine Blister Rust Infection in the Lake States.* USDA For. Serv., Lake States For. Exp. Sta. Paper No. 64. 12 pp.

Kinloch, B. B., Jr., and J. W. Byler. 1981. Relative effectiveness and stability of different resistance mechanisms to white pine blister rust in sugar pine. *Phytopathology* 71:386–391.

Kinloch, B. B., Jr., and M. Comstock. 1980. Cotyledon test for major gene resistance to WPBR in sugar pine. *Can. J. Bot.* 58:1912–1914.

Kinloch, B. B., and M. Comstock. 1981. Race of *Cronartium ribicola* virulent to major gene resistance in sugar pine. *Pl. Dis.* 65:604–605.

Kinloch, B. B., Jr., and J. L. Littlefield. 1976. White pine blister rust: Hypersensitive resistance in sugar pine. *Can. J. Bot.* 55:1148–1155.

Kinloch, B. B., Jr., G. K. Parks, and C. W. Fowler. 1970. White pine blister rust: Simply inherited resistance in sugar pine. *Science* 167:193–195.

Kliejunas, J. T. 1985. Spread and intensification of white pine blister rust in the southern Sierra Nevada California USA. (abst.) *Phytopathology* 75:1367.

Kulhavy, D. L., A. D. Partridge, and R. W. Stark. 1984. Root diseases and blister rust associated with bark beetles (Coleoptera Scolytidae) in western white pine *(Pinus monticola)* in Idaho USA. *Environ. Entomol.* 13:813–817.

Leaphart, C. D., and E. F. Wicker. 1968. The ineffectiveness of cycloheximide and phytoactin as chemical controls of the blister rust disease. *Pl. Dis. Reptr.* 52:6–10.

Lehrer, G. F. 1982. Pathological pruning a useful tool in white pine blister rust control. *Pl. Dis.* 66:1138–1139.

McDonald, G. I., R. J. Hoff, and W. R. Wykoff. 1981. *Computer Simulation of White Pine Blister Rust Epidemics. I. Model Formulation.* USDA For. Serv., Res. Paper INT-258. 136 pp.

Miller, D. R., J. W. Kimmey, and M. E. Fowler. 1959. *White Pine Blister Rust*. USDA For. Serv., For. Pest Leaflet 36. 8 pp.

Moss, V. D. 1957. Acti-dione treatment of blister rust trunk cankers on western white pine. *Pl. Dis. Reptr.* 41:709–714.

Moss, V. D. 1961. Antibiotics for control of blister rust on western white pine. *For. Sci.* 7:380–396.

Pack, C. L. 1933. *White Pine Blister Rust: A Half-Billion Dollar Menace*. The Charles Lathrop Pack Foundation, Washington, D. C. 14 pp.

Patton, R. F. 1961. The effect of age upon susceptibility of eastern white pine to infection by *Cronartium ribicola*. *Phytopathology* 51:429–434.

Patton, R. F., and D. W. Johnson. 1970. Mode of penetration of needles of eastern white pine by *Cronartium ribicola*. *Phytopathology* 60:977–982.

Pennington, L. H. 1925. Relation of weather conditions to the spread of white pine blister rust in the Pacific Northwest. *J. Agric. Res.* 30:593–607.

Phelps, W. R., and R. Weber. 1969. *Characteristics of Blister Rust Cankers on Eastern White Pine*. USDA For. Serv., Res. Note NC-80. 2 pp.

Pierson, R. K., and T. S. Buchanan. 1938. Susceptibility of needles of different ages on *Pinus monticola* seedlings to *Cronartium ribicola* infection. *Phytopathology* 28:833–839.

Powers, H. R., and W. A. Stegall, Jr. 1965. An evaluation of Cycloheximide (acti-dione) for control of white pine blister rust in the Southeast. *Pl. Dis. Reptr.* 49:342–346.

Quick, C. R., and C. H. Lamoureaux. 1967. Field inoculation of white pine blister rust cankers on sugar pine with *Tuberculina maxima*. *Pl. Dis. Reptr.* 51:89–90.

Riker, A. J., T. F. Kouba, W. H. Brener, and L. E. Byam. 1943. White pine selections tested for resistance to blister rust. *J. For.* 41:753–760.

Robbins, K. 1984. *How to Select Planting Sites for Eastern White Pine in the Lake States*. USDA For. Serv., NA-FB/M-8. 7 pp.

Spaulding, P. 1922. *Investigations of the White Pine Blister Rust*. USDA, Bull. No. 957. 100 pp.

Steinhoff, R. J. 1971. *Field Levels of Infection of Progenies of Western White Pine Selected for Blister Rust Resistance*. USDA For. Serv., Res. Note INT-146. 4 pp.

Van Arsdel, E. P. 1961. *Growing White Pine in the Lake States to Avoid Blister Rust*. USDA For. Serv., Lake States For. Expt. Sta. Pap. No. 92. 11 pp.

Van Arsdel, E. P. 1965. *Relationships Between Night Breezes and Blister Rust Spread on Lake States White Pine*. USDA For. Serv., Res. Note LS-60. 4 pp.

Van Arsdel, E. P. 1967. The nocturnal diffusion and transport of spores. *Phytopathology* 57:1221–1229.

Van Arsdel, E. P., A. J. Riker, and R. F. Patton. 1956. The effects of temperature and moisture on the spread of white pine blister rust. *Phytopathology* 46:307–318.

Van Arsdel, E. P., A. J. Riker, T. F. Kouba, V. E. Suomi, and R. A. Bryson. 1961. *Climatic Distribution of Blister Rust on White Pine in Wisconsin*. USDA For. Serv., Paper No. 87. 34 pp.

Weber, R. 1964. *Early Pruning Reduces Blister Rust Mortality in White Pine Plantations*. USDA For. Serv., Res. Note LS-38. 2 pp.

Welch, B. L., and N. E. Martin. 1974. Invasion mechanisms of *Cronartium ribicola* in *Pinus monticola* bark. *Phytopathology* 64:1541–1546.

Wicker, E. F., and J. Y. Woo. 1973. Histology of blister rust cankers parasitized by *Tuberculina maxima*. *Phytopathol. Z.* 76:356–366.

Wicker, E. F., and J. M. Wells. 1970. Incubation period for *Tuberculina maxima* infecting the western white pine blister rust cankers. *Phytopathology* 60:1693.

DISEASE PROFILE

Fusiform Rust

Importance: Fusiform rust is a major obstacle to intensive forest management in the southern United States. Annual losses occur primarily as mortality and unmerchantability resulting from stem infection of young (nursery to 5–8 years old) trees. Stands in high rust-hazard areas are often nearly 100% infected within 10 years from establishment.

Suscepts: This disease alternates between pine and oak. Most important pine suscepts are loblolly, slash, and longleaf pines, although the latter is somewhat resistant. Shortleaf pine is practically immune; needle lesions occur, but galls do not form. Most important natural alternate hosts are water and laurel oak.

Distribution: Fusiform rust occurs from Maryland south to Florida and west to southern Arkansas and Texas.

Causal Agent: The cause of fusiform rust is *Cronartium quercuum* f. sp. *fusiforme* (Uredinales).

Biology: The rust is heteroecious and macrocyclic with pycnial and aecial stages on pine woody tissues and uredinial and telial stages on oak leaves. The fungus overwinters as mycelium in pine gall tissues. Pycnia occur on pine galls in November and December and are followed in early spring by aeciospores, which are wind-disseminated and infect emerging oak leaves. Within 2 weeks, uredinia are produced on infected leaves. The uredinia contain urediniospores, which can infect succulent oak leaves, initiating secondary cycles. Telia are produced shortly after the uredinia. During humid periods, primarily in April and May, telia germinate and produce basidiospores (sporidia), which infect needles and succulent branch and stem tissues of pine. Subsequent colonization by mycelium causes hypertrophy and hyperplasia of cambial initials and ray parenchyma cells and results in the typical fusiform branch and stem galls, visible in 4–6 months. Subsequently, pycnia are formed completing the disease cycle.

Epidemiology: Fusiform rust, rare at the turn of the century, has increased rapidly in large areas of the South since about 1940. Natural populations of loblolly and slash pine are very susceptible, and increases in the amount of succulent, susceptible pine tissue concurrent with intensive management of these species have caused an increased incidence of rust. With decreased use of fire as a management tool, oak abundance has also increased. Also, slash pine, which occurs naturally on wet, oak-free sites, is often planted on dryer sites with abundant oak. Other factors include the planting of infected nursery stock, a decreased proportion of longleaf pine relative to slash and loblolly, and the increased proportion of young stands.

Diagnosis: Symptoms are fusiform (spindle-shaped) galls on pine branches and stems that vary in length from less than 1 cm to 1 m. Often, stems break at these galls. Mortality may also occur when seedlings and young trees are girdled. Inconspicuous, often purple, needle and stem lesions preceed gall formation. Signs include inconspicuous amber pycnial droplets on pine galls and very obvious orange-yellow aeciospores. On oak leaves, orange-yellow uredinial pustules and red-brown hairlike telia occur primarily on the undersurface.

Control Strategy: In nurseries, spray with a protective fungicide during the period of basidiospore production (February–June) and rogue or cull infected seedlings. On plantations, plant rust-resistant species or mixtures of resistant varieties in high-hazard areas. Discriminate against susceptible oak during management practices when possible,

Aecial gall.

and when economically feasible salvage stem-cankered trees (especially those involving more than one half of the stem circumference) prior to stem breakage. In natural stands, salvage stem-cankered trees when feasible. On ornamental and seed orchard trees, prune branch galls near the stem and excise or chemically treat stem galls.

SELECTED REFERENCES

Czabator, F. J. 1971. *Fusiform Rust of Southern Pines—A Critical Review*. USDA For. Serv., Res. Paper SO-65. 39 pp.

15.2.2 Fusiform Rust

Importance

For many years fusiform rust was considered as a seedling disease that occasionally caused losses after reforestation. Since World War II, it has become the most important disease of loblolly and slash pines and is second only to the dwarf mistletoes in national importance. Considering the fact that fusiform rust occurs only in the forests of the southeastern United States, then it must cause tremendous losses indeed where it occurs. The destructive increase of fusiform rust is a direct result of an interesting combination of past land abuse, fire control, the increasing intensity of forest management, and poor forestry practices that, altogether, have favored the rust. The name fusiform rust is widely accepted now. In older literature, it was referred to as southern pine fusiform rust, southern fusiform rust, or cronartium canker. Fusiform rust is native to the southeastern United States.

Fusiform rust attacks the stem and branches of pines and the leaves of oaks. Main stem galls on pines are lethal, while branch infections that do not reach the main stem may kill only the branch and thus cause minimal injury to the tree. On oaks the injury is negligible, and even where the disease is severe, it results only in a slight reduction in the photosynthetic area of the leaves and a corresponding decrease in tree growth. On pines, however, the injury is often severe. In nurseries, fusiform rust increases the cost of seedling production and reduces the annual output of the nurseries. In the past, when seedlings were sorted and graded by hand, diseased individuals were culled, increasing sorting costs. Seedlings are no longer inspected for disease. Today, machines lift, handle, and package seedlings. When diseased seedings die, they leave gaps in the stand. Natural pruning of pines surrounding the gaps is delayed, which reduces quality of the timber produced by these trees. Weed species such as scrub oak or sweetgum may also be released in these gaps.

As infected trees grow older, they seem to survive infection better, and the percentage killed becomes progressively lower. A canker on the stem that does not kill the tree reduces the strength of the stem so that it may be broken at that point by the wind. The deformation of the trunk greatly reduces the value of older trees. This loss in quality and quantity of wood in older trees often exceeds the loss from death of young trees. Fusiform rust may, however, benefit the stand by freeing resources for more rapid growth of the survivors. Some infected trees are badly stunted and deformed. This damage may be slight if they are used as pulpwood, but it is much more serious if they are grown for sawtimber or veneer. The canker may permit entry of wood-decay fungi as the trees mature. Resin-soaked gall tissues contribute to further defect.

Losses due to fusiform rust were estimated in 1952 to be about $10.5 million annually. There are no reliable recent estimates. In one 9-year-old Mississippi plantation, 27% of the trees planted were dead, 67% were alive with one or more stem cankers, 2% had branch cankers only, and 4% were disease-free. In a survey of two Alabama counties, 34–71% of the trees had trunk infections; 60–82% had trunk and/or branch infections. In southern nurseries, losses exceeded 4,000,000 seedlings in 1938; 3,000,000 in 1939; and 1,000,000 in 1940. These losses were almost entirely in slash and loblolly pines. Losses in these same nurseries in longleaf pines rarely exceeded 2–3%, whereas the losses in slash and loblolly ranged from 15 to 35%. Since those data were collected, fusiform rust has spread eastward into Georgia,

South Carolina, northern Florida, and the coastal plain of North Carolina. In 1975 a very conservative estimate of the annual loss was $28,000,000. Based on the changes recorded for all commodities of southern pine in the producer price index, in 1992 a conservative estimate of this loss was $56,000,000.

Suscepts

The fungus that causes fusiform rust is heteroecious, requiring the presence of two unrelated hosts for the completion of its life cycle. Typical of the *Cronartium* rusts, the pycnial and aecial stages occur on various species of pines; the uredinial and telial stages are on oaks. In 1918 the following species of pines were determined to be susceptible to infection: *Pinus heterophylla, P. palustris, P. rigida, P. serotina, P. taeda, P. contorta, P. coulteri, P. halepensis, P. murrayana, P. montana, P. muricata, P. pinea, P. ponderosa, P. radiata,* and *P. sabiniana.* The rust was found in nature only on the first five species, but it was established on the other species by artificial inoculation. Subsequently, the list of susceptible hosts in artificial inoculations of both pines and of oaks has greatly expanded. Many Diploxylon species of pine are susceptible to fusiform rust as well as two species of Haploxylon pines, *P. lambertiana* and *P. aristata.* Some pine and oak species in western North America are rather susceptible. Jeffrey, Monterey, and ponderosa pines are more susceptible to fusiform rust than slash and loblolly pines. This predisposition is cause for concern should the rust be inadvertantly introduced into that region.

Fusiform rust is presently a major problem only in the southeastern United States and only on loblolly and slash pines. It is a lesser problem on longleaf and pond pines. The relative susceptibility of slash and loblolly has been argued, but it seems that there is more difference in susceptibility of individuals within a species than between species. In nurseries, loblolly pine seeds germinate much more slowly and less uniformly than slash pine seeds. Thus losses in slash pine seedlings are usually greater. Slash pine seedlings that were planted at the same time are exposed to one or more infection periods before the loblolly seedlings emerge.

Many species of oaks may serve as a telial host of the fungus. Twenty-two of 28 species of the genus *Quercus* tested were infected following inoculation. Later, sawtooth oak was added as a new host, while four species of *Castanea* including *C. dentata, Castanopsis diversifolia,* and *Lithocarpus densiflorus* were inoculated and infected. The fungus has been observed in nature only on the oaks. Water and willow oaks are highly susceptible. Blackjack, cherrybark, bluejack, running, southern red, northern red, and turkey oaks are moderately to highly susceptible, and scarlet and black oaks are generally hypersensitive. Few telia are produced on the leaves of white oaks. While many oaks are infected, water and laurel oaks are key hosts, because rust incidence in slash pine plantations is strongly correlated with abundance of water oak and in loblolly pine plantations, only with water and laurel oaks.

Distribution

The *Cronartium quercuum* complex has an ancient lineage extending almost as far back as the origins of the pines, perhaps to the early Jurassic period almost 190 million years ago. The prehistoric range of the *C. quercuum* complex in North America is unknown, but, whatever its distribution, the slate was wiped clean during the late

Wisconsin ice age of the Quaternary period, at its maximum about 18,000 years before the present. The southern pines had migrated far south of their present locations and then migrated northward as the glaciers receded. It has been suggested that *C. quercuum* f. sp. *fusiforme* subsequently migrated northward along with loblolly pine out of refugia in Central America and with slash pine from refugia on the Bermuda land mass. Before this time, slash pine and loblolly pine independently coevolved a relationship with the rust that was probably not threatening to either species. This was disturbed when the natural ranges of these two pine species overlapped, perhaps as recently as 4,000 years ago.

Recorded history of fusiform rust began with a description of the disease by Underwood and Earle in 1896. They believed, however, that the spindle-shaped galls on loblolly pine were merely the expression of what was known at that time as the eastern gall rust and later as the pine-oak rust. At that time fusiform rust was relatively scarce, perhaps because of a balanced interaction between the hosts, the pathogen, and the environment. The stability of this natural forest system was disrupted as European settlers entered the region.

This imbalance was worsened by subsequent attempts at forest management. The pine forests of the southeastern United States were a fire subclimax and maintained as such by natural and native-set fires. These fires determined the composition and structure of the pine forests. Fire-resistant longleaf pine was favored on higher and drier sites in a wide band extending along the coastal plain from Virginia to Texas. Less fire-resistant slash pine was confined to low areas and margins of swamps in a narrow strip along the coast from South Carolina to Louisiana. Loblolly pine occurred widely, especially in the upper coastal plain and on moist, inland sites not frequently burned. On most sites susceptible oaks were present in the understory. Blackjack, turkey, and bluejack oaks were prevalent on drier sites, and on moist, well-drained sites, water and willow oaks were present.

After the Revolutionary War land clearing intensified, moving from the lower coastal plain and extending inland into the piedmont and foothills of the Appalachian Mountains. There were several alternating sequences of clearing and abandonment. Abandoned fields were invaded by loblolly and slash pines. The resistant longleaf pine was preferred for lumber until well into the early twentieth century and was harvested; because of presumed regeneration problems, this species was not favored during reforestation. Fire control also favored the establishment of slash and loblolly pines, and lack of fire released the suppressed oak understory.

Beginning in about 1920, the cutover or abandoned lands were planted to restore productivity. Forest tree nurseries at that time produced and distributed diseased seedlings. Most of the diseased seedlings died soon after outplanting, but a few survived long enough to produce a crop of aeciospores, often introducing fusiform rust into areas where the rust was not present.

Regional rust surveys begun after 1930 indicated an increase of fusiform rust in the late 1930s starting in southeast Louisiana and southern Mississippi and Alabama. Ten years later the rust had spread eastward and by 1950 was present in southern South Carolina. The present geographic range of fusiform rust generally coincides with the ranges of slash and loblolly pines. The severity of the rust has varied from less than 1% to approximately 100%. Surveys have been made in several states, and the severity of the rust has been mapped (Fig. 15.10).

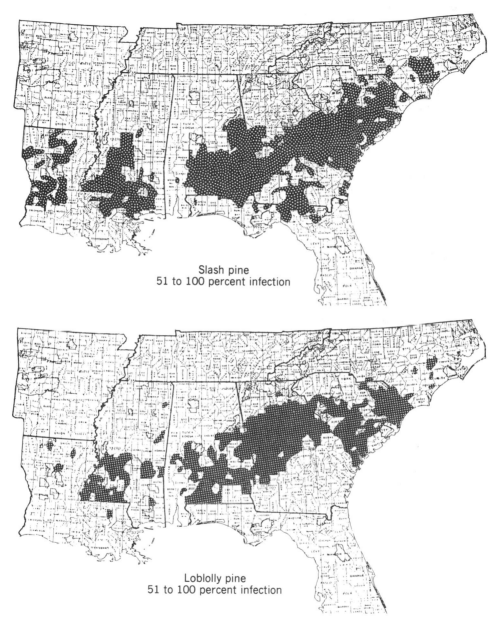

FIGURE 15.10   Areas where 51–100% of 8- to 12-yr-old planted loblolly and slash pines were infected by fusiform rust in 1971–1973. From Powers et al. (1975).

Land-use practices that favored susceptible pine hosts and abundant alternate hosts set the stage for serious rust outbreaks. Introducing the rust on planting stock ensured that the pathogen was present. These factors, in a period of 60–80 years, disrupted the natural equilibrium between host and pathogen that had evolved over thousands of years.

Cause

A form of *Peridermium cerebrum* was described in 1896 as the cause of spindle-shaped galls on loblolly pine. The spindle shape was attributed to host reaction since there were no detectable microscopic differences in this fungus and the well-known *P. cerebrum,* which produced globose galls on shortleaf and Virginia pines. In 1906 *Peridermium fusiforme* was described from loblolly pine from Alabama and Florida, from longleaf pine from Texas and Florida, and from a *Pinus* species from Georgia. This species was separated from *P. cerebrum* largely on the basis of host reaction. In 1914 the connection was proved between the *Peridermium* stage on pine and the *Cronartium* stage on oak. Because the telial stage of this fungus was indistinguishable from the telial stage of the fungus from shortleaf pine, *P. fusiforme* was recognized as a synonym of *P. cerebrum.* The preponderance of the fusiform type of gall in the southeastern United States was attributed to the more vigorous growth of the host following infection rather than to basic differences in the fungi involved. Others also concluded that *P. cerebrum* and *P. fusiforme* were the same fungus because they produced indistinguishable uredinia and telia on oaks. By inoculation tests, however, it was then proved that the fungus causing fusiform galls was a distinct strain or species from that causing the globose galls and produced a constancy of symptoms even when the two pathogens attacked a single pine host species.

After a careful study of *Cronartium cerebrum* and *C. fusiforme,* the two species were separated on the following bases:

1. the constancy of different symptoms produced by the two species even where the host ranges overlapped;
2. the differences in host range;
3. the increased lethal effects of *C. fusiforme;* and
4. several small but distinct differences in microscopic characteristics of the fungi.

Questions, though, continued to be raised concerning the relationship between *C. fusiforme* and *C. quercuum* because of their similar life cycles, same alternate host species, and almost identical morphological characters. Burdsall and Snow (1977) reexamined this controversy and used the designation of special form *(forma specialis)* to describe the two fungi that differ scarcely or not at all from a morphological standpoint but that were adapted to different hosts. *Cronartium fusiforme* was reduced to the rank of *forma specialis* of *C. quercuum,* and three other *formae speciales* were erected based on their pathogenic interactions on different pine species.

The fungus is thus classified as follows:

Class: Basidiomycotina
    Order: Uredinales
      Family: Melampsoraceae
        Genus: *Cronartium*
          Species: *C. quercuum*
          formae speciales:
            C. q. f. sp. *banksianae* (primarily pathogenic on jack pine)
            C. q. f. sp. *virginianae* (primarily pathogenic on Virginia pine)
            C. q. f. sp. *echinatae* (primarily pathogenic on shortleaf pine)
            C. q. f. sp. *fusiforme* (primarily pathogenic on slash and loblolly pine)

This classification has been widely accepted and has helped avoid much confusion in dealing with these respective rusts.

Biology

Cronartium quercuum f. sp. *fusiforme* is a heteroecious rust requiring two unrelated hosts for the completion of its life cycle. The pycnial and aecial stages occur on various species of the genus *Pinus* and the uredinial, telial, and basidial stages on *Quercus* spp. Characteristics of the spore stages are shown in Table 15.2 and the complete life cycle is illustrated in the disease diagram (Fig. 15.11).

The pathogen overwinters as mycelium in the gall on the pine. The aecia, with their masses of powdery yellow aeciospores, develop on these galls in early spring (Fig. 15.12). Aecia are most abundant during February in Mississippi, and they last until early April in the northern part of the range of the fungus. Aeciospores are capable of retaining viability over a long period if stored under cool, dry conditions. Freeze-dried spores are now kept in a spore bank for use in research. Aeciospores are disseminated by wind to young oak leaves that emerge at about the same time. The dispersal of aeciospores is well synchronized with the development of young oak leaves. Aeciospores infect only the leaves of the oak. Occasionally a late frost will injure the oak leaves, and this injury effectively reduces the infections for that year. Because the leaves decrease in susceptibility as they expand and mature, infection on oak can occur only during a few days of the growing season. The aeciospores germinate and penetrate through stomata on the underside of the leaves. A film of water on the surface of the leaf is required for germination and infection. Three self-inhibitors of germination have been isolated and characterized from aeciospores. The role of these self-inhibitors may be to enhance dissemination and dispersal of aeciospores before they germinate.

The mycelium of the fungus is intercellular in the oak leaf, growing in the intercellular spaces. Nourishment is obtained through specialized absorptive structures called haustoria that are inserted into the living cells of the host. Within 15 days or less, a yellow spot appears on the upper surface of the leaf and the orange-yellow uredinia and urediniospores on the lower surface. The urediniospores, like the aeciospores, are one-celled and orange-yellow and contain two nuclei.

TABLE 15.2 Characteristics of the Spores Occurring in the Life Cycle of *Cronartium quercuum* f. sp. *fusiforme*

| Spore | Produced | | Borne by | Function | Nuclei |
	When	Where			
Pycniospore	Fall (Oct–Dec)	On pine	Unknown	Unknown	1 haploid
Aeciospore	Early spring	On pine	Wind	Infects oak	2 haploid
Urediniospore	Mid spring	On oak	Wind	Infects oak	2 haploid
Teliospore[a]	Late spring	On oak		Produces basidiospores	2 haploid 1 diploid
Basidiospore	Late spring	On oak	Wind	Infects pine	1 haploid

[a]The two haploid nuclei fuse in the teliospore. The diploid nucleus that is formed then divides by meiosis to give the four haploid nuclei, which move into the developing basidiospores.

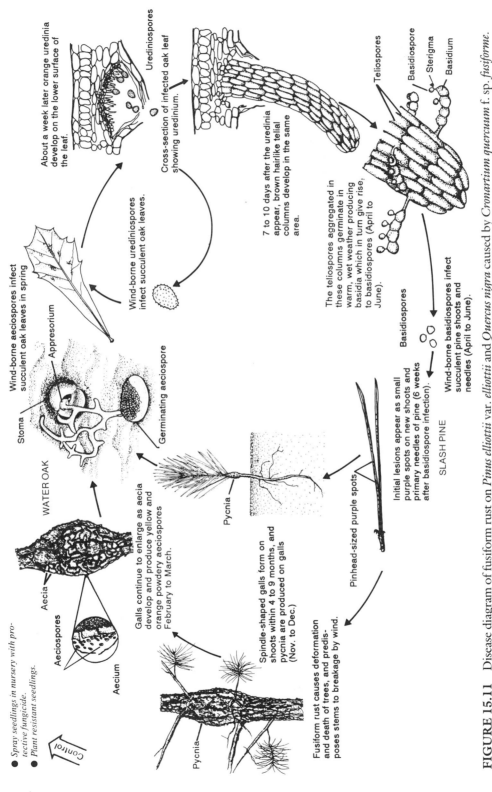

About a week later orange uredinia develop on the lower surface of the leaf.

Urediniospores

Cross-section of infected oak leaf showing uredinium.

Teliospores

Basidiospore

Sterigma

Basidium

7 to 10 days after the uredinia appear, brown hairlike telial columns develop in the same area.

Wind-borne urediniospores infect succulent oak leaves.

Wind-borne aeciospores infect succulent oak leaves in spring

The teliospores aggregated in these columns germinate in warm, wet weather producing basidia which in turn give rise, to basidiospores (April to June).

Basidiospores

Appresorium

Stoma

Wind-borne basidiospores infect succulent pine shoots and needles (April to June).

WATER OAK

Germinating aeciospore

SLASH PINE

Aecia

Aeciospores

Initial lesions appear as small purple spots on new shoots and primary needles of pine (6 weeks after basidiospore infection).

Pycnia

Aecium

Galls continue to enlarge as aecia develop and produce yellow and orange powdery aeciospores February to March.

Pinhead-sized purple spots.

Spindle-shaped galls form on shoots within 4 to 9 months, and pycnia are produced on galls (Nov. to Dec.)

Pycnia

Fusiform rust causes deformation and death of trees, and predisposes stems to breakage by wind.

● Spray seedlings in nursery with protective fungicide.
● Plant resistant seedlings.

Control

526

FIGURE 15.11 Disease diagram of fusiform rust on *Pinus elliottii* var. *elliottii* and *Quercus nigra* caused by *Cronartium quercuum* f. sp. *fusiforme*. Drawn by Valerie Mortensen.

FIGURE 15.12   Aecia-bearing gall of fusiform rust.

The urediniospores also are disseminated by wind and can cause new infections in leaves of oak. These spores are produced in large numbers, mostly in late March and April. In one study, peak production of urediniospores occurred between April 1 and 15, and urediniospore production lasted from 18 to 28 days. The urediniospores can infect only young oak leaves. The spores tend to adhere together in clumps so that their spread by wind from oak to oak is strictly local. The urediniospores increase in number on the oak and increase the amount of inoculum for later infection of the pine. Urediniospores are long-lived, and a small percentage of them may live over winter and serve as primary inoculum on the oak the following spring.

From the same mycelium that gave rise to the uredinia, columns of teliospores are produced as early as February in Florida and as late as June in North Carolina, Virginia, or Maryland. In many cases telia may be produced directly after initial inoculation so that uredinia do not occur. This is especially true later in the spring as the oak leaves mature. The teliospores are one-celled and dark brown and initially contain two nuclei. They are fused both vertically and horizontally into brown, hairlike columns large enough to be visible to the unaided eye. These columns may be so abundant on the lower side of an oak leaf that they have the appearance of velvet. The teliospores aggregated in these columns germinate under favorable weather conditions to produce basidia that in turn give rise to basidiospores. As the basidium forms, the two nuclei in the mature teliospore fuse to form a single diploid nucleus. This nucleus then divides meiotically to give the four haploid nuclei, which migrate into the basidiospores (sporidia) as they develop. Teliospores remain viable for about a month, germinating rapidly when favorable environmental conditions occur. Each teliospore germinates to produce a septate basidium, and on each of the four cells a small, one-celled basidiospore or sporidium is produced. Four basidiospores are produced from each teliospore.

The basidiospores are very delicate and short-lived. They are produced as early as April in Florida and Mississippi and as late as May and June in North Carolina. They are disseminated by wind to the pine host. Basidiospores are dispersed generally at night and early morning, especially during nights preceded by rain during the day or evening. Spores appear when the relative humidity approaches a 100% atmosphere. The time required for release of basidiospores to begin and to reach a maximum rate is inversely related to temperatures from 12 to 19.5°C. The rate of release is related directly to temperature, and no spore release occurs at temperatures less than 8°C. The distance that basidiospores may be carried by wind from oak to pine is unknown (mostly several kilometers), but quartenary sporidia have been observed and spread may be greater than was once thought. Quartenary sporidia are formed when basidiospores land on an unfavorable substrate and germinate to form another basidiospore. Slash pine exposed in a stand of oaks became infected on 7 of 30 days in spring of 1965 and on 10 of 21 days in 1966. Seedlings exposed in an open field were infected only on one day in 1966. The shortest incubation period for slash pine seedlings inoculated with basidiospores was 4 hours, and lesions developed more slowly on seedlings incubated for 4 hours than on those incubated for 6–16 hours.

Infection of the pine normally occurs directly through the cuticle of primary needles or through the cotyledons or succulent stem. Fascicled needles are not susceptible. The basidiospores land on the needle and germinate; then the germ tube penetrates into the mesophyll of the needle. *Cronartium quercuum* f. sp. *fusiforme* can infect fresh wounds on 7-year-old loblolly pines, and thus wounds should be avoided when basidiospores are active. Wounds and the prolific secondary shoots caused by the Nantucket pine tip moth also serve as infection courts. The occurrence of galls on the hypocotyl of pine seedlings below the cotyledon whorl indicates that infection may occur through the epidermis of the succulent stem of young seedlings. In nurseries where burlap bed covers were used, infection of the hypocotyl occurred even before the covers were removed and before the cotyledon whorl appeared aboveground.

From the needle the mycelium grows into the branches and eventually into the cambium of the main stem. The large, septate uninucleate hyphae ramify freely through the mesophyll tissue of the needle into and through the cortex parenchyma of the primary bark, through the ray tissue of the phloem and cambium, and into the xylem rays and parenchyma. They do not penetrate the tracheids. Mycelium spreads tangentially only where the parenchyma cells surrounding a resin duct form a living bridge from ray to ray. In the phloem the hyphae invade the parenchyma and ray cells. Their invasion into the cambium results in rapid cell division and production of wood cells. This abnormal rapid cambial activity results in the formation of the gall. The fungus induces both hyperplasia and hypertrophy. The resulting galls develop more rapidly in younger than older trees. The hyphae of *C. quercuum* f. sp. *fusiforme* in slash pine are intercellular, septate, uninucleate, averaging 3.4 μm wide with unicellular, uninucleate, unbranched or occasionally branched haustoria averaging 13.1 μm long. The fungus can be found in all gall tissues. The height of the rays and the length and width of ray parenchyma cells are greater than normal in galls, and the number of resin ducts is increased. The phloem and cortex of the galls are less affected than the xylem. In some instances the cambium may be killed quickly so that infection results in a sunken canker. Because the pathogen is more lethal to slash pine than to loblolly or longleaf pine, the killing of the cambium and the resulting sunken cankers are more prevalent in this species.

After infection occurs in the spring, no external symptoms of the disease appear until swelling becomes evident 5–6 months later. Progress of the disease is faster on seedlings than on older trees. Galls continue to elongate year after year as long as the stem or branch is living. (Remember that rusts are obligate parasites!) Galls on the branches of slash pine grow toward the trunk at an average rate of about 5 cm/yr, and trunk cankers or galls grow downward at about 13 cm/yr. The rate of spread in loblolly pine is somewhat less rapid than in slash.

During the late summer or fall following infection, the pycnia may appear on the galls (Fig. 15.13). Within the pycnia, numerous tiny, one-celled pycniospores develop. These spores, each of which contains a single haploid nucleus, are exuded from the pycnia in a sticky mass. The pycniospores are presumed to serve as male gametes and fertilize the female flexuous hyphae around the periphery of the pycnium. However, their function is not known for certain. The gall may produce aecia without having produced pycnia. The spring following the appearance of pycnia, or the spring following infection, aecia appear on the gall. Thereafter, the pycnia and aecia usually develop in alternate years as long as the gall remains alive.

Secondary infection on pine has been achieved by inserting aeciospores under the bark of young stems and branches. No pycnia formed on the galls, and the aecia that were produced were apparently identical with those produced on galls developing from natural infection.

Epidemiology

The severity of fusiform rust in any given year is dependent on the weather during the spring of that year. Weather conditions from February through June affect the extent of disease development on both oaks and pines by influencing:

1. the timing and distance of dissemination of aeciospores, urediniospores, and basidiospores;
2. the proportion of basidiospores that remain viable during dissemination (these spores are especially vulnerable to exposure in the atmosphere);
3. the time at which the hosts start to grow in the spring and the rapidity of maturation of leaves on oak and pine (immature tissues of both hosts are more susceptible to infection than mature tissues);
4. the number of spores that germinate, resulting in penetration and establishment in the hosts.

Ideal conditions for infection seem to be when the months of February and March are warm and April and May are cool and wet with several periods of continuous

FIGURE 15.13   Pycnia of fusiform rust on loblolly pine.

moisture lasting 18 hours or more. The warm weather in February and March results in early growth of the oak and pine. Warm March weather results in early appearance of uredinia and early and rapid secondary spread and buildup of the fungus on oaks. Then, cool weather in April and May delays the growth of the oaks so that they remain susceptible longer, thus permitting additional secondary disease cycles on this host and rapid buildup of increased amounts of inoculum. The cool weather in later spring also favors the activity of the fungus.

Weather conditions are particularly influential on germination of teliospores, production and dissemination of basidiospores, and infection of pine. These stages in the disease cycle are favored by protracted periods during which the relative humidity remains at or near 100% with temperatures between 16 and 24°C. Infection can occur when these conditions prevail for as little as 12 hours, as is common on many spring nights. When rain or fog and overcast skies extend the time when these conditions occur during a period that coincides with flushes of new growth on pine, rates of infection can reach epidemic proportions, and severe rust years can result.

Periods of continuous moisture of at least 12 hours are essential for infection of the pine, since teliospore germination, basidiospore production and germination (4–6 hr), and pine infection (4–6 hr) must be a continuous process. Interrupting this process at any point results in rapid death of the basidia and basidiospores. The spread of the disease to pine is thus largely dependent on the number of wet periods of 12 hours or more that occur during April and May when the teliospores on the oak leaves are germinating.

Certain silvicultural practices also influence the severity of the disease. Even-aged planted stands usually are damaged more severely than natural stands. Incidence of fusiform rust in loblolly pine is less when planted under a pine overstory in part due to reduced growth of pine seedlings. Fertilization will increase the number of galls per infected tree. Fertilization causes longer, more succulent shoots, and these shoots are very susceptible to infection. The more intensively a site is prepared for planting (by chopping, cultivating, burning, or fertilizing), the greater will be the disease incidence. All these practices favor oaks. For unknown reasons fusiform rust is of little consequence in some areas where both hosts are present.

Diagnosis

The most characteristic expression of fusiform rust is the spindle-shaped or sunken canker produced on the stem and branches of pines at the point of infection. These galls may vary from a very slight swelling (Fig. 15.14) to a spindle swelling two or three times the normal stem diameter. On older infections, the gall surface is sometimes covered with pitch. On seedlings and young trees this gall normally girdles the tree and kills it within 1–3 years. Rust infection may also stimulate excessive branching at the point of the gall. Nurserymen used to routinely discard slash pine seedlings with basal branches in the belief that this was a symptom of the disease. Basal branching, though, is not necessarily a positive indication of the disease. Older trees are rarely killed but show large sunken cankers, often near the base of the stem (Fig. 15.15). These trees often break near the canker (Fig. 15.16). Although fusiform rust does not always stunt the growth of affected trees, some stands, when affected at an early age, become worthless because of stunting and deformity.

On many trees, the pycnia normally appear from October to December one year after the infection occurs or, on nursery seedlings of slash pine, the same year they are infected. These pycnia are not readily visible themselves, but the sticky matrix in which the pycniospores are exuded is readily seen on the gall surface. Orange-yellow aecia appear on the same galls in the spring following pycnia formation. Occasionally both pycnia and aecia may appear on the same gall in the same season, but in general they appear in alternate years. If they appear on the same gall in the same year, they are on different portions of the gall. The aecia are large and lobed in shape, blisterlike in appearance, and yellow or orange-yellow and are filled with the powdery yellow aeciospores.

In the galls no summerwood is formed; the gall is composed entirely of

FIGURE 15.14   Seedlings of slash pine with swellings that have developed from young fusiform rust infections.

springwood. Xylem tracheids are irregular in shape and much shorter than the tracheids in healthy tissue. In diseased wood the rays are greatly enlarged, both because the individual rays are larger and because the rays contain more cells. The number of resin ducts in diseased wood is approximately double that in healthy wood. Secondary phloem parenchyma is produced at a rate greater than usual, and the individual cells are larger. In the phloem, numerous small cells containing abundant styloid crystals are formed. As the gall ages, the phloem ray cells continue to enlarge until they become isodiametric and lose their identity. The sieve tubes and crystal-bearing cells are soon crushed. The parenchyma cells also become disassociated and appear as spherical bodies in a fungus matrix. Within the phloem, numerous radial resin ducts originate in the rays. As the rays lose their characteristic shape, these canals increase greatly in diameter through repeated division of the epithelial cells. Starch, granular deposits, and fatty globules are abundant in young parenchyma cells within and around diseased tissue in early winter. As the cells age, the stored foods gradually disappear.

On oaks the symptoms are confined to the leaves. Although both uredinia and telia have been reported on the upper as well as the lower surface, the first symptoms to

FIGURE 15.15 Loblolly pine with basal fusiform rust cankers.

appear are orange-colored dots of uredinia usually on the lower surface of the leaves. The uredinial sori on oak leaves are so small and inconspicuous that they are rarely observed unless a special search is made for them. Shortly thereafter the brown, hair-like telial columns appear, also usually on the lower surface. These telial columns are large enough to be readily visible to the unaided eye. In many cases no uredinia are formed, and the telial columns develop directly. On the upper surface of the leaves, a pale yellow spot appears opposite the developing fruiting structures on the lower surface.

Control Strategy

Nurseries. The methods of controlling fusiform rust vary with the value of the crop. It was suggested that the prevalence of galls on the native pines within a 1.2- or 2.5-km radius of the proposed nursery site be used as a measure of risk to the crop and that if 10% of longleaf seedlings are infected, the site is too hazardous for slash or loblolly. Assessment of risk to fusiform rust is a prime consideration today in the southeastern United States when considering a new nursery location. In nurseries the

FIGURE 15.16   Windthrown loblolly pine that broke at the site of a fusiform rust canker.

value of the crop is such that use of fungicide is feasible. In natural stands and plantations, control costs must be minimized, and thus foresters rely primarily on resistant trees.

Early researchers tested a number of fungicides and found several of them useful to control fusiform rust in the nursery. These included Bordeaux mixture, Zineb, Ferbam, and Ziram. It was important to time the sprays so as to precede each predicted wet period during April and May. High-pressure systems centered off the southeastern coast of the United States created circulation necessary to move maritime tropical air over the southeast and produced ideal weather conditions for production of the basidiospores and infection of pine by these spores. It was suggested that nurseries be sprayed weekly as soon as the seedlings appeared aboveground and that the spraying continue until mid June. Where the rust hazard was high, as indicated by the early spring weather, applications were to be twice weekly for the first 3 weeks and then weekly until mid June. These applications were to be moved forward if a wet period was predicted near the end of the weekly spray interval. The spray program was costly and required considerable effort. Daily weather forecasts and spraying only in periods when infection was likely to occur in seed orchards and nurseries reduced the cost and effort without reducing control. Proper spraying reduced nursery rust infections to less than 0.5%.

A major problem in using protectant fungicides like Ferbam was that newly forming leaf and stem tissues are unprotected until the next application. As many as 20 spray applications are needed for effective control. A great boon was realized with the development and release of triadimefon (Bayleton), a systemic fungicide that was more effective than Ferbam and reduced the number of applications needed. Bayleton applied to the soil before planting has prevented galls from forming. Bayleton treatment of seeds protects pine seedlings for at least 4 weeks after planting and, if the seeds are treated, no more than three foliar sprays are needed that growing season.

An early recommendation stressed that infected seedlings should be culled and not shipped from the nursery. Even though most diseased seedlings die before aeciospores are produced, some live to increase the inoculum or introduce the pathogen into a new area. Because few nurseries handle individual seedlings anymore, a more feasible practice is for the purchaser to visually inspect samples of seedlings. If nurseries are properly sprayed with systemic fungicides at timely intervals, inspections should not be necessary.

Eradication of Oaks. Early workers recommended or attempted to eradicate oaks and diseased pines around nursery sites. Because of the rapid regrowth of the oaks, the brevity of the fungus life cycle on oak, and the great difficulty of obtaining complete oak eradication, the procedure was not successful. Cost of oak removal is the determining factor for this practice, which can be greatly reduced if herbicides are used to eradicate the oaks. Eradication of oaks is no longer recommended because of the value of these species for wildlife.

Late Sowing. Some reduction in losses from rust occurred when planting was delayed from mid March to mid April. Losses from *Macrophomina phaseolina*, during the following summer were so great in the late-seeded beds, however, that they more than offset the lower rust infection.

Plantations and Natural Stands. Planting and thinning procedures that favor early self-pruning, maintenance of an overstory, development of uneven-aged stands, planting of alternate-row mixtures of species, and prescribed burning to ignite the resin-soaked surfaces of stem cankers all have been suggested as means to reduce losses due to this disease. Although some of these recommendations may be of value, they frequently cannot be justified economically and have generally not been proven to be that useful.

Dense Planting. The use of a 1.2 × 1.2-m spacing was recommended on high-hazard sites in preference to the more usual 1.8 × 1.8-m or similar wider spacing. The major advantage gained from close spacing is the early natural pruning that prevents many of the branch galls or cankers from reaching the main stem. Dense stands also tend to suppress the growth of oaks. Slash pine can be planted at a closer spacing than loblolly pine because difficulty is often encountered in attempting to get a satisfactory stand of slash pine. Although there is some evidence of a progressive decrease in the percentage of diseased trees as the number of trees planted per area of land increases, the silvicultural disadvantages of dense planting probably outweigh the advantages.

Reduced Thinning. In order to maintain a dense stand and encourage early natural pruning of branches, thinning should be sufficient to prevent stagnation only. Trees with trunk galls or cankers should be removed first; then trees with branch galls or cankers close to the trunk should be removed. Diseased trees should not be removed if their removal will result in excessive thinning. With frequent thinnings, the extra labor cost may be at least partially offset by the use of diseased trees for pulpwood before they are lost. Care must be taken on high-hazard sites for annosum root disease because frequent thinning is likely to result in increased losses. Trees that are not completely girdled by the canker may grow as rapidly as healthy trees and may be salvaged later for pulpwood. Based on theoretical calculations of the strength of wood and the average rate of lateral spread of the canker, one study concluded that a tree with less than 50% of its circumference affected by rust should withstand normal winds in the Charleston, South Carolina, area for at least 5 years. Trees with more than 50% of the circumference affected, trees with a bend at the canker, as well as trees with a deeply sunken canker base are poor risks and should be cut first.

Pruning. Neither early pruning of infected branches nor pruning of all branches to a height of 5.2 m (60% of total height) justified the cost of the operation in slash pine. To remove infected branches in four different years required 26.5 man-hr/ha, while 50.2 man-hr/ha were required to prune all branches up to 5.2 m. The percentage infection in trees pruned in the 4 years was 14.3% compared to 20.3% when none were pruned. Pruning apparently stimulates development on the main stem of tissues that are susceptible to infection. Pruning in seed orchards would be economically feasible. Pruning of large branches near the base of the tree will result in wounds that may serve as infection courts for the entrance of annosum root disease.

Prescribed Burning. There have been attempts to control rust through the elimination of second-year needles by prescribed burning. In addition to the elimination of second-year needles, fire in an 8- to 9-year-old slash pine plantation reduced the number of galls producing aeciospores from 41 to 12%. The fire killed the galled trees much more readily than it killed healthy trees. In younger plantings (3–6 years old), fire killed many of the galled lower branches. Thus fire resulted in an initial reduction in the prevalence of the disease because it reduced the percentage of galls that produced aeciospores, it destroyed many galled branches before the fungus could spread to the trunks, and it reduced the tissue susceptible to attack by killing the second-year needles. On the other hand, fire in many cases results in an early break in the dormancy of the trees, an abundant flush of early spring growth, and an eventual increase in the prevalence of disease due to increased infection of the new wood through the current-year needles. Therefore, prescribed burning cannot be recommended under any circumstances for the control of rust.

Fertilization and Cultivation. In a series of studies, cultivation alone, cultivation and fertilization together, as well as cultivation, fertilization, and interplanting with cotton for the first 3 years after pines were planted in a series of studies resulted in an increase in the severity of fusiform rust. The increased amount of rust in cultivated sites was attributed to the superior growth and greater succulence of seedlings. The increase in disease prevalence was later attributed to the early break in dormancy and early rapid growth resulting from cultivation and fertilization. The increase in growth of slash pines, though, in the fertilized and cultivated plots may be offset by the increase in rust. A decrease in pulpwood yields of about 10% from cultivation and fertilization during the first 3 years were realized when the trees were cut for pulpwood 19 years later. This yield decrease was due to an increase in the percentage of diseased trees from 42% in the untreated stands to 65 and 66% in the cultivated and fertilized and the cultivated, fertilized, and intercropped stands, respectively. There was no relationship between rust incidence and estimated site index, but site preparation increased the incidence of rust. Phosphorus application and rust incidence were related, and phosphorus and potassium applications increased height growth but nitrogen did not. Most authors agree that fertilization and cultivation are not economically feasible where rust hazard is high, because since the increased losses from rust will offset the gains from greater tree growth.

Use of Mycoparasites. Three species of mycoparasitic fungi have been found in association with *C. quercuum* f. sp. *fusiforme* and may exert some degree of natural control. *Tuberculina maxima* Rostr. (Hyphomycetes) parasitizes fusiform rust galls on loblolly and slash pines but may play only a minor role. Sporodochia of *T. maxima* develop in the same area of the rust gall but are separate from the aecia. *Scytalidium uredinicola* Kuhlman, Carmichael, and Miller (Hyphomycetes) was found on fusiform rust galls on loblolly and slash pines. Aeciospore production was reduced 72% (Fig. 15.17). There was also a reduction in germination of aeciospores from heavily parasitized gall areas. *Sphaerellopsis filum* (Biv.-Bern. ex Fr.) Sutton (Coelomycetes) *(Darluca filum)* has been reported on uredinial sori of *C. quercuum* f. sp. *fusiforme* on black, water, and willow oaks. Because of the short, irregular cycle of *C. quercuum* f. sp. *fusiforme* on oak, use of *S. filum* did not appear to be practical as a biological control.

USE OF RESISTANT SPECIES. Where the risk of rust is high, adapted resistant species of pines should be used instead of the very susceptible slash and loblolly. Shortleaf is practically immune to fusiform rust, and longleaf is highly resistant. Shortleaf has largely fallen into disfavor except in the Ozark and Ouachita plateaus of Missouri and Arkansas. Longleaf pine, on the other hand, has enjoyed renewed popularity and, because of brown spot control, is more widely planted than ever before. Loblolly may

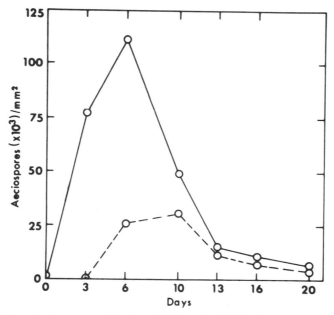

FIGURE 15.17 Aeciospore production of gall surface by fusiform rust in unparasitized areas (solid line) and in areas parasitized by *Scytalidium uredinicola* (dashed line). From Kuhlman (1981).

be favored over slash in high-risk areas because ultimate damage to this species is usually less.

Planting alternate rows of slash and loblolly pines has been suggested. The rapidly growing loblolly pines will tend to suppress the slower growing slash pines initially and thus tend to restrict the development of rust on the suppressed slash. The loblolly may be removed for pulpwood, and the slash must be allowed to grow on to maturity. The effectiveness of rust control by suppressing the initial growth of pines has been questioned, however. In 38 paired plots of loblolly pines, it was found that those pines growing under and partially suppressed by an overstory of older pines had the same amount of rust as similar pines grown in the open.

USE OF LESS SUSCEPTIBLE STRAINS OF SUSCEPTIBLE SPECIES. Selection and hybridization of resistant loblolly and slash pines, and to a limited degree resistant intraspecific hybrids, offer the most promise for reducing losses to fusiform rust. Various investigators have compared the susceptibility of seedlings grown either from selected mother trees or from certain specific locations. A comparison of seedlings grown from seeds collected in Texas, Louisiana, Arkansas, and Georgia found that, when grown in Louisiana, the Georgia strain was more susceptible to rust than the other three strains. The observed differences, though, may have been due to the greater susceptibility of the Georgia pines to the strain of the fungus encountered in Louisiana. When comparing seedlings grown from mother trees selected for resistance to rust, a maximum degree of resistance was observed in the progeny of one mother tree that had an infection rate only 50% of that of the control seedlings.

From a summary of these tests, it may be concluded that variation in susceptibility exists in loblolly and slash pines currently even though the variability in slash pine is not currently considered sufficient for initiating a productive selection program. Apparently natural hybridization between slash and loblolly and loblolly and shortleaf has occurred in the western portion of the loblolly range and accounts for western sources being more resistant than eastern sources. It was also observed that a few individual slash and loblolly pines recovered from infection with fusiform rust. A breakthrough in testing for fusiform rust resistance was made in the 1970s when a technique for inoculating pine seedlings was developed and it was shown that seedling response to artificial inoculation is a reliable index of field performance. This technique allows testing seedlings for resistance to rust using heavy inoculum pressure at an early age. Family screening for resistance can be done on 1-year-old trees at less cost and may be done more accurately than by long-term field testing. Work with interspecific crosses between slash and shortleaf and loblolly and shortleaf signalled the beginning of a new approach to breeding for fusiform rust resistance.

Resistant loblolly and slash pines have been developed and are being planted, often in mixtures to reduce the opportunity for the fungus to produce races that could attack the resistant pines. Experience with other rust fungi has demonstrated that a fungus population will change so as to be more virulent on the resistant host. At one time it was thought that resistance in loblolly pine was related to early growth in spring when weather conditions were more favorable for the fungus. Now it is quite evident that resistance is not necessarily related to time of flushing in spring. Much has been accomplished in developing disease-resistant loblolly pines. In time the sources of seed for both loblolly and slash pine should be resistant trees, not just a single clone but several clones representing a broad base of resistance.

REFERENCES

Arthur, J. C., and F. D. Kern. 1914. North American species of *Peridermium* on pine. *Mycologia* 6:109–138.

Balthis, R. F., and D. A. Anderson. 1944. Effect of cultivation in a young slash pine plantation on the development of Cronartium cankers and forked trees. *J. For.* 42:926–927.

Barber, J. C., K. W. Dorman, and E. Bauer. 1957. Slash pine progeny tests indicate genetic variation in resistance to rust. *USDA For. Serv., Southeast For. Exp. Sta. Note* 104:1–2.

Boggess, W. R., and R. Stahelin. 1948. The incidence of fusiform rust in slash pine plantations receiving cultural treatments. *J. For.* 46:683–685.

Burdsall, H. H., Jr., and G. A. Snow. 1977. Taxonomy of *Cronartium quercuum* and *C. fusiforme. Mycologia* 69:503–508.

Cummins, G. B. 1956. Nomenclature changes for some North American Uredinales. *Mycologia* 48:601–608.

Davis, R. T., and G. A. Snow. 1968. Weather systems related to fusiform rust infection. *Pl. Dis. Reptr.* 52:419–422.

Dinus, R. J. 1974. Knowledge about natural ecosystems as a guide to disease control in managed forests. *Proc. Amer. Phytopath. Soc.* 1:184–190.

Dwinell, L. D. 1974. Susceptibility of southern oaks to *Cronartium fusiforme* and *Cronartium quercuum. Phytopathology* 64:400–403.

Dwinell, L. D., and R. L. Blair. 1974. Sawtooth oak, a new host of *Cronartium fusiforme. Pl. Dis. Reptr.* 58:198–200.

Dwinell, L. D., and H. R. Powers, Jr. 1974. Potential for southern fusiform rust on western pines and oaks. *Pl. Dis. Reptr.* 58:497–500.

Enghardt, H. G., L. F. Smith, and O. O. Wells. 1969. *Pruning to Reduce Fusiform-Rust Damage Not Jusitified on Young Slash Pines.* USDA For. Serv., Res. Note SO-87. 3 pp.

Foster, A. A., and D. W. Krueger. 1961. *Protection of Pine Seed Orchards and Nurseries from Fusiform Rust by Timing Ferbam Sprays to Coincide with Infection Periods.* Ga. For. Res. Pap. No. 1. 4 pp.

Foudin, A. S., P. B. Bush, V. Macko, J. K. Porter, J. D. Robbins, and W. K. Wynn. 1977. Isolation and characterization of three self-inhibitors of germination from aeciospores of *Cronartium fusiforme. Exp. Mycology* 1:128–137.

Gilmore, A. R., and K. W. Livingston. 1958. Cultivating and fertilizing a slash pine plantation: Effects on volume and fusiform rust. *J. For.* 56:481–483.

Goggins, J. F. 1949. *Cronartium fusiforme* on slash and loblolly pine in the Piedmont region of Alabama. *J. For.* 47:978–980.

Goggins, J. F. 1957. Reducing losses from southern fusiform rust. *Alabama Agr. Exp. Sta. Leaflet* 54:1–4.

Hare, R. C., and G. L. Switzer. 1969. *Introgression with Shortleaf Pine May Explain Rust Resistance in Western Loblolly Pine.* USDA For. Serv., Res. Note SO-88. 2 pp.

Hare, R. C., and G. A. Snow. 1983. *Control of Fusiform Rust in Slash Pine with Bayleton (Triadimefon) Seed Treatment.* USDA For. Serv., SO-288. 4 pp.

Hebb, E. A. 1949. A study of the influence of overstory on the incidence of Cronartium cankers on loblolly pine. (abst.) *J. For.* 47:415.

Hedgcock, G. G. 1939. Notes on North American pine–oak species of *Cronartium* on *Castanea, Castanopsis* and *Lithocarpus. Phytopathology* 29:988–1000.

Hedgcock, G. G., and N. R. Hunt. 1918. Notes on *Cronartium cerebrum. Phytopathology* 8:74.

Hedgcock, G. G., and W. H. Long. 1914. Identity of *Peridermium fusiforme* with *Perdimium cerebrum*. *J. Agr. Res.* 2:247–250.

Hedgcock, G. G., and P. V. Siggers. 1949. A comparison of the pine oak rusts. *USDA, Tech. Bull.* 978:1–30.

Henry, B. W. 1953. Fusiform rust. *Amer. For.* 59:28, 45.

Henry, B. W. 1956. Basal branches no symptom of fusiform rust on slash pine seedlings. *Tree Pl. Notes* 24:16.

Jackson, L. W. R., and J. N. Parker. 1958. Anatomy of fusiform rust galls on loblolly pine. *Phytopathology* 48:637–640.

Jewell, F. F. 1960. New pine hosts for southern fusiform rust. *Pl. Dis. Reptr.* 44:673.

Jewell, F. F. 1961. Infection of artificially inoculated shortleaf pine hybrids with fusiform rust. *Pl. Dis. Reptr.* 45:639–640.

Jewell, F. F., and L. N. Eleutcrius. 1963. Amphiphyllous uredia and telia of southern fusiform rust. *Pl. Dis. Reptr.* 47:65.

Jewell, F. F., R. P. True, and S. L. Mallett. 1962. Histology of *Cronartium fusiforme* in slash pine seedlings. *Phytopathology* 52:850–858.

Kais, A. G. 1963. In vitro sporidial germination of *Cronartium fusiforme*. *Phytopathology* 53:987.

Kelley, W. D. 1980. Evaluation of systemic fungicides for control of *Cronartium quercuum* f. sp. *fusiforme* on loblolly pine seedlings. *Pl. Dis.* 64:773–775.

Kinloch, B. B., and R. W. Stonecypher. 1969. Genetic variation in susceptibility to fusiform rust in seedlings from a wild population of loblolly pine. *Phytopathology* 59:1246–1255.

Klawitter, R. A. 1957. Most cankered trees are good risks in loblolly pine sawtimber stands. *USDA For. Serv., SE For. Exp. Sta. Note* 107:1–2.

Kuhlman, E. G. 1981. Mycoparasitic effects of *Scytalidium uredinicola* on aeciospore production and germination of *Cronartium quercuum* f. sp. *fusiforme*. *Phytopathology* 71:186–188.

Kuhlman, E. G., and F. R. Matthews. 1976. Occurrence of *Darluca filum* on *Cronartium strobilinum* and *C. fusiforme* infecting oak. *Phytopathology* 66:1195–1197.

Kuhlman, E. G., and T. Miller. 1976. Occurrence of *Tuberculina maxima* on fusiform rust galls in the southeastern United States. *Pl. Dis. Reptr.* 60:627–629.

Kuhlman, E. G., J. W. Carmichael, and T. Miller. 1976. *Scytalidium uredinicola*, a new mycoparasite of *Cronartium fusiforme* on *Pinus*. *Mycologia* 68:1188–1194.

Kuhlman, E. G., F. R. Matthews, and H. P. Tillerson. 1978. Efficacy of *Darluca filum* for biological control of a *Cronartium fusiforme* and *C. strobilinum*. *Phytopathology* 68:507–511.

Lamb, H. 1937. Rust canker diseases of southern pines. *USDA For. Serv., SE For. Exp. Sta. Occas. Pap.* 72:1–7.

Lamb, H., and B. Sleeth. 1940. Distribution and suggested control measures for the southern pine fusiform rust. *USDA For. Serv., SE For. Exp. Sta. Occas. Pap.* 91:1–5.

Lindgren, R. M. 1948. Thinning pines cankered by fusiform rust. *USDA For. Serv., SE For. Exp. Sta. Notes* 55:1–2.

Lindgren, R. M. 1948. Care in thinning pines with heavy fusiform rust infection. *For. Farmer* 7(12):3.

Meinecke, E. P. 1920. Facultative heteroecism in *Peridermium cerebrum* and *Peridermium harknessii*. *Phytopathology* 10:279–297.

Millar, C. I., and B. B. Kinloch. 1991. Taxonomy, phylogeny, and coevolution of pincs and their stem rusts, pp. 1–38. In Y. Hiratsuka, J. K. Samoil, P. V. Blenis, P. E. Crane, and B. L. Laishley (eds.), *Rusts of Pine*, Proc. IUFRO Rusts of Pine Working Party Conf., Sept. 18–22, 1989, Banff, Alberta, Canada. For. Can. Info. Rept. NOR-X-317. 408 pp.

Miller, T. 1972. Fusiform rust in planted slash pines: Influence of site preparation and spacing. *For. Sci.* 18:70–75.

Miller, T., and F. R. Matthews. 1973. Wounds on loblolly pines as infection courts for *Cronartium fusiforme. USDA For. Serv., SE For. Exp. Sta.* 63:446.

Miller, T., and H. R. Powers. 1983. Fusiform rust resistance in loblolly pine: Artificial inoculation vs. field performance. *Pl. Dis.* 67:33–34.

Muntz, H. H., W. F. Mann, Jr., and N. M. Scarbrough. 1948. Close spacing reduces fusiform rust. *USDA For. Serv., SO For. Exp. Sta., So. For. Notes* 53:1–2.

Powers, H. R. 1962. Fusiform rust and annosus root rot. The South's most serious plantation diseases. Proc. Soc. Amer. For., Atlanta, GA, pp. 34–36.

Powers, H. R. 1974. Breakthrough in testing for fusiform rust resistance. *For. Farmer* 33(4):7–8.

Powers, H. R. 1975. Relative susceptibility of five southern pines to *Cronartium fusiforme. Pl. Dis. Reptr.* 59:312–314.

Powers, H. R. 1984. Control of fusiform rust of southern pines in the USA. *Eur. J. For. Path.* 14:426–431.

Powers, H. R., and R. W. Roncadori. 1966. Teliospore germination and sporidia production by *Cronartium fusiforme. Pl. Dis. Reptr.* 50:432–434.

Powers, H. R., and S. J. Rowan. 1983. Influence of fertilization and ectomycorrhizae on loblolly pine growth and susceptibility to fusiform rust. *So. J. Appl. For.* 7:101–103.

Powers, H. R., J. P. McClure, H. A. Knight, and G. F. Dutrow. 1975. *Fusiform Rust: Forest Survey Incidence Data and Financial Impact in the South.* USDA For. Serv., Res. Pap. SE-127. 15 pp.

Raddi, P., and H. R. Powers. 1982. Relative susceptibility of several European species of pine to fusiform rust. *Eur. J. For. Path.* 12:442–447.

Rhoads, A. S., G. G. Hedgcock, E. Bethel, and C. Hartley. 1918. Host relationships of the North American rusts, other than *Gymnosporangium,* which attack conifers. *Phytopathology* 8:309–352.

Rowan, S. J. 1984. Bayleton seed treatment combined with foliar spray improves fusiform rust control in nurseries. *So. J. Appl. For.* 8:51–54.

Rowan, S. J., W. H. McNab, and E. V. Brender. 1975. *Pine Overstory Reduces Fusiform Rust in Underplanted Loblolly Pine.* USDA For. Serv., SE For. Exp. Sta. Res. Pap. SE-212. 6 pp.

Schmidt, R. A., M. J. Foxe, C. A. Hollis, and W. H. Smith. 1972. Effect of N, P, and K on the incidence of fusiform rust galls on greenhouse-grown seedlings of slash pine. *Phytopathology* 62:788.

Schmidtling, R. C. 1985. Co-evolution of host/pathogen/alternate host systems in fusiform rust of loblolly and slash pines, pp. 13–19. In J. Barrows-Broaddus and H. R. Powers (eds.), *Proc. Rusts of Hard Pines Working Party Conference, S2.06–10.,* Georgia Center for Continuing Education, University of Georgia, Athens. 331 pp.

Sholler, D. L., F. E. Bridgwater, and C. C. Lambeth. 1983. Fusiform rust resistance of select loblolly pine seedlots in the laboratory, nursery and field. *So. J. Appl. For.* 7:198–203.

Sleeth, B. 1943. Fusiform rust control in forest tree nurseries. *Phytopathology* 33:33–44.

Siggers, P. V. 1947. Temperature requirements for germination of spores of *Cronartium fusiforme. Phytopathology* 37:855–864.

Siggers, P. V. 1949. Fire and southern fusiform rust. *For. Farmer* 8(5):16, 21.

Siggers, P. V. 1951. Spray control of the fusiform rust in forest tree nurseries. *J. For.* 49:350–352.

Siggers, P. V. 1955. Control of the fusiform rust of southern pine. *J. For.* 53:442–446.

Siggers, P. V., and R. M. Lindgren. 1947. Cronartium rust—New threat to slash pine. *Naval Stores Rev.* 47(52):13–14.

Siggers, P. V., and R. M. Lindgren. 1947. An old disease—A new problem. *So. Lumberman* 175(220):172–175.

Snow, G. A. 1968. Basidiospore production by *Cronartium fusiforme* as affected by suboptimal temperatures and preconditioning of teliospores. *Phytopathology* 58:1541–1546.

Snow, G. A. 1968. Time required for infection of pine by *Cronartium fusiforme* and effect of field and laboratory exposure after inoculation. *Phytopathology* 58:1547–1550.

Snow, G. A., and A. G. Kais. 1966. Oak leaf development during aecial sporulation of *Cronartium fusiforme*. *Pl. Dis. Reptr.* 50:388–390.

Snow, G. A., and A. G. Kais. 1972. Technique for inoculating pine seedlings with *Cronartium fusiforme*. Biology of rust resistance in forest trees, pp. 325–326. In *Proceedings of a Nato-IUFRO Advanced Study Institute,* USDA, Misc. Pub. 1221. 681 pp.

Snow, G. A., R. J. Dinus, and A. G. Kais. 1975. Variation in pathogenicity of diverse sources of *Cronartium fusiforme* on selected slash pine families. *Phytopathology* 65:170–175.

Snow, G. A., R. C. Froelich, and T. W. Popham. 1968. Weather conditions determining infection of slash pines by *Cronartium fusiforme*. *Phytopathology* 58:1537–1540.

Snow, G. A., F. F. Jewell, and L. N. Eleuterius. 1963. Apparent recovery of slash and loblolly pine seedlings from fusiform rust infection. *Pl. Dis. Reptr.* 47:318–319.

Snow, G. A., S. J. Rowan, J. P. Jones, W. D. Kelley, and J. G. Mexal. 1979. *Using Bayleton (Triadimefon) to Control Fusiform Rust in Pine Tree Nurseries.* USDA For. Serv., Res. Pap. SO-253. 5 pp.

Squillace, A. E., R. J. Dinus, C. A. Hollis, and R. A. Schmidt. 1978. *Relation of Oak Abundance, Deed Source, and Temperature to Geographic Patterns of Fusiform Rust Incidence.* USDA For. Serv., Res. Pap. SE-186. 20 pp.

Tainter, F. H., and R. L. Anderson. 1993. Twenty-six new pine hosts of fusiform rust. *Pl. Dis.* 77:17–20.

Underwood, L. M., and F. S. Earle. 1896. Notes on the pine-inhabiting species of *Peridermium*. *Bul. Torrey Bot. Club* 23:400–405.

Underwood, L. M., and F. S. Earle. 1897. A preliminary list of Alabama fungi, pp. 113–283. In *Ala. Agr. Exp. Sta. Bull. 80.*

Wakeley, P. C. 1944. Geographic source of loblolly pine seed. *J. For.* 42:23–32.

Wenger, K. F. 1950. The mechanical effect of fusiform rust cankers on stems of loblolly pine. *J. For.* 48:331–333.

Westburg, D. L. 1950. *Cultivating Boosts Fusiform Rust Infection.* USDA For. Serv., SO For. Exp. Sta., So. For. Notes 69. 2 pp.

Witcher, W., and R. E. Beach. 1962. *Fomes annosus* infecting through pruned branches of slash pine. *Pl. Dis. Reptr.* 46:64.

DISEASE PROFILE

Comandra Rust

Importance: Comandra rust can cause understocking of ponderosa pine. Lodgepole pine has been severely damaged where the alternate hosts are abundant. In the Lake states, comandra rust is the least damaging of the stem rusts partly because of its restricted distribution. The introduction of this disease to eastern Tennessee and the realization of how susceptible loblolly pine is has caused some concern in the states south and east of Tennessee where loblolly is the primary pine species. In Arkansas some young stands have had as many as 68% of the trees infected.

Suscepts: Comandra rust commonly occurs on 15 species of hard pines, primarily on ponderosa and lodgepole in the western United States and jack pine in the northeastern states. In 1951 the fungus was found on loblolly pine in eastern Tennessee, apparently transported on ponderosa stock received from a western nursery, and it subsequently was reported in Arkansas. Additional pine species are susceptible including shortleaf and Scots pine. The alternate hosts are species of *Comandra,* sometimes called bastard toadflax.

Distribution: Comandra rust occurs from New Brunswick to the Yukon Territory in Canada and south to California, New Mexico, Arkansas, and northern Alabama.

Causal Agents: Comandra rust is caused by the fungus *Cronartium comandrae* (Uredinales).

Biology: Cronartium comandrae forms pycnia from July to October and aecia the following year from May to July. The amount of sporulation varies considerably from year to year. Rodents feeding on the cankers substantially reduce spore production. Infection by aeciospores occurs through stomata, with small chlorotic spots visible in 5 days and uredinia visible in 9 days; urediniospores are dispersed 25–28 days later, and telia are

Aecial gall.

formed 32 days after inoculation. Uredinia are present on the comandra plants from May to late July. Telia form from late June to mid August when leaves absciss. Infection of pine occurs from early July to mid August.

Epidemiology: A saturated atmosphere and moderate temperature for about 20 hours are needed for the basidiospores to form, germinate, and infect the pine host. Aeciospores germinate best at 15°C with some spores germinating after 1-hour incubation. Most aeciospores germinate within 5 hours. Apparently, free water is required for germination. Comandra plants inoculated with urediniospores produced a new generation of uredinia within 9 days.

Diagnosis: Symptoms are cylindrical swellings on seedlings, fusiform swellings on branches, and either circular or elongated cankers on the main stems. Excessive pitch flow is often associated with these cankers. The aeciospores are distinctive because of their pyriform shape.

Control Strategy: As with the other rust diseases in this group, comandra rust can be controlled in the nursery with fungicide sprays. Diseased stock should not be planted as this disease has been moved into new areas on diseased seedlings. Eradication of the alternate host may be feasible. Pruning in plantations helps to reduce losses.

SELECTED REFERENCES

Bergdahl, D. R., and D. W. French. 1976. Epidemiology of comandra rust on jack pine and comandra in Minnesota. *Can. J. For. Res.* 6:326–334.

Powell, J. M. 1974. Environmental factors affecting germination of *Cronartium comandrae* aeciospores. *Can. J. Bot.* 52:659–667.

DISEASE PROFILE

Pine–Sweetfern Rust

Importance: In natural stands, damage is minor unless the alternate host and the fungus are abundant. In Minnesota and Wisconsin some jack pine stands have been seriously damaged. If pines are to be used for any product such as pulp or posts, it is almost impossible to satisfactorily remove the bark from the cankered part of the tree. If the trees are to be grown over a longer rotation, the rust-infected trees will have more decay than healthy trees. Mortality in mature stands may be as great as 43%. Seedlings in nurseries and young trees in plantations often have been severely damaged by this fungus, with losses up to 45%. Jack pine introduced to Minnesota from Canada are more susceptible than jack pine from local seed sources.

Aecia at base of tree. Courtesy of D. W. French, University of Minnesota, St. Paul.

Suscepts: Pine–sweetfern rust occurs on at least 19 species of two-or three-needle pines, sweetfern *(Comptonia asplenifolia),* and sweet gale *(Myrica gale).* In the Lake states, jack pine is the most severely damaged host. In Washington and British Columbia, lodgepole and ponderosa pines are hosts.

Distribution: Sweetfern rust is endemic in the United States and was first reported on nursery seedlings about 1910, soon after seedlings of susceptible pines were produced in quantity in regions where the rust occurs. It is abundant from Maine to Delaware and occurs sporadically as far west as Minnesota. It also is found around Puget Sound, in Washington and British Columbia.

Causal Agents: Sweetfern rust is caused by *Cronartium comptoniae* (Uredinales).

Biology: Pycnia are produced on pine from August to October the year following infection, and aecia are produced the following spring. Successive crops of aecia are borne on the infected stem each spring as long as the fungus remains alive. Aeciospores are wind-disseminated to the leaves of the alternate host where they infect the undersides. In 3–12 days, uredinia are produced and urediniospores are released to infect other sweetfern plants. About 30 days later, telial columns are produced by the same mycelium that previously gave rise to uredinia; these germinate in the summer months, mainly in August, and produce basidiospores that are carried by the wind to pine where they germinate and infect the current year's needles.

Epidemiology: Distribution of the rust is very erratic in natural stands and plantations. The two alternate hosts inhabit opposite extremes in site so the rust is successful in both dry, sandy areas as well as wet, swampy areas. Higher relative humidities may be conducive to infection as cankers are seldom found more than 1.5 m aboveground.

Diagnosis: The fungus attacks the stem of the tree; branches seldom are infected. The infected stems of seedlings become slightly swollen, but on older trees this swelling is less apparent. Often only one side of the bole will be infected. The infection manifests itself as an elongated canker with a flat face surrounded by a series of small galls and ridges of tissue. The orange-colored aecia, which form along the margins of the infected portion of the stem in the spring, are easily recognized, but they are not abundant and are present for a relatively short time. The Zimmerman pine moth is a common inhabitant in and around active infections. The wood decay fungus *Phellinus pini* often sporulates on the canker face. The wood in the rust-infected area is brown, in contrast with the normally light-colored sapwood. Urediniospores on the underside of the leaves are yellow-orange, and the brown telial columns are borne in clusters on the underside of the leaves, but both are inconspicuous.

Control Strategy: When infected stands are thinned, all obviously cankered trees should be removed. Where the disease is prevalent, nurseries should be located away from any sweetfern, or the sweetfern in the adjacent area should be eradicated. It is not known how far basidiospores travel, but apparently they do not travel far. In the nursery, fungicide sprays will reduce the incidence of infection. Resistant varieties are being developed. In Europe *M. gale* grows in association with some susceptible species of pine, and great damage could result if the rust were introduced.

SELECTED REFERENCES

Blanchette, R. A. 1982. *Phellinus (Fomes) pini* decay associated with sweetfern rust in sapwood of jack pine. *Can. J. For. Res.* 12:304–310.

Smeltzer, D. L. K., and D. W. French. 1981. Factors affecting spread of *Cronartium comptoniae* on the sweetfern host. *Can. J. For. Res.* 11:400–408.

DISEASE PROFILE

Pine–Cowwheat Rust

Importance: Pine–cowwheat rust, also known as stalactiform rust, causes rapid mortality in seedlings. On larger trees, elongate cankers form on the main stem, reducing the value of the trees for poles and pilings. Extensive resinosis is usually associated with the cankers. Trees may be predisposed to attack by bark beetles.

Suscepts: Pine–cowwheat rust occurs on hard pines, primarily jack pine in the Lake states and lodgepole pine in western United States. Cowwheat (*Melampyrum* spp.) and Indian paintbrush (*Castilleja* spp.) are alternate hosts. Although Indian paintbrush is susceptible, it is not as important a host for the fungus in the Lake states. Indian paintbrush occurs along roadsides and in open fields where the environment may be less favorable for the fungus. In western states and Canada, Indian paintbrush is an alternate host of consequence along with other species in the Scrophulariaceae including yellow owl's clover (*Orthocarpus luteus*), lousewort (*Pedicularus bracteosa*) and yellow-rattle (*Rhinanthus crista-galli*).

Aecia on lower stem. Courtesy of D. W. French, University of Minnesota, St. Paul.

Distribution: Pine–cowwheat rust is restricted to North America and is found in western
   Canada as far east as Manitoba and in the United States from southern California,
   Colorado, and Utah to Minnesota, Wisconsin, and Michigan.

Causal Agents: This rust disease is caused by *Cronartium coleosporioides* (Uredinales). The
   imperfect name is *Peridermium stalactiforme.*

Biology: Cronartium coleosporioides forms pycnia on pine in August in the year following
   infection and aecia the following spring. The uredinia, telia, and basidiospores are
   produced on the alternate hosts during the summer. Infection of the pine host occurs in
   July and August.

Epidemiology: Little is known of requirements for infection. Artificial inoculation of pines
   and alternate hosts have been successful with either a sprayed spore suspension or by
   dusting spores on the foliage and then incubating at 100% relative humidity for up to 72
   hours at about 20°C. Unlike most stem rusts, *C. coleosporioides* grows vertically from the
   initial infection, spreading into branches of the mid-and upper-crown.

Diagnosis: Cronartium coleosporioides causes elongated swellings and long narrow cankers
   on the main stem of older trees. These cankers resemble mechanical wounds that could
   have resulted from damage caused by a falling tree. The branches in the canker on a rust-
   infested tree are not broken or missing as would be the case with mechanical damage.

Control Strategy: Cowwheat rust can be controlled in the nursery with fungicides such as
   Ferbam. Diseased stock should not be planted. Eradication of alternate hosts adjacent to

nurseries may be economically feasible. Cowwheat requires considerable shade, and if the nursery is located away from stands with the alternate host, incidence of the disease would be reduced. Resistant varieties are possible; however, the disease is rarely of concern in forests.

SELECTED REFERENCES

Anderson, N. A., D. W. French, and R. L. Anderson. 1967. The stalactiform rust on jack pine. *J. For.* 65:398–402.

Hiratsuka, Y., and J. M. Powell. 1976. *Pine Stem Rusts of Canada.* Can. For. Serv., For. Tech. Rep. 4. 83 pp.

Peterson, R. S. 1968. Limb rust of pine: The causal fungi. *Phytopathology* 58:309–315.

DISEASE PROFILE

Limb Rusts

Importance: Limb rusts are so named because they kill branches in the crown without girdling the tree. As the disease progresses and more branches are killed, the crown remaining alive can no longer meet the tree's needs, and the tree will die. More often, however, these weakened trees are attacked by bark beetles. Even though some stands have less than 20% of the trees infected, the aesthetics of the area are severely degraded because the larger trees are usually infected.

Suscepts: Limb rusts occur on Jeffrey and ponderosa pines in the United States and Apache and Montezuma pines in Mexico. One race infests alternate host species of Scrophulariaceae including *Castilleja, Orthocarpus,* and *Pedicularis* spp.

Distribution: Limb rusts occur in all the southwestern United States as well as Oregon, Utah, South Dakota, Mexico, and Guatemala.

Causal Agents: Limb rusts are caused by *Peridermium filamentosum* (Uredinales). There are three races: the Coronado race, which infects species in the Scrophulariaceae, and the Powell and the Inyo races, which do not need an alternate host (autoecious). Because the Coronado race produces a sexual state, it has been named *Cronartium arizonicum.*

Biology: The Coronado race produces uredinia, telia, and basidiospores on the alternate hosts. All the races produce pycnia and aecia on the pine hosts, but the aecia and their method of germination vary with the race. The Coronado race produces long germ tubes, while the other races have short germ tubes. The Powell race has short, swollen, dichotomously branched germ tubes with uninucleate or multinucleate cells in contrast to the binucleate cells of the Coronado race. A third race is called Inyo and can be separated from the Powell race because it has less persistent aecia. The peridium of the Coronado race is only one cell thick, while peridia of the other two races are two or more

Limb rust.

cells thick. In Jeffrey pine stands where both *P. filamentosum* and *P. stalactiforme* occur, the two rusts can be separated because *P. stalactiforme* has lower, more confluent aecia. The thick, smooth areas on aeciospore walls, lack of aecidial fragments on infected twigs from previous year's aecia, and presence of rough bark or cankers are systemic symptoms of the latter.

Epidemiology: Aecia of the Inyo race begin to emerge in early June at low elevations in the Sierra Nevada and then are produced in greater abundance later in June and July. These aecia are intact in August and may retain spores in September. In contrast aecia of *P. stalactiforme* appear in April in southern Coast Ranges and in May in the Sierra Nevada with dissemination usually complete in June. Aecia of the Coronado and Powell races appear in spring. Dehiscence of the Coronado race is usually in June, but it is delayed with the Inyo and Powell races until the advent of abundant rain. Without adequate rain the aecia of the Powell races have remained unbroken until September. Aecia of the Powell race are favored by high temperature (15.5–23°C) more so than the Coronado race, which has an optimum for germination of 13–18°C. Aeciospores of the Powell race may not infect seedlings; however, the Coronado race will infect seedlings.

Diagnosis: The most characteristic symptom is mid-crown mortality with bare branches. Above and below the dead portion of the crown, the living branches may have the remains of the aecia. In some cases, aecia remain in good condition for most of the summer. Aecidial fragments from the previous years' aecia are usually visible with binoculars. The fungi are systemic and thus not limited to a canker or a gall.

Control Strategy: Fungicides will protect against infection of seedlings and eradication of infected hosts around nurseries would be of benefit. In stands, diseased trees can be detected in early stages and eliminated. Shortening cutting cycles in affected stands to 25 years will allow salvage of diseased trees.

SELECTED REFERENCES

Mielke, J. L. 1952. The rust fungus *Cronartium filamentosum* in Rocky Mountain ponderosa pine. *J. For.* 50:365–373.
Peterson, R. S. 1966. *Limb Rust Damage to Pine.* USDA For. Serv., Res. Paper INT-31. 10 pp.
Peterson, R. S. 1968. Limb rust of pine: The causal fungi. *Phytopathology* 58:309–315.

DISEASE PROFILE

Pine–Oak Rust

Importance: Pine–oak rust, also called eastern gall rust, is common in hard pine stands with a red oak component and can be a serious problem in nurseries and young plantations. Galls on the main stem of seedlings and saplings can cause death. Galls on branches are of minor importance.

Suscepts: More than 20 species of hard pines and almost 30 species of oaks are susceptible. In the Lake states region, jack pine is the host most severely damaged. The disease is common on southern pines where it can be confused with fusiform rust.

Distribution: Pine–oak rust is endemic to this country. Its range is practically coincidental

Aecial gall.

with the range of its hosts; it has been found in all parts of the United States and in Canada, Mexico, China, and Japan.

Causal Agents: Pine–oak rust on jack pine is caused by *Cronartium quercuum* f. sp. *banksianae* (Uredinales). On shortleaf pine, it is *C. q.* f.sp. *echinatae,* and on Virginia pine, *C. q.* f.sp. *virginianae.*

Biology: Cronartium quercuum produces pycniospores and aeciospores on pine; one crop of pycniospores is borne, probably about a year after infection, and the following year aecia are produced on the same portion of the gall. In subsequent years, pycnia and aecia are formed in alternate years on new areas of the gall. Aeciospores are carried by the wind to the underside of oak leaves where they germinate and penetrate through the stomata. This mycelium produces urediniospores for a short time in spring or summer depending on geographic location. These spores infect other oaks. Later the same mycelium produces columns of teliospores, which mature in late winter and spring (February to June) in southern states and in the summer (June, July) in the northern states. The teliospores produce basidiospores, which are wind-disseminated to infect pine needles. Galls form during the first or second year following infection and are composed mostly of woody tissue. Their growth is induced by mycelium of the parasite. Conduction of sap is hindered; and if a gall surrounds the stem, the part above the gall is likely to die. Galls on older stems can subject these trees to wind breakage. Blue stain fungi frequently enter the stem through these galls. In oak leaves the mycelium spreads only a few millimeters, and the infected cells die soon after teliospores are produced.

Epidemiology: The time required for teliospores to germinate and produce basidiospores and for the basidiospores to germinate and invade the pine needles is at least 13 hours when the temperature is 12–24°C and the atmosphere is saturated. Conditions necessary for infection of oak leaves by aeciospores or urediniospores are similar. Oak leaves are susceptible when they are less than 3 weeks old.

Diagnosis: On smaller branches spherical galls form and completely surround the stem; on the stems of older trees galls frequently include only a part of the circumference of the stem or may be apparent only as a flat or distorted area. Infection on oak leaves results in small necrotic or chlorotic areas. The urediniospores are yellow to orange, and the telial columns are brown.

Control Strategy: The cutting of scrub oak, with the consequent production of succulent, susceptible sprouts, may tend to increase the disease. The oak in and around nursery sites should be eradicated. It is not known to what distance the basidiospores can infect pine, but it is a greater distance than for some of the other species of *Cronartium.* Pruning is effective, but generally not economically feasible. In nurseries, fungicides can be applied during June and July in northern states and earlier in southern states. Chemicals applied for controlling fusiform rust will be effective against pine–oak rust as well. Infected seedlings should be culled. Individual trees that are resistant have been selected, and eventually this rust may be controlled by planting resistant trees.

SELECTED REFERENCES

Anderson, N. A. 1963. *Eastern Gall Rust.* USDA For. Serv., Pest Leaflet 80. 4 pp.
Nighswander, J. E., and R. F. Patton. 1965. The epidemiology of the jack pine–oak gall rust *(Cronartium quercuum)* in Wisconsin. *Can. J. Bot.* 43:1561–1581.

DISEASE PROFILE

Pine–Pine Rust

Importance: In very susceptible hosts, species stunting and malformation commonly result. Some trees are killed. In western Canada, this is the most common and destructive stem rust of hard pines. Infections in the western states are concentrated in relatively few years and at these times can cause considerable damage. Although somewhat limited in its distribution in the Lake states, it has caused losses in areas where the other rust diseases are not present. It is a serious problem to Christmas tree growers in the northeastern states.

Suscepts: Pine–pine rust, also known as western gall rust, occurs on hard pines, including lodgepole, ponderosa, jack, and Scots pines. Whether or not this rust fungus has an alternate host is unknown.

Distribution: Pine–pine rust occurs across Canada and Alaska. In the United States it is found in eastern states as far south as Virginia, the Lake states, Nebraska, South Dakota, and the western states as well as northern Mexico. In Minnesota this rust dominates north of the Continental Divide with the pine–oak rust being more prevalent south of the Divide.

Causal Agents: Pine–pine rust is caused by *Endocronartium harknessii* (Uredinales). The imperfect stage is *Peridermium harknessii.*

Multiple rust galls. Courtesy of W. Merrill, Pennsylvania State University, University Park.

Biology: *Endocronartium harknessii* is different from the other stem rusts of pines caused by species of *Cronartium* in that it is autoecious with no known alternate host. The aeciospores are really aecidioid teliospores. They have the morphology of aeciospores but function like teliospores. Their germ tubes act as basidia with one to four cells, each containing a single nucleus, and these germ tubes can infect pine in the spring season. Thus there are no basidiospores and no repeating stage. Pycnia are either rare or absent.

Epidemiology: Pine–pine rust has the potential to be a devastating disease. It is apparently limited by unfavorable weather as years of heavy infection occur when climatic conditions favor infection. In artificial inoculations, seedlings of jack pine were successfully infected following incubation at 18°C for 48 hours. There is a maturation related resistance to pine–pine rust in radiata pine.

Diagnosis: Pine–pine rust causes globose galls on seedlings and older trees, and these galls resemble those produced by pine–oak rust. *Cronartium quercuum* produces aeciospore germ tubes about three times as long as those produced by *E. harknessii*. The pine–pine rust often results in large numbers of small-diametered galls in comparison with pine–oak rust, and the pine–pine galls are mainly on the branches rather than on the main stem.

Control Strategy: The pine-pine rust can probably be controlled in the nursery with fungicides. Diseased swellings should be eradicated and infected seedlings should not be shipped into areas where the disease is not known to occur. Because of the extreme susceptibility of some strains of Scots pine, it is important that the fungus not be introduced to other countries outside of North America. Elimination of infected pines around nurseries up to 100 m has been recommended. Resistant varieties will likely be a long-term method of managing the disease. Resistance genes appear to interact multiplicatively.

SELECTED REFERENCES

Burnes, T. A., R. A. Blanchette, C. G. Wang, and D. W. French. 1988. Screening jack pine seedlings for resistance to *Endocronartium harknessii*. *Pl. Dis.* 72:614–616.

van der Kamp, B. J. 1988. Temporal and spatial variation in infection of lodgepole pine by western gall rust. *Pl. Dis.* 72:787–790.

Zagory, D., and W. J. Libby. 1985. Maturation-related resistance of *Pinus radiata* to western gall rust. *Phytopathology* 75:1443–1447.

Ziller, W. G. 1974. *The Tree Rusts of Western Canada*. Can. For. Serv., Pub. 1329. 272 pp.

15.3   CEDAR AND JUNIPER STEM RUSTS

DISEASE PROFILE

Gymnosporangium Rusts

Importance: These rust diseases, common on cedars, junipers, and pomaceous hosts, are important chiefly because of their abundance and because of the damage to pomaceous crops, especially apple. Although branch and tree mortality may occur, forest trees are not usually significantly impared. Ornamental cedars and junipers may be injured or appear unsightly if rust is severe.

Suscepts: Gymnosporangium rusts are host-specific but occur on many host genera, especially in the case of the aecial host. Coniferous genera include *Chamaecyparis, Cupressus, Juniperus,* and *Libocedrus.* Aecial hosts are in four families: Rosaceae, Myricaceae, Hydrangiaceae, and Cupressaceae, with the first being the most important. Aecial host genera include *Amelanchier, Aronia, Chaenomeles, Comptonia, Cotoneaster, Crataegomespilus, Crataegus, Cydonia, Fendlera, Gillenia, Heteromeles, Malus, Mespilus, Myrica, Peraphyllum, Philadelphus, Photinia, Pyrus, Sorbus,* and *Vauquelina.*

Distribution: Disease distribution depends on the host specificity of the pathogen and the range of the hosts. A few occur only in restricted areas of one state, others occur widely over large regions, and still others occur nationwide.

Causal Agents: The fungi that cause these rust diseases belong to the genus *Gymnosporangium* (Uredinales) except for two species that have no known telial stage; these are in the form genus *Uredo.*

Biology: The large genus *Gymnosporangium* is very diverse, but typically pycnia occur on the upper surface and aecia on the lower surface of leaves of the pomaceous hosts, often also on fruits and shoots. Telia occur on the leaves, shoots, and, less frequently, trunks of coniferous hosts. Only *G. nootkatense* produces uredinia and only *G. bermudianum* is autoecious, producing both aecia and telia on the juniperous host. Generally, the pathogen overwinters as perennial mycelium in the conifer host. Telia are produced in the spring and upon germination basidiospores infect the aecial host. Here pycnia and

Pycnia and aecia.

aecia are produced, and spores from aecial pustules infect conifer leaves and green shoots during the summer. Often more than a years' time is required for gall formation following infection of junipers.

Epidemiology: Collectively, the epidemiology of these diseases has been little studied, especially as related to infection of conifers. Aeciospores and basidiospores are wind-disseminated, the latter for at least several kilometers. Basidiospores do not require a dormant period and are produced abundantly in the spring during wet weather, which conditions teliospore germination.

Diagnosis: Symptoms of the disease are easily recognized on the conifer host by the presence of globose, reddish leaf galls (1–5 cm diameter), globoid to fusiform-shaped branch or stem swellings (2 cm to more or less 0.3 m in length), and witches' brooms of branches and stems. On the aecial host, defoliation may result, and infected fruits are often distorted. Signs are conspicuous gelatinous telia (orange-yellow-red-brown) that occur as large horns, cones, cushions, or crests on leaf and branch galls. Telial horns (tendrils) may be greater than 15 cm in length. Teliospores are usually two-celled and borne singly on pedicels. Pycnia are subepidermal and most commonly epiphyllus, globose, and first yellow and later black. Aecia are also subepidermal with a peridium, usually roesteloid, but in a few cases aecidioid. Aeciospores are catenulate and brown and have verrucose walls and numerous germpores. Urediniospores when present are borne singly on pedicels and are catenulate and brownish or yellowish.

Control Strategy. No control is practiced in the forest or in general for conifer hosts. In apple orchards, protective fungicides are used in spray programs. Eradication of cedars and junipers in the vicinity of apple orchards provides additional safeguards and is sometimes regulated by law. Some apple varieties are resistant to rust as are some species of *Juniperus*.

SELECTED REFERENCES

Arthur, J. C. 1934. *Manual of the Rusts in the United States and Canada.* The Science Press Print Co., Lancaster, Pa. 438 pp.

Ziller, W. G. 1974. *The Tree Rusts of Western Canada.* Can. For. Serv., Publ. 1329. 272 pp.

DISEASE PROFILE

Juniper Broom Rust

Importance: Stem swellings and witches' brooms form on junipers. The latter either die while small or grow larger and live for many years. Branch cankers may be associated with infection. Severe infection of alternate hosts may cause leaf lesions and fruit mortality.

Witches' broom.

Juvenile leaves from witches' broom.

Suscepts: Eastern and southern redcedars and prostrate and Rocky Mountain junipers are telial hosts. The pycnial and aecial stages occur on fruits, young stems, and leaves of apple, hawthorn, mountain ash, quince, and serviceberry.

Distribution: Juniper broom rust can be found in North America from coast to coast and as far south as Arizona and Florida.

Causal Agents: Gymnosporangium nidus-avis (Uredinales).

Biology: Infection of junipers is through leaves or green twigs. If the infected area remains alive, fungus growth becomes perennial. Infection of branch tips may lead to formation of a witches' broom, whereas infection of older portions of branches may lead only to swellings. Leaves within the witches' brooms may take on an awl-shaped juvenile form. Cushionlike orange-brown telia form in spring on swollen branches, on branches that form the witches' broom, or from leaf axils. The telia become gelatinized and grow in size. Teliospores germinate in place and produce basidiospores that are wind-disseminated to alternate hosts. They germinate, and the fungus penetrates directly through the cuticle and forms localized colonies. Pycnia form within 14 days. Aecia form approximately 1 month later, producing in midsummer aeciospores that infect junipers.

Epidemiology: As is the case with most tree rusts, the life cycle of *G. nidus-avis* is synchronized with the maturation level of host tissues and rainy periods. The rust overwinters in the inner bark. Telia protrude and germinate after a few hours of rain, producing basidiospores that must intercept susceptible alternate host leaves, germinate, and cause infection before the moist period ends. After aeciospores are produced, their successful germination and infection of the conifer host depends on the frequency and duration of summer rains. It is likely that in much of the range, disease incidence is relatively low because of summer dryness.

Diagnosis: Witches' brooms, containing awl-shaped juvenile leaves, are diagnostic. Branches within the brooms have roughened bark. Support branches and those branches that are infected but not supporting a witches' broom are swollen and may have deep longitudinal fissures. Pycnia appear as small swollen yellow spots on leaves and young twigs. Aecia form on both upper and lower leaf surfaces and will be distinctive when their aecial peridia elongate just before spores are released.

Control Strategy: No control on either host is warranted in forest stands. Junipers and cedars with high individual value may be pruned to remove infected branches.

SELECTED REFERENCES

Sinclair, W. A., H. H. Lyon, and W. T. Johnson. 1987. *Diseases of Trees and Shrubs.* Comstock Publ. Assoc., Cornell Univ. Press, Ithaca. 574 pp.

15.4 FOLIAGE RUSTS

DISEASE PROFILE

Cedar–Apple Rust

Importance: Cedar–apple rust does not damage the juniper host to any extent, but it has caused a loss of 10% of the apple crop in Virginia in one year and ranks fifth in diseases of apples with respect to damage caused, being responsible for an estimated loss of 0.8%

Mature telial gall.

of the total apple crop each year. The loss results from extensive defoliation and deformed fruit.

Suscepts: Cedar–apple rust occurs primarily on genera such as *Malus, Crataegus,* and *Juniperus.*

Distribution: Cedar–apple rust was endemic on wild hosts in North America and was first described in 1822. It has been of economic importance since about 1880. Distribution of the rust is more or less coincidental with the distribution of its hosts in California and central and eastern North America.

Causal Agents: Cedar–apple rust is caused by *Gymnnosporangium juniperi-virginianae* (Uredinales).

Biology: The mycelium is intercellular, and haustoria penetrate the host cells and absorb nourishment. The galls on juniper are of overgrown leaves, composed of both host tissue and mycelium of the parasite. The telial horns or gelatinous tendrils arise from mycelium in the galls; they are made up of teliospores and their stalks, the latter containing a gelatinous material that absorbs water and swells during rains, causing the columns to elongate. The mycelium in apple leaves decomposes the chloroplasts and causes the cells to enlarge and multiply.

Epidemiology: Apple leaves and fruit are infected in spring by basidiospores that land on them, germinate, and enter the host cells. The mycelium spreads locally; pycnia are produced in May and June on the upper side of leaves, and, later in July, aecia form on the underside. Aeciospores are carried by the wind, in the summer and early fall, to cedar leaves, where they germinate, invade the leaves, and stimulate the formation of galls. Teliospores are formed on these galls 19–22 months later and germinate to produce basidiospores, which infect apples. As many as 7.5 million basidiospores may be produced on a single gall. Basidiospores may be carried by the wind to a maximum of about 10 km, but usually not more than 3 km. No urediniospores are produced. Low temperature and high relative humidity are necessary for production of basidiospores and for infection of apples. There apparently are several strains of the fungus that differ in their ability to attack different apple varieties. Conditions for optimal infection of junipers are unknown.

Diagnosis: Globose galls up to 5 cm in diameter are formed on the juniper trees. In spring the surface of these galls is covered with orange-colored columns of teliospores, which elongate and become mucilaginous during moist weather. Infected spots on the apple leaves are yellow in color, a group of pycnia forms in the center of each spot, and groups of aecia are borne on the underside of the leaf, just below the groups of pycnia. Aecia are tubular, and the peridium breaks into characteristic ribbonlike strands. When an immature fruit is infected, growth ceases in the region of the infection, and the fruit becomes distorted. On the fruit, aecia are produced in a ring around the pycnia.

Control Strategy: Cedar–apple rust may be controlled by eradicating the junipers within 1 km of apple orchards (required by law in Virginia, West Virginia, New York, and Arkansas in the apple-growing regions). Another means of control is the use of resistant varieties of apples. Fungicide sprays can be used.

SELECTED REFERENCES

Crowell, I. H. 1934. The hosts, life history and control of the cedar–apple rust fungus, *Gymnosporangium juniperi-virginianae*. *J. Arnold Arboretum* 15:163–232.

DISEASE PROFILE

Pine Needle Rust

Importance: Most conifers are subject to one or more kinds of needle rusts. One of the more common is pine needle rust. This rust can cause locally severe defoliation in nurseries that are located in areas where the alternate hosts are abundant. Even though only the old needles are affected by the disease, the seedlings may be substantially injured.

Suscepts: Hard pines (2- to 3-needle pines), including red pine, are hosts. The major alternate hosts are aster (*Aster* spp.), goldenrod (*Solidago* spp.), and others.

Distribution: Pine needle rust is found wherever the hosts occur, on jack and red pine in Canada and the northern United States; on lodgepole pine in western North America; and on loblolly, pitch, Virginia, and table-mountain pines in the southern and eastern states.

Causal Agents: There are several species of fungi that cause pine needle rust. The major one is caused by *Coleosporium asterum* (Uredinales).

Biology: There are at least two forms of *C. asterum:* one infects goldenrod but not asters; another infects cultivated annual but not cultivated perennial asters or goldenrod. The aeciospores of the western form of *C. asterum* are larger than those that infect goldenrod, which, in turn, are larger than those of the eastern form occurring on asters.

Aecia on needles.

Epidemiology: Aecia develop on pine needles in May and June, and aeciospores are wind-disseminated to the aster or goldenrod hosts, which are infected through the stomata. Uredinia are formed 10–15 days later, and the urediniospores can infect other alternate host plants during the summer. In some regions, the fungus can overwinter, supposedly in the uredinial stage on the rosettes of the goldenrod, and thus maintain itself without the pine host. Infection of the alternate hosts requires 20–25 hr of wet weather. From the same mycelium, in late summer or fall, reddish-brown waxy telia are formed, which are hard when dry and gelatinous when wet. The one-celled teliospores germinate without dormancy into an internal four-celled basidium, with each cell producing a basidiospore and sterigma that projects above the telium. The basidiospores are yellow, and 10–12 hr of wet weather are required for their production. The basidiospores are wind-dispersed to pine needles and cause infection. The fungus can survive in the needles as long as the needles remain alive and produce pycnia and aecia in the spring.

Diagnosis: Needle rusts can be recognized by the white aecia filled with orange-yellow aeciospores that can turn the foliage on a small tree from green to yellow. The uredinia and later the telia are obvious on the underside of the leaves of the alternate hosts. The underside of leaves of goldenrod are often covered with yellow-orange uredinia.

Control Strategy: Not much is known about the dispersal of the basidiospores of these fungi, but based on observations, infection of the pine depends on alternate hosts being relatively close to the pines. Thus eradication of the aster and goldenrod plants within and adjacent to nurseries should alleviate the problem. Fungicides and resistant varieties are possibilities but are not considered economically feasible at this time. There are some 20 species of *Coleosporium*, most of which occur on species in the family Cardnaceae and on pines. Southern pines are the hosts for most of these rust species.

SELECTED REFERENCES

Nicholls, T. H., E. P. Van Arsdel, and R. F. Patton. 1965. *Red Pine Needle Rust Disease in the Lake States.* USDA For. Serv., Res. Note LS-58. 4 pp.

DISEASE PROFILE

Spruce Needle Rust

Importance: There are at least 13 species of spruce needle rusts of which three cause significant injury to the spruce hosts, either by inducing premature defoliation or brooming, both of which result in portions of the crown being killed. Spruce broom rust resulted in an average cull factor of 24% in stands of Engelmann spruce in Colorado. Spruce needle rusts caused major losses for the Christmas tree industry when spruce was used more extensively for that purpose.

Suscepts: The species of spruce subject to these rust diseases include Engelmann, white, black, blue, red, and Sitka spruces. The alternate hosts include species of *Ledum, Chamaedaphne, Rhododendron, Arctostaphylos, Caultheria, Empetrum,* and *Vaccinium.*

Distribution: Spruce-needle rusts occur wherever hosts exist over Alaska, Canada, and northern and western United States.

Causal Agents: The important species, at least those causing injury, include *Chrysomyxa ledicola, C. ledi,* and *C. arctostaphyli* (Uredinales). The other species are of less consequence.

Aecia on current needles. Courtesy of USDA Forest Service.

Biology: Spruce broom rust caused by *C. arctostaphyli* causes perennial systemic witches' brooms. The needles in the broom are etiolated and die as a result of infection.

Epidemiology: The aeciospores of spruce needle rusts develop in the summer and are wind-disseminated to their alternate hosts where they invade the underside of the leaves. Uredinia, and later telia, are produced, and the fungi overwinter in the telial stage. Basidiospores are produced in the spring; they are wind-disseminated and infect the young emerging foliage of the spruce host. Pycnia are followed by the production of aecia. Most of these needle rust species can survive when spruce are not present in the area.

Diagnosis: Spruce needle rusts can be recognized by aecia on the needles early in the growing season. Affected trees often appear yellow and can be detected from some distance because of the discoloration of the foliage. In Canada, defoliation by *C. ledicola* was evident from airplanes. Trees infected by *C. arctostaphyli* can be recognized by the fairly large witches' brooms with only current year needles. Infection on the alternate hosts appears as yellow-brown spots on the upper surface, and their uredinia and telia are present on the underside of the leaves. On some species these spore stages occur on the upper surface of the leaves. A few of the species, including *C. arctostaphyli,* do not produce uredinia.

Control Strategy: The basidiospores of at least some of the spruce rust fungi can infect spruce at considerable distances of more than 2 km from the alternate hosts. Even so, it is advisable to eradicate the alternate hosts near nurseries and plantations where this is feasible. Fungicides are effective but not usually required.

SELECTED REFERENCES

Ziller, W. G. 1974. *The Tree Rusts of Western Canada.* Can. For. Serv., Pub. 1329. 272 pp.

DISEASE PROFILE

Ash Rust

Importance: Ash rust is of limited importance and sporadic occurrence. Severe infections can cause defoliation, which, if repeated for several years, may result in tree mortality.

Suscepts: Ash rust occurs on the alternate hosts of ash and cord or marsh grass (*Spartina* sp.). Of the tree hosts, white ash is most important, but the disease has also been reported on green, red, blue, black, and Carolina ashes, as well as species of *Forestiera.*

Distribution: Ash rust occurs primarily along the eastern and Gulf coasts from Nova Scotia to Florida and west to Texas. However, ash rust has been observed in inland areas of

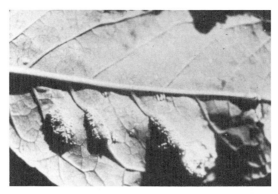

Aecia on ash leaf. Courtesy of D. W. French, University of Minnesota, St. Paul.

New England, the Lake states, the midwest, and also Arkansas, Montana, Colorado, Minnesota, and North and South Dakota.

Causal Agents: Ash rust is caused by *Puccinia sparganioides* (= *P. peridermiospora)* (Uredinales).

Biology: Puccinia sparganioides is a macrocyclic, heteroecious rust with the pycnial and aecial stages on ash and uredinial and telial stages on cord grass. The fungus overwinters in the telial stage, and sporidia, released in early spring, infect the current growth of ash in late spring and summer. Pycnia occur on the upper surface of ash leaves and on petioles and shoots in June. Aecia follow in approximately 2 weeks, occurring on the lower leaf surface and on petioles and shoots. Aecia infect cord grass in May and June, and uredinia produced on cord grass reinfect grass during July and August. Telia occur on cord grass during the fall.

Epidemiology: Ash rust incidence is highest along the coast and diminishes with distance inland. Within marsh areas, disease incidence is inversely correlated with soil water salinity; high concentrations of salt have a fungistatic effect on germination of aeciospores. Several hours of high atmospheric moisture result in teliospore germination and production of sporidia; infection of ash requires 6–8 hours. Optimum temperatures for production of sporidia were 12–21°C in New Hampshire, and the optimum temperature for spore (all types) germination was 25°C in Texas. Dry hot weather in late spring and summer during the ash infection period is thought to reduce the incidence of ash rust.

Diagnosis: Infected leaves, petioles, and shoots of ash are swollen and distorted. Defoliation may occur if rust is severe, and when early defoliation occurs in several consecutive years, trees may die. The most obvious aid to diagnosis is the bright orange-yellow aecial (cluster cup) stage on the undersurface of ash leaves and petioles and shoots. On the latter two plant parts, aecia are aggregated and can be several centimeters in length. The aecia are even more obvious because they occur on the swollen infected areas. Pycnia, which preceed aecia, occur on the upper leaf surface of ash and on petioles and shoots. On the grass host, pale-yellow uredinial pustules and compact black-brown oblong to linear (1–3 mm) telial sori, characteristic of the genus *Puccinia,* are easily seen. Telia contain stalked, two-celled teliospores.

Control Strategy: No control is practical even in forests along the coasts. Protective foliar fungicides used in the spring coincident with infection of ash provide control for shade and ornamental trees. Species other than ash should be used in high hazard-coastal areas.

SELECTED REFERENCES

Van Dyke, C. G., and H. V. Amersan. 1976. Interaction of *Puccinia sparganioides* with smooth cord-
grass *(Spartina alterniflora). Pl. Dis. Reptr.* 60:670–674.

DISEASE PROFILE

Fir Broom Rust

Importance: Witches' brooms form on infected branches. Heavily infected trees may be
stunted and die prematurely. Seedlings and saplings can be killed. The brooms also
provide infection courts for heart decay and can lead to wind or snow injury. The disease

Young witches' broom. Courtesy of D. W. French, University of Minnesota, St. Paul, MN.

is most severe on alpine firs in the northern Rocky Mountains. In general, though, the disease is usually not serious.

Suscepts: Most of the native true firs, including subalpine, balsam, grand, noble, red, Pacific silver, and white, act as hosts. Alternate hosts are chickweed (*Cerastium* spp.) and mouse-ear chickweed (*Stellaria* spp.).

Distribution: Also known as yellow witches' broom, the disease is common across the northern hemisphere wherever its hosts occur. It has been introduced into the Falkland Islands. The disease is more severe in Europe, where it causes open lesions and swellings on the trunks of mature silver firs, resulting in considerable cull.

Causal Agents: Fir broom rust is caused by *Melampsorella caryophyllacearum* (Uredinales).

Biology: Melampsorella caryophyllacearum is a macrocyclic, heteroecious rust, alternating between true firs and members of the chickweed family. An important feature of this fungus is that it becomes systemic and perennial in both hosts and may overwinter in both hosts. Aeciospores are produced on fir and are wind-disseminated to leaves of chickweed. Uredinia develop on both leaf surfaces and appear as tiny orange-red pustules. Urediniospores are released and cause new infections on chickweed. Its systemic and perennial nature allows it to persist in the absence of fir hosts. In the spring, the fungus grows into new shoots and leaves where it produces first telia and later uredinia. Teliospores germinate in place, producing basidiospores that infect fir. Two years are, thus, needed to complete the life cycle. In fir, buds are infected in spring, and the fungus invades the young shoots and eventually induces formation of witches' brooms. Shoots within the broom are thicker and shorter than normal. Needles are also stunted, thickened, and quickly turn brown, pale green, and yellow. They die in autumn, and, thus, only the current needles are present within the broom. Pycnia form on all needles in the spring. Aecia develop in summer along the lines of stomata on the underside of needles. In the fall, needles die and are cast, leaving the broom without foliage over winter.

Epidemiology: Free moisture is required for germination of aeciospores and urediniospores. Both can germinate between 5 and 30°C. The abundance of the rust on fir is favored in cool, moist environments and is influenced by the abundance of alternate hosts. Outbreaks result from favorable seasonal climatic changes.

Diagnosis: In the summer the yellow color of the witches' broom is striking. Close inspection of the foliage will reveal the chlorotic, thickened, and shortened needles in the compact broom. Aecia or aecial scars may be present.

Control Strategy: In recreation areas, brooms may be pruned out to reduce potential breakage and maintain vigor. In forest stands, control is generally not necessary or feasible, but diseased trees can be easily identified and removed during thinning operations.

SELECTED REFERENCES

Scharpf, R. F., (tech. coord.). 1993. *Diseases of Pacific Coast Conifers.* USDA, Agric. Handbook 521. 199 pp.

DISEASE PROFILE

Poplar Rusts

Importance: Poplar rusts have been known for years to cause extensive defoliation of some clones, especially late in the growing season. Recently it was learned that losses in volume of unsprayed trees in relation to those protected by fungicides ranged from 9 to 21% for a resistant clone to 49–190% for a susceptible clone. The amount of loss varied with geographic location.

Suscepts: There are five poplar hosts: eastern cottonwood, quaking aspen, bigtooth aspen, black cottonwood, and balsam poplar. The principal alternate hosts in approximately decreasing order of importance are eastern and western larches, Douglas-fir, and lodgepole and ponderosa pines. The aecial stage is plurivorous, which means occurring on many different hosts.

Distribution: Poplar rusts are found across Canada and the United States and in other countries including Argentina and Japan.

Causal Agents: There are three species of fungi—*Melampsora medusae, M. occidentalis,* and *M. abietus-canadensis* (Uredinales). Closely related species of *Melampsora* cause rusts on willows with alternate hosts in the genera *Abies, Larix,* and *Tsuga.*

Biology: Melampsora medusae, one of the most important and widely distributed of the poplar rusts, produces aecia on larch needles in May, and the spores are wind-dispersed in May and early June to the poplar leaves where infection gives rise to uredinia. The urediniospores are wind-disseminated to other poplars where they can result in the production of a new generation of uredinia. The telia are produced in late summer from the same mycelium that previously produced uredinia. The telia are subepidermal originally and, after overwintering, break through the epidermis when the single-celled

Rust-defoliated plantation. Courtesy of G. Newcombe, Washington State University, Puyallup, Washington.

teliospores produce a basidium with four basidiospores. These spores infect the newly emerging needles of larch. Pycnia are produced, followed by the aecia.

Epidemiology: In the western United States the poplar rusts do not survive on the poplar host. In the central United States, however, *M. medusae* can spread from north to south during the growing season as new generations of urediniospores are produced and blown southward. Maximum aeciospore dispersal occurs at temperatures from 15 to 20°C. The disease usually starts on poplar in the north as a result of infection by aeciospores coming from larch. Few urediniospores are dispersed at temperatures below 10°C. The fungus can overwinter on poplar in the south but soon disappears with the advent of hot weather. Cooler temperatures in September and October are more beneficial to rust development. Rain or overhead sprinkling systems enhance infection of poplars. Seedlings of several conifers are highly susceptible to infection.

Diagnosis: Species of *Melampsora* on conifers can be recognized by the aecia on the needles. On young seedlings, these aecia form on the stem and the cotyledons. The fungus is easily recognized on poplar because of the yellowing of the leaves and premature defoliation. The upper surface of leaves can have large numbers of yellow spots, and the uredinia are very evident on the underside.

Control Strategy: At low levels of incidence, control measures are not necessary. In nurseries and plantations where the disease is causing major losses, fungicides may be applied to poplars at 2-week intervals from June to October. If dates of infection are known, the number of applications could be reduced. Nurseries producing conifers susceptible to poplar rust should be located away from poplars, or these trees should be eradicated. Fungicides applied during the first 3 weeks following bud break should protect against infection. Development of resistant clones of *Populus* is currently receiving much emphasis.

SELECTED REFERENCES

Ziller, W. G. 1974. *The Tree Rusts of Western Canada.* Can. For. Serv., Pub. 1329. 272 pp.

15.5 CONE RUSTS

DISEASE PROFILE

Southern Cone Rust

Importance: Southern cone rust attacks and destroys the developing (first year) cones of slash and longleaf pines. Additional losses to mature (second year) cones result from insect damage, especially that caused by *Dioryctria* spp. These insects are initially attracted to rust-infected first-year cones and then migrate to second-year cones. This disease, which occurs in the natural forest, has become more important with the advent of intensive management of southern pines and the concurrent demand for seed. The economic impact of cone rust has increased because of losses to production of improved quality seeds in seed orchards and seed production areas.

Suscepts: Cone rust occurs on the alternate hosts—pine and oak. The pine hosts are slash and longleaf. The most important alternate hosts are live oak and often evergreen oaks. Water and willow oaks are also hosts.

Distribution: The distribution of southern cone rust is closely associated with the range of live oak. The rust is important in coastal South Carolina and Georgia, south Georgia, Florida, and the Gulf coast of Alabama, Mississippi, and Louisiana.

Causal Agents: Cone rust is caused by *Cronartium strobilinum* (Uredinales).

Biology: Cronartium strobilinum is a macrocyclic, heteroecious rust with pycnial and aecial stages on pine cones and uredinial and telial stages on oak leaves. Basidiospores, from overwintered telia on leaves of evergreen oak, are produced during warm wet weather and infect emerging female flowers (strobili) of slash pine during pollination, usually in December through February. Pycnia form on the conelets in March through June and are followed by aecia, which are wind-disseminated and infect oak leaves. Uredinia are produced on oak leaves throughout the summer, and in the fall and winter telia mature on these leaves. Telia cannot survive the dormant season on dead leaves (the fungus is an obligate parasite) and do not function as inoculum to continue the epidemic. Telia

Cone with aecia. Courtesy of D. W. French, University of Minnesota, St. Paul.

produced on evergreen oaks or perhaps other oaks during mild winters remain viable and produce basidiospores to infect pine flowers in December through February.

Epidemiology: A high incidence of cone rust occurs sporadically and is no doubt conditioned by climatic factors such as favorable temperature and moisture conditions to allow secondary cycles on oak during the spring and summer; winter weather, which allows survival of telia but prevents their germination prior to the appearance of female flowers on pine; and relatively warm temperatures and high atmospheric moisture enhancing telia germination and basidiospore infection during the pollination period.

Diagnosis: Infected conelets swell to three or four times normal size in April. Cone scales in the infected area are swollen and reddish in color. Diseased cones are often attacked by insects and shed prematurely by late summer. Pycnia appear as viscid drops on the infected cone scales in early spring and are soon followed by aecia. This yellow-orange powdery mass of aeciospores, produced from ruptured cone scales, is the most obvious diagnostic feature of cone rust.

Control Strategy: Cone rust is not controlled in the forest nor is control normally recommended for natural seed production areas. However, practical and effective controls are available for seed orchards. Slash and longleaf pine seed orchards should, if possible, be established outside the natural range of live or evergreen oak. Application with protective fungicides during flower emergence and pollination will prevent cone rust. In controlled-pollinated orchards, cone rust is normally prevented by the pollination bags used to enclose and isolate the female flower.

SELECTED REFERENCES

Hepting G. H., and F. R. Matthews. 1970. *Southern Cone Rust.* USDA For. Serv., Pest Leafl. No. 27. 4 pp.
Matthews, F. R. 1964. Some aspects of the biology and the control of southern cone rust. *J. For.* 62:881–884.

15.6 OTHER CONE RUSTS

Southwestern pine cone rust, which occurs in southern Arizona and New Mexico, is caused by *Cronartium conigenum* Hedge. and Hunt with pycnial and aecial stages on cones of Chihuahua pine and uredinial and telial stages on leaves of Emory and white leaf oaks. Spruce cone rust is caused by *Chrysomyxa* spp., which have pycnial and aecial stages on cones of *Picea* spp. and uredinia and telia on wintergreen and woodnymph. Hemlock cone rust is caused by *Melampsora abietis-canadensis* (Farl.) Ludw. with pycnia and aecia on cones and needles of eastern and Carolina hemlock; telia occur on leaves of poplar species. Also there is the autecious rust *M. farlowii* (Arth.) Davis having telia on needles and cones of these same hemlock species.

16

CANKER DISEASES

16.1 INTRODUCTION

Canker diseases are among the most destructive problems in our forests, especially the hardwoods. Perhaps the most famous, and most destructive, is chestnut blight, which has all but eliminated the American chestnut as a commercially important forest tree species.

Cankers are diseases of the bark and cambium. The fungi that cause them, usually ascomycetes and basidiomycetes, grow in the living tissues of the cambium and tolerate the host response. Canker fungi usually kill the cambium, thus girdling the trees. They create an entry court, or a place of entry for decay fungi. Some canker fungi, such as *Hypoxylon mammatum,* the cause of Hypoxylon canker of aspen, can themselves decay wood. Decay reduces the tree's timber value and also predisposes the tree to windsnap. Additional timber losses occur because the cankered part of the tree is culled. Some trees respond to canker fungi by producing layers of callus tissue, while in other cases the canker fungi kill the cambium, resulting in a sunken area. In either case, the tree becomes disfigured. Because the cankered part of the tree is not round, it is difficult to debark and is often cut off and left in the woods. This loss to cull may often be as important as mortality by the canker.

Cankers are classified in several ways. *Diffuse cankers* are those where the fungus grows so quickly that the host forms little or no callus tissue. Diffuse cankers usually girdle infected trees quickly and are thus among the most destructive diseases. In contrast to diffuse cankers, *target-shaped cankers* form when the fungus infects the host. The host responds by forming callus tissue, and then the fungus overcomes the resistant response and overgrows the callus tissue. When this see-saw battle occurs over many years, annual layers of callus tissue are formed, often concentric in shape, giving the appearance of a target from which it is named. Target cankers rarely kill their hosts; rather, decay fungi are usually associated with the canker, and the tree snaps at that point. These cankers, such as Strumella and Eutypella cankers, mostly kill smaller trees; larger trees are usually snapped. When they occur on the butt log, target cankers cause serious losses as they reduce the volume in the most valuable part of the tree.

A separate, but related system of classification considers whether the cankers are annual or perennial. *Annual cankers* usually attack weakened trees. After the tree recovers, the fungus can no longer continue to grow and is gone. This is often the case after drought, when Fusarium cankers are common. *Perennial cankers* are those that can attack vigorous trees and persist many years. The target cankers are classic examples, but also consider that such diffuse cankers as Hypoxylon and chestnut blight are also perennial.

Foresters often attempt to manage forests to minimize the impacts of canker diseases. Sanitation, destroying cankered trees, is often recommended, but is rarely effective. However, cankered trees should be discriminated against during intermediate operations, as they will not provide as much timber as canker-free trees. Burning the cankers to reduce inoculum is of questionable value. Cultural practices that avoid wounding trees can often reduce canker incidence. There is often genetic resistance to cankers, so using resistant plant materials in plantings can minimize canker diseases. Quarantines may be effective against some canker diseases. Regulations have apparently successfully restricted the spread of the European strain of Scleroderris canker but were ineffective in stopping the chestnut blight fungus with its broad host range and profuse sporulation.

16.2 HARDWOOD CANKERS

DISEASE PROFILE

Chestnut Blight

Importance: Chestnut was one of the most valuable of our eastern hardwood tree species. Over a time span of 50 years following its accidental introduction, the chestnut blight virtually eliminated this species from North America.

Suscepts: American and European chestnuts and Allegheny chinkapin are very susceptible. Chinese and Japanese chestnuts are resistant. Some injury is also incurred by post oak, live oak, and eastern chinkapin. The causal organism can also grow as a weak parasite or saprophyte on other species of *Quercus, Acer rubrum, Carya ovata,* and *Rhus typhina.*

Distribution: Chestnut blight is native to the Orient. It was brought to the United States in about 1900 and somewhat later to Europe. By 1940 it had extended across the entire natural range of chestnut in the eastern United States and also appeared in scattered plantings in British Columbia, Oregon, and Washington but was eradicated there following intense control efforts.

Branch/stem canker.

Causal Agents: Chestnut blight is caused by *Cryphonectria (Endothia) parasitica* (Diaporthales).

Biology: Ascospores or conidia germinate in fresh wounds of the living bark. The resulting hyphae form mycelial fans that grow in the inner bark/cambium region. Host cells are killed by chemical and mechanical action. The affected branch or stem is quickly girdled and killed. Pycnidia form in orange to yellow-brown stromata in fissures formed in the bark as the canker advances or on the bark of sunken cankers. Perithecia may also form in the same stromata that produced pycnidia. After the main stem is girdled, adventitious sprouts often form from the roots of American chestnut, but these sprouts are eventually infected and killed. *Cryphonectria parasitica* does not infect roots and so the roots can remain alive for many years, producing repetitious crops of adventitious shoots.

Epidemiology: Conidia are produced in large quantities in a sticky matrix that oozes out of the pycnidia and are carried in splashing rain water and by insects and birds. Ascospores are windborne but can also be carried by insects and birds. *Cryphonectria parasitica* can survive as a saprophyte for many years in dead tissues and continue to produce large quantities of spores, on chestnut as well as on some other plant species.

Diagnosis: Diffuse cankers are formed. They may be either sunken or swollen in relation to the healthy bark and yellow-brown in color. Stromata appear as yellow-orange pustules and are produced abundantly in the invaded bark. Fans of tan mycelium may be found under the bark of cankered areas.

Control Strategy: Early attempts at eradication were unsuccessful. Subsequently, a great effort was expended in searching for resistant strains of American chestnut, but to no avail. Recently, backcrossing has been attempted to improve blight resistance. Hybrids of American and Asiatic chestnuts are backcrossed to American chestnut. If this process is continued through additional generations of backcrosses, resistant chestnuts with growth and form of the American chestnuts will result. Another promising approach for control is the use of hypovirulent strains of *C. parasitica*. These strains contain double-stranded ribonucleic acid, which adversely affects the metabolism of *C. parasitica* to the extent that host resistance mechanisms can become dominant and resist the blight pathogen.

SELECTED REFERENCES

Fulbright, D. W., W. H. Weidlich, K. Z. Haufler, C. S. Thomas, and G. P. Paul. 1983. Chestnut blight and recovering American chestnut trees in Michigan. *Can. J. Bot.* 61:3164–3171.
Hepting, G. H. 1974. Death of the American chestnut. *J. For. Hist.* 18:60–67.

16.2.1 Chestnut Blight

Importance

From the beginning of European settlement until the end of the nineteenth century, chestnut was one of the most valuable of our eastern hardwood species. This species alone dominated the eastern forests amounting to one third to three quarters of the volume in hardwood stands. It is a species that thrived in good and poor soils on mountains and at low elevations. Some trees were 24–30 m tall with diameters of 1 m or greater. One stump was 5.2 m across. Their rapid growth produced sawlog-sized trees in 50 years or so. The wood was used for lumber, ties, posts, and poles. During colonial settlement of North America, chestnut was an indispensible resource (Fig. 16.1). The wood was not especially strong, but the heartwood is naturally resistant to decay, and this and other qualities made the wood valuable for special purposes. High-grade tannin was extracted from chestnut wood and until only recently 50% of the tannin used in the United States was obtained from chestnut. Dead trees are still being cut for tannin extraction. Chestnuts were a major food source for wildlife, and nuts were produced in large numbers every year. The nuts were commonly roasted and consumed by people. With practically all the larger trees dead as a result

FIGURE 16.1 American chestnut was important during the settlement of the United States. With its demise passed the classic chestnut cabin, shingled roof, and fence. From Anagnostakis (1978).

of chestnut blight, only sprouts from the roots of previously killed trees survive throughout most of the commercial range of the species, and these sprouts usually are killed by the fungus before they attain any commercial size. The bark of the trees killed in this way is soon invaded by bark beetles and falls off, and the wood is invaded by insects and secondary fungi. It was once estimated that loss of the chestnut species reduced the value of the hardwood forests in the eastern United States from 15 to 50%. Not only have we almost lost a tree of tremendous value, but it is quite possible that some Lepidopteran moth species that coevolved with the American chestnut are now extinct as a direct result of the blight.

Suscepts

American chestnut *(Castanea dentata)*, European chestnut *(C. sativa)*, and Allegheny chinkapin *(C. pumila)* are very susceptible; *C. japonica* and *C. mollissima*, Oriental chestnut species, are resistant but not immune. Chestnut blight also results in considerable damage to post oak *(Quercus stellata)* and some damage to Eastern Chinkapin *(Castanopsis* sp.). Cankers may also occur on live oak *(Quercus virginiana)*. The causal organism grows as a weak parasite or a saprophyte on other species of *Quercus, Acer rubrum, Carya ovata,* and *Rhus typhina.*

Distribution

Chestnut blight is endemic in the Orient, and it was brought to the United States in about 1900 on imported Asiatic chestnut seedlings. It was first observed in this country in 1904 on chestnut trees growing in the New York Zoological Garden (Fig. 16.2). On the Asiatic chestnuts, it was not considered a serious pathogen, and at first

FIGURE 16.2   A stand of chestnut trees on Long Island, N.Y., in 1907, dying from chestnut blight. From Diller (1965).

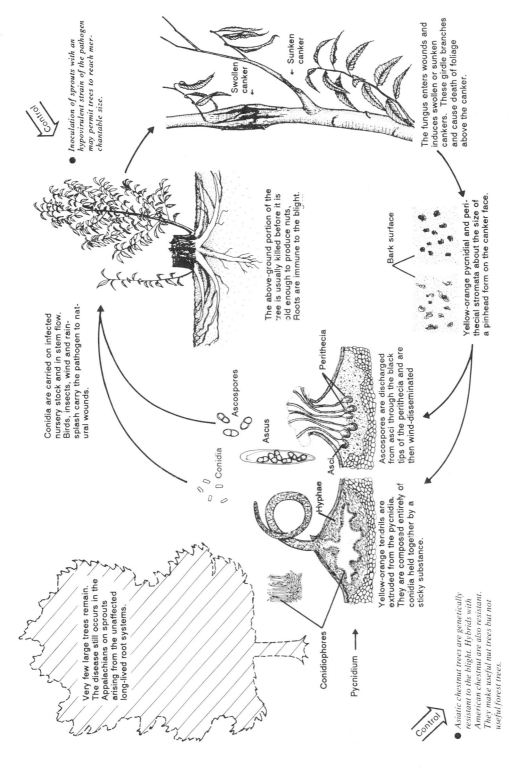

● Inoculation of sprouts with an hypovirulent strain of the pathogen may permit trees to reach merchantable size.

Control

Swollen canker

Sunken canker

The fungus enters wounds and induces swollen or sunken cankers. These girdle branches and cause death of foliage above the canker.

Conidia are carried on infected nursery stock and in stem flow. Birds, insects, wind and rainsplash carry the pathogen to natural wounds.

Ascospores

The above-ground portion of the tree is usually killed before it is old enough to produce nuts. Roots are immune to the blight.

Conidia

Ascus

Perithecia

Ascospores are discharged from asci through the black tips of the perithecia and are then wind-disseminated

Asci

Bark surface

Yellow-orange pycnidial and perithecial stromata about the size of a pinhead form on the canker face.

Hyphae

Very few large trees remain. The disease still occurs in the Appalachians on sprouts arising from the unaffected long-lived root systems.

Yellow-orange tendrils are extruded from the pycnidia. They are composed entirely of conidia held together by a sticky substance.

Conidiophores

Pycnidium

Control

● Asiatic chestnut trees are genetically resistant to the blight. Hybrids with American chestnut are also resistant. They make useful nut trees but not useful forest trees.

FIGURE 16.3 Disease diagram of chestnut blight caused by *Cryphonectria parasitica*. Drawn by Valerie Mortensen.

it was not realized how virulent this fungus was on the American chestnut. The fungus found the American chestnut a very suitable host, and by 1908 the disease extended from Massachusetts to Virginia and was beyond control. By 1940 chestnut blight had invaded almost all the commercial stands of eastern chestnut and killed most of the trees (Fig. 1.4).

Cause

Chestnut blight is caused by *Cryphonectria parasitica* (Murrill) Barr. Until recently the scientific name was *Endothia parasitica*.

Biology

The disease diagram of chestnut blight is shown in Figure 16.3. Infection occurs through wounds, which may be caused by insects, birds, squirrels, rubbing of branches in the wind, or other means. The mycelium grows in the bark and outer sapwood, kills the bark and the cambium, and eventually girdles the tree (Fig. 16.4).

FIGURE 16.4   Front and rear views of a young chestnut blight canker that has just girdled the tree. From Gravatt and Marshall (1926).

The pathogen is a wound parasite, and the germinating spore can live as a saprophyte on necrotic wound debris for some time, but on a susceptible tree the mycelium eventually invades healthy tissue in the inner bark. This invasion is accomplished enzymatically as *C. parasitica* produces several polysaccharide-degrading enzymes that lead to maceration of the cell wall. The enzyme operates most efficiently at a pH of about 5.5, which is roughly equal to the pH of the inner bark. Closer to the mycelium, the pH drops to about 2.8. This is due to the presence of oxalic acid, which is toxic to the host protoplasts. The tree is girdled and killed in from 1 to 10 years. The mycelium remains dormant during cold weather in the winter, resumes growth in the spring, and may grow 25 cm in 1 year. Pycnidia are formed 3–6 weeks after infection. It has been estimated that 100 million conidia may be borne in one pyncidium. These spores are borne in a gelatinous matrix that absorbs water during rain and swells, causing the spores to ooze out in threads (Fig. 16.5). They may be disseminated by rain, insects, and birds. Shortly after chestnut blight was discovered in the United States, woodpeckers were found to carry large numbers of spores on their feet and plumage (Fig. 16.6). Perithecia form 4–8 weeks after infection, and the perithecia produce asci and ascospores over a period of months to years. The mature asci are liberated from the wall of the perithecium, move up the neck, and eject their spores

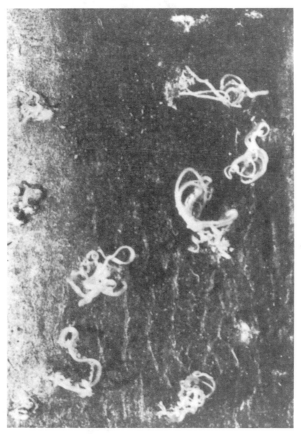

FIGURE 16.5   Spore tendrils produced by pycnidia in infected bark. From Gravatt and Marshal (1926).

FIGURE 16.6   Old chestnut blight canker, showing the work of woodpeckers. From Heald and Studhalter (1914).

in a dry state at the ostiole. The ascospores are disseminated by the wind. The perithecia remain viable for 4–5 years; ascospores from cankers kept in the laboratory for several years have germinated when soaked in water overnight. Both kinds of spores may be profusely produced for a long time on dead trees.

Epidemiology

The success enjoyed by the chestnut blight pathogen is classic among the most devastating tree pathogens introduced into North America. The rapid spread of the disease, 32–40 km/yr, can be attributed to the highly susceptible trees, the millions of spores produced at each canker, and the efficient vectors of the pathogen. In the then almost limitless hardwood forests of the eastern United States, it encountered vast forests composed mainly of the highly susceptible American chestnut. Early control attempts were stymied because rain splash and wind-driven rain efficiently disseminated the sticky conidia of *C. parasitica* to healthy trees adjacent to local infections, and those sticky conidia also were carried long distances by woodpeckers that came to feed on insects colonizing dead trees. The dry, aerially borne ascospores also contributed to spread.

Conidia and ascospores need wounds in order to enter healthy chestnuts, but this is no real obstacle because trees receive many natural wounds during any given time period. At least some of the few remaining disease-free American chestnuts trees that continue to be discovered in small numbers are probably escapes that, when wounded, were not coincidentally challenged with viable spores.

Cryphonectria parasitica also thrives as a saprophyte, especially on dead and decaying chestnut wood and bark but also on a few unrelated woody plant species. Even though chestnut has nearly become extinct, it is unlikely that this pathogen will suffer a similar fate.

Diagnosis

The fungus causes cankers that may become several meters in length and finally surround the branch or trunk on which they occur. The cankered portion of the stem is either sunken or swollen in relation to the healthy bark and is yellow-brown in color. Callus tissue often forms at the border of a canker, sometimes to such an extent that the branch at that point may be twice the normal diameter (Fig. 16.7). Pycnidia and perithecia are produced abundantly in the invaded bark, in stromata that appear as

FIGURE 16.7   Mature chestnut blight cankers. The central portion of the canker on the left is sunken where the fungus has killed the bark. The canker on right is slightly swollen and has the cracks characteristic of swollen cankers. From Anonymous (1954).

FIGURE 16.8   Mycelial fan of chestnut blight exposed by peeling back the outer bark. From Gravatt and Marshall (1926).

yellow-orange pustules. Sometimes fans of tan mycelium may be found under the bark of cankered areas (Fig. 16.8).

Control Strategy

Early in this century, soon after the disease was discovered on American chestnuts, efforts were made to eradicate the disease in Pennsylvania, but they were unsuccessful. In subsequent years a major effort was expended in searching for resistant varieties of the American chestnut, but to date no satisfactory strain has been found, even though isolated disease-free individual trees are still encountered in the Appalachians. The possibility of a resistant variety or individual developing in nature is reduced by the fact that most sprouts are killed before they are old enough to produce nuts, thereby restricting the opportunity for sexual recombination.

Recently an agronomic breeder recognized that resistant American chestnuts might be obtained using the highly successful backcross method, which has been used to improve many agricultural crops. If hybrids of American and Asiatic chestnuts are backcrossed to the American chestnut and this process is continued through additional backcrosses, resistant chestnuts with growth and form of the American chestnut should result.

Another promising approach for control of chestnut blight is the use of hypovirulent strains of *C. parasitica*. Hypovirulent stains of the fungus were first recognized in Italy in the early 1950s where infected trees were healing from old infections. Hypovirulent strains were introduced into chestnut orchards in France, and those orchards have largely recovered their former productivity for nuts. Subsequently, the same phenomenon was recognized in Michigan. The hypovirulent strains are characterized by the presence of double-stranded RNA.

Double-stranded RNAs are probably associated with mycoviruses. Mycoviruses are genetic elements that are associated with many fungi at a very high frequency. While many dsRNAs have no overt effect on their fungal hosts, the dsRNA that infects *C. parasitica* decreases the levels of certain enzymes such as laccase and cellulase as well as the level of metabolities such as oxalate. Infected strains also have altered colony morphology, reduced sporulation, reduced pigmentation, and, most interestingly, reduced virulence. In simpler terms, they are diseased. Virulent strains can be converted to hypovirulent strains, but the two strains must be vegetatively compatible for the conversion to take place. The expectation is that the diseased strain will spread, infecting the fungus in nearby trees or stands. The necessary research to achieve this goal is underway and has met with some success. A major constraint in North America is that strains of *C. parasitica* here have relatively low degrees of vegetative compatibility. European hypovirulent strains experimentally introduced into natural infestations of chestnut blight have not spread the dsRNA at an acceptable rate into the nearby diseased trees.

The dsRNAs associated with hypovirulence from European strains of *C. parasitica* have genetic sequences that are distinct from those of American hypovirulent strains and represent distinct viruslike agents rather than strains of the same agent. Considerable effort is being expended to precisely identify the genetic information contained in the dsRNA responsible for hypovirulence. Hypovirulence may be especially useful for protecting key trees with other desirable traits in a breeding program.

Individual cankers have been halted by applying soil and other materials as a poultice. Chestnuts from infected trees can carry the fungus and thus should not be moved to disease-free areas. Immersion in water at 50°C for 3 minutes shortly after harvesting did not completely eradicate the fungus in nuts.

REFERENCES

Anagnostakis, S. L. 1978. *The American Chestnut: New Hope for a Fallen Giant*. Conn. Agr. Exp. Sta., Bull. 777. 9 pp.

Anagnostakis, S. L. 1979. Hypovirulence conversion in *Endothia parasitica*. *Phytopathology* 69:1226–1229.

Anagnostakis, S. L., and R. A. Jaynes. 1973. Chestnut blight control: Use of hypovirulent cultures. *Pl. Dis. Reptr.* 57:225–226.

Anonymous. 1954. *Chestnut Blight and Resistance Chestnuts*. USDA, Farmer's Bull. 2068. 21 pp.

Baxter, D. V., and L. S. Gill. 1931. *Deterioration of Chestnut in the Southern Appalachians.* USDA, Tech. Bull. 257. 22 pp.

Bramble, W. C. 1936. Reaction of chestnut bark to invasion by *Endothia parasitica. Am. J. Bot.* 23:89–95.

Burnham, C. R. 1988. The restoration of the American chestnut. *Amer. Sci.* 76:478–487.

Clapper, R. B. 1954. Chestnut breeding techniques, and results. *J. Heredity* 45:107–114.

Day, P. R., J. A. Dodds, J. E. Elliston, R. A. Jaynes, and S. L. Anagnostakis. 1977. Double-stranded RNA in *Endothia parasitica. Phytopathology* 67:1393–1396.

Diller, J. 1965. *Chestnut Blight.* USDA For. Serv., For. Pest Leaflet 94. 7 pp.

Elliston, J. E., R. A. Jaynes, P. R. Day, and S. R. Anagnostakis. 1977. A native American hypovirulent strain of *Endothia parasitica.* (abst.) *Proc. Amer. Phytopathol. Soc.* 4:83.

Fulbright, D. W., W. H. Weidlich, K. Z. Haufler, C. S. Thomas, and G. P. Paul. 1983. Chestnut blight and recovering American chestnut trees in Michigan. *Can. J. Bot.* 61:3164–3171.

Gravatt, G. F., and R. P. Marshall. 1926. *Chestnut Blight in the Southern Appalachians.* USDA, Circ. 370. 11 pp.

Griffin, G. J., F. V. Hebard, R. W. Wendt, and J. R. Elkins. 1983. Survival of American chestnut trees: Evaluation of blight resistance and virulence in *Endothia parasitica. Phytopathology* 73:1084–1092.

Heald, F. D., and R. A. Studhalter. 1914. Birds are carriers of the chestnut blight fungus. *J. Agric. Res.* 2:405–422.

Heald, F. D., M. W. Gardner, and R. A. Studhalter. 1915. Air and wind dissemination of ascospores of the chestnut blight fungus. *J. Agric. Res.* 3:493–526.

Hepting, G. H. 1976. The death of the American chestnut. *J. For. Hist.* 18:60–67.

Jaynes, R. A., and N. D. DePalma. 1984. Natural infection of nuts of *Castanea dentata* by *Endothia parasitica. Phytopathology* 74:296–299.

Kuhlman, E. G., H. Bhattacharyya, B. L. Nash, M. L. Double, and W. L. MacDonald. 1984. Identifying hypovirulent isolates of *Cryphonectria parasitica* with broad conversion capacity. *Phytopathology* 74:676–682.

L'Hostis, B., S. T. Hiremath, R. E. Rhoads, and S. A. Ghabrial. 1985. Lack of sequence homology between double-stranded RNA from European and American hypovirulent strains of *Endothia parasitica. J. Gen. Virol.* 66:351–355.

MacDonald, W. L., F. C. Cech, J. Luchok, and C. Smith (eds). 1978. *Proceedings of the American Chestnut Symposium, Jan. 4–5, Morgantown, West Virginia.* 122 pp.

May, C., and R. W. Davidson. 1960. *Endothia parasitica* associated with a canker of live oak. *Pl. Dis. Reptr.* 44:754.

Roane, M. K., G. J. Griffin, and J. R. Elkins. 1986. *Chestnut Blight, Other Endothia Diseases, and the Genus Endothia.* American Phytopathological Society, St. Paul, Minn. 53 pp.

Shapira, R., G. H. Choi, and D. L. Nuss. 1991. Virus-like genetic organization and expression strategy for a double-stranded RNA genetic element associated with biological control of chestnut blight. *The EMBO J.* 10:731–739.

DISEASE PROFILE

Hypoxylon Canker

Importance: Hypoxylon canker is the most destructive disease of aspen and causes about 38% of all tree mortality. Stocking may be reduced to the point that a commercial harvest is no longer feasible.

Suscepts: Trembling aspen is the major host, especially in the Lake States region. Other hosts include bigtooth aspen, balsam poplar, bolleana poplar, European aspen, and some species of willow.

Distribution: The range of Hypoxylon canker is coincidental with the natural range of its hosts in the United States and Canada, except that it is not known to occur in Alaska.

Causal Agents: Hypoxylon canker of aspen is caused by the fungus *Hypoxylon mammatum* (Xylariales).

Hypoxylon Canker. Courtesy of D. W. French, University of Minnesota, St. Paul.

Biology: The infection process is not entirely understood. Apparently, the fungus enters the tree through small branches or stubs. Some infections have been associated with wounds or galls caused by the insects *Saperda inornata* and *Cicada* spp. Oviposition wounds may allow fungus infection of the bark under the green layer, which inhibits spore germination. *Hypoxylon mammatum* also produces a toxin that appears important in the infection process. Once established, the fungus grows rapidly in the tree. Unlike many canker fungi, it has the ability to decay wood as well as colonize the bark. Conidia are produced around the canker margins. Their role is not known, but they may spermatize perithecial initials. In cankers that are 3 years old or older, perithecia form in stromata in older portions of cankers. Trees may be girdled and killed, but often they snap off at the canker.

Epidemiology: There is no relation with site. Insect behavior may greatly influence the epidemiology of *H. mammatum* by producing wounds that are ideal for infection. Canker incidence is greater in mixed stands of low density. *Saperda inornata* is a sun-loving insect, and this propensity may explain why canker incidence is greatest along stand edges or in low-density stands. Ascospores are probably responsible for canker initiation and are discharged for 18 hours after a rain. They are wind-disseminated and dispersed year around, even when air temperatures are below freezing.

Diagnosis: Cankers are of the diffuse type and occur on the main stem and on branches. The affected area becomes yellow, then mottled with darker colors, and finally collapses as the underlying tissues are partially digested. Alternating light- and dark-colored zones are apparent when the bark is sliced open. This feature is useful for diagnosing young cankers. Peglike structures, which bear conidia, form under the outer bark near canker margins. On older cankers, black stromata form in patches. These initially have a whitish covering but soon become covered with the black ascospores and eventually deteriorate into a black, crumbly residue.

Control Strategy: No suitable control measures are known. In areas of high incidence, other tree species should be favored. Resistant varieties that are under development will help reduce losses in the future. The biggest problem will be how to convert millions of hectares of the aspen type to the resistant varieties. Quarantine will help keep *H. mammatum* out of the valuable aspen stands of Alaska.

SELECTED REFERENCES

Anderson, N. A., M. E. Ostry, and G. W. Anderson. 1979. Insect wounds as infection sites for *Hypoxylon mammatum* on trembling aspen. *Phytopathology* 69:476–479.
Manion, P. D., and D. H. Griffin. 1986. Sixty-five years of research on Hypoxylon canker of aspen. *Pl. Dis.* 70:803–808.

16.2.2 Hypoxylon Canker

Importance

Hypoxylon canker is undoubtedly the most destructive canker disease of aspen and one of the most important tree diseases in the Lake states area. A recent study found that about 12% of the living trees are infected and that Hypoxylon canker causes about 38% of all tree mortality. In older stands in Ontario, more than 5% of the trees were cankered in 70% of the aspen stands, and more than 25% of the trees were cankered in 29% of the stands. This mortality reduces stand density below desirable levels; sometimes density is reduced to the point that a commercial harvest is no longer feasible.

Marty (1972) estimated that in areas where aspen harvest exceeded growth, Hypoxylon canker reduced harvest value by $4.4 million/yr. The 200 million ft^3 estimated lost in the Lake states was approximately equal to the annual growth. Annual growth and utilization could be doubled if Hypoxylon canker were eliminated. In much of Minnesota, the upper peninsula of Michigan, and Wisconsin, areas where growth exceeded harvest, disease impact was considered minor. This fact probably is no longer true, since the establishment of a forest products industry that requires aspen as the raw material for producing various types of reconstituted particle and fiber board products and pulp. Hypoxylon canker is favorable for browsing animal species such as deer because infections cause small openings in the stand.

Suscepts

Hypoxylon canker is of primary concern on trembling aspen *(Populus tremuloides)*. Other hosts include bigtooth aspen *(P. grandidentata)*, balsam poplar *(P. balsamifera)*, bolleana poplar *(P. alba* var. *nivea)*, and European aspen *(P. tremula)*. Hypoxylon cankers are also found on species of *Salix,* the willows. Hybrid poplars may also be cankered.

Distribution

Hypoxylon canker is widely distributed, occurring throughout the natural range of its hosts in the United States and Canada, except that this disease has not been reported from Alaska. Although present in the western United States, Hypoxylon canker is not common.

Cause

Hypoxylon canker on aspen is caused by *Hypoxylon mammatum* (Xylariales). Earlier this fungus was known as *H. pruinatum*.

Biology

Despite extensive research on this disease, forest pathologists still do not clearly understand the infection process. What we do know about the disease cycle is summarized in Figure 16.9. Apparently, the fungus enters the tree through or around branch stubs, as more than 90% of the cankers are associated with branch stubs. In some stands, new cankers have been associated with wounds or galls caused by *Saperda inornata* and *Cicada* spp. *Hypoxylon mammatum* produces a toxin that appears important in the infection process. When the toxin (fungus-free) is applied to host tissues, it kills them. Perhaps the fungus and toxin are introduced into the tree by insects. The fungus invades the wood first and then the bark tissues. Thus the mycelium

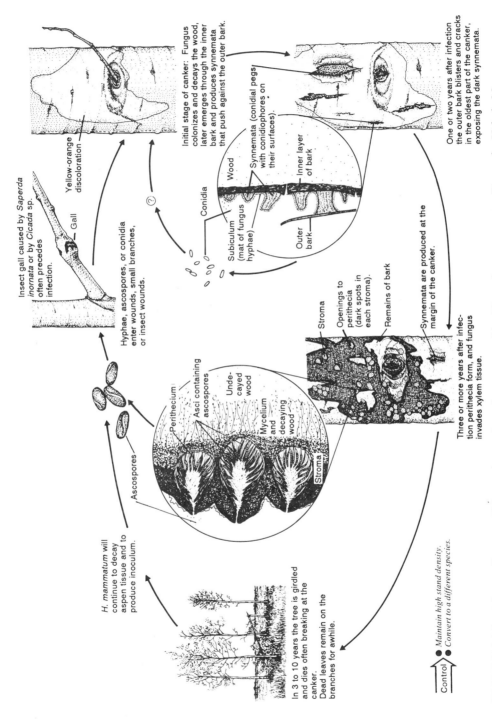

Yellow-orange discoloration

Insect gall caused by *Saperda inornata* or by *Cicada* sp. often precedes infection.

Gall

Hyphae, ascospores, or conidia enter wounds, small branches, or insect wounds.

Initial stage of canker: Fungus colonizes and decays the wood, later emerges through the inner bark and produces synnemata that push against the outer bark.

One or two years after infection the outer bark blisters and cracks in the oldest part of the canker, exposing the dark synnemata.

Synnemata (conidial pegs with conidiophores on their surfaces).

Wood

Inner layer of bark

Conidia

Subiculum (mat of fungus hyphae)

Outer bark

Synnemata are produced at the margin of the canker.

Stroma

Openings to perithecia (dark spots in each stroma).

Remains of bark

Three or more years after infection perithecia form, and fungus invades xylem tissue.

Perithecium

Asci containing ascospores

Undecayed wood

Mycelium and decaying wood

Stroma

Ascospores

H. mammatum will continue to decay aspen tissue and to produce inoculum.

In 3 to 10 years the tree is girdled and dies often breaking at the canker. Dead leaves remain on the branches for awhile.

Control
Maintain high stand density.
Convert to a different species.

FIGURE 16.9 Disease diagram of Hypoxylon canker of aspen caused by *Hypoxylon mammatum.* Drawn by Valerie Mortensen.

in the wood extends beyond where it is present in the bark. The green layer of the bark inhibits spore germination. Pyrocatechol and two glycosides inhibitory to *H. mammatum* have been also found in aspen bark, so aspen bark may not be a favorable substrate for growth of *H. mammatum*. Perhaps the toxin produced by the fungus in the wood kills the bark and permits fungal growth there.

Once established, *H. mammatum* grows rapidly in the tree. Cankers may be 1 m long and halfway around a 13-cm tree in 3 years. The trees may be girdled and thus killed. *Hypoxylon mammatum*'s rather unique ability to decay wood causes some trees, especially larger ones, to snap at the point of the canker before they are killed.

Conidia are formed around the margins of all but the youngest cankers. The synnemata on which the conidia are produced separate and split open the bark, allowing these spores to be wind-disseminated (Fig. 16.10). Perithecia form in stromata on cankers 3 years or older, and new stromata form each year (Fig. 16.11). Perithecia fill with asci and ascospores during wet weather and discharge spores for 18 hours after a rain. Ascospores are wind-disseminated. Their dispersal is so effective that they have been collected in isolated stands from which all known cankers had been removed. Spores are produced and dispersed from cankers on dead trees. On felled trees, spore production was not significantly reduced until 2 years after felling. Ascospores are dispersed throughout the growing season and during the winter as well.

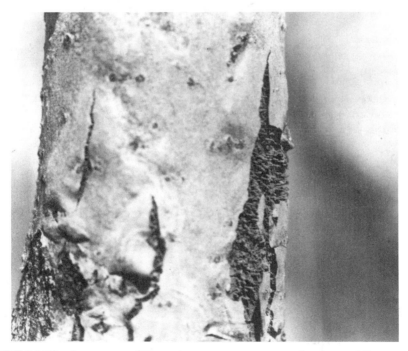

FIGURE 16.10 Synnemata of *Hypoxylon mammatum* on a canker on trembling aspen. These synnemata are formed under the bark and push their way through it to expose the conidia to the wind. Courtesy of D. W. French, University of Minnesota, St. Paul.

FIGURE 16.11   Perithecial stromata of *Hypoxylon mammatum* on an older section of a canker. Courtesy of D. W. French, University of Minnesota, St. Paul.

Epidemiology

The incidence of cankers is related to stand density and other unknown factors. There are more cankers in low-density and mixed-species stands. Significantly more cankers occur at stand edges than 20–40 m into the same stand. Incidence varies with geographic location. Incidence is low in extreme northern Minnesota and high in parts of Wisconsin. A relationship between canker incidence and site quality has not been established, but evidence from greenhouse studies suggests that trees subjected to water stress are more susceptible.

Insect behavior may greatly influence the epidemiology of Hypoxylon canker. *Saperda inornata* makes wounds that appear to be ideal infection sites for *H. mammatum*. Eggs are laid in these wounds, and larval activity stimulates the host to produce a gall. Larvae maintain an open tunnel to the outside of the branch to eject frass from the gall. These tunnels are below the tissue which is toxic to *H. mammatum*. Ascospores can germinate in these wounds and lead to canker development. Evidence that ascospores are responsible for canker initiation is accumulating. Egg-laying activity takes place in branches and stems usually less than 1.5 cm in diameter, which might explain why cankers are more common in smaller trees and in the tops of larger trees. Hypoxylon cankers colonizing these small branches may be more likely to move into the stems. The original infected branch is killed and breaks off, leaving the branch stub so typical of Hypoxylon cankers. *Saperda inornata* is a sun-loving insect and prefers open-growing aspen and branches of aspen growing along the stand edge. This preference may explain the greater incidence of cankers in these situations. Further investigation of this insect may provide a better understanding of this disease.

Diagnosis

Hypoxylon cankers are usually of the diffuse type and on larger trees may be as long as 3 m. They occur on the main stem, where they are most important, and on branches. The cankers are usually associated with a dead branch or branch stub. Affected areas usually become yellow and then mottled yellow and darker colors. A few weeks after invasion, the bark surface collapses irregularly due to partial digestion of host tissues. Alternating light- and dark-colored zones are apparent when the bark is sliced open (Fig. 16.12); the dark layers are masses of mycelium, and the light layers are decayed host tissues. This symptom is very useful in detecting young cankers. Synnemata that will produce conidia form under the outer bark near the margins of the canker. Synnemata can easily be seen with the naked eye or a hand lens. On

FIGURE 16.12   Alternating light- and dark-colored zones under the bark of a Hypoxylon canker. The dark layers are masses of mycelium, and the light layers are decayed host tissues. Courtesy of D. W. French, University of Minnesota, St. Paul.

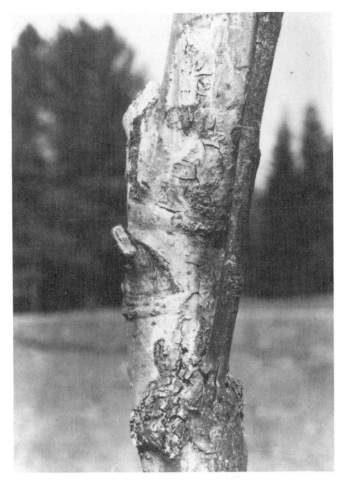

FIGURE 16.13   Hypoxylon canker on trembling aspen showing the black, crumbly appearance of the canker. Courtesy of D. W. French, University of Minnesota, St. Paul.

cankers at least 3 years old, stromata will be present. These stromata become covered with black ascospores, and the host tissues are further deteriorated, giving the canker a black, crumbling appearance (Fig. 16.13). Callus tissue sometimes forms at the canker margin (Fig. 16.14), especially on hosts other than trembling aspen.

Control Strategy

Suitable control measures are not known. High-density stands with a minimum of other tree species will have smaller losses to Hypoxylon canker. Where the incidence of Hypoxylon canker is characteristically high, other species should be grown if possible. Shorter (pathological) rotations can minimize losses.

In the future, resistant aspens may become available. There are several problems, however. So far, hybrids involving *P. tremula* and other species, and selections from *P. tremuloides* have not resulted in resistant trees with good form. There is considerable variation in resistance to *H. mammatum* within clones of *P. tremuloides,* and there

FIGURE 16.14 Hypoxylon canker on resistant variety of aspen, showing callus growth on canker margins. Courtesy of D. W. French, University of Minnesota, St. Paul.

are clones known to recover from infections and close cankers. Even if a resistant aspen is developed, it will be a major task to convert millions of hectares of susceptible aspen forest to the resistant host because of this species' ability to sucker.

Understanding the role of insects may illuminate opportunities for managing Hypoxylon canker. Managing stands to minimize their susceptibility to insects such as *Saperda* may be beneficial. Direct control of insects would probably not be feasible.

SELECTED REFERENCES

Adams, R. D. (ed.). 1990. *Aspen Symposium '89, Proceedings.* Serv., Gen. Tech. Rept. NC-140. 348 pp.

Anderson, N. A., and M. E. Ostry. 1983. Galleries of *Saperda inornata* as infection courts of *Hypoxylon mammatum* on trembling aspen. *Phytopathology* 73:836.

Anderson, N. A., M. E. Ostry, and G. W. Anderson. 1979. Insect wounds as infection sites for *Hypoxylon mammatum* on trembling aspen. *Phytopathology* 69:476–479.

Bier, J. E. 1940. *Studies in Forest Pathology. III. Hypoxylon Canker of Poplar.* Can. Dept. Agri., Pub. 691. Tech. Bull. 27:1–40.

Griffin, D. H., K. E. Quinn, G. S. Gilbert, C. J. Wang, and S. Rosemarin. 1992. The role of ascospores and conidia as propagules in the disease cycle of *Hypoxylon mammatum. Phytopathology* 82:114–119.

Manion, P. D., and D. H. Griffin. 1986. Sixty-five years of research on Hypoxylon canker of aspen. *Pl. Dis.* 70:803–808.

Marty, R. 1972. The economic impact of Hypoxylon canker on the Lake States resource, pp. 21–26. In *Aspen Symp. Proc.* USDA For. Serv., Gen. Tech. Rep. NC-1. 154 pp.

Ostry, M. E., and N. A. Anderson. 1983. Infection of trembling aspen by *Hypoxylon mammatum* through cicada oviposition wounds. *Phytopathology* 73:1092–1096.

DISEASE PROFILE

Hypoxylon Dieback

Importance: A common cankerlike disease of red and black oaks that have been stressed by drought or suppressed by competition.

Suscepts: Primarily on oaks of the *Erythrobalanus* subgenus, but also on white oak, beech, hickories, basswood, ironwood, and maples.

Distribution: Hypoxylon dieback is found from the great plains eastward, wherever susceptible hosts occur, and especially where the trees are stressed.

Causal Agents: Hypoxylon dieback is caused by *Hypoxylon atropunctatum* (Xylariales), but other species such as *H. mediterraneum, H. punctulatum,* and *H. rubiginosum* may be locally prevalent. *Hypoxylon atropunctatum* is more abundant in the southeastern states from Virginia south to Florida and then westward to central Texas.

Biology: Hypoxylon atropunctatum is known to latently infect living bark tissue of oak seedlings. As the seedlings mature into trees and eventually become stressed, the fungus is in a positional advantage to colonize host tissues in advance of other organisms. Growth of the fungus into living oak tissues is stimulated when their relative water content declines to 60–70%. *Hypoxylon atropunctatum* quickly colonizes the sapwood, uses cell contents, and also produces a soft rot type of decay of the wood cell walls.

Epidemiology: Several weeks after colonizing the sapwood of a tree stem, the fungus produces large conidial stromata that loosen the bark and produce large numbers of conidia. Their function is unknown. Four to six weeks later, perithecia are produced in the same stromata. They are ejected following rains and are presumed to be responsible for infection.

Diagnosis: Dieback of one or more branches in the upper crown is a symptom of stress. *Hypoxylon atropunctatum* produces little evidence of a canker. A positive sign of infection is the stromata beneath sloughing bark, which may be produced at any time of the year. These symptoms may appear first on larger dying branches but will soon be evident on the main stem as the entire tree dies. The single-celled conidia are spherical,

Stroma of *Hypoxylon atropunctatum*.

3–4 μm in diameter, and may be from black to grayish-black or brown. Later, as the stroma thickens and perithecia form, the surface of the stroma often appears bluish-gray. Ascospores are dark brown, unicellular, and football-shaped. Decayed sapwood is light yellow with irregular black zone lines. Optimum growth rate of the fungus in culture is more than 30°C.

Control Strategy: Early experimental infections have not been confirmed. The current belief is that *H. atropunctatum* is not a pathogen but, rather, an extremely successful initial invader of recently dead or dying sapwood tissues. The present high incidence of Hypoxylon dieback in forest stands is related to a large exent to past land abuses such as land clearing for agriculture and charcoal production, subsequent abandonment, poor forestry practices, and introduction of exotic forest pests, in addition to more recent but also poorly understood perturbations such as acidic precipitation and ozone pollution. Minimization of stressful conditions would be the major control strategy. Match the tree species with the site. In ornamental plantings, fertilization, mulching, pest control, and avoidance of injury should be the goals.

SELECTED REFERENCES

Bassett, E. N., and P. Fenn. 1984. Latent colonization and pathogenicity of *Hypoxylon atropunctatum* on oaks. *Pl. Dis.* 68:317–319.

Starkey, D. A., S. W. Oak, G. W. Ryan, F. H. Tainter, C. Redmond, and H. D. Brown. 1989. *Evaluation of Oak Decline Areas in the South.* USDA For. Serv., Prot. Rept. R8–PR17. 36 pp.

Tainter, F. H., and W. D. Gubler. 1974. Effect of invasion by *Hypoxylon* and other microorganisms on carbohydrate reserves of oak-wilted trees. *For. Sci.* 20:337–342.

DISEASE PROFILE

Beech Bark Disease

Importance: Epidemic development of the beech scale–nectria complex from 1920 to 1950 aroused only mild concern because American beech then was considered to be low-valued. Solution of beech utilization problems in the 1950s, however, has created new markets whereby demands for beech timber have focused new attention upon the disease.

Suscepts: All species and varieties of European and American beech are susceptible in varying degrees. Varietal differences in American beech are known, namely red beech (most susceptible), white beech (intermediate), and gray beech (least susceptible).

Distribution: In Europe the disease has been known since the 1840s and is now found wherever beech is grown. In North America the disease was first observed in 1920 in Nova Scotia, about 30 years after the scale insect (a European import) was known to be

Tree infested with beech scale. From Shigo (1971).

present in the area of Halifax. Current distribution of disease components is characterized by three stages: (1) the advance front, that is, scale present and fungus scarce (east-central West Virginia and northern Virginia); (2) the killing front, that is, both agents present and synergistic on beech (central New York and northern Pennsylvania); and (3) the aftermath where only beech survivors, that is, disease escapes, remain (Vermont, eastern New York, Maine, and the Canadian Maritimes east of the St. Lawrence River).

Causal Agents: Beech bark disease is caused by *Cryptococcus fagi* (Eriococcidae), *Nectria coccinea* var. *faginata* (Hypocreales). The imperfect stage is *Cylindrocarpon* sp. (Hyphomycetes).

Biology: The scale insect *C. fagi* creates a minute feeding wound by stylet penetration of bark parenchyma. Subsequent drying and shrinkage of tissue opens the wound further and provides access for the windborne deposition of ascospores of *N. coccinea* whereby a canker 1–2 cm in diameter results. Tree injury is determined by the number of cankers, their coalescence, and the ultimate girdling effect, and is directly correlated with the intensity of scale infestation.

Epidemiology: Disease spread and intensification is largely a function of population dynamics of the insect and suscept predisposition. Although the scale is flightless, its small size (only 0.5–1 mm long at maturity) suggests various modes of local and long-distance dispersal, such as wind and birds, respectively. Severe winters may limit the scale somewhat in that air temperatures of $-37°C$ or lower are lethal.

Diagnosis: The scale ordinarily is sedentary, except for the motile crawler stage in August and September, and is generally hidden from view by a white, cottony-waxy covering, which is seen as white patches en masse if the attack is at all serious. With the advent of the fungus, the scale population declines as its food supply is destroyed. The fungus forms white sporodochia in the spring and clustered, red, lemon-shaped perithecia in the fall.

Control Strategy: The scale and fungus may be contained, respectively, by a predator, *Chilocorus stigma* (the twice-stabbed ladybird beetle) and a mycoparasite, *Gonatorrhodiella highlei.* Eradication by salvage is practiced, but the sapwood of dying trees is so quickly invaded and stain degraded by wood saprophytes, such as *Hypoxylon* spp., that timing is critical.

SELECTED REFERENCES

Burns, B. S., and D. R. Houston. 1987. Managing beech bark disease: Evaluating defects and reducing losses. *Nor. J. Appl. For.* 4:28–33.

Houston, D. R., E. J. Parker, and D. Lonsdale. 1979. Beech bark disease: Patterns of spread and development of the initiating agent *Cryptococcus fagisuga. Can. J. For. Res.* 9:336–344.

Shigo, A. L. 1971. The beech bark disease spreads southward. *South. Lumberman.* Dec. 15. (unnumbered).

16.2.3 Beech Bark Disease

Importance

Beech bark disease is a devastating disease of beech trees in the northeastern United States. During the 1930s there was great concern over the disease because it was new and spectacular. Then the concern waned rapidly because beech was not very valuable from a commercial standpoint. Beech wood is difficult to dry properly and was not used. As the wood industry learned to dry and use beech profitably, however, they found that little sound beech remained. The disease continues as an epidemic on American beech and has caused considerable mortality over a large area. Several organisms interact with the principal disease agents and have roles in the spread of the disease, its symptoms, and the mortality of beech. As a timber species, and as an important component of urban/suburban forests, beech has recently experienced a renewed interest. As beech bark disease has continued to spread south and west, it has also attracted attention.

Suscepts

Beech bark disease is a disease complex that occurs on American beech *(Fagus grandifolia),* its southern variety *(F. grandifolia* var. *caroliniana),* and European beech *(F. sylvatica).* It results from the sequential attack by the beech scale insect *Cryptococcus fagisuga* and fungi in the genus *Nectria.*

Distribution

Reports of beech bark disease in Europe date back to 1849, but the fungus was not associated with the disease until 1914. The causal fungus *N. coccinea* var. *faginata* appears less virulent in Europe, where the disease is not considered to be as serious as it once was.

In North America the first recorded outbreak of beech bark disease occurred in Nova Scotia about 1920, some 30 years after the first known appearance of the insect on ornamental European beech in the Halifax public gardens. By 1930 the disease had become so common throughout Nova Scotia that hardly a beech stand could be found in which there had not already been some mortality. It was first found in the United States in Massachusetts in 1929. Once established, the disease spread rapidly through the forests at an average rate of 10–15 km/yr.

By the 1930s beech bark disease was distributed generally throughout eastern Maine and in isolated localities in the rest of the state, on the eastern slopes of the White Mountains in New Hampshire, and in the Catskill Mountains of New England to eastern and central New York and to northern Pennsylvania. By the early 1970s the disease had spread throughout Massachusetts and Connecticut and into eastern Pennsylvania. During this time the scale insect had preceded the disease complex into new areas.

In 1981 an isolated outbreak of scale was discovered on the Monongahela National Forest in West Virginia. In 1983 the scale was discovered on the George Washington National Forest in Virginia, near the West Virginia infestation. This area was about 400 km away from the advancing front in Pennsylvania and indicates the possible long-range spread of the scale. In 1984 the scale was discovered in Ohio. In 1993 beech bark disease was discovered in the Great Smokey's National Park in Tennessee (Fig. 16.15). Figure 1.4 shows the approximate presently known distribution of beech bark disease.

The patterns of insect spread and the subsequent occurrence of *Nectria* infection and tree death have led to a time-space classification of the stages of disease development. There are three zones: first is the advancing front containing large, older beech and building populations of the scale; second is the killing front, with large popula-

FIGURE 16.15 The chronological spread of beech bark disease.

tions of scale and severe outbreaks of *Nectria* infection; the third, or aftermath zone, shows evidence of prior mortality, few older trees, and beech stands consisting of small trees of sprout origin. In the second stage (Fig. 16.16) the disease is most spectacular.

Cause

In North America the major fungus involved in the beech bark disease complex is *Nectria coccinea* var. *faginata* and sometimes *N. galligena* or *N. ochroleuca*. In Europe the fungus is *N. coccinea*.

FIGURE 16.16 Mountainside with many beech trees killed by the beech bark disease. From Shigo (1963).

Biology

The disease diagram of the beech bark disease complex is shown in Figure 16.17. *Cryptococcus fagisuga* reproduces parthenogenically. No males exist. Eggs are deposited on the bark during midsummer; the mature females then die. The eggs hatch from late summer to early winter. The larvae (also known as crawlers or nymphs) either

1. remain underneath the bodies of the previous generation for protection;
2. migrate to cracks or other protected areas;
3. are washed, blown, or fall off the tree and die; or
4. are carried by wind, birds, or ladybugs (*Chilocorus stigma* feeds on the scale insect) to other trees where they establish colonies.

The first larval instar is the only phase of *C. fagisuga* with legs; thus it is the principal means for the spread of infestations. Once established in the feeding position, the scale insects remain there the rest of their lives. The nymphs insert a 2-mm-long stylet into the bark and begin feeding. Once feeding begins, the first stage nymphs molt into second instar nymphs. The scale overwinters as the second instar nymph (Fig. 16.18). In the spring, the second instar nymphs pupate, mature as females, and then lay eggs.

The adults are yellow, elliptical (almost globose), pubescent, with red eyes and diminished antennae and legs. They are from 0.5 to 1.0 mm long and have numerous glands on their bodies from which they excrete fine strands of wax

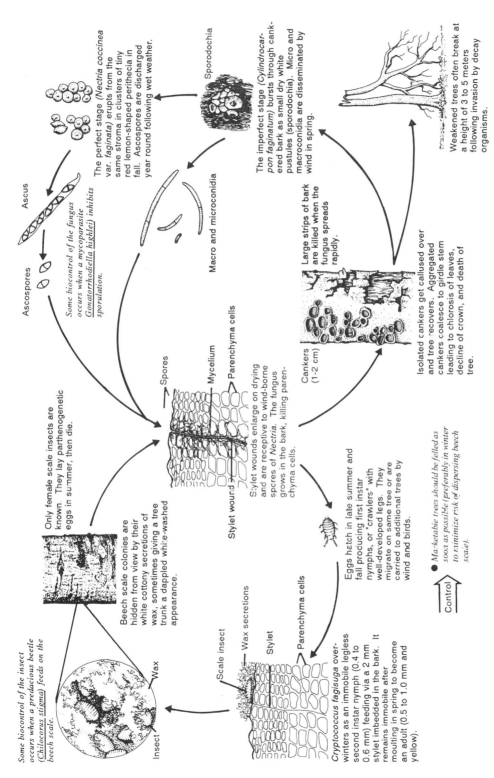

Some biocontrol of the insect occurs when a predacious beetle (Chilocorus stigma) feeds on the beech scale.

Beech scale colonies are hidden from view by their white cottony secretions of wax, sometimes giving a tree trunk a dappled white-washed appearance.

Only female scale insects are known. They lay parthenogenetic eggs in summer, then die.

Ascus

Ascospores

Some biocontrol of the fungus occurs when a mycoparasite Gonatorhodiella highlei) inhibits sporulation.

The perfect stage (Nectria coccinea var. faginata) erupts from the same stroma in clusters of tiny red lemon-shaped perithecia in fall. Ascospores are discharged year round following wet weather.

Sporodochia

Macro and microconidia

The imperfect stage (Cylindrocarpon faginatum) bursts through cankered bark as small dry white pustules (sporodochia). Micro and macroconidia are disseminated by wind in spring.

Weakened trees often break at a height of 3 to 5 meters following invasion by decay organisms.

Spores

Mycelium

Parenchyma cells

Stylet wounds enlarge on drying and are receptive to wind-borne spores of Nectria. The fungus grows in the bark, killing parenchyma cells.

Stylet wound

Large strips of bark are killed when the fungus spreads rapidly.

Cankers (1-2 cm)

Isolated cankers get callused over and tree "recovers". Aggregated cankers coalesce to girdle stem leading to chlorosis of leaves, decline of crown, and death of tree.

Eggs hatch in late summer and fall producing first instar nymphs, or "crawlers" with well-developed legs. They migrate on same tree or are carried to additional trees by wind and birds.

Cryptococcus fagisuga overwinters as an immobile legless second instar nymph (0.4 to 0.6 mm) feeding via a 2 mm stylet imbedded in the bark. It remains immobile after moulting in spring to become an adult (0.5 to 1.0 mm and yellow).

Parenchyma cells

Stylet

Scale insect

Wax secretions

Wax

Insect

Marketable trees should be felled as soon as possible (preferably in winter to minimize risk of dispersing beech scale).

Control

FIGURE 16.17 Disease diagram of beech bark disease caused by *Cryptococcus fagisuga* and *Nectria coccinea* var. *faginata*. Drawn by Valerie Mortensen.

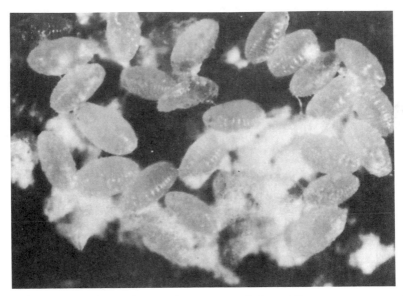

FIGURE 16.18 Second stage larvae of *Cryptococcus fagisuga*. The wax was removed before photographing. The larvae are 0.4 to 0.6 mm long. From Shigo (1970).

(Fig. 16.19). Infestations contain countless individuals. Each individual adds to the waxy secretion, and after several generations, the wooly mass provides excellent protection from the elements. The densest masses occur in the lenticular crevices, under mosses and lichens, and in abaxillary angles of lower branches. *Cryptococcus fagisuga* infestations usually spread to every beech in an area, but with differences in severity. Vigorous trees in vigorous stands are less damaged.

Most bark necrosis is due to *Nectria* species that infect and kill bark stressed by scale infestation. Following wound entry, the fungal mycelium develops stromata and ruptures the bark. The stromata then form white pustules that contain sporodochia. On these, macroconidia and microconidia develop in dense white clusters resembling the waxy secretions of the scale colonies. This is the *Cylindrocarpon* stage. After the conidia are disseminated by wind, the perithecia develop on the same stroma. The perithecia appear as tiny red lemons (Fig. 16.20). Each ascus within the perithecia contains 8 two-celled ascospores, which are also dispersed by wind.

A single infection spreads, causing increased necrosis and weakening the host. Spores multiply the number of infections and the lesions may coalesce and girdle the tree. At the same time infection of nearby trees intensifies. Large populations of scale and *Nectria*, and abundant mortality, typify advanced infections. Mild infections on vigorous hosts may be sealed off by callus tissue, resulting in a pitted and gnarled appearance (Fig. 16.21).

Epidemiology

Beech bark is not a sporadic disease. The patterns of relentless spread over time are typical of those resulting from the introduction of a lethal disease into a susceptible host population. The generally high degree of infestation that occurs whenever and

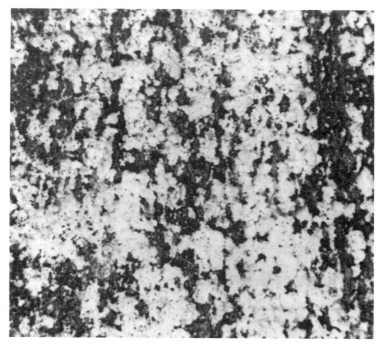

FIGURE 16.19   A section of beech bark covered with the white wax secretion of the beech scale. From Shigo (1970).

wherever *C. fagisuga* spreads suggests also that environmental predisposition to attack by the insect is not necessary. Therefore, the spread and development of the beech bark disease is governed largely by the spread and development of *C. fagisuga*.

Spread of *C. fagisuga* takes place during the period when it is capable of becoming established on a fresh substratum, that is as eggs and as nymphs. Because there are no

FIGURE 16.20   Perithecia of *Nectria coccinea* var. *faginata*. They may be in clusters up to 40 or in a straight line. The perithecia are about 0.3 × 0.25 mm. From Shigo (1970).

FIGURE 16.21   Beech trees with moderate injury to beech bark disease. The craterlike scars indicate where small isolated Nectria cankers were walled off by callus tissue. From Shigo (1970).

winged forms, dissemination is effected by the wandering of the crawlers (nymphs) and by agents that carry eggs and nymphs. The crawlers are active and can travel up to 2 m on a sunny afternoon. The first attacks on nearby trees begin at the butt and proceed upward.

Wind is assumed to carry eggs and nymphs, and birds and mammals may also play a role. Wind can undoubtedly carry eggs and nymphs for many kilometers and lead to the establishment of infestations in distant areas. Glue-covered glass slides exposed in the air catch great numbers of nymphs, far more than slides exposed near the ground.

Man and his vehicles of transportation may also carry infested beech stock or a bit of wood with eggs or nymphs on it and scatter them on a suitable substrate. This mode was probably the source of the first outbreaks in Nova Scotia, Virginia, and Tennessee.

Environmental and host conditions influence development of the scale insect. The spread and intensity of localized infestations are mainly influenced by wind, moisture, and temperature. Sharp declines in scale populations were noted in the Adirondack Mountains following extremely low winter temperatures in 1979–1980, and it was also thought that similar population crashes in northeastern Pennsylvania in the early 1970s were associated with unusually heavy autumn rainfall.

The scale insect alone can affect tree vigor. After 3–5 years, the scale has wounded the tree sufficiently so that the Nectria fungus can enter. The fungus may girdle the beech. In some instances, areas on the trunk where the fungus is abundant may look like cankers.

More than one species of *Nectria* follows infestations by *C. fagisuga*. In Europe the species is *N. coccinea,* while in the northeastern United States the species is *N. coccinea* var. *faginata*. The species recorded in West Virginia and Tennessee was *N. galligena*. *Nectria ochroleuca* was recorded in the westerly limit of the disease. All species have a life history similar to many other ascomycetes. Copious discharge of ascospores occurs following rainy periods. Both ascospores and conidia are carried by air currents. Becuase of the need for wounds, fungus infection patterns are closely related to infestation patterns of the scale insect.

Diagnosis

The first symptom of beech bark disease is a wooly-appearing, white, waxy material caused by the secretions of the scale insect that appears on infested twigs or bark. The scale accumulates in heavy, localized colonies that can cover entire trees, resulting in a snowy appearance.

Early symptoms of *Nectria* infection include a reddish-brown fluid oozing from the wound site. This is followed by the appearance of sunken, necrotic areas on the bark. Small, white cushions (sporodochia) producing asexual spores (conidia) then frequently push through these cracks. These conidia vary in shape from single-celled ovals to eight-celled sickle shapes and are produced in dry heads, well suited for wind-dissemination. The imperfect stage can be found from midsummer until fall and can be easily mistaken for small isolated colonies of the scale insect.

In the fall, perithecia in clusters form on the bark. They are conspicuous and are bright red and lemon-shaped. Ascospores are discharged only when the perithecia are moistened sufficiently. When ascospores dry, they appear as white crusts on the tips of perithecia. Perithecia continue to produce viable spores the following year.

Ascospore length is useful to differentiate between species. If a sample of 25 ascospores has a mean length greater than 14.3 μm, it is *N. galligena;* if it is less than 13.3 μm, then it is *N. coccinea* var. *faginata*. If the mean is between 13.3 and 14.3 μm, then other characteristics must be examined.

The fungi may colonize large areas on some trees, at times completely girdling them. On such trees the perithecia can redden large areas of the bark. Frequently the fungus invades only narrow strips on the bole, and the subsequent symptoms differ from those on trees that have been girdled. Callus tissue forms around these strips and the bark becomes roughened.

Foliage of infected trees often becomes chlorotic, small, and sparse. Later, the leaves turn yellow and usually remain on the tree during the summer. Trees with weak, thin crowns may persist for several years before they die. Sometimes, however, trees may succumb before these foliar symptoms appear. Even while trees are still alive, they may be attacked by Ambrosia beetles and decayed by fungi such as *Armillaria* and *Hypoxylon* spp.

Control Strategy

Nursery inspection and prohibition of the transportation of nursery stock or other materials likely to harbor the scale might slow its spread into regions isolated by natural barriers. When trees are protected against infestation or disinfested within a year after attack by the scale insect, fungal infection does not generally follow. Insecticides can effectively protect ornamental and landscape beeches exposed to infestation and eradicate the insect from infested trees.

Dying and dead trees deteriorate rapidly and are soon unfit even for firewood. Timely salvage operations are necessay to secure the greatest possible return on a rapidly depreciating investment and to prevent the secondary infection of other uncankered trees by Ambrosia beetles. The outlying outbreak in West Virginia alerted forest managers to favor the harvest of older beech in their management of northern hardwoods.

Some trees resist the disease, even when in close proximity to heavy infestations, and some relic old beeches remain in the aftermath zone. There is no complete explanation for this resistance. Because the *Nectria* fungus is weakly parasitic and depends on wound entry, most investigations of resistance have centered on *C. fagisuga* infestation (even though the insect cannot kill the tree). Some trees remain free of the scale insect and the fungus. Although uncommon, these trees are often in small groups, arising as ramets of a clone from root sprouts or as half- or full-sibs. This observation suggests some form of genetic resistance.

Other organisms affect host resistance. *Xylococcus betulae,* another scale insect, infests beech; it roughens the bark and appears to predispose trees to infestation by *C. fagisuga.* A common bark-inhabiting fungus *Ascodichaena rugosa* is negatively associated with *C. fagisuga* infestations. When *A. rugosa* is abundant, *C. fagisuga* has difficulty becoming established. The mycoparasite *Nematogonum ferrugineum (Gonatorrhodiella highlei)* often parasitizes *Nectria* species. In culture it inhibits sporulation of *N. coccinea* var. *faginata.* The buff-brown mycelium and spores of the mycoparasite may be diagnostically useful where red perithecia cannot be seen. Alhough detrimental to *N. coccinea* var. *faginata, N. ferrugineum* does not seem to influence the beech bark disease. A ladybug, *Chilocorus stigma,* and a lampyrid, *Lusidota corrusca,* feed on the scale insect. These two insects seem to have little effect on scale populations but may be very important by carrying the crawlers on their bodies as they fly or are blown by wind to new beech stands.

The silvicultural objective of managing stands damaged by the beech bark disease is to reduce the amount of diseased and disease-susceptible beech and to maintain or increase the amount of beech resistant to infestation by beech scale. After harvesting beech, root sprouts will develop and form dense thickets. Beech thickets that develop from root sprouts are genetically the same as the parent stock. Susceptible parents produce susceptible sprouts. The increase in susceptible beech may lead to a new and more serious outbreak of beech bark disease. Therefore, prior to harvesting, resistant and susceptible beech must be identified. This procedure will determine if beech can be managed or if it should be discouraged. Within the silvicultural guidelines, careful selection of harvesting equipment, timing, layout, and control can effectively result in the future reduction of beech bark disease. Herbicides may be needed to control dense, advance reproduction of beech and sprouting from roots and stumps of susceptible beech.

Knowledge of the patterns of insect dissemination and intensification can be useful in anticipating where beech bark disease will occur and in predicting the course of its development. Because the patterns of infestation are only indirect measures of patterns of dissemination, trapping techniques should be used to monitor the aerial transport of the insect within and above the canopy of susceptible stands.

REFERENCES

Cotter, H. V. T., and R. O. Blanchard. 1980. Identification of the two *Nectria* taxa causing bole cankers on American beech. *Pl. Dis.* 65:332–334.

Erlich, J. 1934. The beech bark disease, a *Nectria* disease of *Fagus* following *Cryptococcus fagi* (Baer). *Can. J. Res.* 10:593–692.

Houston, D. R. 1975. Beech bark disease—The aftermath forests are structured for a new outbreak. *J. For.* 73:660–663.

Houston, D. R., and D. B. Houston. 1990. Genetic mosaics in American beech: Patterns of resistance and susceptibility to beech bark disease. *Phytopathology* 80:119.

Houston, D. R., and E. M. Mahoney. 1987. Beech bark disease: Association of *Nectria ochroleuca* in W. VA, PA and Ontario. *Phytopathology* 77:1615.

Houston, D. R., E. J. Parker, and D. Lonsdale. 1979. Beech bark disease: patterns of spread and development of the initiating agent *Cryptococcus fagisuga. Can. J. For. Res.* 9:336–343.

Houston, D. R., E. J. Parker, R. Perrin, and K. J. Lang. 1979. Beech bark disease: A comparison of the disease in North America, Great Britain, France, and Germany. *Eur. J. For. Path.* 9:199–211.

Houston, D. R., and H. T. Valentine. 1988. Beech bark disease: The temporal pattern of cankering in aftermath forests in Maine. *Can. J. For. Res.* 18:38–42.

Lonsdale, D. 1980. Nectria infection of beech bark in relation to infestation by *Cryptococcus fagisuga* Lindiger. *Eur. J. For. Path.* 10:161–168.

Mielke, M. E., C. Haynes, and W. L. MacDonald. 1982. Beech scale and *Nectria galligena* on beech in the Monongahela National Forest, West Virginia. *Pl. Dis.* 66:851–852.

Mielke, M. E., D. B. Houston, and D. R. Houston. 1985. First report of *Cryptococcus fagisuga,* initiator of beech bark disease, in Virginia and Ohio. *Pl. Dis.* 69:905.

Ostrofsky, W. D., and M. L. McCormack, Jr. 1986. Silvicultural management of beech and the beech bark disease. *Nor. J. Appl. For.* 3:89–91.

Shigo, A. L. 1963. *Beech Bark Disease.* For. Notes. Summer Issue. 4 pp.

Shigo, A. L. 1964. Organism interactions in the beech bark disease. *Phytopathology* 54:263–269.

Shigo, A. L. 1970. *Beech Bark Disease.* USDA For. Serv., For Pest Leaflet 75. 8 pp.

Shigo, A. L. 1972. The beech bark disease today in the northeastern U. S. *J. For.* 70:286–289.

Towers, B. T., B. A. Ammerman, J. F. Conner, and A. W. Reinke. 1974. 1973 beech bark disease status in Pennsylvania. *Pl. Dis. Reptr.* 58:718–719.

DISEASE PROFILE

Nectria Canker

Importance: When Nectria cankers develop on the main stem, tree value is reduced, and these trees are subject to wind breakage. Although Nectria cankers result in the death of only a small portion of the trees affected, these trees occupy space and thus reduce the stand productivity. The canker can serve as an opening for the entrance of wood-decay fungi, and the discoloration associated with the canker degrades the value of the wood for pulp and lumber. Nectria canker has caused serious losses in shade trees, especially in nurseries.

Suscepts: Many species of hardwood trees are subject to Nectria canker. The principal hosts are trembling aspen, bigtooth aspen, white and yellow birches, basswood, black walnut, American elm, red and sugar maples, red and white oaks, apples, locusts, and other species.

Distribution: Nectria cankers are found wherever hosts occur.

Causal Agents: Nectria canker is caused by *Nectria galligena* (Hypocreales). The imperfect stage is *Cylindrocarpon heteronemum* (Hyphomycetes). Other species of *Nectria*, such as *N. cinnabarina,* are responsible for similar cankers.

Biology: Infection occurs through wounds that must reach the cambium. The fungus kills host tissues in advance of its mycelium but does not grow very far each year. A typical target-shaped canker that consists of a series of callus ridges develops. Each ridge of host tissue is invaded during the dormant season. The fungus produces indoleacetic acid, which may stimulate the marginal callus growth.

Nectria canker. Courtesy of D. W. French, University of Minnesota, St. Paul.

Epidemiology: Infection occurs through openings around branch stubs or wounds of various kinds. In nurseries, pruning wounds are a common entry point. Conidia are produced on small white sporodochia soon after infection. Conidia, which are pinkish and produced in sticky masses, need free water and temperatures from 18 to 30°C for germination. Conidia can infect leaf scars up to 10 days after abscission, and the amount of infection increases with increased number of conidia. In succeeding years, red lemon-shaped perithecia are produced in the summer and fall, and the unequally two-celled ascospores are formed and dispersed during periods of wet weather. The conidia are probably more common early in the growing season, while ascospores are dispersed abundantly in August and September and at other times of the year as well. The fungus can overwinter as perithecia, which give rise to ascospores, which are dispersed early in the next growing season.

Diagnosis: Cankers are usually oval or elliptical and normally are associated with branch stubs. Generally the cankers are open, but on aspen the face of the canker is often covered with bark and tends to be irregular in shape. Because the fungus grows at about the same rate as the host that produces the callus, the ridges of old callus are quite uniform and result in a targetlike appearance.

Control Strategy: It is impossible to eradicate all the diseased trees in stand improvement work, but they should be discriminated against because they are likely to be short-lived or produce trees of low merchantable value. Species of trees on which infection is most common should be discouraged. It may even be preferable to convert heavily cankered hardwood stands to conifers, either by planting or by release cuttings, rather than to try to eradicate the diseased trees. *Nectria cinnabarina* on maples and locusts can be controlled in part by eradicating diseased trees and by avoiding pruning during wet weather. This species produces abundant sporodochia and perithecia on downed material, the latter during the dormant season.

SELECTED REFERENCES

Bedker, P. J., R. A. Blanchette, and D. W. French. 1982. *Nectria cinnabarina:* The cause of a canker disease of honey locust in Minnesota. *Pl. Dis.* 66:1066–1067.

Bedker, P. J., and R. A. Blanchette. 1984. Identification and control of cankers caused by *Nectria cinnabarina* of honey locust. *J. Arboric.* 10:33–39.

Brandt, R. W. 1964. *Nectria Canker of Hardwoods.* USDA For. Serv., Pest Leafl. 84. 7 pp.

DISEASE PROFILE

Thyronectria/Tubercularia Canker

Importance: Both *Thyronectria* and *Tubercularia* cause dieback and cankers on hosts that are drought-stressed or weakened by an adverse environment, such as freezing.

Thyronectria canker. From Jacobi and Riffle (1986).

Suscepts: Thyronectria canker is aggressive on honeylocust and, to a lesser extent, on Japanese honeylocust and mimosa. Tubercularia canker affects alder buckthorn, Russian olive, Siberian elm, and winged euonymous.

Distribution: Thyronectria canker occurs from Massachusetts to Colorado to the Gulf states but is most destructive in the great plains. Tubercularia canker is destructive only in the great plains.

Causal Agents: For Thyronectria canker, *Thyronectria austroamericana* (Hypocreales) is the source, with *Gyrostroma* sp. as the imperfect state. *Tubercularia ulmea* (Hyphomycetes) is the pathogen causing Tubercularia canker and dieback.

Biology: Both pathogens cause formation of branch and stem cankers, which often lead to dieback. On mimosa *T. austroamericana* systemically invades the sapwood and causes wilt symptoms similar to those resulting from mimosa wilt, with entire tree death the usual result. Fruiting structures (pycnidial stromata with Thyronectria and sporodochia with Tubercularia) form in the bark of the killed canker area. Clusters of perithecia form later in the season, occasionally with Tubercularia and usually with Thyronectria. The former has bright red perithecia that are indistinguishable from those of *Nectria cinnabarina*. The latter perithecia are easily recognizable by their yellow-brown bodies with black tips. Ascospores of the latter are yellowish, elliptical in shape, and divided into many cells.

Epidemiology: For Thyronectria, conidia are released after wetting and may be extruded in cirrhi. They are likely to be disseminated by rain-splash and possibly by insects. Ascospores are forcibly discharged and are wind-disseminated. Only fresh wounds are susceptible to infection by either spore. For Tubercularia, conidia are dispersed by rain-splash and in artificial inoculations have caused cankers to form within 10 days.

Diagnosis: Presence of girdling cankers and associated dieback on these hosts growing under stress are strongly suggestive of either of these two diseases. Examination of sporulating fruiting structures will confirm identification.

Control Strategy: Avoid the use of honeylocust outside of its native habitat. Avoid pruning wounds and sunscald. Select resistant cultivars. Avoid prevention of normal cold hardiness. This strategy is important because the affected species are among only a few that can survive the harsh winters on the great plains.

SELECTED REFERENCES

Carter, J. C. 1947. Tubercularia canker and dieback of Siberian elm (*Ulmus pumila* L.). *Phytopathology* 37:243–246.

Crowe, F., D. Starkey, and V. Lengkeek. 1982. Honeylocust canker in Kansas caused by *Thyronectria austro-americana. Pl. Dis.* 66:155–158.

Jacobi, W. R. 1989. Resistance of honeylocust cultivars to *Thyronectria austro-americana. Pl. Dis.* 73:805–807.

Jacobi, W. R., and J. W. Riffle. 1986. Thyronectria canker of honeylocusts, pp. 60–61. In J. W. Riffle and G. W. Peterson, (eds.), *Diseases of Trees of the Great Plains.* USDA For. Serv., Gen. Tech. Rept. RM-129. 149 pp.

DISEASE PROFILE

Strumella Canker

Importance: This perennial, usually nongirdling lower stem canker of hardwoods is common on red oaks in the Appalachian region but seldom exceeds an incidence of 2–3%. Infected trees may break at the canker or live to maturity. Such trees are doomed to be culls in terms of timber value and they occupy growing space.

Suscepts: Seven species of white oaks and five species of red oaks known to be susceptible. Beech, hickory, red maple, ironwood, basswood, and black gum are occasional suscepts, as was American chestnut before its demise from chestnut blight.

Distribution: The disease is cosmopolitan in the east from New Hampshire to North Carolina and occurs to a lesser extent in the central and north-central states. The causal fungus, as a saprophyte only, has also been reported from Missouri, Oregon, and Ontario.

Causal Agents: The causal fungus is *Urnula craterium* (Pezizales). The imperfect stage is *Conoplea globosa* [*Strumella coryneoidea* (Hyphomycetes)].

Biology: Mycelium overwinters in cankered tissue and in oak debris. On the latter, apothecia develop over a few weeks in early spring to release windborne ascospores that germinate readily. Despite their limited availability, ascospores are the likeliest source of inoculum. Conidia are also produced but only after an occasional cankered tree dies. These spores have never been germinated so their causative role is uncertain. Because canker etiology has been shown only by mycelial inoculation, the mode of infection is

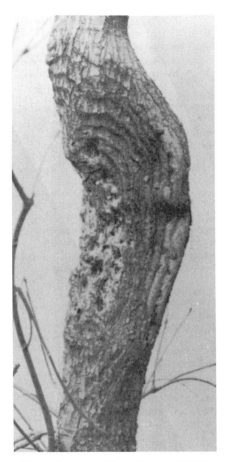

Strumella canker. From Houston (1966).

also unknown. Cankers invariably center on branch stubs, the assumption being that spore inoculum establishes penetration either parasitically via wounds in living branches or saprophytically in branches dead or dying from suppression. The cankers are generally initiated before age 25 and are usually localized to the lower 3 m of stem. Young trees, up to 10 cm in diameter, may be girdled and killed by rapid and diffuse cankering. Typically though, perennial, nongirdling cankers develop and persist for years as marginal callusing is countered annually by fungal colonization and necrosis. In time, *Urnula* decays the decorticated canker face, and stem breakage may occur. The cycle is completed with the production of conidia and ascospores following tree death and upon windthrow, respectively.

Epidemiology: The potential for spread is limited by stand age (25 years) and oak composition, particularly of the more susceptible red oaks, but it is not related to tree vigor.

Diagnosis: Canker symptoms, which start as yellowish bark discoloration, may ultimately range from diffuse to target types in development. Diagnosis is best typified, however, by a perennial, elliptical canker that in time displays decortication and decay of the canker face. Sterile sporodochia may form on the cankered bark. They are dark brown, mounded pustules, 1–3 mm in diameter, that sporulate and multiply saprophytically at and beyond the canker after the tree dies. This so-called gunpowder stage in sporodochial development marks the production of abundant, globose to pyriform, brown conidia, (6.7–8.1 × 4.7–5.8 μm) borne at the junctures of sinuous, spiculose conidiophores, all features that characterize the genus. The apothecial stage forms in the spring on fallen diseased trees or on oak slash. Occurring in stalked clusters, the black, leathery, urn-shaped cups (3–4 cm in diameter and 4–6 cm deep) flare open at maturity to display a ragged margin. The asci are narrowly cylindrical with 8 one-celled, hyaline ascospores, 25–35 × 12–14 μm.

Control Strategy: Infected trees should be felled at a young age in timber stand improvement, and, because of the *Urnula* stage, the cankered portion must be removed and destroyed or raised off the ground to dry rapidly.

SELECTED REFERENCE

Houston, D. R. 1966. *Strumella Canker of Oaks.* USDA For. Serv., For. Pest Leafl. 101. 8 pp.

DISEASE PROFILE

Eutypella Canker

Importance: The importance of this disease is not known; although in some areas such as northeastern Wisconsin, it is common on hard maple. If present to any extent, it can cause serious loss by deforming the butt log. Small trees are suppressed and killed by the fungus.

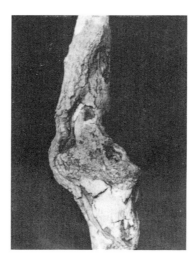

Eutypella canker. From Kessler and Hadfield (1972).

Suscepts: Eutypella canker occurs most commonly on hard or sugar maple; it is also found on red and Norway maples as well as boxelder and sycamore.

Distribution: Eutypella canker is found over much of the range of the hosts.

Causal Agents: Eutypella canker is caused by *Eutypella parasitica* (Diatrypales). The imperfect stage is *Libertella* spp. (Coelomycetes).

Biology: A dead branch stub is usually present in the center of each canker, indicating that the fungus probably enters at this point. During the first few years, forming cankers are depressed with the bark tightly attached. Older cankers have a much roughened appearance and usually are outlined with a large callus ridge. Mycelial fans grow in the inner bark just under the expanding canker margin. A callus and cork layer forms each year; it is then breached during the next year. *Eutypella parasitica* produces a brown decay in the sapwood beneath the canker. This decay weakens the stem and often leads to wind breakage.

Epidemiology: Conidia are produced, but their importance is unknown. Ascospores are ejected coincident with rain and are wind-dispersed. They infect small branches that have been broken or wounded. The fungus is easy to isolate from diseased bark or wood.

Diagnosis: The cankers usually occur on the main stem from the ground up to about 3 m. As a general rule, the callus growth is extensive and often will be evident from the opposite side of the tree. Patches of bark will adhere to the surface of the canker, and white to tan mycelial fans occur under the bark at the extremities of the canker. The centers of old cankers contain numerous embedded black perithecia. Their black, crusty necks protrude from the canker face.

Control Strategy: The only control measure that can be recommended is to discriminate against infected trees when thinnings and improvement cuttings are made. In some areas where a high percentage of the hard maples are infected, it might be advisable to discriminate against the species.

SELECTED REFERENCES

Johnson, D. W., and J. E. Kuntz. 1979. Eutypella canker of maple: Ascospore discharge and dissemination. *Phytopathology* 69:130–135.

Kessler, K. J., and J. S. Hadfield. 1972. *Eutypella Canker of Maple.* USDA For. Serv., For. Pest Leaflet 136. 6 pp.

DISEASE PROFILE

Fusarium Canker of Hardwoods

Importance: First observed on yellow-poplar in Ohio in 1956 and on sugar maple in Pennsylvania in 1959, this annual canker has since sporadically occurred in eastern North America. Affected trees are seldom killed directly; however, stems degrade from

Fusarium canker on oak stem. From Toole (1966).

cankering, and stain is often compounded by decay establishment and, in time, stem breakage and mortality.

Suscepts: As judged by disease prevalence, yellow-poplar and sugar maple are most susceptible. Natural infection also has occurred on black poplar in Wisconsin; hybrid poplars in Ohio; eastern cottonwood in Arkansas, Mississippi, and Quebec; nursery-grown western cottonwood in British Columbia; white ash, black cherry, and red oak in Pennsylvania; swamp tupelo in Louisiana; and bottomland oaks in Mississippi.

Distribution: The collective host ranges suggest an occurrence throughout eastern North America.

Causal Agents: Fusarium cankers on hardwood are caused by *Fusarium* spp. but especially *F. solani* (Hyphomycetes).

Biology: Inoculum has been detected in the air and soil and on bark of healthy trees. Stem infection is largely wound-dependent during suscept dormancy and is conditioned by temperature. Spore germination is not inhibited by cortical tissue extracts. Canker development in yellow-poplar is marked by complete degradation of starch and partial degradation of pectin and cellulose, while lignin is unaffected; canker extension and fungus growth cease within 18 months following inoculation as limited by lignification of a barrier zone of cells in the cortex, phloem, and rays of the bark. Thus host defenses limit the fungus to a single year of pathogenesis influenced by temperature.

Epidemiology: Sugar maple canker in Pennsylvania occurs primarily on soils with poor internal drainage and during springs when cambial injury from sudden freezes is suspected to provide wound courts for pathogen entry. The epidemic began in the early 1930s, peaked during the 1950s when disease incidence doubled every 4 years, and began to subside in the early 1960s. In northeastern Minnesota, where nearly 80% of the cankers originated from 1963 to 1972, distribution of the disease was not correlated with any specific factor, climatic or otherwise. Predisposition of yellow-poplar to cankering in Ohio has been attributed to adverse sites, specifically spoil banks and dry soils, and to growth suppression from close spacing of plantations, all aggravated by periodically dry growing seasons. The linking of canker severity with upland sites and pine mixtures in

North Carolina did not hold in South Carolina where only high soil potassium was an important indicator of canker severity. Occasional cankering of bottomland hardwoods along the Mississippi River is associated with flooding and wounding from floating debris. In this case, wound inoculation may occur during or after flooding from water-or airborne inoculum, respectively.

Diagnosis: Annual cankering is common to all suscepts. Cankers on yellow-poplar initially appear as single elliptical depressions in the bark, which are best delineated by shaving the margin to expose the blackened inner tissue. The cankers may range from a few centimeters to more than a meter in length as they coalesce. Because girdling seldom occurs, affected trees usually survive. In a severely degraded condition, decortication exposes sapwood to secondary invaders, which may stain and decay the wood locally behind the canker face. Ultimately callus closes the wound. The cankers on maple differ primarily in bark features, which may range from a hob-nail effect to larger bark plates. Diagnosis is easily confirmed by tissue isolations from the canker margin.

Control Strategy: Disease avoidance should be practiced by planting or otherwise favoring natural regeneration of yellow-poplar and sugar maple only on loamy, moderately moist, well-drained soils. Exclusion is not likely to have any effect, because species and even subspecies of *Fusarium* are so broadly indigenous. Neither protection nor eradication are economically feasible; however, the latter should theoretically have merit when thinning in young, high-value stands.

SELECTED REFERENCES

Dochinger, L. S., and C. E. Seliskar. 1962. Fusarium canker found on yellow-poplar. *J. For.* 60:331–333.
Toole, E. R. 1963. Cottonwood canker caused by *Fusarium solani. Pl. Dis. Reptr.* 47:1032–1035.
Weidensaul, T. C., and F. A. Wood. 1974. Analysis of a maple canker epidemic in Pennsylvania. *Phytopathology* 64:1024–1027.

DISEASE PROFILE

Ceratocystis Canker

Importance: In Colorado, Ceratocystis canker was found in 71% of the plots and 4.4% of the aspen were infected. In the eastern United States, severe outbreaks have been caused by pruning or other human activities. In California on prune trees, the disease reached epidemic proportions in 1965 with 6,000 ha of trees affected. In some orchards, 100% of the trees had cankers.

Suscepts: Ceratocystis cankers occur on species of *Populus,* especially trembling aspen; on the London plane (sycamore), coffee, cacao, and rubber; and on fruit trees such as almond, prune, peach and apricot.

Distribution: This disease is found worldwide over much of the range of the hosts.

Causal Agents: Ceratocystis cankers are caused by *Ceratocystis fimbriata* (Ophiostomatales), varieties of this species, and very possibly by other species of the same genus.

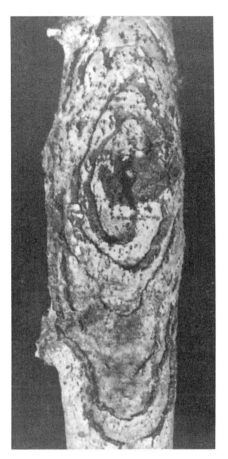

Ceratocystis canker. From Wood and French (1963). Reprinted from *Forest Science*, Vol. 9, No. 2, published by the Society of American Foresters, 5400 Grosvenor Lane, Bethesda, MD 20814-2198. Not for further reproduction.

Ceratocystis fimbriata f. *platani* causes sycamore canker. Other species that may have potential for inducing cankers are frequently found in cankers including those supposedly caused by species of *Nectria*.

Biology: Infection occurs through wounds and branches. Once in the host the hyphae colonize the phloem and cambium. Brown gummy materials are produced in response to the fungus, and these plus dead cells restrict the spread of the fungus. Although the fungus can invade xylem and pith, it is restricted, and in fruit trees it seldom penetrates further than one growth ring. Perithecia with hat-shaped ascospores and conidia (hyaline and brown) are produced on the cankered portions of the hosts and, as is characteristic for species in the genus *Ceratocystis*, these spores are insect-disseminated. Among possible insect vectors are a nitidulid beetle and a drosophilid fly. Passage of spores through the intestinal tract of the former did not result in loss of viability. The fungus remained with the beetles up to 8 days after they stopped feeding on the fungus. Contaminated larvae in sterile soil produced adults with the fungus.

Epidemiology: Local dissemination of spores is by rain-splash. Insects are attracted to cankers by the fruity odor emitted by the fungus. They become contaminated with the sticky conidia and ascospores and transmit these when they visit fresh wounds.

Diagnosis: Ceratocystis cankers are more or less target-shaped with or without bark adhering to the surface. In fruit trees the wood behind the canker becomes stained. In sycamore, discoloration of wood is very evident. In most hosts, the fungus grows slowly enough so that the tree can form a new layer of callus tissue each year, thus the target shape.

Control Strategy: Epidemics on London plane trees in the northeastern states prior to World War II were eventually controlled by proper sanitation practices. In fruit trees the incidence of the disease has increased with the greater use of mechanical tree shakers. Thus the disease can be reduced very likely by avoiding wounds. If wounds do occur, the infected tissues can be removed and the area treated with a fungicide such as benomyl or an antiseptic wound dressing. Tree species and varieties within species vary in their resistance to Ceratocystis cankers. In aspen and other forest tree species, no control measures are known other than minimizing logging wounds.

SELECTED REFERENCES

DeVay, J. C., F. L. Lukezic, H. English, E. E. Trujillo, and W. J. Moller. 1968. Ceratocystis canker of deciduous fruit trees. *Phytopathology* 58:949–950.

Hinds, T. E. 1964. Distribution of aspen cankers in Colorado. *Pl. Dis. Reptr.* 48:610–614.

Wood, F. A., and D. W. French. 1963. *Ceratocystis fimbriata,* the cause of a stem canker of quaking aspen. *For. Sci.* 9:232–235.

DISEASE PROFILE

Septoria Canker

Importance: As a leaf-spotting organism, the *Septoria* fungus is of no consequence; but on many fast growing hybrid poplars, it is a limiting disease.

Suscepts: Hybrids such as *Populus rasumowskyana, P. petrowskyana,* and *P. berolinensis* and the Northwest (native hybrid) are all susceptible. In mixed plantations the susceptible trees are gradually eliminated.

Distribution: Many species of poplars present across North America are probably susceptible to leaf spotting caused by the same fungus that causes cankers on some of the hybrids.

Causal Agents: Septoria canker and leaf spot is caused by *Mycosphaerella populorum* (Dothidiales); the imperfect stage is *Septoria musiva* (Coelomycetes).

Biology: The fungus usually enters through wounds. In eastern cottonwood, unwounded first-year stems can be invaded. Leaf spotting may occur first, and the spores responsible for cankers can come from these leaf infections. Cankers do not usually continue to develop, but new infections may continue to occur, often leading to wind breakage.

Epidemiology: Infection in the spring comes from ascospores, produced in perithecia on dead leaves or from conidia produced in pycnidia on cankers.

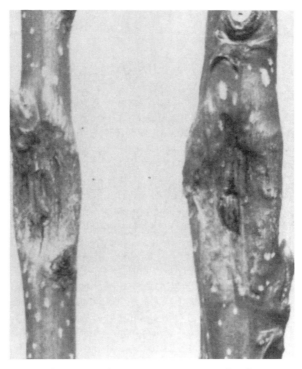

Septoria cankers. From Waterman (undated).

Diagnosis: Irregularly shaped, gray to black spots occur in the leaves; and in some cases, the spots may involve most of a leaf's surface. Pycnidia, which ooze pink spore tendrils, develop in the spots. Cankers appear first as discolored, slightly depressed areas in the bark, later turning dark brown as the tissues dry. Callus tissues form and often seal over the canker.

Control Strategy: The only successful control measure is the use of resistant varieties, which necessitates a selection program for each part of the country. Surface sterilization of cuttings should be mandatory to avoid introducing new more virulent strains. Wider spacing in plantations to reduce relative humidity might be of some benefit.

SELECTED REFERENCES

Bier, J. E. 1939. Septoria canker of introduced and native hybrid poplars. *Can. J. Res.* 17:195–204.
Filer, T. H., F. I. McCracken, C. A. Mohn, and W. K. Randall. 1971. Septoria canker on nursery stock of *Populus deltoides*. *Pl. Dis. Reptr.* 55:460–463.
Waterman, A. M. undated. *Septoria Canker.* USDA Circ. No. 947. 24 pp.

DISEASE PROFILE

Fire Blight

Importance: Since the last century, fire blight has caused devastating losses in pear and
apple orchards in the eastern United States, and, early in this century, all but eliminated
pear trees in some California counties. Major losses have also occurred in nurseries
raising mountain ash, cotoneaster, and pyracantha. Fire blight is a major concern in
European fruit-producing regions.

Suscepts: Along with the genera *Malus* and *Pyrus,* 37 other genera involving 129 species are
susceptible to fire blight. Additional seriously affected species are in the genera
Cotoneaster, Crataegus, Cydonia, Pyracantha, and *Sorbus.*

Distribution: Fire blight has been spread as far as Chile and New Zealand and is now in a
total of 14 countries. In the United States it was first found in New York in 1780 and
during the next century spread across the country, being found in California in 1888. In
the late 1940s, fire blight was found in Great Britain. Next it appeared in Poland,
Netherlands, Denmark, Germany, and, in the early 1970s, France.

Causal Agents: Fire blight is caused by the bacterium *Erwinia amylovora.*

Biology: Erwinia amylovora can overwinter on holdover cankers or as resident bacteria on
buds and small shoots. Bacterial cells overwinter only in a portion of the cankers, and
these cells account for only a portion of the infections. Bacterial cells causing the primary
infection in spring are disseminated by insects, wind, or rain, with insects probably the
primary vector. Secondary infections are far more numerous and are responsible for the
major injury to hosts. Inoculum for secondary infection is carried by rain, wind, insects,
birds, and people. Pruning equipment and shipments of fruits can cause further spread.
Most bacteria in bark tissues late in the growing season die as the host tissues die. The

Blight-killed branch. Courtesy of D. W. French, University of Minnesota, St. Paul.

most common entry point for the bacterium is the flowers, but infection can also occur in any succulent tissue. Bacteria enter through natural openings, multiply in the intercellular spaces with collapse of host cells, followed by discoloration and dying of host tissues.

Epidemiology: Incidence is favored by succulent tissue, weather, soil conditions, and certain cultural practices. Succulent tissues are likely to be low in carbohydrate and more susceptible. Slow but fairly vigorous growth results in more resistant tissues. Increased amounts of nitrogen result in prolonged succulence of tissues and more infection. Warm wet weather, followed by periods of high humidity, favors infection. Moderately high temperatures favor blight. High relative humidity reduces sugar concentration in succulent tissues, rendering them more susceptible. Injury as a result of wind and hail provides more openings for entrance of bacteria. Blight is likely to be more common on wet than well-drained sites. Cultivation and severe pruning increase blight incidence.

Diagnosis: Shoots turn brown or black as though affected by fire. Cankers form on the branches or even the main stem. The cankers are reddish brown, often with a small dead twig in the center. Bacterial ooze on pear can form streaks on and below the canker. Apples turn black with red margins around dead tissues. Pears turn black with dark green margins. Bacterial ooze will occur on fruit. Severely cankered trees are killed.

Control Strategy: In orchards, reduction of excessive succulent tissue can be minimized by reducing the amount of nitrogen and by pruning less severely. Locating orchards in sites that are not excessively wet will reduce disease incidence. Growing highly susceptible varieties, especially of pears, will increase incidence in the rest of the orchard. In establishing orchards, select a reliable source of disease-free trees. Pruning, if properly done, will help eradicate sources of inoculum. Pruning equipment should be sterilized between cuts. Chemical control is possible, but not totally satisfactory. Copper sulfate plus lime causes fruit russeting. Antibiotics, particularly streptomycin, have had some success, but tolerance has developed, reducing its effectiveness. Biological control using insects to carry nonpathogenic bacterial competitors of *E. amylovora* has shown promise.

SELECTED REFERENCE

Van Der Zuet, T., and H. L. Keil. 1979. *Fire Blight, a Bacterial Disease of Rosaceous Plants.* USDA, Agric. Handbook No. 510. 200 pp.

DISEASE PROFILE

Butternut Canker

Importance: Butternut had been assumed to be reasonably free of diseases. Butternut canker was reported in Wisconsin in 1967; however, it apparently had been active for several years prior to this time. In southwestern Wisconsin, many stands have sustained substantial losses with 80% of the trees infected and 32% dead. In North Carolina and Virginia the population of butternuts fell from 7.5 million in 1966 to 2.5 million in 1986.

Butternut canker. Courtesy of D. W. French, University of Minnesota, St. Paul.

Suscepts: Butternut is the primary host. Cankers occur infrequently on black walnut.

Distribution: Butternut canker has been found throughout the range of butternut in eastern North America, but the disease has not been reported from more northerly regions where butternuts occur.

Causal Agents: The fungus causing the canker disease is *Sirococcus clavigignenti-juglandacearum* (Coelomycetes).

Biology: Sirococcus clavigignenti-juglandacearum enters its host through leaf scars, buds, bark wounds including insect wounds, and splits in the bark. Young cankers may not be detected unless the bark is removed; older cankers are exposed and may exist for many years with annual layers of callus. Trees girdled by cankers may produce sprouts that are quickly infected and killed. Butternut seedlings are killed within 2 weeks at favorable temperatures, 16–20°C. Cankers in black walnut expand slowly, and callus layers sometimes limit spread. The fungus produces hyphal pegs and conidia in pycnidia, which form at the base of hyphal pegs. Presumably conidia are dispersed by rain and carried further by insects.

Epidemiology: Conidia are extruded in cirrhi, with individual spores being disseminated by rain-splash. Conidia are short-lived unless they are shaded and in high humidity. The microenvironment associated with lower branches tends to retain favorable moisture

Butternut canker with bark removed. From Nicholls et al. (1978).

conditions for spore germination longer than in upper branches. Sporulation may continue for up to 20 months on dead host tissues.

Diagnosis: Butternut cankers are elliptical to fusiform-shaped, whether on branches, stems, or roots. They are perennial, covered when young and later either open or partially covered with deteriorating bark. The wood under cankers is dark brown. During spring a black exudate oozes from fissures in cankers.

Control Strategy: Control measures are not known. Black walnuts should not be planted near diseased stands of butternut.

SELECTED REFERENCES

Nair, V. M. G., C. J. Kostichka, and J. E. Kuntz. 1979. *Sirococcus clavigignenti-juglandacearum* an unde-scribed species causing canker on butternut. *Mycologia* 71:641–646.

Nichols, T. H., K. J. Kessler, Jr., and J. E. Kuntz. 1978. *How to Identify Butternut Canker.* USDA For. Serv., Leaflet. 1 pp.

Tisserat, N., and J. E. Kuntz. 1984. Butternut canker: Development on individual trees and increase within a plantation. *Pl. Dis.* 68:613–616.

DISEASE PROFILE

Diffuse Cankers

Importance: There are two major diffuse cankers, Cryptosphaeria canker and sooty-bark canker, both of aspen.

Suscepts: Both are serious pathogens on quaking aspen but may also occur on other species of Populus.

Distribution: Cryptosphaeria canker is found from Arizona into Alaska and all across northern United States and southern Canada. It is especially common in Colorado. Sooty-bark canker is a common disease in the Rocky Mountains and in the eastern United States and Canada.

Causal Agents: Cryptosphaeria canker is caused by *Cryptosphaeria lignyota* (Diatrypales); the imperfect stage is *Cytosporina* sp. (Coelomycetes). Sooty-bark canker is caused by *Encoelia pruinosa* (Helotiales) *(Cenangium singulare).*

Biology: Cankers are elongated and form on the main stem, becoming brownish (Cryptosphaeria) or blackish (sooty bark). Canker growth is very rapid; with sooty bark, the fungus can grow 1 m in 1 year.

Epidemiology: With Cryptosphaeria, light orange acervuli develop near the canker margin. After the bark is dead for at least 1 year, single perithecia develop in a grayish pseudostroma. Both spore types are presumed to infect wounds. With sooty-bark, large numbers of apothecia develop on the blackened bark, producing ascospores that are forcibly ejected.

Diagnosis: Cryptosphaeria cankers are elongate, with some bleeding at the margins. Large numbers of highly visible black perithecia form in the bark. There may be extensive internal stain and decay in the stem behind the canker. Sooty-bark cankers often have arcs of blackened bark tissue as a result of fungus attack. Apothecia that form on the black surface, are silver gray in color.

Control Strategy: Until more is known about these diseases, the only control strategies that can be suggested is to eradicate infected trees and change species if the diseases continue to cause significant losses.

SELECTED REFERENCES

Hinds, T. E. 1981. Cryptosphaeria canker and Libertella decay of aspen. *Phytopathology* 71:1137–1145.

Juzwik, J., and D. W. French. 1989. Cryptosphaeria canker on *Populus tremuloides* in Minnesota. *Can. J. Bot.* 68:2044–2045.

Juzwik, J., D. W. French, and T. E. Hinds. 1986. *Encoelia pruinosa* on *Populus tremuloides* in Minnesota: occurrence, pathogenicity, and comparison with Colorado isolates. *Can. J. Bot.* 64:2728–2731.

16.3 CONIFERS

DISEASE PROFILE

Scleroderris Canker

Importance: Recognized in 1964, and possibly present as early as 1942, Scleroderris canker has caused serious mortality of red pine and jack pine in plantations and nurseries in the Lake states. The so-called European strain has caused serious losses in red and Scots pine stands in New York and Vermont.

Suscepts: In Europe, Scleroderris canker occurs on pine and spruce species. In the northeastern United States, all eastern conifers are susceptible except for balsam fir and

Xylem discoloration. From Skilling and O'Brien (1979).

northern white cedar. Pines, however, are most susceptible, and both seedlings and large trees succumb to the fungus. Another species in the western United States attacks white, grand, Pacific silver, alpine, and red firs.

Distribution: Scleroderris canker can be a problem in young spruce and pine stands in eastern Canada and in northern Michigan, Minnesota, and Ontario in Canada. The European strain is present in 12 northeastern New York counties, all of northern Vermont, Maine, southern Quebec, and New Brunswick.

Causal Agents: The primary fungus is *Ascocalyx abietina* (Helotiales) *(Gremmeniella abietina, Scleroderris lagerbergii);* a related fungus, *Grovesiella abieticola (Scleroderris abieticola),* attacks subalpine fir in Colorado and other firs in the western states.

Biology: Buds and needles of new shoots are infected from early spring to late summer. These die within a few months. Small cankers form on infected branches. Conidia are produced in pycnidia; oozed out in pink spore tendrils, and dispersed by splashing water and possibly insects. Ascospores are produced in apothecia, and these are dispersed by wind. The perfect stage of the European strain occurs infrequently in North America.

Epidemiology: Ascocalyx abietina is favored by low temperatures, about 15–20°C. In the Lake states, the disease is most severe in low areas where cold air can accumulate.

Diagnosis: The most characteristic symptom other than the partially dead crowns is a yellow-green color that develops beneath the bark. In April and May the bases of needles on affected trees turn brown, and the needles fall from the tree. Stem cankers may not always be obvious. Small black pycnidia on needles and the pink spore tendrils they produce, and groups of apothecia, are evidence of infection by *A. abietina.* There is seldom any evidence of infection during the year in which the tree became infected.

Control Strategy: In nurseries, Scleroderris canker is easily controlled with fungicide sprays. On high-risk sites, susceptible species should not be planted. Spring plantings may be desirable as frost injuries are avoided. Planting under partial overstories results in less disease than completely open areas because of less frost injury in the former. Caution should be used with eradication, and the removed material should probably be burned or buried as the fungus has survived for 10 months in dead branches. Some resistance has been observed.

SELECTED REFERENCES

Dorworth, C. E. 1981. Status of pathogenic and physiologic races of *Gremmeniella abietina. Pl. Dis.* 65:927–931.

Skilling, D. D., and J. T. O'Brien. 1979. *How to Identify Scleroderris Canker.* USDA For. Serv., Leaflet. 1 pp.

Skilling, D. D., B. Schneider, and D. Fasking. 1986. *Biology and Control of Sceroderris Canker in North America.* USDA For. Serv., Res. Pap. NC-275. 19 pp.

16.3.1 Scleroderris Canker

Importance

Scleroderris canker, caused by the so-called European strain of the fungus, by 1979 involved 20,000 ha of red and Scots pine stands in northeastern New York (Fig. 16.22). Infected trees are defoliated and killed. The fungus has caused serious losses in Vermont. The North American strain of the fungus caused serious losses in limited areas in Michigan, Wisconsin, Minnesota, and adjacent regions of Canada. Mortality

FIGURE 16.22 Pole-sized red pine dying from *Ascocalyx abietina* infection in northern New York.

in red pine plantations has been as high as 40% and in jack pine as high as 39%. These losses have been reduced in recent years as a result of the development and use of control methods. If the European strain spreads west, the potential losses in our pine forests across the northern United States could be devastating.

Suscepts

Scleroderris canker occurs on pine and spruce species in Europe, where the disease was first described by Brunchorst in 1888. Although there is rather good evidence that the disease was in the United States at least as early as 1942, it was not recognized as such until 1964 (1962 in Canada). More recent studies have demonstrated that almost all eastern conifers are susceptible to the European strain except for balsam fir (Fraser fir is susceptible) and northern white cedar. The pines, however, are most susceptible, and both seedlings and large trees succumb to the fungus.

Another related species of the fungus causes a similar disease on white, grand, Pacific silver, alpine, and red firs.

Distribution

At the present time, it is a problem in young spruce and pine stands in eastern Canada and in young pines in northern Michigan, Minnesota, and Ontario in Canada (Fig. 16.23). Scleroderris canker caused by a different species was found in subalpine fir in Colorado in 1964, and the disease was dated back to 1956. The European strain is

FIGURE 16.23 Distribution of *Ascocalyx abietina* in North America; shaded area is of the North American strain; solid area is of the European strain. Drawn from data presented in Skilling (1981), Myron and Davis (1986), and Laflamme and Lachance (1987).

now known in 12 northeastern New York counties, all of northern Vermont, Maine, southern Quebec, and New Brunswick.

Cause

The primary fungus involved is *Ascocalyx abietina* (Lagerberg) Schläpfer *(Gremmeniella abietina, Scleroderris lagerbergii)*; *S. abieticola* is the species on subalpine fir in Colorado and other firs in the western states.

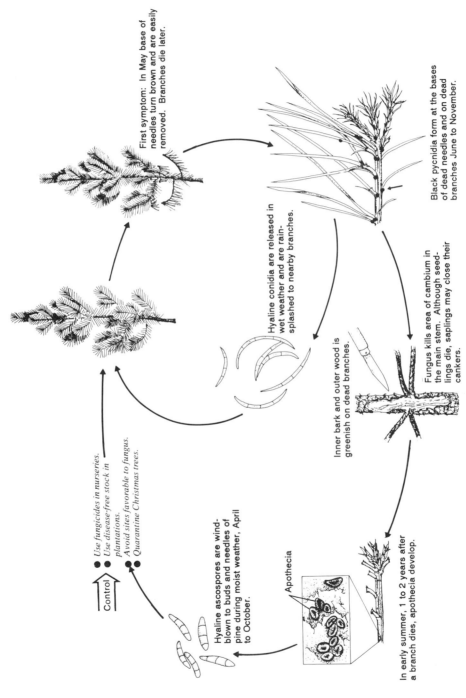

First symptom: In May base of needles turn brown and are easily removed. Branches die later.

Black pycnidia form at the bases of dead needles and on dead branches June to November.

Hyaline conidia are released in wet weather and are rain-splashed to nearby branches.

Inner bark and outer wood is greenish on dead branches.

Fungus kills area of cambium in the main stem. Although seed-lings die, saplings may close their cankers.

Control

Use fungicides in nurseries.
Use disease-free stock in plantations.
Avoid sites favorable to fungus.
Quarantine Christmas trees.

Hyaline ascospores are wind-blown to buds and needles of pine during moist weather, April to October.

Apothecia

In early summer, 1 to 2 years after a branch dies, apothecia develop.

FIGURE 16.24 Disease diagram of Scleroderris canker on pine caused by *Ascocalyx abietina*. Drawn by Valerie Mortensen.

Biology

The disease diagram for Scleroderris canker is shown in Figure 16.24. *Ascocalyx abietina* infects both buds and needles of new shoots of susceptible hosts from early spring to late summer. Infected shoots die within a few months. Entire trees die within a few years. Small cankers form on infected branches. Two kinds of spores are produced—conidia and ascospores. The conidia, produced in pycnidia, are oozed out in pink spore tendrils and dispersed during wet periods by splashing and possibly by insects. These conidia are produced one year after infection. The ascospores produced in apothecia are wind blown to new hosts. Apothecia form 2 years after infection and mature in spring; maximum numbers of spores are available between May 15 and September 3. Apothecia on spruce develop in the latter part of the summer, and spore dispersal occurs from the middle of July to the middle of October. The perfect stage of the European strain occurs infrequently in New York and other New England states. Movement of diseased trees is potentially the most important means of spread

FIGURE 16.25   Wilting and drooping of dead needles killed by *Ascocalyx abietina*. From Skilling and Nicholls (1975).

because of the possibility of moving greater distances. Infection of newly planted trees can occur from larger infected trees adjacent to the plantation.

Epidemiology

Ascocalyx abietina is favored by low temperatures. The optimum temperature for spore discharge and mycelium development ranges from 15 to 20°C, although some isolates are favored by temperatures below 15°C.

Diagnosis

The most characteristic symptom other than the partially dead crowns is the yellow-green color that develops beneath the bark. In April and May the bases of the needles on affected trees turn brown, and the needles fall from the tree (Fig. 16.25). Stem cankers are another symptom but are not always obvious or necessarily indicative of Scleroderris canker. There is no evidence of infection at the end of the growing season during which the tree becomes infected. The presence of the small black pycnidia on infected needles (Fig. 16.26), the pink spore tendrils produced from these pycnidia, and the groups of apothecia are evidence of the presence of *A. abietina* (Fig. 16.27).

Management

In nurseries, Scleroderris canker has been controlled with Maneb or chlorothalonil applications at biweekly intervals from May through September. The number of applications could be reduced if the fungicide is applied prior to when infection might occur. Seven applications of chlorothalonil from late May to mid August at 3.6 gm/l water can be used. Additional sprays may be required in wet years. Rogueing infected trees is not very practical. If eradication is attempted, it is important to realize that the fungus can survive in dead branches for 10 months and thus they must be burned

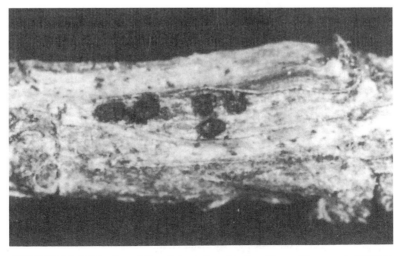

FIGURE 16.26 Pycnidia of *Ascocalyx abietina*. From Dorworth (1972).

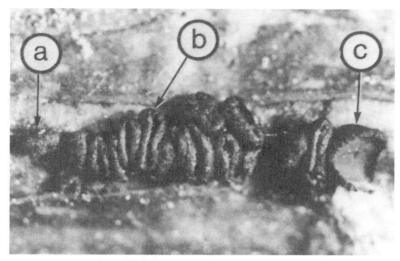

FIGURE 16.27   *Ascocalyx abietina* on stem of jack pine with (a) immature, (b) mature closed, and (c) mature expanded apothecia. From Dorworth (1972).

to destroy the source of inoculum. On sites where the disease has resulted in significant losses, susceptible species should not be planted. In northern Michigan, for example, this would mean using species other than red and jack pines. Spring plantings have been twice as successful as fall plantings, probably because frost injuries are avoided. Planting where partial overstories exist will result in less disease than completely open areas because of less frost damage in the former. Frost pockets favor infection. Resistant jack pine have been found in Canada.

SELECTED REFERENCES

Cordell, C. E., D. D. Skilling, and J. W. Benzie. 1968. Susceptibility of three pine species to *Scleroderris lagerbergii* in Upper Michigan. *Pl. Dis. Reptr.* 52:37–39.

Dorworth, C. E. 1972. Longevity of *Scleroderris lagerbergii* Gremmen in pine slash. *Bimonthly Res. Notes* 28(1):5.

Laflamme, G., and D. Lachance. 1987. Large infection center of Scleroderris canker (European race) in Quebec Province. *Pl. Dis.* 71:1041–1043.

Myren, D. T., and C. N. Davis. 1986. European race of Scleroderris canker found in Ontario. *Pl. Dis.* 70:475.

Skilling, D. D. 1971. Epidemiology of *Scleroderris lagerbergii*. *Eur. J. For. Path.* 2:16–21.

Skilling, D. D. 1981. Scleroderris canker—The stituation in 1980. *J. For.* 79:95–97.

Skilling, D. D., and T. H. Nicholls. 1975. Shoot blight and Scleroderris canker—A new disease of pine plantations. *Am. Christmas Tree J.* 19:7–8.

Skilling, D. D., and C. D. Waddell. 1974. Fungicides for control of Scleroderris canker. *Pl. Dis. Reptr.* 58:1097–1100.

Skilling, D. D., B. S. Schneider, and J. A. Sullivan. 1977. Scleroderris canker on Austrian and ponderosa pine in New York. *Pl. Dis. Reptr.* 61:707–708.

DISEASE PROFILE

Pitch Canker

Importance: Pitch canker, first noticed in 1945 on Virginia pine in North Carolina, occurred sporadically until 1974, when incidence and severity increased dramatically on large areas in Florida. In central Florida, slash pine plantations experienced severe damage, as have slash and loblolly pine seed orchards in Florida and other southern states. In 1986 pitch canker was found on Monterey pine in California. Since then, it has been found in several areas of California, in Mexico, and on nine additional hosts.

Suscepts: Important suscepts are slash, South Florida slash, longleaf, Virginia, and loblolly pines. Other naturally infected pines include shortleaf, Scots, pitch, table-mountain, and eastern white pine and *Pinus occidentalis*. In California, Monterey pine is the most important host, as it has a very limited natural range. Other suscepts include Monterey × knobcone, Bishop, Coulter, Torrey, shore, aleppo, Italian stone, and Canary Island pines. Recently, Douglas-fir and ponderosa pine have become infected. These latter two commercially important timber species have caused considerable concern as they have large, contiguous ranges throughout the western United States.

Distribution: Pitch canker occurs from Virginia to south Florida and west to Mississippi and Tennessee. As of 1992 the disease was found in 10 counties of California.

Causal Agents: The fungal causal agent is *Fusarium subglutinans* (Wollenweb. & Reinking) P.E.Nelson, T.A.Tousoun & Marasas (Hyphomycetes). The fungus was previously known as *F. moniliforme* J.Sheld. var. *subglutinans* Wollenweb. & Reinking. A perfect stage is unknown.

Biology: Many aspects of the disease cycle are not yet known. Infections appear on branches formed during the current season, especially the terminal leader. Mycelium

Courtesy of T. D. Waldrop, USDA Forest Service, Clemson, S. C.

Known range of pitch canker.

usually girdles the branches and upper parts of the main stem. In less susceptible hosts such as slash pine, the main stem is less frequently girdled, and these trees usually recover. Wounds appear to be required for infection. In Florida, *Ryacionia* spp. (tip moths) and *Pissodes nemorensis* (deodar weevil) have been associated with pitch canker. In California, insect associations appear to be evolving with the recent introduction of the pathogen. The dry twig and cone beetle, four twig beetles in the genus *Pityophthorus,* a cone beetle, three species of *Ips,* and the red turpentine beetle have been found carrying the pitch canker fungus.

Epidemiology: Pitch canker occurs sporadically throughout the South and is common on slash pine plantations and seed orchards in Florida. In 1974, extensive outbreaks of pitch canker occurred on large areas of 15- to 20-year-old planted slash pine in central Florida. Disease incidence reached 100% in some stands, and there was significant growth loss and mortality. These unique outbreaks may have been associated with overstocked stands and drought conditions during times when the populations of deodar weevils were large and the fungus was sporulating abundantly on small cankered branches. As we learn more about the developing infestations in California, it may become obvious that the epidemiology of this disease is regulated by insect behaviors. *Ips* beetle populations may increase in areas of infestations. These beetles may carry the fungus into uninfested brood material, thereby increasing the reservoir of inoculum, which may further increase disease incidence.

Diagnosis: Symptoms include the typical flagging of girdled branches, often the terminal leader; a sunken, perhaps inconspicuous canker; copius resin flow from the cankered area; and pitch-soaked wood in the cankered area. Trees may recover, with a crooked stem resulting when an uninfested lateral assumes dominance. During outbreaks, many cankers may occur on current season's branches, and entire crowns may die back. The sporodochia typical of the genus *Fusarium* may be abundant on small infected branches.

Control Strategy: In nurseries and seed orchards, fungicides and insecticides can reduce pitch-canker incidence. These chemicals, however, have not been demonstrated to be effective on ornamental trees or in forest stands and are not registered for use in preventing this disease. Minimizing damage during cone harvesting can reduce wounding, thereby reducing incidence. Infected branch tips may be pruned to remove the infection from a tree. Cut branches and trees should be destroyed to minimize brood material for bark beetle vectors of the fungus. In California a quarantine to restrict movement of bark-intact wood from infested counties could restrict the spread of pitch canker.

SELECTED REFERENCES

Correll, J. C., T. R. Gordon, A. H. McCain, J. W. Fox, C. S. Koehler, D. L. Wood, and M. E. Schultz. 1991. Pitch canker disease in California: Pathogencity, distribution, and canker development on Monterey pine *(Pinus radiata)*. *Pl. Dis.* 75:676–682.

Dwinell, L. D., J. B. Barrows-Broaddus, and E. G. Kuhlman. 1985. Pitch canker: a disease complex of southern pine. *Pl. Dis.* 69:270–276.

Fox, J. W., D. L. Wood, and C. S. Koehler. 1991. Engraver beetles (Scolytidae:*Ips* species) as vectors of the pitch canker fungus, *Fusarium subglutinans. Can. Ent.* 123:1355–1367.

DISEASE PROFILE

Larch Canker

Importance: Larch canker in Europe has caused extensive losses in some larch stands and only minor damage in others. The introduction of the fungus causing larch canker to the United States caused a great deal of concern for fear that it would spread to our native larch species and cause serious losses. The disease was found in 1935 and again in 1952 after all diseased trees had been removed in 1927. In 1980 the disease was found in New Brunswick and Nova Scotia on eastern larch and a year later in Maine. Age of oldest

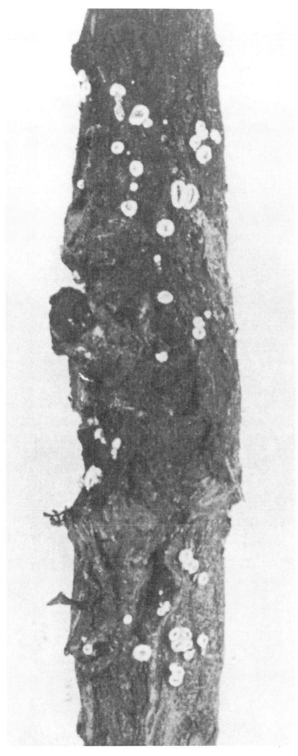

Canker with apothecia. From Hahn and Ayers (1936). Reprinted from *Journal of Forestry,* Vol. 34, No. 10, published by the Society of American Foresters, 5400 Grosvenor Lane, Bethesda, MD 20814-2198. Not for further reproduction.

cankers indicated that the disease had been present since at least 1958 in New Brunswick. The disease is not considered important in this country.

Suscepts: Larch canker has been found only on species of *Larix* and *Pseudolarix.*

Distribution: Larch canker is found almost everywhere that European larch is found. In the mountains of central Europe, the natural range of European larch, larch canker is not important. In northern Europe, however, since the middle of the nineteenth century, the larch canker has been responsible for failures of European larch plantations. Larch canker was introduced to the United States, specifically Massachusetts, in 1904 and 1907. It has been in Japan for at least 100 years.

Causal Agents: Larch canker is caused by *Trichoscyphella* (= *Dasyscypha) willkommii* (Helotiales).

Biology: Trichoscyphella willkommii enters its host through wounds, especially frost openings. Young trees less than 10 years of age (chiefly 2- to 5-year-old shoots) are more susceptible than older trees. The fungus causes cankers on which pycnidia are formed on cushions of white mycelium. Later apothecia are produced throughout the growing season. The apothecia are white with a short stalk. The ascospores are wind-disseminated.

Epidemiology: Susceptibility to larch canker apparently decreases with age. Weather can be a factor. Oceanic climate or cool and humid weather have been reported to increase disease incidence, but this is not always the case. Some reports suggest that suppressed trees are more susceptible. Mixed species stands appear to be more resistant.

Diagnosis: Cankers initially are elliptic or circular and sunken, with resin exudation. Callus forms around the canker unless the branch is girdled and killed and the callus layers are evident. In old cankers the wood is exposed, and the surrounding bark is dark in color.

Control Strategy: In places where the disease is not causing serious losses, control is not necessary. Using resistant species or more suitable seed sources of European larch will result in insignificant losses. Planting larch on more suitable sites is of benefit. In spite of the limited losses to larch canker, it is important to keep the fungus out of regions where it does not occur now. Once the canker is introduced, locating all the infections and eliminating the disease as evidenced by the need for repeated eradications from the United States is difficult.

SELECTED REFERENCES

Hahn, G. G., and T. T. Ayers. 1936. The European larch canker and its relation to certain other cankers of conifers in the United States. *J. For.* 34:898–908.

McComb, A. L. 1955. The European larch: Its races, site requirements and characteristics. *For. Sci.* 1:298–318.

Ostaff, D. P. 1985. Age distribution of European larch canker in New Brunswick. *Pl. Dis.* 69:796–798.

Cytospora Canker

Importance: Cytospora canker injures some trees growing outside their normal range or under unfavorable conditions. During droughts, shade and windbreak trees may be affected. Severely pruned trees, especially topped trees, and cuttings in storage are particularly susceptible. Losses in propagating beds have been as high as 75% or more.

Suscepts: Cytospora canker has been found on more than 70 species of hardwood trees and shrubs as well as some conifers. It is important primarily on fruit trees, poplars, willows, and Colorado blue spruce.

Distribution: Cytospora cankers occur worldwide.

Causal Agents: On hardwoods, Cytospora canker is caused by *Cytospora chrysosperma* (Coelomycetes); the perfect stage is *Valsa sordida* (Diaporthales), which is not common. Pycnidia have many connecting locules and one pore through which spores are exuded. On spruce, Cytospora canker is caused by *C. kunzei. Cytospora abietis* causes a canker of true firs in the western United States.

Biology: Species of *Cytospora* are probably omnipresent. On predisposed hosts, the fungus enters through wounds or dying tissues and grows in the bark and outer sapwood. The fungus may produce windblown ascospores, but more commonly, hyaline, single-celled spores are produced in gelatinous tendrils during and after rains. These spores are spread by rain, insects, and people.

Epidemiology: The development of the disease is related to drought or other stress. A

Lower branch dieback due to Cytospora.

relationship between decreased bark moisture content and susceptibility to fungi that cause cankers has been suggested. Dwarf mistletoe predisposes true firs to infection by *C. abietis*.

Diagnosis: When *Cytospora* parasitizes a living branch, the bark in the invaded area is killed, and callus tissue frequently forms at the edge of the invaded region. The canker appears as a dead, sunken area, sometimes discolored, and may extend 1 m or more along a branch or the trunk of a tree. A canker frequently starts at a wound or dead branch stub. Pycnidia are often very abundant; forming in dead bark to protrude as small, dark pimples. During moist weather, threads of hyaline spores are exuded in an orangish, gelatinous matrix. The threads often dry out, harden, and remain on the canker for weeks, providing a good sign by which the fungus can be recognized. This does not necessarily mean that *Cytospora* has killed the branch. Young Colorado blue spruce planted outside their range may be free of cankers for their first 5–15 years. Branches at the base of the tree are killed first. Although trees are rarely killed, branch mortality throughout the crown makes the trees unsightly and leads to their removal. Excessive pitch exudation on dying branches is a good indication that Cytospora canker is present.

Control Strategy: Selecting locally grown plant materials from local seed sources will ensure use of materials adapted to the site. Planting these trees in good soil on a proper site is perhaps the best method to avoid Cytospora cankers. Plants should be provided with adequate water and fertilizer to keep them vigorous and resistant to attack. Avoid excessive pruning. Cuttings for propagation should be stored no warmer than 2°C to prevent the fungus from developing during storage. Blue spruce should be planted on protected sites. Avoid southwest-facing slopes and sandy or gravelly soils.

SELECTED REFERENCES

Bloomberg, W. V. 1962. Cytospora canker of poplars: factors influencing the development of the disease. *Can. J. Bot.* 40:1271.

Scharpf, R. F. 1983. Temperature-influenced growth and pathogenicity of *Cytospora abietis* on white fir. *Pl. Dis.* 67:137–139.

Schoeneweiss, D. F. 1983. Drought predisposition to Cytospora canker in blue spruce. *Pl. Dis.* 67:383–385.

DISEASE PROFILE

Basal Canker of White Pine

Importance: Basal canker of white pine occurs on nearly 400,000 ha of the Tug Hill plateau in north-central New York. Trees 8- to 10-years-old are girdled and killed in less than 1 year, creating openings in the young plantations.

Suscepts: Eastern white pine is the host.

Distribution: This unique site/insect/disease complex is restricted to the Tug Hill plateau in New York.

Multiple cankers. From Houston (1969). Reprinted from *Forest Science,* Vol. 15, No. 1, Published by the Society of American Foresters, 5400 Grosvenor Lane, Bethesda, MD 20814-2198. Not for further reproduction.

Causal Agents: *Pragmopara pithya* (Helotiales) causes perennial cankers and is isolated from about 50% of the basal cankers.

Biology: *Pragmopara pithya* enters the host through injuries made by the black meadow ant and by snow and ice. The ant chews the lenticels of the young stems and then sprays the wounds with formic acid, causing a small lesion. This action is believed to be a survival mechanism designed to kill vegetation that eventually would shade and cool the mound, preventing egg hatch. Multiple attacks may girdle small stems. Initially the lesions are free of fungi, but eventually fungi, including *P. pithya* and *Fusarium* sp., invade the wounds. These fungi also invade branch axil-wounds created when low branches are weighted down by snow.

Epidemiology: The Tug Hill plateau rises from Lake Ontario in the west to the Black River valley. Very heavy snows occur, and the short growing seasons contributed to late settlement and early abandonment of fields that were subsequently planted to pine.

Diagnosis: Basal cankers or cankers associated with branch-axil wounds on white pine in plantations on the Tug Hill plateau. At first, this canker was thought to be blister rust because the region is a high-hazard zone for that disease.

Control Strategy: Suggested control measures include chemical control of ants where they exist in large numbers. Basal canker is most prevalent on the sheltered side of north–

south hedgerows and stone piles, and in depressions. Species other than white pine should be planted on these sites.

SELECTED REFERENCES

Houston, D. R. 1969. Basal canker of white pine. *For. Sci.* 15:66–83.

DISEASE PROFILE

Sirococcus Shoot Blight

Importance: Sirococcus shoot blight has caused serious losses in some years in some locations. In the Lake states, not only were overstory trees affected, but the fungus was also infecting large percentages of the reproduction under these trees, especially in red pine stands. Seedlings die soon after they are infected, while large infected trees survive for several years and may never be killed. The disease has caused extensive damage on western species but, potentially, may be more destructive in nurseries than in forest stands where a degree of thinning is desirable. In Alaska on western hemlock, 19% of the branches have been killed.

Branch dieback. From O'Brien (1972).

Suscepts: Hosts other than red pine include western hemlock; Jeffrey, Coulter, ponderosa, lodgepole, and jack pines; Douglas-fir; white fir; and species of spruce.

Distribution: Sirococcus blight occurs in the Lake states and in New York and eastern Canada. It has also been found from California to Alaska. In Europe the disease occurs on spruce, fir, and pines.

Causal Agents: Sirococcus shoot blight is caused by *Sirococcus conigenus* (Coelomycetes).

Biology: Infection of red pine occurs in May and June from conidia produced in pycnidia, which develop within a few weeks after the infected shoots die in July and August. The spores are disseminated probably by rain and insects. Infection occurs in May and June, and the severity may vary with region and weather. Adventitious buds form at the base of the dead shoots and these buds develop the following year and often are infected and yellow. Repeated infections result in deformed or dead trees.

Epidemiology: Sirococcus conigenus conidia can be splashed or carried by rain from infected overstory trees to understory trees. How the fungus infects overstory trees is not known. The disease appears to reach epidemic proportions and then subside, suggesting that its occurrence is related to some special weather condition. The fungus is favored by mild temperatures, wet weather, and low light intensities.

Diagnosis: The current shoots are invaded and killed by the fungus. Tiny pitch droplets occur on current-year shoots and needles in the early stage of infection. Small, black pycnidia, which are light brown or tan when first formed, develop on the dead shoots and under the sheath of dead needles.

Control Strategy: Eradication of infected overstory trees reduces the amount of infection in understory trees. Removal of infected shoots also reduces disease incidence, but removal would be a feasible approach to control only in recreation sites and valuable plantations. In nurseries and on Christmas trees, fungicides should be applied once every 2 weeks during the infection period. The disease could be avoided by growing containerized stock inside with sufficient ventilation and supplemental lighting. In natural stands, control may not be warranted. Sirococcus blight on western hemlock reduces height growth, but, if these stands are thinned, the poor trees can be removed and subsequently conditions are less favorable for the disease.

SELECTED REFERENCES

O'Brien, J. T. 1972. *Distribution of "Deerskin Droop" in the Lake States (1971).* USDA For. Serv., FPM Rept. No. S-72–2. 3 pp.

Wicker, E. F., T. H. Laurent, and S. Iraelson. 1978. *Sirococcus Shoot Blight Damage to Western Hemlock Regeneration at Thomas Bay, Alaska.* USDA For. Serv., Res. Pub. INT-198. 11 pp.

Wall, R. E., and L. P. Magasi. 1976. Environmental factors affect *Sirococcus* shoot blight of black spruce. *Can. J. For. Res.* 6:448–452.

DISEASE PROFILE

Diplodia Canker

Importance: Diplodia canker, while seldom a problem in natural pine forests, causes considerable damage to pine windbreaks and landscapes. Pine outside their normal range are especially subject to damage, such as red pine in Missouri.

Suscepts: This disease is most common on Austrian pine but occurs on Scots, red, ponderosa, mugo, and Monterey pines and others.

Distribution: In the United States, Diplodia canker is of greatest concern in the central part of the country and to some extent in the eastern states. It is a problem as well in California, Hawaii, New Zealand, Australia, and South Africa.

Causal Agents: Diplodia canker is caused by *Sphaeropsis sapinea (Diplodia pinea)* (Coelomycetes).

Biology and Epidemiology: The fungus can infect unwounded new shoots and enter through wounds in current-year and older tissues. Hail and insect wounds serve as infection courts as do pruning wounds. Trees under stress such as drought may be more susceptible. Cones are susceptible their second year. Pycnidia form on needles, sheaths, scales of cones, and bark. Pycnidia may develop abundantly with high rainfall in late summer but normally occur more commonly in spring. Conidia are dispersed from March through October. Infection occurs following rainy periods. High relative humidity is required for infection. New shoots are primarily susceptible for about 2 weeks following bud opening. Second-year cones are usually infected in late May or early June.

Diagnosis: Stunted shoots with short, brown needles or dead shoots sometimes occur

Branch tip dieback. From Peterson and Johnson (1986).

throughout the crown but often are more evident in the lower crown. If not controlled, repeated infections result in reduced growth, deformation, and finally death of the tree. The small black pycnidia form on needles, needle sheaths, and scales of second-year cones. Cankers are often present on the main stem of trees infected in previous years.

Control Strategy: Protective fungicides should be applied at 1-week intervals with the first application at bud break. However, this will not prevent infection of the 1-year-old cones. Pine should not be sheared for at least 2 weeks after bud break. Seedlings should not be planted near infected trees. Susceptible windbreak trees should be removed from around pine nurseries. In nurseries, three applications of systemic fungicide applied at 2-week intervals when buds form in July through August resulted in little infection of 1–0 seedlings. In 2–0 stock with moderate to severe infection, fungicide needs to be applied at 2-week intervals during bud break and early shoot development.

SELECTED REFERENCES

Peterson, G. W. 1977. Infection, epidemiology, and control of Diplodia blight of Austrian, ponderosa and Scots pines. *Phytopathology* 67:511–514.

Peterson, G. W., and D. W. Johnson. 1986. Diplodia blight of pines, pp. 128–129. In J. W. Riffle and G. W. Peterson, (eds.), *Diseases of Trees in the Great Plains.* USDA For. Serv., Gen. Tech. Rept. RM-129 149 pp.

DISEASE PROFILE

Atropellis Canker

Importance: Atropellis canker is generally only of scattered occurrence, but occasionally reaches epidemic proportions in lodgepole and ponderosa pines. Trunk deformities and early mortality comprise the major losses in the western United States. Cankered stems are difficult to debark. In the eastern United States, economic losses also occur in Christmas tree plantations.

Suscepts: Many native and introduced hard pines, especially lodgepole, ponderosa, loblolly, shortleaf, Virginia, and Austrian pines.

Distribution: It can be found across North America wherever hosts are growing.

Causal Agents: Atropellis piniphila (Helotiales) is the most important species in western North America. *Atropellis pinicola* is most prevalent on the Pacific coast. *Atropellis apiculata* is known only in North Carolina and Virginia. *Atropellis tingens* is the most common species in eastern North America.

Biology: Infection starts at a branch axil. Colonization proceeds slowly, forming elongated cankers. Resinosus occurs at the canker margin as long as it is active. A wedge-shaped bluish black stain forms under the canker face. Pycnidial stromata, and then black apothecia, form on killed bark each year as the canker enlarges.

Epidemiology: Epidemics have been documented for *A. piniphila* and *A. tingens,* the former in stressed stands of lodgepole and ponderosa pine. Slower-growing trees are generally

Atropellis canker.

more damaged by cankers. The role of the conidia is unknown. Ascospores are dispersed by wind, mainly in summer.

Diagnosis: Cankers appear as elongated depressions covered with bark. Multiple cankers may be present, and these may be connected by the characteristic discolored wood. Apothecia are black with a brown interior, measuring 2–4 mm across when moist. They are present year around. Species are separated based on ascospore morphology.

Control Strategy: Because dense stands are at higher risk to infection, any silvicultural treatment that encourages trees to grow normally should reduce losses. Apparently some lodgepole pines are genetically resistant. In Christmas trees infected branches should be removed and destroyed. Trees with stem cankers may be deformed but may be harvested before the stem is girdled.

SELECTED REFERENCES

Hopkins, J. C. 1963. Atropellis canker of lodgepole pine: Etiology, symptoms, and canker development rates. *Can. J. Bot.* 41:1535–1545.

Sinclair, W. A., H. H. Lyon, and W. T. Johnson. 1987. *Diseases of Trees and Shrubs.* Comstock Publishing Assoc., Ithaca, N.Y. 574 pp.

17

STEM PATHOLOGY: SYSTEMIC DISEASES

17.1 VASCULAR WILTS AND STAINS

Vascular wilt diseases are among the most devastating of tree diseases. As the name implies, they involve the vascular system of the tree, and the primary symptom produced is wilting. During a sunny day a healthy tree may transpire several hundred liters of water. An important function of this transpirational water is the cooling effect it has as it evaporates from the leaf surfaces. The transpirational water within the tree is under extreme tension during a sunny day, and any break in this column of water causes formation of air embolisms that effectively stop the flow of transpirational water in the particular vessels affected. A healthy tree reacts by bypassing these vessels, and there is no net loss of transpirational flow. Infection by a vascular wilt pathogen initiates a series of events that quickly blocks all or most of the transpirational water flow, and the tree is unable to bypass this blockage. The result is usually a dramatic rapid wilt. Two major vascular wilt diseases are the Dutch elm disease and oak wilt. The former was introduced and has devastated native elms. The latter is a native disease that could potentially be just as devastating but lacks an efficient vector. A major problem facing the vascular wilt pathogens is that of escaping the confines of the diseased tree. This problem has led to some interesting pathogen/vector relationships, including not only insects but humans as well.

DISEASE PROFILE

Dutch Elm Disease

Importance: This disease was responsible for eliminating the majority of elm trees in Europe and North America within a few decades after its accidental introduction. It was especially devastating on elm shade trees. Elm species were the predominant shade trees in many urban areas on both continents. Because of their growth form and tolerance to adverse conditions, some elm species were particularly suited for planting along canals, roads, and boulevards. There are no replacement species that have quite the same desirable traits.

Suscepts: All species of elm native to North America and most European species are susceptible, although there are considerable varietal differences in susceptibility even within the very susceptible species. Siberian elm, often used in windbreaks and shelterbelt plantings, and certain other Asiatic species are resistant, but they are not immune.

Distribution: First found in the Netherlands in 1919, the disease spread rapidly across European countries. The disease is probably native to Asia. It was first identified in North America in 1930. The disease is now in southeastern Canada and in every state except possibly Arizona and Florida. In 1967 a more virulent form of the pathogen from North America was introduced into England and subsequently caused a second epidemic to move across Europe.

Causal Agents: Dutch elm disease is caused by the fungus *Ophiostoma (Ceratocystis) ulmi* (Ophiostomatales). The smaller European elm bark beetle and the native elm bark beetle are important vectors in North America. In Europe the larger European elm bark beetle is an important vector.

Biology: The pathogen is spread either by insects or through root grafts. Once inside the

Removing diseased elm. From Anonymous (1967).

host, the fungus grows and is carried through the cells of the outer sapwood. The tree reacts by producing gums and tyloses, which in turn plug the vascular system of the tree. A toxin, cerato-ulmin, is produced by the fungus and may play a role in early stages of pathogenesis. Susceptibility to insect-vectored spores is greatest in spring when the springwood vessels are being formed. After the tree dies, the pathogen colonizes the remaining outer sapwood and inner bark. Special asexual sporophores called synnemata are formed in the walls of insect galleries, and the spores are disseminated by emerging bark beetles. Sexual spores, formed in perithecia, may also be produced if both mating types of the fungus are present.

Epidemiology: The success of this pathogen depends to some extent on the aggressiveness of its insect vectors. Spread through root grafting is also important but only between closely spaced trees. A mycovirus, also known as a double-stranded RNA, may have been responsible for the loss of virulence during the first European epidemic.

Diagnosis: Wilt symptoms are first evident in June and become most prominent in July and August as soil moisture becomes limiting. The leaves turn yellow, curl, dry out, and fall. When the infection has been initiated by an insect, a branch may exhibit these symptoms forming a "flag." Then the rest of the tree wilts. A brownish ring of discoloration forms in the outermost band of sapwood.

Control Strategy: Culturing may be needed to distinguish this disease from elm yellows or native elm wilt; however, any tree with a progressive dieback, regardless of cause, provides suitable brood material for elm bark beetles, and such trees must be dealt with in the control program. Sanitation or eradication of dead and dying elm wood is a necessity. Prevention of spread through root grafts is especially important for trees growing close together. Insecticides are no longer recommended because of danger to the environment and nontarget organisms. Systemic fungicides may be effective if properly applied. Resistant elms are now available. All these strategies are most effective if they are incorporated in an integrated approach or as part of a community control program.

SELECTED REFERENCES

Anonymous. 1967. *Death of a Giant.* USDA Agri. Res. Serv., Picture No. 202. 4 pp.
Stipes, R. J., and R. J. Campana. 1981. *Compendium of Elm Diseases.* The American Phytopathological Society, St. Paul, MN. 96 pp.

17.2   DUTCH ELM DISEASE

Importance

Dutch elm disease (DED) is of greatest importance on shade trees, the value of which cannot be estimated easily. In some areas of Europe, 60–70% of the trees were killed. Considering the prevalence of elms in the cities of the United States, losses from this disease have been very great (Fig. 17.1). In 13 years in Connecticut, 25% of the elms were infected or killed. As of 1963, Detroit, Michigan, had lost 10,733 trees valued at approximately $200 each or a total loss for one city of over $2 million. In 1964 the state of Michigan spent an estimated $8,750,000 on control measures and lost trees with an aesthetic value placed at $30,000,000. The province of Quebec has lost more than 600,000 trees; assuming a very moderate value of $100 for their removal, the cost of eliminating these dead trees would be $60,000,000. Over 90% of the elms in Champaign-Urbana and Bloomington, Illinois, succumbed to Dutch elm disease

FIGURE 17.1   The University of Illinois boardwalk in Champaign-Urbana: (above) lined with stately American elms and (below) after the area had been replanted with other species of deciduous trees. From Carter (1967).

within just 10 years. Champaign-Urbana had approximately 14,000 elms in 1944 and only 40 survived in 1972. In 1974 it was estimated that the state of Illinois had lost over 90% of its elms. Similar disasters can occur in several more states. With the introduction of a more virulent strain of the fungus from Canada to England, the losses in England by 1974 had reached 4.5 million of the total 23 million elms.

In contrast to these extensive losses, some cities have greatly reduced the losses to

DED, avoided or delayed the cost of removing large numbers of elms, and prolonged the life and utility of their elms. For example, Winnipeg, Manitoba, Canada, has an aggressive program to protect its elms. Since 1975, the annual loss of elms is less than 2% of the population per year, a figure that includes elms removed for reasons other than DED. Managing DED is not limited by knowledge of effective control measures; it is constrained by a lack of commitment to implement and maintain effective disease management and urban forestry programs.

Suscepts

All species of elm native to North America and most European species are susceptible, although there are considerable varietal differences in susceptibility even within the very susceptible species. Some of the European selections are somewhat resistant. Siberian elm, introduced to the United States as a fast-growing tree for use in windbreaks and shelterbelt plantings, and certain other Asiatic species are resistant, but not immune.

Distribution

Dutch elm disease, which was first found in the Netherlands in 1919, is probably native to either Europe or Asia. Dying elms had been observed in France and other countries prior to 1919. The disease spread rapidly through Europe, and by 1934 it was found in practically all European countries and in the British Isles. In 1930 a few diseased trees were found in Ohio, and the fungus was identified by Dr. Curtis May. The disease had been introduced into the United States from Europe in logs that contained both the fungus and European elm bark beetles (these same beetles were in Massachusetts as early as 1909). The beetles escaped at different places along the route of travel from New York to the veneer factories in Ohio and Indiana, hence the sporadic infections in these areas. The beetles and fungus are still gradually spreading from this beginning. By 1974 Dutch elm disease had spread through southeastern Canada, south to Georgia, Mississippi (1968), Alabama (1969), Texas (1970), and west to Colorado, Idaho, and Oregon (1974). Dutch elm disease is in California, Montana, Utah, and Washington, as well. This leaves only Florida, Louisiana, New Mexico, Arizona, and Nevada (Fig. 11.4). The European elm bark beetle is found in every state except Arizona and Florida.

Cause

Dutch elm disease is caused by the Ascomycete *Ophiostoma ulmi* (Buisman) Nannfeldt. This fungus was formerly known as *Ceratocystis ulmi* (Buisman) C. Moreau. The asexual stage producing synnemata (or coremia) is *Pesotum ulmi* (Schwartz) Crane & Schoknecht. A yeastlike stage once known as the *Cephalosporium* stage is now called the *Sporothrix* stage.

Biology

Ophiostoma ulmi enters the tree either through wounds made by bark beetles or moves from infected to healthy trees through grafted roots. Apparently for infection to occur, only relatively few spores need to be introduced into the tree. When 4- or 8-year-old trees have been inoculated artificially, 100 spores per tree resulted in as severe wilt as in trees inoculated with more spores. Once inside the host, the fungus grows and is carried through the cells of the outer sapwood. The tree reacts to the fungus by producing gums and tyloses, which in turn plug the vascular system of the tree. Fungal structures may also contribute to the plugging. The tree wilts because of reduced water available to the foliage. A toxin may also be involved in wilting. The degree of foliar symptoms depends on the location of the vascular plugging. Blockage in the main stem causes less severe symptoms than blockage in the small branches.

The susceptibility of elms is high in spring, somewhat less in July, and then intermediate in August. This general statement varies with geographic location. In Quebec, Canada, susceptibility was high from May 15 to late July, and maximum susceptibility was from the end of May to late June. Apparently, season of susceptibility is related in part to the production of the large xylem vessels. Formation begins in the top of the tree and progresses to the base of the main stem. Large vessels start to form in the twigs early in May at bud break. By first leaf expansion near the end of May, large vessels are contiguous from twigs to the base of main stem. It has been suggested that not only are the large springwood vessels needed for initial entry of the fungus, but elms with larger, more contiguous vessels are more susceptible.

Weather conditions may influence the ability of *O. ulmi* to infect elms. Disease incidence presumably has been greater in years when soil moisture was high. Although elms under moisture stress are less susceptible, trees in poor health may attract breeding bark beetles, which introduce *O. ulmi* during colonization. Ultimately, bark beetles may introduce the fungus more frequently into stressed trees during breeding attempts than while feeding in vigorous trees.

Once established in a host, the fungus grows or is moved through the cells of the outer sapwood. The fungus has moved down twigs of young elms at 1.4–2.5 cm/day. At this rate, weeks and months might pass before external symptoms appear. Extensive defoliation and death occur in the following year. Many twig inoculations remain localized, and less than 6% of the twigs have additional symptoms the following year. Small conidia are produced on the mycelium or by budding in the vessels and are carried to other parts of the tree. After the fungus is well established, it may produce another kind of conidium beneath the bark, especially in insect galleries. The special fungus structures that produce these conidia in the tree and also form readily in culture are called synnemata or coremia (Fig. 17.2). *Synnemata* are black stalked structures 1 mm high at the top of which the spores are borne in a gelatinous material adapted to insect dissemination. These spores are carried to other trees on emerging bark beetles. If both A and B mating types of *O. ulmi* are present, ascospores are produced in perithecia with long black necks from which the spores ooze in a sticky matrix. Only one mating type needs to be present to cause disease. All three types of spores produced by this fungus are adapted to transport by insects.

The elm bark beetles are the primary insect vectors. In the United States, they are the smaller European elm bark beetle *(Scolytus multistriatus)* and the native elm bark beetle *(Hylurgopinus rufipes)* (Fig. 17.3). In Europe *Scolytus scolytus,* the larger European elm bark beetle, is an important vector. Occasionally, other insects have been suspected as vectors. In areas where both beetles occur, *S. multistriatus* is usually more important. This insect attacks brood material more aggressively and will limit the number of *H. rufipes.* It feeds in the thinner-barked portions of trees and is more

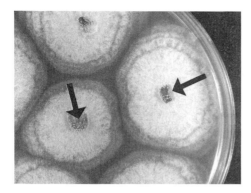

FIGURE 17.2 (Left) Culture of *Ophiostoma ulmi* showing synnemata (arrows). (Right) Closeup of perithecia of *Ophiostoma ulmi*. Courtesy of D. W. French, University of Minnesota, St. Paul.

likely to introduce the fungus. However, *H. rufipes* occurs over the entire range of elm, while *S. multistriatus* is less able to survive cold winters; thus its northward movement is limited.

The fungus produces conidia in the larval tunnels and pupal chambers of these beetles, and when the beetles emerge, they carry the spores on their bodies. The adult beetles usually travel less than 100 m before they feed on the small twigs of healthy elms and inoculate them. Beetles prefer to feed in trees having twigs with more acute angles; also the fungus is more likely to be introduced when beetles feed on the side of a branch rather than directly in a branch axil.

Usually there are at least two broods of smaller European elm bark beetles each year. The overwintering brood emerges in May or June and reportedly is responsible for most of the spread. The second brood emerges in August. Under optimal conditions, the bark beetles can lay eggs and the larvae emerge in a minimum of 28 days. In mild climates or in warm years a third or even fourth generation may occur. After feeding, the beetles may travel more than 3 km to find a breeding place in dead or dying elms. During breeding, the elm bark beetles may introduce *O. ulmi* into elm material that previously did not have DED. Any suitable breeding sites may be colonized by *O. ulmi*, regardless of the cause of the elm's demise, and the beetles would emerge carrying the fungus. Although we usually think of dead or dying trees being attacked by breeding elm bark beetles, elms stressed by pruning or drought have become diseased as a result of elm bark beetle breeding attempts. In Minnesota a large number of elms were inoculated in September and October when the European beetles attempted to breed in living trees that presumably were stressed by a severe drought.

The native elm bark beetle differs from the European beetle in several respects. The native beetle has 1.5 generations per season and thus overwinters as newly emerged adults or as larvae. The adults emerging in late summer and fall enter the base of healthy trees where they survive the winter protected by the thick bark. These beetles emerge in April, about a month earlier in the spring than adults of *S. multistriatus*. Native elm bark beetles feed on larger living branches and then search out dead and dying elms for breeding. The native elm bark beetles may also introduce *O. ulmi* when they breed. Thus the insects leaving this brood material will carry *O. ulmi*. The larvae

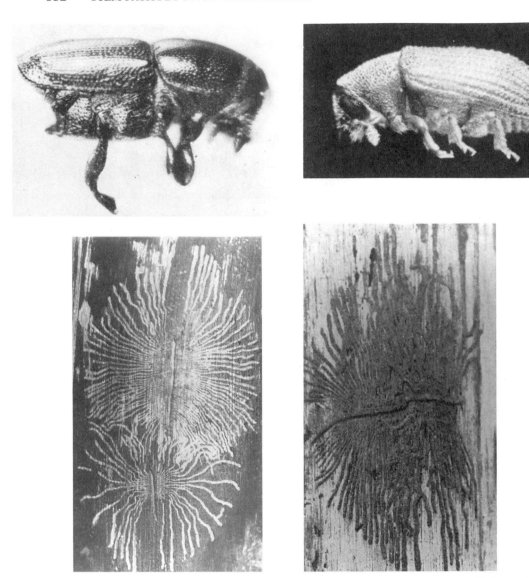

FIGURE 17.3   (Left) Adult of the smaller European elm bark beetle, and egg galleries and larval mines. From Carter (1967). (Right) Adult of the native elm bark beetle, and egg galleries and larval mines. From Whitten (1966).

FIGURE 17.4   Roots of two elms grafted together. Bark is removed to show vascular discolorations. From McKenzie and Becker (1937).

from these galleries emerge in the summer and in turn lay eggs, which overwinter as larvae.

In addition to spread by insects, the fungus can pass through the common vascular system of roots, which have become fused (Fig. 17.4). As many as 35–80% of the infections have occurred by this means.

Tyloses (outgrowths from living cells adjacent to vessels) develop in the large vessels of the current year's growth in infected elms. Large accumulations of tannin along with the tyloses account for the discoloration that can be seen in cross sections of branches. The fungus can penetrate through pits between vessels and can penetrate cell walls as well. In advanced infections, the middle lamella is broken down by the fungus. Lignin and pectic materials plug vessels, suggesting that *O. ulmi* possesses pectolytic enzymes. Sugar concentrations in the sap of wilting trees were up to seven times those in the sap of healthy elms. The concentration of nitrogenous materials was several times greater in diseased than in healthy sap after inoculation during spring when elms were susceptible, but it was lower after inoculations in late July. As a result of the presence of the fungus, which may be widespread in the tree, polyphenols are formed, oxidized, and accumulated in the xylem elements. A disease diagram outlining the disease is shown in Fig. 17.5.

Epidemiology

The incidence of DED is related to the populations of beetles, the weather, and, to some degree, the distribution of the host trees and virulence of the fungus. The usual pattern for DED after its introduction to a new area is to start slowly with only a few trees becoming infected. Dutch elm disease was first found in St. Paul, Minnesota, in 1961, and in the next 7 years only 30 cases of the disease were reported. After this

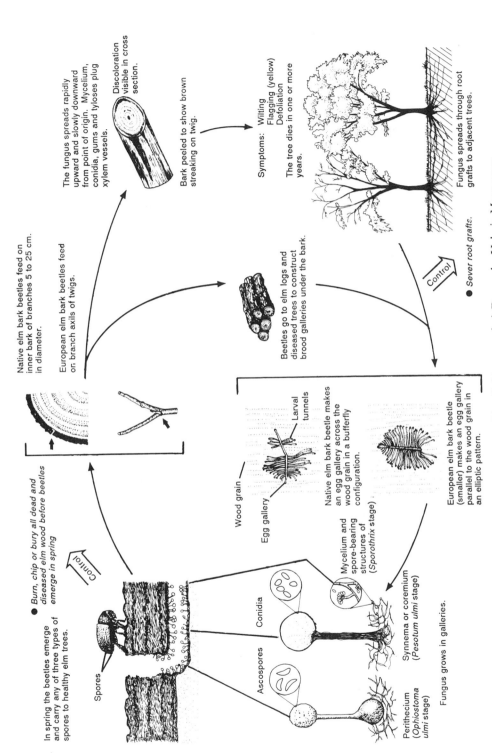

The fungus spreads rapidly upward and slowly downward from point of origin. Mycelium, conidia, gums and tyloses plug xylem vessels.

Discoloration visible in cross section.

Bark peeled to show brown streaking on twig.

Symptoms: Wilting
Flagging (yellow)
Defoliation

The tree dies in one or more years.

Fungus spreads through root grafts to adjacent trees.

Control

• Sever root grafts.

Native elm bark beetles feed on inner bark of branches 5 to 25 cm. in diameter.

European elm bark beetles feed on branch axils of twigs.

Beetles go to elm logs and diseased trees to construct brood galleries under the bark.

• Burn, chip or bury all dead and diseased elm wood before beetles emerge in spring

Control

In spring the beetles emerge and carry any of three types of spores to healthy elm trees.

Spores

Wood grain

Egg gallery

Larval tunnels

Native elm bark beetle makes an egg gallery across the wood grain in a butterfly configuration.

European elm bark beetle (smaller) makes an egg gallery parallel to the wood grain in an elliptic pattern.

Mycelium and spore-bearing structures of (Sporothrix stage).

Conidia

Ascospores

Synnema or coremium (Pesotum ulmi stage)

Fungus grows in galleries.

Perithecium (Ophiostoma ulmi stage)

FIGURE 17.5 Disease diagram of Dutch elm disease caused by *Ophiostoma ulmi*. Drawn by Valerie Mortensen.

initial slow pace, the inoculum level increased more rapidly and numbers of diseased trees increased greatly. In 1977 St. Paul lost over 50,000 elms. The rapid increase in the numbers of trees lost can be attributed to an increase in the number of bark beetles because of more dead elms and an abundance of elm firewood. Unless the sanitation program eliminates at least 90% of the beetle-infested material, the epidemic will continue.

Weather can affect the survival of elm bark beetles and thus the incidence of disease. Cool weather during the growing season slows beetle development. On the other hand, hot, dry weather not only favors the development of beetles but also stresses living trees. Stress may cause trees to be more attractive to breeding beetles, which can result in increased disease incidence the following season. Severe winter weather can reduce the beetle population. Mortality of bark beetles above the snow line is greatly increased when temperatures remain below $-34°C$ for 48 hours.

Midwestern states suffered greater loss of trees than eastern states because of the extensive use of elms for shade and boulevard trees. In many communities, elm was the only species used and it was planted at a very close spacing, sometimes as little as 4.6 m. This spacing provided maximum opportunities for DED: an abundant source of food and brood material for bark beetles and a high frequency of root grafting. Siberian elms used for windbreaks and fast-growing shade trees provided additional breeding sites for beetles. The Siberian elms, although resistant to DED, frequently suffered winter injury and were attacked by bark beetles carrying spores of *O. ulmi*.

There are two strains of *O. ulmi*, known as aggressive and nonaggressive. These are distinguished by their pathogenicity on elms with moderate resistance and by their morphology and growth rate in culture. Aggressive isolates, as their name suggests, are more pathogenic and in culture grow more rapidly and produce more aerial mycelium. There is some information to suggest that nonaggressive strains may contain double-stranded ribonucleic acid (dsRNA), characteristic of a fungal virus.

Diagnosis

Elms, especially if small and infected early in the season, may die the same season. Other trees may survive for a year and sometimes longer. Leaf symptoms are first evident in June and most prominent in July and August. The leaves turn yellow or dull green, curl, dry out, and fall. Usually an entire branch or section of the tree wilts and this is called flagging, followed by wilting of the rest of the tree (Fig. 17.6). In the outer sapwood, a continuous or disrupted brown ring forms, usually confined to a narrow band in the spring wood (Fig. 17.7). Trees infected with other wilt-inducing pathogens such as *Verticillium albo-atrum* or *Dothiorella ulmi* may have similar symptoms and can be distinguished from DED for certain only by culturing the fungus. Keep in mind, however, that if the elms provide suitable breeding habitat, bark beetles may attack the tree and introduce *O. ulmi*. For this reason, any elm that will provide suitable brood material is a hazard and should be eradicated regardless of the cause of death. Culturing to identify the pathogen is unnecessary if the elm is wilting progressively.

Control Strategy

Dutch elm disease has changed, or in those areas with abundant elms has the potential to change, the characteristics of the landscape. While disease management pro-

FIGURE 17.6   Advanced wilting symptoms of Dutch elm disease.

grams may be instituted to protect the elm resource, perhaps a more easily justified rationale for an effective program is financial. Several studies have shown that an effective DED management program is expensive, but much less so than the removal of the elms within a short period. Thus effective DED management programs not only save elms, but they also save money. And, by prolonging the life of the elms, communities have time to plant other species and let them grow to a useful size. These communities will have an effective shade tree population when the elms finally succumb.

Originally, attempts to control Dutch elm disease involved eradicating the infected trees. The disease was estimated to be moving at almost 8 km/yr, and, with the backing of federal money, infected trees were located and destroyed. This program was all but discontinued during World War II, and DED spread rapidly. Quarantine regulations were established to prohibit the import of more infested elm material from Europe, but this measure was too late. Quarantines established within the United States might have slowed movement of the disease, especially in areas beyond the range of native elms. We assume that introductions of Dutch elm disease to many western states were the result of man's activities. In Minnesota the disease appeared first in areas over 160 km from known infection centers. Man probably carried the fungus to these advanced locations, most likely in firewood containing contaminated beetles. In spite of the lesson learned by the United States in the 1930s, rock elm logs containing a more aggressive strain of the fungus were shipped to England, and in 1967 a new epidemic of Dutch elm disease spread across the English countryside from three ports of entry.

Once established, the most important methods of controlling DED are sanitation and disruption of common root systems. Other control methods often receive more

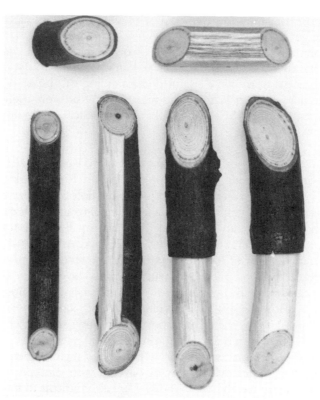

FIGURE 17.7   Vascular discolorations in current year's xylem of branch pieces taken from a tree infected with Dutch elm disease.

attention but are ineffective unless disease incidence is low. Often overlooked, however, is detection. Diseased trees must be located before management can take place. Many programs failed because of inadequate detection. They failed to realize that detection is the cornerstone of Dutch elm disease management.

The population of elms to be protected, and those in nearby surrounding areas, must be examined at least monthly. An inspection at 2-week intervals is better. Consider the following situation, and remember that elm bark beetles can emerge within 28 days. During a monthly detection survey, an infected elm is identified and has already been attacked by the smaller European elm bark beetle. The homeowner is notified to remove the tree. During the next inspection, 1 month later, the inspector finds the tree still standing. Where are the bark beetles? Off feeding in nearby trees? Or consider a tree that is diseased but is not detected until the next inspection. If the bark beetles enter the tree soon after the first inspection, there will be very little time after detection to remove the tree before the beetles emerge. Or, consider another example where a tree is suspected of being infected. Samples are taken for culturing but prove negative. During the next inspection, samples are taken again, and this time they are positive for DED. Where will the beetles be by the time the tree is removed? These examples may seem trivial, but they do occur, and they illustrate some of the real problems in dealing with DED. Frequent inspection provides more opportunity to achieve the goals of sanitation: removal of bark beetle brood material before the bark beetles emerge.

Ground surveys are usually done on foot or on a bicycle. Elms can be examined from a vehicle, but there should be a driver and an observer for safety. Flying above the community can also provide a useful understanding of disease hotspots. Aerial photography has not been satisfactory, although it might be more useful with further study. The best results used a helicopter and a 35mm camera to obtain oblique, true-color photography.

Sanitation. Sanitation is the most important control measure for DED. Its aim is to reduce or eliminate bark beetle vectors, not the fungus. If elm bark beetles do not carry the fungus between trees and root grafts are disrupted, *O. ulmi* cannot spread. Sanitation consists of destroying any dead and dying elms in which elm bark beetles can breed. Both the native elm bark beetle, and the smaller European bark beetle breed in any dead or dying elm material with intact bark. As many as 25,000 beetles have emerged from 929 cm^2 of bark, illustrating the tremendous reproductive potential of these insects. Any potential brood material should be eliminated. This includes elms with other diseases or those damaged by construction, lightning, and the like. Bark beetle brood material must be detected, removed, and disposed of before the bark beetles emerge. Perhaps the most important breeding sites for elm bark beetles, and often the most difficult to manage, are elm firewood piles.

Special effort should be made to eradicate elm brood material before bark beetles emerge in spring. Woodpiles, which may be covered with snow, offer the beetles a protected place to overwinter. These woodpiles are also difficult to find, but they are an important source of overwintering bark beetles. Firewood is easy to dispose of, as it can be burned or debarked. Both bark beetles breed between the bark and wood of dead or dying elms. Debarking will expose the larvae, and they dry out and die. Thus debarked elm material is not suitable brood material, and neither the bark nor the wood is hazardous. Where DED is epidemic, however, the extremely large volume of infected trees poses serious problems for disposal. In the past, these trees could be burned or buried, but now, with the advent of more intense concern about environmental quality, this is very costly. It is more reasonable to use this material, if possible. The tops and small-diametered material can be chipped and used for mulch. These chips will not support bark beetles. The larger material can be chipped for use in wood fiber products or sawn into a variety of items such as railroad ties or pallets. Finding an outlet for this material and keeping up with the large volumes of elm coming into the disposal site is not always possible. Perhaps more important is the high percentage of iron and other hardware ingrown in the wood. This is often difficult to detect and is damaging to saw blades and planer knives.

Where many elms are dying or if the trees cannot be used promptly, bark beetle colonization can be prevented in standing elms by applying potassium iodide (1.8 km/4.2 l of water) in an ax frill. Elms must be treated in an early stage of wilt to keep beetles out. Pressure injection at 330 km/6.45 cm^2 with cacodylic acid (hydroxydimethylarsine oxide) using 200–5,000 ml per tree of a solution of the acid (144 g/l) almost completely protected diseased elms from beetle invasion. These measures should be considered only for use in wild areas, because of the difficulty of identifying treated trees in a community.

Pentachlorophenol (1% in fuel oil) applied to elm logs has prevented bark beetle establishment and is a method that could be used after the trees have been felled. Chlorpyrifos is also labeled for use on elm wood at elm disposal sites. Insecticides may kill attacking insects or prevent attack until the elm wood dries out and is no longer suitable brood material. Once the bark falls off the material, the bark beetles will no longer attempt to breed there. These chemicals can be used only if they are registered for that use by the Environmental Protection Agency, and they must be used according to the label.

Root Grafts. Prevention of spread through root grafts is the other major component of successful control programs. In Illinois and presumably elsewhere, 93% of all infections occurring through common root systems occur within 9.1 m of previously infected trees. The fungus, however, has moved to trees as far as 18 m away, even though 15 m of this distance were under a paved street. Generally, however, trees more than 9–11 m away from infected trees are not likely to become infected in this manner.

Severing common root systems mechanically or chemically will prevent further spread of the fungus. Mechanical trenchers, or a vibratory plow, trenching 0.9–1.1 m deep will disrupt most root grafts. Mechanical trenching is effective, but in urban areas, underground utilities often restrict its use. SMDC (sodium N-methyl-dithiocarbamate), with tradenames Vapam or VPM, is a less expensive and less effective procedure for isolating an infected tree. Depending on the size of the tree, 50–200 ml of SMDC, mixed one part of the chemical to three parts of water, are poured into each of a series of holes 1.9 cm in diameter, 0.45 m deep and placed at 15-cm intervals approximately halfway between infected and healthy trees. SMDC should not be used within 2–3 m of healthy trees. Roots can graft under concrete and other paved surfaces, providing avenues for spread if they are not severed. To prevent this spread, root grafts must be severed completely around infected trees, which means drilling through sidewalks and streets. Holes may be made through sidewalk expansion joints or where the curb and gutter meets the pavement. Each hole must be closed immediately after the chemical has been added.

A second barrier can be placed outside of the first group of healthy trees in the event that the fungus has moved into one of these trees but is not yet evident. The infected tree should not be removed until 2 weeks after the chemical barrier has been established; the reason for this is to isolate the infected tree before the adjacent trees draw fungus spores through these root grafts. If the tree already has been attacked by breeding elm bark beetles and an SMDC barrier is to be used, it may be difficult to allow time for the chemical to work and yet remove the trees before the bark beetles emerge. Thus it is necessary to detect these trees as soon as possible; monthly surveys will not permit effective use of chemical root barriers.

If trees are detected early enough, they can be removed before the fungus enters the root system. This approach is perhaps the most effective way to prevent root graft spread. Frequent detection surveys provide time to remove trees before the fungus spreads into the root systems and into adjacent trees.

Pruning. Elms inoculated by beetles and in an early stage of disease development can be saved by pruning out the affected portion of the tree. Pruning will not remove infections that entered the tree through grafted roots. Pruning must be done as soon as possible after the fungus is detected. With a good detection program at frequent intervals, trees that might be saved by pruning can be identified and pruned to remove the fungus before it reaches the main stem of the tree. It has been suggested that at least 2–3 m of the branch removed should be free of vascular discoloration. However, to be certain of saving the tree, remove the branch at the main stem. Pruning in this manner has saved trees and saved the community the cost of tree removal.

Bark beetles are attracted to fresh wounds. In Michigan, trees trimmed during late July, August, or early September seemed more susceptible than trees pruned at other times. Nevertheless, trimming any time during the growing season increased disease incidence. The increased incidence is not of a magnitude that should discourage pruning to remove dead wood, hanging limbs or other hazards, or infected branches. Wounds made when bark beetles are active should be covered with a wound dressing to make the wounds less attractive to the insects. Some recommend that pruning equipment be sterilized between cuts. There is, however, little evidence that *O. ulmi* can be transmitted to healthy elms on pruning tools.

Systemic Fungicides. Systemic fungicides, such as Lignasan BLP (methyl-2-benzimidazolecarbamate phosphate) and Arbotect 20-S [2-(4-thiazolyl benzimidazole hypophosphite)], have shown some promise for controlling DED. Systemic fungicides that prove to have some beneficial effect should be considered only as an addition to the major control measures, namely sanitation and disruption of common root systems. The cost of injection prevents treatment of large portions of the elm population. Thus elm injection should be considered only for key landscape elms with little risk of becoming infected through root grafts. Injection is effective only for American elms with DBH greater than 25 cm. Smaller trees, or other species of elms, may be damaged by the treatment.

Lignasan BLP moves throughout the tree more quickly than Arbotect 20-S but is difficult to detect in branches after about a month, while Arbotect 20-S will remain in a tree for longer periods depending on the amount used. There is a great deal of confusion about the rates of chemicals to use. Early labels prescribed amounts that were not effective. Consequently, there are terms like "3X rate" or "3 year dose." The effective rate of Arbotect 20-S is 336 g/13 cm of tree diameter measured at breast height. Lignasan, because of its short persistence, is no longer recommended for use.

No systemic fungicide is effective unless properly applied to a tree that can be saved. Elms in advanced stages of wilt or those infected through common roots should not be treated. Therapeutic injections are not recommended when the symptoms appear before July because the fungus often affects more of the tree than we would expect from crown symptoms. Once *O. ulmi* is well established in the main stem, that tree cannot be saved. Elms in an early stage of the disease, preferably less than 5% of the crown affected, or trees not yet infected are suitable for injection.

The soil must be removed from the base of the tree to expose the root flares. Injecting into the root flares distributes the chemical throughout a tree with less damage. The injection points should be no more than 10–20 cm apart, providing the equivalent of two points of injection for each 2.5 cm of tree diameter 1.4 m above the ground. Trees should not be injected before they reach full leaf because the xylem elements that carry the fungicide are not yet complete from top to bottom of the tree. Longer protection is gained with injections later in the growing season.

Originally pressure systems providing 13.6 km/6.45 cm^2 or more were recommended, but 4.5 km/6.45 cm^2 is adequate. Arbotect 20-S will remain in the tree for two years and possibly into the third season following proper injection. The injection holes should not be plugged with wooden dowels, which are susceptible to decay. Injection damages the tree, causing extensive wood discoloration. Injection damage must be considered, especially if the tree is being retreated.

Many other chemicals have been evaluated to varying degrees, including a large number of antibiotics, but none of these chemicals has proven to be effective.

Insecticides. Insecticides have been used to prevent bark beetles from feeding on healthy elms. In the past DDT was used, but it cannot be used now because of its hazard to warm-blooded animals, especially birds. Methoxychlor is less persistent and must be applied in late winter or early spring prior to bud expansion. Success of methoxychlor depends on complete coverage of bark surfaces. Helicopter applications have been successful but require an experienced operator. As a general rule, the trees should be sprayed when the temperature is above freezing, and there is little wind. In many places the days with these weather conditions are few. Consequently, it may be necessary to add methanol to depress the freezing point so that the insecticide can be applied when the temperature is slightly below freezing.

Based on studies in five Illinois cities, losses to Dutch elm disease averaged 1.9% of the residual elm population with sanitation alone and 0.6% with sanitation and spraying. In two communities that discontinued their spray program, the incidence that had averaged 0.8% for 2 years when trees were sprayed increased to almost 3.0% in the second season after halting the spray program.

Chlorpyrifos (tradename Dursban), applied to the main stem of elms up to a height of about 3 m, has reduced populations of the native elm bark beetle adults overwintering under the thick bark. Thus in the northern parts of the United States and in Canada where this species is the important vector, Dursban applied to the elms in September, or in the spring prior to emergence, will kill beetles present in the bark. All elms, public and private, must be treated to reduce the bark beetle population. This treatment is not effective against larvae of either elm bark beetle, as they overwinter throughout dead and dying trees.

Systemic insecticides have been tried, especially a compound called Bidrin. Theoretically, these compounds are distributed throughout the tree and discourage or limit the feeding of the bark beetles before the fungus has been introduced. Bidrin and all the other systemic insecticides have not been satisfactory for several reasons. First, it is difficult to inject enough insecticide to limit bark beetle activity and yet not injure the tree. The treatment must be repeated every year and, around each point of injection, extensive discoloration develops. Bidrin, as is the case with other systemic insecticides, is very toxic to warm-blooded animals including man; thus these compounds must be handled with great care.

Bark beetles are attracted to elm trees and stimulated to feed by compounds produced by the tree. Pheromones produced by virgin female elm bark beetles are far more potent and draw both male and female beetles to the site. Although not as attractive as the virgin female beetles, similar compounds produced in the laboratory will attract bark beetles. If the attractant is exposed with a sticky substance, bark beetles can be trapped and kept from feeding on healthy elms. These traps have been placed around communities, usually on utility poles, to intercept beetles entering and also to draw bark beetles from within the community. As many as three lines of traps have been used to intercept a maximum number of beetles. The traps are excellent monitors of beetle activities but have had little impact on the incidence of DED.

An analogous concept is the use of trap trees to remove bark beetles. Bark beetles are encouraged to attack (for breeding) diseased or unwanted elms either naturally or by using pheromones, and then the trees are removed and destroyed before the larvae mature. While trap trees may reduce bark beetle populations where brood material is limiting, trap trees have not proven effective in areas with large populations of elms.

Resistant Species. Resistant elms are a possibility for the future. Resistant elm selections must also be resistant to pathogens other than *O. ulmi* and must be suited to the climate where they are planted. Another possibility, of course, is to substitute other tree species, but there is no real replacement for the American elm, and many alternative trees have their own disease problems.

Species, and individuals within a species, differ in their resistance to *O. ulmi*. Resistant elms have smaller-diametered vessels and fewer contiguous vessels, and the vessel groups are smaller than susceptible elms. This relationship is shown in Table 17.1 developed by McNabb et al. (1970).

In branches of resistant trees, the interval from inoculation to the appearance of symptoms was longer, fewer shoots wilted, and less extensive invasion occurred than in branches of susceptible trees. There are additional reasons for greater resistance of one tree over another, including more rapid secretion of gums. It has been noted that five times as many native elm bark beetles are attracted to American elm than rock elm in northern Wisconsin.

Among species present in North America, Siberian elms are more resistant than the native elms, but they lack the growth form of American elm, and in the northern United States they are subject to winter injury. Siberian elms with winter injury can harbor large numbers of bark beetles and thus add to the problem of controlling the disease. Siberian elms have 28 chromosomes, and American elms have 56. By treating Siberian elm with colchicine, the chromosomes split, and the cells do not divide. This process yields a Siberian elm with 56 chromosomes that can hybridize with the American elm. Thus it has been possible to produce some hybrids that incorporate the resistance of Siberian elm. Two varieties that have proven to be resistant to DED and are reasonably hardy are the Urban elm and the Soporo Autumn Gold elm. The Urban elm combines genes from *Ulmus carpinifolia* and *U. pumila*.

TABLE 17.1 Anatomical Characteristics of Resistant Elm Trees

Tree Species	Resistance Rating[a]	Vessel Diameter (rel. units)	Number of Contiguous Vessels	Size of Vessel Group (rel. units)
Celtis	7	4.7	1.4	6.5
Ulmus pumila	5	5.2	2.7	4.0
"Christine Buisman" elm	4	2.6	9.2	23.9
Ulmus americana	0	8.0	9.1	72.9

[a] Resistance rating: 0—highly susceptible to 7—immune.

Control. Biological control is another possibility. Parasites of bark beetles such as a wasp, *Dendrosoter protuberans,* have reduced bark beetle populations in experiments, but in nature they have had no appreciable effect on beetle numbers. This parasitic wasp can be reared rather easily in the laboratory and has survived for more than one season in Michigan. One of the disadvantages of relying on *D. protuberans* is that the species prefers to parasitize beetles under thin-barked portions of the tree. Possibly only 5–10% of the beetles survive the winter season in the northern United States, and these beetles are concentrated in the lower portion of the trunk where the bark is thick.

Several fungi can also reduce the success of bark beetle reproduction. These fungi include *Phomopsis oblonga, Beauvaria bassiana, Trichoderma harzianum, T. polysnorum,* and *Scytalidium lignicola.* Bark beetles may attack logs colonized by some of these fungi less frequently than noncolonized logs, or the larvae develop abnormally and rarely mature. These fungi, however, have not been used in wide-scale attempts to reduce bark beetle reproduction.

Community Control Programs. Sanitation is essential to the success of any program designed to manage Dutch elm disease. Without sanitation, no other approach is worth the expense. Sanitation, in the broadest view, includes detection and removal of dead and dying elms and disruption of common roots between elms. Programs will fail if large amounts of beetle-containing material are missed. Prompt elimination of all dead and dying elm material, including firewood piles, is essential. The public relations aspects of managing firewood piles pose an especially difficult problem for community foresters. Finding them involves entering private property, which citizens may not view favorably. Brood material in Siberian elms damaged as a result of unfavorable winter weather must also be eliminated.

Disruption of common root systems is a form of sanitation and must be included in the program. Mechanical trenching is difficult in an urban environment with underground utilities. VAPAM is the best alternative but is not always effective. Severing root grafts under pavement may require difficult negotiations between the city forester and the city street department. Reducing native elm bark beetle populations with Dursban is a form of sanitation, and, in northern areas where this beetle is an important vector, it should be used.

The other control measures are of far less importance and may fit only trees of high value. Systemic fungicides are expensive, limited only to healthy or recently infected trees, and are temporary protection at best. Pruning is more effective than systemic fungicides but requires effective management. Insecticides such as methoxychlor help but are expensive and not always effective. Resistant trees may provide a future solution but are of little use in protecting the magnificent elms we have today.

Implementing these control measures is a difficult task. Many of the problems encountered are social or political in nature, and yet biologists are responsible for the results. The best vehicle for managing DED, and other urban pest problems, is an effective urban forestry program. Dutch elm disease is perhaps responsible for the implementation of urban forestry programs in much of the country. Regardless of the scale of the urban forestry program, it consolidates and institutionalizes responsibility for the shade trees. If communities can maintain their urban forestry program, and an effective sanitation program in the face of budgetary crises, DED will have minimal impact on their elm population.

SELECTED REFERENCES

Andersen, J. L., R. J. Campana, A. L. Shigo, and W. C. Shortle. 1985. Wound response of *Ulmus americana*. 1: Results of chemical injection in attempts to control Dutch elm disease. *J. Arboric.* 11:137–142.

Anderson, D. E., N. F. Sloan, and A. R. Hastings. 1980. *Control of Overwintering Native Elm Bark Beetles by Application of Dursban 4E to Elm Trees.* USDA For. Serv., For. Ins. and Dis. Mgmt. NA-FB/P6. 8 pp.

Baker, F. A., and D. W. French. 1985. Economic effectiveness of operational therapeutic pruning for control of Dutch elm disease. *J. Arboric.* 11:247–249.

Baughman, N. J. 1985. Economics of Dutch elm disease control: Model and case study. *J. For.* 83:554–557.

Birch, N. C., T. D. Paine, and J. C. Miller. 1981. Effectiveness of pheromone mass-trapping of the smaller European elm bark beetle. *Calif. Agric.* 35:6–7.

Cannon, W. N., Jr., J. H. Barger, and C. J. Kostichka. 1986. *Time and Materials Needed to Survey, Inject Systemic Fungicides, and Install Root-Graft Barriers for Dutch Elm Disease Management.* USDA For. Serv., Res. Pap. NE-585. 6 pp.

Cannon, W. N., Jr., and D. P. Worley. 1980. *Dutch Elm Disease Control: Performance and Costs.* USDA For. Serv., Res. Pap. NE-457. 8 pp.

Carter, J. C. 1967. *Dutch Elm Disease in Illinois.* Ill. Nat. Hist. Surv., Circ. 53. 20 pp.

Cuthbert, R.A., W.N. Cannon, Jr., and J.W. Peacock. 1975. *Relative Importance of Root Grafts and Bark Beetles to the Spread of Dutch Elm Disease.* USDA For. Serv., Res. Note. NE-206. 4 pp.

Gardiner, L. M. 1981. Seasonal activity of the native elm bark beetle, *Hylurgopinus rufipes,* in central Ontario (Coleoptera:Scolytidae). *Can. Ent.* 113:341–348.

Gemma, J. N., G. C. Hartmann, and S. S. Wasti. 1984. Inhibitory interactions between *Ceratocystis ulmi* and several species of entomogenous fungi. *Mycologia* 76:256–260.

Gibbs, J. N., D. R. Houston, and E. B. Smalley. 1979. Aggressive and non-aggressive strains of *Ceratocystis ulmi* in North America. *Phytopathology* 69:1215–1219.

Gregory, G. F., and J. R. Allison. 1979. The comparative effectiveness of pruning versus pruning plus injection of trunk and/or limb for therapy of Dutch elm disease in American elms. *J. Arboric.* 5:1–4.

Hart, J. H., W. E. Wallner, M. R. Caris, and G. K. Dennis. 1967. Increase in Dutch elm disease associated with summer trimming. *Pl. Dis. Reptr.* 51:476–479.

Jassim, H. K., H. A. Foster, and C. P. Fairhurst. 1990. Biological control of Dutch elm disease: Larvicidal activity of *Trichoderma harzianum, T. polysporum* and *Scytalidium lignicola* in *Scolytus scolytus* and *S. multistriatus* reared in artificial culture. *Ann. Appl. Biol.* 117:187–196.

Lanier, G. N. 1987. Fungicides for Dutch elm disease: comparative evaluation of commercial products. *J. Arboric.* 13:189–195.

Lanier, G. N., J. F. Sherman, R. J. Rabaglia, and A. H. Jones. 1984. Insecticides for control of bark beetles that spread Dutch elm disease. *J. Arboric.* 10:265–272.

Lillesand, T. M., D. E. Meisner, D. W. French, and W. L. Johnson. 1981. Evaluation of digital photographic enhancement for Dutch elm disease detection. *Photogramm. Eng.* 48:1581–1592.

McKenzie, M. A., and W. B. Becker. 1937. *The Dutch Elm Disease—A New Threat to the Elm.* Mass. Agri. Exp. Sta., Bull. No. 343. 16 pp.

McNabb, H. S., H. M. Heybroek, and W. L. MacDonald. 1970. Anatomical factors in resistance to Dutch elm disease. *Neth. J. Pl. Path.* 76:196–204.

Neely, D. 1984. Dutch elm disease control in Illinois municipalities. *Pl. Dis.* 68:302–303.

Phillipsen, W. J., M. E. Ascerno, and V. R. Landwehr. 1986. Colonization, emergence and survival of *Hylurgopinus rufipes,* and *Scolytus multistriatus* (Coleoptera:Scolytidae) in insecticide-treated elm wood. *J. Econ. Ent.* 79:1347–1350.

Pusey, P. L., and C. L. Wilson. 1982. Detection of double-stranded RNA in *Ceratocystis ulmi. Phytopathology* 72:423–428.

Richards, W. C., and S. Takai. 1988. Production of cerato-ulmin in white elm following artificial inoculation with *Ceratocystis ulmi. Physiol. and Molec. Plant Pathol.* 33:279–285.

Sherwood, S. C., and D. R. Betters. 1981. Benefit–cost analysis of municipal Dutch elm disease control programs in Colorado. *J. Arboric.* 7:291–298.

Shigo, A. L., and R. Campana. 1977. Discolored and decayed wood associated with injection wounds in American elm. *J. Arboric.* 3:230–235.

Sinclair, W. A., and R. J. Campana. 1978. Dutch elm disease perspectives after 60 years. *Northeast Regional Publ. Search Agriculture* 8(5):1–52.

Weber, J. 1981. A natural biological control of Dutch elm disease. *Nature* 292:449–451.

Westwood, A. R. 1991. *A Cost Benefit Analysis of Manitoba's Integrated Dutch Elm Disease Management Program 1975–1990.* Forestry Branch, Manitoba Natural Resources, Winnipeg, Manitoba, Canada. 22 pp.

Whitten, R. R. 1966. *Elm Bark Beetles.* USDA, Circ. 185, 8 pp.

DISEASE PROFILE

Elm Yellows

Importance: Elm yellows (once called elm phloem necrosis) has devastated elm populations in the eastern United States. It would have undoubtedly been more important had not the Dutch elm disease killed such a large proportion of the elm population.

Suscepts: Native elm species are susceptible, especially American, red, and winged elms.

Distribution: Elm yellows is now present in more than 20 of the eastern and midwestern states and in Utah.

Causal Agents: Prior to the mid 1960s, a virus was assumed to be the pathogen. Now it is recognized that elm yellows is caused by a mycoplasmalike organism (MLO).

Biology: The MLO is disseminated by the white-banded leafhopper. The MLO colonizes the phloem. Both nymphs and adults feed on phloem tissues and can acquire the MLO. A year may pass before symptoms appear, and several years may pass before the tree dies.

Epidemiology: Conditions favorable for the leafhoppers generally favor high incidence of the disease. An epidemic may spread 5–8 km/yr with single tree or satellite infections occurring some distance from the killing front.

Diagnosis: Elm yellows may be confused with Dutch elm disease except that with the former all the branches are affected at once. Infected phloem tissues turn a tan or brown color. Pieces of the inner bark may be placed in a closed vial for a few minutes and will

Wilt symptoms. From Tucker (1945).

Discolored wood. From Gibson et al. (1981).

produce a wintergreen odor. In red elm, yellows causes small witches' brooms to form, but the wintergreen odor and discolored phloem are not present. Unlike fungi and bacteria, MLOs cannot be grown on culture media.

Control Strategy: Some suppression of symptoms has been accomplished by injection with tetracycline antibiotics. There is little evidence that insecticides to control the leafhoppers have been effective. Transmission through common root systems can be prevented as recommended for Dutch elm disease. Asian and European elms appear to be resistant to elm yellows.

SELECTED REFERENCES

Braun, E. J., and W. A. Sinclair. 1979. Phloem necrosis of elms: Symptoms and histopathological observations in tolerant hosts. *Phytopathology* 69:354–358.

Gibson, L. P., A. R. Hastings, and L. A. LaMadeleine. 1981. *How to Differentiate Dutch Elm Disease from Elm Phloem Necrosis*. USDA For. Serv., NA-FB/P-11. 1 pp.

Sinclair, W. A., and T. H. Filer. 1974. Diagnostic features of elm phloem necrosis. *Arborists' News* 39:145–149.

Tucker, C. M. 1945. *Phloem Necrosis, A Destructive Disease of the American Elm*. Univ. Missouri, Agri. Exp. Sta., Circ. 305. 15 pp.

17.3 ELM YELLOWS

Importance

Elm yellows, also known as elm phloem necrosis, has devastated elm populations in the northeastern and midwestern United States and in the South, such as Stoneville and Greenville, Mississippi. In Syracuse, New York, 58% of the elm population was killed in just 4 years. Elm yellows can increase losses to Dutch elm disease by providing more dead trees in which bark beetles can increase their populations.

Suscepts

Elm yellows is an important disease of American, red, and winged elm. It also occurs on cedar and September elm. Susceptibility of rock elm is unknown.

Distribution

Elm yellows is now present in at least 20 states but is concentrated in the central part of the eastern United States and New York (Fig. 17.8). Only recently was it found in Minnesota, Wisconsin, and Utah.

Causal Agents

Elm yellows is caused by an MLO that belongs to a group of microorganisms discovered in the mid 1960s. Prior to this time it was assumed that a virus was responsible.

Biology

The disease cycle is shown in Figure 17.9. The MLO is disseminated to healthy trees by the white-banded leafhopper *(Scaphoideus luteolus)*. Both nymphs and adults of this insect use their tubular mouthparts to feed on phloem. During this feeding they acquire the MLO from or introduce it into the phloem. The MLO colonizes the phloem tissues. A rapid deposition of callose occurs within sieve tubes, followed by collapse of the sieve elements and companion cells. The cambium then produces replacement phloem with greatly reduced sieve elements. Starch accumulates in diseased phloem. Roots are affected first and may die as a result. The disease does not become apparent for a year or longer after infection, but trees usually die during the first growing season that symptoms appear. Some trees survive 2 or even 3 years after symptoms appear.

Scaphoideus luteolus is 5 mm long. Adults lay eggs in the outer layer of elm bark, and after overwintering they hatch in April and May. Adults are present during the

FIGURE 17.8   The distribution of elm yellows.

last half of the summer. There is only one generation per year. Incubation of MLO in the insect ranges from 5 to 33 days.

Epidemiology

It was originally assumed that elm yellows would be restricted by the distribution of the vector and that the northern and western states would not have to be concerned. While this may be generally true, the elm leafhoppers have inoculated elms in New York, Michigan, Minnesota, and probably Wisconsin. Conditions favorable for the leafhoppers undoubtedly favor high incidence of the disease. Elm yellows may exist

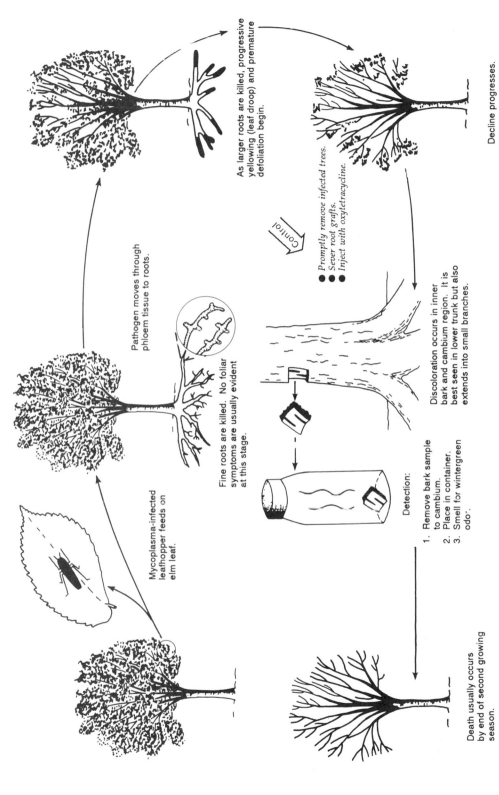

As larger roots are killed, progressive yellowing (leaf droop) and premature defoliation begin.

Decline progresses.

Pathogen moves through phloem tissue to roots.

Control

● Promptly remove infected trees.
●● Sever root grafts.
● Inject with oxytetracycline.

Fine roots are killed. No foliar symptoms are usually evident at this stage.

Mycoplasma-infected leafhopper feeds on elm leaf.

Discoloration occurs in inner bark and cambium region. It is best seen in lower trunk but also extends into small branches.

Detection:

1. Remove bark sample to cambium.
2. Place in container.
3. Smell for wintergreen odor.

Death usually occurs by end of second growing season.

FIGURE 17.9 Disease cycle of elm yellows. Redrawn from Tatter (1978). Courtesy of Academic Press and T. A. Tatter.

669

at a low incidence in an area for years before the disease becomes prevalent. Once an epiphytotic occurs, in many situations most of the elms in the area are usually killed.

Elm yellows has spread 5–8 km/yr. Occasionally single trees or small satellite infections are found outside of contiguous infested areas. These sites are presumed to result from long distance transport of leafhoppers by wind.

Diagnosis

The symptoms vary with the host. In American elm epinasty, yellowing, premature defoliation, or wilting are usually preceded by extensive necrosis of feeder roots. The MLO affects the living phloem tissues, and they turn tan or brown. This color change is first apparent in the inner phloem region of the roots or base of the tree. The inner bark has a wintergreen odor that can best be detected by placing pieces of inner phloem into a closed vial for a few minutes. The most obvious, but not the most specific, symptom is yellowing of the foliage, which might be confused with Dutch elm disease. In contrast to Dutch elm disease, which causes branch flagging, elm yellows affects all the branches at once. Wilting is most likely to occur during hot weather in midsummer.

In red elm, yellows causes witchs' brooms, and, because the tree usually dies within that year, the brooms are only a few centimeters long. They can be 1 m long if the tree survives longer than 1 year. Wintergreen odor and discolored phloem do not occur. If an odor is present, it resembles maple syrup. Wilted leaves appear to have more red in them than is the case in American elm.

Unlike fungi and bacteria, MLOs cannot be grown on cultural media. Their involvement in disease is confirmed by electron microscopy or indirectly by administering tetracycline antibiotics and observing remission of symptoms.

Control Strategy

Very little can be suggested for reducing losses to elm yellows. Eradication of infected trees has been suggested, but even intensive sanitation efforts to combat Dutch elm disease have had little impact on the epidemic of elm yellows. Insecticides have been recommended for controlling leafhoppers, but there is little evidence that such treatments were effective. Transmission through common root systems can be prevented as recommended for Dutch elm disease. Asian and European elms appear to be resistant to elm yellows, and using these species in breeding for resistance to Dutch elm disease may also provide resistance to elm yellows. Some suppression of symptoms has been accomplished by injection with tetracycline antibiotics. Where both Dutch elm disease and elm yellows are present, efforts and investments in disease management to preserve elms in addition to sanitation are probably not wise.

SELECTED REFERENCES

Braun, E. J., and W. A. Sinclair. 1976. Histopathology of phloem necrosis in *Ulmus americana*. *Phytopathology* 66:598–607.

Braun, E. J., and W. A. Sinclair. 1979. Phloem necrosis of elms: Symptoms and histopathological observations in tolerant hosts. *Phytopathology* 69:354–358.

Lanier, G. N., D. C. Schubert, and P. D. Manion. 1988. Dutch elm disease and elm yellows in central New York. *Pl. Dis.* 72:189–194.

Sinclair, W. A. 1981. Elm yellows, pp. 25–31. In R. J. Stipes and R. J. Campana (eds.), *Compendium of Elm Diseases.* American Phytopathological Society, St. Paul, Minn. 96 pp.

Sinclair, W. A., E. J. Braun, and A. O. Larsen. 1976. Update on phloem necrosis of elms. *J. Arboric.* 2:106–113.

Sinclair, W. A., and T. H. Filer. 1974. Diagnostic features of elm phloem necrosis. *Arborists' News* 39:145–149.

Swingle, R. U., R. R. Whitten, and H. C. Young. 1949. *The Identification and Control of Elm Phloem Necrosis and Dutch Elm Disease.* Ohio Agr. Expt. Sta., Special Circ. 80. 11 pp.

Tattar, T. A. 1978. *Diseases of Shade Trees.* Academic Press, New York. 384 pp.

Wilson, C. L., C. E. Seliskar, and C. R. Krause. 1972. Mycoplasma-like bodies associated with elm phloem necrosis. *Phytopathology* 62:140–143.

DISEASE PROFILE

Oak Wilt

Importance: Oak wilt is an important disease of oaks in the eastern United States. It is especially destructive in the north-central states and is becoming increasingly important in Texas.

Suscepts: Thirty-three species and varieties of oak are susceptible. The red or black oaks are very susceptible; the white oaks generally exhibit greater resistance.

Distribution: Oak wilt is believed to have been present in the United States for many decades but was originally found only in Iowa, Minnesota, and Wisconsin. It is now found from Minnesota to Pennsylvania, south to South Carolina, west to northern Arkansas and eastern Oklahoma, and north through eastern Kansas and Nebraska. Oak wilt also occurs in much of northeastern and central Texas.

Causal Agents: The pathogen is a fungus, *Ceratocystis fagacearum* (Ophiostomatales).

Biology: Ceratocystis fagacearum can spread through root grafts or by spores transmitted by insects. Spread through root grafts, or through common root systems in the case of Texas live oak, accounts for the irregular circular patches of dead trees in infection centers. Conidia and, to a lesser extent, ascospores are produced on mats and pads of specialized mycelium, which develop between the bark and cambium. The pads increase in size, forcing a break in the bark. Insects are attracted by a fruity odor produced by the fungus and enter through the crack to feed on the mats. Sap-feeding beetles are the major vectors of *C. fagacearum.* After feeding on the mats and becoming infested with spores, they must feed between late May to early June on sap draining from fresh wounds of healthy trees in order to transmit the pathogen successfully. Dates may vary with geographic location. Once inside the tree, the fungus can spread rapidly through the vascular system. The tree is stimulated to produce gums and tyloses that plug the water-conducting tissues and cause the tree to wilt. After the tree dies and the cambium turns brown, pads and mats may be produced.

Wilt symptoms. Courtesy of D. W. French, University of Minnesota, St. Paul.

Epidemiology: Oak wilt has been especially serious in coppice-reproduced stands where there are large numbers of oaks with few other kinds of trees present. In mixed stands, movement is restricted because of the lack of root grafts. A major reason for the unpredictability of rapid spread in many instances is that known vectors are relatively inefficient. In Europe a potentially effective vector, the oak bark beetle, has caused much concern about the destructive potential of oak wilt, should it be introduced to that continent.

Diagnosis: With red and black oaks, leaves turn a dull green or bronze color, which rapidly progresses from the margins to the base, shortly after which the leaf is cast. With white oaks, leaf symptoms tend to be more localized, and the balance of the crown may remain healthy. Vascular discolorations may be present, being more common in white oaks in the north and common in red or blacks in the south. White oaks may survive for years, whereas red or black oaks die within weeks.

Control Strategy: In urban areas, spread through common root systems is the major cause of new cases of oak wilt. Disruption of those root grafts is a most essential part of any control program. Restricting pruning and other wounding activities during the time when springwood vessels are being formed eliminates much insect spread. This approach varies with location; in Minnesota avoid pruning during May and June. Injection with systemic fungicides such as propiconizole may be effective if used as protectants or in the very earliest stages of infection. Eradication of diseased oaks in forest stands is not practiced much anymore but can be effective in urban forests if done before spores form.

SELECTED REFERENCE

Gibbs, J. N., and D. W. French. 1980. *The Transmission of Oak Wilt.* USDA For. Serv., Res. Pap. NC-185. 17 pp.

17.4 OAK WILT

Importance

Oak wilt is an important disease in portions of the north-central United States, where the oak species make up about 80% of the woodland trees and in many residential areas where oaks are the predominant species. This disease is becoming increasingly important in Texas. Oak wilt is especially serious in coppice-reproduced stands where there are large numbers of oaks per acre with few other kinds of trees present. In the central United States the disease is important in that the oaks account for over half of the lumber production. If, for some reason, the disease continues to increase in intensity to the east and south, the major oak resources of the country will be threatened. An evaluation of the impact of oak wilt in Wisconsin and Minnesota indicated that if oak wilt were eliminated, there would be an increase in net growth of 11%.

Suscepts

All 33 species and varieties of *Quercus* tested are susceptible, as are three species of *Castanea,* one species of *Castanopsis,* and one species of *Lithocarpus.* The red or black oaks are very susceptible; the white oaks range from highly resistant to quite susceptible. *Quercus alba* is resistant, while varieties of *Q. macrocarpa* range from resistant to susceptible.

Distribution

Oak wilt has been present in the United States for many decades, but the mortality of oak was attributed to several factors other than this fungus until 1941. Dr. Carl Hartley, in a letter written in November 1928 and in reference to his notes of 1912, described mortality in scarlet oak stands in Minnesota and Wisconsin, which undoubtedly was due to oak wilt. These communications suggest that the disease was present for many years prior to its formal recognition. Originally it was found in Iowa, Minnesota, and Wisconsin. It is now distributed from Minnesota east to Pennsylvania, south to South Carolina and Tennessee, west to the northern half of Arkansas and eastern Oklahoma, north through eastern Kansas and Nebraska to Minnesota (Fig. 17.10). Oak wilt also occurs in northeastern and central Texas.

Cause

The fungus causing oak wilt was first described as *Chalara quercina* (Fungi Imperfecti); later the perfect stage was found and named *Endoconidiophora fagacearum* T. W. Bretz (Ascomycetes). Now it is called *Ceratocystis fagacearum* (T. W. Bretz) J. Hunt. Both Bretz and Hunt met untimely deaths. Ted Bretz routinely used mercuric chloride as a surface disinfectant in his laboratory and in the middle of a very produc-

FIGURE 17.10   The distribution of oak wilt in North America.

tive career died as a result of mercury toxicity. John Hunt had recently finished his doctoral requirements, which included a revision of the genus *Ceratocystis,* and was driving across the northern United States on the way to his new employment. Along the way he gave a ride to a hitchhiker and was murdered. Thus his promising career was snuffed out before he had a chance to pursue it.

Biology

Ceratocystis fagacearum can spread in two ways: through root systems of trees grafted together and by spores transmitted by insects. This movement is shown in the disease diagram in Figure 17.11. Spread of the fungus through common root systems ac-

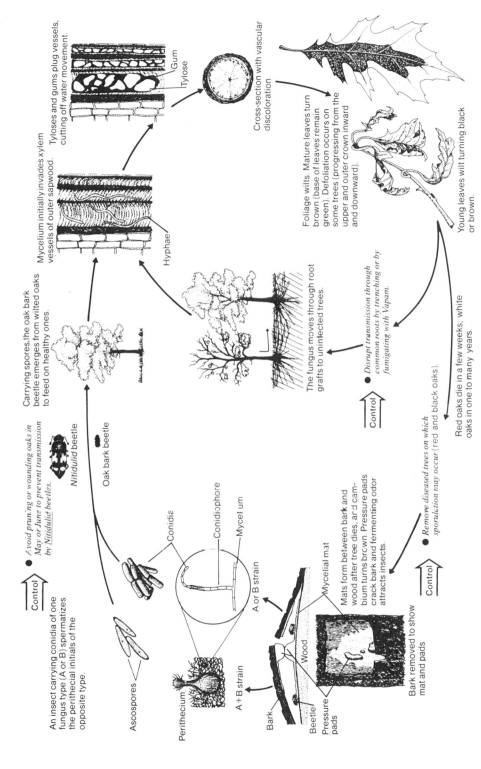

Control
• Avoid pruning or wounding oaks in May or June to prevent transmission by Nitidulid beetles.

An insect carrying conidia of one fungus type (A or B) spermatizes the perithecial initials of the opposite type.

Ascospores

Perithecium

A + B strain

Conidia

Conidiophore

Mycelium

A or B strain

Bark

Beetle

Wood

Pressure pads

Mycelial mat

Mats form between bark and wood after tree dies, and cambium turns brown. Pressure pads crack bark and fermenting odor attracts insects.

Bark removed to show mat and pads

Control
• Remove diseased trees on which sporulation may occur (red and black oaks).

Nitidulid beetle

Oak bark beetle

Carrying spores, the oak bark beetle emerges from wilted oaks to feed on healthy ones.

Mycelium initially invades xylem vessels of outer sapwood.

Hyphae

Tyloses and gums plug vessels, cutting off water movement.

Gum

Tylose

Cross-section with vascular discoloration

The fungus moves through root grafts to uninfected trees.

Control
• Disrupt transmission through common roots by trenching or by fumigating with Vapam.

Foliage wilts. Mature leaves turn brown (base of leaves remain green). Defoliation occurs on some trees (progressing from the upper and outer crown inward and downward).

Young leaves wilt turning black or brown.

Red oaks die in a few weeks; white oaks in one to many years.

FIGURE 17.11   Disease diagram of oak wilt caused by Ceratocystis fagacearum. Drawn by Valerie Mortensen.

675

counts for the irregular circular patches of dead trees. These mortality centers gradually increase in size. The fungus produces two types of spores: conidia, and less frequently ascospores in perithecia. Both types of spores are produced on spore mats underneath the bark of red and black oaks (Fig. 17.12). Spore mats range in size from about 2.5–20 cm long by 1.3–10 cm wide, although mats up to 33 × 18 cm have been found. The center portion of the mat consists of specialized mycelium and one or more pressure pads that, in connection with opposing pads attached to the inside of bark, develop sufficient pressure to split open the bark. Conidia are produced in abundance. Perithecia and ascospores may be present on mats if the two compatible strains, called A and B, are present in the tree or brought together by insects.

Conidia and ascospores are spread by insects and possibly to a minor extent by other agents such as squirrels. Sap-feeding beetles, in the family *Nitidulidae,* and oak bark beetles such as species of *Pseudopityophthorus* transmit both kinds of spores. The spores apparently must be introduced to healthy trees through wounds, especially fresh wounds made in May and June. At least this is the case with the *Nitidulidae*. In red oaks, at least those species investigated, wounds are susceptible for up to 8 days, while wounds are susceptible for 18–24 hours in bur oaks, depending on the weather.

FIGURE 17.12   Spore mats on red oak, with Nitidulid beetles. Courtesy of D. W. French, University of Minnesota, St. Paul.

TABLE 17.2 Time of Spore Mat Formation on Red and Black Oaks

Tree Wilts	Spores Form
June	Fall of same year
July, August	May and June of following year
September	Late June of following year, if at all

If certain other fungi, such as *Ceratocystis piceae*, are present in the wound prior to introduction of the oak wilt fungus, the tree will not be infected. Thus wounds only 24 hours old may not be receptive to infection. Infection seems to occur through fresh wounds made during a very limited period in late May or early June. Often this period may last only 2 weeks. To be certain of avoiding infection and to allow for variation in weather, we consider that infection can occur at any time during May and June.

Once inside the tree, the fungus can spread rapidly through the vascular system of the tree. Substances produced by the fungus stimulate the formation of tyloses by the host. These tyloses along with gummy materials are formed in the water-conducting tissues of the tree and plug the vessels causing the trees to wilt. The oak wilt fungus first invades the xylem vessels in the outer sapwood. Later, hyphae move into adjacent xylem parenchyma through the pits. The hyphae grow intercellularly, intracellularly, and within the cell wall itself. The fungus degrades the walls of infected sapwood cells, progressing from the lumen into the cell wall.

After the trees die and the cambium turns brown, the fungus produces spore mats. The maximum production of spore mats occurs in the fall and spring in relation to when the tree wilts. On trees wilting in July and August, mats will be produced the same fall and then again the following spring (Table 17.2). On trees that wilt in June, mat production usually is completed during the fall of the same year, and for trees that wilt in September, mats, if formed at all, will develop the following June. Spore mat formation is apparently a function of sapwood moisture content. Mats form sooner in small- than in large-diametered trees. Sapwood moisture content remains high in a wilting tree, higher than it would be if the tree was not infected. Mats form when sapwood moisture content ranges from 37 to 53% (based on green weight). Sapwood moisture contents in white oaks are less than this, which may explain why few mats are found on the white oaks. Patterns of mat formation may vary across the country where oak wilt occurs. In Missouri and Arkansas, few mats are found. This tendency is not due to the fungus strains because mats form when oaks in the northern United States are inoculated with the southern isolates of the fungus.

Epidemiology

The incidence and rate of spread of oak wilt depends in part on the availability of susceptible hosts. In the northern United States, most new cases of oak wilt result from spread through common root systems with only an occasional incidence of overland spread. Root grafting occurs only between oaks of the same species. Rates of spread have been reported to range from 1–7 m/yr to 11–16 m/yr in live oak stands in Texas. In mixed stands, movement is restricted because of the lack of root grafts.

With the exception of live oaks, root grafts are less frequent in the more southerly portion of the range. If the fungus is present in both red and white oaks, then both groups of species will be affected in that stand. On occasion, the fungus is only in the red oaks, and the white oaks will survive.

Spread of the fungus by insects may vary considerably from year to year, depending on availability of fresh wounds, abundance of inoculum and populations of Nitidulid beetles. Also, some circumstantial evidence suggests that overland spread does not occur in more northerly locations such as northern Minnesota because the timing or some other factor is not suitable. In the more southerly areas, the fungus may be restricted by high temperatures that develop in wilted branches and stems. Other than in Texas, the distribution of the disease has been quite static, and spread of any significance is more likely to occur when aided by people than when left to natural processes. The fungus stops spreading on occasion even though more oaks are near.

Diagnosis

With red and black oaks, the leaves turn a dull green or bronze color, which progresses toward the base of the leaf, the blade tissue along the petiole being the last to change color (Fig. 17.13). The symptoms progress very rapidly over the tree, usually beginning at or near the top. The fungus is capable of changing a seemingly healthy tree into a completely dead tree in a few weeks. Brown to black discolorations may be found in the sapwood, usually in the most recent ring. This symptom is more evident in the white oaks than in the red or black oaks. No red oak has been known to recover once infected.

The white oaks are more resistant. The leaves turn brown from apex toward the

FIGURE 17.13   Red oak leaf symptoms. Courtesy of D. W. French, University of Minnesota, St. Paul.

FIGURE 17.14   Live oak leaf symptoms. Figure on right, courtesy of D. W. French, University of Minnesota, St. Paul.

base of the leaf and the discoloration resembles fall coloration. Infected branches may be found scattered through the crown, while the balance of the tree remains healthy. Vascular discoloration is more evident in these species than in the red oaks. The trees usually die very slowly, surviving for one to several years.

On live oaks, flagging or browning of leaves on individual limbs occurs during spring months. Later, leaves on infected limbs exhibit more specific patterns of chlorosis and necrosis (Fig. 17.14), typically this is dead tissue along the veins and tip of the leaf. Complete defoliation is common within 30 days after infection. Although most infected live oaks die within a given infection center, 10% or more of the infected trees may survive for 1 year or more or recover completely.

Control Strategy

In many areas of the country, most new cases of oak wilt are the result of infection through common root systems; thus disruption of those root grafts is a most essential part of any control program. Short distance spread, through root systems that are grafted together, can be prevented by severing root systems either chemically or mechanically. Chemically, this is done by killing a portion of the root systems halfway between infected and surrounding healthy trees. As with Dutch elm disease, the best chemical for this purpose is SMDC (sodium N-methyldithiocarbamate), known by the trade name Vapam. One part of SMDC to three parts of water should be poured into a series of holes 15 cm apart and 38–46 cm deep. Use 50–200 ml in each hole depending on the size of the trees. Each hole should be closed immediately, and the chemical should not be placed within 2.4–3.0 m of healthy trees. The infected tree should be left standing for 2 weeks after the application of SMDC. A second barrier outside the first set of healthy appearing oaks provides an extra measure of protection in case the fungus has moved into the seemingly healthy trees prior to placement of the first barrier. Mechanical trenches must be at least 1.3 m deep to prevent spread of the fungus through common root systems. A vibratory plow with a blade at least 1.5 m long is the most efficient means of placing barriers (Fig. 17.15). In most stands the plow moves at about 21 m/min and does not require the extra step of filling as with

FIGURE 17.15   Vibratory plow, with blade removed from soil. Courtesy of D. W. French, University of Minnesota, St. Paul.

trenches. Where underground obstacles, such as utilities and rockiness permit, mechanical trenching is less expensive, quicker, and more certain to prevent further spread than chemical treatment.

Long distance or overland dissemination, as a result of insects carrying spores, is more difficult to prevent. The amount of inoculum can be reduced by eradicating infected oaks before spores form. If we assume that overland spread occurs only in May and June, then those trees on which spores will occur need to be destroyed before the spring season. To be certain, we recommend destruction before April. Spores will not form on oaks dead for longer than 1 year. Spore mats form mainly on red or black oaks, and primarily only those wilting in July or August will have spore mats during the susceptible period. Thus only these trees need to be eliminated by burying, burning, or using them. Removing the bark prevents spore mat formation, but bark removal must be complete. Using the logs from infected trees for lumber, railroad ties or pallets or as firewood is an acceptable means of eliminating this source of inoculum.

In forest stands where many trees are involved, sporulation can be prevented or limited by girdling the infected oaks as they begin to wilt. Deep girdling (removal of bark and sapwood) near the base of the tree or chemical girdling with sodium arsenite accelerates drying of the tree. More rapid drying causes the fungus to produce spores earlier when they are not likely to cause infection, spore production to extend over a shorter interval, fewer spores to be produced per mat, and the pressure pads to be unable to crack open the bark. Sodium arsenite is hazardous to wildlife and, thus, must be used with care. Spore production can be prevented or the spores rendered useless by wrapping logs in 4-mil or heavier plastic for that period in spring during which infection by insect-carried spores can occur (May and June!). The plastic can be removed in July, and the wood can be used for firewood. The conditions under the plastic are highly unfavorable for *C. fagacearum*. Usually, the spore mats are overrun by *Trichoderma* sp. Other fungi such as species of *Trichothecium* and *Hypoxylon* (especially *H. atropunctatum*) compete with the oak wilt fungus and may prevent spore production. These fungi occur naturally in oaks and show some promise for biological control.

In eradication programs the white oaks should not be removed because the fungus seldom sporulates on these trees. If these trees are felled, the fungus is more likely to be drawn into the adjacent healthy oaks. Stumps of infected trees can be poisoned with ammate to reduce the time the oak wilt fungus survives in the root systems of such trees. *Ceratocystis fagacearum* can survive for 3–4 years in untreated root systems, while in ammate-treated stumps, the fungus survived in 17% of the trees for 1 year and in none after 2 years. The fungus seldom survived more than a year in root systems of trees that had been girdled to the heartwood. Only trees that will have spore mats in May and June need be treated. They are only the red and black oaks that wilt in July and August.

Oak should not be wounded during the spring season, when trees are susceptible to invasion by *C. fagacearum*. If oaks are damaged or pruning is necessary, cover the exposed surfaces immediately with a tree wound dressing to discourage insects.

Overland spread occurs infrequently, and yet this is the only way in which new infection centers can occur. It is true that occasionally a large number of trees will become infected at one time because of line-clearing or tree-pruning operations, but if such unnecessary wounding in the spring can be avoided, new infections are relatively rare. If insect dissemination was more of a factor, we doubt that there would be many oaks left in the northern United States. Until we know more about insect dissemination of *C. fagacearum*, control of the disease by limiting insect populations does not seem feasible.

Recent studies suggest that chemical injections into oaks may prevent sporulation. Some success was obtained with copper sulfate, but this chemical is very corrosive and difficult to inject. Cacodylic acid appears to be the best of all the compounds evaluated to date.

In spite of some efforts in locating and developing resistant oaks, none has been forthcoming as yet. Oak trees can be used with little risk in metropolitan areas, however, because if these trees are mixed with other species and are at least 15 m apart, the fungus would have difficulty moving to healthy trees. Systemic fungicides such as Lignasan BLP and Arbotect 20-S have been evaluated for treatment of diseased white oaks. The white oaks are resistant and, with the assistance of a single injection, may be able to overcome the fungus and survive for many years without further treatment.

Trees within an oak wilt disease area can be injected with the systemic fungicide Alamo to prevent tree death. This treatment must be applied before symptoms are observed on more than 30% of the tree's canopy; consequently, it is difficult to apply to red oaks because symptom progression takes place in 3–5 weeks. For live oaks, though, treatment is feasible because symptom progression may take 3–6 months.

The sterol-inhibiting fungicide propiconazole (Alamo) is applied at the rate of 2–3 ml/l of water/2.54 cm of trunk diameter by injection at low pressure (ca. 1.4 k/cm^2) into the root flares. The lower rate is used as a preventive treatment, and the higher rate is for trees that are symptomatic of wilt or that are next to diseased trees. Holes approximately 8–10 cm apart are drilled into the lower trunk and root flares and to a depth of 1.3–2.54 cm. The holes must be drilled with a power drill using an 8-mm bit. A 12- to 16-1 pressure sprayer may be used as a reservoir for fungicide suspension. The reservoir is connected to the tree with plastic tubing and specially designed injection ports.

The oak wilt fungus has been used as a selective silvicide in Minnesota, Arkansas, and other states. It effectively eradicated undesirable oaks. Obviously, it must be used only where *C. fagacearum* occurs naturally and must be handled in such a way as to not endanger valuable oaks. In Minnesota where this fungus has been used as a selective silvicide, there is no evidence of increased incidence of oak wilt in surrounding stands. Even when provided an opportunity to move into adjacent oak stands, the fungus had not done so.

SELECTED REFERENCES

Appel, D. N., C. F. Drees, and J. Johnson. 1985. An extended range for oak wilt and *Ceratocystis fagacearum* compatibility types in the United States. *Can. J. Bot.* 1325–1328.

Appel, D. N., R. C. Maggio, E. L. Nelson, and M. J. Jeger. 1989. Measurement of expanding oak wilt centers in live oak. *Phytopathology* 79:1318–1322.

Campbell, R. N., and D. W. French. 1955. Moisture content of oaks and mat formation by the oak wilt fungus. *For. Sci.* 1:265–270.

French, D. W., and D. B. Schroeder. 1969. Oak wilt fungus, *Ceratocystis fagacearum,* as a selective silvicide. *For. Sci.* 15:198–203.

French, D. W., and W. C. Stienstra. 1978. *Oak Wilt.* Agr. Ext. Service, Univ. of Minnesota, Extension Folder 310. 6 pp.

Gibbs, J. N. 1980. Role of *Ceratocystis piceae* in preventing infection by *Ceratocystis fagacearum* in Minnesota. *Trans. Br. Mycol. Soc.* 74:171–174.

Gibbs, J. N., and D. W. French. 1980. *The Transmission of Oak Wilt.* USDA For. Serv., Res. Pap. NC-185. 17 pp.

Juzwik, J., and D. W. French. 1983. *Ceratocystis fagacearum* and *C. piceae* on the surfaces of free-flying and fungus-mat-inhabiting nitidulids. *Phytopathology* 73:1164–1168.

Juzwik, J., D. W. French, and J. Jeresek. 1985. Overland spread of the oak wilt fungus in Minnesota. *J. Arboric.* 11:323–327.

Lewis, R., Jr., and A. R. Brook. 1985. An evaluation of Arbotect and Lignasan trunk injections as potential treatments for oak wilt in live oaks. *J. Arboric.* 11:125–129.

Mistretta, P. A., R. L. Anderson, W. L. MacDonald, and R. Lewis, Jr. 1984. *Annotated Bibliography of Oak Wilt 1943–1980.* USDA For. Serv., Gen. Tech. Rep. WO-45. 132 pp.

Osrerbauer, N. K., T. Salisbury, and D. W. French. 1994. Propiconazole as a treatment for oak wilt in *Quercus alba* and *Q. macrocarpa. J. Arboric.* 20:202.

Schmidt, E. L. 1983. Minimum temperature for methyl bromide eradication of *Ceratocystis fagacearum* in red oak log pieces. *Pl. Dis.* 67:1338–1339.

Tainter, F. H. 1986. Growth, sporulation and mucilage production by *Ceratocystis fagacearum* at high temperatures. *Pl. Dis.* 70:339–342.

Ulliman, J. J., and D. W. French. 1977. Detection of oak wilt with color IR aerial photography. *Photogramm. Eng. and Remote Sensing* 43:1267–1272.

DISEASE PROFILE

Conifer Wilt (*Ophiostoma* and *Ceratocystis* spp.)

Importance: Conifer wilt is not a clearly defined group of diseases and is certainly not a single disease. For this reason its importance is not known. The combination of conifer wilt and the attack by bark beetles has resulted in significant losses to many tree species. It is very difficult, however, to separate tree mortality caused by conifer wilt from the attacks by bark beetles.

Suscepts: Conifer wilts occur on many species of pine and other coniferous species, and very likely every conifer species is susceptible to some degree. The disease has been observed and studied on shortleaf, loblolly, and eastern white pine in the eastern United States as well as on the western white, lodgepole, ponderosa, and sugar pine in the western states.

Distribution: The extent of conifer wilt is likely to correspond with the distribution of its hosts and is worldwide in extent.

Causal Agents: The causal fungi include *Ophiostoma piliferum* (Ophiostomatales), *O. minus, O. ips,* and other species of this genus as well as species of *Ceratocystis.*

Biology: Bark beetle species of *Dendroctonus* and *Ips* are the vectors that carry the fungi. When beetle populations are present in large numbers, they can attack reasonably healthy trees. If the hosts are predisposed by drought or some other unfavorable environmental condition, they are more subject to attack. Attacking adult beetles chew into the phloem and inoculate the tissues with spores that germinate. The resultant mycelium then begins to colonize the phloem and xylem. If host resistance (resinosus) is not sufficient to repel the beetles, the fungi quickly kill the host cells and further resinosus ceases. Fungal invasion is assisted by beetle gallery construction. Fungal growth is most abundant in the rays but eventually extends throughout the sapwood, producing a blue stain of the wood.

Epidemiology: Because of the complexity of the conifer wilts, no one set of factors determines the amount of the disease. The incidence of wilt is undoubtedly related to bark beetle populations and their activities. The fungi involved are ubiquitous and always associated with the bark beetles.

Diagnosis: Affected trees wilt and die. The symptoms are similar to those caused by drought. Moisture content of the sapwood in the infected portion of the tree is reduced, and the wood may be blue stained. A high percentage of the pits are aspirated.

Control Strategy: Reducing beetle populations and keeping the trees in good condition are the best approaches for dealing with this disease/insect complex. Overly dense stands reduce tree vigor and create conditions very conducive for beetle attack. These stressed trees are, furthermore, less able to withstand attack after it has begun.

SELECTED REFERENCES

Basham, H. G. 1970. Wilt of loblolly pine inoculated with blue-stain fungi of the genus *Ceratocystis.* *Phytopathology* 60:750–754.

Lorio, P. L., Jr., and J. D. Hodges. 1985. Theories of interactions among bark beetles, associated microorganisms, and host trees, pp. 485–492. In *Proc. Third Biennial South. Silv. Res. Conf.* USDA For. Serv., Gen. Tech. Rep. SO-54.

DISEASE PROFILE

Black-Stain Root Disease

Importance: Black-stain root disease is becoming a serious disease in commercial forests. Foresters are concerned because black stain increases with disturbance.

Suscepts: All pines tested are susceptible. The disease occurs on Jeffrey, knobcone, lodgepole, ponderosa, pinyon, singleleaf pinyon, western white, and sugar pines. It is also important on Douglas-fir and western hemlock.

Distribution: Black-stain root disease exists only in the western United States and Canada, occurring on ponderosa and Jeffrey pines in California; pinyon pines in California, Nevada, Arizona, Utah, and Colorado; and primarily on Douglas-fir in the Pacific Northwest.

Causal Agents: Black-stain root disease is caused by the imperfect fungus *Leptographium (Verticicladiella) wageneri.* The sexual form, *Ophiostoma (Ceratocystis) wageneri* (Pyrenomycetes, Ophiostomatales), has only been found once.

Biology: Leptographium wageneri affects seedlings and older trees. It can enter through nonwounded roots ≤5 mm in diameter, through root grafts, and through insect-feeding wounds. It is confined to mature xylem tracheids and moves through bordered pits. It cannot penetrate cell walls directly. Xylem and phloem function are disrupted. Mycelium in the tracheids and gums serve to plug the xylem. As mycelial growth and xylem plugging increase, trees succumb to moisture stress. The black stain precedes the fungus and, in susceptible pine seedlings, especially ponderosa, can be within 1–2 cm of the apical meristem. *Leptographium wageneri* spreads through root grafts, killing patches of trees. It also can grow short distances through soil and across root contacts. In addition, several root-feeding insects spread *L. wageneri,* both in long- and short-distance dispersal. The insects breed in fresh stumps or in the soil, and emerging adults feed on nearby healthy roots. Mortality centers in ponderosa pine enlarge an average of 1 m/yr in radius.

Wilt symptoms. From Harrington and Cobb (1988). Courtesy of F. W. Cobb, Jr.

There are three variants of *L. wageneri,* existing primarily on pinyons, Douglas-fir, or hard pines. The hard-pine variant also affects white pines when they grow nearby. Although these variants may affect other species, crossover infection is rare.

Epidemiology: Considered to be native to western forests, this disease has increased greatly during the past 30 years due to accelerated logging of old-growth stands and increasingly intensive management of young stands. The disease is associated with road building, tractor logging, and especially precommerical thinning possibly because of the attraction of insect vectors to stressed trees and freshly cut stumps. Infection is greater at soil temperatures of 16°C than at temperatures exceeding 21°C. *Leptographium wageneri* infects and colonizes more seedlings in moist soils, and spread is greater on moist sites. This often means that black stain may be most severe on better sites. *Armillaria* spp., mountain pine beetle, and the red turpentine beetle often attack trees weakened by black stain.

Diagnosis: In large trees the symptoms include reduced growth of the leader, chlorotic needles, and loss of foliage. Trees die in 1–8 years. The dark brown or black stain from which this disease gets its name occurs in the sapwood above infected roots and may extend to breast height. This stain differs from blue stain in that it is not wedge-shaped. Affected wood is also resinous.

Control Strategy: Trenching has been used to sever continuous root systems. Trees have been felled around disease centers to take advantage of the slow spread of the fungus along roots and its short persistence in stumps. Both of these measures have failed. Alternating species can be of use. Douglas-firs are not susceptible to hard-pine pathovars, and pines are not susceptible to the Douglas-fir pathovars. Unfortunately, the nature of the site may limit the use of other species. Studies in Oregon suggest that thinning in May, June, and July compared to thinning in September through January reduces activity of insect vectors in Douglas-fir. Similar studies in California, however, found as many of these vectors in stumps thinned the previous year after the peak of vector activity as in freshly thinned stumps. If black stain is a threat, avoid precommercial thinning.

SELECTED REFERENCES

Harrington, T. C., and F. W. Cobb, Jr. (eds.). 1988. *Leptographium Root Diseases on Conifers.* APS Press, St. Paul, MN. 149 pp.

Hunt, R. S., and D. J. Morrison. 1986. Black-stain root disease on lodgepole pine in British Columbia. *Can. J. For. Res.* 16:996–999.

17.5   BLACK-STAIN ROOT DISEASE

Importance

Black-stain root disease is not considered a serious disease in commercial forests, but it does have the potential of being of considerable consequence. The disease is currently present in more than 20 areas in the western United States. The largest infection center, in southern California, involves several thousand hectares of singleleaf pinyon pine, and, although not seriously affecting commercial stands, heavy losses have been incurred on recreation lands. The disease results in predisposition of the trees to attack by bark beetles and an increase in their population.

Suscepts

All pines tested are susceptible to black-stain root disease. The disease has been reported on Jeffrey, knobcone, lodgepole, ponderosa, pinyon, singleleaf pinyon, eastern white, western white, and sugar pine. It occurs on Douglas-fir as well.

Distribution

Black-stain root disease is found only in the western United States and southwestern Canada (Fig. 17.16). It occurs on ponderosa and Jeffrey pines in northern California, on singleleaf pinyon pine in southern California, and on other hosts in Arizona, Colorado, Montana, Nevada, Oregon, Utah, and Washington.

FIGURE 17.16   Distribution of black-stain root disease in North America.

Cause

The black-stain disease is caused by *Ophiostoma (Ceratocystis) wageneri* (Goheen and Cobb) Harrington. However, the perfect stage is rare; hence, the causal fungus is usually referred to by its imperfect form, *Leptographium wageneri* (Kendrick) Wingfield, formerly known as *Vertcicladiella wageneri* Kendrick.

Biology

Leptographium wageneri affects both seedlings and older trees and can enter through nonwounded roots. When the fungus invades the tree, it passes through the phloem and cambium; otherwise it is confined to the xylem and mature tracheids. The fungus moves from one cell to another through bordered pits and is not able to penetrate directly through cell walls (Fig. 17.17). Tyloses are formed in invaded cells. The stain preceeds the fungus and in susceptible pines, especially ponderosa, can be within 1–2 cm of the apical meristem. The fungus has been isolated from all parts of inoculated ponderosa seedlings including juvenile needles. The hyphae in the tree often are surrounded by a sheath that together with the hypha (averaging 5.5 μm) will fill the tracheid.

Much like the vascular wilt fungi, *L. wageneri* probably spreads along roots in contact with each other, and thus patches of trees are killed. Insects may be responsible for long distance dispersal of the fungus. Susceptible trees may not recover once infected, and rate of mortality depends on host resistance and age of tree. In addition, several root-feeding insects spread *L. wageneri*. These insects include *Hylastes nigrinus*, *Steremnius carinatus*, and *Pissodes fasiculatus*. They breed in freshly cut stumps or in the soil, and the adults emerge to feed on nearby healthy roots. While these insects play a role in long-distance dispersal of *L. wageneri*, they are also important in the spread of the fungus within a root-disease center. The rate of spread has been estimated at 1 m/yr in radius (range 0–7 m). The optimum temperature for the fungus is about 15°C, and soil temperatures and rainfall, as it affects soil temperature, may limit disease spread. Bark beetles attack trees weakened by *L. wageneri*, making it difficult to determine the true cause of death. Thus the losses are likely to be more extensive than originally thought.

There are three variants of *L. wageneri*, existing primarily on pinyons, on Douglas-fir, or hard pines. The hard-pine variant also affects white pines when they grow near diseased hard pines. Although these variants may affect other species, crossover infection is rare.

Epidemiology

Environment does not seem to be restrictive, as *L. wageneri* is found both in the very hot semiarid regions of the southwestern United States to areas in the Northwest with 1.52 m of annual rainfall. Soil conditions where the disease occurs vary from shallow and rocky to deep clay loams. The disease is found at sea level and up to elevations of 2,700 m.

This disease is considered to be native to western forests, but it has increased greatly during the past 30 years. There is some speculation that this is due to accelerated logging of old growth stands and increasingly intensive management of young stands. Black-stain root disease is associated with human disturbance such as road

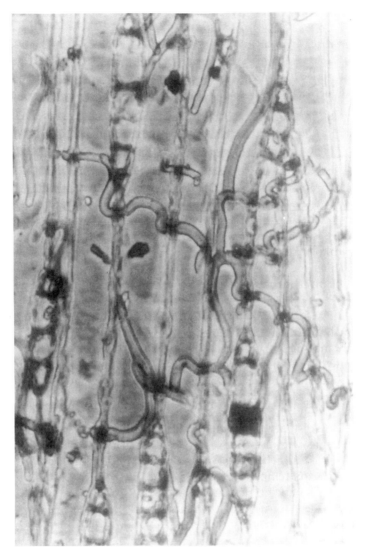

FIGURE 17.17   Hyphae of *L. wageneri* in tracheids of Douglas-fir. From Harrington and Cobb (1988). Courtesy of F. W. Cobb, Jr.

building, tractor logging, and especially precommercial thinning. This relationship may be due to the attraction of the insect vectors to stressed trees and freshly cut stumps.

The soil environment can influence the development of black-stain root disease. Infection is greater at soil temperatures of 16°C than at temperatures exceeding 21°C. *Leptographium wageneri* infects and colonizes more seedlings in moist soils, and spread is greater on moist sites. Consequently, black stain may be most severe on better sites.

Other diseases and insects often attack trees with black stain. *Armillaria* spp. often attack infected trees at the root collar. Mountain pine beetle *(Dendroctonus ponderosae)* and the red turpentine beetle *(D. valens)* often attack trees weakened by black stain. These other pests may affect the rate at which the trees express symptoms or even die.

FIGURE 17.18 Partial cross section of ponderosa pine showing black stain in the outer growth rings. Compare with the blue stain in lower left and right corners. From Harrington and Cobb (1988). Courtesy of F. W. Cobb, Jr.

Diagnosis

The first symptoms of the disease are reduced growth of leader, shortened and chlorotic needles, and loss of needles. The tree appears generally sick, not different from root rot-affected trees. Trees die in 1–8 years time. The dark brown to black stain, from which the fungus gets its name, in the main stem differs from blue stain in that it is not wedge-shaped (Fig. 17.18). The discolored wood is resinous. Bark beetles are often present.

Control Strategy

Other than substituting resistant species such as true firs or junipers, there is little in the way of sound controls that can be recommended. Trenching has been used to sever continuous root systems. Trees have been felled around disease centers to take advantage of the slow spread of the fungus along roots and its short persistence in stumps. Both of these measures have failed, perhaps due to insects carrying the fungus beyond the barrier.

Alternating species can be of use. Douglas-firs are not susceptible to hard-pine pathovars, and pines are not susceptible to the Douglas-fir pathovars. Unfortunately, in many cases the site limits the use of other species.

Forest pathologists and entomologists are working together to understand this disease problem. Studies in Oregon suggest that thinning in May, June, and July reduced activity of *H. nigrinus* and *P. fasciatus* in Douglas-fir compared to thinning in September through January. Similar studies in California, however, found as many of these vectors in stumps thinned the previous year after the peak of vector activity as in freshly thinned stumps. If black stain is a threat, avoiding precommercial thinning should be considered.

SELECTED REFERENCES

Cobb, F. W., Jr., G. W. Slaughter, D. L. Rowney, and C. J. DeMars. 1982. Rate of spread of *Ceratocystis wageneri* in ponderosa pine stands in the central Sierra Nevada. *Phytopathology* 72:1359–1362.

Harrington, T. C., and F. W. Cobb, Jr. 1983. Pathogenicity of *Leptographium* and *Verticicladiella* spp. isolated from roots of western North American conifers. *Phytopathology* 73:596–599.

Harrington, T. C., and F. W. Cobb, Jr. (eds.). 1988. *Leptographium Root Disease on Conifers.* APS Press, St. Paul, Minn. 149 pp.

Harrington, T. C., F. W. Cobb, Jr., and J. W. Lownsbery. 1985. Activity of *Hylastes nigrinus,* a vector *Verticicladiella wageneri* in thinned stands of Douglas-fir. *Can. J. For. Res.* 15:519–523.

Hessburg, P. F., and E. M. Hansen. 1986. Mechanisms of intertree transmission of *Ceratocystis wageneri* in young Douglas-fir. *Can. J. For. Res.* 16:1250–1254.

Hunt, R. S., and D. J. Morrison. 1986. Black-stain root disease on lodgepole pine in British Columbia. *Can. J. For. Res.* 16:996–999.

Witcosky, J. J., T. D. Schowalter, and E. M. Hansen. 1986. The influence of time of precommercial thinning on the colonization of Douglas-fir by three species of root-colonizing insects. *Can. J. For. Res.* 16:745–749.

Witcosky, J. J., T. D. Schowalter, and E. M. Hansen. 1986. *Hylastes nigrinus* (Coleoptera:Scolytidae), *Pissodes fasciatus,* and *Steremnius carinatus* (Coleoptera:Curculionidae) as vectors of black stain root disease of Douglas-fir. *Environ. Entomol.* 15:1090–1095.

Zambino, P. J., and T. C. Harrington. 1989. Isozyme variation within and among host-specialized varieties of *Leptographium wageneri. Mycologia* 81:122–133.

DISEASE PROFILE

Verticillium Wilt

Importance: Verticillium wilt has been known since 1914, but the extent of losses is unknown. Nurseries have experienced serious losses when large percentages of some maple varieties are affected. Every year in many parts of the country, although not causing overwhelming losses, many ornamental maples are lost to this disease.

Suscepts: Sugar, Norway, and red maples and their cultivars are susceptible. Some cultivars of Norway maple may be most susceptible. Other tree species that are subject to the disease include elm, catalpa, and ash. Verticillium wilt occurs on a wide range of hosts

Wilt symptoms.

Discolored wood.

other than trees, including such commonly grown crops as potatoes and tomatoes. It is not known to be a common pathogen of forests, but recently it has occurred on *Ceanothus* on a logged site in California.

Distribution: Verticillium wilt occurs wherever the hosts occur and has been reported commonly on trees in Europe and North America.

Causal Agents: Verticillium albo-atrum (Hyphomycetes) is the causal organism. *Verticillium dahliae* is a closely related species capable of causing a similar disease.

Biology: Verticillium albo-atrum enters its host through wounds, usually root wounds. Wounds in sugar maple seedlings were susceptible for 16 days, and then, as the wounds aged, fewer plants became infected. Once in the host, the fungus rapidly colonizes the cortex, endodermis, and vessel members. Conidia are produced within 8 days. On resistant hosts the cortex is colonized, but few hyphae develop in the vessel members; conidia are not present in the vascular system. Microsclerotia form in both cortex and vascular systems within 14 days of infection.

Epidemiology: Availability of inoculum very likely is an important factor in the incidence of Verticillium wilt. The fungus can survive in soils separate from the host for 2 years and longer, and, if susceptible trees are planted in infected soil, wilt will frequently result. The disease incidence is greater in wet soils, but expression of wilt symptoms is greater in dry soils.

Diagnosis: The first symptom is usually wilting of one or more branches, often near the top of the tree. The entire crown of small trees may wilt, but in large trees usually only portions of the crown are affected. A diagnostic symptom is a green discoloration in the wood of maples. In elms this discoloration is brown. Small branches may not be discolored; branches 2 cm in diameter or larger usually are. Other discolorations associated with frost crack or other factors should not be confused with Verticillium wilt. Although large trees may survive for many years, they are not likely to recover. Positive identification must be made by culturing the fungus and observing its characteristic verticillate conidiophores.

Control Strategy: No practical control measures are known. The life of the tree may be prolonged by pruning infected branches from the tree. Fertilizing and watering have been suggested and may aid the tree in compartmentalizing the infection. Fungicides applied as soil drenches have reduced symptom development, but none have provided total control. When replacing a tree killed by Verticillium, do not use a susceptible species. Avoid planting maples, especially sugar and Norway maples, in soils known to have been associated with Verticillium wilt. If a susceptible species must be used, use a resistant variety or cultivar.

SELECTED REFERENCES

Harrington, T. C., and F. W. Cobb. 1984. *Verticillium albo-atrum* on *Ceanothus* in a California forest. *Pl. Dis.* 68:1012.

Himelick, E. B. 1969. *Tree and Shrub Hosts for Verticillium albo-atrum.* Ill. Nat. Hist. Surv., Biol. Notes 66. 8 pp.

DISEASE PROFILE

Mimosa Wilt

Importance: Discovered in Tryon, North Carolina, in 1935, this destructive vascular wilt disease has since dominated the pathology of mimosa, particularly in the southeastern states. There, the tree, also known as silktree, is a favored ornamental around homes and a common natural escape on a variety of forested sites. It is native to Asia and, since

Wilt symptoms.

1831, has been planted extensively from New York to Florida and west to Texas and California. The disease is now a common sight in most southern communities.

Suscepts: In general, mimosa is very susceptible; however, a few wilt-resistant selections have been made and propagated to a limited extent.

Distribution: The wilt has appeared over most of mimosa's planted range, from New York and New Jersey to Florida and west to Mississippi. The disease is also known in Russia, Argentina, and Puerto Rico.

Causal Agents: Mimosa wilt is caused by *Fusarium oxysporum* f. sp. *perniciosum* (Hyphomycetes).

Biology: Spores from lenticellular sporodochia form on wilt-killed trees and are dispersed by rain-splash and stem flow to enter the soil where they transform into chlamydospores in very large concentrations. Infection through the roots is optimal at 28°C and may be enhanced by root exudates. The failure of trunk inoculations to produce disease is attributed to either an active defense mechanism in stem tissues or a root infection threshold requisite for stem invasion. The fungus progresses into the stem via vessels of the current growth ring as microconidia but not mycelium, thus suggesting passive transport in the transpiration stream. Gum deposits form first in pit apertures and then increase to occlude eventually the xylem vessels at the time of foliage wilt. Unlike other wilt diseases, tyloses do not develop in mimosa wilt. Affected trees usually wilt, defoliate, and die within 1 year of symptom onset, and there is no recovery. During the later stages of wilt, the fungus proliferates by mycelial growth into all parts of branches, twigs, and even seeds to a limited extent via the ray parenchyma system.

Epidemiology: The disease occurs under such a wide variety of circumstances that no soil, pH, or topographic situation appears limiting. By 1938 mimosa mortality in Tryon, North Carolina, had approached the 70% level. The current wide distribution and local persistence of the disease points to the movement of infested soil, water, planting stock, and seeds (ca. 6% transmission to progeny), either naturally or by man's activities. Synergism between the wilt fungus and root knot nematodes has also been demonstrated.

Diagnosis: Host specificity and distinctive symptoms facilitate diagnosis. The principal symptoms are wilting and defoliation with concurrent brown streaking of sapwood in the current annual ring of roots, stem, and branches. Salmon- or orange-colored sporodochia become evident after defoliation and through the succeeding year. The fungus is readily isolated from recently wilted branches. Macroconidia are distinctive for species identification; however, validation of forma *perniciosum* requires suscept inoculation and proof of pathogenicity.

Control Strategy: First control attempts were aimed at selecting and propagating resistant cultivars. Control directed at protection and possible chemotherapy have attained some success with soil drenches of systemic fungicides. In terms of disease avoidance, *Fusarium*-infested soils may be too ubiquitous to expect practical disease escape, but certainly soils surrounding the root zone of wilt-killed trees should be avoided. Eradication by destroying infected trees, preferably on a community-wide scale, should reap practical results through inoculum reduction.

SELECTED REFERENCES

Crandal, B. S., and W. L. Baker. 1950. The wilt disease of American persimmon caused by *Cephalosporium diospyri*. *Phytopathology* 40:307–325.

Hepting, G. H. 1939. *A Vascular Wilt of the Mimosa Tree (Albizia julibrissin)*. USDA, Circ. 535. 10 pp.

Phipps, P. M., and R. J. Stipes. 1976. Histopathology of mimosa infected with *Fusarium oxysporum* f. sp. *perniciosum*. *Phytopathology* 66:839–843.

DISEASE PROFILE

Persimmon Wilt

Importance: Persimmon wilt has been a devastating disease and has eliminated extensive acreages of this tree in the central basin of Tennessee and in areas reaching into Florida and west to Oklahoma where persimmon was once a common species. Persimmon is considered a weed in some regions where farmers have difficulty eradicating the species from cultivated fields. On the positive side, persimmon helps control erosion, supplies food for wildlife and the wood, because of its hardness, was valuable for such items as heads of golf clubs.

Suscepts: Persimmon wilt occurs on the American persimmon, which grows over much of the southeastern United States, north to a line from New Jersey to Iowa and west from Oklahoma and Texas. Other species of persimmon *(Diospyros)* range from highly

Wilt symptoms. From Crandal and Baker (1950).

susceptible to immune. *Diospyros lotus* and *D. kaki*, two introduced species, are highly resistant, but the wilt pathogen can grow in them. If they are grafted onto American persimmon, the fungus will move into the roots and kill the trees.

Distribution: Persimmon wilt was known to exist in central Tennessee as early as 1931, but where the fungus originated and exactly when is unknown. The disease is now present in 10 states from Tennessee and North Carolina, south to the Gulf, and west to Arkansas, Oklahoma, and Texas.

Causal Agents: Persimmon wilt is caused by *Acromonium (Cephalosporium) diospyri* (Hyphomycetes). The fungus was described and identified in 1945.

Biology: Acromonium diospyri enters the host through wounds and, if entry occurs early in the growing season, it spreads through the entire tree within a month. If inoculated late in the season, it remains more or less inactive until the host starts growing the following season. The trees wilt rapidly as the fungus spreads through the vessels. Gums and tyloses form as a result of the infection. As soon as the tree or parts of it die, spores are produced at the cambium, from August until cold weather occurs in the fall. They remain viable until the following spring.

Epidemiology: Incidence may be determined to a large degree by the availability of wounds either caused by cattle or by insects. In grazing lands, cattle injuries are probably more important than insect wounds. Wind may be a factor in causing wounds. Two insects that may be primary factors are the powder-post beetle and the twig girdler.

Diagnosis: The most striking symptom is abrupt wilting of the foliage, usually starting at the top of the tree. Often the tree is completely defoliated and dead within 2 months after being infected. Trees infected late in the season will have small chlorotic leaves the following spring, and these trees may survive for 1–2 years. Brown to black streaks occur in the outer rings. This discoloration is present throughout the tree when the first foliar symptoms appear. At the end of the growing season, the fungus produces pinkish spore masses beneath the bark. In young portions of the stem, the spores are in red blisters, which rupture to release spores. Spore masses under older bark are not released until the bark becomes loose.

Control Strategy: It has been suggested that it is beneficial to protect healthy persimmon from wounding. This means reducing populations of possible insect vectors, which requires eliminating dead persimmon and destroying all branches and twigs from these trees. This method is very difficult to accomplish, especially since both insect species inhabit other tree species as well as persimmon. Probably most important is the eradication of diseased persimmon to eliminate sources of spores. Substitution of introduced species that are immune is a possibility, but it is not done to any degree. Resistance in *D. virginiana* has not been reported. Where persimmon has been a troublesome weed tree, *A. diospyri* has been used as a selective silvicide.

SELECTED REFERENCE

Crandall, B. S., and W. L. Baker. 1950. The wilt disease of American persimmon caused by *Cephalosporium diospyri. Phytopathology* 40:307–325.

18

DIEBACKS AND DECLINES

18.1 DIEBACKS AND DECLINES

Most of the diseases we have discussed thus far result from either single biotic or abiotic causes or have simple multiple causes.

Chapter 3 outlined a large number of abiotic factors that can either directly or indirectly lead to tree disease. Many times the symptoms are of such a nature that the exact abiotic cause can be determined following a little detective work. Sometimes the exact cause may be difficult to uncover because of infestation by secondary organisms. This chapter outlines several diseases of abiotic cause that were very difficult to explain at the time they were recognized. The series of diagnostic steps that forest pathologists have used to unravel the causes of the following examples were best summarized by Manion (1991) in his presentation of the decline syndrome.

Certain long-term or slowly changing *predisposing factors* alter the trees' ability to withstand or respond to. Global warming, a drier climate, or abused soils would be examples of these predisposing factors. A second group of factors, called *inciting factors* are of short duration and would include actions such as short-term drought, insect defoliation, late frost, or exposure to a pollutant. These factors, if acting upon a tree already stressed from exposure to the predisposing factors, further respond by some degree of twig, branch, or root-tip dieback. Such affected trees usually do not store sufficient food reserves to withstand continued stressful events and are receptive to the third group of factors called *contributing factors*. This group includes not only environmental factors but also biotic factors such as fungi and insects that are attracted to these stressed trees. Their appearance may confound the original predisposing and inciting factors and make these cases very difficult to diagnose.

REFERENCE

Manion, P. D. 1991. *Tree Disease Concepts.* 2nd ed. Prentice-Hall, Englewood Cliffs, N.J. 402 pp.

DISEASE PROFILE

Birch Dieback

Importance: This disease caused extensive growth loss and mortality to birch on thousands of square kilometers of forests from 1930 to 1950. Birch dieback has historical significance in that it was the first of several dieback–decline diseases to receive intensive research that eventually has led to our present understanding of these important forest tree disease complexes.

Suscepts: White or paper and yellow birch were affected.

Distribution: Birch dieback occurred in New Brunswick, Nova Scotia, and Quebec provinces of Canada and Maine, New Hampshire, Vermont, and New York.

Causal Agents: Intensive investigations of climatic, edaphic, and biotic factors failed to establish a cause of birch dieback. Edaphic factors associated with rootlet mortality in shallow-rooted trees were most suspect. Increased soil temperature associated with drought conditions killed birch rootlets in artificial tests, but there was no evidence that droughts occurred during the period of extensive dieback. It was suggested that root damage associated with soil freezing during winters without snow cover may have been responsible for birch dieback.

Biology: Mortality of rootlets and a failure of affected trees to generate sufficient new rootlets appeared as the initial reaction. This condition led to extensive crown dieback and mortality. Weakened trees were attacked and often killed by secondary root-decaying fungi and insects, especially the bronze birch borer.

Epidemiology: Birch are subject to deterioration associated with site disturbances, and birch dieback in particular was associated with shallow-rooted trees subject to unfavorable edaphic factors. Rootlet damage may be initiated by severe soil freezing, which induces physiological drought and frost heaving. Such conditions were thought to occur during several winters with little or no snow cover. Increased soil temperature and drought also induced rootlet mortality, but there is little evidence to suggest that these conditions actually occurred during the period of extensive dieback. There is little evidence for contageous spread of the disease but rather it appeared to intensify within local areas. By 1950 birch dieback had abated; young trees grew normally, and older trees that were not severely diseased recovered.

Diagnosis: Symptoms typical of dieback–decline diseases occurred and reflect an overall host response to an adverse physical environment. Leaves became chlorotic, foliage was sparse, and extensive crown dieback was evident as shoot tips and branches died. Bunching of foliage occurred in the living lower crown. These outward symptoms were preceded by rootlet mortality and followed by reduced annual increment. When these conditions persisted, mortality occurred in 3–5 years.

Control Strategy: There was no control for this disease of unknown cause. Losses were minimized by proper salvage cuttings. Because residual birch is subject to post-logging decadence, diseased stands should be either clear-cut or only lightly thinned. Yellow birch, being tolerant and uneven-aged, is more easily managed by thinning than is the short-lived, intolerant, even-aged white birch.

Other Similar Diseases: Residual birch left on a site during harvesting are subject to deterioration. This problem, which is referred to as post-logging decadence, is the result of an abrupt increase of light and soil temperature associated with stand opening. Initially, this problem was confounded with birch dieback. While there may be some commonality of causal factors, these problems differ in that birch dieback occurred on undisturbed sites. Nevertheless, care must be taken not to thin birch stands so severely that residual birch are affected by post-logging decadence.

SELECTED REFERENCES

Clark, J., and G. W. Barter. 1958. Growth and climate in relation to dieback of yellow birch. *For. Sci.* 4:343–364.

Pomerleau, R., and M. Lortie. 1962. Relationship of dieback to the rooting depth of white birch. *For. Sci.* 8:219–224.

DISEASE PROFILE

Pole Blight of Western White Pine

Importance: Pole blight is one of the most serious diseases of western white pine. Of the 300,000 ha of pole-size stands of western white pine, nearly 43,000 ha (one eighth) are affected by pole blight. According to a 1955 survey, 57–81% of the basal area is affected in severely blighted areas; of this a large proportion is dead. The impact of these losses is all the more serious because they occur in stands of large trees of which there is a deficiency to meet future needs for white pine lumber.

Suscepts: Pole blight affects western white pine. Typically, 35- to 150-year-old (pole-size) trees in natural forests are attacked, although in 1973 the disease was found in plantations of younger trees.

Distribution: Pole blight occurs throughout the northern portion of the western white pine type in northern Idaho, western Montana, eastern Washington, and British Columbia.

Causal Agents: Presently, the best evidence suggests that unfavorable edaphic factors initiate the disease. Pole blight occurred on sites with low soil-moisture-holding capacities during the drought period of 1916–1940. Fungi appear to be involved only as secondary invaders.

Biology: Presumably rootlet mortality induced by drought results in crown dieback and tree death. Blighted trees may be invaded by secondary invaders that hasten their death.

Epidemiology: Tree ring records of western white pine indicate that a period of unfavorable growth conditioned by an extended drought occurred from 1916 to 1940. On sites with

From Gill et al. (1951).

soils having a low moisture-holding capacity (soil <46 cm deep having <10 cm of water-storage capacity in the top 1 m), moisture stress during this drought period appears to have conditioned blight. Spread of the disease does not appear to be of a contagious pattern, but within affected areas intensification usually occurs.

Diagnosis: Symptoms are chlorotic, dwarfed, often tufted needles in the upper crown, followed by leader and branch dieback. Symptoms progress down the crown, and top dieback may preceed by several years death of the tree, which usually occurs 5–10 years following initial decline. Long, narrow necrotic lesions may occur on the bole, most often on the lower half. The wood beneath these lesions may be blue-stained and/or resin-soaked. These outward symptoms are preceeded by rootlet mortality and accompanied by reduced radial growth increment.

Control Strategy: There is no control for pole blight. Thinnings and salvage cuttings in affected stands have not retarded the progress of the disease. Salvage cuttings are recommended when economically feasible, especially because affected trees rarely recover. During regeneration, consideration should be given to avoiding high-risk sites with soils of low-moisture capacity.

SELECTED REFERENCES

Gill, L. S., C. D. Leaphart, and S. R. Andrews. 1951. Preliminary results of inoculations with a species of *Leptographium* on western white pine. USDA, For. Path. Rel. No. 35. 14 pp.

Leaphart, C. D., O. L. Copeland, Jr., and D. P. Graham. 1957. *Pole Blight of Western White Pine.* USDA For. Serv., Pest Leafl. No. 16. 4 pp.

Leaphart, C. D., and A. R. Stage. 1971. Climate: A factor in the origin of the pole blight disease of *Pinus monticola* Dougl. *Ecology* 52:229–239.

DISEASE PROFILE

Sweetgum Blight

Importance: Sweetgum blight, first described in 1948, was responsible for widespread deterioration of sweetgum in the 1950s. A survey in 1954 indicated that 35% of the sweetgums and 40% of the total merchantable wood volume were affected by blight. Only 1% of the trees were killed, but growth loss was considerable. In local areas where the disease was severe, a higher percentage of trees died. Blight was most severe in pole- and log-sized trees.

Suscepts: Sweetgum, both natural and planted, were affected by this disease. Generally, pole- and log-size trees (>15 years of age) were most affected.

Distribution: Sweetgum blight occurred throughout the natural range of sweetgum and was most prevalent in the coastal states from Delaware to Texas. Sweetgum in Maryland was most affected, although blight was locally severe in South Carolina, Florida, Alabama, Mississippi, and Louisiana.

Sweetgum blight. From Miller and Gravatt (1952).

Causal Agents: No biotic causal agents were identified with sweetgum blight, and results from several studies indicate that drought was the primary causal factor.

Biology: Rootlet mortality was followed by crown deterioration. Many trees recovered provided crown dieback did not exceed 50%.

Epidemiology: Disease incidence and severity was highest among older trees and on sites with soils having a water deficiency (both upland and floodplain slack water soils). Likewise, irrigation reversed the dieback process and trees recovered. Although the incidence of sweetgum blight increased 4–5%/yr within affected areas, there was no evidence for contagious spread.

Diagnosis: Symptoms were typical of dieback–decline diseases. Leaves were chlorotic, dwarfed, and sparse and became prematurely colored in the fall. Crown dieback, evident as twigs and branches died, progressed from the top to bottom. In older trees a "stag-head" appearance was evident; otherwise, there was general crown deterioration. These symptoms were preceeded by rootlet mortality.

Control Strategy: There was no control for sweetgum blight with the exceptions of making salvage cuttings to remove trees with extensive crown dieback and avoiding planting sweetgum for saw-log rotations on drought-prone soils.

SELECTED REFERENCES

Hepting, G. H. 1955. A southwide survey for sweetgum blight. *Pl. Dis. Reptr.* 39:261–265.
Miller, P. R., and G. F. Gravatt. 1952. The sweetgum blight. *Pl. Dis. Reptr.* 36:247–252.
Toole, E. R. 1959. *Sweetgum Blight.* USDA For. Serv., Pest Leafl. No. 37. 4 pp.
Toole, E. R. 1959. Sweetgum blight as related to alluvial soils of the Mississippi River Floodplain. *For. Sci.* 5:2–9.

DISEASE PROFILE

Oak Decline

Importance: Oak mortality is common to eastern forests, especially in the Appalachian Mountains, and a high incidence was reported in 1912–1920, 1925–1932, 1953–1956, 1958–1960, 1964–1966, 1969–1971, and 1980–1986. Mortality and growth losses (40–85%) are widespread; and with the exception of chestnut blight, oak mortality may cause overall the greatest impact on these eastern hardwood forests where oak species, which replaced chestnut, predominate. This typical dieback–decline disease also demonstrates the role of predisposing biotic factors (other than fungi) in disease development.

Suscepts: Species in both the red and white oak groups are affected, especially northern red, scarlet, white, and chestnut oak. Of these species scarlet oak is perhaps the most severely injured.

Distribution: Oak mortality is widespread in the natural range of oak, especially in the Appalachian Mountain region of Pennsylvania, West Virginia, and Virginia. It also occurs in New York, New Jersey, Connecticut, Missouri, Tennessee, and Arkansas.

Causal Agents: Oak mortality results from initial injury by biotic, climatic, or edaphic predisposing factors followed by invasion of weakened trees by secondary fungi and insects. Both drought and early spring frosts play a role as predisposing factors in certain areas and years. However, by far the most important predisposing agents are defoliating insects. Numerous (> 20) oak-defoliating insects have been identified, but those primarily associated with oak mortality are oak leaf tier, fall canker worm, forest tent caterpillar, and gypsy moth. Secondary invaders include the two-lined chestnut borer, Armillaria root decay, and various stem and branch decay fungi.

Declined oak.

Biology: Trees attacked by leaf-feeding insects may succumb after several years of defoliation, or they may be weakened by one severe or several partial defoliations. Droughts, early spring frosts, or ice and hail damage may have similar effects. Trees thus weakened show depleted carbohydrate reserves, reduced increment, root mortality, and crown dieback and are subject to attack by secondary invaders that often provide the coup de grace.

Epidemiology: The epidemiology of this disease is closely associated with the time and location of the initiating factors, that is, defoliating insects and, on occasion, drought and other extremes of climate. Mortality follows in the path of insects, and latent losses occur for several years thereafter. Trees not severely injured by predisposing factors or secondary invaders often recover during normal growth periods.

Diagnosis: Extensive crown dieback and associated mortality are readily observed. Of course, in the case of insect attack, dieback symptoms are preceeded by defoliation. If drought-induced, then gradual crown dieback is preceeded by root mortality; in some instances stem cankers are evident.

Control Strategy: Direct control of the defoliating insects is not economically feasible in forests. Because scarlet oak is severely affected and is subject to decay after 80 years, this species should be removed in salvage cuts and thinnings. White pine, yellow-poplar, and red maple should be favored on appropriate sites as compared with red oaks. Favor white oak on good sites in place of chestnut oak. If plantations are established, dry sites should be avoided. Species composition seems to be the only feasible means of avoiding oak mortality, the incidence of which has increased coincident with the increase of phytofagus insects on oak. Arrival of the gypsy moth to the eastern oak forests has complicated the control of oak decline.

SELECTED REFERENCES

Staley, J. M. 1965. Decline and mortality of red and scarlet oaks. *For. Sci.* 11:2–17.
Tainter, F. H., W. A. Retzlaff, D. A. Starkey, and S. W. Oak. 1990. Decline of radial growth in red oaks is associated with short-term changes in climate. *Eur. J. For. Path.* 20:95–105.

18.2 OAK DECLINE

Oak decline is a term used to decribe oaks that are not growing as they should and implies that some expected level of health or productivity is not being achieved. Oak decline does not usually refer to decline or death of single or scattered trees unless their incidence is unacceptably high over a large area. Usually, many trees over a large area are affected, exhibiting symptoms of crown dieback, reduced radial growth, and premature death. Oak decline may cause significant losses to forest managers because it results in lower yields of wood or fiber. For wildlife managers the loss of mast production may be harmful to some wildlife species, but the formation of snags may be beneficial as wildlife habitat. There is a general concern, though, that, if oak decline continues, many benefits of the oak forests of the eastern United States may be lost, and these sites will convert to less versatile and less valuable species mixtures.

The eastern oak forest has been in a state of flux since retreat of the glacial ice fields of 18,000 years ago. Pollen records show that oaks were present in the late glacial forest, but today these species are present only near the southern margin of the boreal forest. Prior repeated glaciations may have reduced the genetic variability of the oak population and made them more sensitive to a changing environment during the warming period following glaciation.

Information derived from polar ice cores suggests that the average water and air temperature during the last 850,000 years was 15°C. The warming trend realized after the glaciers began to recede was the last of only three short-term warming periods observed during this entire time. Concurrent with this warming trend, a change in orbital oscillation 9,000 years ago caused the climate to become warmer and drier in what is now the southeastern United States. As a result, oaks and hickories became prevalent. Several thousand years ago a tropical maritime air mass from the Gulf of Mexico became dominant, and this initiated a high frequency of lightning-caused fires that favored pine at the expense of the oak–hickory forest.

This warming trend accelerated in the 1880s, and the northern temperate zone grew markedly warmer, with the period from 1875 to 1975 as one of the warmest in about 4,000 years. Global average temperatures increased by about 0.3–0.6°C, and the period from 1910 to 1940 experienced a nearly straight-line increase from −0.1 to +0.3°C above average in the southern Appalachians. These high temperatures and prolonged regional droughts beginning in the early 1950s caused depressed radial growth and acted as a set of predisposing factors which subsequently resulted in decline and premature death of oaks.

But, accumulating evidence suggests that the initiation of the oak decline syndrome was far more complex than just a simple response to drought and high temperature. The natural distribution of oak forests would have been much different had it not been for human interference, first by pre-Columbian inhabitants and later by European settlers.

The influence of fire increased dramatically about 10,000 years ago with the advent of aboriginal humans. Fire was used to drive game, improve habitat, and make food gathering easier. Regular burning created large, open meadows with widely spaced trees. Regular burning was continued by European settlers to improve forage for livestock. On upland and Piedmont sites repeated burning killed young hardwoods and injured older trees. Because other species lacked the ability to sprout from the root collar after being top-killed, recurrent fires may have favored a high proportion of oaks in the current stands. Subsequent and effective fire control by the middle of the twentieth century led to the development of dense canopies and subcanopies. This decreased the development of large advance reproduction of oaks and has led to difficulty in creating new oak stands.

Following the invention of the cotton gin in 1794, a period of significant land clearing and abandonment was initiated, and virtually all the original forests of the Appalachian Piedmont and many of those of the adjacent uplands were cleared. By the time the Great Depression forced many of the most recent farmers out of business, severe sheet and gulley erosion had drastically reduced soil fertility. Oaks that revegetated these sites produced scrubby growth and restricted root systems as reactions to their attempts at colonizing the heavy subsoils.

Lands in the forested uplands and Appalachian Mountains that escaped clearing for agriculture suffered from the logger. The original dense forests served as resources for the development of large regional markets for product mixes ranging from charcoal to veneer logs. By the first few decades of the 1900s, repeated logging of these sites left only the poorest trees. The increased use of heavy logging machinery severely scarred the sites and contributed to erosion on the steeper slopes. Bole and root injur-

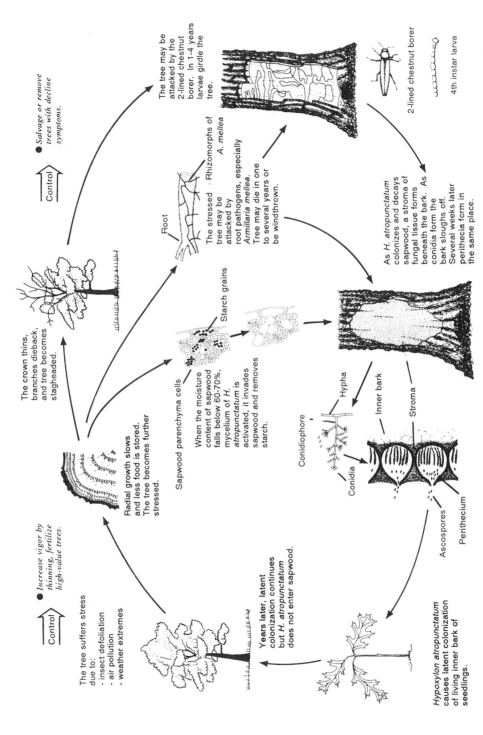

FIGURE 18.1 Disease diagram of oak decline, caused by a combination of abiotic and biotic factors.

ies on surviving trees served as infection courts for a variety of root- and butt-infecting pathogens that contribute to decline.

The loss of the American chestnut from vast areas in the southern Appalachian Mountains resulted in their replacement by various species of oaks which were released from suppression as the chestnuts died. These species may have been less well adapted for growth on stressful sites than the chestnut they replaced; the stands they formed are among those that have subsequently suffered from oak decline.

All the preceding examples illustrate the multitude of predisposing environmental factors (Manion 1991) which have served to make many of the oak forests susceptible to decline and premature death. Leaf defoliators and cambial borers serve as examples of the second group of factors, the inciting factors (Manion 1991). These factors have a physiological action that is short in duration, but this action further reduces the tree's ability to respond to additional stressful events. Single-event defoliations such as the oak leaf tier may not cause significant stress, but if defoliation occurs in consecutive years or is associated with other stresses, irreversible injury can result.

The stage is now set for the action of the third set of factors, which can lead to decline and death—the contributing factors (Manion 1991). Their combined actions are shown in the disease diagram in Figure 18.1. These are usually a series of biotic agents, any or all of which are attracted to the stressed trees and have a quicker, fatal effect. A serious insect is the two-lined chestnut borer, which lays eggs in the branches of stressed trees. The larvae feed in the phloem and kill the branches. This action further reduces the vigor of the affected tree but is only a prelude to a more severe effect. This insect then moves down the tree and eventually kills it by girdling the main stem. Weakly pathogenic fungi, such as *Armillaria* spp. and *Hypoxylon atropunctatum,* also invade and colonize the tissues of these stressed trees. These organisms can take advantage of the food resources in the outer sapwood of the stems and roots. Stressed trees have an altered nitrogen metabolism that leads to higher concentrations of certain nutrients in the roots. These nutrients tend to make roots especially attractive to attack by secondary root pathogens such as *Armillaria* spp. Stressed oaks may not be able to produce sufficient polyphenolics and wound periderm to stop the invading pathogen. As more roots are invaded, the tree is further reduced in vigor, and death often results. The outer stem of a stressed oak contains nitrogen, sugar, and starch in relatively large amounts. As the moisture content of the outer sapwood declines to 60–70%, *H. atropunctatum,* which has heretofore survived as a latent inhabitant of the bark since the tree was a seedling, is stimulated to colonize the sapwood quickly. It metabolizes the carbohydrate cell contents and then produces a soft-rot type of decay of the woody cell walls. *Hypoxylon atropunctatum* is often the last of the contributing factors acting upon stressed oaks and, because it produces highly visible stromata which slough off the bark of the main stem and larger branches, signals the beginning of an intense biodegradation of these trees by a host of other decay fungi, insects, woodpeckers, predators, and other wildlife forms.

REFERENCES

Abrams, M. D. 1992. Fire and the development of oak forests. *Bioscience* 42:346–353.

Ammon, V., T. E. Nebeker, T. H. Filer, F. I. McCracken, J. D. Solomon, and H. E. Kennedy. 1989. *Oak Decline.* Miss. Agri. and For. Exp. Sta., Tech. Bull. 161. 15 pp.

Bartuska, A. M. 1990. Air pollution impacts on forests in North America, pp. 141–154. In W. Grodzinski, E. Cowling, and A. Breymeyer (eds.), *Ecological Risks, Perspectives From Poland and the United States*. National Academy Press, Washington, D.C.

Bassett, E. N., and P. Fenn. 1984. Latent colonization and pathogenicity of *Hypoxylon atropunctatum* on oaks: *Quercus alba, Quercus marylandica, Quercus velutina. Pl. Dis.* 68:317–319.

Biocca, M., F. H. Tainter, D. A. Starkey, S. W. Oak, and J. G. Williams. 1993. The persistence of oak decline in the western North Carolina Nantahala Mountains. *Castanea* 58:178–184.

Burnett, H. 1987. The great drought of 1986. *Am. For.* 93(May/June):22–25.

Delcourt, H. R. 1979. Late Quaternary vegetation history of the eastern highland rim and adjacent Cumberland Plateau of Tennessee. *Ecol. Monographs* 49:255–280.

Frothingham, E. H. 1931. *Timber Growing and Logging Practices in the Southern Appalachian Region*. USDA, Tech. Bull. 250. 93 pp.

Manion, P. D. 1991. *Tree Disease Concepts.* 2nd ed. Prentice-Hall, Englewood Cliffs, N.J. 402 pp.

Matthews, S. W. 1976. What's happening to our climate? *Nat. Geog.* 150:576–615.

Millers, I., D. S. Shriner, and D. Rizzo. 1989. *History of Hardwood Decline in the Eastern United States*. USDA For. Serv., Gen. Tech. Rept. NE-125. 75 pp.

Mueller-Dumbois, D., J. E. Canfield, R. A. Holt, and G. P. Buelow. 1983. Tree-group death in North American and Hawaiian forests: a pathological problem or a new problem for vegetation ecology? *Phytocoenologia* 11:117–137.

Nelson, T. C. 1955. Chestnut replacement in the Southern Highlands. *Ecology* 36:352–353.

Oak, S. W., C. M. Huber, and R. M. Sheffield. 1991. *Incidence and Impact of Oak Decline in Western Virginia, 1986*. USDA For. Serv., Res. Bull. SE-123. 16 pp.

Sayers, W. B. 1971. The king is dead, long live the king. *Am. For.* 77:20–23, 40–41.

Shaw, C. G., III, and G. A. Kile. 1991. *Armillaria Root Disease*. USDA For. Serv., Agri. Handbook 691. 233 pp.

Shigo, A. L. 1985. Wounded forests, starving trees. *J. For.* 83:668–673.

Tainter, F. H., W. A. Retzlaff, D. A. Starkey, and S. W. Oak. 1990. Decline of radial growth in red oaks is associated with short-term changes in climate. *Eur. J. For. Path.* 20:95–105.

Whitehead, D. R. 1981. Late-Pleistocene vegetational changes in northeastern North Carolina. *Ecol. Monographs* 51:451–472.

DISEASE PROFILE

Ash Dieback

Importance: White ash is a valuable species and an important component of the hardwood forests of the northeastern United States. Ash dieback, first noticed in 1930 affecting roadside trees, was widespread on a variety of sites during the period 1955–1965. A survey in New York in 1960–1961 indicated that approximately one third of the ash were affected and approximately 5% were dead.

Ash dieback. From Ross (1966).

Suscepts: White and green ash ages 15–150 years were affected on a variety of sites including hedgerows, forests, and ornamental plantings.

Distribution: Ash dieback occurred from Maine south and west to Pennsylvania and New York, including New Hampshire, Vermont, Massachusetts, Rhode Island, and Connecticut. Dieback was prevalent in New York and especially in the Hudson River Valley.

Causal Agents: Ash dieback was thought to be induced by drought conditions. Weakened trees were invaded by secondary canker fungi that hastened tree death. Two such fungi, *Cytophoma pruinosa* and *Fusicoccum* sp. (Coelomycetes), cause severe stem and branch cankers on weakened trees.

Biology: During periods of drought, ash trees suffered a loss of vigor and subsequent decline. Root and crown deterioration were often accompanied by cankers formed in response to secondary fungi. These facultative parasites developed best during the dormant season unless the tree was in an advanced stage of decline. In the case of *C. pruinosa,* pycnidia formed in the bark on dead areas adjacent to the canker.

Epidemiology: The occurrence of ash dieback was related to rainfall, and disease incidence increased during periods of drought. Examination of plots in New York in 1968 indicated that the disease had not spread or intensified. Secondary canker fungi developed slowly except during very dry periods or when trees were in advanced stages of decline. The sticky spores produced by *C. pruinosa* were thought to be rain-splashed.

Diagnosis: Symptoms were typical of tree decline (i.e., chlorotic, dwarfed, and sparse foliage often tufted on branches, branch and crown dieback, reduced increment and tree mortality). Small trees died in 2–3 years and larger ones in 5–8 years. Stem and branch cankers were yellow to green-brown, slightly sunken with irregular margins. Water sprouts often occurred below the canker. Signs of the canker fungus *C. pruinosa* are pycnidia, which occurred in the bark on dead areas adjacent to the canker.

Control Strategy: Salvage cuttings to use valuable trees are feasible but must be accomplished before or very soon after tree death as deterioration caused by insects and decay fungi occurs rapidly. Thinning may reduce root competition in dense forest stands and irrigation and fertilization of ornamental trees in early stages of decline may be helpful.

SELECTED REFERENCE

Ross, E. W. 1966. *Ash Dieback—Etiological and Developmental Studies.* State Univ. Coll. of For. at Syracuse Univ. N.Y., Tech. Publ. No. 88. 80 pp.

DISEASE PROFILE

Maple Blight

Importance: While declines of sugar maples are common and widespread throughout the hardwood forest of the central and northeastern states, maple blight has special significance. This disease, initiated in 1957 and culminated within a few years, occurred in only a few small areas. Maple blight is of primary importance as an example of the sequence of events (causal agents) that can condition dieback—decline diseases. The present epiphytotic of maple decline in the northeastern states likewise has a complex etiology and is discussed at greater length later in this chapter.

Suscepts: Sugar maple was the species affected by maple blight.

Distribution: Maple blight first occurred in several small areas in northeastern Wisconsin.

Causal Agents: The primary cause of maple blight was defoliation by the maple webworm. However, the sequence of events leading to the blight and culminating in tree mortality is both complex and instructive. Stand composition, climate factors, several leaf-rolling insects, the maple webworm, and a root-decay fungus were involved.

Biology: Abundant light and increased temperatures in open stands with a high proportion of sugar maples provided favorable conditions for the buildup of leaf-rolling insects. Two such insects, *Sparganothus acerivorana* and *Acleris chalyeana* and to a lesser degree *Gracilaria* sp., caused leaf-rolling and defoliation. However, the primary defoliator was the maple webworm. This insect depended on the rolled leaves for a niche in which to lay eggs, and the population of webworms increased because of the leaf rollers. The period of defoliation by the webworm was critical. When insect defoliation occurred in July and early August, new leaves were formed and prior to hardening were subsequently killed by fall frosts. Trees thus weakened by two such defoliations suffered crown decline and attack by *Armillaria* spp. and often died within several years.

Epidemiology: Maple blight occurred in intensely managed, selectively cut, understocked stands with a high percentage of sugar maple. Increased light and temperature in opened stands favored the buildup of leaf rollers, which in turn favored the increase of the webworm defoliator. Blight occurred only in areas that were severely defoliated in 1957. There was no evidence of spread or intensification, and recovery was evident after 1957, although some trees continued to decline due to severe crown dieback and secondary

Maple blight. From Anonymous (1964).

invaders. Neither site, age, nor size of tree was associated with maple blight, although very young trees were not affected.

Diagnosis: Crown dieback from the tip toward the interior was the primary symptom. Water sprouts on the living portion of affected branches were evident, and in some cases limb and stem lesions occurred. Of course, these symptoms were preceded by defoliation caused by insects and fall frosts. Because of the suddenness of the defoliation, there was no evidence of a slow reduction in increment or prolonged rootlet mortality which often occurs in slowly developing declines.

Control Strategy: There are two silvicultural opportunities to prevent maple blight. These are control stand density to prevent understocked conditions that allow the increase of defoliating insects and control species composition to ensure that stands do not contain a high proportion of sugar maple conducive to a buildup of maple leaf-feeding insects.

Other Maple Declines: There are abundant records of various maple declines and diebacks, and these, while of somewhat similar etiology, should not be confused with the maple blight described here.

SELECTED REFERENCES

Anonymous. 1964. *The Causes of Maple Blight in the Lake States*. USDA For. Serv., Res. Pap. LS-10. 15 pp.

Giese, R. L., D. R. Houston, D. M. Benjamin, J. R. Kuntz, J. E. Kapler, and D. D. Skilling. 1964. *Studies of Maple Blight*. Univ. of Wisc., Res. Bull. 250. 128 pp.

DISEASE PROFILE

Pine Wood Nematode

Importance: The pine wood nematode is a serious problem in Japan, apparently introduced from the United States. Losses are extensive in Japanese black and red pines. In North America, injury to native conifers is minor; however, a complex of organisms is involved in the demise of both native and exotic stressed trees. The pine wood nematode is not known to occur in Europe, and the European Plant Protection Organization has banned import of softwood products except kiln-dried lumber from North America. This embargo has a substantial impact on timber exports from North America.

Suscepts: The nematode affects seedlings of almost all pines tested; larch, spruce, firs, and cedars are also susceptible. From an economic standpoint, the most important host in the United States is Scots pine, which is often used as an ornamental.

Wilt symptoms in white pine.

Distribution: Associated organisms are probably present throughout the natural range of conifers in North America, but losses occur only where mean summer temperature exceeds 20°C for 6–8 weeks.

Causal Agents: The disease is associated with the nematode *Bursaphelenchus xylophilus*. In Japan, where the disease had been attributed to Cerambycid activity, the nematode was previously described as *B. lignicolus*. Sawyer beetles in the genus *Monochamus* (Cerambycidae) are largely responsible for spread of the nematode. In Japan *M. alternatus* is the primary vector; in the United States *M. carolinensis, M. scutellatus,* and *M. titillator* are the major vectors.

Biology: The nematode is intimately associated with bark beetles. Emerging sawyer beetles carry the nematode in their tracheae and then exhibit one of two behaviors, both of which influence the fate of the nematode. In the phytophagous phase, some sawyers fly off to healthy trees and feed on the tender bark of young twigs. A dispersal stage of the nematode leaves the insect through the spiracles and enters the feeding wound. Once in the wood the nematodes moult to become adults and then mate and reproduce rapidly in resin canals. They feed on parenchyma and epithelial cells and may secrete a toxin that kills cells in advance of their activity. Upon emergence some sawyers are attracted to recently dead, dying, or stressed trees and may introduce the nematode during oviposition, so that if dying trees do not contain the nematode, it will be introduced. This is called the mycophagous phase because the nematodes feed on fungi in the wood, including the blue-stain fungi, which are introduced by other bark beetles. A dying tree may contain millions of nematodes. Larvae of the sawyer beetles develop in these trees, and when they pupate, they stimulate the nematode to produce the dispersal stage, also known as dauerlarvae. As the pupae mature, the dauerlarvae enter the beetle through the spiracles and leave the host tree with the beetle. Pine wood nematodes will readily kill seedlings. When inoculated into larger, vigorous trees, however, the nematodes do not incite a disease. The stresses affecting an introduced plant such as Scots pine may play a role in conditioning their susceptibility to the nematode.

Epidemiology: Understanding the epidemiology of this disease is difficult because of the two different phases of the life cycle. A tree cannot be considered to be diseased just because the nematodes are present. Infested trees must be examined for other primary causes of death or stress. In Japan, drought and high summer temperatures are factors in disease development.

Diagnosis: The symptoms are similar to those attributed to drought: affected trees wilt and die within 40–50 days. Secondary organisms often obscure the true cause of death.

Control Strategy: Insecticides have been used to prevent attack by the Cerambycids in Japanese forests with some success, but such applications have not been used in the United States. Perhaps the best management practice is to plant on suitable sites. If Scots pine is to be used as an ornamental in the warmer areas of the United States, adequate water must be provided.

SELECTED REFERENCES

Dwinell, L. D. 1986. *Ecology of the Pinewood Nematode in Southern Pine Chip Piles*. USDA For. Serv., Res. Pap. SE-258. 14 pp.

Rutherford, T. A., Y. Mamiya, and J. M. Webster. 1990. Nematode-induced pine wilt disease: Factors influencing its occurrence and distribution. *For. Sci.* 36:145–155.

18.3 PINE WOOD NEMATODE

Importance

The pine wood nematode causes a disease known as pine wilt and is a serious problem in Japan, apparently introduced from the United States. Losses are extensive in forests of Japanese black pine and Japanese red pine. These forests are extremely vulnerable because they are composed almost entirely of these two species. The causal organism is also present in North America, but damage to native conifers is minor. However, this complex of organisms is involved in the demise of stressed trees, both native and exotic, and has been indirectly damaging to the United States and Canada because of perceived problems. The pine wood nematode is not known to occur in Europe; because the Europeans do not want it, the European Plant Protection Organization has banned import of softwood products except kiln-dried lumber from North America. This embargo will have a substantial impact on timber exports from North America. The province of British Columbia alone stands to lose an annual market worth $600 million CAN; the southern United States will lose $20 million annually.

Suscepts

The pine wood nematode affects seedlings of almost all pines tested; larch, spruce, firs, and cedars are also susceptible. Large trees become diseased only where the mean summer temperature exceeds 20°C for 6–8 weeks. Slash pine is the only native species that has died when inoculated. From an economic standpoint, the most important host in the United States is the exotic Scots pine, which is often used as an ornamental. Japanese black pine and Japanese red pine also wilt when inoculated with the nematodes. Other exotic pines are probably also affected.

Distribution

Although a map has been compiled of the known distribution of the pine wood nematode in North America (Fig. 18.2), the nematode probably occurs throughout the natural range of conifers in North America. The pine wood nematode, however, as mentioned before, occurs only where the mean summer temperature exceeds 20°C for 6–8 weeks.

Cause

The pine wood nematode is *Bursaphelenchus xylophilus* (Steiner & Buhrer) Nickle. In Japan, where the disease had been atributed to Cerambycid activity, the nematode was previously described as *B. lignicolus* (Mamiya & Kizohara). Nematodes from Japan produce fertile offspring when mated with nematodes from the United States, and thus the two nematodes are considered to be of the same species (conspecific). A morphologically identical nematode was discovered in preserved insect specimens collected in the United States. This nematode had been described in 1934 as *Aphlenchoides xylophilus*. Because of the rules of nomenclature, the pine wood nematode is known as *Bursaphelenchus xylophilus*.

Insects, primarily the sawyer beetles in the genus *Monochamus* (Cerambycidae), are

FIGURE 18.2 The approximate known distribution of pine wood nematode in North America.

responsible for the spread of the nematode. In Japan *M. alternatus* is the primary vector; in the United States *M. carolinensis, M. scutellatus,* and *M. titillator* are the major vectors.

Biology

Members of the genus *Bursaphelenchus* are intimately associated with bark beetles. The sawyer beetles emerge carrying the nematode in their tracheae and then exhibit one of two behaviors, both of which influence the fate of the nematode. Some sawyers fly off to healthy trees and feed on the tender bark of young twigs, creating a wound.

During this feeding, a dispersal stage of the nematode leaves the insect through the spiracles and enters the feeding wound. Once in the wood, the nematodes moult to become adults; they mate and reproduce rapidly in the resin canals of the host. The nematodes feed on parenchyma and epithelial cells and may secrete a toxin that kills cells in advance of their activity. This part of the life cycle is called the phytophagous phase, because the nematodes are feeding on plant cells.

Upon emergence, or after maturation feeding, some sawyers are attracted to recently dead, dying, or stressed trees to oviposit in them. The insects may introduce the nematode into trees during oviposition, so that, if dying trees do not contain the pine wood nematode, it will be introduced. This is called the mycophagous phase, because the nematodes feed on fungi in the wood. These fungi include the blue-stain fungi, which are introduced into the tree by other bark beetles.

A dying tree may contain millions of nematodes. Larvae of the sawyer beetles develop in these trees, and when they pupate, they stimulate the nematode to produce the dispersal stage, also known as dauerlarvae. As the pupae matures, the dauerlarvae enter the beetle through the spiracles and leave the host tree with the beetle.

In most native forests of North America, the pine wood nematode is probably one of the early colonizers of dead trees or dead parts of trees. The nematode is carried there by pine sawyers using the dead tree for brood material. Other common associates of the pine wood nematode include species of *Ips,* blue-stain fungi, and mites and bacteria carried by colonizing bark beetles. This complex of organisms occurs in trees killed recently by many primary diseases including white pine blister rust, *Dothistroma pini, Armillaria,* dwarf mistletoe, and *Sphaeropsis sapinea,* and insects such as the southern pine beetle. Thus it is easy to understand the confusion about whether the pine wood nematodes are pathogenic. Any conditions that favor attack by members of the secondary complex will encourage attacks by other members, and the pine wood nematodes will be introduced into the tree. With their rapid reproductive capacity, there will be millions of nematodes in a tree shortly after their introduction. We could easily attribute the mortality to the nematode. Many have done so erroneously.

This mistake poses an interesting question for pest managers: When is a tree killed? We have no way to monitor "brain waves" or other vital signs. Trees may continue to function long after they have received a fatal injury. For example, a lodgepole pine girdled in Utah may remain green for at least 2 years after girdling, yet we know it will die. Pathologists can use only experience to guide them. We know that when certain pathogens (and insects) attack a tree, it will most certainly die. Other fungi and insects appear only on weakened trees; they rarely attack vigorous trees. Most consider them secondary invaders, yet they are colonizing living trees. As a student of pest management, you must first understand what is normal for a tree. Only then can you see deviations from the normal. Keep in mind that a tree may be stressed by factors that cause no symptoms and that may not affect us. Abnormal dry periods of a few days or weeks may not trigger a reaction in forest managers but can do so in forest trees.

To determine if the pine wood nematode is a pathogen, Koch's Rules of Proof must be repeated. That is, the nematodes must be constantly associated with declining trees; they must be isolated in pure culture; the pure culture must be placed into living trees and reproduce the disease; and the nematodes must be reisolated from the diseased trees. Because of the ease of obtaining and working with seedlings, they are often used for inoculations. The pine wood nematodes will readily kill seedlings.

When inoculated into larger, vigorous trees, however, the pine wood nematodes do not incite a disease. Only two instances are known where inoculation of pine wood nematodes have resulted in mortality of large trees: one with Scots pine in Iowa, and the other with slash pine in Florida. The disease seems to be most important on Scots pine, an exotic species widely used as an ornamental tree. The stresses affecting an introduced tree species such as Scots pine may play a role in conditioning their susceptibility to pests such as the pine wood nematode.

Epidemiology

Understanding the epidemiology of this disease is difficult because of the two different phases of the life cycle. A tree cannot be considered to have pine wilt just because the nematodes are present in the tree; they may have been introduced during the mycophagous phase by breeding insects. Because these insects prefer to breed in stressed or dying trees, infested trees must be examined for other primary causes of death or stress. Drought conditions have been associated with increase in pine wilt, perhaps because the population of Cerambycid vectors increases and because the trees are more susceptible to the nematodes. Even in Japan, drought and high summer temperatures are considered factors in disease development.

Diagnosis

The symptoms are similar to those attributed to drought: affected trees wilt and die within 40–50 days (Fig. 18.3). Secondary organisms may attack affected trees, often obscuring the true cause of death.

FIGURE 18.3 Drooping needles on white pine, often associated with infestation by pine wood nematodes.

Control Strategy

Insecticides have been used to prevent attack by the Cerambycids in Japanese forests with some success, but such applications have not been used in the United States. Perhaps the best management practice is to plant species on suitable sites. If Scots pine is to be used as an ornamental in the warmer areas of the United States, adequate water must be provided.

SELECTED REFERENCES

Appleby, J. E., and R. B. Malek (eds.). 1982. Proc. 1982 National Pine Wilt Disease Workshop. Univ. Ill. 137 pp.

Bedker, P. J., and R. A. Blanchette. 1988. Mortality of Scots pine following inoculation with the pinewood nematode, *Bursaphelenchus xylophilus*. *Can. J. For. Res.* 18:574–580.

Dwinell, L. D. 1986. *Ecology of the Pinewood Nematode in Southern Pine Chip Piles*. USDA For. Serv., Res. Pap. SE-258. 14 pp.

Dwinell, L. D., and W. R. Nickle. 1989. *An Overview of the Pine Wood Nematode Ban in North America*. USDA For. Serv., Gen. Tech. Rep. SE-55. 13 pp.

Luzzi, M. A., R. C. Wilkinson, and A. C. Tarjan. 1984. Transmission of the pinewood nematode, *Bursaphelenchus xylophilus*, to slash pine trees and log bolts by a Cerambycid beetle, *Monochamus titillator*, in Florida. *J. Nematol.* 16:37–40.

Mamiya, Y. 1983. Pathology of the pine wilt disease caused by *Bursaphelenchus xylophilus*. *Annu. Rev. Phytopathol.* 21:201–220.

Rutherford, T. A., Y. Mamiya, and J. M. Webster. 1990. Nematode-induced pine wilt disease: Factors influencing its occurrence and distribution. *For. Sci.* 36:145–155.

Wingfield, M. J. (ed.). 1987. *Pathogenicity of the Pine Wood Nematode*. Amer. Phytopathol. Soc. Symp. Series, APS Press. St. Paul, MN. 122 pp.

Wingfield, M. J., R. A. Blanchette, and T. N. Nicholls. 1984. Is the pine wood nematode an important pathogen in the United States? *J. For.* 82:232–235.

Wingfield, M. J., R. A. Blanchette, T. N. Nicholls, and K. Robbins. 1982. Association of the pine wood nematode with stressed trees in Minnesota, Iowa, and Wisconsin. *Pl. Dis.* 66:934–937.

DISEASE PROFILE

Sapstreak

Importance: The amount of loss caused by sapstreak is not known, although locally and at times it has caused substantial degrade and mortality. At one time in the early 1960s, 10% of the sugar maple in one stand in the Upper Peninsula of Michigan were infected.

Wilt symptoms. From Kessler (1972).

Stained wood. From Skelly (undated).

These infected trees were either dominants or codominants. Sapstreak reduced the value of saw logs 32% and lumber 57%.

Suscepts: The sapstreak disease of sugar maple may be the best known of the diseases in this classification, but other hosts have similar problems. Yellow-poplar is affected. In Russia, oaks and birch are affected by a similar fungus. It is very reasonable to assume that there are other diseases caused by related fungi that have not been identified.

Distribution: Sapstreak of maple has been reported in North Carolina (about 1935), Michigan, Wisconsin, and Vermont. It undoubtedly occurs in other states as well. Sapstreak of yellow-poplar is scattered throughout North Carolina and Tennessee. The disease on oaks and birch occurs in Russia.

Causal Agents: Sapstreak in maple and yellow-poplar is caused by *Ceratocystis virescens (C. coerulescens)* (Ophiostomatales). The Russian fungus is *C. roboris.*

Biology: The incidence of sapstreak is related to some extent with logging injuries, and *C. virescens* enters through these wounds at the base of the tree or in the roots. How the fungus kills the tree is not understood, but presumably it interferes with the metabolism of the host by depleting or converting necessary nutrients. Some plugging of the xylem tissues may occur. The fungus readily sporulates on wood surfaces, producing both conidia and ascospores. As with other species in this genus, the spores are sticky and well adapted to spread by insects.

Epidemiology: The disease incidence is probably related to availability of inoculum, suitable wounds in the trees, and adequate vectors. The fungus usually enters at the base of the tree, but it has been found as high as 17 m and in one case had developed from a wound 13 m aboveground.

Diagnosis: In the early stages of the disease, the foliage is dwarfed in portions of the crown, and this may spread to the balance of the crown in subsequent years. Dieback then occurs, and after 3 or 4 years the tree may die. The sapstreaking is very apparent when the trees are felled and may occur over most of the cross section of the tree. The tips of discoloration extending toward the cambium can be greenish in color in contrast to the general grayish color in the center. The fungus will develop profusely over the exposed surface, sporulate abundantly, and produce an isobutyl acetate (banana oil) odor that is quite distinctive.

Control Strategy: Until more is known about this complex of diseases and the fungus or fungi involved, there is little to recommend in the way of control measures. Undoubtedly it is important to avoid wounding sugar maple trees. This is important not only to prevent sapstreak but also to restrict entry of all other stains and decay organisms. Recommended control measures, which include removing infected trees and painting stump surfaces to prevent sporulation, have no basis in fact at the present time.

SELECTED REFERENCES

Houston, D. R. 1993. *Recognizing and Managing Sapstreak Disease of Sugar Maple.* USDA For. Serv., Res. Pap. NE-675. 11 pp.

Kessler, K. J., Jr. 1972. *Sapstreak Disease of Sugar Maple.* USDA For. Serv., For. Pest Leaflet 128. 4 pp.

Skelly, J. M., ed. Undated. *Diagnosing Injury to Eastern Forest Trees.* Nat. Acid Precip. Assess. Program. 122 pp.

Wilson, C. L. 1967. Vascular mycosis of oak in Russia. *Pl. Dis.Rep.* 51:739–741.

19

MISTLETOES

19.1 BIOLOGY OF MISTLETOES

Most parasitic flowering plants are contained within the following families: the mistletoes and dwarf mistletoes (Loranthaceae and Viscaceae); the sandalwoods (Santalaceae, Olacaceae, and Myzodendraceae); the broom-rapes, parasitic figworts, and witchweeds (Orobanchaceae and Scrophulariaceae); the dodders (Convolvulaceae, Lauraceae, Lennoaceae, and Krameriaceae); and Rafflesiaceae, Hydnoraceae, and Balanophoraceae.

Dodder, leafy mistletoes, and dwarf mistletoes are among the most important parasitic flowering plants causing economically important damage to forest trees in North America. These parasites cause injury primarily by diverting nutrients and water from their hosts.

19.2 DWARF MISTLETOES

Of all the parasitic flowering plants, the dwarf mistletoes are of the greatest concern in the North American conifer forests and are recognized as the single most important disease problem in the western conifer forests of the United States. As the older timber has been harvested, dwarf mistletoe has become an even more important problem than heart decays (Table 10.1).

Importance

Dwarf mistletoe is a major problem on ponderosa pine but also causes serious losses on lodgepole pine, Douglas-fir, true firs, western larch, spruces, and other western conifers. In Oregon and Washington diseases cause an annual reduction in wood produced of 3,133 million bd ft or 37 million m³, which amounts to about 13% of the total wood produced each year. The losses consist of 15 million m³ of growth loss, 12 million m³ of mortality, and 10 million m³ of cull. Growth loss is caused primarily by the dwarf mistletoes and root diseases (7 million and 6 million m³, respectively). Mortality is due mainly to dwarf mistletoes and root diseases (7 million and 5 million m³). Heart decay, or rots, cause all but 0.2 million m³ of the cull.

Black spruce is the primary dwarf mistletoe host in the Lake states. In Minnesota the area out of production due to dwarf mistletoe has been estimated at as much as 11%, which would involve approximately 62,000 ha of commercial black spruce type. In Arizona and New Mexico 36% of the ponderosa pine stands are infected, and the parasite is present on at least 1 million ha of the type, involving an annual volume loss of 55 million to 75 million bd ft and an additional 20 million to 27 million bd ft in Douglas-fir. Larch dwarf mistletoe was found in 64% of the stands sampled in the Coeur d'Alene National Forest. In the Clark Fork unit in western Montana, dwarf mistletoes were present in 23% of the sample plots. Surveys in the Colville National Forest and adjacent private lands in northeastern Washington revealed widespread, severe infection in Douglas-fir and western larch stands. By species, 80% of the Douglas-fir volume and 92% of the larch volume were in infected trees. Dwarf mistletoe was found in approximately 70% of all larch stands sampled in the Kootenai National Forest, Montana. In the central Rocky Mountains, dwarf mistletoe was present in 51% of the area surveyed, and the volume in healthy stands was 1.5 times that in infected stands. Red fir has been eliminated in some locations in California following heavy infections.

The losses from dwarf mistletoe are not only measured by mortality but also growth reduction, poorer quality, less seed production, wind breakage, and predisposition to insects and decay. Infected trees do not die rapidly, but rather death results from a gradual suppression of growth. Diameter and height growth loss in lodgepole pine averaged 0.7%/yr since the time of infection. Lightly infected western hemlock

FIGURE 19.1 Dwarf mistletoe infection of black spruce.

trees had 41% greater volume growth and 84% greater height growth than severely infected trees. Western hemlock lost nearly 5.5 m³/hc/yr. The quality of wood from infected trees is lowered because of more large knots, abnormal grain, and spongy, pitchy wood.

Diagnosis

The portion of the tree invaded by the parasite becomes swollen and, as the result of excessive branching, witches' brooms develop (Fig. 19.1). In black spruce, infected branches are stimulated to develop as new leaders. Occasionally this happens several times on the same tree. In contrast to the temporarily more rapid growth of the infected portion of the tree, the growth rate in the balance of the tree is drastically reduced, and the original leader dies. Eventually the entire tree dies. Infection centers containing several infected trees appear as irregular circular patches that can be detected rather easily on aerial photographs. Once the aerial shoots are produced, dwarf mistletoe is obvious, and these shoots indicate the species of *Arceuthobium*.

Suscepts

There are approximately 40 species of dwarf mistletoes. Most of these species occur in North America (Fig. 19.2). Areas containing the highest concentrations of dif-

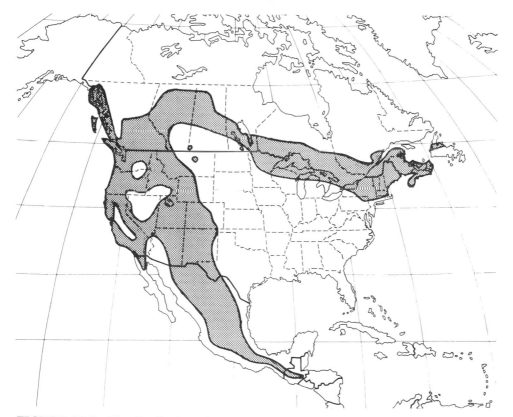

FIGURE 19.2 The distribution of *Arceuthobium* in North America. Redrawn from Hawksworth and Wiens (1972).

ferent species are in Durango, Mexico, and in the Mt. Shasta area of northern California.

All species of *Arceuthobium* and their principal hosts follow. The more important species are marked with an asterisk.

Arceuthobium	*Principal Host*
A. abietinum f. sp. concoloris*	Abies concolor, Abies grandis
A. abietinum f. sp. magnificae	Abies procera, Abies magnifica
A. abietis-religosae	Abies religiosae
A. americanum*	Pinus banksiana, Pinus contorta subsp. latifolia, Pinus contorta subsp. murrayana
A. apachecum	Pinus strobiformis
A. bicarinatum	Pinus occidentalis
A. blumeri	Pinus strobiformis
A. californicum	Pinus lambertiana
A. campylopodum	Pinus ponderosa, Pinus jeffreyi, Pinus attenuata
A. cyanocarpum	Pinus flexilis, Pinus aristata
A. divaricatum	Pinus edulis, Pinus monophylla, Pinus quadrifolia, Pinus cembroides
A. douglasii*	Pseudotsuga menziesii
A. gillii subsp. gillii	Pinus leiophylla chihuahuana
A. gillii subsp. nigrum	Pinus leiophylla leiphylla, Pinus leiophylla chihuahuana, Pinus lumholtzii, Pinus teocote
A. globosum	Pinus cooperi, Pinus douglasiana, Pinus durangensis, Pinus engelmannii, Pinus hartwegii, Pinus lawsonii, Pinus michoacana, Pinus montezumae, Pinus pringlei, Pinus pseudostrobus, Pinus rudis, Pinus tenuifolia
A. hondurense	Pinus oocarpa
A. laricis*	Larix occidentalis
A. microcarpum	Picea engelmannii, Picea pungens
A. occidentale	Pinus sabiana, Pinus radiata, Pinus muricata
A. pusillum*	Picea mariana, Picea glauca
A. rubrum	Pinus cooperi, Pinus durangensis, Pinus engelmannii, Pinus teocote
A. strictum	Pinus leiophylla chihauhuana
A. tsugense*	Tsuga heterophylla, Tsuga mertensiana, the interior form of Pseudotsuga menziesii
A. vaginatum subsp. vaginatum*	Pinus cooperi, Pinus engelmannii, Pinus hartwegii, Pinus montezumae, Pinus ponderosa arizonica, Pinus ponderosa scopulorum, Pinus rudis, Pinus lawsonii
A. vaginatum subsp. cryptopodum	Pinus ponderosa scopulorum, Pinus ponderosa arizonica, Pinus engelmannii
A. vaginatum subsp. durangense	Pinus montezumae, Pinus durangensis
A. verticilliflorum	Pinus cooperi, Pinus engelmannii

Some of the preceding species were formerly subdivided into form species or varieties that are specific for certain hosts. Thus *A. abietinum* f. sp. *concoloris (A. campylopodum)* was found only on white fir, and *A. abietinum* f. sp. *magnifica* could infect only red fir.

Biology

The disease diagram for the eastern dwarf mistletoe is shown in Figure 19.3. Species of *Arceuthobium* are dioecious. At least 25 species of insects visited pistillate and staminate flowers of *A. americanum* suggesting that this and other species are insect pollinated (Fig. 19.4, left). The fruit is a one-seeded berry (Fig. 19.4, right), in which the seed is surrounded by a mucilaginous pulp called viscin.

The seeds of the dwarf mistletoe plants are expelled forcibly for an average distance of about 4.6–9.1 m, though maximal distances of 14.6 m or more have been reported. Seeds shot from branches of trees may have a dispersal radius of 10 m or more depending on height of origin and, to a minor extent, wind. New infection centers probably result from seed dissemination by birds. The seeds adhere to objects they strike because of their mucilaginous coating of viscin. The seeds may begin to germinate within a few weeks, or, as is true of most species, germination is delayed until the following spring. Apparently dwarf mistletoe seeds can photosynthetically fix carbon dioxide, and this capability may help seeds to overwinter in a viable condition. After germinating, the emerging radicle comes in contact with a needle fascicle or other irregularity in the bark surface where a holdfast forms, and the primary haustorium penetrates the bark.

Once established in the cortex of its host, the parasite develops a more extensive absorption system, a portion of which is gradually included in the xylem of the host as a result of the xylem growing around structures called sinkers (Fig. 19.5). Vessel elements in these sinkers connect with tracheids in the host xylem, and the parasite derives most of its nourishment from the host. The mistletoe plants contain abundant chloroplasts, but the amount of carbon dioxide they fix is relatively small. In ponderosa pine a minimum of 4 years and usually 5 years or more are required for completion of a cycle from seed dispersal to the production of new plants with seed. *Arceuthobium americanum* on lodgepole pine will produce fruit 6 years after seed is deposited on the host. Aerial shoots may bear one crop of seeds or many, depending upon the species; after the aerial shoots are shed, a persistent basal cup remains.

The endophytic system of *Arceuthobium pusillum*, which causes witches' brooms to form on black spruce, completely encircles most of the twigs of the broom for their full length, just outside the cambium (Fig. 19.4). New mistletoe shoots may grow out from this tissue. The endophytic systems of the mistletoe plants is perennial and usually remains alive until the host dies.

Supposedly, dwarf mistletoes require considerable light to develop well; therefore, they are most prevalent in open stands or on upper branches. The spread of the parasite is very slow, depending on several factors. It is more rapid from overstory trees to an understory, moving no more than an average of 0.3–0.6 m/yr in even-aged stands. In more open stands, dwarf mistletoe will move about 1/3 more rapidly than in dense stands. No relationship with site productivity is evident. Dwarf mistletoe occurs more commonly on ridges and least commonly on bottom lands. On the plateau adjacent to the Grand Canyon in Arizona, the dwarf mistletoe is restricted to a 1.3-km-wide belt

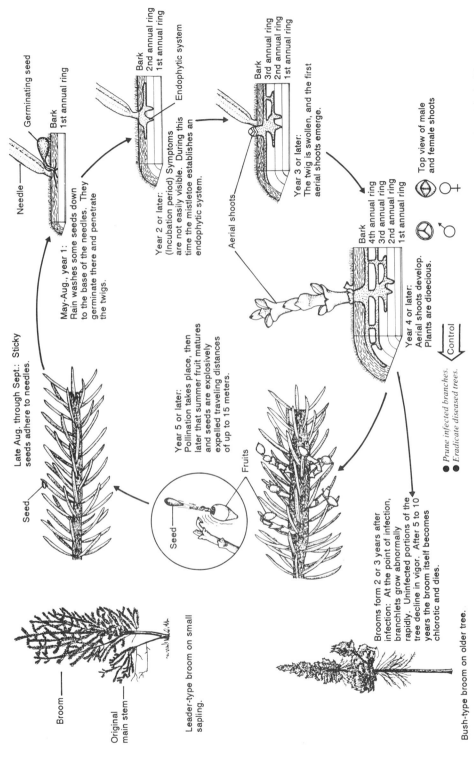

Needle

Germinating seed

Bark

1st annual ring

May-Aug., year 1:
Rain washes some seeds down to the base of the needles. They germinate there and penetrate the twigs.

Bark
2nd annual ring
1st annual ring

Endophytic system

Year 2 or later:
(Incubation period) Symptoms are not easily visible. During this time the mistletoe establishes an endophytic system.

Bark
3rd annual ring
2nd annual ring
1st annual ring

Year 3 or later:
The twig is swollen, and the first aerial shoots emerge.

Top view of male and female shoots

Aerial shoots

Bark
4th annual ring
3rd annual ring
2nd annual ring
1st annual ring

Year 4 or later:
Aerial shoots develop. Plants are dioecious.

Control

● *Prune infected branches.*
● *Eradicate diseased trees.*

Late Aug. through Sept.: Sticky seeds adhere to needles.

Seed

Year 5 or later:
Pollination takes place, then later that summer fruit matures and seeds are explosively expelled traveling distances of up to 15 meters.

Fruits

Seed

Brooms form 2 or 3 years after infection: At the point of infection, branchlets grow abnormally rapidly. Uninfected portions of the tree decline in vigor. After 5 to 10 years the broom itself becomes chlorotic and dies.

Broom

Original main stem

Leader-type broom on small sapling.

Bush-type broom on older tree.

FIGURE 19.3   Disease diagram of eastern dwarf mistletoe on black spruce caused by *Arceuthobium pusillum*. Drawn by Valerie Mortensen.

FIGURE 19.4   *Arceuthobium pusillum,* the eastern dwarf mistletoe: (left) male flowers and (right) mature fruits.

paralleling the south rim of the Canyon, even though ponderosa pine extends for several kilometers beyond this. Apparently dwarf mistletoe increases at higher elevations though in some locations it is most abundant at the median altitude.

Control Strategy

In the past, fire has been the major factor in reducing the incidence of dwarf mistletoe. With improved fire control in more recent times, the dwarf mistletoes have been allowed to cause ever-increasing losses. Eradication is one possible method of reducing losses caused by dwarf mistletoe. Because dwarf mistletoes are obligate parasites, they die when their host is killed. The most practical means of eradication is to clear-cut infection centers as part of a logging operation followed by prescribed burning to eliminate latent and undetected infections. In black spruce in Minnesota, clear-cutting infected trees 20–40 m into the surrounding healthy trees is of questionable value. Clear-cutting followed by fire will totally eradicate black spruce and its parasite and

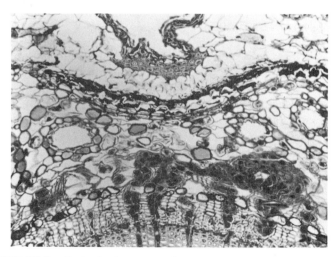

FIGURE 19.5   Endophytic system of *Arceuthobium pusillum* in black spruce.

provide a very suitable seed bed for regenerating the spruce. For large infection centers where it may not be possible to burn all the infected trees, it may be necessary to plant a nonsusceptible host between residual infected trees and the newly cleared area. Initially, fuel oil was used to ensure an adequate fire, but subsequently effective fires were obtained without the use of fuel oil. More recently, cutover areas have been burned efficiently using a helicopter with a drip torch or ignited fuel dropped in ping pong ball-sized capsules. Thus a very efficient economical method for controlling dwarf mistletoe in black spruce is now available. An essential part of any management program, especially when dealing with extensive relatively inaccessible acreages as is the case with black spruce, is detection and mapping of the infection centers. Infection centers can be detected with infrared 35mm photography.

When dwarf mistletoes cannot be eradicated from a stand, as is most often the case in western states, the goal of dwarf mistletoe management is to reduce the dwarf mistletoe rating (DMR) such that it will be lower at the next stand entry. This is possible because dwarf mistletoes are slow-moving pathogens; the infestation in the stand took many years to become severe, and it will change slowly in the ensuing years. When possible, extend the treatment area 20–40 m beyond obvious infection. This will subject to treatment those trees with infections in their incubation period. Using natural barriers to the dwarf mistletoe will prevent susceptible regeneration from becoming infested. Stands of nonhost species, ridge tops, streams, or roads can prevent the further spread of the parasite.

If a susceptible species is to be regenerated in the understory, heavily infected trees in the overstory must be removed. Trees with DMR of 4 or less may be left for seed trees, but these must be removed as soon as the regeneration is established. Trees with DMR of 5 or 6 are poor seed trees for two reasons: they produce few viable seeds because they have been weakened by the dwarf mistletoe, and they produce abundant dwarf mistletoe seeds that will threaten regeneration. Heavily infested understory trees should be removed in future entries; infested advanced regeneration may be discriminated against.

In recreation sites, dwarf mistletoe can be managed by pruning. Pruning dwarf mistletoe brooms is especially effective in reducing the impact of the parasite on the host and restoring tree vigor. However, pruning must be repeated at 3- to-5 year intervals, because the dwarf mistletoe plants will end their incubation and become visible.

Trees with bole infections should be discriminated against during stand entries because of their cull factor. These infections pose little threat of spreading the disease to adjacent trees. Bole infections may be lethal to trees less than 5 cm in diameter, but have little impact on larger trees.

Chemical and biological controls are of only limited value at present but have some promise for the future. Chemical controls such as the use of herbicides have effectively killed aerial portions of dwarf mistletoe plants but have not killed the entire parasite without killing the host. Some fungi that are parasitic on the mistletoe itself, as well as insects, birds, and mammals, reduce the amount of seed available for dispersal, but their total effect is localized and usually not significant. No serious attempt has been made to use some of the fungi, such as *Colletotrichum gloeosporioides, Septogloeum gillii,* and *Wallrothiella arceuthobii,* in controlling dwarf mistletoe. Among the many insects that attack mistletoe plants, those that may reduce the amount of seed are a spittlebug

(Clastoptera obtusa), a plant bug *(Neoborella tumida)*, and larvae of certain species of *Lepidoptera.*

In western United States and Canada, eradication of infection centers using fire is not always acceptable. Fire has been recommended for use in lodgepole pine in Alberta, at least for eliminating the nonmerchantable infected trees. Another and more common method of managing the dwarf mistletoes in the West is to take that alternative action, which predicts maximum yield for that site and the current disease situation. An example provided by Hawksworth (1973) follows. A lodgepole pine stand is 50 years old and heavily infected; the stand could be left as is, or the stand can be salvaged immediately and a new stand started. If nothing is done, the present 42 m^3/ha would increase to 68 m^3 in 30 years and then decline as trees die. If, however, a new stand is generated, volume would increase and after 50 years surpass that of the untreated stand and in 100 years be six times that in the untreated stand.

It is important to protect young stands from infection, and young stands of lodgepole pine up to 15 years of age should have priority because disease intensity is lowest in such stands. Old, infected residual trees bordering these young stands should be cut first. Eradication in the young stand can be accomplished by tree removal or pruning. Two treatments at 3-year intervals are suggested. In stands 16–40 years old, all trees should be removed up to a distance of 15 m beyond the infection center. An additional 12 m should have diseased trees or diseased branches removed. Stands over 40 years old, if lightly infected, can be left, and if heavily infected, removed as soon as reasonable.

Resistant varieties is one approach to the management of the dwarf mistletoes that has not been fully explored. Roth (1966) has offered some explanations for the lack of resistance in ponderosa pine. Limited seed dispersal has prevented general epiphytotics exerting widespread selection pressure. Infection centers are subject to fire, and these areas are reseeded by adjacent susceptible trees. An exception does exist in ponderosa pine, however, and especially in Oregon's Rogue River Valley where a variety of ponderosa pine with drooping needles is common. On these trees the mistletoe seed, rather than sliding down the needle toward the stem, slides off the end of the needle. Dwarf mistletoe is almost absent in the Rogue River Valley.

For high-value trees in recreation sites, an ethylene-releasing chemical "cerone" when applied from the ground has caused abscission of mistletoe plants but does not eliminate the endophytic system. Eliminating seed production will prevent further spread of the mistletoe for a year or two. In these stands, pruning would permanently free the trees from dwarf mistletoe, and improve their vigor.

SELECTED REFERENCES

Acciavatti, R. E., and M. J. Weiss. 1974. Evaluation of dwarf mistletoe on Engelmann spruce, Fort Apache Indian Reservation. *Pl. Dis. Reptr.* 58:418–419.

Alexander, M. E., and F. G. Hawksworth. 1975. *Wildland Fires and Dwarf Mistletoes: A Literature Review of Ecology and Prescribed Burning.* USDA For. Serv., Gen. Tech. Rept. RM-14. 12 pp.

Barrett, J. W., and L. F. Roth. 1985. *Response of Dwarf Mistletoe-Infested Ponderosa Pine to Thinning: 1. Sapling Growth.* USDA For. Serv., Res. Pap. PNW-330. 15 pp.

French, D. W., M. P. Meyer, and R. L. Anderson. 1968. Control of dwarf mistletoe in black spruce. *J. For.* 66:359–360.

Gill, L. S. 1935. *Arceuthobium* in the United States. *Trans. Conn. Acad. Arts. Sci.* 32:111–245.

Grego, S., D. Weins, R. E. Stevens, and F. G. Hawksworth. 1974. Pollination studies of *Arceuthobium americanum* in Utah and Colorado. *Southwest. Nat.* 19:65–73.

Hawksworth, F. G. 1961. *Dwarf Mistletoes of Ponderosa Pine in the Southwest.* USDA, Tech. Bull. 1246:1–112.

Hawksworth, F. G. 1977. *The 6–Class Dwarf Mistletoe Rating System.* USDA For. Serv., Gen. Tech. Rept. RM-48. 7 pp.

Hawksworth F. G., and D. W. Johnson. 1989. *Biology and Management of Dwarf Mistletoe in Lodgepole Pine in the Rocky Mountains.* USDA For. Serv., Gen. Tech. Rept. RM-169. 38 pp.

Hawksworth, F. G., and D. Weins. 1972. *Biology and Classification of Dwarf Mistletoes (Arceuthobium).* USDA For. Serv., Agri. Handbook 469. 49 pp.

Hudler, G. T., T. Nicholls, D. W. French, and G. Warner. 1974. Dissemination of seeds of the eastern dwarf mistletoe by birds. *Can. J. For. Res.* 4:409–412.

Irving, F. D., and D. W. French. 1971. Control by fire of dwarf mistletoe in black spruce. *J. For.* 69:28–30.

Kimmey, J. W. 1957. *Dwarf Mistletoes of California and Their Control.* USDA, Calif. For. and Range Exp. Sta., Tech. Pap. No. 19. 12 pp.

Knutson, D. M. 1974. Infection techniques and seedling response to dwarf mistletoe. *Pl. Dis. Reptr.* 58:235–238.

Kuijt, J. 1955. Dwarf mistletoes. *Bot. Rev.* 21:569–627.

Kuijt, J. 1969. *The Biology of Parasitic Flowering Plants.* University of California Press, Berkeley. 246 pp.

Leonard, O. A., and R. J. Hull. 1965. Translocation relationships in and between mistletoes and their hosts. *Hilgardia* 37:115–153.

Mark, W. F., and F. G. Hawksworth. 1974. How important are bole infections in spread of ponderosa pine dwarf mistletoe. *J. For.* 71:146–147.

Ostry, M. E., T. H. Nicholls, and D. W. French. 1983. *Animal Vectors of Eastern Dwarf Mistletoe of Black Spruce.* USDA For. Serv., Res. Pap. NC-232. 16 pp.

Parmeter, J. R., and R. F. Scharpf. 1963. Dwarf mistletoe on red fir and white fir in California. *J. For.* 61:371–374.

Roth, L. F. 1966. Foliar habit of ponderosa pine as a heritable basis for resistance to dwarf mistletoe, pp. 221–228. In H. D. Gerhold et al. (eds.). *Breeding Pest-Resistant Trees.* Pergamon Press. Oxford, New York. 505 pp.

Smith, R. B. 1969. Assessing dwarf mistletoe on western hemlock. *For. Sci.* 15:277–285.

Tainter, F. H. 1971. The ultrastructure of *Arceuthobium pusillum. Can. J. Bot.* 49:1615–1622.

Wicker, E. F. 1974. *Ecology of Dwarf Mistletoe Seed.* USDA, Res. Pap. INT-154. 28 pp.

DISEASE PROFILE

Southwestern Dwarf Mistletoe

Importance: In the southwestern United States, ponderosa pine comprises 90% of the annual timber harvest. The southwestern dwarf mistletoe is recognized as the most important pathogen in these forests, reducing height and diameter growth and eventually killing the trees.

Suscepts: Southwestern dwarf mistletoe commonly occurs on ponderosa pine, varieties *scopulorum* and *arizonica,* and Apache pine. Lodgepole pine and bristlecone pine are occasional hosts.

Distribution: Southwestern dwarf mistletoe is common throughout the ponderosa pine forests of Colorado, Arizona, Utah, New Mexico, west Texas, and Mexico.

Causal Agents: Southwestern dwarf mistletoe is the common name for *Arceuthobium vaginatum* subsp. *cryptopodum,* a parasitic flowering plant in the family Viscaceae.

Biology: Fruits of *A. vaginatum* subsp. *cryptopodum* mature in July and August. Seeds are forcibly discharged with most landing within 15 m of the source. Some land on twigs; others land on needles and are washed onto the twig by rain. Only a small portion germinates on young twigs, which takes place in August and September, approximately 1 month after dispersal. The hypocotyl grows along the host bark until it exhausts its food supply or until it encounters a needle sheath, a bud, or a needle scar. It then forms a mound of tissue called a holdfast. A haustorium develops from this holdfast and penetrates the host cortex to initiate formation of the endophytic system of the parasite within the host. During the next 2 years, the infection is latent. Fusiform swellings are

Witches' broom on ponderosa pine.

usually the first symptom produced, but they may be absent. Aerial shoots of the dwarf mistletoe may appear as early as 2 years after inoculation, but more commonly after 3 or 4 years. The first male and female flowers appear in May and June in the fifth year after inoculation. Pollination is considered to be effected by both wind and insects. Fruits develop over the next year, maturing in July and August. Thus for this species the minimum generation time is 6 years.

Epidemiology: Southwestern dwarf mistletoe spreads through young, evenly aged, open stands at 0.52 m/yr and, in more dense stands, at 0.37 m/yr. Maximal spread from overstory to understory can average 12.8 m/yr. Birds may infrequently carry seeds for greater distances. The effects of infection on tree growth and vigor are correlated with the amount of dwarf mistletoe in the tree, determined by using the six-class dwarf mistletoe rating (DMR) system. Trees with DMRs of 5 and 6 had an average volume of 0.255 and 0.216 m^3, respectively, as compared with an average of 0.325 m^3 for trees in classes 0, 1, and 2 in the same stand. This difference is primarily attributed to a reduction in merchantable height, although a slight reduction in total diameter may be observed. These effects are more pronounced when witches' brooms are formed. Mortality is also an important cause of volume loss, as approximately 2% of the infected trees die each year.

Diagnosis: The first visible symptom of dwarf mistletoe infection is an asymetrical twig swelling that later becomes fusiform-shaped. These infections may remain localized or may stimulate the host to form witches' brooms. There is usually no visible decline in tree vigor until the DMR reaches 4 or 5.

Control Strategy: The goal of dwarf mistletoe management is to eliminate the parasite or reduce it to a tolerable level. When a stand is clear-cut, the cutting should extend 20–40 m beyond apparently affected trees, and all infected trees should be removed before the regeneration is 5 years old to prevent spread to the understory. In young stands, sanitation cuttings and thinnings are considered. Sanitation cuttings involve cutting all visibly infected trees. It should be noted that this cutting does not eliminate dwarf mistletoe, but only reduces it in the stand temporarily because many infected trees cannot be detected. Sanitation thinning is similar, but the emphasis is on spacing in the stand and removing heavily infected trees. Thinning, however, is not recommended in stands where the average DMR is greater than 3.0. Removing dwarf mistletoe infections can increase the growth and vigor of the tree, but because of the cost, pruning can be used only to protect high-value trees, such as in campgrounds.

SELECTED REFERENCE

Hawksworth, F. G. 1961. *Dwarf Mistletoe of Ponderosa Pine in the Southwest.* USDA For. Serv., Tech. Bull. 1246. 112 pp.

DISEASE PROFILE

Douglas-fir Dwarf Mistletoe

Importance: The Douglas-fir dwarf mistletoe causes growth loss, deformity, and mortality.

Suscepts: Douglas-fir is the only commonly infected host of this parasite, although white, grand, and subalpine firs and Engelmann and blue spruces may be infected when they occur in mixture with infected Douglas-fir.

Distribution: The Douglas-fir dwarf mistletoe occurs throughout the range of Douglas-fir, except that it is not of any concern west of the Cascades in British Columbia, Washington, or Oregon.

Causal Agents: Douglas-fir dwarf mistletoe is caused by *Arceuthobium douglasii,* a parasitic flowering plant in the family Viscaceae.

Biology: The aerial shoots of *A. douglasii* average 2 cm in length but may reach 7 cm. Flowering occurs in March and April, and fruits mature in September of the year following pollination. Seeds are forcibly ejected from the fruits and intercepted by host needles. Germination occurs in March. Two to three years later, aerial shoots appear, thus this parasite may complete its life cycle in a minimum of 5 years.

Epidemiology: Douglas-fir dwarf mistletoe is thought to behave like other dwarf mistletoes: lateral spread occurs at about 0.6 m/yr, somewhat greater in open stands. Spread from overstory to understory is much more rapid than from overstory to overstory trees.

Diagnosis: The most noticeable symptom of dwarf mistletoe infection is the witches' broom. Although most infections begin as a localized fusiform swelling, the endophytic or root

Witches' brooms on Douglas-fir. From Kimmey and Graham (1960).

system will invade a dormant bud, stimulating it to grow and form a systemic witches' broom. Signs include the aerial shoots and the basal cups, which persist after the aerial shoots die.

Control Strategy: Douglas-fir dwarf mistletoe can be most effectively managed by preventing the establishment of dwarf mistletoe in regeneration. Management can be accomplished by establishing cutting boundaries in healthy stands or along natural barriers—roads, rivers or resistant species, and clear cutting the stand. Infected residuals and susceptible advanced regeneration should be removed, using prescribed burning, scarification, or hand felling. If eradication of infected residuals is not feasible, future disease incidence may be reduced by encouraging resistant species, for example, ponderosa pine. In stands less than 40 years old sanitation cuts may reduce losses. The cost of sanitation may be greatly reduced when combined with spacing operations. Discriminate against heavily infected trees. Lightly infected trees can be killed or pruned. In infested stands older than 40 years, sanitation is difficult. Because losses increase as the disease intensifies, early removal of these stands is suggested to obtain maximum production on the site.

SELECTED REFERENCES

Hawksworth, F. G., and D. Wiens. 1972. *Biology and Classification of Dwarf Mistletoes (Arceuthobium).* USDA, Agric. Handbk. 401. 234 pp.

Kimmey, J. W., and D. P. Graham. 1960. *Dwarf Mistletoes of the Intermountain and Northern Rocky Mountain Regions and Suggestions for Control.* USDA For. Serv., Intermountain For. and Range Exp. Sta. Res. Pap. 60. 19 pp.

DISEASE PROFILE

Eastern Dwarf Mistletoe

Importance: The eastern dwarf mistletoe has long been known as the major disease of black spruce, with 11–26% of stands in national forests in Michigan, Wisconsin, and Minnesota infested.

Suscepts: Black spruce is the primary host. White spruce is an important host when growing in pure stands. Red spruce is also a host. This dwarf mistletoe occasionally occurs on eastern larch and rarely on jack pine but only when these species are growing in association with infected black spruce.

Distribution: Eastern dwarf mistletoe occurs throughout the commercial range of black spruce from eastern Saskatchewan through southern Manitoba, Ontario, Quebec, the Maritime Provinces, and Newfoundland in Canada. In the United States it is found in northern Minnesota, Wisconsin, and Michigan, and also in Pennsylvania, New Jersey, and the northern New England states.

Causal Agents: The pathogen is *Arceuthobium pusillum,* a parasitic flowering plant in the Viscaceae.

Numerous witches' brooms on black spruce.

Biology: Fruits mature in August or September and discharge seeds to a maximum horizontal distance of 17 m, although the average distance is 1–2 m. The viscin cells that coat the seed serve to bind the seeds in place. Seeds that have not been attacked by fungi or insects germinate between June and August with host penetration complete by September. For 2 years or longer, the infection is invisible. Aerial shoots appear 4 years after seed dispersal. During their first year, aerial shoots produce flower buds, and flowering occurs the following spring. The aerial shoots are dioecious. Pollen apparently is transported by insects, although wind pollination may occur. Fruits mature during the summer, and seeds are dispersed in August or September. The minimum generation time for *A. pusillum* is 5 years, although most infections require 7 years or more to produce fruit. *Arceuthobium pusillum* causes large witches' brooms on black spruce. These brooms alter tree physiology and cause rapid tree death, which, in turn, often creates circular mortality centers.

Epidemiology: Long-distance dissemination by birds occurs infrequently, and thus eastern dwarf mistletoe is a slow-moving pathogen, spreading 0.7 m/yr. Only 54% of the female aerial shoots that flower persist until August. Each aerial shoot has an average of 4.9 flowers, but only an average of 2.4 fruits are produced per aerial shoot. In some years discharge of fruits is drastically reduced or eliminated by fall frosts. A majority of surviving fruits discharge seeds that are not intercepted by suitable infection sites. The viscin cells that allow the seed to move to the twig may also allow the seed to be washed from the host. As many as 78% of the seeds placed on needles in September were lost by June the following year. Fungi and insects take their toll. Only 5% of the seeds placed on spruce needles germinated. Yet, this parasite thrives.

Diagnosis: Witches' brooms and new leaders are the most evident indication of infection along with centers of dead trees. Spruce in advanced stages lose their color and appear less thrifty. Infected portions of the trees usually have dwarf mistletoe plants, which are evident on close examination. Trees can be infected but not have external symptoms.

Control Strategy: Dwarf mistletoe can best be managed in black spruce by eradicating infected trees. Mortality can be detected using aerial photography with color infrared film. Infection centers can be clear-cut, including 20–40 m into the surrounding healthy stand. The commercial timber is harvested and the cleared area burned if sufficient fuel is available. If a tree-length logging system is used, residuals must be eradicated during or after the harvest. A computer simulation can predict losses over the rotation age of the stand and thus indicate whether control is necessary and the advantages to be gained by burning or other treatment. There is no tree species that can be used in place of black spruce as these are usually wet sites and are not suitable for other species, except possibly for eastern larch, which is less desirable and suffers from its own pests. Resistant varieties have not been found.

SELECTED REFERENCE

Baker, F. A., D. W. French, and G. W. Hudler. 1981. Development of *Arceuthobium pusillum* on inoculated black spruce. *For. Sci.* 27:203–205.

DISEASE PROFILE

Lodgepole Pine Dwarf Mistletoe

Importance: Lodgepole pine occupies about 6.4 million ha in the western United States, as well as extensive areas in western Canada. At least half of the lodgepole pine type in the United States is infested. In Alberta the annual loss to this parasite is estimated at 270,000 m^3/yr. The parasite reduces yield of lodgepole pine in Wyoming, Utah, Nevada, and Idaho by more than 480,000 m^3/yr. In Manitoba it is the most serious disease of jack pine, infesting 9% of the commercial jack pine forest area and reducing total jack pine volume by 4–8%.

Suscepts: Lodgepole pine mistletoe occurs principally on lodgepole pine and jack pine. Ponderosa pine is attacked, usually where mixed with lodgepole pine, but occasionally in pure stands. Other species attacked are bristlecone, whitebark, and limber pine; white, Engelmann, and blue spruce; and Douglas-fir.

Distribution: Lodgepole pine dwarf mistletoe is more widely distributed than any of the other dwarf mistletoes, occurring throughout the range of lodgepole pine, from California north to British Columbia eastward in Idaho, Wyoming, Utah, and Colorado, and in Canada. This dwarf mistletoe extends through Alberta into Saskatchewan, Manitoba, and Ontario on jack pine. The infestation center in Ontario may have been eradicated by wild fires.

Causal Agents: Lodgepole pine dwarf mistletoe is caused by *Arceuthobium americanum,* a parasitic flowering plant in the family Viscaceae.

Biology: The aerial shoots of *A. americanum* average 6 cm in size but may grow to 30 cm. These shoots flower in April and May, each infection producing flowers of only one sex.

Witches' brooms in lower crown. From Gottfried and Embry (1977).

Fruits mature in August or September of the year following pollination. As is typical of this genus, seeds are forcibly discharged and intercepted by needles of a potential host. Seeds washed to twigs by rains begin to germinate in May of the year following dispersal. The following year a swelling may appear at the point of infection and shoots may be visible in the third year after dispersal. Flowering occurs the fifth year after dispersal, with seed maturation and dispersal the following year. Thus *A. americanum* requires 6 years to complete its life cycle. Aerial shoots of this species may persist for many years.

Epidemiology: Lodgepole pine dwarf mistletoe spreads through stands at about 0.5 m/yr, and at 0.7 m/yr in stands in which the canopy has not closed. Once a tree is infected, the disease intensifies in its crown. Only when trees become heavily infected (dwarf mistletoe rating greater than 3) can a volume loss be detected. This volume loss is due to a reduction in height and diameter and increases with the time since infection. Mortality also contributes to volume loss but is not a major factor in lodgepole pine. In jack pine, however, mortality is much more important than growth loss.

Diagnosis: The first visible symptom of infection by *A. americanum* is a fusiform swelling at the point of penetration. The main stem of the host, if infected, may also have a fusiform swelling, as well as aerial shoots. Large systemic brooms may form; they allow easy detection.

Control Strategy: Lodgepole pine stands less than 40 years old, with less than 40% infection, may be thinned to specified growing stock levels, discriminating against infected trees. Lightly infected trees may be left to maintain desired spacing. Complete sanitation in immature stands does not appear feasible. The effects of thinning to various levels can be examined using the Forest Vegetation Simulator (FVS). Eradication during or immediately after harvesting is more acceptable. This is achieved by felling residuals, shearing, prescribed burning, or scarification with chains or shark-finned barrels.

SELECTED REFERENCES

Gottfried, G. J., and R. S. Embry. 1977. *Distribution of Douglas-Fir and Ponderosa Pine Dwarf Mistletoe in a Virgin Arizona mixed Conifer Stand.* USDA For. Serv., Res. Pap. RM-192. 16 pp.
Hawksworth, F. G., and D. W. Johnson. 1989. *Biology and Management of Dwarf Mistletoe in Lodgepole Pine in the Rocky Mountains.* USDA For. Serv., Gen. Tech. Rep. RM-169. 38 pp.

DISEASE PROFILE

Hemlock Dwarf Mistletoe

Importance: In recent years the hemlock dwarf mistletoe has been recognized as an important cause of losses in forest stands. In Washington and Oregon the annual loss attributed to growth reduction and defect is estimated at over 28,330 m^3. In British Columbia this loss is estimated to be 1.7 million m^3/yr.

Suscepts: Principal hosts of this parasite are western and mountain hemlocks, subalpine, Pacific silver, grand, and noble firs. Lodgepole, whitebark, and western white pine are also hosts, especially when associated with infected hemlocks.

Distribution: Hemlock dwarf mistletoe ranges from Alaska to central California along coastal regions and the west slope of the Cascades.

Causal Agents: Hemlock dwarf mistletoe is caused by *Arceuthobium tsugense,* a parasitic flowering plant in the family Viscaceae.

Biology: Arceuthobium tsugense, like all dwarf mistletoes, is dioecious, producing male and female flowers on separate infections. Flowering takes place in July and August, and the

Entire crown is a mass of witches' brooms. From Weir (1916).

fruits mature in late September and October of the following year. Seeds are forcibly discharged from the fruits to distances of 16 m. Seeds may germinate in February in the milder weather of the coastal areas and several months later on interior sites. Penetration of the twig occurs during the summer after germination. After 1–2 years, a small, fusiform swelling develops at the point of infection. In another 1–2 years, aerial shoots may appear. These shoots range from greenish to reddish yellow in color and from 5 to 12 cm long. These aerial shoots are perennial and may produce a crop of flowers and fruits for several years. Although hemlock dwarf mistletoe may occasionally complete its life cycle in 4 years, 5 or 6 years is more common.

Epidemiology: Hemlock dwarf mistletoe, as with other dwarf mistletoes, spreads more rapidly from infected overstory to understory trees affecting those in a 10- to 16-m radius almost immediately. Thus less than 22 evenly distributed infected trees could infect all susceptible regeneration on a hectare. Spread through even-aged stands is usually less than 0.6 m/yr. Spread through open stands is about 1.5 times more rapid, due to more vigorous dwarf mistletoe plants and longer seed flights. The vigor of the dwarf mistletoe plants appears to be related to the amount of light the plants receive. Although plants may remain alive when shaded, they produce few if any aerial shoots and thus do not contribute to spread and intensification of the parasite. Intensification within a tree usually occurs rapidly. Disease intensity estimated with the DMR system increases by 1 unit every 10–20 years. The vertical rate of spread within a crown is about 0.6 m/yr, and this rate is also greater in open than in closed stands.

Diagnosis: The spindle-shaped swellings are characteristic. Even though restricted to the swelling, the parasite may stimulate the host to form witches' brooms. In addition to the aerial shoots, an important sign of infection are the basal cups, which persist after an aerial shoot dies. When the main stem is invaded, large swellings or cankers may form.

Control Strategy: Mature infested hemlock stands should be clear-cut, with boundaries located in uninfested stands or along natural barriers. Infected residual trees or advanced regeneration should be eliminated during or after harvesting. Broadcast burning slash, or where terrain permits, mechanical scarification with a drag scarifier or drum chopper have been effective in removing these sources of inoculum. In immature stands, sanitation may be necessary to attain an acceptable yield from an infested stand. Infected residual trees and infected bordering stands should be cut first. Young trees with stem infections should be killed. Trees with branch infections may be killed, or pruned, to remove infections if this is feasible. When combined with juvenile spacing, sanitation can be accomplished with little additional cost. Intermediate cuttings in infected stands should be avoided, as opening the canopy will stimulate the dwarf mistletoe.

SELECTED REFERENCES

Baranyay, J. A., and R. B. Smith. 1972. *Dwarf Mistletoes in British Columbia and Recommendations for Their Control.* Pac. For. Res. Cent. Rept. BC-X-72. 18 pp.

Weir, J. R. 1916. *Mistletoe Injury to Conifers in the Northwest.* USDA, Bull. No. 360. 39 pp.

DISEASE PROFILE

Leafy Mistletoes

Importance: The common "Christmas" or "leafy" mistletoes are usually considered to be curiosities or desired plants for Christmas festivities, but they can be serious tree killers in local areas. *Phoradendron* occurs principally on hardwood trees, but some western species also occur on conifers *(Juniperus, Cupressus, Abies)*. There are 14 taxa of *Phoradendron* in the United States, all but 2 of which are western. Generally, *Phoradendron* are more serious in ornamental trees and in orchards than in forest situations. The leafy mistletoes are essentially water parasites; and as a general rule they have less impact on their hosts than do the dwarf mistletoes.

Suscepts: Fourteen species of leafy mistletoes are found in the United States. Five of these species occur on conifers, the remainder occur on hardwoods (Table 19.1).

Distribution: In the United States the leafy mistletoes do not extend north of a line drawn across the country from approximately New Jersey to Oregon. Their northward extension is probably limited by cold temperature. Leafy mistletoes do not occur in Canada. The majority of leafy mistletoes occur in the southwestern states, with two occurring in the eastern states. There are many more in Mexico and Central America.

Causal Agents: Leafy mistletoes are parasitic flowering plants in the genus *Phoradendron*.

Biology: Most species have well-developed leaves, and in one species the stems are up to 38 cm in diameter. For most species the stems are less than 2.5 cm in diameter.

Water oak in winter, with numerous infections.

TABLE 19.1 *Phoradendron* in the United States

Species	Principal Hosts	Distribution
On Hardwoods		
1. *P. californicum*	*Prosopis, Cercidum*	California, Nevada, Arizona
2. *P. macrophyllum*	Many hardwoods—esp. *Populus, Fraxinus,* and *Salix.* Known from about 60 species of 30 genera.	West Texas to North California
3. *P. rubrum*	*Swietenia*	South Florida
4. *P. serotinum* (= *P. flavescens*)	Many hardwoods. Known from about 110 species of 50 genera.	Eastern United States; New Jersey, Ohio, Indiana, Missouri, south to Gulf
5. *P. tomentosum*	*Celtis, Prosopis*	West Texas, Oklahoma
6. *P. villosum*		
a. subsp. *villosum*	*Quercus*	California, Oregon
b. subsp. *coryae*	*Quercus*	Arizona, New Mexico, Texas
On Conifers		
7. *P. densum*	*Cupressus, Juniperus*	Oregon, California, Arizona
8. *P. pauciflorum*	*Abies concolor*	California, Arizona
9. *P. capitallatum*	*Juniperus*	South Arizona, South New Mexico
10. *P.* sp. (undescribed species)	*Juniperus*	South New Mexico
11. *P. juniperinum*	*Juniperus*	Colorado to Oregon, south to Mexico
12. *P. libocedri*	*Libocedrus decurrens*	Oregon, California

Pollination in most species of *Phoradendron* is believed to be effected by birds and insects. The mature fruits are eaten by birds, but the seeds are passed intact. The outer coat of the seed is covered with a sticky substance (viscin), which binds the seed in place on the host twig. There, it germinates and penetrates the host tissue. Subsequently, aerial shoots appear. These shoots persist after the host normally drops its leaves allowing for easy detection during winter months. These species cause overgrowth in the host. The infected portion of the tree may be swollen, and a broom may result. Portions of the tree beyond the point of mistletoe infection may become deformed and die, but in many instances the host will live for years with only minor reduction in growth rate.

Epidemiology: The epidemiology of this group of parasites has not been studied in view of the small amount of damage they cause, but birds are proven to be a major vector for the seeds.

Diagnosis: Before the aerial shoots of the leafy mistletoes appear, a swelling of the twig at the point of infection may be evident. Eventually the parasite may form a witches' broom at this point. Branches beyond the infection may die.

Control Strategy: Control of *Phoradendron* is not warranted in forest stands. In some areas, the leafy mistletoes are highly desirable and are collected and sold during the Christmas season. In shade trees or other high-value trees, annually pruning infected branches provides effective control. The branches should be cut at least 0.3 m proximal to the mistletoe shoots closest to the bole. In walnut orchards in California, applications of phenoxy herbicides during the dormant season have been effective in ridding the trees of this parasite.

SELECTED REFERENCES

Gill, L. S., and F. G. Hawksworth. 1961. *The Mistletoes, a Literature Review.* USDA For. Serv., Tech. Bull. No. 1242. 87 pp.

Hawksworth, F. G. 1979. Mistletoes and their role in North American forestry, pp. 13–23. In *Proc. Second Symp. Parasitic Weeds.*

20

WOOD DECAY

20.1 DISCOLORATION AND DECAY IN LIVING TREES

Decay of wood is caused by fungi that use wood cell wall constituents as a source of nutrition. These unique organisms have both beneficial and damaging influences on forest stands. They play a virtually irreplaceable role in the earth's carbon cycle by returning to the atmosphere billions of tons of carbon each year. As shown by the following general equations, if wood decay were to cease while photosynthesis continued at its usual rate, life as we know it on earth would stagnate for lack of atmospheric carbon dioxide. It has been predicted that this would occur in a little less than 40 years.

$$\text{Energy} + 6\,CO_2 + 5\,H_2O \underset{\text{Decay (respiration)}}{\overset{\text{Photosynthesis}}{\rightleftharpoons}} \underset{\text{wood}}{C_6H_{10}O_5} + 6\,O_2$$

Decay fungi break down the xylem tissues in branches, roots, and stems of trees. This breakdown is important in ridding forest lands of wood from trees that are killed by diseases, insects, fire, windthrow, and such, or suppressed by competition within forest stands. These fungi also aid in the natural pruning of branches of many species of trees and thus improve the quality of timber available for harvest.

Wood-destroying fungi also contribute to soil fertility and to natural soil-forming processes. They add organic matter with highly desirable ion-exchange properties to the soil and contribute to the cycling of essential mineral elements. They also aid in aeration of the soil by decaying roots that are broken or sloughed off by trees. In this way, new channels for exchange of gases are formed within forest soils.

By ridding forest lands of woody debris, wood-decaying fungi also decrease the hazard for wildfire in the forest and make it a much more desirable habitat for wildlife and for man. Without wood-decaying fungi, the forest floor would soon become an impenetrable thicket of dead and broken trees and other organic debris.

Wood-destroying fungi also cause very important damaging influences in forest stands. As already described in Chapter 13, certain fungi that decay structural roots can kill trees. Such fungi can interfere with the transport of water and essential mineral nutrients through roots. They can also predispose trees to windthrow. As discussed in Chapter 16, certain of the canker-forming fungi also decay the dead or weakened sapwood surrounding cankers of both hardwoods and conifers. As we shall now learn, a host of additional decay fungi can invade and destroy, discolor, or weaken the stemwood of living trees and of dead standing timber. These fungi make the stems of trees less useful, and in some cases completely useless, for manufacture into lumber, plywood, or even into pulp and paper products.

Taxonomy of Wood-Destroying Fungi

An essential prerequisite to effective forest management is accurate identification of the cause of disease. In the case of most decay diseases of trees, sporocarps of the fungi themselves often provide reliable clues to identification of the causal organisms. With the identity established, the forester will know something about the extent of the decay in a tree, the condition of the wood, and the potential for further loss if the infected tree is left standing. A wildlife manager may choose to manage in favor of a particular decay condition to produce habitat suitable for nesting or feeding of wildlife.

The classification of wood-decay sporophores, or fruit bodies, has undergone a great deal of change during the last two decades. This change has resulted in two names, old and new, for many species. The new names should be used; however, much of the literature will have only the old names. Because the new classification is still in a state of flux, the classification system in this book will use the most important features of the older system but will list the decay fungi by using both names. The old name is in parentheses. The keys in the following partial list are useful aids for the identification of causal organisms.

REFERENCES

Christensen, C. M. 1965. *Common Fleshy Fungi.* Burgess Publishing Company, Minneapolis, Minn. 237 pp.

Fergus, C. L. 1960. *Illustrated Genera of Wood Decay Fungi.* Burgess Publishing Company, Minneapolis, Minn. 132 pp.

Gilbertson, R. L. 1974. *Fungi That Decay Ponderosa Pine.* University of Arizona Press, Tucson, Ariz. 197 pp.

Gilbertson, R. L. 1980. Wood-rotting fungi of North America. *Mycologia* 72:1–49.

Larsen, M. J. 1974. A contribution to the taxonomy of the genus *Tomentella. Mycologia Memoir.* 4. 145 pp.

Lentz, P. L. 1955. *Stereum and Allied Genera of Fungi in the Upper Mississippi Valley.* USDA, Agri. Monograph 24. 90 pp.

Lowe, J. L. 1957. *Polyporaceae of North America. The Genus Fomes.* State Univ. Forestry at Syracuse, Bull. No. 80. 97 pp.

Overholts, L. O. 1953. *The Polyporaceae of the United States, Alaska, and Canada.* University of Michigan Press, Ann Arbor, Mich. 466 pp.

Partridge, A. D., and D. L. Miller. 1974. *Major Wood Decays in the Inland Northwest*. Idaho Res. Foundation., University of Idaho. 125 pp.

Sinclair, W. A. , H. H. Lyon, and W. T. Johnson. 1987. *Diseases of Trees and Shrubs*. Cornell University Press. 574 pp.

Slysh, A. R. 1960. *The Genus Peniophora in New York State and Adjacent Regions*. State Univ. Forestry at Syracuse, Tech. Bull. 83. 95 pp.

Readers are urged to review the summary classification of the fungi found in Chapter 4. The most important wood-destroying fungal taxa that can be pathogenic on living trees or that cause damage to primary wood products are listed in the following abbreviated scheme.

Classification
    Basidiomycotina
        Aphyllophorales
            Thelephoraceae—hymenium borne on a smooth surface
            Hydnaceae—hymenium borne on the outside of teeth
            Polyporaceae—hymenium borne on the inside of pores
        Agaricales
            Agaricaceae—hymenium borne on the surface of gills
        Ascomycotina
            Xylariaceae—ascospores borne in perithecia

Thelephoraceae. The fungi in this family are characterized by the hymenium being borne on a smooth surface. The important species follow:

1. *Chondrostereum (Stereum) purpureum* is a parasite of numerous fruit trees and causes considerable damage to apples and plums. The fungus enters through wounds and decays the interior of the trunk. The effect that this fungus has on a tree is sometimes called silver leaf when the foliage is off-color because of decay in that part of the tree.

2. *Cystostereum (Stereum) murrayi* causes a white or light brown, sweet-smelling heart decay of living hardwoods, especially yellow birch.

3. *Peniophora rufa (Stereum rufum)* causes a sap decay on fallen aspen.

4. *Phanerochaete (Peniophora) gigantea* is one of the most important decay fungi in coniferous logs and in the South is a major cause of decay in stored pulpwood.

5. *Stereum frustulosum* causes a white pocket decay of oaks (Fig. 20.1).

6. *Stereum gausapatum* causes a white mottled heart decay of sprout oaks in the eastern and central states, entering through the heartwood connecting the old stump and the sprout.

7. *Stereum sanguinolentum* causes a red top decay of balsam fir throughout the range of this tree, and it is second in importance only to *Perenniporia subacida* in the decay of balsam fir. The fungus enters through branch stubs and infects chiefly that part of the tree corresponding to the heartwood. The infected wood soon becomes red or light brown. In cross section the infection may be patchy at first, but later it is more or less solid. The wood is completely disintegrated in small areas, forming pockets that are about 3 mm long. Even in an advanced

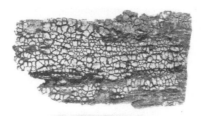

FIGURE 20.1   Fruiting bodies of *Stereum frustulosum*.

stage of decay the wood is fairly solid. See the Disease Profile on Red Heart Decay.

Hydnaceae. In this family the hymenium is borne upon the outside of tapering tooth-like structures that are formed on the underside of the pileus and project downward. The genera most important in the decay of trees and logs are *Climacodon, Echinodontium, Hericium,* and *Irpex.*

Teeth in early stage of development forming poroid or lamelloid structures—*Echinodontium* (fruit bodies perennial, hoof shaped) and *Irpex* (fruit bodies annual, usually resupinate).

FIGURE 20.2   Fruiting bodies of *Climacodon septentrionalis.* Courtesy of C. M. Christensen, University of Minnesota, St. Paul.

Teeth long and awl-shaped, fruit bodies stalked, shelf-like or coralloid—*Climacodon* and *Hericium*.

1. *Climacodon septentrionalis (Hydnum septentrionale)* causes a light-colored heart decay of living maple, oak, and birch. Although it has not been studied sufficiently, its prevalence in some areas indicates that it may be responsible for considerable decay in these species (Fig. 20.2).

2. *Echinodontium tinctorium,* the only species of this genus in North America, causes a stringy red heart decay of many western conifers, especially western hemlock, white fir, and Douglas-fir. In the Blue Mountain region of Oregon, overmature white firs are so thoroughly infected with this fungus that a common cruising rule is to cull completely all trees over 56 cm at 1.4 m height and reduce the estimate of gross increment by 25% for trees less than 56 cm. Some stands of western hemlock in the Inland Empire are 97% infected, with a cull loss of 20% or more. In trees with normal red heartwood, the early stages of the decay are not very apparent, even though the wood may be considerably weakened. The fungus may be present in the wood several meters in advance of the visibly infected area. Fruit bodies are formed only after decay has become rather extensive in the tree. See the Disease Profile on Brown Stringy Decay.

3. *Hericium (Hydnum) erinaceus* causes a white heart decay of oaks (Fig. 20.3).

4. *Irpex* spp. occur commonly on dead branches and on slash, especially that of hardwood trees (Fig. 20.4). None are known to cause much damage in the forest.

Polyporaceae. The distinguishing characteristic of this family is that the basidia are borne upon the walls of pores or tubes. These tubes are arranged vertically on the under surface of the fruit bodies and are open at the bottom so that, when the basidiospores are discharged, they fall until they are out of the tubes and then are blown away by the wind. The orientation of these tubes is critical in spore dispersal and there may be a 30% reduction if pores are 3° from the perpendicular.

The following key includes the most important genera of this family.

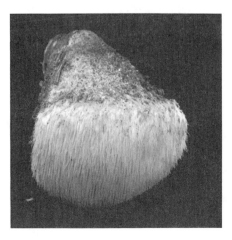

FIGURE 20.3   Fruiting body of *Hericium erinaceus.*

FIGURE 20.4 Fruiting bodies of *Irpex tulipifera*.

Fruit bodies annual
    Fruit bodies resupinate
        Tubes in the form of shallow depressions between irregular ridges—*Merulius*
        Length of tubes greater than the diameter—*Poria*
    Fruit bodies effused-reflexed, sessile, or stalked
        Tubes round to angular
            Tubes equally sunken in context—*Polyporus*
            Tubes unequally sunken in context—*Trametes*
        Tubes daedaloid—*Daedalia*
Fruit bodies perennial—*Fomes*

Merulius and *Poria*

1. *Ceriporiopsis rivulosa (Poria albipellucida)* causes a white ring decay in redwood and is especially prevalent in the northern part of the redwood region.
2. *Perenniporia (Poria) subacida* causes a yellow, feathery root and butt decay of balsam fir throughout the range of this tree and on firs in western forests. The fungus apparently lives in the soil, enters through wounds, and decays

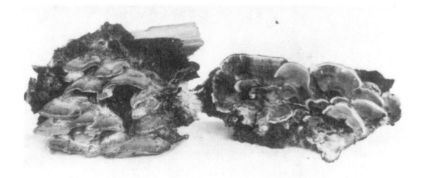

FIGURE 20.5 Fruiting bodies of *Bjerkandera adusta*. Courtesy of C. M. Christensen, University of Minnesota, St. Paul.

the roots and central portion of the lower stem. It causes extensive cull losses in some stands of balsam fir and predisposes the trees to windthrow.

3. *Phellinus (Poria) weirii* (See Chapter 13 on laminated root disease.)
4. *Postia (Poria) sequoiae* causes a brown cubical decay in redwood averaging 24% cull in this species.
5. *Serpula lacrimans (Merulius lacrymans)* and *Meriliporia (Poria) incrassata* cause decay in buildings and stored lumber.

Polyporus

1. *Bjerkandera adusta (Polyporus adustus)* is found on the logs of many species of hardwoods, especially aspen, basswood, elm, and maple (Fig. 20.5).
2. *Dichomitus squalens (Polyporus anceps)* causes a red heart and sap decay of the trunk of several conifers throughout the country (Fig. 20.6). It is the most important fungus causing decay of ponderosa pine in the Southwest. It is of value in that it decays slash, but it also causes heart decay and is especially damaging in older stands. In other regions it is common on ponderosa and other pines, as well as spruce. Fruit bodies are uncommon

FIGURE 20.6   Fruiting bodies of *Dichomitus squalens*. Courtesy of C. M. Christensen, University of Minnesota, St. Paul.

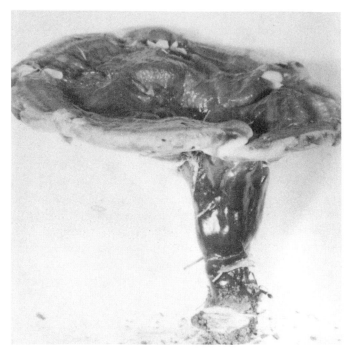

FIGURE 20.7   Fruiting body of *Ganoderma tsugae*. Courtesy of C. M. Christensen, University of Minnesota, St. Paul.

and are produced only on dead trees or logs. It enters trees through the sapwood of dead branches and then converts to a heart decayer when it reaches the main stem.

3. *Ganoderma (Polyporus) tsugae* is very common on eastern hemlock logs and stumps (Fig. 20.7).

FIGURE 20.8   Fruiting bodies of *Inonotus circinatus.* Courtesy of C. M. Christensen, University of Minnesota, St. Paul.

FIGURE 20.9   Fruiting bodies of *Ischnoderma resinosum*. Courtesy of C. M. Christensen, University of Minnesota, St. Paul.

4. *Inonotus (Polyporus) circinatus* causes a white pocket decay of the roots and butt of conifers including pine, spruce, and fir (Fig. 20.8). It supposedly enters through wounds at the base of the tree or in the roots and grows upward in the heartwood and outward into the sapwood. Aside from the actual damage caused by decay, the wood becomes red, and later white pockets are formed; the decay is thus similar to that caused by *Phellinus pini*. This decay has not been studied extensively, but the fruit bodies of the fungus, which occur on the ground around the infected tree, are very common in some areas; thus it may be of importance. This fungus is easily confused with *I. tomentosus*.

FIGURE 20.10   Fruiting bodies of *Laetiporus sulphureus*.

FIGURE 20.11   Fruiting bodies of *Phellinus gilvus*.

5. *Inonotus (Polyporus) tomentosus* is the same as or closely related to *Inonotus circinatus* and is responsible for killing small patches of spruce (Chapter 13.5).

6. *Inonotus (Polyporus) hispidus* causes a spongy white decay of oaks and other deciduous species (See the Disease Profile on Canker Decays of Hardwoods). The decay caused by this fungus often results in irregularly shaped cankers on the main stem.

7. *Ischnoderma resinosum (Polyporus resinosus)* causes a straw-colored decay of both sapwood and heartwood of deciduous trees (Fig. 20.9).

FIGURE 20.12   Fruiting bodies of *Polyporus badius*. Courtesy of C. M. Christensen, University of Minnesota, St. Paul.

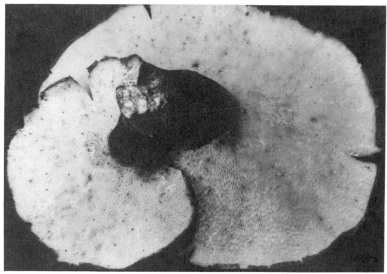

FIGURE 20.13   Fruiting body of *Polyporus squamosus*. Courtesy of C. M. Christensen, University of Minnesota, St. Paul.

8. *Laetiporus (Polyporus) sulphureus* causes a brown cubical decay of the trunk and is found on living and dead deciduous trees and rarely on conifers; it is common on oak (Fig. 20.10). It enters through wounds.

9. *Phaeolus (Polyporus) schweinitzii* causes a brown cubical decay of roots and butt of all conifers except cedars, cypress, and junipers (See the Disease Profile on Schweinitzii Root and Butt Disease).

10. *Phellinus (Polyporus) gilvus* is found on dead standing trees and logs of deciduous species and is particularly common on oak (Fig. 20.11).

11. *Polyporus badius (picipes)* is common on hardwood logs in the later stages of decay; it is common on basswood (Fig. 20.12).

12. *Polyporus squamosus* causes a white heart decay of living and dead deciduous trees; it is common on elm (Fig. 20.13).

FIGURE 20.14   Fruiting bodies of *Pycnoporus cinnabarinus*.

FIGURE 20.15   Fruiting bodies of *Spongipellis unicolor*. Courtesy of C. M. Christensen, University of Minnesota, St. Paul.

13. *Postia (Polyporus) amarus* causes an important heart decay of incense cedar.

14. *Pycnoporus (Polyporus) cinnabarinus* is found on deciduous trees and is common on oak (Fig. 20.14).

15. *Spongipellis unicolor (Polyporus obtusus)* causes a spongy white decay of living oaks, entering principally through fire scars (Fig. 20.15).

16. *Trametes hirsuta (Polyporus hirsutus)* is found on dead wood and trees of deciduous species and rarely on coniferous species.

17. *Trametes (Polyporus) versicolor* is found on dead trees and wood of deciduous species and rarely on coniferous wood (Fig. 20.16).

18. *Trichaptum biforme (Polyporus pargamenus)* is found on deciduous trees, especially aspen and birch (Fig. 20.17). This fungus enters the tree soon

FIGURE 20.16   Fruiting bodies of *Trametes versicolor*. Courtesy of C. M. Christensen, University of Minnesota, St. Paul.

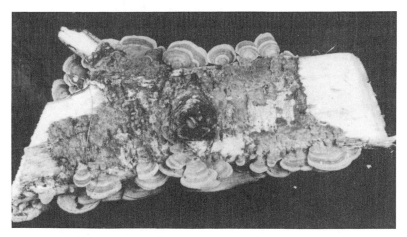

FIGURE 20.17   Fruiting bodies of *Trichaptum biforme*. Courtesy of C. M. Christensen, University of Minnesota, St. Paul.

after it is dead and causes a white pocket decay of the sapwood and heartwood.

Daedalia
1. *Cerrena (Daedalia) unicolor* causes a white sap and heart decay of deciduous trees (Fig. 20.18). It is one of the first fungi to attack trees killed by other causes and although it has not been studied much, its abundance, especially on birch and aspen, indicates that it is important. It has been found occasionally causing a decay of living trees.

2. *Daedaleopsis (Daedalia) confragosa* is very common, especially on dead branches of birch, aspen, willow, and alder that are lying on the ground; it is of no economic importance except that it decays slash (Fig. 20.19).

3. *Daedalia quercina* is found chiefly on the stumps and old logs of deciduous species, especially oak, and occasionally is found on ties and posts (Fig. 20.20).

Fomes
1. *Fomes fomentarius* causes a spongy white sap and heart decay of living and dead birch and may occasionally be found on aspen (Fig. 20.21). It enters through dead tops and branches and progresses throughout the tree. It rarely fruits on living trees, but fruit bodies are formed abundantly on standing and downed dead trees. It is not known how much damage is caused by this fungus.

2. *Fomitopsis (Fomes) pinicola* causes a brown cubical sap and heart decay of conifer and hardwood trees; it has been reported on 37 hardwood and 54 conifer species in Europe and North America (Fig. 20.22). It grows in living trees, dead standing trees, and logs, and it is most destructive in standing trees that have been killed by other causes. The fungus enters through fire scars or other wounds, beetle tunnels, and branch stubs.

FIGURE 20.18   Fruiting bodies of *Cerrena unicolor*. Courtesy of C. M. Christensen, University of Minnesota, St. Paul.

3. *Ganoderma (Fomes) applanatum* causes a mottled root and butt decay of a large number of hardwoods; the decay is rather similar to that caused by *Phellinus igniarius* (Fig. 20.23). It is not known how this fungus enters the trees nor how much damage it causes in our forests, but the prevalence of fruit bodies in hardwood stands indicates that it is responsible for

FIGURE 20.19   Fruiting bodies of *Daedaleopsis confragosa*.

FIGURE 20.20   Fruiting bodies of *Daedalia quercina*.

considerable decay, especially in overmature trees. It may be a major cause of windthrow of aspen in Colorado.

4. *Heterobasidion annosum (Fomes annosus)* is discussed in Chapter 13.2 Annosum Root Disease.

5. *Oxyporus populinus (Fomes connatus)* causes a white heart decay of deciduous trees, especially sugar maple, apparently entering through wounds. The fruit bodies often occur in the wound where the fungus probably entered the tree. This fungus is found fruiting in Eutypella cankers.

6. *Phellinus (Fomes) igniarius* causes a white heart and sap decay of many species of living deciduous trees, especially aspen, birch, maple, and ironwood, and it is one of the most important fungi on these species. Fruit bodies are formed on aspen; they are typically hoof-shaped and appear just below the branch stubs. The fruit bodies on other hosts are shelflike. See the Disease Profile on White Trunk Decay.

7. *Phellinus (Fomes) pini* is described in the Disease Profile on Red Ring Decay.

8. *Phellinus (Fomes) robustus* causes a white decay in sapwood and heartwood of living red oaks as well as other hardwood species (Fig. 20.24). Invasion by the fungus often results in cankers on the main stem. This species is found on a wide range of hosts (including conifers) as a sap-decaying fungus.

Agaricaceae. The distinguishing character of the fungi in this family is that the hymenium is borne on the sides of gills. Relatively few of the species in this large family cause wood decay or are parasitic on trees.

FIGURE 20.21   Fruiting body of *Fomes fomentarius*.

FIGURE 20.22   Fruiting body of *Fomitopsis pinicola*.

Armillaria spp. This important genus is discussed in Chapter 13.4 Armillaria Root Disease.

Lenzites
1. *Gloeophyllum sepiarium (Lenzites sepiaria)* causes a brown cubical decay chiefly in coniferous wood; it is fairly common on logs, stumps, ties, poles and posts (Fig. 20.25).

FIGURE 20.23   Fruiting body of *Ganoderma applanatum*. Courtesy of C. M. Christensen, University of Minnesota, St. Paul.

FIGURE 20.24   Fruiting body of *Phellinus robustus*. Courtesy of C. M. Christensen, University of Minnesota, St. Paul.

2. *Lenzites betulina* is common on slash up to several centimeters in diameter, especially that of birch, although it decays the slash of a number of hardwood species (Fig. 20.26).

Other genera of importance in the Agaricaceae follow:

Lentinus—*Neolentinus (Lentinus) lepideus* causes brown cubical decay of conifers and is most important in decaying slash and wood products (Fig. 20.27). It is more tolerant of creosote than many other fungi.

FIGURE 20.25   Fruiting bodies of *Gloeophyllum sepiarium*.

FIGURE 20.26   Fruiting bodies of *Lenzites betulina*. Courtesy of C. M. Christensen, University of Minnesota, St. Paul.

Pleurotus—Species of *Pleurotus* cause white decay in many deciduous species (Fig. 20.28). The fruit bodies of some of the species are good to eat.

Schizophyllum—*Schizophyllum commune* causes white decay in many deciduous species (Fig. 20.29). This species has been used for studying the genetics of the Agaricales.

Xylariaceae. Species of fungi in the genera *Daldinia, Hypoxylon,* and *Xylaria* can cause decay in deciduous species. The rate of decay may be somewhat slower than it is for fungi in the Basidiomycetes, but undoubtedly they cause substantial losses, both in standing trees and in slash. They cause a white decay often with black zone lines and for some species the decay occurs primarily at or near the surface and is favored by high moisture content. In standing trees, decay caused by these fungi is usually associated with wounds or cankers.

20.2   HEART DECAY

Heart Decay in Living Trees

Decay in the central stemwood of living trees is commonly called heart decay or heart rot. This is the most damaging of all types of tree diseases. Losses due to heart decay are greater—in fact they are more than two times greater—than those due to all other diseases taken together. This is true in both hardwoods and conifers. The volume of usable timber destroyed or decreased in value by heart decay fungi in the United States is about 30% as much as the total harvested for use by the people of this country. Heart decays are also very important causes of loss in Canadian forests.

In both hardwoods and conifers, heart decays are much more important in mature and overmature stands than in young stands. This fact is the basis of one of the most important management techniques in forest pathology—the concept of a pathological

FIGURE 20.27   Fruiting body of *Neolentinus lepideus*. Courtesy of C. M. Christensen, University of Minnesota, St. Paul.

rotation. This concept is useful for diseases that become important after a certain relatively dependable number of years in the life of the stand. Under these conditions, losses due to heart decay can be minimized simply be harvesting the timber before disease losses become economically important. In the early 1900s, heart decay in the southern pines averaged about 20% of the volume of harvested sawtimber; at present, losses due to heart decay are less than 1%. Over this span of time, the average age of southern pine trees at time of harvest decreased from more than 80 years to less than 40 years. Heart decay is rarely important before age 60.

As this simple principle of pathological rotation is applied to more and more species of trees, losses due to heart decay will continue to decrease in importance. But so long as we try to maintain large volumes of mature and overmature timber for such uses as wildlife habitat—as for example in some of the conifer forests of the United States and Canada—losses due to heart decay will continue to be important.

Wounds that break through the protective layer of bark and expose the wood beneath the cambium are major pathways for entrance of many heart decay fungi.

FIGURE 20.28   Fruiting bodies of *Pleurotus ulmarius*. Courtesy of C. M. Christensen, University of Minnesota, St. Paul.

Management practices such as control of wild fires and more careful logging operations will diminish future losses due to heart decay because they decrease the number of entrance points for decay fungi.

In earlier times, the term *heart decay* was used to imply decay of the heartwood portion of tree stems. We now know that certain diffuse-porous species of trees (notably maple, birch, beech, aspen, and sweetgum) do not form a true heartwood. By *heartwood* we mean a cylinder of wood in the center of tree stems that contains no functional conductive or storage (parenchyma) cells. Heartwood serves merely to provide mechanical support for the crown of the tree. For this reason, the term *heart decay* should be used only to indicate the position of decay within tree stems rather then the type of tissue affected.

Wood that is decayed by heart decay fungi is commonly referred to as *cull*. The volume of decayed wood in a tree or forest stand is called *cull volume*. If the amount of cull is expressed as a percentage of the total volume of wood in an tree or in a forest stand, the term *percent cull* is commonly used.

Heart decay develops in progressive stages. Early stages are called *incipient decay* or *incipient rot;* in later stages the terms *advanced decay* or *advanced rot* are appropriate. In its final stages, when the tissue is totally destroyed, we find just a hole in the tree stem. This condition is especially damaging in veneer bolts because the bolt frequently

FIGURE 20.29 Fruiting bodies (lower surface) of *Schizophyllum commune*. Courtesy of C. M. Christensen, University of Minnesota, St. Paul.

cannot be held securely in a veneer lathe. As a result, otherwise veneer-quality logs must be used for less-valuable products.

The longitudinal portion of tree stems affected by various heart decay fungi varies with both the species of fungi and species of trees involved. Thus some fungi are called root and butt decayers while other are called top decayers, and still others are called trunk decayers depending on the portion of the tree stem usually affected. Some species of decay fungi cause root and butt decay in one species and a trunk decay in other tree species. Fungi that cause both cankers and decay of stemwood are called *canker decayers* or *canker rotters*.

Some of the common fungi that cause decay in various conifers and hardwoods are listed in Table 20.1. The relative prevalence of many of these fungi varies with locality and sometimes with the site within a given locality in which the trees are grown. Note in the table that some species of trees are attacked by a large number of different heart-decay fungi, while others are attacked by comparatively few. Because heartwood of redwood and incense cedar is rich in very fungitoxic extractives, it is understandable that very few species of decay fungi are sufficiently tolerant of these substances to be able to decay the naturally durable heartwood of these trees (see Chapter 12 for further information about natural decay resistance). The prevalence of heart decay caused by *Stereum sanguinolentum* in balsam fir is due to the unusual tolerance of this fungus to extractives in the heartwood of the tree. By contrast, the extremely high

TABLE 20.1 Major Heart Decay Fungi in Certain Conifers and Hardwoods

Balsam fir

Stereum sanguinolentum, Perenniporia (Poria) subacida, Scytinostroma (Corticium) galactinum, Coniophora puteana

Douglas-fir

Phellinus (Fomes) pini, Fomitopsis (Fomes) officinalis, Fomitopsis (Fomes) cajanderi (Fomes subroseus), Phaeolus schweinitizii, Stereum sanguinolentum, Echinodontium tinctorium, Fomitopsis rosea (Fomes roseus), Pholiota adiposa, Dichomitus squalens (Polyporus anceps)

Incense cedar

Postia amara (Polyporus amarus)

Redwood

Postia (Poria) sequoiae, Ceriporiopsis rivulosa (Poria albipellucida)

Western hemlock

Heterobasidion annosum (Fomes annosus), Stereum sanguinolentum, Phellinus (Poria) weirii, Phellinus (Fomes) pini, Echinodontium tinctorium, Perenniporia subacida, Pholiota adiposa, Armillaria spp., Fomitopsis (Fomes) pinicola, Phaeolus schweinitizii, Laetiporus (Polyporus) sulphureus, Ganoderma (Fomes) applanatum

Aspen

Phellinus (Fomes) igniarius

Appalachian oaks

Stereum frustulosum, Hericium (Hydnum) erinaceus, Laetiporus (Polyporus) sulphureus, Bondarzewia (Polyporus) berkeleyi, Stereum gausapatum, Phellinus (Poria) spiculosa, Inonotus (Polyporus) hispidus

Sugar maple

Armillaria spp., Hypochnicium (Corticium) vellereum, Oxyporus (Fomes) connatus, Phellinus (Fomes) igniarius, Inonotus (Polyporus) glomeratus, Ustilina deusta (vulgaris)

Sweetgum

Rigidoporus (Fomes) geotropus, Lentinus (Panus) tigrinus, Pleurotus dryinus (corticatus), Pleurotus ostreatus, Tyromyces (Polyporus) fissilis, Hericium erinacius

Yellow birch

Phellinus laevigatus (Fomes igniarius var. laevigatus), Pholiota adiposa, Inonotus (Polyporus) hispidus, Inonotus obliquus (Poria obliqua), Cystosporium (Stereum) murrayi, Armillaria spp.

---

incidence of *Phellinus igniarius* in aspen has never been explained adequately. The wood of this tree is essentially devoid of toxic extractives. It is also extremely susceptible to decay both in living trees and in wood products after harvest. Thus we can only conclude that *P. igniarius* appears to be uniquely adapted to the particular conditions that exist in living trees of aspen.

It is beyond the scope of this brief introduction to include detailed information on all the various types of fungi that cause heart decay in living trees. The representative heart decays discussed here have been selected to include fungi affecting the major trees in various geographical regions and to illustrate the variety of ways in which decay fungi are known to enter and develop in living trees.

Although decay in trees has been a primary concern of forest pathologists since the time of Hartig, many new ideas are currently being developed that relate to our understanding of the processes of infection and development of decay fungi. In the future these continuing researches will alter and adjust the concepts described in these few pages. This emphasizes the importance of obtaining the most current information before developing management plans for heart decays and other tree diseases.

Heart Decay in Douglas-fir

Heart decays are the most damaging of all diseases in old-growth stands of Douglas-fir. *Phellinus pini* causes the greatest amount of cull although substantial losses also are caused by *Fomitopsis officinalis, Fomitopsis rosea* and *Phaeolus schweinitizii*. In some localities other fungi have been reported as indicated in Table 20.1.

Amounts of cull in Douglas-fir increase with age of the trees approximately as follows:

Tree Age	Volume of Cull (%)
<40 years	0
40–80 years	2
80–120 years	4
120–160 years	9
160–200 years	34
<200 years	85

The decay caused by *P. pini* is commonly called red heart or red ring rot because of the reddish color of the wood in incipient stages of decay. In advanced stages the decay is called white pocket rot. The distinctive white pockets appear as spindle-shaped cavities that are oriented parallel to the grain and contain soft white fibers separated by zones of apparently sound wood. The decayed wood sometimes is filled with resin. In the incipient (red) stage, the strength of the wood is not significantly affected. In the white-pocket stage, however, the amount of strength loss is roughly proportional to the abundance of white pockets.

Decay by *P. pini* usually is confined to the heartwood, but in some trees in which decay of the heartwood is very advanced, sapwood is also attacked. This attack indicates that *P. pini* must be considered a pathogenic fungus as well as a saprophytic destroyer of dead heartwood. Boyce (1961) indicates that *P. pini* may continue to develop in trees after felling but is of no importance in the destruction of logging slash. The decay has never been found to continue to develop in wood products after harvest. Thus products made from partially decayed wood will be just as stable as they are when they were manufactured.

Sporocarps and swollen knots or punk knots are the most reliable external indicators of heart decay caused by *P. pini*. The sporocarps vary substantially in appearance from thin, shell-shaped bracket types of structures to substantial, hoof-shaped sporocarps with deep furrows on the upper side. They also vary in size from 2 to 5 cm to as much as 0.3 m in width. The spore-bearing pores are visible to the naked eye on the lower side of the sporocarps (see "Taxonomy of Wood-Destroying Fungi"). Punk knots are usually formed at branches when the tree has partially or completely overgrown an old sporocarp. The punky material of the knot consists of a large mass of hyphae. Sometimes that knot develops sufficient pressure beneath the bark to produce distinctive swollen knots. The column of decay usually extends 2–9 m above and below each sporocarp depending on the age of the trees. Decay extends about half this far above and below swollen knots. If a tree bears many sporocarps, it is often a total loss.

Management to minimize losses due to *P. pini* involves three general approaches:

1. harvesting the trees before losses become excessive;
2. in intensively managed stands, pruning branches before they exceed 5 cm in diameter; and
3. using partially decayed wood.

As indicated previously, heart-decay losses begin about age 40 but become greatly accelerated after age 160. From this standpoint, harvesting should be scheduled before age 160. First priority in harvesting should be given to removal of the oldest trees with the largest number of sporocarps and other indications of defect.

In the western United States and Canada, pruning is a very uncommon practice at present but may become more widely used in the future.

In the past several years, the technical utilization requirements for Douglas-fir have been adjusted to permit the utilization of white-pocket logs. Plywood containing white-pocket veneers are used as decorative paneling, in packaging crates, sheathing, and other uses where adequate structural integrity and/or attractiveness are desired rather than maximum strength. Use of both sound and partially decayed veneer sheets in the same plywood composite is also permitted. These specifications by the Douglas-Fir Plywood Association were a major boon to use of logs infected with *P. pini*. Pulp and paper products made from wood partially decayed by *P. pini* are also satisfactory for many purposes. Reasonable yields of good-quality pulp can be obtained from logs that contain wood too decayed for use as plywood. The heart decay caused by *P. pini* well illustrates that economic and technological as well as biological factors must be considered in formulating management strategies to minimize disease losses.

Echinodontium tinctorium Heart Decay

This fungus, which is commonly called the Indian Paint fungus, occurs only in western North America. It is most damaging on the true firs but attacks western hemlock in certain localities and causes occasional heart decay in Douglas-fir, western redcedar, and Engelmann spruce. Volume losses in overmature western hemlock can be as great as 30% or more in the interior region of its range, but losses in coastal stands rarely exceed 5%. Losses in the true firs average 15–20% in mature stands. Little damage is observed in stands of less than 150 years, although heart decay has been detected in trees as young as 72 years of age.

Echinodontium tinctorium causes a brown stringy type of decay that is restricted to the heartwood of affected trees. The stringy characteristic of the decay is due to greater attack of the springwood portion of the annual rings. The decay frequently extends the entire length of the heartwood in stems and often extends into both branches and roots. This fungus functions only in living trees; it causes no important decay of dead trees or logging slash and does not continue to decay wood in use.

Sporocarps of this fungus are hard, woody, and hoof-shaped. They can be up to 0.3 m or more in width. They usually develop on the underside of dead branches. The interior of the sporocarps is a characteristic rusty red due to a pigment that was commonly used by Indians—hence the common name. Sporocarps and characteristic punky knots are formed only after the heart decay has become extensive. This additional decay complicates the problem of accurately estimating amount of cull on the basis of external indicators.

The process of infection by *E. tinctorium* is better understood than perhaps any other heart decay of conifers. For this reason, this aspect of the life history of the fungus will be emphasized in the discussion that follows. For many years it was presumed that dead branch stubs were the primary avenue of entrance of this fungus. However, Etheridge (1972) demonstrated that various Ascomycetes and Fungi Imperfecti colonize the wood exposed in larger broken branch stubs so rapidly that they provide an effective barrier against infection by *E. tinctorium*. Initial infection by *E. tinctorium* does occur, however, through tiny branchlets only 0.5 to 1.5 mm in diameter that are broken off from the upper side of main branches near the stem. Branchlets larger than 1.5 mm and smaller than 0.5 mm apparently are not suitable for infection. Hyphae of the fungus spread along the pith of the tiny branchlet and the main branch on which these branchlets occur. Some of these infections extend into the central wood of the main branches where they become inactive. Here the fungus can remain in a quiescent (dormant) condition for as long as 20, 50, or even 100 years. When the main branch dies or breaks off (or for other reasons conditions become favorable for reactiviation of the fungus), the fungus becomes active again, and decay of the heartwood portion of the main branches begins. Infections from various branches frequently coalesce in the main stem of the tree to form a continuous column of decayed heartwood.

Heart decay caused by *E. tinctorium* is controlled by two management approaches:

1. harvesting trees at an economically determined pathological rotation age, that is, before net losses due to heart decay become economically important;
2. preventing the occurrence of branch infections in young stands by silvicultural procedures.

According to Etheridge (1972), such procedures should be aimed at regulating the amount of light reaching the lower crown so as to limit the diameter of secondary branchlets to below 0.5 or above 1.5 mm or induce the natural pruning of main branches before heartwood forms in them (between 35 and 45 years). Artificial pruning before heartwood forms in them also may be effective although expensive.

Use of wood decayed by *E. tinctorium* is not feasible even for pulping purposes. The decay is of the brown cubical type in which the wood loses essentially all its strength in a very early stage of the decay process.

White Trunk Decay of Aspen

Phellinus igniarius is the most damaging of all heart-decay fungi in hardwoods. It is destructive in beech, birch, and maple but can be catastrophic in its effect on aspen. In the Lakes states region of the United States and Canada, *P. igniarius* is the principle reason that the economic life of aspen trees is limited to about 40 or 50 years. The fungus, which is sometimes called the false tinder fungus, occurs throughout the natural range of aspen. Three major strains of the fungus are known. All three develop in aspen, but the most common and damaging is called *P. igniarius* var. *populinus*. The severity of heart decay in aspen varies with location, but this apparently is due more to genetically controlled variation in susceptibility of the trees in various localities than to site factors that influence infection or decay processes. Vigorously growing trees within a given locality tend to contain less heart decay than trees in less vigorous stands.

The decay caused by *P. igniarius* is a uniform white trunk decay. In advanced stages the decayed wood contains fine black zone lines that extend in a characteristic, almost random fashion through the decayed wood. In the earliest stages of decay, the wood shows light yellowish-white streaks. In advanced stages, there is often an abrupt transition from sound to decayed wood. This transition zone frequently shows as a dark yellow-green to brown zone.

The decay caused by *P. igniarius* can occur throughout aspen stems but more frequently occurs in the middle rather than in the butt or top of the stem. The fungus commonly continues to cause decay in dead trees and in logging slash but does not cause decay of wood in service.

Sporocarps, swollen punky knots, and so-called sterile conks (actually a mass of sporocarplike tissue that does not form spores) are the most reliable external indicators of heart decay caused by *P. igniarius*. Swollen knots usually develop before sporocarps are formed, usually at branch stubs. The sporocarps are usually hoof-shaped, grey to black on top with conspicuous cracks when mature. A new hymenial layer is formed each year. This is a characteristic of the genus *Fomes* (see "Taxonomy of Wood-Destroying Fungi"). Each sporophore indicates a decay column 1.5–3 m in length; thus a tree with even a few sporophores usually is a total loss. Branch stubs, frost cracks, and other wounds that expose the wood beneath the bark are presumed to be the most common avenue of entrance for *P. igniarius*.

Management of *P. igniarius* heart decay is achieved mainly by harvesting the trees before the decay becomes excessive. Rotation ages to avoid decay vary with the region in which the tree is growing and may be 40, 60, 90, or even 130 years. Because the decay is of the white variety, wood in the incipient and intermediate stages of decay can be used for production of pulp and decorative paneling. Felling badly infected trees to prevent development of sporocarps or production of spores has been recommended but is not worthwhile because the fungi continue to disperse spores from sporocarps on these fallen trees.

Heart Decay in Southern Hardwoods

A great variety of fungi cause heart decay in the mixed hardwood forests of the southern United States (see Table 20.1). Two general types of forest are recognized within this region: Appalachian hardwoods that include several species of oak, yellow-poplar, basswood, ash, sugar maple, black cherry, and black walnut and coastal plain and piedmont hardwoods that include various species of oak, sweetgum, various hickories, yellow-poplar, black gum, elms, red maple, river birch, beech and sycamore. Most of these species form a true heartwood. Both types of forest have been abused by a combination of uncontrolled wildfires, grazing by livestock, careless logging, an opportunistic system of harvesting called high grading (repeatedly cutting the best and leaving the worst trees), abandonment of agricultural land, and a general lack of informed forest management. These conditions have contributed to the development of large areas of overstocked, unthrifty stands of poor-quality hardwoods in which heart decay is the greatest single cause of loss.

Very extensive studies have been made of the association between heart decay in various species of southern hardwoods and the wounds through which the causal fungi apparently enter the tree stems. These wounds include basal injuries such as fire scars and logging injuries, as well as top injuries that include broken branches, cankers, insect borings, and broken tops. Fire scars have been by far the most common entrance court for decay fungi.

In contrast with heart decay in conifers, the production of sporocarps is not a reliable indication of the amount, or in some cases even the type, of decay in hardwoods. Thus other external indications of interior defects have had to be developed. Various types of statistical tables have been constructed to show the relationship between amount of cull and such factors as age and species of tree, type and size of wound (whether basal, midtrunk, or top and percentage circumference of the stem involved), and growth habit of the tree (whether of seedling or sprout origin). These tables are very useful for predicting either the current amount of cull in standing trees or the amount of cull that can be expected to develop at various periods of time in the future.

The extensive studies described here provide the basis for the following generalizations concerning heart decay in southern hardwoods:

1. The amount of heart decay increases directly with the age of the stand. Thus the older that stand, the greater the amount of heart decay.
2. Trees of seedling origin generally contain less decay than trees of sprout origin.
3. Wounded trees, particularly trees with basal stem injuries, contain more decay than trees without visible wounds.
4. Trees with basal stem injuries usually contain more decay than trees with broken branch stubs or broken tops.
5. The larger the wound (whether it be at the base, midtrunk, or top of the tree), the greater the amount of heart decay.
6. The longer the time since wounding, the greater the amount of decay.
7. The heavier the logging machinery and the more carelessly that machinery is used, the greater the amount of heart decay in the residual trees. Logging with light machinery or horses usually results in many fewer basal stem injuries than logging with heavy machinery.

All these generalizations except 2 and 5 apply equally well to northern and southern hardwoods (see "Discoloration and Decay in Northern Hardwoods").

Many hardwood trees originate as sprouts that develop from cut stumps. Trees of sprout origin frequently contain more decay than trees of seedling origin because decay spreads from the heartwood of the parent stump into the heartwood of the developing sprouts.

Roth and Sleeth (1939) have recommended the following procedures to minimize losses due to heart decay in stands of sprout oak. For stands up to 8 cm in diameter and not over 20 years old:

1. Seedlings or seedling sprouts should be favored as crop trees.
2. Sprouts from small stumps (preferably less than 8 or 10 cm and not over 15 cm in diameter at ground level) should be favored over those from larger stumps.
3. Wherever possible, sprouts originating low on the parent stumps (near ground level or below) should be favored over those of higher origin. Sprouts from

low-cut stumps should be favored over those from high-cut stumps. Controlled fire after logging operations can sometimes be used to kill buds high on cut stumps and thus favor lower stem sprouts and root sprouts.

4. In the selection and thinning of fused sprouts, the procedure recommended is to cut flush at the crotch, or as nearly flush as can be done without injury to the favored sprout, so that rapid sealing may take place.

Recommendations for stands more than 8 cm at breast height and generally over 20 years old follow:

1. Single sprouts are preferable to fused sprouts, particularly if the latter require thinning. Single-stemmed trees usually develop from smaller stumps than twin trees.
2. Sprouts with open stump wounds or with enlarged butts should be cut.
3. Clumps of large sprouts that are fused for some distance above ground level or that have low V-type crotches should be treated as a unit and either cut or left intact.

A certain amount of discolored wood is usually associated with the heart decay that develops after wounding of most southern hardwoods. In some species such as oaks, this zone of discolored wood is usually narrow and inconspicuous. In other cases, most notably in yellow-poplar and tupelo, the volume of discolored wood may be very substantial. This is true especially in the years immediately after wounding before extensive decay has developed. In yellow-poplar the colors are often spectacular—shades of dark yellowish green, pink, red, blue, purple and black; black is also very common in tupelo. These discolorations are important factors influencing the use of dimension stock and veneers used for furniture production, cabinet work, and so on. They can also increase the requirements for bleaching chemicals in pulping of hardwoods.

Hepting and Blaisdel (1936) reported a dark cylindrical zone of protection wood surrounding central columns of decay in wounded sweetgum trees. This zone marked the position of the cambium at the time of injury and appeared to confine the discoloration and decay within this zone. This process of barrier zone formation (as it is now called) is a general reaction of many hardwoods to wounding. This is discussed more fully in "Discoloration and Decay in Northern Hardwoods."

In the future, heart decay in southern hardwoods should be expected to become a progressively less important source of loss. This decrease in importance occurs for a number of reasons.

1. Increasing control of wild fire should decrease the incidence of fire scars.
2. Average age at time of felling of hardwood sawlogs will probably continue to decrease. This decrease will give less time for development of heart decay.
3. The frequency of logging injuries on trees remaining after selective logging should continue to decline because of two recent technological changes. Low-quality and injured hardwoods can now be used for production of pulp and paper products. Machinery and pulping technology are available for conversion of whole hardwood trees (branches, bark, and leaves) into chips and ultimately

into pulp and paper products. Both of these innovations provide economic incentives for complete clear-felling of hardwood stands as opposed to high grading.

4. Increasing use of site-preparation techniques, including use of controlled fire and/on rolling choppers, potentially should decrease the frequency of sprouts originating high on residual stumps. This practice should diminish the vulnerability of sprout oaks to heart decay originating from parent stumps.

Two recent developments related to item 3 include increasing use of very heavy (and therefore very damaging) logging equipment and the use of a harvesting system called economic clear-cutting instead of complete clear-felling. This practice not only results in a substantial number of badly wounded trees left on the land but also provides sufficient shading of the area to inhibit the development of seedlings and sprouts of the more desirable light-demanding species of hardwoods.

Discoloration and Decay in Northern Hardwoods

In the mixed hardwood forests of the northeastern United States and eastern Canada, discoloration and decay associated with wounds are a major cause of loss in the quality of hardwood lumber, dimension stock, and veneer. The predominant hardwoods of this region include beech, paper birch, yellow birch, sugar maple, red maple, aspen, and white ash. In contrast with most southern hardwoods, many of these species produce diffuse-porus wood and do not form a true heartwood. This means that the wood of beech, birch, and maple trees potentially should be uniform in texture and light in color from cambium to pith. But such trees are relatively rare because most trees sustain one or more major wounds that lead to discoloration and sometimes to decay of the stem wood before they mature. Because the processes of discoloration and decay and the resistance of the trees to these processes have been studied more intensively in maple, birch, and beech than in other hardwoods, knowledge developed with these species will be emphasized in the discussion that follows.

Discoloration and decay in northern hardwoods are initiated by aboveground wounds that break through the protective layer of the bark and expose the wood beneath the cambium. These wounds include branches and stems that are broken by wind, animals, or accumulations of ice or snow; they also include fire scars and a variety of mechanical injuries including scrapes by logging equipment and deeper holes made by insects, birds, small boys, and even foresters using increment borers. These wounds injure and kill some cells directly. They also expose other cells to the atmosphere. Initial discoloration results from chemical reactions that in part are analogous with the passive browning of freshly cut peach or apple tissue and in part involve the wound-induced formation of new substances that are synthesized by the injured tissues of the tree.

As soon as the wound occurs, the newly exposed tissues at the surface of the wound are exposed to a myriad of microorganisms that are carried by the wounding agent itself, are borne in the air or in rain that washes over the surface of the wound, or are carried by insects, birds, and other animals that visit the wound surface. In addition, any microorganisms that may already be present in the wood prior to injury may be activated as in the case of the dormant infections by *Echinodontium tinctorium* described earlier in this chapter. Depending on the vigor of the tree, the severity of the wound, and the competitive interactions that occur both among the microorganisms present and between these organisms and the cells of the tree, the processes of discoloration and development of microorganisms may stop with only very superficial evidence of the injury, or the process may proceed further.

Soon after the wound occurs, the cambium of the tree reacts by forming a layer of unique cells in a barrier zone that frequently extends over the entire surface of the cambium of the tree at the time of injury. This barrier zone confines the developing discoloration and microorganisms to tissues that existed in the tree at the time of injury. Both the discoloration and associated microorganisms develop radially toward the pith, tangentially in both directions around the stem and longitudinally both upward and downward in the stem; but they do not develop outward into the tissue formed after wounding because the barrier zone is impervious to penetration by microorganisms. In addition to forming a barrier zone at the cambium, a layer of callus tissue is formed at the margin of the wound. If the callus spreads rapidly and the wound is small, the wound soon may be completely sealed off. If so, further development of discoloration and microorganisms is inhibited and the process stops.

If the wound remains open, however, a spatial succession of microorganisms is established in the wood of the tree. This succession includes bacteria and certain Ascomycetes and Fungi Imperfecti as well as decay fungi. The Ascomycetes and Fungi Imperfecti are usually members of the genera *Phialophora, Fusarium, Cephalosporium, Margocinomyces,* and some others. They are called pioneer organisms because they can be detected first near the wound and usually can be found in tissues that surround the decay fungi. This succession of organisms develops most rapidly along the grain of the wood, less rapidly in a radial direction inward and toward the pith, and least rapidly around the stem in both tangential directions from the wound.

As the zone occupied by the bacteria and other pioneer organisms expands, the acidity of the tissue changes from about pH 4 or 5, as is usually found in normal sapwood, to pH 6, 7, or even higher. The acidity of the zone occupied by the decay fungi, by contrast, is usually pH 4 or less. In addition, the moisture and mineral contents of the affected tissues frequently increase. This increase in mineral content is important because it provides a basis for detecting decay electronically. This will be discussed in the next section. Also, the color of the tissue is altered by compounds produced by the living parenchyma cells of the wood, the bacteria, the fungi, or perhaps by all three acting together. Some of the compounds produced by the parenchyma cells of the wood are toxic to the pioneer organisms and decay fungi and tend to inhibit their further spread and development. This defense reaction is called reaction zone formation. Its effectiveness depends on the physiological status of the tree.

The wound-sealing processes of the margin of the wound continue while the preceding succession of microorganisms is spreading within the tree stem. If the wound becomes completely sealed over, all these processes—discoloration, microbial development, and decay—may be stopped before significant heart decay develops. If not, the processes continue on, and a larger and larger volume of tissue becomes affected.

As time goes on, the column of discolored tissue continues to expand both radially and tangentially until all the tissues inside the barrier zone laid down by the cambium may become affected in the vicinity of the wound. The column also continues to extend longitudinally, first near the wound and then at greater and greater distances away. The tissues may pass through incipient, intermediate, and advanced stages of decay. The continuing linear extension of the column of discolored and decayed wood is often inhibited by substances produced by the parenchyma cells of the wood in the reaction zone that develops at the advancing margin of the column. The effectiveness of this reaction zone in limiting the linear extension of the column is determined by the physiological status of the tree in relation to the aggressiveness of the advancing microorganisms.

As each year passes following the initial wound, new annual increments of nondiscolored sapwood are laid down outside the original barrier zone. This tissue remains free of discoloration, microorganisms, and decay unless the tree is wounded again. In such cases, a new series of events is initiated including formation of a new barrier zone that compartmentalizes the stem again. The discoloration and organisms associated with the subsequent wound may be similar or very different from those associated with the first wound.

The processes discussed previously can be summarized as four general phenomena:

1. cylindrical compartmentalization of stems by barrier zones of abnormal cells formed by the cambium so that tissues formed after wounding remain free of discoloration and decay;
2. development of a spatial succession of microorganisms inside the barrier zone—this succession includes bacteria, certain Ascomycetes and Deuteromycetes, and decay fungi;
3. formation of a reaction zone that inhibits longitudinal spread of the discoloration and advancing microorganisms; and
4. complete arrest of the discoloration, microbial succession, and decay processes when wound sealing becomes complete.

Recent research results suggest that these four general phenomena occur in both southern and northern hardwoods including both ring-porous and diffuse-porous species as well as those that do and do not develop a true heartwood. One difference in the ring-porous species, however, is that wounding often interferes with normal heartwood formation.

Management of northern hardwoods so as to minimize losses due to discoloration and decay involves the following general approaches.

1. In so far as possible avoid major wounds above ground by using clear-felling in preference to selective cutting, removing injured trees as the last stage of selective cutting operations, and artificially pruning of lower branches early so as to obtain as rapid sealing of branch stubs as possible.
2. Harvest trees at an age before discoloration and decay become economically important.
3. Thin excessive stems in sprout stands as early as possible and favor sprouts as widely separated on the same stump as possible.

Prevention of Heart Decay in Shade and Other Ornamental Trees

Because individual shade and other ornamental trees have substantially greater value than individual forest trees, much greater care and expense can be applied to prevent damage by heart decay and other diseases.

In planting trees for ornamental purposes, it is imperative to select trees that are well-adapted to the local conditions of climate, soil type, and the like. This is just as important in the prevention of heart decay as it is in many other diseases. Recommendations for the prevention and control of heart decay in ornamental trees follow:

1. Throughout the life of the tree, check soil fertility periodically by soil analysis and fertilize as necessary to maintain tree vigor. Watering in periods of extreme drought and drainage of water from flooded areas may also be desirable.

2. Avoid burning of trash that could produce injurious fumes or cause damage to stems or branches by excessive heating especially in freezing weather.

3. Make as few changes as possible in the soil grade and/or drainage patterns in the vicinity of the tree. These changes often interfere with normal root respiration and thus diminish the vigor of the tree.

4. In selecting trees of species that form a true heartwood, especially oaks, favor trees of seedling origin rather than of sprout origin.

5. Avoid wounds that break through the bark on roots, stems, and branches during construction of buildings, walkways, driveways, and roadways.

6. If wounds are inflicted, take the following precautions to minimize their potential as courts of infection for heart-decay fungi:

    Broken stems. Trim the stem below the broken top so that water will drain off rather than collect on the wound surface.

    Broken branches. Trim the broken branch as nearly vertically and as nearly flush to the bark of the main stem as possible while remaining outside the branch collar ridge (Fig. 20.30). Undercut branches before severing them so that the bark below the branch does not peel away as the branch falls.

    Superficial scrapes. Trim the bark with a sharp knife in a rounded elliptical area around the scrape so that sealing of the wood (covering by wound callus) will be encouraged.

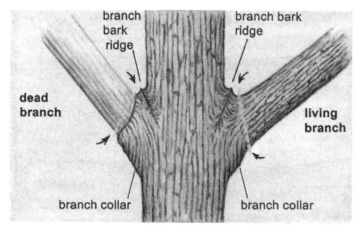

FIGURE 20.30   The proper way to target the cut when pruning a branch. From Bedker et al. (1995).

Roots. Trim the severed end of the root so that a blunt rather than a splintered end is exposed. An asphaltic wound dressing applied to the cut end of the root will avoid constant contact between the soil and the exposed wood.

In all pruning operations, take care to avoid tearing the bark away from the wood of the root, stem, or branch. Pruning broken branches, stems, and roots should be done as soon after wounding as possible. Wound dressings and use of toxic chemicals such as ethyl alcohol, pentachlorophenol, or creosote are of questionable value for treatment of branches and superficial scrapes because they do not prevent invasion of the wood by decay fungi. The barrier zone formed by the cambium will effectively confine the decay within the tissues present at the time the tree was wounded. Wound dressings might be of value if they are used to fill or cover a sunken area that otherwise would collect water or to prevent contact between soil and exposed wood.

7. Take precautions to avoid later wounds that may be serious. For example, broadly forked stems sometimes split and provide a huge entrance point for heart-decay fungi. Using cable or steel rods to tie the separate stems or heavy branches together may avoid later splits. Also, if it is known that certain branches will need to be removed later in the development of the tree, early pruning will provide a smaller wound that will heal over more rapidly than the larger ones that will be necessary later in the life of the tree. Precautionary pruning is best done in the fall or winter to avoid too rapid drying of the exposed bark-wood interface. Pruning in late spring and summer frequently leads to separation of wood and bark around pruning wounds.

8. If heart decay has already begun in trees, its presence can be detected electronically. As decay develops, various mineral elements including potassium, sodium, calcium, and magnesium collect in the decayed wood. The presence of these ions can be detected by drilling a very fine hole into the suspected zone and measuring the change in resistance to a pulsed electric current as the probe is inserted deeper and deeper into the wound. If an abrupt decrease in electrical resistance is detected, the wood is decayed. A commercial instrument is available to make such measurements. It is called a Shigometer in honor of Dr. Alex Shigo who has devoted much of his life to the study of discolorations and decays in living trees.

If heart decay has already begun in branches but has not reached the stem, trim off the branch as nearly flush with the stem as possible, but outside the branch bark collar ridge. If the decay is already established in the stem, remove as much of the decayed wood as possible to allow good ventilation of the cavity. If rain water will collect in the cavity, however, either plug the cavity with plastic foam, concrete, or wound dressing or provide a drainage hole at the bottom of the cavity. In removing decayed wood be careful not to break through the barrier zone that was formed around the column of decayed wood that is being removed.

REFERENCES

Aho, P. E. 1971. *Decay of Engelmann Spruce in the Blue Mountains of Oregon and Washington.* USDA For. Serv., Res. Pap. PNW-116. 16 pp.

Aho, P. E. 1974. *Defect Estimation for Grand Fir in the Blue Mountains of Oregon and Washington.* USDA For. Serv., Res. Pap. PNW-175. 12 pp.

Andrews, S. R. 1955. *Red Rot of Ponderosa Pine.* USDA For. Serv., Agri. Monograph No. 23. 34 pp.

Basham, J. T. 1960. *The Effects of Decay on the Production of Trembling Aspen Pulpwood in the Upper Pic Region of Ontario.* Can. Dept. Agri., Publ. 1060. 25 pp.

Basham, J. T. 1973. Heart rot of black spruce in Ontario. II. The mycoflora in defective and normal wood of living trees. *Can. J. Bot.* 51:1379–1392.

Bedker, P. J., J. G. O'Brien, and M. E. Mielke. 1995. *How to Prune Trees.* USDA For. Serv., NA-FR-01–95. 30 pp.

Berry, F. H., and J. A. Beaton. 1972. *Decay in Oak in the Central Hardwood Region.* USDA For. Serv., Res. Pap. NE-242. 11 pp.

Boyce, J. S. 1961. *Forest Pathology,* 3rd ed. McGraw-Hill, New York. 572 pp.

Davidson, A. G. 1957. Decay of balsam fir, *Abies balsamea* (L.) Mill., in the Atlantic Provinces. *Can. J. Bot.* 35:857–874.

Davidson, A. G., and D. E. Etheridge. 1963. Infection of balsam fir, *Abies balsamea* (L.) Mill., by *Stereum sanguinolentum* (Alb. and Schw. ex Fr.) Fr. *Can. J. Bot.* 41:759–765.

Etheridge, D. E. 1972. *True Heartrots of British Columbia.* For. Insect & Dis. Surv., Pest Leaflet No. 55. 14 pp.

Etheridge, D. E., and H. M. Craig. 1976. Factors influencing infection and initiation of decay by Indian paint fungus in western hemlock. *Can. J. Bot.* 6:299–318.

Farr, W. A., V. J. LaBau, and T. H. Lament. 1976. *Estimation of Decay in Old-Growth Western Hemlock and Sitka Spruce in Southeast Alaska.* USDA For. Serv., Res. Pap. PNW-204. 24 pp.

Foster, R. E., J. E. Browne, and A. T. Foster. 1958. *Decay of Western Hemlock and Amabilis Fir in the Kitmat Region of British Columbia.* Can. Dept. Agri., Publ. 1029. 37 pp.

Hepting, G. H. 1935. *Decay Following Fire in Young Mississippi Delta Hardwoods.* USDA, Tech. Bull. 494. 32 pp.

Hepting, G. H., and D. J. Blaisdell. 1936. A protective zone in red gum fire scars. *Phytopathology* 26:62–67.

Hinds, T. E. 1963. *Extent of Decay Associated with Fomes igniarius Sporophores in Colorado Aspen.* USDA For. Serv., Res. Note RN-4. 4 pp.

Kimmey, J. W., and P. C. Lightle. 1955. Fungi associated with cull in redwood. *For. Sci.* 1:104–110.

Landis, T. D., and A. K. Evans. 1974. A relationship between *Fomes applanatus* and aspen windthrow. *Pl. Dis. Reptr.* 58:110-113.

Lightle, P. C., and S. R. Andrews. 1968. *Red Rot in Residual Ponderosa Pine Stands on the Navajo Indian Reservation.* USDA For. Serv., Res. Pap. RM-37. 12 pp.

Maloy, O. C. 1973. Reducing decay in grand fir. *J. For.* 71:706–707.

Manion, P. D., and D. W. French. 1968. Inoculation of living aspen trees with basidiospores of *Fomes igniarius* var. *populinus. Phytopathology* 58:1302–1304.

Manion, P. D., and D. W. French. 1969. The role of glucose in stimulating germination of *Fomes igniarius* var. *populinus* basidiospores. *Phytopathology* 59:293–296.

McCracken, F. I., and E. R. Toole. 1974. Felling infected oaks in natural stands reduces dissemination of *Polyporus hispidus* spores. *Phytopathology* 64:265–266.

Merrill, W. 1970. Spore germination and host penetration by heartrotting Hymenomycetes. *Annu. Rev. Phytopath.* 8:281–300.

Palmer, J. G., and C. May. 1970. Additives, durability, and expansion of a urethane foam useful in tree cavity fills. *Pl. Dis. Reptr.* 54:858–862.

Roth, E. R. 1956. Decay following thinning of sprout oak clumps. *J. For.* 54:26–30.

Roth, E. R., and B. Sleeth. 1939. *Butt Rot in Unburned Sprout Oak Stands.* USDA, Tech. Bull. 684. 42 pp.

Shigo, A. L. 1965. *Organism Interactions in Decay and Discoloration in Beech, Birch, and Maple.* USDA For. Serv., Res. Pap. NE-43. 23 pp.

Shigo, A. L. 1974. Relative abilities of *Phialophora melinii, Fomes connatus,* and *F. igniarius* to invade freshly wounded tissues of *Acer rubrum. Phytopathology* 64:708–710.

Shigo, A. L. 1989. *A New Tree Biology: Facts, Photos, and Philosophies on Trees and Their Problems and Proper Care,* 2nd. ed. Shigo and Trees, Associates, Durham, N.H. 618 pp.

Shigo, A. L., and E. H. Larsen. 1969. *A Photo Guide to the Patterns of Discoloration and Decay in Living Northern Hardwood Trees.* USDA For. Serv., Res. Pap. NE-127. 100 pp.

Shigo, A. L., H. R. Skutt, and R. A. Lessard. 1972. Detection of discolored and decayed wood in living trees using a pulsed electric current. *Can. J. For. Res.* 2:54–56.

Thomas, G. P. 1958. *The Occurrence of the Indian Paint Fungus Echinodontium tinctorium E. & E., in British Columbia.* Can. Dept. Agri., Publ. 1041. 30 pp.

van der Kamp, B. J. 1975. The distribution of microorganisms associated with decay of western red cedar. *Can. J. For. Res.* 5:61–67.

Wagener, W. W., and R. W. Davidson. 1954. Heart rots in living trees. *Bot. Rev.* 20:61–134.

Wagener, W. W., and R. V. Bega. 1958. *Heart Rots of Incense Cedar.* USDA For. Serv., For. Pest Leaflet 30. 7 pp.

Wilcox, W. W. 1970. Tolerance of *Polyporus amarus* to extractives from incense cedar heartwood. *Phytopathology* 60:919–923.

DISEASE PROFILE

Red Heart Decay

Importance: This disease causes a severe heart rot of living balsam fir in North America and in living trees and stored pulpwood of Norway spruce in Scandinavia. It has also caused a serious decay problem in exotic pine species in highland plantations in Kenya. In the western United States it causes death of planted conifer saplings and is known as mottled bark disease. It has caused heart decay of pruned eastern white pine.

Suscepts: Conifers, especially balsam fir and spruces, serve as hosts.

Distribution: Red-heart decay is widespread in temperate forests of the northern hemisphere as a saprophyte on coniferous slash. On susceptible hosts it causes heart decay.

Causal Agents: Red heart decay is caused by the fungus *Stereum sanguinolentum* (Aphyllophorales).

Sporocarps. From Boyce (1961).

Biology: Infection of living saplings probably occurs through basal wounds. In large trees the causal fungus likely enters the heartwood after colonizing dead branches or dead sapwood around wounds. The decay occurs throughout the aerial portion of the bole but rarely invades the butt.

Epidemiology: Stereum sanguinolentum may be widespread as a decayer of coniferous slash. The prevalence of this decay fungus in living balsam fir has been attributed to its tolerance to extractives in the heartwood of the tree. Low temperatures favor infection by *S. sanguinolentum* over competitors. Optimal daily temperatures for infection were between 7 and 13°C. Above 15.5°C there was intense competition for the wood surface by such fungi as *Peniophora cinerea, Alternaria alternata (tenuis),* and *Ophiostoma (Ceratocystis) piceae.*

Diagnosis: In diseased saplings the inner bark surface on roots, root collar, and butt has a white, mottled appearance. In advanced stages the underlying sapwood becomes soft and spongy. Small sporophores may cover the bark of the butt and exposed roots. They are small and crustlike, with a pale olive-buff upper surface. Incipient decay in heartwood of balsam fir appears water-soaked yet firm and is reddish-brown. Rays of decay extend outward from the main body of decay, which may be in a solid large mass or in irregular patches. Thin mycelial belts form early during the decay process. Wood with advanced decay appears light brown and dry and is friable. Sporophores form in profusion on dead wood but do not appear on living trees.

Control Strategy: Consider harvesting stands of fir and spruce that have branch breakage due to wind or ice as these may provide points of entry for decay.

Decayed wood. From Kaufert (1935).

SELECTED REFERENCES

Boyce, J. S. 1961. *Forest Pathology,* 3rd ed. McGraw-Hill, New York. 572 pp.

Davidson, A. G., and O. E. Etheridge. 1963. Infection of balsam fir, *Abies balsamea* (L.) Mill, by *Stereum sanguinolentum* (Alb. and Schw. ex Fr.) Fr. *Can. J. Bot.* 41:761–765.

Griffin, H. D. 1969. Studies on *Stereum sanguinolentum* in Kenya pine plantations. *Can. J. Bot.* 47:761–771.

Kaufert, F. 1935. Heart rot of balsam fir in the Lake states, with special reference to forest management. Univ. Minn. Agri. Exp. Sta., Tech. Bull. No. 110. 27 pp.

DISEASE PROFILE

Brown Stringy Decay

Importance: This damaging heart decay occurs only in western North America where it is important in old-growth stands of conifers.

Sporocarp.

Decayed wood. From Boyce (1930).

Suscepts: It is most damaging on the true firs but attacks western hemlock in certain localities and causes occasional heart rot in Douglas-fir, western redcedar, and Engelmann spruce.

Distribution: Most prevalent in western Montana, Idaho, Oregon, and Washington.

Causal Agents: Echinodontium tinctorium (Aphyllophorales), also known as the Indian paint fungus, is the source of brown stringy decay.

Biology: This fungus is active only in living trees. Infection occurs through tiny branchlets (0.5–1.5 mm diameter) that are broken off from the upper side of main branches near the stem. Hyphae spread into the pith of the main branch where they remain quiescent for many years. When the branch dies or breaks off, the fungus becomes active again and begins to decay the heartwood portion of the main branch. As these multiple infections coalesce, a continuous column of decayed heartwood may form in the main stem. Advanced decay frequently extends the entire length of the heartwood, extending into the roots and larger branches. The decay will frequently remove all the heartwood but does not extend into the living sapwood.

Epidemiology: Infection does not occur through large broken branch stubs as earlier suspected. Those sites are colonized so rapidly by various other fungi that they provide an initial barrier against infection by *E. tinctorium*. Its ability to remain quiescent for decades in the pith of tiny branchlets has apparently given it a competitive edge.

Diagnosis: The perennial sporophores are hard, woody, and hoof-shaped, with large teeth on the underside. The interior is a striking red. Sporocarps and punky knots are formed only after heart rot has become extensive. A single sporophore usually indicates extensive decay. Incipient decay in white fir forms light-brown or golden-tan spots in the light-colored heartwood. Next, rusty-reddish streaks appear following the grain. At this stage the wood is badly weakened. As decay progresses, the color intensifies and the wood tends to separate between the annual rings. Advanced decay is brown, fibrous, and stringy.

Control Strategy: Echinodontium tinctorium is controlled by harvesting trees at an economically determined pathological rotation age (i.e., before net losses due to heart decay become economically important) and maintaining dense stands to prune branches naturally.

SELECTED REFERENCES

Boyce, J. S. 1930. *Decay in Pacific Northwest Conifers.* Yale Univ., Bull. No. 1. 51 pp.
Etheridge, D. E., and H. M. Craig. 1976. Factors influencing infection and initiation of decay by the Indian paint fungus in western hemlock. *Can. J. Bot.* 6:299–318.

DISEASE PROFILE

White Trunk Decay

Importance: This disease limits the economic life of aspen trees in the Lake states region.

Suscepts: The hosts are primarily aspen, but also beech, birch, maple, and other hardwood species.

Distribution: White trunk decay is found in the temperate northern hemisphere and especially in the Lake states region of Canada and the United States.

Causal Agents: Phellinus (Fomes) igniarius (Aphyllophorales), sometimes called the false tinder fungus.

Biology: Infection occurs through branch stubs, frost cracks, and other wounds that expose wood beneath the bark. Glucose is necessary for germination of basidiospores, and they are stimulated to germinate on wounds less than a few days old. In living trees the decay is confined to the main portion of the heartwood, which it eventually decays. After tree death, decay continues into the sapwood.

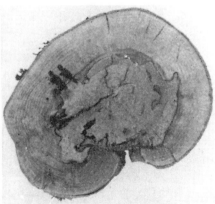

Decayed wood. From Boyce (1961).

Epidemiology: Abundant sporocarp production on diseased living trees as well as on down or dead trees means that basidiospores are likely plentiful. Fresh wounds are necessary for infection, but some studies have found that the wood needs to be altered in some way before it can be invaded. Spores germinated when placed on wounds in July but not in April or May, suggesting the need for a higher temperature for spore germination. The fungus remains active in dead trees and slash but does not invade aspen logs in storage.

Diagnosis: Sporocarps and swollen knots are the most reliable indicators. Even a few sporocarps on a tree indicate extensive internal decay. Wood in early stages of decay contains light yellowish-white streaks. Advanced decay is of the white decay type and may be surrounded by black zone lines. There is often an abrupt transition from sound to decayed wood.

Control Strategy: Losses due to *P. igniarius* are minimized mainly by harvesting the trees before the decay becomes excessive. Rotation ages may be no more than 40–50 years where the disease is severe. Vigorously growing trees tend to have less heart decay than do less vigorous trees. Because the decay is a white decay, decayed wood in early or intermediate stages can be used for pulp.

SELECTED REFERENCES

Boyce, J. S. 1961. *Forest Pathology,* 3rd ed. McGraw-Hill, New York. 572 pp.

Manion, P. D., and D. W. French. 1969. The role of glucose in stimulating germination of *Fomes igniarius* var. *populinus* basidiospores. *Phytopathology* 59:293–296.

Shigo, A. L. 1974. Relative abilities of *Phialophora melinii, Fomes connatus* and *F. igniarius* to invade freshly wounded tissues of *Acer rubrum. Phytopathology* 64:708–710.

DISEASE PROFILE

Red Ring Decay

Importance: Red ring decay, also known as red ring rot, red heart, conk rot, ring scale, or white speck, is a decay of the heartwood of living conifers.

Suscepts: Most standing coniferous species are susceptible, especially Douglas-fir, larches, pines, and spruces. Incidence is greater in older trees.

Distribution: Red ring decay is found throughout the temperate zone in the northern hemisphere.

Causal Agents: Phellinus (Fomes) pini (Aphyllophorales) is the pathogen. This terminology is correct because this fungus can attack living sapwood.

Biology: Infections occur through dead branch stubs, rarely through open wounds. Incipient decay appears as a discoloration of the heartwood, with different hosts exhibiting characteristic colors. During incipient decay, hyphae are found only in latewood. Lignin in the middle lamella between trachieds is selectively degraded. Advanced decay, also known as white pocket decay, appears as many elongated pockets or cavities parallel to the grain and separated by apparently sound wood. These pockets contain mainly cellulose. Eventually the pockets merge so that the decayed wood is a mass of white fibers. Decay may continue for a time after an infected tree dies or is cut but *P. pini* is of no importance as a decayer of slash.

Epidemiology: Perennial sporophores may be produced soon after the decay has been initiated. The basidiospores are wind-disseminated. Spore numbers are greatest in spring and late fall and especially abundant when the temperature rises following cool but not freezing weather.

(upper) Sporophore. (middle) Decayed log. (lower) Closeup of decayed wood. From Boyce (1930).

Diagnosis: Sporophores are the best external indicators of decay but they are extremely variable in size and shape. On some hosts, *P. pini* does not sporulate until the tree dies. They may be shell-shaped, bracketlike, hoof-shaped, or even resupinate. The lower surface is rather diagnostic, with the pore mouths varying from small and almost circular to large and irregular or daedaloid. Punk knots are also good indicators of decay. They are filled with the same yellowish-brown-colored mycelium as is the interior of a sporophore. Swollen punk knots result when the tree attempts to overgrow the forming punk knots. Incipient decay in Douglas-fir is reddish-or olive-purple; in white and red spruce it is light purplish-gray deepening to a reddish-brown; in western white, ponderosa, and southern pines it is pink to reddish-brown. The wood is firm, tough, and resinous in this stage. In advanced decay the white pockets are distinctive.

Control Strategy: Red ring decay was largely a problem in old-growth forests. As these have been harvested in recent decades, its importance has declined. At one time red ring decay was so prevalent that decayed wood with the distinctive white pockets was marketed for its aesthetic qualities and for products such as studs where strength reduction was not critical. In southern yellow pine forests where the name red heart is more commonly used, the endangered red-cockaded woodpecker may preferentially select old, infected trees in which to construct nest and roost cavities. These mature pine forests are being harvested and replaced with plantations of much shorter rotations. On federal forest lands the lengthening of rotations has been mandated, which may permit decays to flourish.

SELECTED REFERENCES

Blanchette, R. A. 1980. Wood decomposition by *Phellinus (Fomes) pini:* a scanning microscope study. *Can. J. Bot.* 58:1496–1503.

Boyce, J. S. 1930. *Decay in Pacific Northwest conifers.* Yale Univ., Bull. No. 1. 51 pp.

DISEASE PROFILE

Decay of Southern Hardwoods

Importance: Previous land abuses have contributed to the development of large areas of overstocked, unthrifty stands of poor-quality hardwoods in which heart decay is the greatest single cause of loss.

Suscepts: A great variety of hardwood species are affected including: oaks, yellow-poplar, sweetgum, sycamore, black gum, elms, cottonwood, hackberry, ash, persimmon, maples, black cherry, black walnut, hickories, and river birch.

Distribution: The mixed hardwood forests of the southern United States are divided into three general types: Appalachian hardwoods, coastal plain and piedmont hardwoods, and southern hardwoods. All have been abused to varying degrees by wild fires, grazing by livestock, careless logging, high-grading, and a general lack of good forest management.

Butt decay. Courtesy of J. S. Boyce, Jr., USDA Forest Service.

Causal Agents: A great variety of fungi cause heart decay in the southern hardwoods.

Biology: Heart decay is associated with wounds through which the causal fungi enter the tree. These include fire scars, logging injuries, broken branches, cankers, insect wounds, and broken tops. In coppice stands that reproduce by sprouts, decay frequently spreads from the heart decay of the stump into the developing sprout.

Epidemiology: Spores of heart decay fungi are apparently plentiful in the air. Therefore, the incidence of decay is proportional to wounding. In general, older trees have more wounds and therefore more decay; basal stem injuries usually lead to more decay than do top injuries; wider wounds lead to more decay; the longer the time since wounding, the greater the amount of decay; and use of heavier logging machinery leads to more decay in the residual trees.

Diagnosis: In contrast with heart decay in conifers, the production of sporocarps is not a reliable indication of the amount of decay in hardwoods. Various tables showing statistical relationships between the amount of decay and age, site index, tree species, degree of wounding, and whether seedling or sprout origin are useful for predicting both the current amount of decay and the expected amount at various time periods in the future.

Control Strategy: In addition to minimizing the decay-conducive factors listed in the section on epidemiology, rules of thumb have been developed to minimize decay originating from parent stumps in stands of sprout origin. In young stands, seedlings should be favored as crop trees, sprouts from small stumps should be favored over those from large stumps, low sprouts should be favored over those of higher origin, and sprouts from low-cut stumps should be favored over those from high-cut stumps.

SELECTED REFERENCES

Hepting, G. H. 1935. *Decay Following Fire in Young Mississippi Delta Hardwoods*. USDA, Tech. Bull. 494. 32 pp.
Roth, E. R. 1956. Decay following thinning of sprout oak clumps. *J. For.* 54:26–30.

DISEASE PROFILE

Discoloration and Decay in Northern Hardwoods

Importance: In northern hardwoods, discoloration and decay associated with wounds are a major cause of loss in the quality of hardwood lumber, dimension stock, and veneer.

Suscepts: The predominant hardwoods affected include beech, paper birch, yellow birch, sugar maple, red maple, aspen, and white ash.

Distribution: Discoloration and decay can be found in the mixed hardwood forests of eastern Canada and northeastern United States.

Causal Agents: Reaction of the host to wounding, and subsequent colonization of the wound by a succession of microorganisms and, lastly, wood decayers causes discoloration and decay that are responsible for the losses.

Biology: A wound that exposes the cambium causes a host reaction that is visible as a discoloration that frequently extends over the entire surface of the cambium forming a barrier zone. Contents of the wounded cells are digested by a myriad of microorganisms that are either airborne to the site or arrive in some other manner. Next, bacteria and certain Ascomycetes and Fungi Imperfecti invade the wound, increasing the moisture content of the surrounding wood and raising its pH to 6, 7, or even higher. As time goes on, the column of discolored tissue continues to expand within the tree stem inside the volume delimited by the barrier zone. As it expands it is followed by the succession of early invaders. As long as the wound does not become completely sealed over, this process continues and a larger volume of tissue is affected. Eventually the open wound surface is colonized by decay fungi. If the wound becomes completely sealed over, all these processes, including decay, may be stopped.

Epidemiology: The barrier zone effectively prevents the outward spread of discoloration and decay. If the cambium is injured later, the entire process begins over again.

Diagnosis: Visible wounds are a good indication of the presence of discoloration and decay in the standing tree. Vigorous callus growth over the wound will ensure that there will be a minimum of discoloration and probably no decay.

Discoloration following a wound. Courtesy of D. W. French, University of Minnesota, St. Paul.

Control Strategy: Avoid major wounds to tree stems and roots, maintain stand vigor as high as possible, harvest trees before discoloration and decay become economically important, and thin excessive stems in sprout stands as soon as possible.

SELECTED REFERENCES

Shigo, A. L. 1965. *Organism Interaction in Decay and Discoloration in Beech, Birch, and Maple.* USDA For. Serv., NE Res. Pap. No. 43. 23 pp.
Shigo, A. L. 1969. The death and decay of trees. *Nat. Hist.* 78(3):43–47.

DISEASE PROFILE

Canker Decays of Hardwoods

Importance: Canker-decay fungi cause serious degrade and cull.

Suscepts: Important hosts are red oaks, but also hickory, honeylocust, white oaks, and other hardwoods.

Distribution: Canker decays are abundant across the southeastern United States. They are common in the piedmont hardwood areas of North and South Carolina and Georgia, as well as in the bottomlands of the Mississippi delta.

Causal Agents: Hispidus canker is caused by *Inonotus (Polyporus) hispidus* (Aphyllophorales); spiculosa canker is caused by *Phellinus (Poria) spiculosus),* a similar

Hispidus canker.

Spiculosa canker. From Toole (1959).

Irpex canker.

canker is caused by *Phellinus laevigatus (Poria laevigata);* and irpex canker is caused by *Spongipellis pachyodon (Irpex mollis).*

Biology: Like most wood-decaying fungi, those causing canker decays produce sporocarps that produce basidiospores, which are wind-disseminated to wounds on susceptible hosts. These fungi not only decay the heartwood, but they also kill the cambium and decay the sapwood above and below the entrance point on the tree. Their ability to cause cankers distinguishes them from the heartwood decay fungi.

Epidemiology: The canker-decay fungi produce large numbers of basidiospores for several weeks after basidiocarp formation. The cankers caused by the canker-decay fungi tend to be conspicuous. This is because the canker-decay fungi, while able to kill the cambium, can do so only when host growth is limited. During the spring a ridge of callus tissue is formed by the host in an attempt to seal over the canker. The callus tissue is invaded by the canker-decay fungus. These pathogens are seldom able to grow completely through the callus layer. As additional callus layers are formed and killed, the trees may develop a fusiform-shaped swelling (hispidus).

Diagnosis: The basidiocarps are distinctive for each canker decay but in the case of spiculosa canker, do not grow on living, infected trees but only develop on decayed logs and snags (dead trees). To identify spiculosa canker on living trees, an axe cut can be made into the base of a suspected branch trace, revealing a mass of brown mycelium, although actual decay is a white decay. Hispidus sporocarps are 5–30 cm wide, spongy, hairy, stalkless, and yellowish-brown to rusty red. Within a few months after their formation, they dry to a blackened mass and fall to the ground. This decay is also a white decay, and the decayed wood becomes soft and pale yellow. Irpex cankers are usually smaller than hispidus cankers. The sporocarps are small and creamy white with short, jagged teeth on the lower surface.

Control Strategy: Cankered trees cull very rapidly. Although quick removal should be considered, depending on management objectives, these trees may be used by wildlife.

SELECTED REFERENCES

Campbell, W. A. 1942. A species of *Poria* causing rot and cankers of hickory and oak. *Mycologia* 34:17–26.

McCracken, F. I., and E. R. Toole. 1974. Felling infected oaks in natural stands reduces dissemination of *Polyporus hispidus. Phytopathology* 64:265–266.

Roth, E. R. 1950. Cankers and decay of oaks associated with *Irpex mollis. Pl. Dis. Reptr.* 34:347–348.

Toole, E. R. 1959. *Canker-Rot in Southern Hardwoods.* USDA For. Serv., For. Pest Leafl. 33. 4 pp.

DISEASE PROFILE

Wetwood

Importance: Wetwood prevents wounds from properly healing, increasing the tree's susceptibility to wind breakage. In wood products, especially of aspen, wetwood causes excessive shrinkage and other internal defects. It is a major constraint in the production of high-quality oak lumber.

Bleached bark from bacterial ooze.

Suscepts: Practically all hardwoods and some softwoods are susceptible in varying degrees. Economic losses have occurred in true firs, aspen and cottonwoods, willow, redwood, red oaks, and elm. Internationally, wetwood is severe in some species of *Eucalyptus* wherever they are planted and in *Nothofagus* in native forests of Chile and Argentina. Wetwood in oak is known as bacterial wetwood.

Distribution: Wetwood can be found worldwide, probably wherever susceptible hosts are grown.

Causal Agents: The bacterium *Erwinia nimipressuralis* has been identified as the cause of wetwood in elm. Other anaerobic heterotrophic bacteria isolated from wetwood-affected trees include *Clostridium, Bacteroides, Edwardsiella, Klebsiella,* and *Lactobacillus* species.

Biology: The bacteria associated with wetwood can enter the tree in some way, possibly through roots, and, once inside, they live on materials present in the wood, probably cell contents. Decomposition of complex carbohydrates, but not cellulose, in the tree has some similarities with digestion in rumenants. In some cases the bacteria produce gas, and this is forced out through openings in the wood. In elms a toxin that may cause dieback is produced and will kill the cambium if allowed to soak through the bark.

Epidemiology: Little is known of how or why wetwood develops. Some species, such as elm, seem to be predisposed, and almost every tree is infected. Wetwood in other species, such as red oak or aspen, may be influenced by site quality or tree vigor.

Diagnosis: Diseased elms will have discolored or blackened areas on the bark where the bacterium has forced ooze out of the tree. The ooze soon oxidizes to a brown, foul-smelling material and attracts insects that are commonly found living in it. The wood has a water-soaked appearance. Elm wood has a brown to gray discoloration, aspen is often colorless but will excessively shrink when dried, red oak will have a sharp, acidlike odor and may exhibit internal honeycombing when dried.

Control Strategy: For wetwood of elm, the only control measure that has been recommended for many years is to relieve the tree from the toxic ooze. This is done by installing a drainage pipe immediately below the point of oozing. The damage done by wounding may be more harmful than doing nothing. In wood of aspen, oak, and other species, wetwood-affected pieces must be segregated out and dried by slower drying schedules. A reliable means for identifying logs and green lumber products for presence of wetwood would be a big boon for the wood-working industry. An accurate system has been developed for presorting of bacterial oak lumber. In *Eucalyptus* and *Nothofagus* there is no internal checking, or honeycombing, so the wood can be rather severely kiln dried and then the original volume of the collapsed wood recovered by steam reconditioning.

SELECTED REFERENCES

Schink, B., J. C. Ward, and J. G. Zeikus. 1981. Microbiology of wetwood: Role of anaerobic bacterial populations in living trees. *J. Gen. Micro.* 123:313–322.

Ward, J. C., and D. A. Groom. 1983. Bacterial oak: Drying problems. *For. Prod. J.* 33:57–65.

Glossary

Abiotic disease. Disease resulting from nonliving agents.

Acervulus (pl. acervuli). An open saucerlike fruiting structure bearing conidia.

Aeciospore. Spore borne inside an aecium.

Aecium (pl. aecia). A cuplike structure bearing aeciospores in the rust fungi.

Aerobic. Requiring oxygen to grow.

Agar. A substance, from certain red algae, used in media on and in which microorganisms will grow.

Anaerobic. Requiring no oxygen for growth.

Annulus (pl. annuli). A ring on the stipe of a mushroom; the remains of a veil that covered the gills.

Antheridium (pl. antheridia). A male gametangium.

Antibiotic. A substance that is produced by a living organism and that can inhibit another living organism.

Antibody. Substance produced in the blood of an animal in response to the introduction of an antigen.

Antigen. Any proteinaceous substance that when placed into an animal stimulates the production of antibodies.

Apothecium (pl. apothecia). A saucerlike ascocarp in which the asci are borne.

Ascocarp. A structure in or on which asci are produced in the Ascomycotina.

Ascogonium (pl. ascogonia). The female gemetangium of the Ascomycotina.

Ascospore. A sexual spore borne in an ascus.

Ascus. A saclike hypha containing usually eight ascospores, the sexual spores of the Ascomycotina.

Aseptate. Without crosswalls.

Asexual. Reproduction not involving the fusion of two nuclei.

Atrophy. Lack of growth.

Autoecious. A characteristic of fungi that complete their life cycle on one host, such as certain of the rust fungi.

Avirulence. Inability to parasitize a given host.

Azygospore. A spore that is morphologically like a zygospore but that forms parthenogenetically.

Bacteriophage. A virus associated with bacteria.

Bacterium. A unicellular, chlorophylless microorganism that reproduces by fission.

Basidiospore. A sexual spore produced on a basidium.

Basidium (pl. basidia). A hyphal structure in which fusion of compatible nuclei and reduction division usually occur, and on which the sexual spores of the Basidiomycotina are borne.

Bifurcate. To divide into two branches.

Biotic disease. Disease caused by living organisms.

Biotype. A population of individuals that are identical genetically.

Blight. A disease symptom characterized by rapid discoloration and death of all parts of a plant.

Canker. A visible dead area, usually of limited extent, in the cortex or bark of a plant.

Capillitium. Sterile, threadlike structures with the spores in the sporangia of the Myxomycota.

Capsule. A sheath of polysaccharide gum surrounding a bacterium.

Cardinal temperatures. The minimum, optimum, and maximum temperatures at which an organism will grow.

Carrier. A plant that harbors a virus without apparent symptoms.

Cellulose. A linear polymer of glucose molecules linked by 1,4-beta-glycosidic bonds. The linear cellulose chains in wood cell walls are bound laterally into a linear, partially crystalline structure.

Chlamydospore. A thick-walled resting spore, asexual or sexual, usually found in an intercalary position.

Chlorosis. The fading of green leaves due to failure of chlorophyll to form.

Clamp connection. A hyphal outgrowth that, at cell division, makes a connection between the resulting two cells by fusion, common in the Basidiomycotina.

Cleistothecium. An ascocarp without any opening.

Coenocyte. A unicellular thallus containing many nuclei.

Columella. A sterile structure in a sporangium or other fructification; often a swelling at the terminal end of the sporophore.

Conidiophore. A hypha that bears conidia.

Conidium (pl. conidia). A deciduous asexual spore usually borne at the tip of a special hypha called a conidiophore.

Context. The sterile portion above the hymenium in the fleshy fruit bodies of the Basidiomycotina.

Cull. Wood in a living tree that is not useful because of decay.

Cystidium (pl. cystidia). A large sterile hyaline structure in the hymenium of Basidiomycotina.

Dieback. Death of the terminal portions of the crown of a tree, usually a gradual process.

Dikaryotic. A cell with two haploid nuclei that are not fused.

Dioecious. Having the male reproductive organs in one individual, the female in another.

Diploid. A nucleus with 2n chromosomes.

Disease. Any deviation from the normal that deleteriously affects a plant or any part of it or its economic value.

Echinulate. Spiny.

Endemic disease. A disease constantly infecting a few plants throughout an area.

Endogenous. Borne within.

Endospore. A spore formed within a cell.

Enzyme. Enzymes are biochemical catalysts produced by living cells that facilitate reactions necessary for physiological processes in living cells. Enzymes are in general heat labile, water-soluble protein molecules that increase the rate of a biochemical reaction without being permanently altered in the process. They are also active in very low concentrations.

Epidemic disease. A disease sporadically infecting a large number of hosts in an area and causing considerable loss.

Epiphytotic disease. A disease sporadically infecting a large number of plants in an area and causing considerable loss.

Etiolation. Paleness due to nondevelopment of chlorophyll in part or all of a plant.

Etiology. The study of the causes of disease.

Exogenous. Borne outside of.

Facultative parasites. Organisms that are ordinarily saprophytic but under certain conditions may become parasitic.

Facultative saprophytes. Organisms that usually live as parasites but under certain conditions are capable of saprophytic growth.

Fasciated. Broadened and flattened.

Fiber saturation point. That water content at which the cell walls of wood are saturated with water but no free water occurs in the lumens.

Flagellum (pl. flagella). A slender extrusion of protoplasm in bacteria and certain fungi used as an organ of locomotion.

Flagging. A symptom of a disease characterized by a dead or dying portion of a tree's crown.

Fruit body. A special structure composed of mycelium in which or on which the spore-producing organs are borne; a structure bearing spores.

Fungicide. A substance causing destruction of fungi.

Fungistatic. The state in which a fungus is living but not growing.

Fungus (pl. fungi). A plant without chlorophyll, having a vegetative body called a thallus and composed of hyphae, and reproducing by spores.

Gametangium (pl. gametangia). A structure containing gametes or nuclei that fuse and produce sexual spores.

Gamete. A male or female reproductive cell, or the nuclei within a gametangium that fuse and produce sexual spores.

Geotropism. Growth in response to gravity.

Germination. In the fungi, a change from a dormant state (such as a spore) to an actively growing state.

Germ tube. Initial hypha arising from a spore or other resting structure.

Gills. The lamellae on the under-side of the pileus of a mushroom on which the hymenium is borne.

Haploid. A nucleus with 1n chromosomes.

Haustorium (pl. haustoria). A specialized hypha that enters a host cell and absorbs food for the rest of the fungus colony.

Heliotropism. A growth in response to light.

Heteroecism. The growth of rust fungi for part of their life cycle on one species of host and for the remainder on another, usually only distantly related to the first.

Heterogamy. Condition in which the gametangia that fuse can be morphologically differentiated into male and female elements.

Heterokaryotic. Containing nuclei of different strains.

Heterothallism. The production of sexual spores by a fungus only as a result of fusion of two thalli developing from different spores of opposite sex.

Homothallism. The production of sexual spores by a fungus on a thallus developing from one spore.

Hormone (plant). An organic substance that controls growth, is active in minute quantities, and is capable of acting at a site removed from where it is found.

Host. An organism upon which an organism of a different species grows and from which all or most of its food is derived.

Hyaline. Without color.

Hymenium. An aggregation of asci or basidia in a layer.

Hyperplasia. Pathologic overgrowth due to an abnormally large number of cells.

Hypertrophy. In general, pathologic overgrowth of a part of a plant; specifically, pathologic overgrowth due to abnormally large cells.

Hypha (pl. hyphae). A single branch of a fungus thallus.

Hypoplasia. Pathologic underdevelopment of a part of a plant.

Immunity. Totally nonsusceptible.

Incitant. Agent that causes the disease. Same as *causal organism*.

Incubation. The time between inoculation and the appearance of symptoms of a disease.

Infection. The process of a pathogen obtaining food from a host, after having established itself as a parasite.

Inoculation. The act of bringing the disease-producing organism in contact with the host, where infection may occur.

Inoculum. Any part of a pathogen, such as a spore or bit of mycelium, that is able to grow and cause infection of a host.

Inoculum potential. The total potential of a pathogen to cause infection; it is related to the number of spores, amount of mycelium, or the total propagules as well as the ability of the pathogen to invade the host.

Intercellular. Between cells.

Intracellular. Within cells.

Isogamy. Condition in which the gametangia that fuse are not morphologically distinguishable.

Karyogamy. The fusion of two nuclei.

Lesion. Localized necrosis or diseased area.

Macroconidium (pl. macroconidia). A large conidium as compared to a microconidium.

Macrocyclic. Rust fungi that have four or five types of spores as compared to a microcyclic rust fungus with less than four types of spores.

Medium (pl. media). Substrate on which fungi and other organisms will grow.

Meiosis. Reduction division of the chromosomes.

Microconidium (pl. microconidia). A small conidium.

Micron. A unit of measurement, 1/1,000 millimeter, 1/25,400 inch.

Mitosis. Indirect cell division.

Morphology. The study of shape, form, and structure.

Mosaic. A virus disease that causes the affected plants to become mottled.

Mushroom. Loosely, any fleshy fungus; strictly, the gill fungi.

Mutualism. A symbiotic relationship beneficial to both organisms.

Mycelium. A mass of hyphae composing the thallus of a fungus.

Mycology. The science dealing with fungi.

Mycorrhiza (pl. mycorrhizae). The compound structure resulting from the growth of mycelium of a fungus on and within the short roots of higher plants, the two elements being constantly arranged in an orderly manner.

Necrosis. Death of host tissue.

Nucleic acid. The substances out of which chromosomes and certain other cell constituents that control cellular functions are made. Two forms are found in living cells—DNA (deoxyribonucleic acid) and RNA (ribonucleic acid).

Nucleus. The part of the cell that is made up chiefly of chromosomes or of chromatin material.

Obligate parasite. An organism that can grow only parasitically; it must grow only on or in and derive its food from another living organism.

Obligate saprophyte. An organism that can grow only saprophytically.

Oogonium. A female gametangium.

Oospore. A sexual spore borne in an oogonium.

Operculum. A hinged cap on a sporangium or ascus.

Organelles. Discrete structural units in the cytoplasm of all living cells such as chloroplasts, ribosomes, and mitochondria.

Ostiole. A porelike opening in a perithecium or pycnidium through which the spores escape from the fruit body.

Parameter. A defined variable.

Paraphysis (pl. paraphyses). Sterile hyphae between asci or basidia in a hymenium.

Parasite. An organism that grows part or all of the time on or within another organism of a different species and derives part or all of its food from it.

Parthenogenesis. Development of sexual reproduction from female gametes alone.

Pathogen. A parasite that causes disease.

Pathogenesis. That phase of the life cycle of a pathogen when it is associated with the living tissues of its host.

Pathogenicity. Ability of an organism to cause disease.

Pathologic disease. Disease caused by living organisms.

Pathological rotation. Age of a tree species at which time the heart decay fungi will decay more wood than is produced by the tree.

Peridium (pl. peridia). The covering or layer of mycelium over aeciospores and other spores.

Perithecium (pl. perithecia). A spherical ascocarp opening by a pore.

pH. A measure of acidity or hydrogen ion concentration; it is expressed in a figure representing the log of the reciprocal of the normality of a solution with respect to active hydrogen ions.

Phylogeny. A study of the history of a species and the interrelationships of different species.

Physiologic form. One of a group of forms alike in morphology but unlike in certain cultural, physiological, biochemical, pathological or other characters.

Pileus. The hymenium supporting part of a fruit body, usually referring to the stalked Agarics. The cap of a mushroom, including context and gills.

Plasmodium (pl. plasmodia). Naked mass of protoplasm in the Myxomycota.

Pleomorphism. The ability of a fungus to produce more than one kind of spore.

Pollard. To prune back a plant, usually implying drastic pruning or removal of all branches.

Predisposition. The influence of environment on the susceptibility of a plant or animal to disease.

Promycelium. The name given to the basidium in the fungi that cause rusts and smuts.

Protein. Chemical compounds that are involved in the metabolism of all living cells. Proteins include all enzymes, certain hormones, a portion of virus particles, and certain structural units of cell walls. They are made up of 22 different amino acids that are bound together in a complex helical structure.

Pustule. A small swelling on a host, usually with a mass of spores.

Pycnidiospore. An asexual spore borne in a pycnidium; one kind of conidium.

Pycnidium. A spherical fruit body in the Ascomycotina and Deuteromycetes that opens by a pore and that bears conidia.

Pycniospore. The spore that functions as a gamete in the rust fungi; borne in a pycnium.

Pycnium (pl. pycnia). A spherical fruit body opening by a pore in which the spores are borne; it functions as gametes in the rust fungi.

Race. A subdivision of a variety based on physiological rather than morphological characteristics.

Resupinate. Lying flat on the substrate.

Reticulate. Having a netted appearance.

Rhizomorph. A specialized form of mycelium consisting of several strands of hyphae twisted together so as to be rootlike.

Roguing. The eradication of diseased plants in a stand.

Rosette. Crowded or clustered foliage as a result of less than normal elongation of internodes.

Saprophyte. An organism that lives on dead organic matter.

Sclerotium (pl. sclerotia). A small hard body of densely interwoven hyphae often becoming detached from the rest of the mycelium and capable of remaining alive in a dormant state for a long time and enduring unfavorable conditions.

Septate. With cross walls.

Seta (pl. setae). A large sterile black structure in the hymenium of Basidiomycotina.

Sign. The manifestation of disease by the presence of structures of the causal agent.

Slime mold. A chlorophylless thallophyte in which the thallus consists of a plasmodium, a naked mass of protoplasm, and that reproduces by means of spores.

Somatic. The vegetative or mycelial stage of a fungus.

Sorus (pl. sori). A fruiting structure that consists of a mass of spores in or on host tissue.

Species. A unit of classification of closely related individuals resembling one another in certain inherited characteristics and capable of mating and producing fertile offspring.

Spermatium (pl. spermatia). A nonmotile sporelike structure that acts as a male gamete.

Sporangiophore. A hypha that bears a sporangium.

Sporangiospore. An asexual spore borne within a sporangium; this is the most common kind of asexual spore in the Mastigomycotina.

Sporangium. A fruit body producing endogenous asexual spores in the Mastigomycotina.

Spore. The unit of reproduction in the fungi, as well as certain other plants; a one- to many-celled structure that is liberated from the mycelium and can live in a dormant state, each cell being capable of germinating and reproducing the organism.

Sporidium (pl. sporidia). The name given to the basidiospores in the rust and smut fungi.

Sporodochium (pl. sporodochia). A cushion-shaped fruiting structure with conidia.

Sporophore. A spore-producing or spore-supporting structure.

Sterigma (pl. sterigmata). A slender stalk arising from a basidium, the tip of which expands, becomes abstricted, and thus forms a spore, such as a basidiospore.

Stipe. The stem of a mushroom.

Strain. A subdivision within a species.

Stroma (pl. stromata). A pseudoparenchymatous mass of mycelium on which sporophores are borne or in which fruit bodies are contained.

Symbiosis. The living together of two taxonomically different organisms, with organic union.

Symptom. Any perceptible change in host structure caused by disease.

Synnema (pl. synnemata). A group of branched upright hyphae that produce spores.

Systemic infection. An infection in which the parasite is found generally throughout the host tissue.

Taxonomy. The science of classification of organisms; it is chiefly a study of relationships as evidenced by morphologic characters.

Teliospore. In the rust fungi, the spore in which the male and female nuclei fuse, and from which the promycelium or basidium arises.

Thallus (pl. thalli). The vegetative body of a thallophyte.

Tolerance. Ability of the plant to endure the development and survive invasion of a parasite with minimum symptoms of disease.

Trichogyne. A projection from an ascogonium through which the male nucleus enters the ascogonium.

Tropism. A definite response to a certain stimulus.

Urediniospore. Asexual dikaryotic spore produced in a uredinium.

Uredinium (pl. uredinia). In the rust fungi, a fruit body producing urediniospores.

Variety. A subdivision in a species.

Vector. An agent, such as an insect, that may transmit a fungus or other microorganisms.

Virulence. The degree or measure of pathogenicity.

Viruliferous. Capable of transmitting a virus.

Virus. A submicroscopic disease-producing agent that multiplies only in living tissues.

Volva. A cuplike structure at the base of the stipe of a mushroom, the remains of a veil that covered the entire mushroom.

Witches' broom. The structure resulting from prolific branching, usually induced by a parasite.

Zoospore. A flagellated sporangiospore in the Mastigomycotina.

Zygospore. A sexual spore resulting directly from the fusion of morphologically similar gametangia.

INDEX *

* *Note:* Pages listed in boldface refer to major coverage of that topic.

801